UNSOLVED PROBLEMS OF NOISE AND FLUCTUATIONS

Previous Proceedings in the Series of International Conferences on Unsolved Problems of Noise and Fluctuations

	Year	Held in	Publisher	ISBN
2nd	1999	Adelaide, Australia	AIP Conference Proceedings vol. 511	1-56396-826-6
1st	1996	Szeged, Hungary	World Scientific	981-02-3199-7

Other Related Titles from AIP Conference Proceedings

661 Modeling of Complex Systems: Seventh Granada Lectures
Edited by Pedro L. Garrido and Joaquín Marro, April 2003, 0-7354-0121-7

658 Modern Challenges in Statistical Mechanics: Patterns, Noise, and the Interplay of Nonlinearity and Complexity; Pan American Studies Institute
Edited by V. M. Kenkre and K. Lindenberg, March 2003, 0-7354-0118-7

622 Experimental Chaos: 6th Experimental Chaos Conference
Edited by S. Boccaletti, B. J. Gluckman, J. Kurths, L. M. Pecora, and M. L. Spano, June 2002, 0-7354-0071-7

597 Nonequilibrium and Nonlinear Dynamics in Nuclear and Other Finite Systems: International Conference
Edited by Zhuxia Li, Ke Wu, Xizhen Wu, enguang Zhao, and Fumihiko Sakata, November 2001, 0-7354-0041-5

560 Nanoscale Linear and Nonlinear Optics: International School on Quantum Electronics
Edited by Mario Bertolotti, Charles Bowden, and Concita Sibilia, April 2001, 1-56396-993-9

502 Stochastic and Chaotic Dynamics in the Lakes: STOCHAOS
Edited by David S. Broomhead, Elena A. Luchinskaya, Peter V. E. McClintock, and Tom Mullin, February 2000, 1-56396-915-7

501 Stochastic Dynamics and Pattern Formation in Biological and Complex Systems: The APCTP Conference
Edited by Seunghwan Kim, Kyoung J. Lee, and Wokyung Sung, January 2000, 1-56396-914-9

466 Quantum 1/f Noise and Other Low Frequency Fluctuations in Electronic Devices: Seventh Symposium
Edited by Peter H. Handel and Alma L. Chung, March 1999, 1-56396-854-1

411 Applied Nonlinear Dynamics and Stochastic Systems Near the Millennium
Edited by James B. Kadtke and Adi Bulsara, November 1997, 1-56396-736-7

UNSOLVED PROBLEMS OF NOISE AND FLUCTUATIONS

UPoN 2002: Third International Conference on Unsolved Problems of Noise and Fluctuations in Physics, Biology, and High Technology

Washington, DC 3–6 September 2002

EDITOR
Sergey M. Bezrukov
NICHD, National Institutes of Health
Bethesda, Maryland

SPONSORING ORGANIZATIONS
National Institute of Child Health and Human Development
National Institute of Mental Health
National Institute of Neurological Disorders and Stroke
National Institute of Deafness and Other Communication Disorders

AMERICAN INSTITUTE OF PHYSICS

Melville, New York, 2003
AIP CONFERENCE PROCEEDINGS ■ VOLUME 665

Editor:

Sergey M. Bezrukov
Lab. Physical and Structural Biology
Building 9, Room 1N-124B
NICHD, National Institutes of Health
Bethesda, MD 20892-0924
USA

E-mail: bezrukov@helix.nih.gov

L.C. Catalog Card No. 2003105095
ISBN 0-7354-0127-6
ISSN 0094-243X
Printed in the United States of America

CONTENTS

SENSORY SYSTEMS

NEURONS AND NEURONAL NETWORKS

CENTRAL NERVOUS SYSTEM

ION CHANNELS AND MOTORS

SINGLE MOLECULES

GENETICS

NOISE AND NOISE ANALYSIS FOR MEDICAL APPLICATIONS

NONLINEAR DYNAMICS

MATERIALS, DEVICES

WAVES, HYDRODYNAMICS, AND MECHANICS

GENERAL THEORY AND EXPERIMENT

PREFACE

The third international conference on Unsolved Problems of Noise and fluctuations in physics, biology, and high technology (UPoN'2002) took place at the National Institutes of Health in Bethesda (Washington, DC metro area) on September 3 – September 6, 2002. In order to promote the penetration of concepts and quantitative methods of the physics of fluctuations into biological sciences and medicine, the conference brought together scientists from physics, biophysics, biomedical engineering, biology, and medicine. The conference was sponsored by four NIH institutes: National Institute of Child Health and Human Development, National Institute of Mental Health, National Institute of Neurological Disorders and Stroke, and National Institute on Deafness and Other Communication Disorders.

The third of the series, this conference adhered to well established UPoN traditions including double-blind refereeing of all conference contributions. The first conference, pioneered by Laszlo Kish, held in Szeged, Hungary in 1996, was mostly devoted to high technology devices. The second one, organized by Derek Abbott, was held in Adelaide, Australia in 1999 and focused largely on mathematical aspects and paradoxes in noise and fluctuation research.

The purview of this third conference was shifted toward biology and medicine. Among the topics covered were: the constructive role of noise in the central nervous system, neuronal networks, and sensory transduction (hearing in humans, photo- and electroreception in marine animals), encoding of information into nerve pulse trains, single molecules and noise (including single molecule detection and characterization by nanopores – molecular 'Coulter counting'), concepts of noise in neurophysiology (randomness and order in brain and heart electrical activities under normal conditions and in pathology), the role of noise in genetic regulation and gene expression, biosensors, etc.

Judging by the participants' response and the number of 'thank you' letters, I infer that the conference was memorable, useful, and successful. I would like to thank the scientific directors of the four NIH institutes, Dr. Owen Rennert, Dr. Robert Desimone, Dr. Story Landis, and Dr. Robert Wenthold, for their vision and support that made this conference possible. I am also grateful to the members of the scientific committee for their valuable advice and help in peer review of the submitted contributions.

The compilation of papers that you hold in your hands reflects the spirit of the conference. As the conference itself did, this volume represents the best research in the field of noise and fluctuations in biologically relevant fields of science. I hope that for this reason it will be interesting to a wide and diverse audience of researchers around the world.

Sergey Bezrukov,
UPoN'2002 chair

Scientific Committee

Derek Abbott (*Adelaide University, Australia*)
Robert Austin (*Cornell University, USA*)
Dean Astumian (*University of Maine, USA*)
Gijs Bosman (*University of Florida, USA*)
Adi Bulsara (*SPAWAR Systems Center, USA*)
Jim Collins (*Boston University, USA*)
David Deamer (*University of California, USA*)
Louis DeFelice (*Vanderbilt University, USA*)
Jamal Deen (*McMaster University, Canada*)
Charlie Doering (*University of Michigan, USA*)
Peter Hanggi (*University of Augsburg, Germany*)
Laszlo Kish (*Texas A&M University, USA*)
Mikhail Levinshtein (*AF Ioffe Physical Technical Institute, Russia*)
Andre Longtin (*University of Ottawa, Canada*)
Peter McClintock (*University of Lancaster, England*)
Frank Moss (*University of Missouri, USA*)
Ralph Nossal (*National Institutes of Health, USA*)
Lino Reggiani (*University of Lecce, Italy*)
Barry Richmond (*National Institutes of Health, USA*)
Lutz Schimansky-Geier (*Humboldt University, Germany*)
Michael Shlesinger (*Office of Naval Research, USA*)
Gene Stanley (*Boston University, USA*)
Charles Stevens (*Salk Institute for Biological Studies, USA*)
Nico Van Kampen (*Institute of Theor. Physics, The Netherlands*)
Lode Vandamme (*University of Eindhoven, The Netherlands*)
Michael Weissman (*University of Illinois, USA*)
Kurt Wiesenfeld (*Georgia Institute of Technology, USA*)

Sponsors

National Institute of Child Health and Human Development,
Dr. Owen Rennert, Scientific Director

National Institute of Mental Health,
Dr. Robert Desimone, Scientific Director

National Institute of Neurological Disorders and Stroke,
Dr. Story Landis, Scientific Director

National Institute on Deafness and Other Communication Disorders,
Dr. Robert Wenthold, Scientific Director

OPENING ADDRESSES BY FORMER UPoN CHAIRS

Unsolved Problems of Noise: at the Frontiers of Research

Laszlo B. Kish[1]

Department of Electrical Engineering, Texas A&M University, College Station TX 77843-3128, USA

The research of noise is going through a revolution. More and more scientists realize that the noise is not only a kind of rubbish that has to be removed from the signal but it also carries a vast amount of information and, in many cases of biology and physics, is functionally involved in the system operation. The trend can be well seen in Table 1. The relative portion of papers where noise(s) or fluctuation(s) keywords produce a hit in the SCI database has grown by almost a factor of six during the last decade. It is important to note that acoustical noise papers make up only about 10% of these numbers.

Year	Total Number of Papers (SCI)	Noise/Fluctuation Papers	Ratio
1970	314347	1066	0.3%
1980	554615	1722	0.3%
1990	689629	2221	0.3%
2001	999620	**16550**	**1.7%**

TABLE 1. Evolution of the field of noise and fluctuations

The UPoN conference at the National Institutes of Health was certainly the crowning of the UPoN idea and this conference series. This would not have been possible without the hard and meticulous work of the organizers, especially of Sergey Bezrukov. The selection and the quality of the presented papers made this meeting a unique one in the history of noise research.

[1] Until 1999, his family name was Kiss

CP665, *Unsolved Problems of Noise and Fluctuations: UPoN 2002: Third International Conference*, edited by S. M. Bezrukov

It is an appropriate moment to cite the names of some great scientists in the history of noise and fluctuation research, such as James Maxwell and Ludwig Boltzmann (gas molecule velocity thermal noise in kinetic gas theory), Max Planck (thermal noise of electromagnetic field, the black body radiation), Albert Einstein (thermal noise of the velocity of small particles in a fluid, the Brownian motion; spontaneous and induced emission), and Claude Shannon (noise and bandwidth, the ultimate limitations of information transfer rate). All these scientists have realized and solved important unsolved problems of noise and, by doing that, catalyzed the development of the field to its present magnificent state.

I hope that we all join the effort so that the UPoN conference series will keep on evolving at the frontiers of scientific research.

Address on Behalf of the Committee

Derek Abbott

Centre for Biomedical Engineering (CBME), EEE Dept.,
University of Adelaide, SA 5005, Australia

I am delighted and honored to address so many distinguished researchers in the area of noise and fluctuations, here at the National Institutes of Health in Washington, DC.

In 1996, first UPoN conference, held in Szeged, Hungary, had a theme in the area of materials. In 1999, the second UPoN was held in Adelaide, Australia, and had a theme centered around fundamental issues in noise. Now in 2002, there is a strong theme on fluctuations in biological systems and thus is it highly appropriate that our host is the NIH.

The study of fluctuations is perhaps one of the most difficult areas of science to master, with many pitfalls for the unwary – it takes a rather special breed of scientist to boldly tackle the many subtleties and intricacies. But the rewards are great, because noise and fluctuations are so fundamental to all systems and crucial to understanding how the world operates at the microscopic level.

In this generation we are witnessing huge leaps in biophysics and biomedical engineering. We are moving into an age of bioinformatics. The study of noise and fluctuations in these systems is perhaps the most challenging and exciting – we are therefore honored that we have the splendid leadership of Sergey Bezrukov in chairing this meeting, as he is very much at the forefront and cutting edge of this field.

So, on behalf of the committee, I would like to extend our thanks to Sergey for his vision, his leadership and many late hours spent putting this conference together. He has truly put a superb program together – one that will surely make a lasting impact on the field for years to come.

The committee would also like to thank Laszlo Kish, the founder of the UPoN series, also for his tremendous vision and energy, who has made this conference series so popular and significant.

A special thanks is due to the hosting institution, the NIH, who have made available the splendid and stimulating setting for this conference.

Finally, last but not least, a many thanks must go to the speakers for coming here from distant parts of the world, bringing their enthusiasm and creativity for us all to learn.

CP665, *Unsolved Problems of Noise and Fluctuations: UPoN 2002: Third International Conference*,
edited by S. M. Bezrukov

KEYNOTE TALK

Forces and Fluctuations: Nourished by Noise

V. Adrian Parsegian

Laboratory of Physical and Structural Biology, NICHD, National Institutes of Health, Bethesda, MD 20892-0924

As you all know, the idea of "noise as nuisance" evolved only slowly into the idea of noise as a source of information. Static on the radio taught us about charges dancing in hot metals. Today, the noise in currents through ionic channels reveals dynamics of biomolecules – "noise as news"!

What I will describe here is "noise as nourishment." The noise that is intrinsic to our surroundings creates intermolecular forces. Specifically, I am going to talk about forces between bodies in terms of the fluctuations of the charges and electromagnetic fields within them, so-called van der Waals forces. We know that neutral materials are made up of plus and minus charges. At any given instant transient positions and velocities of the charges will deviate from their smeared average to create electromagnetic fields. These motions will correlate between the bodies so as to create attractive forces between like bodies. The essence of this attraction is that natural motions of charges within materials show up not only in forces but in "noise." Conversely, by looking at the spontaneous motions revealed as noise, we learn how materials convert their natural motions into intermolecular forces.

When we think of noise and spontaneous motion of charges we usually think of thermal fluctuations. With van der Waals forces most of the important frequencies at which these interactions occur are those whose photon energies are far greater than thermal energy kT. In fact, it is not the temperature that drives most of the contributing fluctuations. Most interactions come from motions described by the uncertainty principle, spontaneous motion even at the very coldest temperatures, so called zero-point fluctuations. The modern view is that we are bathed in an infinity of vacuum electromagnetic-field fluctuations. These fields can be imagined as coming from all the universe. Basically interacting bodies feel those fluctuations, filter them by their absorption spectra, and soak up certain of those zero-point fluctuations to drive the charge configurations that create interaction. That is a mouthful to say. It is hard to take in at the outset, but I hope that within the next twelve or fourteen minutes you'll see what this means and how it connects with what you do.

The story begins with van der Waals' 1873 Ph.D. thesis wherein he changed Boyle's ideal gas law in two important ways:

One, the particles in a gas are of finite size, so that the volume available to them is the volume of the whole box minus the volume of the particles themselves.

CP665, *Unsolved Problems of Noise and Fluctuations: UPoN 2002: Third International Conference*, edited by S. M. Bezrukov
2003 American Institute of Physics 0-7354-0127-6

Two, where the Boyle's Law had pressure, one must write pressure plus another term that varies with the volume. The pressure is actually less than in Boyle's law; it is less by an amount that changes with the volume of the box. From this second point it became clear that there are attractive forces between the molecules, forces that change with the distance between them.

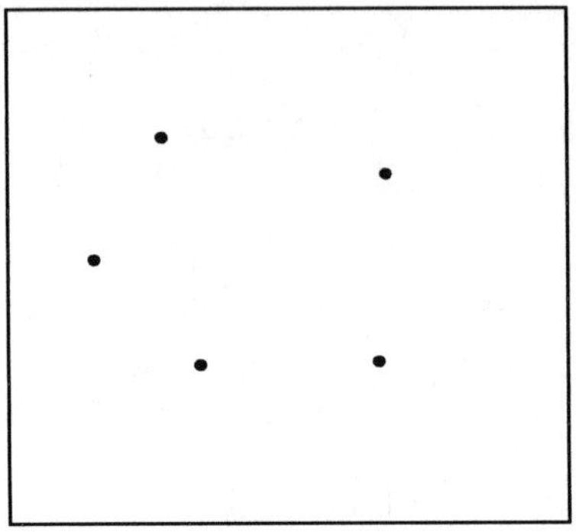

1660 Boyle's Law: pV constant

Nowadays : $p_{\text{Boyle}}V=NkT$, N number of particles, k the Boltzmann constant, T absolute temperature, p pressure, V volume of box.

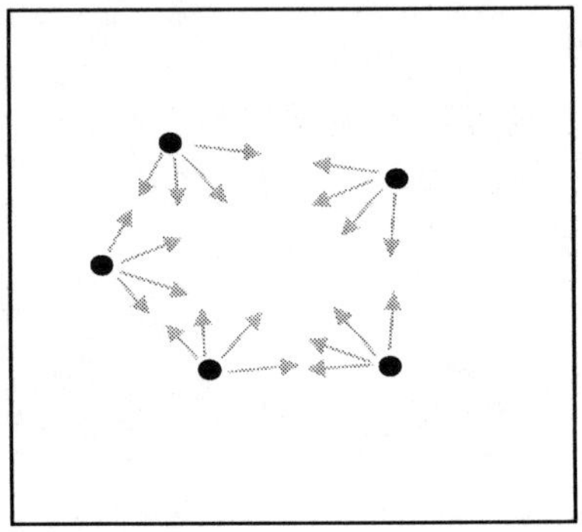

1873 van der Waals

$$\left(p_{\text{vdW}}+\frac{\text{a}}{V^2}\right)(V-\text{b})=NkT$$

Coefficient **a** ≥ 0, $p_{vdW} \leq p_{Boyle}$ due to attractive forces; total volume of particles, **b** ≥ 0.

Even more important, van der Waals' thesis was on the "continuity of the gas and the liquid states." What he was pointing out was that these attractive forces seen in the gas are, as we know well today, interactions that create liquids and solids, condensed media. The attractive force seen in the gas shows up as an important feature of condensed materials. What was the source of this attraction?

Remember, van der Waals was working at a time when the electromagnetic theory had recently been codified by Maxwell; Hertz was soon to develop the principles of receivers and transmitters, proving that current fluctuations created electromagnetic waves that could be detected and converted back into currents at a distant place; and in 1894 a graduate student named Lebedev was writing a description of intermolecular forces in his Ph.D. thesis. Lebedev's insight is the germ of the noise/forces idea I am describing today. He said it so clearly that the best I can do is to quote (Lebedev's ideas are instructively summarized by Derjaguin [1]):

Hidden in Hertz's research, in the interpretation of light oscillations as electromagnetic processes, is still another as yet undealt with question, that of the source of light emission of the processes which take place in the molecular vibrator at the time when it gives up light energy to the surrounding space; such a problem leads us, on the one hand, into the region of spectral analysis (absorption spectra), *and on the other hand, quite unexpectedly as it were, to one of the most complicated problems of modern physics -- the study of molecular forces.*

The latter circumstance follows from the following considerations: Adopting the point of view of the electromagnetic theory of light, we must state that between two radiating molecules, just as between two vibrators in which electromagnetic oscillations are excited, there exist ponderomotive (physical body) *forces: They are due to the electromagnetic interaction between the alternating electric current in the molecules .... or the alternating charges in them ...; we must therefore state that there exist between the molecules in such a case molecular forces whose cause is inseparably linked with the radiation processes.*

Of greatest interest and of greatest difficulty is the case of a physical body in which many molecules act simultaneously on one another, the vibrations of the latter not being independent owing to their close proximity.

What I think Lebedev was saying is that if two molecules are signaling each other via electromagnetic waves, then the other molecules around those two are necessarily responding and interacting too. He recognized the formidable many-body problem that was not to be solved for another 60 years. Although still not appreciated by many colloid scientists even today, the problem is pretty obvious. A typical electromagnetic wavelength is vastly greater than the distance between atoms in a condensed material. Any electromagnetic field must simultaneously act on a vast set of atoms. Except for a dilute gas, we cannot think of pairs of individual atoms or molecules seeing each other independent of interactions with all their neighbors.

What happened to this note in Lebedev's Ph.D. thesis? First of all, like much good work, it was ignored. Not long after Lebedev's thesis, Planck was addressing the problem of black body radiation. What does it take to heat up empty space? It was commonly understood that, just as when you excite certain modes in playing a musical instrument, there are electromagnetic modes in the box. As you heat, the energy of those electromagnetic oscillators increases. The price of heating is the heat capacity of the box. Remember Planck's great discovery. He found that he could not fit the heating energy unless he treated changes in oscillator energies as discrete units, what Einstein soon recognized as photons. Only by discretization of electromagnetic energy could he talk about how much heat it took to warm empty space. This was the birth of quantum theory. It also provided the germ of an idea by Casimir 45 years later. What if we take Planck's black box but now change the distance between two of its parallel conducting walls? What happens to the energy of those modes in there? Casimir found that there was an attractive force, that is to say, that the energy went down as the walls moved together.

Now, the problem is this. (I am changing the history a little bit to make the science clear.) If you look at the frequencies at which the important interactions are occurring, their photons have energies far higher than those that can be excited by thermal energy. What could be the source of electromagnetic fluctuations that are too costly to be from thermal energy, the kind of fluctuation that we usually associate with "noise"? Casimir explained later [2]:

Here is what happened. During a visit I paid to Copenhagen, it must have been in 1946 or 1947, Bohr asked me what I had been doing and I explained our work on van der Waals forces. "That is nice," he said, "that is something new." I then explained I should like to find a simple and elegant derivation of my results. Bohr thought this over, then mumbled something like "must have something to do with the zero-point energy." That was all, but in retrospect I have to admit that I owe much to this remark.

The idea that emerged is that there are, as I stated in the beginning, fluctuations that are intrinsic to the nature of open space. These fluctuations occur at zero temperature; the forces between bodies occur because the bodies, the walls of the box, are perturbing those fluctuations that flow through all space. There is physical reality to the vacuum. (See the lucid comment by Kleppner [3]). If you have two metal plates, then there is all this electromagnetic noise of the universe shining on them. If they absorb some waves, then there are fewer or weaker electromagnetic fields between the plates than outside. That inequality, in itself, creates an attractive force. (see, e.g., ref [4]).

Imagine two boats floating near each other on rough seas. The boats damp some of the water waves that hit them so that the water between is relatively still. The net effect on each boat is that there are waves coming at it from all directions, except those from the other boat. The net force will be to push the two boats together. From hearing that much, you can begin to see that the degree to which the boats absorb the waves makes a different to the force between them. All the material properties, how they harness this "noise" should show up in the force. The analogy to ideally conducting plates is boats that absorb all waves. Real boats absorb only some waves; real materials screen out only some electromagnetic fluctuations, the fluctuations that occur at their material absorption frequencies. This last point is the key to seeing how material properties couple to forces.

Only five years after Casimir made his formulation (now the renowned "Casimir Effect"), Lifshitz was able to generalize it to recognize real material properties. He solved Lebedev's problem. How to formulate the forces between large bodies? Look at their electromagnetic absorption spectra. Convert their natural measured motions into estimated interaction.

Now back to the roots of your subject, noise. First we can see the Nyquist theorem in a new light. Then we can see how changes in material properties change not only the "noise" that comes out of the theorem but also change intermolecular forces.

How do you noise-niks look at material? How do you learn from it? You put it between the plates of a capacitor, apply a voltage V and observe a current I.

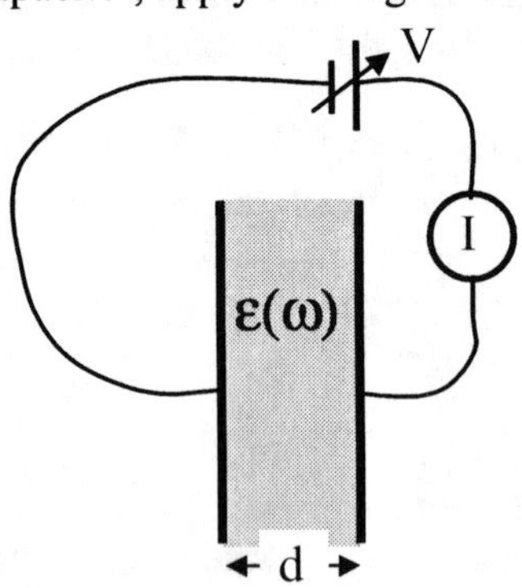

The impedance of a capacitor against voltage, sometimes called "reactance," $-4\pi d / i\omega\varepsilon(\omega)$, involves the dielectric response $\varepsilon(\omega)$, which I have written as a function of the radial frequency ω (radians/second) = $2\pi f$, with the usual frequency f (Hertz). This response $\varepsilon(\omega) = \varepsilon'(\omega) + i\varepsilon''(\omega)$, has two parts. One is springiness, the storage part $\varepsilon'(\omega)$. The other part is dissipation $\varepsilon''(\omega)$. In response to the applied oscillating field, the material can act like a spring – store the electrical energy during one part of the cycle then feed it back to you during the other part. The material can also dissipate energy, create heat out of the applied electrical energy.

And it works the other way. Apply no voltage, just take a current meter and watch how the current flickers creates noise as a function of frequency. The famous 1928 Nyquist formula tells us $\left(\mathrm{I}^2\right)_f = 4kT/R$ with the material resistance R [5]. The hotter it gets, the more noise, inversely proportional to resistance.

The modern view of this is to look at the same frequency fluctuations to find that the mean square current at radial frequency ω is driven by dissipative part $\varepsilon''(\omega)$ of $\varepsilon(\omega)$.

$$\left(\mathrm{I}^2\right)_\omega = \frac{\hbar\omega}{2\pi}\coth\left(\frac{\hbar\omega}{2kT}\right)\mathrm{Re}\left[\frac{1}{Z_\omega}\right] = \frac{\hbar\omega}{2\pi}\coth\left(\frac{\hbar\omega}{2kT}\right)\frac{\omega\varepsilon''(\omega)}{4\pi d}$$

Here $\hbar\omega$ is the photon energy and kT is thermal energy.

When it is hot, so $\hbar\omega \ll kT$, we are in the classical limit, $\frac{\hbar\omega}{2\pi}\coth\left(\frac{\hbar\omega}{2kT}\right) \to \frac{kT}{\pi}$, and current fluctuations are proportional to temperature as in Nyquist's $\left(\mathrm{I}^2\right)_f = 4kT/R$.

In the quantum limit $\hbar\omega \gg kT$, $\frac{\hbar\omega}{2\pi}\coth\left(\frac{\hbar\omega}{2kT}\right) \to \frac{\hbar\omega}{2\pi}$. Temperature is gone; current fluctuations are proportional to photon energy. As T goes to zero, there still something left. There is still noise in this circuit, but now it is not due to the thermal fluctuations. Rather it is due simply to the so-called uncertainty principle. If something can move it does. It was motions at zero temperature that Bohr pointed out to Casimir in 1946 or 1947.

Learning from the static of 1920's radios, Nyquist had already seen the connection to material properties. His 1928 paper observes:

> *Ordinary electric conductors are sources of spontaneous fluctuations of voltage which can be measured with sufficiently sensitive instruments. This property of conductors appears to be the result of thermal agitation of the electric charges in the material of the conductor,*

and ends with a heroic:

> *the limit to the smallness of voltage which can be usefully amplified is often set, not by the vacuum tube, but by the very matter of which electrical circuits are built.*

And that's the key. If we want to know about charge fluctuation forces between bodies A and B across medium m, we look to the "noise" these materials make in an idealized capacitor.

Imagine flat bodies A and B separated at separation l. In the 1950's Lifshitz, and then Dzyaloshinski, Lifshitz and Pitaevskii, generalized Casimir's result for the interaction of two parallel conductors. They included material dielectric properties in the form of dielectric susceptibilities ε_A, ε_m, and ε_B [6-8].

The historical sequence of events looks like this:

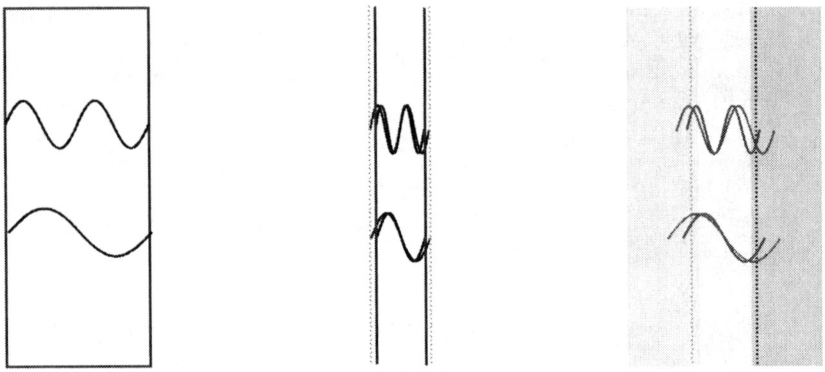

(a) Planck: hollow "black" box with opaque or conducting walls. Heat capacity shows exchange of energy in quanta, photons.

(b) Casimir: two parallel flat ideally conducting surfaces. Electrodynamic force between the two plates across a vacuum.

(c) Lifshitz, Dzyaloshinskii and Pitaevski: any two flat surfaces of any materials facing across a third medium.

The simplest form of this Lifshitz interaction is that there is an inverse square dependence, $-\mathrm{H}/12\pi l^2$ of the energy per unit area. The Hamaker coefficient is given by

$$\mathrm{H} = \mathrm{H}(l) = \frac{3kT}{2} \sum_{\substack{\text{sampling} \\ \text{frequencies } \xi_n}} \left(\frac{\varepsilon_A - \varepsilon_m}{\varepsilon_A + \varepsilon_m} \right) \left(\frac{\varepsilon_B - \varepsilon_m}{\varepsilon_B + \varepsilon_m} \right) \times Rel(l).$$

The essential feature is that differences in the responses of materials, evaluated at a set of discrete sampling frequencies stretching over the entire spectrum, creates the force. The coefficient H itself has distance dependence. This material-dependent contribution is modulated by a "relativistic" factor $Rel(l) \leq 1$ that takes into account the finite time needed for an electromagnetic signal to fly across the gap l.

How do we find the needed information about these ε's? First we use a form to express how materials respond to the exponentially decaying fields of spontaneous fluctuations, at so called imaginary frequencies. The germ of this exponential response is in the same dissipation term that created noise in Nyquist's ideal capacitor.

Use the Kramers-Kronig transform $\varepsilon(i\xi) \equiv 1 + \frac{2}{\pi}\int_0^\infty \frac{\omega\varepsilon''(\omega)}{\omega^2+\xi^2}d\omega$ where we recall $\omega\varepsilon''(\omega)$ from dissipation term $\varepsilon''(\omega)$ in $\left(\mathrm{I}^2\right)_\omega = \frac{\hbar\omega}{2\pi}\coth\left(\frac{\hbar\omega}{2kT}\right)\frac{\omega\varepsilon''(\omega)}{4\pi d}$. Through a transform of the natural frequencies of the materials, we see how the same property that drives current and voltage fluctuations feeds fluctuation forces.

We usually think of thermally driven charge fluctuations as in the original $\left(\mathrm{I}^2\right)_f = 4kT/R$ Nyquist formula. In fact, as I have now reminded you several times, most of the sampling frequencies in the summation for coefficient H are at frequencies whose corresponding photon energies are much greater than thermal energy kT. The important fluctuations are often in the quantum limit, the zero-point fluctuations that "*must have something to do with the zero-point energy.*"

The capacitor-dielectric picture automatically solves another problem, Lebedev's "*Of greatest interest and of greatest difficulty is the case of a physical body in which many molecules act simultaneously on one another, the vibrations of the latter not being independent owing to their close proximity.*"

The material is treated whole rather than as a set of individual molecules. Their response, in the capacitor and in the A-m-B configuration is such that they "*act simultaneously on one another.*" The noisome many-body problem is solved by Nature herself. All we humans have to do is to pay attention to the spontaneous noise from a fully composed material.

The fact that this noise itself indicates the energy absorption by the material in the capacitor lets us step outside the materials themselves. Yes, by thinking interior thoughts we can imagine that their constituent charges fluctuate. Yes, this can be at frequencies so high that they are not thermally excitable. How about exterior thoughts?

If we take seriously the idea of space and the reality of the vacuum, then there are charge fluctuations all around us. Back to the boat-on-rough-seas picture. Imagine bodies A and B in m and all of it bathed by electromagnetic waves coming from the cosmos. The extent of response to these outside waves is proportional to the extent of absorption of materials at all frequencies. Cosmic noise harnessed by material charges and turned into forces.

Students of "noise" should appreciate the power of their concepts in other branches of science. In our own lab we are changing the assembly of membranes and macromolecules by adding solutes to the intervening water; from the perspective of noise and dielectrics, a solution is a different medium from a pure solvent. We learn this as children when we put salt and sugar in water to see the pretty, light-distorting schlieren of refractive index. These days absorption spectra are being used to create transistors and ceramics [9], to control aerosols [10] and colloids [11].

Van der Waals forces have even been in the newspapers. Researchers now believe that geckoes use these forces to crawl on walls [12]. Talk about physics in biology! It looks as though there is plenty of energy in these interactions. Geckoes walk on microscopic feet pads or setae. They can change interactions as they walk, control the stickiness by peeling off their gecko feet as we peel adhesive tape. It seems that geckoes learned about noise and quantum electrodynamics a very long time ago. Next time you see a gecko, remember that it is buoyed by the vacuum-infinity electromagnetic-fluctuation noise of the universe; its feet are filtering out some of those waves because of their absorption spectra. The noise of the universe creates the forces that allow him to walk on a wall. A pretty neat trick. Maybe it is time that we learn it too.

REFERENCES

1. Derjaguin, B. V., P. N. Lebedev's ideas on the nature of molecular forces, *Soviet Physics Uspekhi*, **10**, 108 (1967).
2. The Casimir Effect 50 Years Later, Proceedings of the Fourth Workshop on Quantum Field Theory under the Influence of External Conditions, Ed. Michael Bordag, World Scientific, Singapore, 1999.
3. Kleppner, D., With apologies to Casimir, *Physics Today*, **43**, 9 (1990).
4. Yam, P., Exploiting zero point energy, *Scientific American*, **277**, 82 (1997).
5. Nyquist, H., Thermal agitation of electric charge in conductors, *Physical Review* **32**,110 (1928).
6. Lifshitz, E. M., *Dokl. Akad. Nauk. SSSR*, **97**, 643 (1954); ibid., **100**, 879 (1955); *Zh. Eksp. Teor. Fiz.*, **29**, 94 (1955).
7. Lifshitz E. M. and L. P. Pitaevskii, Statistical Physics, Part 2, Chapter VIII, in Landau and Lifshitz Course of Theoretical Physics, Volume 9, Pergamon Press, 1991.
8. Dzyaloshinskii, I. E., E. M. Lifshitz, and L.P. Pitaevskii, *Advances in Physics*, **10**,165 (1961).
9. French, R. H., Origins and applications of London dispersion forces and Hamaker constants in ceramics", *J. Amer. Ceram. Soc.*, **83**, 2117 (2000).
10. Arunachalam, V. R., R. Lucchese, and W. H. Marlow, Simulations of aerosol aggregation including long-range interactions, *Phys. Rev. E*, **60**, 2051 (1999).
11. Prieve, D. C., Measurement of colloidal forces with TIRM, *Adv. Colloid Interface Sci.*, **82**:93 (1999).
12. Autumn, K., M. Sitti, Y. A. Liang, An. M. Peattie, W. R. Hansen, S. Sponberg, T.W. Kenny, R. Fearing, J. N. Israelachvili, and R. J. Full, Evidence for van der Waals adhesion in gecko setai, *Proc. Natl. Acad. Sci. USA*, **99**, 12252 (2002).

CONSTRUCTIVE ROLE OF NOISE

Non-Markovian Effects in Pattern Forming Chemical Systems

J. M. Sancho*, Miguel A. Santos†, S. Alonso** and F. Sagués**

*Departament d'Estructura i Constituents de la Matèria, Facultat de Física, Universitat de Barcelona, Diagonal 647, E-08028 Barcelona, Spain
e-mail: jmsancho@ecm.ub.es
†Fraunhofer Institute for Autonomous Intelligent Systems Schloss Birlinghoven, 53754 Sankt Augustin, Germany
e-mail:Miguel.Santos@ais.FhG.de
**Departament de Química Física, Universitat de Barcelona, Diagonal 647, E-08028 Barcelona, Spain
e-mail:s.alonso@qf.ub.es;f.sagues@qf.ub.es

Abstract. Active media are a suitable benchmark to study the constructive role of random fluctuations in pattern forming systems. The mathematical framework to model these pattern forming systems is by using Partial Differential Equations. Due to the presence of realistic noises in the experiments we have to deal with Non-Markovian Stochastic Partial Differential Equations whose treatment is an *unsolved problem* in the theory of Stochastic Processes. A preliminary analytical approach to this topic is presented. Quantitative and qualitative agreement is found between theoretical and numerical results.

INTRODUCTION

It is nowadays accepted as a promising paradigm of nonlinear science that random fluctuations can generate order or other counterintuitive effects. This constructive role of noise is under very active research [1]. The different physical mechanisms of such ordering role of noise are studied by using simplified modelizations and appropriate selected experimental setups. There is a common agreement that the most relevant ingredient of such unexpected phenomenon is the cooperative interaction between fluctuations, nonlinearities and spatial coupling.

Active media are one of the best scenario to understand this phenomenon due to its experimental access and the advanced status of their deterministic studies [2, 3, 4, 5]. These systems present a dynamical regime of wave propagation phenomenology and show a large diversity of spatio-temporal patterns, ranging from planar waves to much more complicated structures such as targets or scroll waves.

A theoretical modelizations based in Partial Differential Equations (PDE) is the most common tool for the description and study of wave dynamics [5]. Nevertheless a much more difficult task appears when random fluctuations are incorporated, because then we have to deal then with Stochastic Partial Differential Equations (SPDE). This is a subject of active rigorous mathematical studies with unclear results with respect to the needs of a more applied practitioner. Moreover if, trying to simplify the problem, we perform a

CP665, *Unsolved Problems of Noise and Fluctuations: UPoN 2002: Third International Conference*, edited by S. M. Bezrukov

Markovian approximation we have to decide which kind of stochastic calculus to follow, either Ito or Stratonovich, but in any case formulated to cover spatio-temporal variables or fields.

But this is not the last difficulty. In most of the experiments, random fluctuations have a well controlled spatio-temporal structure, so a Markovian approximation seems to be indeed far from what we need to face. In these circumstances our problem is intrinsically non-Markovian and as a consequence we have to deal with Non-Markovian SPDE.

The rigorous treatment of Non Markovian SPDE's is an *unsolved problem* in the theory of Stochastic Processes and as a consequence new mathematical tools have to be developed to deal with such a problem. Here we will present a preliminary theoretical approach to Non Markovian SPDE's which will focus on its predictive strength [1, 6]. Our analytical, although non exact, calculations will lead to systematic predictions on the possible effects of random fluctuations. Numerical simulations implemented will be used also as a test of the reliability and accuracy of our analytical treatment. In doing so, our main objectives are: (i) To present practical analytical and numerical tools to deal with such problems, and (ii) to apply these techniques to realistic modelizations of chemical excitable systems.

This paper is organized as follows. In Section 2 we present the main aspects of our analytical and numerical methodology. Section 3 contains the explicit results for a very simple modelization of an excitable system. Finally we end up with some conclusions and perspectives.

THEORETICAL APPROACH

Here we will review the most relevant points of the methodology starting, for comparison purposes, with the analysis of the Markovian case, following with the non Markovian problem and ending with the description of the simulation algorithms.

Instead of starting with a one field model we will introduce a generic two field scheme widely used in the studies of active media. The simplest description for wave propagation is based on a model with two variables called *activator*, $u(\mathbf{x},t)$, and *inhibitor*, $v(\mathbf{x},t)$, with different time scales. These equations have the general form,

$$\partial_t u = D\nabla^2 u + f_u(u,v,b); \qquad \partial_t v = f_v(u,v), \tag{1}$$

where f_i are the reactive terms and the inhibitor diffusion is discarded. The conditions $f_i = 0$ determine the deterministic null-clines whose intersections points give the homogeneous stationary states of the system. The relative stability or nature of these points can be obtained by a simple linear analysis. b is an external parameter which controls the excitability properties of the medium. In the experiments on the photosensitive Belousov-Zhabotinsky reaction (see later), the parameter b will stand for the intensity of the incident light [7, 8, 9, 10]. Controlled fluctuations can be introduced through it as, $b \to b + \eta(\mathbf{x},t)$. Then one can arrive to the following generic system of stochastic partial differential equations (SPDE), with multiplicative noise coupled only with variable u,

$$\partial_t u = D\nabla^2 u + f_u(u,v,b) + g(u)_u \eta(\mathbf{x},t); \qquad \partial_t v = f_v(u,v), \tag{2}$$

where $\eta(\mathbf{x},t)$ is a Gaussian temporal white noise with zero mean and a correlation,

$$\langle \eta(\mathbf{x}',t')\eta(\mathbf{x},t)\rangle = C(\mathbf{x}-\mathbf{x}')\delta(t-t'). \tag{3}$$

In the Markovian case we do not need to assume that the noise is also white in space.

Now depending on the modelization involved in a particular system, chemical, biological, neuronal, etc., different reactive functions f_i and g_i, are proposed. For the sake of simplicity, and because it appears in this way in the model we will study below, we have considered that the function g_u depends only on u. Our methodology can be straightforwardly adapted to the general case of g_u depending on both variables u and v, as well as to the case of several noise sources.

The set of equations (2)–(3) with appropriate boundary and initial conditions define the problem we want to study, whose exact solution is impossible within present knowledge, but for which analytical information, even approximate, is highly valuable. All doubts about the interpretation of any component of these equations are dilucidated transforming them into a set of ordinary stochastic differential equations in a d-dimensional lattice [1].

Here we will outline the methodology to understand the main effects of the random fluctuations. The argument is the following: Past experimental and numerical evidences tell us that fluctuations not only have a dispersive or disordering effects but also systematic ones. If it is so, one can try to guess where they come from. A preliminary analysis of (2) shows that the noisy term $g(u)_u\eta(\mathbf{x},t)$ has a non zero mean value when the correlation of the noise is delta correlated in time and the Stratonovich calculus is used to deal with the multiplicative noise. This is the usual interpretation when we face more realistic noises as is the aim of this paper.

In spite of the fact that the noise has zero average, explicit calculations for white noise in time gives [1, 6],

$$< g_u(u)\eta(\mathbf{x},t) >= C(0) < g_u(u)g_u'(u) >\equiv \langle \Phi_0(u)\rangle \neq 0, \tag{4}$$

The above contribution will be the source of the systematic and quite often ordering effects of the noise [1]. By adding and substracting $\Phi_0(u)$ to the equation of motion for u in (2), we arrive at,

$$\partial_t u = D\nabla^2 u + f_u(u,v,b) + \Phi_0(u) + R(u,t); \qquad \partial_t v = f_v(u,v), \tag{5}$$

and $R(u,t) = g(u)\eta(\mathbf{x},t) - \Phi_0(u)$ is a more complicated random term, but which has now a zero mean value and whose fluctuating and non systematic effects can be in principle discarded for non very large noise intensities. Then one arrives to the so called *effective deterministic model*,

$$\partial_t u = D\nabla^2 u + f_u(u,v,b) + \Phi_0(u); \qquad \partial_t v = f_v(u,v). \tag{6}$$

Thus our analytical approach relays on the evaluation of $\Phi_0(u)$. This is a simple task for a white noises in time as expressed by (4) [1], but it is much more involved for spatio-temporal structured noises [6]. In the particular model we will study in the next section, the new reactive term $\Phi(u)$ can be absorbed in the old one $f_u(u,v,b)$ through a renormalization of the model parameters in such a way that its systematic influence can be evaluated using already known results.

In the case of distributed spatio-temporal noise we prescribe a correlation,

$$\langle \eta(\mathbf{x}',t')\eta(\mathbf{x},t)\rangle = C(\frac{\mathbf{x}-\mathbf{x}'}{\lambda})\gamma(\frac{t-t'}{\tau}), \tag{7}$$

where τ and λ are the noise parameters, correlation time, and characteristic length respectively. We have also assumed that the spatial and time dependences are independent quantities with $C(\mathbf{x})$ and $\gamma(t)$ being the spatial and temporal correlations functions,

$$C(r/\lambda) = \frac{\sigma^2}{(\sqrt{2\pi})^d\,\lambda^d}e^{-\frac{r^2}{2\lambda^2}}; \qquad \gamma(s/\tau) = \frac{1}{\tau}e^{-\frac{s}{\tau}}, \tag{8}$$

where σ^2 is the noise intensity.

Following the approach described before, we need to evaluate the quantity $\Phi(u)$. We will proceed by extending previous techniques developed for 0-dimensional variables [11], recently generalized to include extended systems [1, 6]. Thus we will assume that the parameter controlling the non-Markovicity τ, is small enough to perform an expansion keeping only the first contribution.

$$\Phi(u) = \Phi_0(u) + \tau\Phi_1(u) + \ldots, \tag{9}$$

$\Phi_0(u)$ is the Markovian contribution already evaluated in Eq. (4), and $\Phi_1(u)$ is the first correction. For large values of τ we will make here a simple regularization. For some particular cases it is possible to obtain a explicit expressions for higher order terms. However, the reliability of these contributions is still to be assessed.

In order to see how is the phenomenology of a system described by SPDE's such as Eqs. (2) and (3) and also to test the accuracy of the theoretical predictions we can proceed with the numerical simulation of this type of equations. We will follow the Heun method (a second order Runge-Kutta algorithm) extended for SPDE's [1, 12]. The set of Eqs. (2) and (3) are discretised in a *d–dimensional lattice* with N squared cells, each one of linear size Δx. The fields $u(\mathbf{x},t)$ and $v(\mathbf{x},t)$ transform into N time dependent variables $u_i(t)$ and $v_i(t)$ where i is the position index. In the discrete lattice we have N couples of ordinary stochastic differential equations. The simulations have been done choosing appropriate initial conditions, either in a one or two dimensional lattices with appropriate Δx and Δt indicated in the figure captions corresponding to the numerical results.

A MODEL OF EXCITABLE CHEMICAL REACTION

Our theoretical approach can be applied to any model to describe active systems. For the sake of simplicity we will use Barkley modelization for a generic active media [13]. This model has simple polynomial reactive terms,

$$f_u(u,v,b) = \frac{1}{\varepsilon}u(1-u)(u-\frac{v+b}{a}); \qquad f_v(u,v) = u - v, \tag{10}$$

where ε controls the time scale of the activator and a is another system parameter. Including the fluctuations in b, the multiplicative noise reactive term is,

$$g_u(u) = \frac{1}{a\varepsilon} u(1-u). \tag{11}$$

Explicit calculations of the stochastic average defined in (4), in the Markovian approximation, gives for this model,

$$\Phi(u) = \frac{C(0)}{a^2\varepsilon^2} u(u-1)(2u-1). \tag{12}$$

This systematic contribution is incorporated easily into the activator equation as a new reactive term. Fortunately the whole reactive term can be reorganized and the effective reactive term is of the same form as the original deterministic one, but with the following new renormalized parameters,

$$a' = a - \frac{2C(0)}{a\varepsilon}, \qquad b' = b - \frac{C(0)}{a\varepsilon}, \qquad \varepsilon' = \frac{\varepsilon}{1 - 2C(0)/a^2\varepsilon}. \tag{13}$$

As the parameter b controls the excitability of the system one can expect important changes. In particular the decrease of b as the noise intensity increases will drive the system to more excitable regimes and larger wave velocities. These aspects will appear clearly in our simulations.

This same approach has been applied to the more complicated Oregonator model with similar success [10], although much more involved calculations.

For model (10), the explicit evaluation of the first non Markovian correction to $\Phi(u)$ is [12],

$$\Phi_1(u)\,\tau = -\frac{C(0)u(1-u)(1-2u)}{a^2\varepsilon^2}\frac{\tau}{\tau_0}, \tag{14}$$

where we have introduced a new time scale τ_0 defined as,

$$\tau_0^{-1}(\lambda) = -\frac{DC''(0)}{C(0)} - \frac{\beta}{\varepsilon} + 2D\gamma. \tag{15}$$

This is the relevant parameter of our analysis. It contains three terms each one being an inverse characteristic time. The most important for us is the first one because it keeps the dependence on the noise correlation length λ, being independent of σ^2 and τ. The other two terms are associated with the time scale of the reaction and diffusion respectively. The parameters β and γ are obtained from the model [12].

Using this result we arrive to an effective version of Barkley's model (10), with the following renormalized and regularized parameter redefinitions,

$$a' = a\left(1 - \frac{2C_R(0)}{\varepsilon a^2}\right); \qquad b' = b - \frac{C_R(0)}{\varepsilon a}, \tag{16}$$

$$\varepsilon' = \frac{\varepsilon}{\left(1 - \frac{2C_R(0)}{\varepsilon a^2}\right)}; \qquad C_R(0) = \frac{C(0)}{(1+\tau/\tau_0(\lambda))}, \tag{17}$$

where $C_R(0)$ is a new parameter which renormalizes $C(0)$ being in this way the relevant effective parameter in our analytical approach keeping the non Markovian information. It can be understood as a renormalization of the noise intensity which incorporates the influence of the three parameters of the noise (remember that $C(0) \sim \sigma^2/\lambda^d$). The white noise intensity is trivially recovered for $\tau \to 0$.

The main effects of the noise come from the quantity Φ_0, and they have been studied in Ref. [10]. There the focus was on the transitions between the different excitable regimes as one increases the intensity σ^2 of the noise. The influence of the parameters τ and λ are not so strong and they affect the transitions points quantitatively in the direction indicated by the renormalized values of the parameters.

A more refined test of the pure non Markovian effects is to look at the behaviour of the wave velocity in the excitable regime as a function of the noise parameters. We do it in two steps: first we consider only temporal correlations and next we introduce also spatial correlations.

Assuming a delta spatial correlation function, $C(\mathbf{x}-\mathbf{x}') = \sigma^2\delta(\mathbf{x}-\mathbf{x}')$, then the quantities $C(0)$ and $C''(0)$ evaluated in a lattice are,

$$C(0) = \frac{\sigma^2}{\Delta x^d}, \qquad C''(0) = -\frac{2^d\sigma^2}{\Delta x^{2+d}}. \tag{18}$$

Quantitative results are summarized in Fig.(1), which compare results from the simulation of the whole stochastic equations with those obtained simulating only the deterministic effective model system Eqs. (10) and (16)-(17).

In this case the main contribution to τ_0 on Eq.(15) arises from the first term which is one order of magnitude bigger than the others ($\tau_0 \approx 0.0047$ $t.u.$ for the system parameters of Fig.(1)). Noting that the horizontal axis of this figure is logarithmic, one can see that the increase of the propagation speed when τ approaches the new temporal scale introduced by the noise τ_0.

Let us take a further step by adding the parameter of spatial correlation length λ. The explicit expressions for $C(0)$ and $C''(0)$, obtained from Eq.(8), are,

$$C(0) = \frac{\sigma^2}{\pi\lambda^2}, \qquad C''(0) = -\frac{\sigma^2}{\pi\lambda^4}. \tag{19}$$

Keeping the noise intensity (σ^2) constant, increasing either the temporal (τ) or length (λ) correlation parameters, the velocity of wave propagation decreases, as seen in Fig.(2). This is better explained as a decrease in the effective intensity of the fluctuations (see Eqs.(17, 19)). Differently to the previous calculation, we change the new temporal scale introduced by the noise (τ_0) when we modify the correlation length of the noise. In particular the first term of Eq.(15) becomes increasingly irrelevant when λ increases and thus its effects are small (see caption of Fig.(2)). These are the so called non trivial effects of λ. Strong effects, although trivial, are those coming form the intensity of the noise, $C(0) \sim \sigma^2/\lambda^d$, which can be absorbed as a renormalized intensity in a white noise assumption (Eq.(17)).

These figures clearly show how the noise parameters τ and λ affect quantitatively the dynamics of chemical waves in excitable media under external fluctuations, and how these results are explained by our renormalized approach.

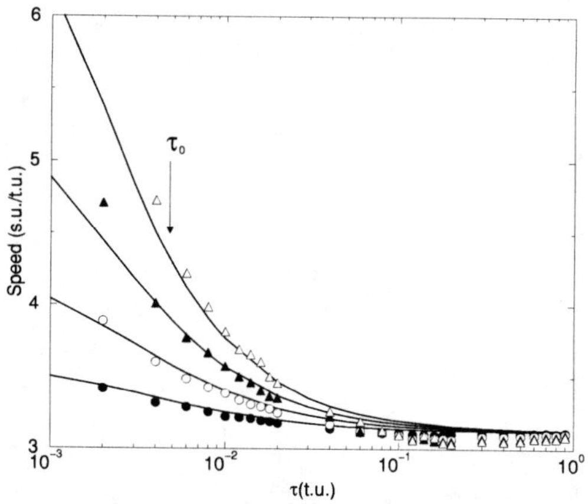

FIGURE 1. Mean velocity, averaged over 25 realizations, of noisy traveling waves in an one dimensional excitable media vs. the correlation time of the noise (τ) for different noise intensities: $\sigma^2 = 0.001$ (filled circles), $\sigma^2 = 0.002$ (circles), $\sigma^2 = 0.003$ (filled triangles) and $\sigma^2 = 0.004$ (triangles). Velocity of traveling waves for the effective deterministic model using the renormalized values of the parameters Eqs.(16-17) (solid lines). Parameters of the model and of the simulations are: $a = 0.7$, $b = 0.1$, $\varepsilon = 0.02$, $\Delta x = 0.1\ s.u.$, $\Delta t = 0.001\ t.u.$, $L = 400 s.u.$

CONCLUSIONS AND PERSPECTIVES

We have presented a methodology to deal with non Markovian Stochastic Partial Differential Equations. This is of special relevance in the studies of experimental systems where fluctuations are necessarily non white.

Moreover, there are physically realizable situations in which there is no possibility of a Markovian approximation as a starting point. An example is the study of front dynamics in smectic-C* liquid crystals in the presence of stochastic electric and magnetic fields [14]. In this case the noise enters quadratically ξ^2 and the white noise assumption cannot be made.

Our main theoretical result is the renormalization of the parameters of the deterministic model which allows us to predict and understand the noise effects by using only deterministic calculations and arguments. Numerical simulations of these equations support our theoretical results with a high degree of accuracy.

Although much more theoretical resources need to be invested in this topic, the practical recipes presented here, open the possibility to grasp what type of consequences can be expected when Markovicity cannot be assumed.

ACKNOWLEDGMENTS

We acknowledge financial support from the Dirección General de Enseñanza Superior (Spain) under projects BFM2000-0624 and BXX2000-0638.

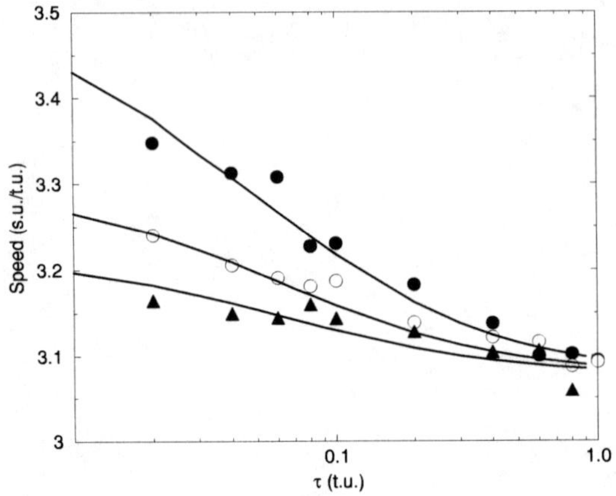

FIGURE 2. Mean velocity of noisy traveling waves in a one dimensional excitable media vs. the correlation time of the noise (τ), averaged over 25 realizations, for the same noise intensity ($\sigma^2 = 0.001$) and for different correlation lengths: $\lambda = 3\Delta x\ s.u.$ (filled circles) ($\tau_0 = 0.057\ t.u.$), $\lambda = 4\Delta x\ s.u.$ (circles) ($\tau_0 = 0.061\ t.u.$) and $\lambda = 5\Delta x\ s.u.$ (filled triangles) ($\tau_0 = 0.064\ t.u.$). Velocity of traveling waves for the effective deterministic model using the renormalized values of the parameters Eqs.(16-17) (solid lines). Parameters of the model and of the simulations are: $a = 0.7$, $b = 0.1$, $\varepsilon = 0.02$, $\Delta x = 0.2\ s.u.$, $\Delta t = 0.005\ t.u.$.

REFERENCES

1. García-Ojalvo, J., and Sancho, J. M., *Noise in Spatially Extended Systems*, Springer-Verlag, New York, 1999.
2. Zaikin, A. N., Zhabotinskii, A. M., *Nature* **225**, 535-537 (1970).
3. *Chemical waves and Patterns* edited by Kapral, R., and Showalter, K., Kluwer, Dordrecht 1995.
4. Goldbeter, A., *Biochemical Oscillations and Cellular Rhythms* Cambridge Univ. Press, Cambridge 1996.
5. Mikhailov, A. S., *Phys. Rep.* **184**, 307-374 (1989).
6. Santos, M. A., and Sancho, J.M., *Phys. Rev. E* **64**, 016129(11) (2001).
7. Kádár, S., Wang, J., Showalter, k., *Nature* **391**, 770 (1998).
8. Sendinha-Nadal, I., et al. *Phys. Rev. Lett.* **80**, 5437 (1998)
9. Sendinha-Nadal, I., et al. *Phys. Rev. Lett.* **84**, 2734 (2000)
10. Alonso, S., et al. *Phys. Rev. Lett.* **87**, 078302 (2001).
11. Sancho, J. M., San Miguel, M., Katz, S., and Gunton, J.D., *Phys. Rev. A* **26**, 1589 (1992).
12. Alonso, S., Sagués, F., Sancho, J. M., *Phys. Rev. E* **65**, 066107 (2002).
13. Barkley, D., Kenss, M., Tuckerman, L. S., *Phys. Rev. A* **42** 2489 (1990).
14. Santos, M. A., Sancho, J. M., and Lamberti, P. W., *European Physical Journal B* **24**, 135 (2001).

Stochastic Resonance with Delay and Traffic

Toru Ohira

Sony Computer Science Laboratories, Shinagawa, Tokyo 141-0022, Japan

Abstract. The phenomena of stochastic resonance has gained more research attentions recently. In this phenomena, noise, which is normally considered an obstacle to information processing, is treated to have an constructive effect causing a resonance with external signals. A particular model which causes "resonance" with noise and delay is focused. We see how regular dynamical patterns appear in this context. We then discuss possible directions where such stochastic resonance could be useful. We proceed by considering simple concrete models for computer network traffic and pedestrian traffic.

INTRODUCTION

Many natural and artificial systems are associated with noise or fluctuation. Issues relating to noise have consequently been a major topic in a variety of fields (see, e.g., [1, 2]). From the point of view of information processing, noise has been considered an obstacle; however, it was recently found or suggested that noise can actually be an integral part of biological information processing, leading to an active research area named as "Stochastic Resonance" [3, 4, 5, 6, 7]. In this paper, we present a rather non-standard examples of stochastic resonance. First we present a models to show "Delayed Stochastic Resonance". This is a "resonant" phenomena using noise and delay. We then try to connect stochastic resonance to traffic problems. Preliminary investigations in that direction are presented for a computer network traffic model and a pedestrian traffic model.

DELAYED STOCHASTIC RESONANCE

In this section, we present particular models showing resonance with noise and delay, delayed stochastic resonance, are discussed.

Delay is a factor which is also important in traffic problems as well as in other fields[8, 9, 10, 11, 12, 13]. Here we give more detailed account for a special case of stochastic resonance. In these models, we do not have an incoming signals. Instead, delay in feedback or interaction is used. We term this phenomena as "delayed stochastic resonance".

Let us first discuss a very simple single unit model[14], which is defined by three parameters, p, q, and τ (Figure 1(A)). $X(t)$ is the state at time steps t and the formal definition is given by the conditional probabilities as follows:

CP665, *Unsolved Problems of Noise and Fluctuations: UPoN 2002: Third International Conference,* edited by S. M. Bezrukov

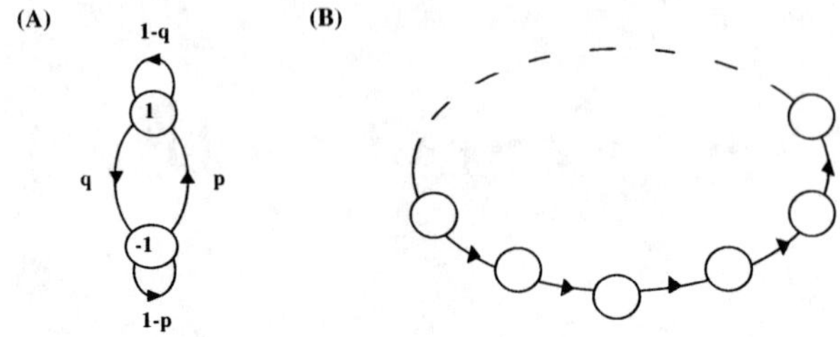

FIGURE 1. (A) Schematic view of binary model. (B) A ring network

$$
\begin{aligned}
P(X(t+1) = +1|X(t-\tau) = -1) &= p \\
P(X(t+1) = -1|X(t-\tau) = -1) &= 1-p \\
P(X(t+1) = +1|X(t-\tau) = +1) &= 1-q \\
P(X(t+1) = -1|X(t-\tau) = +1) &= q
\end{aligned}
$$

Hence, the transition probability of the model depends on its state at τ steps in the past and is a special case of delayed random walks [15, 16].

We randomly generate $X(t)$ for the interval $t = (-\tau, 0)$. Simulations are performed in which parameters are varied and $X(t)$ is recorded up to 10^6 steps. From the trajectory $X(t)$, we construct a residence time histogram $h(u)$ for the system to be in the state -1 for u consecutive steps. Some examples of histograms and $X(t)$ are shown in Figure 2 ($q = 1-q = 0.5$, $\tau = 10$). We note that with $p << 0.5$, as in Figure 2(A), the model has a tendency to switch or spike to the $X = 1$ state after the time step interval of τ. But the spike trains do not last long and result in a small peak in the histogram. For the case of Figure 2(C) where p is closer to 0.5, we observe less regular transitions and the peak height is again small. With appropriate p as in Figure 2(B), spikes tend to appear at interval τ more frequently, resulting in higher peaks in the histogram. This change of peak height in histograms which reaches maximum (Figure 2(D)) at an appropriate noise level is one way to characterize stochastic resonance. It turns out that this histogram $h(u)$ can be obtained exactly in the steady state.

$$
\begin{aligned}
h(u) &= (\frac{q}{p+q})^u(\frac{p}{p+q})^2 \quad (1 \leq u < \tau) & (1) \\
&= (\frac{q}{p+q})^\tau(\frac{p}{p+q})(1-q) \quad (u = \tau) & (2) \\
&= (\frac{q}{p+q})^u(p)^2 \quad (u > \tau) & (3)
\end{aligned}
$$

Also, by differentiating equation (2) with respect to p, we can derive the resonant condition for the peak to reach maximum height as

$$q = p\tau \tag{4}$$

Choosing an appropriate p is equivalent to "tuning" noise with other parameters appropriately fixed. In this sense, our model shares a feature of the stochastic resonance.

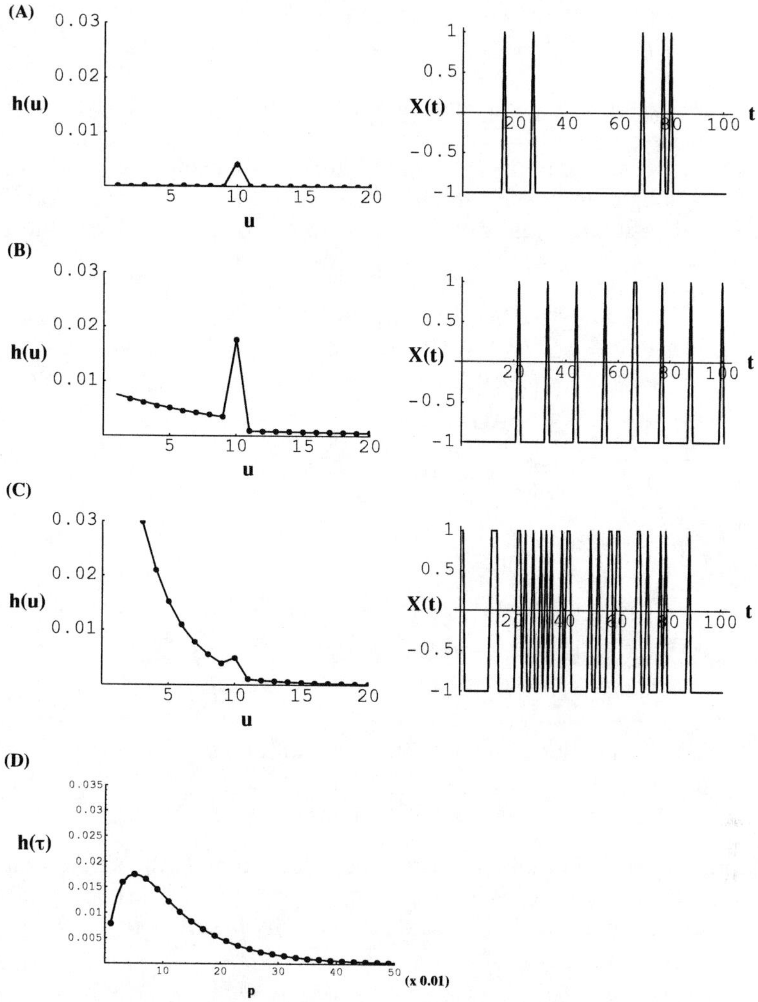

FIGURE 2. Residence time histogram and dynamics of $X(t)$ as we change p. The values of p are (A) $p = 0.005$, (B) $p = 0.05$, (C) $p = 0.2$. The solid line in the histogram is from the analytical expression given in equations (1-3). Also, in (D) we show a plot of peak height by varying p. The solid line is from equation (2). The parameters are $\tau = 10$, $q = 0.5$.

The next model we consider is a 1-D ring network of M binary units (Figure 1(b)). The state $X_i(t)$ of the ith unit at time step t is either -1 or 1. The interaction is assumed to be only one way, and the conditional probability of transition for the ith unit depends on the state of $i-1$th unit at τ steps earlier

$$\begin{aligned} P(X_i(t+1) = +1 | X_{i-1}(t-\tau) = -1) &= p \\ P(X_i(t+1) = -1 | X_{i-1}(t-\tau) = -1) &= 1-p \end{aligned}$$

$$
\begin{aligned}
P(X_i(t+1) = +1 | X_{i-1}(t-\tau) = +1) &= 1-q \\
P(X_i(t+1) = -1 | X_{i-1}(t-\tau) = +1) &= q.
\end{aligned} \tag{5}
$$

where p and q are probability parameters and the Mth unit is connected to the 1st unit (a ring) (Figure 1(B)).

Typical behavior of a unit in the network is similar to the single unit model discussed above. As in the single unit case, we can treat this model analytically. The "resonant interspike interval" histogram show peaks occurs at $s = M(\tau+1) - 1$, and the histogram is given as follows:

$$
\begin{aligned}
h(u) &= \alpha^2\beta^u \qquad (0 \leq u < s) \\
&= Q_{++}\beta^s \qquad (u = s) \\
&= Q_{+-}^2 Q_{--}^{(u-s-1)} \beta^{(2s-u)} \qquad (s+1 \leq u \leq 2s) \\
&= Q_{+-}^2 Q_{--}^{(u-s-1)} \qquad (2s+1 \leq u)
\end{aligned} \tag{6}
$$

where $\alpha = \frac{p}{p+q}$, $\beta = \frac{q}{p+q}$ and

$$
\begin{aligned}
Q_{++} &= \alpha^2 + (1-p-q)^M(\alpha\beta) \\
Q_{+-} &= \alpha\beta - (1-p-q)^M(\alpha\beta) \\
Q_{--} &= \beta^2 + (1-p-q)^M(\alpha\beta)
\end{aligned} \tag{7}
$$

From this we can again derive the resonance condition as

$$
q = M(\tau+1)p \tag{8}
$$

The main notable point with this network model as well as in the single unit model is that we do not need to have explicit oscillator to produce a rhythmic dynamical behavior. The delay in the interaction with the tuned stochasticity is sufficient.

TRAFFIC MODELS

We now briefly discuss our preliminary results of stochastic resonance in the context of traffic models.

Computer Traffic Network Model

The phenomenology of the nature of computer network traffic has commanded much attention recently. Analysis, simulations and experiments based on such concepts as "phase transitions" and "self–similarity" are active research topics (see e.g. [17, 18, 19]). This paper also addresses the phase transition nature of computer network traffic [20]. Our main focus, however, is not so much on understanding the nature of traffic itself,

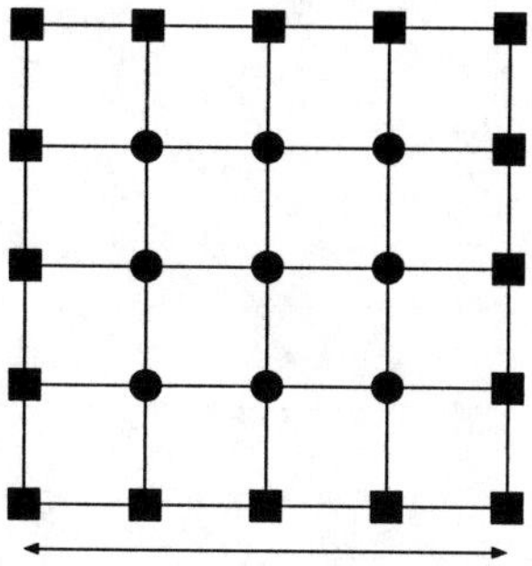

N nodes

FIGURE 3. Network architecture

rather we propose an autonomously adaptive probabilistic routing strategy to "ease" the phase transition point of the network traffic into a congestion state.

The network architecture considered in the model consists of nodes placed as a two-dimensional lattice (Figure 3). It is a square with N nodes (routers) on each side and N^2 nodes as a whole. Packets are generated and destroyed on nodes on the boundary of the lattice (squares in the figure), but not on inner nodes (circles). Inner nodes only forward packets received from neighbor nodes. A node on the boundary generates a packet according to the Poisson arrival with λ and sends it to a destination node selected randomly among the nodes on the boundary (including itself). Each node has a receiving queue of unlimited length through which packets are forwarded to the destination and then destroyed.

During each unit time, each node goes through the following process in order to forward packets. It picks a packet from the head of its queue, decides to which node among its neighbors it will forward the packet, and then puts it at the end of the queue of the selected node. The next node is selected so that the packet is delivered to the destination along the shortest path. If more than one candidate of the next node exist, a strategy is needed to select the recipient. In our simulation we consider two strategies. One of them makes the decision deterministically, which we call "deterministic routing", and the other does so probabilistically, which we call "probabilistic routing". With deterministic routing, the node among the candidates to which the least number of packets has been forwarded so far is selected as the next node.

The probabilistic routing strategy we propose and compare with this deterministic routing is given by introducing a particular form of "randomness". When we have multiple routes A and B based on the destination address, we assign the probability to choose a route A or B by the following equation:

$$P(A) = \frac{e^{-\beta X_A}}{e^{-\beta X_A} + e^{-\beta X_B}} \tag{9}$$

$$P(B) = \frac{e^{-\beta X_B}}{e^{-\beta X_A} + e^{-\beta X_B}} \tag{10}$$

$$1 = P(A) + P(B) \tag{11}$$

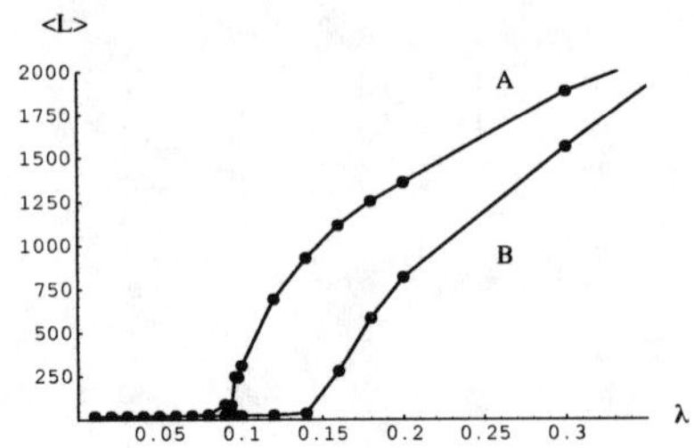

FIGURE 4. Comparison of the phase transition behavior of the deterministic (A) and the probabilistic ($\beta = 0.008$)(B) routing for $\langle L \rangle$. The system size is $N = 20$.

where β is a parameter. X_A and X_B are the number of packets the router has already sent in the direction A and B.

The characteristics of these routing strategies are "autonomous" and "adaptive". They are autonomous as their routing choice is made without any reference to or interaction with other routers. And they are adaptive as a routers routing choice changes based on its past choices and hence time.

We now note a couple of points about this probability function. First, we note that the main difference with the deterministic routing is that even if $X_A > X_B$, there is some probability of choosing and sending a packet to A. Second, we can recover the deterministic model by letting $\beta \to \infty$. Also, if we set $\beta = 0$, we have a completely random choice of A and B, i.e., A and B will be chosen with equal probability of 0.5 regardless of X_A and X_B. Hence, β is a control parameter which determines the degree of randomness of choice. This is analogous to the concept of "temperature" in statistical physics [21]. (To be more precise, β corresponds to the inverse temperature.)

We quantify network traffic congestion by the average "lifetime" of a packet $\langle L \rangle$, which is the average time between the sending and receiving of a packet. (Averages are taken over packets.)

In Figure 4, we show a typical behavior of $\langle L \rangle$ as we change the creation rate λ of the packet using the deterministic routing. The simulation is performed with various system sizes of N, and the system is run up to time step 10,000. The phase transition behavior is clearly observable as λ increases beyond a "critical rate", λ_c. This transition into the congestion phase is sharper with increasing size as in other physical systems showing phase transitions.

We now compare the deterministic routing with the proposed probabilistic routing. We can see from Figure 4 that the onset of phase transition is moved to higher λ showing that the model network with probabilistic routing can tolerate more packets before going into a congestion phase.

To examine the effect of randomness for this shift of phase transition point, we plot in Figure 5 the phase transition points as a function of β. We see that we need to choose β appropriately (≈ 0.01 in Figure 5) to have a desired phase point shift. In this sense, this is a type of stochastic resonance.

For individual routers, the deterministic routing appears to be the most balanced way of sending packets to the next router. Use of the probabilistic routing strategy means that

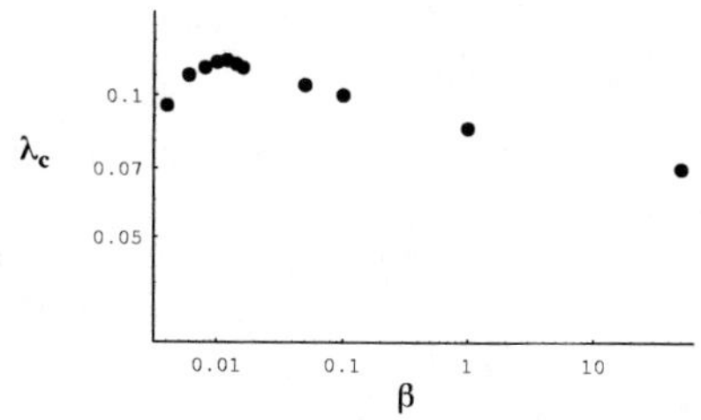

FIGURE 5. Change of "critical rate", λ_c as β is varied for the probabilistic routing. The system size is $N = 20$.

this balance is sometimes intentionally upset. The fact that easing of the phase transition point takes place nonetheless means that an emergent collective behavior of routers is playing a crucial role in deciding the congestion nature of the network.

Pedestrian Model

Among the pedestrian models, which have gained much attention recently, a counter flow model[22] is studied here. Our result is very preliminary, but shows a indication of stochastic resonance of flow in the reverse sense. We mean by "reverse sense" that with tuned "noise" the pedestrian flow shows the worst case of total grid lock.

Pedestrians are randomly placed on a two dimensional rectangular lattice of size $(W \times L) = (150 \times 400)$. The boundary condition is periodic on all sides (torus). We consider two sets of $N = 9000$ pedestrians. One set tries to move to the right and the other to the left on the lattice. We set (S_r, S_l) as parameters for them to step sideways. At each time step, a pedestrian is chosen to move one step to one of its four neighboring site by the following rules.

- If no one is in front of you in the direction you are heading, he moves one step forward.
- Otherwise, he looks at both his right and left sides.
- If only right (left) side is open and $S_r > 0$ ($S_l > 0$), he moves to the open site.
- If both sides are open, he moves to the right and left site with the probabilities $P_r = \frac{S_r}{S_r + S_l}$ and $P_l == \frac{S_l}{S_r + S_l}$
- If neither site is open, he does not move.

With this setting, we performed computer simulations. We obtained the following preliminary results.

- $(S_r, S_l) = (1.0, 0.0)$. This is the situation where only forward or to the right moves are allowed. We typically observe low flow, but not a grid lock, of pedestrians.
- $(S_r, S_l) = (1.0, 0.3)$. The left moves are now allowed. However, we typically observe a grid lock situation with no flow.

- $(S_r, S_l) = (1.0, 0.8)$. The more left moves are taken. Now, we do not see any congestion and they are in free flow.

The above result is an indication that we have a stochastic resonance in the reverse sense: with tuned probability of side stepping, the worst case of no flow situation appeared. More investigations are needed for understanding of this phenomena.

DISCUSSION

We have consider here a possible connection of stochastic resonance and traffic models. This type of effort is just at the beginning state. We expect, however, that such approach can be useful to investigate congestion problems in various traffic systems.

REFERENCES

1. R. L. Stratonovich: *Topics in the Theory of Random Noise* (Gordon and Breach, New York, 1963)
2. N. G. van Kampen: *Stochastic Processes in Physics and Chemistry* (North-Holland, Amsterdam 1992)
3. A. Longtin, A. Bulsara, F. Moss: Phys. Rev. Lett. **67**, 656, (1991)
4. K. Wiesenfeld, F. Moss: Nature, **373**, 33 (1995)
5. J. J. Collins, C. C. Chow, T. T. Imhoff: Nature, **376**, 236 (1995)
6. A. R. Bulsara, L. Gammaitoni: Physics Today, **49**, 3, 39 (1996)
7. L. Gammaitoni, P. Hänggi, P. Jung, F. Marchesoni: Rev. Mod. Phys. **70**, 223 (1998)
8. M. Bando, K. Hasebe, A. Nakayama, A. Shibata, Y. Sugiyama: Phys. Rev. E. **51**, 1035 (1995)
9. K. Nakanishi, K. Itoh, Y. Igarashi, M. Bando: Phys. Rev. E. **55**, 6519 (1997)
10. M. Konishi: Neural Computation **3**, 1 (1991)
11. M. C. Mackey, L. Glass: Science **197**, 287 (1977)
12. K.L. Cooke, Z.Grossman: J. Math. Analysis and Applications **86**, 592 (1982)
13. A. Longtin, J. Milton: Biol. Cybern. **61**, 51 (1989)
14. T. Ohira, Y. Sato: Phys. Rev. Lett. **82** 2811 (1999)
15. T. Ohira, J. G. Milton: Phys. Rev. E. **52**, 3277 (1995)
16. T. Ohira, T. Yamane: Phys. Rev. E. **61**, 1247 (2000)
17. W. E. Leland, M.S. Taqqu, W. Willinger, D. V. Wilson: Communication Review, **23**, 183 (1993)
18. M. Takayasu, H. Takayasu, T. Sato: Physica A, **233**, 824 (1996)
19. M. S. Taqqu, W. Willinger, R. Sherman: ACM/SIGCOMM Computer Communication Review, **27**, 5 (1997)
20. T. Ohira and R. Sawatari: Phys. Rev. E. **58** 193 (1998)
21. R. K. Pathria: *Statistical Mechanics* (Pergamon Press, Oxford 1972)
22. M. Muramatsu, T. Irie, T. Nagatani: Physica A, **267**, 487 (1999)

Pulse Propagation in Bistable Neural Models via Noise-Induced Resetting

J. García-Ojalvo*†, R. Báscones*, F. Sagués**, J.M. Sancho‡ and L. Schimansky–Geier§

*Departament de Física i Enginyeria Nuclear, Universitat Politècnica de Catalunya, Colom 11, E-08222 Terrassa, Spain
†E-mail: jordi.g.ojalvo@upc.es
**Departament de Química Física, Universitat de Barcelona, Diagonal 647, E–08028 Barcelona, Spain
‡Departament d'Estructura i Constituents de la Matèria, Universitat de Barcelona, Diagonal 647, E–08028 Barcelona, Spain
§Institut für Physik, Humboldt–Universität zu Berlin, Invalidenstr. 110, D–10115 Berlin, Germany

Abstract. It is usually assumed that the propagation of action potentials through neural media requires local excitability. Here we review recent investigations showing that action potentials can also propagate when the medium is locally bistable, provided a fast-slow activator-inhibitor dynamics exists, and a suitable source of noise is acting upon the system. We show that this phenomenon is generic to different types of neural models and noise sources, and describe experimental observations in electronic neurons that confirm these predictions. The possible relevance of these results to signal propagation in real neural systems is still an open question.

INTRODUCTION

Active propagation of electrical signals in neurons is usually associated with the excitable properties of these cells. Excitable neurons are characterised by a membrane potential with a single equilibrium state, which they abandon when subject to a sufficiently strong applied current. When this happens, the membrane potential goes through a large excursion before eventually returning to its steady state. This pulsing response, called an action potential, is the most important information-bearing signal that living systems use to control their functions [1].

Even though most neurons are excitable, recent investigations have shown that mammalian thalamocortical neurons exhibit bistable dynamics under certain physiological conditions, a conclusion that can be expected to apply to other excitable cells with similar biophysical properties [2]. Usually, bistable media allow only for the propagation of fronts, in the form of switching waves between the two stable steady states of the system [3]. Bistable neurons, in the absence of a resetting mechanism that takes the system back to the initial steady state after excitation, would in principle be unable to generate a pulsed action potential. However, a suitable source of noise may provide the necessary resetting mechanism and allow propagation of action potentials even in bistable situations. In this communication we review recent work in this direction, describing several types of neural models and noise sources that exhibit noise-induced resetting and ex-

CP665, *Unsolved Problems of Noise and Fluctuations: UPoN 2002: Third International Conference,* edited by S. M. Bezrukov

citability, together with experimental verification in arrays of electronic circuits whose dynamics mimics that of neurons. Two different neural models are used, both of which provide a reduced two-variable, fast-slow description of neural dynamics (i.e. they are FitzHugh-Nagumo-type systems): the Barkley model and the cubic-nullcline model. In order to show that the reported phenomena are not specific of the type of noise chosen, we demonstrate that they arise when the noise is added to either the fast or the slow variable of the model.

BARKLEY MODEL WITH NOISE IN THE ACTIVATOR VARIABLE

A computationally convenient model of neural dynamics is given by the one-dimensional Barkley model, which reads [4]:

$$\frac{du_i}{dt} = \frac{1}{\varepsilon} u_i (1 - u_i) \left(u_i - \frac{v_i + b}{a} \right) + \sum_{j \in N(i)} c_{ij} u_j \tag{1}$$

$$\frac{dv_i}{dt} = \gamma u_i - v_i \,, \tag{2}$$

where $u_i(t)$ is the activating variable representing the membrane potential, and $v_i(t)$ acts as an inhibitor related to the time-dependent conductance of the potassium channels [5]. The time scale ratio verifies $\varepsilon \ll 1$, so that the activator evolves much faster than the inhibitor. The sum term in Eq. (1) represents diffusion of the activator among the neighboring elements of cell i, $N(i)$, which are separated a distance Δx. The weight coefficients c_{ij} correspond to the second-order discretization of the Laplacian operator [6]. The constant parameters a, b and γ determine the type of local dynamics exhibited by the system, which can vary among excitable, bistable, and oscillatory. The excitability parameter b controls the transition between the first two regimes, which are represented in Fig. 1 in the phase plane (u, v) and in the absence of diffusion. In that figure, solid lines represent the activator nullclines, defined by $f(u,v) = 0$, and dashed lines correspond to the inhibitor nullcline $g(u,v) = 0$.

We now consider that the system operates in a deterministically bistable regime, namely that represented in Fig. 1(a). In that situation, the response of the medium to an initial perturbation of the u-variable at one boundary of the one-dimensional medium is shown in Fig. 2(a). The perturbation taking the system to the excited state 1 (plotted in black in the figure) propagates towards the right, until the whole medium ends up in that state. Hence, an action potential cannot be transmitted in this bistable regime by an initial condition like the one described here. We will now show that this property can be recovered in the presence of noise. Let us consider that the parameter b fluctuates around a mean value, $b \to b + \eta_i(t)$, with $\eta_i(t)$ being a zero-mean Gaussian white noise with correlation:

$$\langle \eta_i(t) \eta_j(t') \rangle = 2 \frac{\sigma^2}{\Delta x} \delta_{ij} \delta(t - t') \,. \tag{3}$$

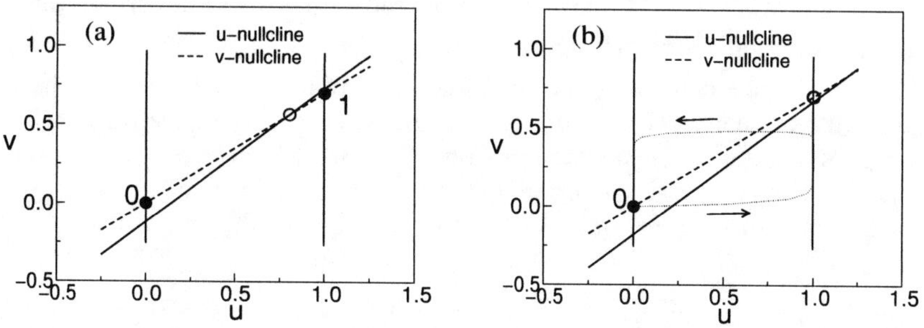

FIGURE 1. Nullclines of the Barkley model in the (a) bistable [$b = 0.12$] and (b) excitable [$b = 0.18$] regimes. Other parameters are $a = 0.85$ and $\gamma = 0.7$. Solid (empty) circles denote stable (unstable) fixed points.

The effect of a small amount of these fluctuations on the spatiotemporal response of the system to a boundary perturbation is shown in Fig. 2(b). The figure shows that noise induces, in an organized way, a decay of the excited state that resets the system after the first wavefront passes through. The result is that the medium behaves as if it was excitable, allowing the propagation of action potentials even in bistable conditions [7] (we note that activator-inhibitor bistable media allow pulse propagation for carefully chosen initial conditions [8], but these are not realistic in neural systems).

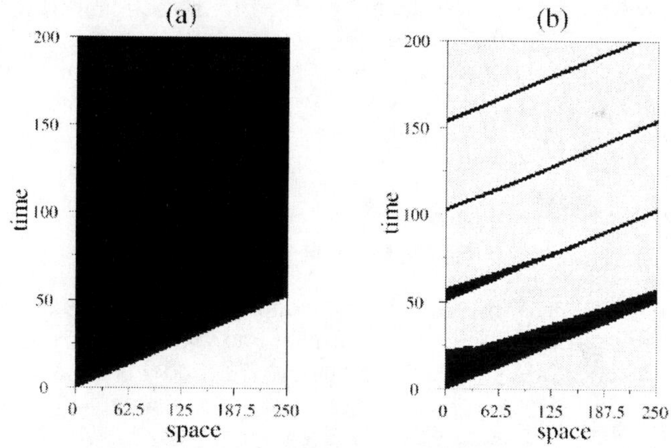

FIGURE 2. Spatiotemporal evolution of a one-dimensional bistable medium governed by model (1)-(2) without (a) and with (b) multiplicative noise. The profile of the activator u is plotted in greyscale along the 1-d system (horizontally) at increasing times (vertically). Black coding corresponds to $u = 1$, and white to $u = 0$. Parameters used are those of Fig. 1, plus $\varepsilon = 0.01$ and $\Delta x = 0.25$. The lattice size is 1000 cells. In plot (b) $\sigma^2 = 10^{-4}$, and periodic boundary conditions are introduced at $t = 30$ to trap the pulse in a ring.

The role of noise in this system can be understood taking into account that, in the Stratonovich interpretation which is being used here, the stochastic term has a nonzero average that makes a systematic contribution to the system dynamics [9]. Consequently, one can write down an effective model where this systematic contribution is made explicit by a deterministic drift [7]. The resulting model has the same form as the original one, but with renormalized parameters $a' = a - \frac{2\sigma^2}{\varepsilon a \Delta x}$, $b' = b - \frac{\sigma^2}{\varepsilon a \Delta x}$, and $\varepsilon' = \frac{\varepsilon}{1-2\sigma^2/\varepsilon a^2 \Delta x}$. According to this, one can easily plot the phase diagram of the model, which is shown in Fig. 3(a). It can be clearly seen that increasing the intensity of the noise reduces the bistable region and enhances the excitable one. The two different excitable zones of the phase diagram correspond to each one of the two states 0 and 1 of Fig. 1(a) being the rest state. Finally, for large enough values of the noise intensity an oscillatory regime is reached [7].

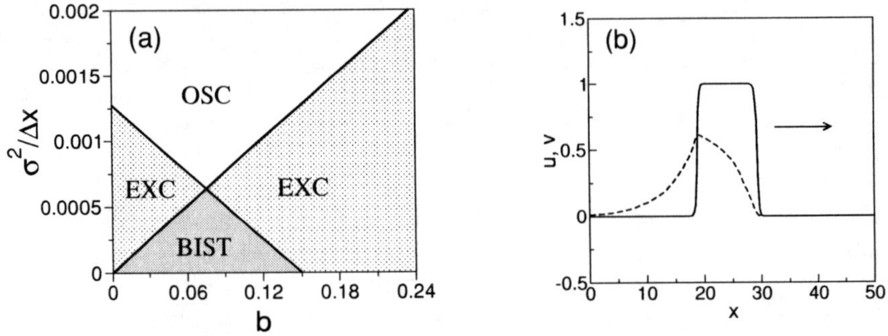

FIGURE 3. Left: phase diagram of the Barkley model in the σ^2-b plane. Parameters are those of Fig. 2. Right: spatial profile of the u and v variables (solid and dashed lines, respectively) for a steady-state pulse in the situation of Fig. 2(b).

Figure 3(b) shows the steady-state spatial profile of a noise-induced pulse generated under the conditions of Fig. 2(b). The influence of noise is not evident in the profiles of neither the activator nor the inhibitor, since according to Eq. (1) fluctuations in b are zero when u is either 1 or 0 (i.e. in the two steady states of the system).

CUBIC NULLCLINE MODEL WITH NOISE IN THE ACTIVATOR VARIABLE

The phenomena described in the previous section are not particular of the Barkley model. We now present a second FitzHugh-Nagumo type model, namely the cubic nullcline model, which exhibits the same behavior. The model is defined by:

$$\frac{\partial u_i}{\partial t} = \frac{1}{\varepsilon}\left[u_i(1-u_i)(u_i-b) - v\right] + \sum_{j\in N(i)} c_{ij} u_j \tag{4}$$

$$\frac{\partial v}{\partial t} = cu - v, \tag{5}$$

where now the nullcline of the activator is not piecewise linear, but cubic. As in the Barkley model, the parameter b controls the excitability of the system, with the medium becoming bistable for $b < 1 - 2\sqrt{c}$. Also similarly to what was described in the previous section, when noise is added to the parameter b, i.e. $b \to b + \eta_i(t)$, it provides a resetting mechanism that allows the propagation of pulses through the medium even in bistable conditions. The situation is illustrated in Fig. 4, which shows the propagation of an action potential through the bistable medium after a boundary perturbation, and its spatial profile for both the activator and the inhibitor variables. In this case fluctuations are manifest in the activator profile, which is the variable directly affected by the noise, given that now the fluctuations are not zero neither at the excited nor at the refractory (i.e. at the back tail) states.

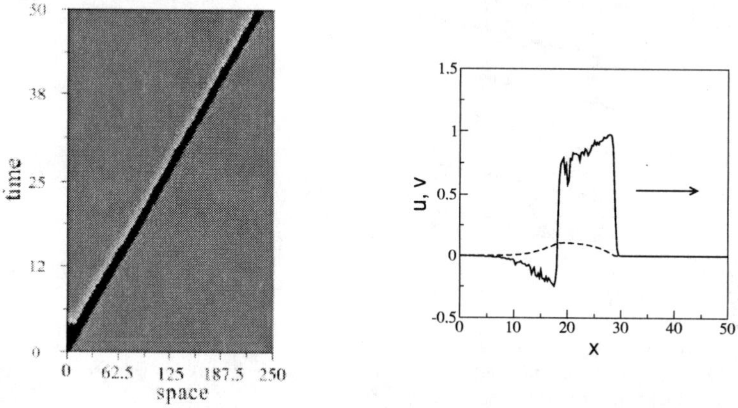

FIGURE 4. Left: spatiotemporal evolution of the cubic nullcline model in a deterministically bistable regime for $\sigma^2 = 5 \times 10^{-5}$, $b = c = 0.15$, $\varepsilon = 0.01$ and $\Delta x = 0.25$. Right: spatial profile of the u and v variables (solid and dashed lines, respectively) for the corresponding steady-state pulse.

BARKLEY MODEL WITH NOISE IN THE INHIBITOR VARIABLE

We have seen that the noise-induced resetting mechanism is not related to the particular features of the Barkley model, since a similar behavior can be found in another type of FitzHugh-Nagumo model. We will now show that this phenomenon is also generic with respect to the type of noise considered. In the previous two sections, noise was assumed to act in a similar way (i.e. on the parameter controlling the vertical position of the positive-slope part of the u-nullcline, and hence the excitability of the system) on the equation of the fast activator variable. We now consider again the Barkley model (1)-(2), but now assuming that it is the parameter γ the one that fluctuates, $\gamma \to \gamma + \eta_i(t)$, with the spatiotemporal zero-mean Gaussian white noise $\eta_i(t)$ defined again by (3). In this case the noise produces random fluctuations of the slope of the v-nullcline [see for instance Fig. 1(a)], and hence one can expect again that a bistable regime will be turned to excitable due to noise (the reverse is also true, producing for instance spiral breakup in an excitable medium due to locally and instantaneously induced bistability [10]). Figure 5(a) shows an example of the propagation of an action potential through a bistable medium in the presence of a small amount of noise in γ [11]. The corresponding spatial profile of the propagating pulse is shown in Fig. 6. It can be seen that fluctuations affect the profile of the inhibitor, as expected. When noise intensity becomes large enough the dynamics of pulse propagation becomes more complex, and backfiring events start to appear, as shown in Fig. 5(b).

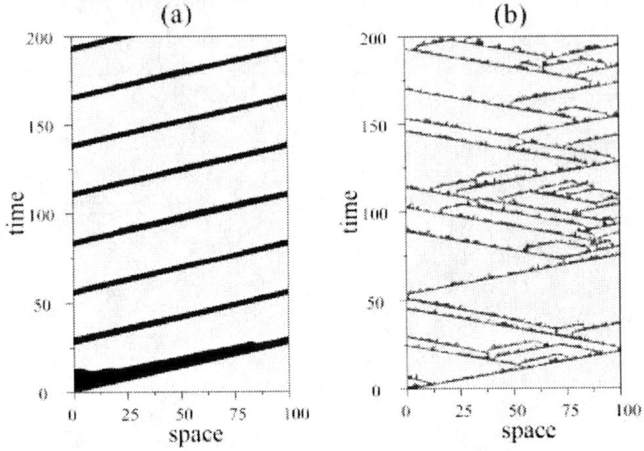

FIGURE 5. Spatiotemporal evolution of the Barkley model in a bistable regime with noise in the inhibitor equation. The parameters used are $a = 0.85$, $b = 0.1$, $\gamma = 0.7$, $\varepsilon = 0.01$ and $\Delta x = 0.25$. The lattice has 400 cells. Left: $\sigma^2 = 0.0015$; right: $\sigma^2 = 0.02$.

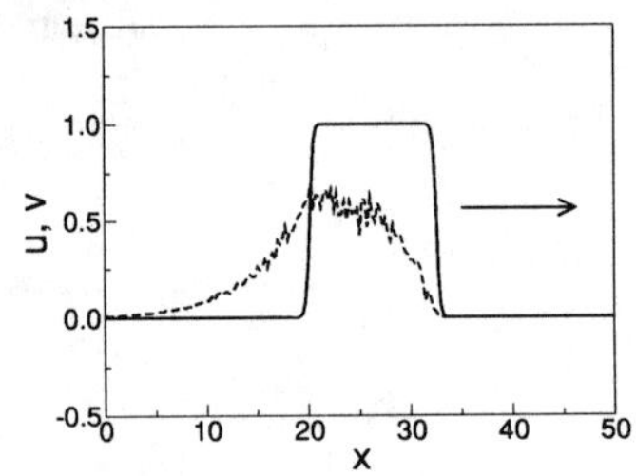

FIGURE 6. Spatial profile of the u and v variables (solid and dashed lines, respectively) for a steady-state pulse in a deterministically bistable regime. Parameters are those of Fig. 5, with $\sigma^2 = 2 \times 10^{-4}$.

LINEAR ARRAY OF COUPLED ELECTRONIC NEURONS

In order to verify the generic predictions made above, we have performed experiments on an array of nonlinear electronic circuits with dynamical behavior similar to the models described in the previous section. In particular, the circuits we consider display a transition between an excitable and a bistable regime controlled by a feeding voltage. A schematic description of the circuits used is given in the left plot of Fig. 7. The boxed region in the figure corresponds to the nonlinear resistance giving rise to excitability/bistability. The parameter controlling the transition between the two regimes is the feeding voltage V_+ of the operational amplifier. For large values of this parameter the circuit operates in a bistable mode, as shown in the nullcline plot of Fig. 8(a), where the reduced phase plane V_1-V_2 is used [12]. On the other hand, when that feeding voltage is low enough the system behaves in an excitable way [Fig. 8(b)].

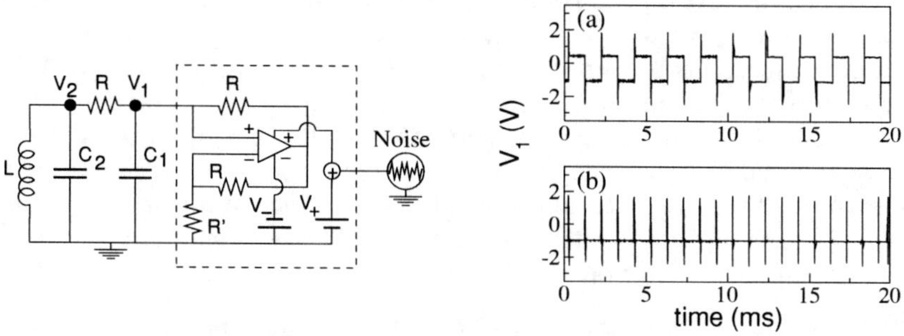

FIGURE 7. Left: Nonlinear electronic circuit with bistable/excitable regime. The values of the elements are $R = 270\,\Omega$, $L = 10\,mH$, $C_1 = 1\,nF$, $C_2 = 10\,nF$, $R' = 220\,\Omega$, $V_+ = 2\,V$ and $V_- = 5\,V$. Right: Response of the last circuit of the chain to a periodic train of pulses applied to the first circuit, when the array is placed in the deterministically bistable regime. (a) $V_{noise} = 0.30\,V$; (b) $V_{noise} = 0.65\,V$. The noise intensity is computed as the root mean square of the random time series provided by the noise generator.

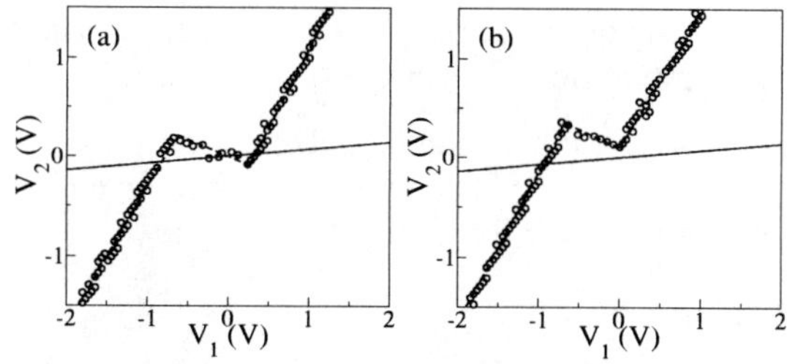

FIGURE 8. Nulclines of the isolated circuit in the reduced plane V_1-V_2 for two different values of the positive supply voltage V_+: (a) $V_+ = 2\,V$, bistable dynamics; (b) $V_+ = 0\,V$, excitable dynamics. The solid line corresponds to the V_2-nulcline, and the circles to the V_1-nulcline.

We built a linear array of twenty circuits of the type described above in the bistable regime, unidirectionally coupled by means of an operational amplifier in series with a resistor [12]. A common source of electronic noise (generated by amplifying the shot noise of a pn junction [13] and acting as indicated in the left plot of Fig. 7), is applied to all circuits of the chain. When the noise is small (or zero) and a signal composed of a train of spikes is introduced into the first circuit of the array, it induces a periodic alternation of switches between the two stable steady states of the system, as shown in Fig. 7(a). This figure plots the response of the last circuit of the chain, and indicates that in these conditions a train of fronts has propagated through the medium. If, however, a large enough amount of noise is added to the system, the transmitted signal becomes a train of pulses, as shown in Fig. 7(b). This observation confirms the predictions made in the previous sections, in the sense that noise induces resetting and allows the propagation of excitable pulses in the bistable regime, instead of fronts. We have also observed [12] that noise reduces the region of propagation failure (i.e. propagation occurs for coupling values lower than without noise), in agreement with previous studies of noise-enhanced propagation on diode oscillator arrays [13].

CONCLUSIONS AND OPEN QUESTIONS

In conclusion, the investigations reviewed in this communication show that parametric noise may induce propagation of excitable pulses through activator-inhibitor media in a bistable regime. This property is observed in different types of FitzHugh-Nagumo models and different types of noise, and has been confirmed by experimental observations in an array of nonlinear electronic circuits. The ubiquity of noise in neural tissue and the recent observation of bistability in certain types of neurons [2, 14] highlights the importance of the topic studied here, although the relevance of the present results to the propagation of action potentials through real neural tissue is still an open problem. In particular, the following questions need to be answered:

1. First, it should be assessed whether a clear distinction between the bistable and excitable states of neurons on an experimental basis can be made.
2. Experimental measurements should be made in order to establish whether action potentials propagate in bistable neural tissue. If this is the case, the present results indicate that noise could play a relevant role in this process.
3. The noise sources considered in this paper are all parametric, i.e. originated by fluctuations in one of the model parameters. It needs to be established whether this is a realistic assumption in neural tissue.

The need to answer these questions is undeniable, since noise cannot be neglected in neural systems. Whether fluctuations have mainly a destructive or a constructive role in neural dynamics, is an important open problem.

ACKNOWLEDGMENTS

This work was supported by the MCyT (Spain), under projects BFM2000-0624, BXX2000-0638, BFM2001-2159 and BFM2002-04369, and by the Sfb 555 (Germany).

REFERENCES

1. J. Keener and J. Snyder, *Mathematical Physiology* (Springer, New York, 1998).
2. S.W. Hughes, D.W. Cope, T.I. Tóth, S.R. Williams, and V. Crunelli, *J. Physiol.* **517**, 805 (1999).
3. A. Mikhailov, *Foundations of Synergetics I*, 2nd. Ed. (Springer, Berlin, 1994).
4. D. Barkley, M. Kness, and L.S. Tuckerman, *Phys. Rev. A*, **42**, 2489 (1990).
5. J.D. Murray, *Mathematical Biology*, 2nd edition (Springer, Berlin, 1993).
6. M. Abramowitz and I.A. Stegun, *Handbook of Mathematical Functions* (Dover, New York, 1972).
7. J. García-Ojalvo, F. Sagués, J.M. Sancho, and L. Schimansky-Geier, *Phys. Rev. E* **65**, 011105 (2002).
8. Pulses are obtained if an initial local activator perturbation is surrounded by inhibitor in only one side, which produces pulse propagation in the opposite direction.
9. J. García-Ojalvo and J.M. Sancho, *Noise in Spatially Extended Systems* (Springer, New York, 1999).
10. J. García-Ojalvo and L. Schimansky-Geier, Europhys. Lett. **47**, 298 (1999).
11. J. García-Ojalvo and L. Schimansky-Geier, J. Stat. Phys. **101**, 473 (2000).
12. R. Báscones, J. García-Ojalvo, and J.M. Sancho, *Phys. Rev. E* **65**, 061108 (2002).
13. M. Löcher, D. Cigna, and E.R. Hunt, Phys. Rev. Lett. **80**, 5212 (1998).
14. P. Heyward, M. Ennis, A. Keller, and M.T. Shipley, J. Neurosci. **21**, 5311 (2001).

Illusions and *Ghost* Resonances: How We Could See What Isn't There.

Dante R. Chialvo

Department of Physiology, University of California, Los Angeles, California 90024, USA
Email: dchialvo@ucla.edu. Web:http://www.bol.ucla.edu/~dchialvo

Abstract. To judge with certainty which information our senses are paying attention to is a hard problem, because often what we see, hear, or feel is not out there. In this paper, objective explanations are discussed for these tantalizing illusions demonstrating how relatively simple nonlinear stochastic transformations make the solution possible.

INTRODUCTION

Despite important advances, is still not well understood what a sensory neuron encodes. How to be certain that a given spike-train from a sensory neuron is encoding a particular aspect of the physical world? In this contribution we choose to discuss this aspect by way of an example in which the brain reads something that objectively isn't there ... The unsolved problem is how the brain extracts the pitch of complex sounds, and the solution proposed is a stochastic nonlinear mechanism which is biologically plausible. Work in this problem can be traced back to Pythagoras [1], who believed that music was more than entertainment. For him music was the expression of a divine principle *-harmonia-* that is able to bring order to chaos. Pitch is the subjective attribute of a sound sensation by which it may be ordered on a scale from high to low. In the simplest case, when the sound is composed of a single tone, like one made by a tuning fork, the pitch equals the frequency of the tone. However, natural sounds more often are complex, i.e., are comprised of the superposition of many tones. Pythagoras was interested in understanding the production of sounds with different pitch, and looked for the physical mechanism involved in the generation of complex tones. It is said that he was passing by a blacksmith's shop and heard the sound of hammers striking on a piece of iron. He noticed that, although the sounds made by the different hammers were all different, they were all in harmony with each other, except one. After toying around with other possibilities, he discovered that the pleasant harmony was related with certain ratios of the weight of the hammers, which were twelve, nine, eight and six pounds. He noted that striking together those with weights ratios of 1/2 -the six and twelve pound ones-, produced the intervals of an octave, using those with ratios of 2/3 (the eight and twelve pound) produced the sound of a fifth, and using a ratio of 3/4 -the nine and twelve pound hammers- produced the sound of a fourth. These were pleasant tones, derived all from the integers 1, 2, 3, and 4, which were

CP665, *Unsolved Problems of Noise and Fluctuations: UPoN 2002: Third International Conference,* edited by S. M. Bezrukov

later replicated by Pythagoras using strings stretched by similar weights. At that moment in history, by his understanding of how to produce consistent musical consonant intervals, the world of sensations stopped being magic and became physics.

As often happens, one happy discovery reveals another puzzling mystery. Pythagoras' discovery opened the yet unresolved mystery of the location of the "consonance detector", i.e., which brain area finds the harmony by computing these pleasant magic ratios? Furthermore, how is it that even untrained individuals, from disparate cultures, all share the same consonance appreciation? Finding universality in this problem, belongs to the yet unwritten pages in the theory of music, awaiting to be formally solved as any other complicated problem.

The scope of these pages is to discuss how the brain could extract the pitch of say just one of Pythagoras' hammers. Of course, it will also bear relevance into the larger problem of harmony, but we will leave that discussion out for the moment.

Sometimes what is heard is not what is being listen to

Natural sounds, as birdsongs, speech or those produced by a hammer striking on a piece of iron are complex, in the sense that they are formed by several discrete spectral lines. As it was said above, if pitch is a unitary attribute of a sensory experience, then how is it that a complex sound can be perceived as a unique entity? In other words, what is the neural process by which the brain fuses all the sound's frequency components into the perceived single pitch?

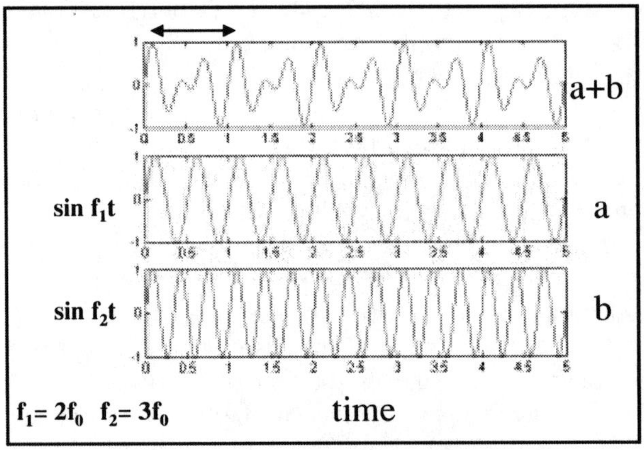

Figure 1: A complex tone (top trace) is constructed by adding two sinusoids (a and b) which are the first and second harmonics of f_0, respectively. When played, at relatively low volume, the resulting time series produces an f_0 pitch (the double-headed arrow indicates the peaks spaced by $1/f_0$).

Some inspiration came from simple inspection of complex tones. Consider the example in Fig. 1 where two sinusoids (with frequencies multiples of a third one) are added. It is well known, that when the resulting complex tone is played, the third frequency, though absent in the sound, is heard. The heard frequency corresponds to the intervals between the peaks of the time series, indicated by the double-headed arrow in the figure. Because the missing frequency is also, from the Fourier viewpoint, the fundamental, this psychoacoustic phenomenon is often called the *"missing fundamental"* illusion [2-4]. Only if the tones are harmonic, like in the example of Fig. 1, (i.e., multiples of f_0), the fundamental and the difference between the tones are the same. In any case the peaks are produced by constructive interference between the tones, a process first demonstrated by Thomas Young in 1800 when explained the origin of Newton' rings [5] and set the basis of the wave nature of light: two periodic processes will add up around their common subharmonic frequency.

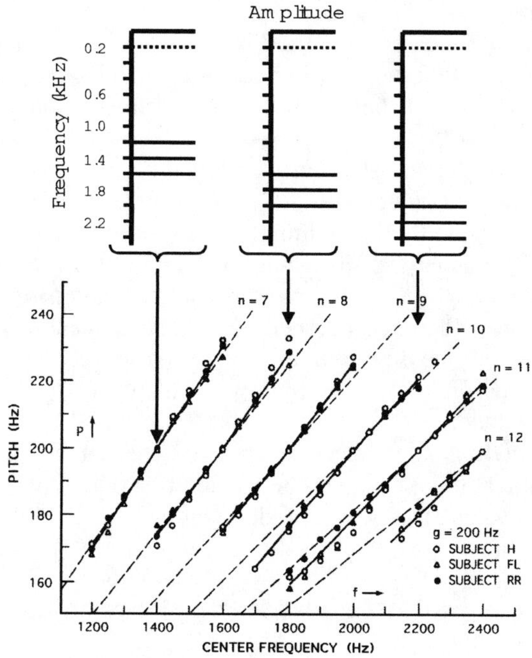

Figure 2: Results from Schouten's pitch shift experiments [4] using three-tones sounds with center frequency from 1.2 up to 2.4 kHz. The top diagrams depict the line amplitude spectra for three examples of the complex sounds listened by the subjects. Symbols (circles and triangles) in the bottom graph indicate the pitch heard by each of the three subjects for each complex sound. (Redrawn and modified from ref. [4]) The dotted lines (in the original publication) were meant to convey visually the idea that a 1/n function consistently underestimates the linear relationship between frequency and pitch shift. (g=200 Hz is "f_0" and "n" is "k" in our notation)

It was assumed, for a long time, that what it is heard is the *frequency difference* between the tones and that the brain could compute that simply by passing the complex tones trough a simple nonlinearity. But Schouten et al [4], four decades ago, has clearly demonstrated that this is not so, by frequency shifting an initially harmonic complex stimulus. Specifically, the complex tones they used were composed by three frequencies:

$$f_1 = kf_0 + \Delta f, \quad f_2 = (k+1)f_0 + \Delta f, \quad f3 = (k+2)f_0 + \Delta f \qquad (1)$$

If the brain computes differences then, with this stimulus, one should heard a constant pitch since this experimental paradigm preserved the equal spacing between tones. Instead, as shown in Figure 2, perception was found to shift linearly with the shift Δf, in the region explored (between 1.2 and 2.4 kHz). Another important observation these authors made was to note that the slope of the linear relation between Δf and perceived shift decreases for increasing k. In addition, they noted that a 1/n function was underestimating the experimental data (see Fig. 2). Finally, a qualitative, but not less important, observation was made: there was a peculiar ambiguity in the pitch judgment. As illustrated in the plot, there were always between two to four perceived pitches for any given sound.

Schouten et al. [4] findings became the litmus test for theories of pitch perception and despite an extensive literature on the subject no satisfactory quantitative theory exists to explain the origins of these scaling relationship between complex tones and perceived pitch.

The specific nature of the neural activity involved in pitch processing has long been the subject of discussions (for reviews see [6-8]). One view proposes that pitch of complex sounds is extracted by the auditory system by first deriving a spectral profile from a (tonotopic) frequency-specific auditory input, followed by some pattern-matching mechanisms [9-10]. Others have suggested mechanisms based on the timing of the auditory nerve fiber activity, irrespective of any frequency organization [11-14]. A serious drawback in deciding the merits of these theories is the absence of quantitative predictions and/or the ambiguity as to where in the brain a proposed computation is being done.

Departing from previous approaches, in the next section, a conjecture is discussed that interprets complex tones perception as the byproduct of a *linear* process of interference (or coincidence) of waves and a *nonlinear* detection of the interference peaks by a noisy threshold.

Robust ghost stochastic resonance

The main idea is that the illusion we hear is uniquely determined by the peaks of the complex tone (as the example in the top time series of Figure 1). Although it is a trivial fact that neurons can easily detect those peaks, it was not anticipated how well the quantitative aspects of the psycho acoustic phenomenon can be reproduced by this simple assumption, as was discussed in ref [15-19]. In that work, a simple toy model of a sensory neuron driven by complex tones was considered. Briefly, the system was a non-dynamical threshold device [16] that reduces the most elementary neuron dynamics to a set of rules comparing the input signal x(t) with a threshold U_{th} like: x(t) > U_{th} or < U_{th}. Anytime x(t) crosses the fixed threshold $U_{th}=1$, a "spike" is emitted. The only quantities of relevance for the problem here are the inter-spike intervals (ISI). The model is driven with combinations of sinusoidal of various frequencies, i.e., simulating complex sounds used in experiments [4,13,14]:

$$x(t) = A\left(\sin(f_1)2\pi t + \sin(f_2)2\pi t + \ldots \sin(f_n)2\pi t\right)\frac{1}{n} + \varepsilon(t) \qquad (2)$$

where $f_1 = kf_0$, $f_2=(k+1)f_0$,... $f_n=(k+n-1)f_0$ and $k > 1$. The term $\varepsilon(t)$ is white noise with zero mean and Gaussian distribution with variance σ . The region of parameters of concern here is A small (a rule of thumb is A=0.9U_{th}) such that the deterministic forcing alone is not enough to cross the threshold and fire a spike. The range of σ values need to be explored on the region where the timing of the spikes is expected that will become more coherent with the input signal. Notice that the deterministic terms in Eq. 2 represent the complex tones where the fundamental f_0 is absent. Since, as was commented above, sounds of this form are perceived with a pitch equal to f_0, i.e., the "missing" fundamental, numerical experiments were aimed to investigate the response of the model to these signals.

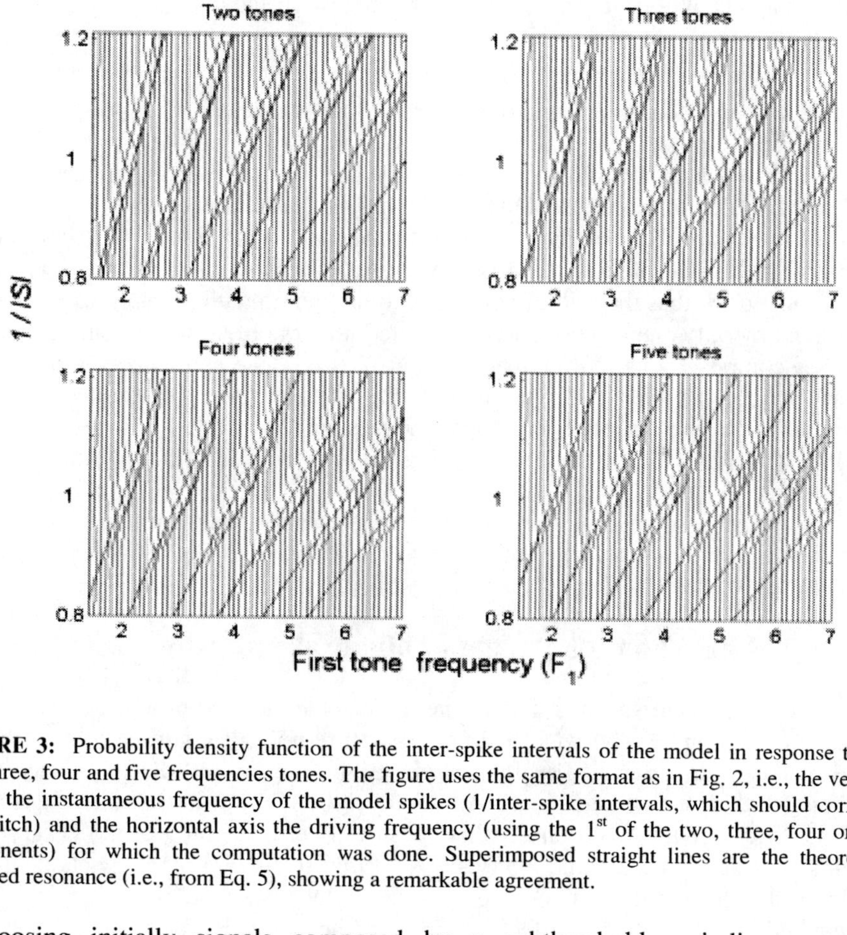

FIGURE 3: Probability density function of the inter-spike intervals of the model in response to the two, three, four and five frequencies tones. The figure uses the same format as in Fig. 2, i.e., the vertical axis is the instantaneous frequency of the model spikes (1/inter-spike intervals, which should correlate with pitch) and the horizontal axis the driving frequency (using the 1st of the two, three, four or five components) for which the computation was done. Superimposed straight lines are the theoretical expected resonance (i.e., from Eq. 5), showing a remarkable agreement.

Choosing initially signals composed by n subthreshold periodic terms and increasing noise it was found: I) there is a wide range of noise for which the neuron spikes are spaced preferably ~ $1/f_0$ [15-19]. Because there is an optimum noise intensity for which the system emits spikes at a frequency that is not in the input we also call this phenomenon "ghost" stochastic resonance [17-19]. Thus, analogous to the commented psycho-acoustic experiments the neuron's strongest resonance occurs for the frequency missing in the input. II) This resonance to the "missing" f_0 is not analogous to the difference f_2 - f_1. This was found by studying the response to inharmonic signals constructed by shifting all components of the harmonic signal by the same amount Δf :

$$f_1 = kf_0 + \Delta f\,,\quad f_2 = (k+1)f_0 + \Delta f\,,\quad ...fn = (k+n-1)f_0 + \Delta f \qquad (3)$$

In this case, despite that the difference $f_n - f_{n-1}$ remains constant, the location of the strongest resonance was found to shift linearly as $f_p = f_0 + \Delta f/(k+1/2)$ for n=2 and $f_p = f_0 + \Delta f/(k+1\,)$for n=3. Furthermore, it is possible to generalize predicting the response of the neuron to stimuli composed of N sinusoidal signals of frequencies:

$$kf_0 + \Delta f\,,\qquad (k+1)f_0 + \Delta f\,,\quad ...(k+N-1)f_0 + \Delta f \qquad (4)$$

and the resonance will be expected to occur at frequencies given by:

$$f_r = f_0 + \frac{\Delta f}{k+(N-1)/2} \qquad (5)$$

Figure 3 shows four representative examples of numerical results using complex tones composed of 2, 3, 4 and 5 sinusoids (using noise variance $\sigma = 0.1U_{th}$). The graphs represent the probability of observing spikes with a given instantaneous firing frequency f_r (1/ISI in the ordinate) as a function of the frequency (f_1, abscissa) of the lowest of two (three, four or five) components of the driving signal. The lines are the theoretical prediction of Eq. 5 that over-imposes exactly with the simulation results. Note, from Eq. 5, that the neural responses to any combinations of sinusoids will be represented by only one of two diagrams shifted horizontally by an amount equal to f_0, i.e., tones composed of *even* numbers of sinusoids produce spikes spaced at intervals with statistics as the ones in the left panels; tones with *odd* numbers produce responses like those in the right panels. Thus, the responses will always fall on lines with slopes 2/3, 2/5, 2/7, 2/9, 2/11, etc for even number of tones, and 1/2,1/3, 1/4,1/5,1/6, etc for odd numbers.

In conclusion, Schouten et al. observations were in the right direction: the function best explaining their data wasn't of the form 1/n; now, from Eq. 5, we know that it is 1/(n+1) [15,19].

What about phase, about binaural, and why noise?

The main argument sustains that all the illusions are related to a linear interference process that is nonlinearly detected by a noisy threshold. It could be argued that the shape of the time series is sensitive to phase, and unless the detection mechanism accounts for that, the spike timings can be affected. While it is true that the shape of the time series changes with phase, this doesn't affect the timing of the peaks, which remains separated by the same intervals. Note that for inharmonic signals phases are continuously changing, relative to each other. The only case in which the phase relationships are fixed is in perfectly harmonic signals, which hardly exist outside of the computer. Coincidentally it should be noted that in reality the phenomenon of the missing fundamental *is* phase insensitive.

A similar (although weaker) missing fundamental illusion exists as well when each of the two tones enters trough a different ear. The neural substrate of this binaural phenomenon must be at some central structure. As discussed in [19] the same mechanism and arguments hold to explain this illusion. In brief, in analogy with the scenario described for sinusoids, the two spike trains coming from each cochlear nerve *add and interfere* to be subsequently *nonlinearly detected* by the noisy threshold of some neuron. The only difference is that spikes, instead of sinusoids, are added but the rest of the dynamics are equivalent, with a "coincidence detection" flavor.

Finally, why noise? After all, one could design a deterministic peak detector that works as well as the noisy threshold discussed here. The answer is that a deterministic mechanism requires additional parameters to control at which value a spike is generated. Empirically, we found that it can be an important problem. Instead, a noisy mechanism warrants that the largest peaks of the interference have always the largest probability to induce spikes. This is important, such that even in the cases where the amplitude of the signal fluctuates, or when the shape of the signal is distorted by the phase effects commented above, a "hierarchy" is safely maintained: the biggest peak produced by the interference process will always be the most likely to produce a spike. Paradoxically, by adding noise, and without extra neuronal mechanisms, detecting in this way the missing fundamental is a robust process.

CONCLUSIONS

The views discussed in this paper emphasize simple nonlinear transformations by which sensory neurons can fabricate illusions. In line with previous ideas, is the assumption that the pitch information is in the temporal envelope of the signal, but in contrast, previous proposals have suggested mechanisms using relatively sophisticated structures. These include delay lines [11]; oscillators in combination with integration circuits [20], neural networks [12], and timing nets [7]; but the search for such neural structures has been unsuccessful. Others, [21,22] have instead proposed more abstract alternatives for this problem. Insofar as the mechanism we propose is concerned, there is nothing on it that is peculiar of the auditory system. Then, there is the possibility of identical phenomenology in other sensory modalities such as visual, touch, etc. [23,24], as well as outside biology, where ghost resonance should be an ubiquitous dynamics of driven nonlinear systems, as in lasers [18] and electronic circuits as well [17].

Finally, these illusions stress the difficulty in judging which information our senses are paying attention to when what we see, hear, or feel is not out there. This can be of some importance for the design of optimum sensory implants. Consider, for instance that the calculation of the most usual Shannon information transmission measures, between neuronal input (complex tones) and its output (spike trains) will reflect, for the examples discussed in this paper, the wrong conclusion that no information is being encoded (is just an illusion after all!). The solution reveals itself, as soon as we abandon the wrong assumption that trough our senses we asymptotically access a *linear* picture of the physical world.

ACKNOWLEDGMENTS

This work was supported in part by grants from the National Institute of Health and by the Veterans Administration Greater Los Angeles Healthcare System.

REFERENCES

1. Strohmeier J. and P. Westbrook, *Divine Harmony. The life and Teachings of Phytagoras.* (1999), Berkeley, CA. Berkeley Hills Books. 159.
2. von Helmholtz H. (1895) *On the Sensations of Tone as a Physiological Basis for the Theory of Music.* trans. Ellis AJ. Longmans, Green and Co., New York.
3. de Boer E., *On the "residue" and auditory pitch perception.* In *Handbook of Sensory Physiology* (1976), eds. Keidel W.D., Neff W.D. Springer-Verlag, Berlin.
4. Schouten J.F., R.J. Ritsma and BL Cardozo, *Pitch of the residue.* J. Acoust. Soc. Am. **34**, 1418-1424 (1962).
5. Gamow G. (1961). *The great physicists from Galileo to Einstein*, Mineola, NY Dover Publications, Inc.
6. Greenberg S., J.T. Marsh, W.S. Brown and J.C. Smith, *Neural temporal coding of low pitch. I. Human frequency following responses to complex tones.* Hearing Res. **25**, 91–114 (1987).
7. Cariani P.A., *Temporal coding of periodicity pitch in the auditory system: an overview.* Neural Plast., **6**(4):147-172 (1999).
8. Tramo M.J., P.A. Cariani, B. Delgutte and L.D. Braida, *Neurobiological foundations for the theory of harmony in western tonal music.* Ann N Y Acad Sci **930**:92-116, (2001).
9. Goldstein J., *An optimum processor theory for the central formation of pitch of complex tones.* J. Acoust. Soc. Am. **54**, 1496–1516 (1973).
10. Cohen M., S. Grossberg and L. Wyse, *A spectral network model of pitch perception.* J. Acoust. Soc. Am. **98**, 862–879 (1995).
11. Licklider, J.C.R. *A duplex theory of pitch perception.* Experientia **7**, 128–134 (1951).
12. Meddis, R. and J. Hewitt, *Virtual pitch and phase sensitivity of a computer model of the auditory periphery I: pitch identification.* J. Acoust. Soc. Am. **89**, 2866–2882 (1991).
13. Cariani P.A., and B. Delgutte, *Neural correlates of the pitch of complex tones. I. Pitch and pitch salience.* J. Neurophysiol. **76**(3): 1698-1716, (1996).
14. Cariani P.A. and B. Delgutte, *Neural correlates of the pitch of complex tones. II. Pitch shift, pitch ambiguity, phase invariance, pitch circularity, rate pitch, and the dominance region for pitch.* J. Neurophysiol. **76**(3): 1717-1734, (1996).
15. Chialvo D.R., O. Calvo, D.L. Gonzalez, O. Piro and G.V. Savino, *Subharmonic resonance and synchronization in neuronal systems.* Phys. Rev. E, **65**, 050902 (2002).
16. Gingl Z., L.B. Kiss and F. Moss, *Non-dynamical stochastic resonance - theory and experiments with white and arbitrarily colored noise.* Europhys. Lett., **29**, 3, 191-196 (1995).
17. Calvo O. and D.R. Chialvo, *Ghost stochastic resonance on an electronic circuit.* Submitted to Electronics Letters (2002).
18. Buldu J.M, D.R. Chialvo, C.R. Mirasso, M.C. Torrent and J. Garcia-Ojalvo, *Ghost Resonance in a semiconductor laser with optical feedback.* Submitted to Phys. Rev. Letters, (2002).
19. Chialvo D.R., *A neural mechanism for the missing fundamental illusion and the perception of pitch.* Submitted to Nature, (2002).
20. Langner G., *Neural Processing and Representation of Periodicity Pitch.* Acta Otolaryngol (Stockh); Suppl **532**: 68-76 (1997).
21. Cartwright J.H., D.L. Gonzalez and O. Piro, *Pitch perception: a dynamical-systems perspective.* Proc. Natl. Acad. Sci. U.S.A., **98**, 9, 4855-4859 (2001).
22. Julicher F., D. Andor and T. Duke, *Physical basis of two-tone interference in hearing.* Proc. Natl. Acad. Sci. U. S.A. **98**,16, 9080-9085. (2001).
23. Kingdom F.A. and D.R. Simmons, *The missing-fundamental illusion at isoluminance.* Perception, **27**, 12, 1451-1460 (1998).
24. Fujii K., S. Kita, T. Matsushima, and Y. Ando, *The missing fundamental phenomenon in temporal vision.* Psychol Res. **64**(2), 149-154 (2000).

Open Questions For Suprathreshold Stochastic Resonance In Sensory Neural Models for Motion Detection Using Artificial Insect Vision

Mark D. McDonnell* and Derek Abbott*

*Centre for Biomedical Engineering (CBME) and
Department of Electrical & Electronic Engineering
The University of Adelaide, SA 5005, Australia

Abstract. Stochastic Resonance (SR) occurs when the presence of noise in a nonlinear system can induce an optimal output from that system, and has been observed in a diverse range of physical and biological systems, including neurons. Despite this widespread observation of SR, to date very few engineering applications inspired by SR have been proposed, and one of the goals of our research is to explore possible new practical applications designed to replicate the benefits of SR. In particular, since about 1991, our group has designed and implemented a number of motion detection VLSI chips based on insect vision. We are currently investigating the possibility of replicating the benefits of SR in artificial insect-vision based motion detection systems, in particular a newly described form of SR called Suprathreshold Stochastic Resonance (SSR). The current paper is intended to review and identify the key open questions and avenues for future research relating to SR and SSR in such systems.

INTRODUCTION AND MOTIVATION

There are many tasks that are performed routinely by biological organisms that either cannot be replicated at all by artificial means (that is, designed systems, whether software, electronics or mechanical systems), or cannot be replicated to anywhere near the same efficiency or with much inferior performance. Such tasks include obvious ones like abstract and creative thinking, pattern recognition, and intelligence in human beings. But there are also other seemingly lesser, more trivial tasks that are routinely performed by creatures considered much less intelligent – that is, insects. Insects such as a common house fly or honey bee, achieve motion detection tasks, for example avoiding the fly swatter, and velocity estimation and visual tracking of an object, for example a bee landing on a leaf swaying in the wind. Despite much research into robotics, artificial intelligence, Uninhabited Aerial Vehicles (UAVs) and the like no one has devised a microrobot that can do all the mechanical and visual processing tasks that a bee can do, in such a small volume.

Other examples include the other senses – artificial noses are only now starting to look possible [1], the human ear is very good at distinguishing faint signals from noise, or even listening to two conversations at once, and the entire surface area of an animal is sensitive to a wide range of heat, pressure and textures.

Hence, an important idea is to study how biology achieves these tasks, in order to

CP665, *Unsolved Problems of Noise and Fluctuations: UPoN 2002: Third International Conference*,
edited by S. M. Bezrukov

learn from it and perhaps re-engineer systems according to this new knowledge. Such a philosophy is not uncommon – there are fields of research known as Biomimetic Engineering and Biomorphic Engineering. Although the engineering described by these terms is often at the molecular level, the concepts are just as relevant at the system level, given the fact that whole biological systems often outperform artificial systems. It is well known in the noise and fluctuation community that sensory neurons are very noisy when compared with what are acceptable levels of noise in electronic engineering [2]. Since the brain (which consists of a system of neurons) seems to function extremely well in this noise, it seems likely that either the noise is being intentionally used by the brain, or that the brain has come up with a good way around the noise.

This is a widely recognized fact, and there has been much research based around the idea, showing that the presence of noise in neurons – whether in real biological experiments, or in mathematical models of neurons, can improve some figure of merit at the output of the neuron, when compared to that figure of merit in the absence of noise. Such a phenomenon is known as Stochastic Resonance (SR).

STOCHASTIC RESONANCE

The term "Stochastic Resonance" [3, 4, 5, 6, 7] was first coined by Roberto Benzi in 1980 [8], as a name for the mechanism he suggested was behind the periodic behavior of the earth's ice ages [9]. Since this time, SR has usually been loosely defined as occurring when an increase in input noise leads to an increase in output signal to noise ratio in a nonlinear system driven by a periodic force, although, in more recent times the context of SR has been extended to include aperiodic signals, and many other measures other than SNR have been used as the figure of merit, such as cross-correlation coefficient [10] and transmitted information [11, 12]. There are many non-linear systems in which SR has been observed, such as electronic devices [13], ring lasers [14], SQUIDS (super conducting quantum interference devices) [15] and in biological sensory neurons [16, 17] and ion channels [18].

NOISE IN NEURONS

Although neurons are often modelled by complicated mathematical equations (usually coupled differential equations), for example the FitzHugh-Nagumo model and the Hodgkin-Huxley model, the essential qualitative feature of a neuron is that the output of a neuron produces a spike when the input signal increases above a threshold. Following the spike, there is a refractory period during which the neuron cannot spike again. The feature we are interested in here is the main nonlinearity involved, that is, the threshold.

SR in single threshold systems

Originally, it was thought that SR could only occur in dynamical systems obeying some differential equation. However later work showed that SR could occur in non-dynamical statistical systems, where the only components required were a threshold, noise and a subthreshold signal [19]. SR in threshold systems [20], such as a neuron, is a very simple concept to demonstrate: when noise is added to an input signal that is too small to cause the neuron to spike without noise, output spikes occur which are correlated with the amplitude of the input signal. The addition of random noise is equivalent to random changes in the values of the neuron's threshold, hence ensuring that the originally subthreshold signal becomes suprathreshold occasionally.

Although this phenomena has been extensively studied by physicists under the name of SR, the essence of it has been known to engineers for decades, by the name of dithering. Dithering has applications in the processing of noisy images and also in Analog to Digital Converters (ADCs), especially in audio systems. The addition of noise to a continuous audio signal at the input of an ADC increases the dynamic range, and decreases the output distortion, by whitening the output quantization noise, at the expense of a small increase in the output noise floor.

Some authors have recognized this fact [21, 22, 23], and either suggested that SR in threshold systems should be called "noise assisted threshold crossing" [21, 22], or that SR should be the name for naturally observed noise assisted behavior, where noise is inherently present, and dithering should be the name for artificially introduced noise, where it would not normally be present [24]. Accordingly, it seems appropriate to call the phenomenon SR in neurons, since noise seems to be unavoidable. On the other hand, perhaps evolution has evolved the brain so that the noise is there by design. Regardless of whether we call the phenomenon SR or dithering, it seems indisputable that noise in neurons causes threshold crossings and hence output spikes, that would not have occurred otherwise. Is this for the same reasons as dithering in ADC's? We believe this is worth studying. Perhaps some insights for new types of dithering may be found by the study of SR in neurons and other types of SR.

SSR – A new type of SR occurring in arrays of neurons

Recently, a series of papers by Stocks has brought to light a new form of stochastic resonance, which he has called Suprathreshold Stochastic Resonance [12]. This occurs in an array of threshold devices subject to the same input signal, but independent additive noise. The output from each device is then summed to give an overall output. Such a configuration is shown in Fig. 1. Previous work on noise in a parallel arrays of neurons [25] did not show the existence of SR in a spike timing precision measure for suprathreshold signals. Additionally, exact expressions for the correlation coefficient in parallel arrays of devices have previously been derived using linear response theory, both theoretically [26], and specifically for a parallel array of threshold devices [27]. In the latter case, however, the signal was always subthreshold, and the noise was not independent in each device.

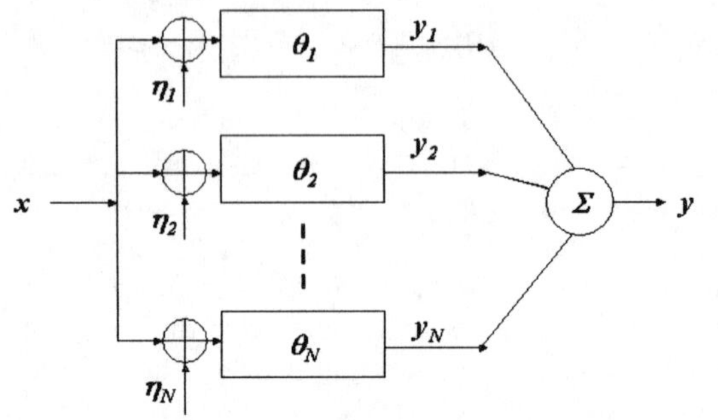

FIGURE 1. Array of N threshold devices. All devices receive the same signal, x and additive independent noise, η_i. The overall output, y, is the sum of the outputs from each device, y_i.

Stocks has used information theory to show how stochastic resonance can occur in such a system, for signals of arbitrary magnitude. If the array is considered as a semi-continuous channel [28], then the transmitted information through such a channel is given by

$$I = -\sum_{n=0}^{N} Q(n)\log_2 Q(n) - \left(-\int_{-\infty}^{\infty} P(x)\sum_{n=0}^{N} P(n|x)\log_2 P(n|x)dx\right), \tag{1}$$

where $P(x)$ is the probability density of the input signal x, $Q(n) = \int_{-\infty}^{\infty} P(n|x)P(x)dx$ is the probability of the output signal y being equal to n ($n = 0, 1, \ldots, N$) and $P(n|x)$ the conditional probability that the output is n given the input is x [12]. $P(n|x)$ can be calculated from knowledge of the threshold settings and the noise probability density function. When all thresholds are set equal to the signal mean, Stocks has shown that the maximum transmitted information given by Eq. (1) occurs for a nonzero noise intensity for both simple Heavyside style threshold devices [12, 29, 30], as well as in an array of more realistic neuron models – FitzHugh-Nagamo neurons [31]. A typical plot of the SSR effect is shown in Fig. 2, where both the input signal, and the noise at each device are Gaussian random variables.

Analysis of SSR

The concept of threshold crossing due to noise mentioned previously is of the form such that in a single neuron a signal, which would not be otherwise detected, is detected in the presence of noise. This is a binary process – the neuron output is either a spike

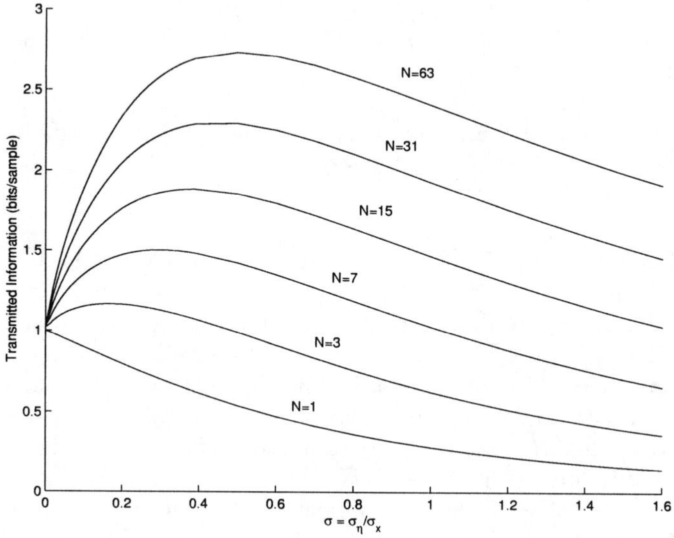

FIGURE 2. Plot of the transmitted information against the ratio of signal standard deviation to noise standard deviation, for Gaussian signal and noise, for various values of N

or not a spike. However if we look at an ensemble of more than one neuron, we are no longer restricted to a binary process.

Consider an array of N identical neurons (all with the same threshold value) in parallel as in Fig. 1. In the absence of noise, (or the same noise at each neuron) the overall output can only be 0 or N, indicating that the signal is either less than or greater than the threshold. However if independent noise is added to the signal at each neuron (or equivalently, each threshold varied randomly), then the output can take on any of $N+1$ values; that is, the number of neurons giving an output spike. What does this number represent? It is a scale that rates by how much the input signal exceeds or is below the threshold. Qualitatively, if the output is $N/2$, then half the thresholds have spiked and half have not spiked. If the noise is symmetrically distributed with the same mean value as the signal, then on average, if $N/2$ neurons spike, the signal must have the same value as the threshold. If N neurons spike, then on average, the signal must be much larger than the threshold, since no value of noise is negative enough to decrease the input below the threshold. If no neurons spike, then on average the signal must be much smaller than the threshold, since no value of noise is positive enough to increase the input above the threshold. All other values of the output are the intermediate values of input signal, between very large and very small compared to the threshold.

Indeed, as mentioned, Stocks showed that the transmitted information through such an array is maximized by a nonzero value of noise intensity. He has coined the term SSR to describe this effect, to differentiate it from SR in single threshold systems, where the signal must always be subthreshold for SR to occur. In the array the size of the signal is irrelevant when compared to the threshold, since the output is an indication of the magnitude of the signal.

The remarkable thing about this system is it has effectively performed an analog to digital conversion – an array of neurons can do ADC! Perhaps this is a case of engineers mimicking nature without realizing it.

Of course the distribution and power spectrum of the signal and noise affect how well the array works, but it is still effectively an ADC, which is potentially more robust to changes in the dynamic range of the input signal than conventional engineering designs.

SR IN INSECT VISION BASED MOTION DETECTION

In 1997, Moini *et al* published a paper describing the design and architecture of a VLSI motion detection chip [32]. This chip is based on the Horridge *template model* of the insect vision system [33] and is based on a (nonlinear) neuronal model. Blackwell has noted that "because of the non-linearities in the visual system, it is not unreasonable to expect that SR would be observed," [34] and indeed has "demonstrated that contrast detection thresholds of luminance sinusoids is improved when the sinusoid is embedded in low contrast noise." Therefore it is anticipated that as in real neurons, SR could be beneficially incorporated into the design of insect vision based chips.

SR in the Template Model

The reason that SR can occur in the Horridge template model is simply due to the presence of a threshold. The template model, a model based on the insect visual system [33], works by the spatio-temporal tracking of leading and trailing edges of moving objects. A large number of spatially separated parallel receivers are required. Due to this spatial separation, each receiver's input signal (the luminance) is slightly different. At each receiver, the input signal is bandpass filtered, thresholded and sampled. The input signal is then classified according to whether it is increasing in luminance, decreasing in luminance or has no change. A two by two template is then formed from this classification of the temporal rate of change of the input signal at two adjacent time samples and between each pair of adjacent receivers. This two by two template thus has 81 different possibilities. It has been shown that 8 of these are robust enough to indicate motion [33].

In the VLSI implementations of the template model designed by Moini *et al* [32], at each receiver the input signal is time differentiated and then thresholded. Note that since we require a case of no change, there are in fact two thresholds required. Since any approaching edge signal is at first subthreshold to the detector but becomes suprathreshold, the situation becomes equivalent to that of a subthreshold signal in the basic threshold systems mentioned earlier. Hence, the presence of noise in the system should presumably cause stochastic resonance to occur when signals are subthreshold.

Recent work by Harmer *et al* [35] has investigated the effects of noise in three different models of motion detection; the Reichardt correlation model [36], the directionally sensitive local inhibitory motion detector (DSLIMD) [37] and the Horridge template model [33]. For the template model, Harmer showed that under certain conditions both

SR and SSR occurred in a network of such motion detectors [38]. He concluded that SR can be of benefit by increasing the dynamic range of the system in low contrast conditions, for subthreshold signals, due to the same mechanism as SR in threshold based systems described earlier. However, by simulating an array situation within the template model, he also concluded that SSR could provide enhancement of suprathreshold signals in high contrast conditions [35].

FURTHER OPEN QUESTIONS

One of the open questions arising from Harmer's work is to investigate whether SR or SSR can occur in motion detection models other than the template model. The two other models he considered, although nonlinear, do not possess a threshold, so it is not immediately obvious whether SR could occur. Hence, further investigation into these models and others would be significant. Of particular interest is an elaborated version of the Reichardt correlator [39]. Although many physiological and behavioral experiments support the theory that Reichardt correlators form the basis for motion detection in insects, as well as in human vision [40, 41], the basic Reichardt correlator does not function as a velocity estimator, a function that both insects and humans are capable of. The elaborated Reichardt correlator has been shown by Dror *et al* to give improved velocity estimation performance in response to a variety of stimuli [40]. Our research aims to further investigate the performance of such elaborated Reichardt correlators in the presence of noise, and to search for conditions under which SR may occur.

A further investigation of SSR in the context of artificial vision is in the area of automatic gain control and logarithmic compression. It is known that both insect [42] and human visual systems [30] respond over a very wide dynamic range of light intensities without saturating. This is achieved by the adaptation of sensory neuron threshold levels to the mean value of the stimulus. Stocks *et al* noted that this phenomenon is analogous to the situation of SSR, where maximum noise-enhanced information transmission occurs precisely when neurons adapt to the mean value of the input signal [31]. Further research to explore these open questions surrounding the possible exploitation of this observation, in the design of artificial motion detectors, is of great interest.

ACKNOWLEDGMENTS

This work was funded by the Australian Research Council and the Sir Ross and Sir Keith Smith fund.

REFERENCES

1. Solis, J.L., Kish, L.B., Vajtai, R., Granqvist, C.G., Olsson, J., Schnürer, J., and Lantto, V., *Sensors and Actuators B: Chemical* **77**, 312–315 (2001).
2. Bialek, W., DeWeese, M., Rieke, F., and Warland, D., *Physica A* **200**, 581–593 (1993).

3. Dykman, M.I., Luchinsky, D.G., Manella, R., McClintock, P.V.E., Stein, N.D., and Stocks, N.G., *Il Nuovo Cimento* **17**, 661–683 (1995).
4. L. Gammaitoni, P. H., and Jung, P., *Reviews of Modern Physics* **70**, 223–287 (1998).
5. Wiesenfeld, K., and Jaramillo, F., *Chaos* **8**, 539–548 (1998).
6. Abbott, D., *Chaos* **11**, 526–538 (2001).
7. Harmer, G.P., Davis, B.R., and Abbott, D., *IEEE Transactions on Instrumentation and Measurement* **51**, 299–309 (2002).
8. Benzi, R., *at the NATO International School of Climatology, Ettore Majorana Center for Scientific Culture, Erice, Italy*, (unpublished), 1980.
9. Benzi, R., Sutera, A., and Vulpiani, A., *Journal of Physics A: Mathematical and General* **14**, L453–L457 (1981).
10. Collins, J.J., Chow, C.C., and Imhoff, T.T., *Physical Review E* **52**, R3321–R3324 (1995).
11. Heneghan, C., Chow, C.C., Collins, J.J., Imhoff, T.T., Lowen, S.B., and Teich, M.C., *Physical Review E* **54**, R2228–R2231 (1996).
12. Stocks, N.G., *Physical Review Letters* **84**, 2310–2313 (2000).
13. Melnikov, V.I., *Physical Review E* **48**, 2481–2489 (1993).
14. McNamara, B., Kiesenfeld, K., and Roy, R., *Physical Review Letters* **60**, 2626–2629 (1988).
15. Wiesenfeld, K., and Moss, F., *Nature* **373**, 33–36 (1995).
16. Douglas, J.K., Wilkens, L., Pantazelou, E., and Moss, F., *Nature* **365**, 337–339 (1993).
17. Wiesenfeld, K., Pantazelou, E., Dames, C., Pierson, D., and Moss, F., *Physical Review Letters* **72**, 2125–2129 (1994).
18. Bezrukov, S.M., and Vodyanoy, I., *Nature* **378**, 362–364 (1995).
19. Gingl, Z., Kiss, L.B., and Moss, F., *EuroPhysics Letters* **29**, 191–196 (1995).
20. Jung, P., and Mayer-Kress, G., *Il Nuovo Cimento* **17**, 827–834 (1995).
21. Gammaitoni, L., *Physical Review E* **52**, 4691–4698 (1995).
22. Gammaitoni, L., *Physics Letters A* **208**, 315–322 (1995).
23. Wannamaker, R.A., Lipshitz, S.P., and Vanderkooy, J., *Physical Review E* **61**, 233–236 (2000).
24. Ando, B., and Graziani, S., *IEEE Instrumentation and Measurement Magazine* **4**, 24–31 (2001).
25. Pei, X., L.Wilkens, and Moss, F., *Physical Review Letters* **77**, 4679–4682 (1996).
26. Neiman, A., Schimansky-Geier, L., and Moss, F., *Physical Review E* **56**, R9–R12 (1997).
27. Gailey, P.C., Neiman, A., Collins, J.J., and Moss, F., *Physical Review Letters* **79**, 4701–4704 (1997).
28. Feinstein, A., *Foundations of Information Theory*, McGraw-Hill, New York, 1958.
29. Stocks, N.G., *Physics Letters A* **279**, 308–312 (2001).
30. Stocks, N.G., *Physical Review E* **63**, 041114 (2001).
31. Stocks, N.G., and Manella, R., *Physical Review E* **64**, 030902(R) (2001).
32. Moini, A., Bouzerdoum, A., Eshraghian, K., Yakovleff, A., Nguyen, X.T., Blanksby, A., Beare, R., Abbott, D., and Bogner, R.E., *IEEE Journal of Solid-State Circuits* **32**, 279–283 (1997).
33. Horridge, G.A., and Sobey, P., *International Journal of Optoelectronics* **6**, 177–193 (1991).
34. Blackwell, K.T., *Vision Research* **38**, 267–280 (1998).
35. Harmer, G.P., and Abbott, D., *Microelectronics Journal* **32**, 959–967 (2001).
36. Hassenstein, B., and Reichardt, W., "Structure of a mechanism of perception of optical movement," in *Proceedings of the 1st International Conference on Cybernetics*, 1956, pp. 797–801.
37. Beare, R.J., *Image segmentation based on local motion detection*, Ph.D. thesis, Department of Electrical and Electronic Engineering, Adelaide University, Australia (1997).
38. Harmer, G.P., *Stochastic Processing for Enhancement of Artificial Insect Vision*, Ph.D. thesis, Department of Electrical and Electronic Engineering, Adelaide University, Australia (2001).
39. Dror, R.O., *Accuracy of visual velocity estimation by* Reichardt *correlators*, Ph.D. thesis, Churchill College, University of Cambridge (1998).
40. Dror, R.O., O'Carroll, D.C., and Laughlin, S.B., *Journal of the Optical Society of America. A, Optics and Image Science* **18**, 241–252 (2001).
41. Morgan, M.J., and Chubb, C., *Vision Research* **39**, 4217–4231 (1999).
42. Nguyen, X.T., *Smart* VLSI *micro-sensors for velocity estimation inspired by insect vision*, Ph.D. thesis, Department of Electrical and Electronic Engineering, Adelaide University, Australia (1996).

Unsolved Problems of Intracellular Noise

Johan Paulsson

Department of Molecular Biology, Princeton University
Washington Road, Princeton, NJ 08544-1014
Fax: 609-258-3980, Email: paulsson@princeton.edu

Abstract. Many molecules are present at so low numbers per cell that significant fluctuations arise spontaneously. Such 'noise' can randomize developmental pathways, disrupt cell cycle control or force metabolites away from their optimal levels. It can also be exploited for non-genetic individuality or, surprisingly, for more reliable and deterministic control. However, in spite of the mechanistic and evolutionary significance of noise, both explicit modeling and implicit verbal reasoning in molecular biology are completely dominated by macroscopic kinetics. Here I discuss some particularly under-addressed issues of noise in genetic and metabolic networks: 1) relations between systematic macro- and mesoscopic approaches; 2) order and disorder in gene expression; 3) autorepression for checking fluctuations; 4) noise suppression by noise; 5) phase-transitions in metabolic systems; 6) effects of cell growth and division; and 7) mono- and bistable bimodal switches.

INTRODUCTION

Intracellular processes operate far from equilibrium and most chemical species are present in low numbers per cell. Random fluctuations can then be anything from nonexistent to dominant depending on the kinetic design. The last decade has experienced a renaissance in stochastic modeling, but by focusing on Monte-Carlo simulations of complicated systems rather than analytical treatments of simple systems, a few central principles have permeated the literature without being recognized as such. This has left the basic problems not only unsolved, but largely unidentified. Considering the pervasive influence of randomness in genetic and metabolic networks, and recent advances in single-cell experimental techniques, this puts integrative stochastic approaches among the highest priorities of quantitative molecular biology.

UNSOLVED PROBLEMS

This study briefly reviews a number of systems without going into mathematical detail. More quantitative analytical descriptions of each problem are either referenced or will be published elsewhere.

CP665, *Unsolved Problems of Noise and Fluctuations: UPoN 2002: Third International Conference*, edited by S. M. Bezrukov

Systematic Macro- and Mesoscopic Descriptions

Macroscopic studies of intracellular kinetics have been greatly influenced by Biochemical Systems Analysis [1] (BSA) and Metabolic Control Analysis [2] (MCA). Both these schools approximate reaction rates using logarithmic gains that measure the normalized change in a response f to a normalized change in a signal s, $\partial\ln(f)/\partial\ln(s)$. A similar approach could be useful when studying intrinsic fluctuations. In van Kampen's Ω-expansion [3], the zeroth order approximation of the master equation gives the macroscopic limit, and the first order approximation gives a linear Fokker-Planck – the Linear Noise Approximation. In the stationary state, this results in matrix equation [4]:

$$AC + CA^T + \Omega B = 0 \tag{1}$$

where C is the sought covariance matrix, A is the Jacobian of the macroscopic system evaluated at steady state, Ω is the system volume, and B is a diffusion matrix that depends on the stoichiometry and steady state rates of the involved reactions. The connection to BSA and MCA can be made by interpreting A and ΩB in terms of three types of parameters: 1) steady state average lifetimes; 2) steady state average copy numbers; and 3) steady state logarithmic sensitivities for how reaction rates depend on concentrations. The simplest example is the one-dimensional birth-and-death process with intensities $J_+(x)$ and $J_-(x)$ for reactions $x \rightarrow x + 1$ and $x \rightarrow x - 1$, that can be approximated by an Ornstein-Uhlenbeck process:

$$\partial_t p(x) = -\frac{\overline{H}}{\overline{\tau}} \times \partial_x [(x - \langle \overline{x} \rangle) \times p(x)] + \frac{\langle \overline{x} \rangle}{\overline{\tau}} \times \partial_x^2 p(x) \tag{2}$$

where over-bar denotes steady state, τ is the average lifetime, $\langle x \rangle$ is the average copy number and $H = \partial\ln(j_-/j_+)/\partial\ln y$, where j is the deterministic limit of J. Parameter H determines the strength of the linearised drift while the stationary average determines the diffusion. With σ for standard deviations, relative x-fluctuations around a stable fix-point then follow the standard:

$$\frac{\overline{\sigma}_x^2}{\langle \overline{x} \rangle^2} = \frac{1}{\langle \overline{x} \rangle} \times \frac{1}{\overline{H}}. \tag{3}$$

Linear noise approximations are well understood in fluctuation theory and the problem is mainly pedagogic, trying to link interpretations of stochastic models to established traditions in biochemistry. In fact, the two-dimensional version of the procedure above can be used to connect and generalize the vast majority of published stochastic cell studies.

Order and Disorder in Random Gene Expression

Most analysis of random gene expression have focused on intrinsic fluctuations in gene activity [5,6] or mRNA levels [7,8,9,10,11,12]. In gene models, independent copies randomly turn ON and OFF with rates λ and μ as in a random telegraph process. The number of active genes (y) then displays a stationary Binomial with correlation time $\tau_y = 1/(\lambda + \mu)$. In mRNA models, mRNAs have a constant birth intensity and a death intensity $1/\tau_y$ per copy. The number of mRNAs (also denoted y) then displays a stationary Poissonian with correlation time τ_y. Assuming that proteins (x) have a birth intensity proportional to y and a death intensity $1/\tau_x$ per copy, leads to the exact [11]:

$$\frac{\overline{\sigma}_x^2}{\langle \overline{x} \rangle^2} = \frac{1}{\langle \overline{x} \rangle} + \frac{\overline{\sigma}_y^2}{\langle \overline{y} \rangle^2} \frac{\tau_y}{\tau_y + \tau_x} \tag{4}$$

The seemingly different published analyses [5-13] are thus identical or close to identical, and y could also represent other sources of external noise [13]. That does not mean that analyses of gene expression have come to closure. For instance, it is not clear that genes, mRNAs and proteins really can be treated as single chemical species. This relates to a central controversy in molecular biology. On one hand there is a widespread notion that intracellular noise is reduced simply because the involved processes are built up by numerous reaction steps. On the other, explicit analyses suggest that the underlying reactions just add external noise, as in Eq. (4).

Measurements of mRNA stability in yeast [14] indicate that exponential decay is the rare exception and that most mRNAs instead are degraded in a multistep process resulting in narrowly distributed lifetimes. Such clock-like decay decreases the correlation time for a given average lifetime and can thereby reduce protein fluctuations. Another example of clock-like kinetics is replication of bacterial plasmids [15, 11]. Some plasmids ensure that the synthesis reaction is divided into very many irreversible steps. This reduces the variation in the time between subsequent replications and can greatly reduce the probability of going down to zero copies [15, 11]. However, it does not change the macroscopic behavior and is invisible in rate equation descriptions [15, 11].

Even the simplest cell processes tend to be composed of many energetically driven reaction steps, but exploiting multistep kinetics to suppress copy number fluctuations is not straightforward. This is perhaps the most central and least understood problem of noise in molecular biology: when are intracellular processes ordered versus disordered.

Autorepression

Autorepression has been suggested to suppress internal noise by speeding up adjustments. However, the parameters that affect the drift for returning to steady state may also affect the rate of diffusion towards further deviations. For the adjustment rate H/τ in Eq. (2), autorepression thus suppresses internal noise by increasing H rather than decreasing τ [11]. Another commonly presumed effect of autorepression is that it suppresses external noise by reducing the susceptibility to fluctuations in the underlying parameters, i.e., by making the steady state levels more robust to disorder.

But there are also many reasons why autorepression should increase noise: 1) It decreases average copy numbers. 2) In a dynamically disordered environment, it expedites adjustments to meandering quasi-steady-states and thus prevents time-averaging. 3) In some systems it makes the quasi-steady-state *more* susceptible to changes in external parameters [11]. 4) It can involve slowly fluctuating signal molecules that induce more disorder.

The consensus that autorepression automatically suppresses noise is thus merely mythological and the requirements for suppressing or amplifying internal or external noise are not well understood in real biochemical systems.

Noise Suppression by Noise

Consider the bimolecular reaction rate $r = x(x-1) \approx \langle x \rangle^2$ where the protein copy number x is subject to both internal and external noise as in Eq. (4). A change in the underlying parameters can then increase the average $\langle x \rangle$, but at the same decrease the variance σ_x^2 so much that $\langle r \rangle = \langle x(x-1) \rangle = \langle x \rangle^2 - \langle x \rangle + \sigma_x^2 \approx \langle x \rangle^2 + \sigma_x^2$ actually decreases. If we insist on studying $\langle r \rangle$ in terms of $\langle x \rangle$, even if $\langle x \rangle$-values do not map to unique $\langle r \rangle$-values, random x-fluctuations can then lift all restrictions imposed by the reaction mechanism, and $\langle r \rangle$ can mimic e.g. $\langle x \rangle^4$ or $\langle x \rangle^{-1}$. This is possible for all nonlinear kinetic mechanisms: Because the average of the function does not coincide with the function of the average, the effective nonlinearity depends both on the nonlinearity of the transfer function and on the nonlinearity of how the probability mass of a distribution depends on the average. As a direct consequence, cells can exploit small-number fluctuations to suppress other small-number fluctuations. For instance, many bacterial plasmids (x) regulate their own replication using plasmid-encoded inhibitors (y). In some cases the inhibitor birth and death intensities seem to be $I_+ = k_y x$ and $I_- = y/\tau_y$, while the plasmid birth and death intensities seem to be $J_+ = r(y)x$ and $J_- = x/\tau_x$ where r is a monotonically decreasing function of y. The birth intensity J_+ is proportional to x because plasmids are DNA molecules and therefore templates for their own synthesis. Since $\tau_x >> \tau_y$, the two-dimensional process for x and y can be replaced by a one-dimensional process for x and a conditional equation for the probability $q(y|x)$ of y given x. But because r is nonlinear in y, the idealization that $\langle r|x \rangle = r_{\langle y|x \rangle}$ can be very misleading even if $\langle y|x \rangle \sim x$. In particular, some plasmids are mechanistically limited to $r(y) = k_x(K + y)^{-1}$ so that in the absence of fluctuations $H<1$ and $\bar{\sigma}_x^2/\langle \bar{x} \rangle > 1$ in Eqs. (2)-(3). However, computing the real average:

$$\langle r|x \rangle = \sum_{y=0}^{\infty} r(y) q(y|x) = k_x e^{-k_y x \tau_y} \sum_{y=0}^{\infty} \frac{(k_y x \tau_y)^y}{(K+y)\, y!}, \tag{5}$$

shows that y-noise can greatly increase the sensitivity of $\langle r|x \rangle$ to x, a phenomenon termed Stochastic Focusing [10, 16, 17]. An intuitive reason for the effect is that the probability that y randomly moves from the average $k_{yx} x \tau_y$ to low values – from which $\langle r|x \rangle$ receives a disproportional contribution – depends on the birth intensity $k_{yx} x$ at every step. The probability mass in the tail of $q(y|x)$ then responds sensitively to x. This makes it possible to have:

$$H_{effective} = -\frac{\partial \ln \left(\langle r|x \rangle \right)}{\partial \ln (x)} >> 1 . \tag{6}$$

Significant conditional y-fluctuations can thus reduce x-fluctuations indefinitely. This remains to be demonstrated experimentally and the general requirements on noise properties, transfer functions and time-scales need to be clarified.

Phase Transitions

Many smaller cell compounds, such as nucleotides and amino acids, are present in large copy numbers per cell. This is often thought to imply insignificant internal noise due to random births and deaths of individual molecules. However, numerous simple intracellular processes appear to operate close to phase transitions where internal noise can be enormous at any average. For instance, with $\varnothing$ as a constant source and sink of molecules, consider the reaction scheme [4]:

$$\begin{array}{ll} X \longrightarrow \varnothing & \text{(a)} \\ \varnothing \longrightarrow X & \text{(b)} \\ Y \longrightarrow \varnothing & \text{(c)} \\ \varnothing \longrightarrow Y & \text{(d)} \\ X + Y \longrightarrow C & \text{(e)} \end{array} \tag{7}$$

where X and Y are formed at constant rates, have intrinsic half-lives and irreversibly form complex C in a bimolecular reaction. This illustrates simple anabolic reactions where small building blocks form higher-level compounds, and also applies to the many stable heterodimers in cells. In both cases, cells avoid wasteful turnover of X and Y by ensuring that reactions (a) and (c) are rare compared to (e). However, in this economic regime, the system displays characteristics of operating near a non-equilibrium phase transition [4]: 1) steady state levels are ultrasensitive to changes in parameters; 2) random fluctuations are extreme; 3) relaxation rates become critically slow; and 4) X and Y are negatively correlated in such a way that the formation rate of C fluctuates insignificantly in spite of large fluctuations in X and Y. This is much like the board game Scrabble where letters are picked randomly from a pool and then used to form words. Because words tend to have both vowels and consonants, a temporary shortage of one or the other tends to linger. In anabolic reactions, the limiting building block is similarly kept at low levels until – by chance – another building block becomes limiting and takes its place. This spontaneously leads to enormous fluctuations in the dominating building block and a random succession of different shortages.

Near-critical behavior is not limited to anabolic reactions but have been suggested in dynamic instability of microtubules [18], replication of unregulated plasmids [11], positive feedback and autocatalysis [10], phosphorylation-dephosphorylation reactions [19, 20] and simple systems with Michaelis-Menten elimination [10]. One of the open questions of phase transitions in anabolic reactions is to what extent they can be attenuated by regulatory mechanisms – such as negative feedback – without simultaneously reducing cell efficiency [4].

Growth and Division

In a growing cell population, a dividing mother cell has on average twice as much of all components as a newborn daughter. Rather than converging to a stationary distribution, cell components then approach a cyclostationary state that repeats itself generation after generation. With a Poisson process for synthesis during the cell cycle and binomial partitioning of molecules at cell division, the cyclostationary distribution is a Poissonian with an average proportional to the cell age, but for most other mechanisms, cyclostationary distributions are harder to calculate.

In the deterministic limit, exponential cell growth with rate μ simply contributes a dilution term $-\mu x_i$ to each concentration rate equation dx_i/dt, thus mimicking first-order degradation. A similar trick is often used in stochastic models where cyclical growth and division is replaced by an exponential decay term [11, 13, 4]. This simplifies the analysis greatly and has been shown to be valid or approximately valid in some simple cases [11,13]. However, in contrast to the macroscopic limit, it is not clear when or how well the idealization works in stochastic models.

A related question is whether or not binomial partitioning increases or decreases overall fluctuations. It is often assumed that since the daughter cells each receive a random sample of the mother cell, they must display larger fluctuations. However, since also the average changes, this depends on the measure of dispersion. Furthermore, since on average half of the copies segregate to each daughter, cells whose mothers had more copies at cell division have also typically lost more copies to the sister cell. Random partitioning at cell division does thus not only add diffusion, but also first-order drift. In some cases, e.g. when molecules are born at constant rates and degraded by saturated enzymes, this drift can dominate the restoring force and be essential to suppress fluctuations [4].

Mono- and Bistable Bimodality

Many cell systems use positive feedback loops to obtain bistability. Due to intrinsic fluctuations around either fix-point, this is thought to lead to pseudo-exponential escape rates and allow a clonal population of cells to exploit distinctly different behaviors for the same environmental input. However, it is far from clear that cells need bistability for this effect. They could also switch randomly between very different states by filtering fluctuations around a single fix-point. In other words, since cells operate at the mesoscopic scale where genes, mRNAs and proteins often are present in one or a few copies per cell, they do not need macroscopically bistable mechanisms to obtain bimodal behavior. The reason why most suggestions for bimodal mechanisms also are bistable may instead reflect that bistability is easier to identify in reaction schemes.

Questions that remain to be answered are 1) Do bistable systems occupy both states due to intrinsic fluctuations or due to external disturbances in the parameters? 2) What simple macroscopically monostable systems are bistable mesoscopically? 3) Does filtering fluctuations around a mesoscopically monostable fix-point lead to an upper limit in apparent escape times due to the limited average lifetimes of the molecular hardware? 4) Does filtering fluctuations around a mesoscopically monostable fix-point lead to a flickering transition between states, i.e., alter the distribution of sojourn times?

CONCLUSIONS

The last few years' explosion in single-cell [21-24] and single-molecule [25] experiments provide an unprecedented opportunity for stochastic modeling. There is also a great need for convergence in the basic perspectives. Stochastic biokinetics has had a long history and it is time for it to also have a tradition. The major challenge is to develop a consistent point of view.

ACKNOWLEDGMENTS

This work was supported by a Lewis-Thomas fellowship from Princeton University. I am grateful to C.N. Peterson for comments on the manuscript.

REFERENCES

1 Savageau, M.A., *Biochemical systems analysis.* Addison-Weasley (1976)
2 Fell, D., *Understanding the control of metabolism* Portland Press (1996).
3 van Kampen, N.G., *Stochastic Processes in Physics and Chemistry.* North-Holland (1992).
4 Elf, J., Paulsson, J., Berg, O.G., and Ehrenberg, M., *Biophys. J.* **84**, 154-170 (2003).
5 Peccoud, J. and Ycart, B., *Theor. pop. Biol.* **48**, 222-34 (1995).
6 Kepler T.B. and Elston, T.C., *Biophys. J.* **81** 3116-36 (2001).
7 Rigney, D.R., *J. theor. Biol.* **69**, 761-6 (1977).
8 Berg, O.G., *J. theor. Biol.* **173**, 307-20 (1978).
9 McAdams, H.H., and Arkin, A., *Proc. natl. Acad. Sci. USA* **94**, 814-19 (1997)
10 Paulsson, J. Berg, O.G., and Ehrenberg, M., *Proc. natl. Acad. Sci.* **97**, 7148-53 (2000)
11 Paulsson, J., and Ehrenberg, M., *Quart. Rev. Biophys.* **34**, 1-59 (2001).
12 Thattai, M., and van Oudenaarden, A., *Proc. natl. Acad. Sci. USA* **98**, 8614-9 (2001).
13 Swain, P.S., Elowitz, M.B., and Siggia, E.D., *Proc Natl Acad Sci U S A.* **99**, 12795-800 (2002)
14 Decker, C.J., and Parker, R., *Genes Dev.* **7** 1632-43 (1993).
15 Paulsson J., and Ehrenberg M., *J. Mol. Biol.* **297** 179 (2000)
16 Paulsson J., and Ehrenberg M., *Phys Rev Lett* **23**, 5447 (2000)
17 Berg, O.G., Paulsson, J., and Ehrenberg, M., *Biophys J.* **79**, 2944-53 (2000)

18 Dogterom, M., and Leibler, S., *Phys. Rev. Lett.* **70**, 1347-50 (1993).
19 Goldbeter, A., and Koshland, D.E., *Proc. natl. Acad. Sci. USA* **78**, 6840-4 (1981)
20 Berg, O.G., Paulsson, J., and Ehrenberg, M., *Biophys. J.* **79**, 1228-1236 (2000)
21 Lobner-Olesen, A., *EMBO J* **18**, 1712-1721 (1999).
22 Becskei A., and Serrano L., *Nature*, **405**, 590-593 (2000).
23 Ozbudak E.M., Thattai, M., Kurtser, I., Grossman, A.D., and van Oudenaarden, A., *Nat. Genet.* **1**, 69-73 (2002).
24 Elowitz M.B., Levine A.J., Siggia E.D., and Swain P.S., *Science* **297**, 183-6 (2002)
25 Xie, X.S., *Single Molecule* **4**, 2, 229 (2001).

Noise-induced Hypersensitivity and Stochastic Resonance: Can Living Systems Use Them at a Molecular Level?

O. V. Gerashchenko, S. L. Ginzburg, and M. A. Pustovoit

*B.P.Konstantinov Petersburg Nuclear Physics Institute,
Gatchina Leningrad district 188350 Russia
Email: gerashch@pnpi.spb.ru*

Abstract. Living organisms are known to receive and transduce useful signals in a very noisy environment. A natural question whether they can use this noise constructively is still lacking solid answer. Some recent works demonstrated that one of the phenomena where noise plays a constructive role, the stochastic resonance (SR), occurs in biology at system and cellular levels of organization. Its existence at molecular level, in ionic channels of cell membranes, is now a subject for intensive, though mainly theoretical, studies. In the present work we study a simple two-state model of ionic channel together with its continuous equivalent, the asymmetrical Kramers oscillator, with external periodic signal and dichotomous external noise. We found that the new kind of SR appears for adiabatically slow external noise, when its amplitude becomes equal to static bias, thus dynamically recovering the system symmetry. These findings are confirmed with analog circuit simulations. The next promising effect is the noise-induced hypersensitivity to small signals. The phenomenon arises in stochastic systems with on-off intermittency. Such a system at optimal external noise intensity and correlation time becomes sensitive to an ultrasmall signal, amplifying it by many orders of magnitude. Such a hypersensitivity is often robust to additive thermal noise. We speculate that a similar effect might occur in various sensory systems that are known to be very noisy and to display unique sensitivity to environmental signals. A simple model system of two neurons with common strong noise source is hypersensitive to small differential signal only in the ideal case of absence of internal noise.

INTRODUCTION

Stochastic resonance is presently a well-recognized example of constructive role of noise in nature. The essence of this phenomenon is the ability of noise, whether internal or external, to optimize signal transduction in a nonlinear system functioning in an otherwise non-optimal regime. Since such non-optimal regimes occur more frequently in biological systems rather than in man-made ones, substantial efforts were performed to demonstrate the use of SR in biology. It was indeed found in various biological systems, from crayfish mechanoreceptor to human visual perception and fish search for prey (see [1] for review). Now the existence of SR at system and cellular levels of biological organization can be considered as proven. However, at the molecular level much work is still to be done; the unique experiment in an ensemble of ionic channels in artificial membrane supported by several theoretical studies [2]

CP665, *Unsolved Problems of Noise and Fluctuations: UPoN 2002: Third International Conference*,
edited by S. M. Bezrukov

had given rise to the concept of threshold-free SR applicable to a wide class of systems, but left questions on the role of SR in real ion channels. In particular, there are some doubts in the literature whether a single ion channel is able to demonstrate SR [3].

SR IN A SINGLE ION CHANNEL

In the present work we take a simple two-state model of ion channel with asymmetric stimulus-dependent rates $\alpha(V)$ and $\beta(V)$ of random transitions between open and closed state. We take two forms of these dependences (see Fig.1): the Goychuk-Hanggi (GH) one used in [3]

$$\alpha(V)=\frac{a_2(V+V_0)}{1-\exp\{-b_2(V+V_0)\}},\quad \beta(V)=a_1\exp(-b_1V),$$

and the one of Hodgkin-Huxley (HH) type, which is used widely in ion channel biophysics [4]:

$$\alpha(V)=a_2\exp\left[b_2(V+V_0)\right],\quad \beta(V)=a_1.$$

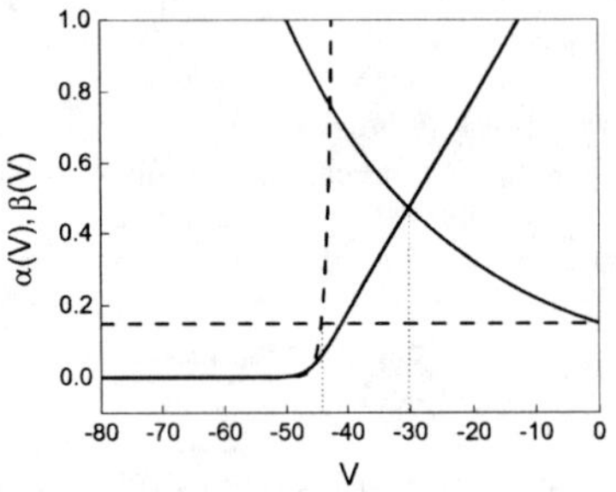

FIGURE 1. Transition rates versus voltage for GH (solid) and HH (dashed) models of ion channel.

The external stimulus V consists from the static voltage bias, the control noise, and the periodic signal. There are two kinds of stochasticity in our model, the internal one related to transitions between two states, and the external one caused by external noise. Besides, there are three characteristic times in the problem: the time of switching between states, equal to a_i^{-1}, the time of correlation of dichotomous noise γ^{-1}, and the period of signal T. We take $T \gg \gamma^{-1}, a_i^{-1}$, however allowing the ratio between γ and a_i to vary. We consider two cases, the adiabatic one $T \gg \gamma^{-1} \gg a_i^{-1}$, and the fast noise one $T \gg a_i^{-1} \gg \gamma^{-1}$. In both these limits it is possible, for dichotomous external noise and square-wave periodic signal, to explicitly calculate output signal and noise power as functions of parameters of the stimulus (the somewhat bulky formulas are given in [5]).
The signal-to-noise ratio (SNR) dependences on the bias p and the noise amplitude u are presented in Fig.2 for fast and slow noise cases.

We see that SR of new kind appears in this initially highly asymmetrical system: when the noise amplitude becomes close to bias value, the interstate transition probabilities become equal (the noise dynamically recovers the system symmetry), and the signal-to-noise ratio passes through a maximum. The occurrence of such resonance is universal and requires only the presence of a point where the curves of dependences of transition rates on external potential intersect. This kind of SR differs from that was

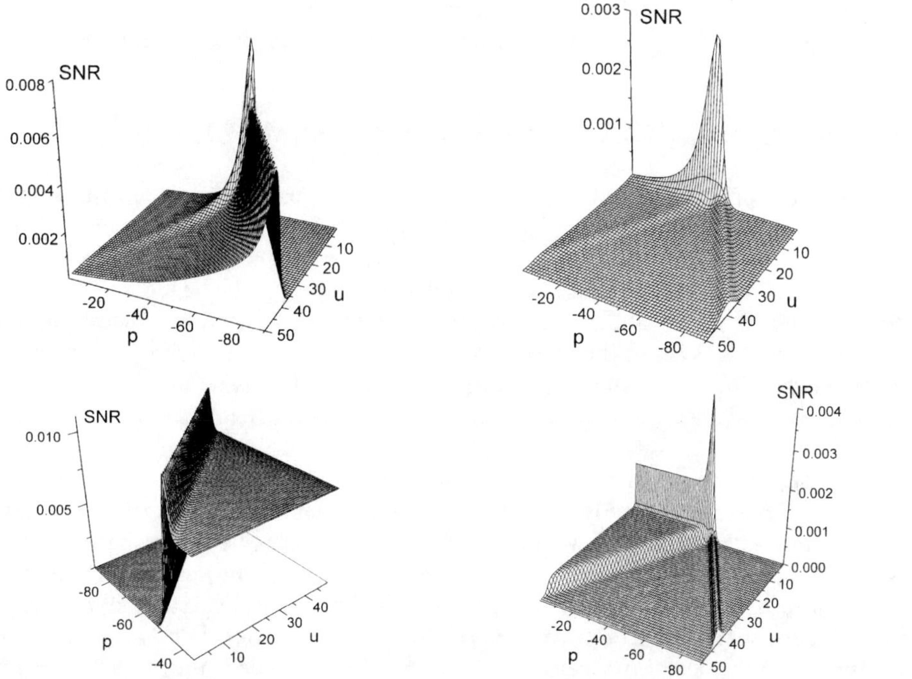

FIGURE 2. Signal-to-noise ratio in our model for the values of parameters $V_0 = 46, a_2 = 0.03,\ b_2 = 1, a_1 = 0.015, b_1 = 0.038,\ \mathrm{v} = 1$, versus the bias p and the noise amplitude u, for fast(left) and slow (right) noise cases. The upper graphs are drawn for GH transition rates, the lower for HH ones. All voltages are expressed in mV. SR is observed at $p < -V_0$.

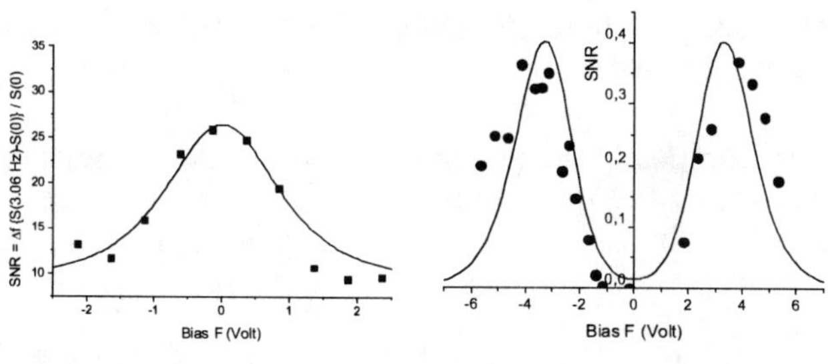

FIGURE 3. Results of electronic circuit simulations of an overdamped nonlinear oscillator under action of the white "internal" noise, the periodic square-wave signal with frequency 3.06 Hz, the bias voltage F and the dichotomous noise: fast (correlation time much less than Kramers time, left graph) and slow (the opposite case, right graph). In the second case, the two maxima appear at the absolute value of bias voltage equal to the amplitude of dichotomous noise, like in Fig.2. Solid lines represent theoretical predictions.

found by Bezrukov and Vodyanoy [2], since the last is observed in the range $\alpha \ll \gamma \ll \beta$ (rare short pulses). The analytical studies are confirmed by analog circuit simulation of the continuous equivalent of our model, the asymmetrical Kramers oscillator (Fig.3). Therefore, we confirm the possibility of appearance of SR in a single ion channel and found the conditions of its existence. The phenomenon is well fitted in a general SR framework of optimizing signal transduction in a non-optimal system by noise.

NOISE-INDUCED SENSITIVITY TO SMALL SIGNAL

The second phenomenon that is promising as a possible mechanism of noise utilization in biology is the noise-induced hypersensitivity to small signals [6]. This effect appears in systems that display on-off intermittency: a parametric noise drives the system back and forth through bifurcation point, altering stochastically the type of system dynamics. As a result, the bursts with a power-law amplitude distribution are generated. Such a system becomes hypersensitive to an asymmetry caused by ultrasmall slow alternating signal, and responds by large bursts whose sign is identical with that of a signal. Therefore the signal gain factor reaches many orders of magnitude.

We investigate the phenomenon in a nonlinear overdamped oscillator with external multiplicative noise. Physically, the system can be imagined as a Brownian particle moving in time-dependent potential that randomly alternates between single-well and double-well form. The multiplicative noise drives the system through bifurcation point and therefore induces *on-off* intermittency. The latter means that the probability density function (PDF) of dynamic variable obeys a power law, i.e., very large fluctuations can appear with reasonable probability. Besides, the system becomes sensitive to the sign of external exponentially small signal responding by large bursts of the same sign (Fig.4).

The system is described by the following Stratonovich SDE:

$$\frac{dx}{dt} = \lambda x - f(x) + \beta\, \xi(t) x + AE(t) + \sigma \varphi(t),$$

where $f(x)$ is the nonlinear function, $\xi(t)$ and $\varphi(t)$ are the mutually uncorrelated noise sources, and $E(t)$ is the zero-mean square-wave signal of unit amplitude, $\lambda, \beta, A, \sigma$ are constants.

We investigate, theoretically, numerically and by analog simulations, the system with various nonlinearities $f(x)$, as well as with various noise sources: white, colored Gaussian and dichotomous noise. For all these cases we obtain the power-law PDF:

$$F(x,t) \sim |x|^{\alpha-1}\, sign(E(t)x).$$

The hypersensitivity occurs when α (the hypersensitivity index) is small. The exact form of α is different for different systems. For white and dichotomous noise we get, correspondingly:

$$\alpha = \frac{2\lambda}{\beta^2}, \quad \alpha = \frac{\lambda}{\tau(\Delta^2 - \lambda^2)},$$

where Δ is the amplitude and τ is the correlation time of dichotomous noise.

The signal gain factor, easily obtained from PDF, is:

$$I = \frac{\langle x(t) \rangle}{AE(t)} \sim \frac{1}{A^{1-|\alpha|}}.$$

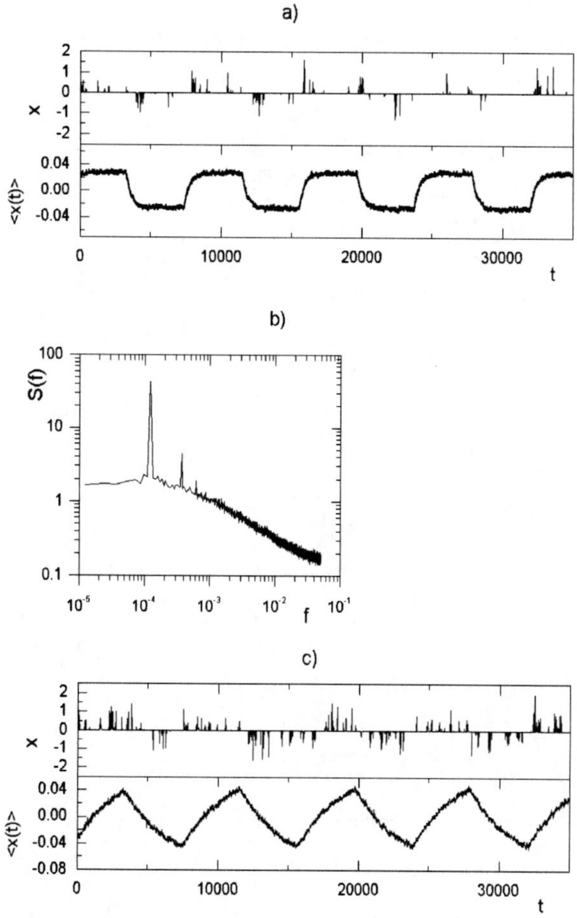

FIGURE 4. (a) Time series of raw output signal x and its ensemble average $\langle x(t) \rangle$ over 4100 samples with identical phase of input signal for $\beta = 1.0, \lambda = -0.01$; (b) power spectrum of x for the signal (a), averaged over 200 samples of random phase; (c) the same as for (a) but for $\lambda = -0.03$. The input signal amplitude $A = 10^{-11}$, period $T = 8192$. The multiplicative noise is white.

It can be very large for small α and A, and passes through a maximum in an optimum range of noise intensities and correlation times (Fig.5).

So, we demonstrate that for various systems with *on-off* intermittency a universal scaling index of hypersensitivity exists. The condition of hypersensitivity, determined by the value of this index, is the same for different systems with *on-off* intermittency, and the system specifics reveals itself only in dependence of this factor from system parameters. We hope that the observed phenomenon could be used in new types of detecting devices, and, probably, shed a light on the unique ability of biological systems to detect weak signals in noisy environment.

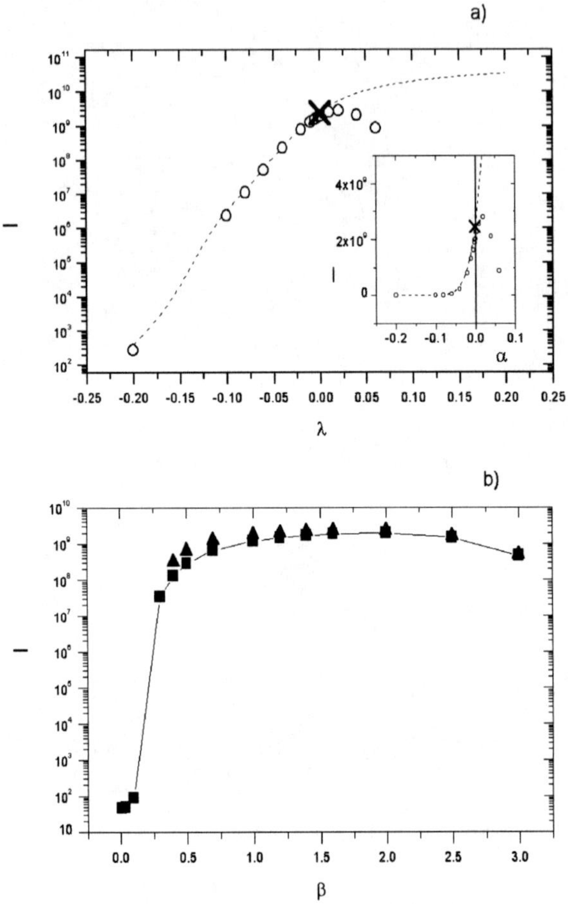

FIGURE 5. (a) Gain factor I versus parameter λ (main graph) and vs. "hypersensitivity index" α (inset) for fixed intensity of white multiplicative noise. Dashed line represents simulations with a constant input signal. The cross displays an estimate given in the text. (b) The gain factor I versus noise intensity β for $\lambda = -0.01$ (squares) and 0.01 (triangles). The input signal is the same as in Fig. 4.

Despite of systems with multiplicative noise, on-off intermittency was also found in identical chaotic maps synchronized by common noise [7]. Therefore it is natural to examine whether two sensory cells or neurons with a common noise source (provided, for example, by third neuron) are able to display sensitivity to small environmental signals. Using two identical model neurons biased slightly below spiking threshold and synchronized by strong common synaptic noise, we show that a very small slow differential voltage signal desynchronizes the neurons, so the difference of their membrane potentials, considered as output signal, exceeds the input voltage by six orders of magnitude. However, unlike the multiplicative noise case, the system is not robust to internal noise or even small nonidentity of neurons, and therefore this particular class of hypersensitive systems is unlikely to serve as a model of remarkable sensitivity of living organisms to environmental stimuli.

CONCLUSIONS AND OPEN QUESTIONS

We have investigated the two phenomena that appear in nonlinear stochastic systems: stochastic resonance in a doubly stochastic system and noise induced sensitivity to ultrasmall signal. We speculate that similar effects might occur in various sensory systems that are known to be very noisy and to display unique sensitivity to environmental signals. For SR in a biased ionic channel, our model is realistic enough to be tested further in a real single-channel experiment. But for noise-induced sensitivity, a relevant biophysical model is still not found. We ruled out the simplest one, the couple of neurons synchronized by common noise.

ACKNOWLEDGMENTS

This work was supported by the RFBR (grant 02-02-16979) and by the State Programs “Quantum macrophysics” and “Investigations of collective and quantum effects in condensed matter”.

REFERENCES

1. L. Gammaitoni, P. Hanggi, P. Jung, and F. Marchesoni, Rev. Mod. Phys. **70**, 223 (1998); V.S. Anishchenko, A.B. Neiman, F. Moss, and L.Shimansky-Geier, Sov. Phys. Usp. **42**, 7 (1999).
2. S.M. Bezrukov and I.Vodyanoy, Nature **378**, 362 (1995); S.M. Bezrukov and I. Vodyanoy, Nature **385**, 319 (1997); S.M. Bezrukov, Phys. Lett. A, **248**, 29 (1998).
3. I. Goychuk and P. Hanggi, Phys. Rev. E **61**, 4272 (2000); G. Schmid, I. Goychuk, and P. Hanggi, Europhys. Lett. **56**, 22 (2001).
4. B. Hille, Ionic Channels of Excitable Membranes, 2nd ed. (Sinauer Associates, Sunderland, MA, 1992).
5. S.L. Ginzburg and M.A. Pustovoit, Phys. Rev. E. **66**, 021107 (2002).
6. S.L. Ginzburg and M.A. Pustovoit, Phys. Rev. Lett. **80**, 4840 (1998); O.V. Gerashchenko, S.L. Ginzburg and M.A. Pustovoit, Eur. Phys. J. B **19**, 101 (2001).
7. H.L. Yang, E.J. Ding, Phys. Rev. E **50**, R3295 (1994).

Stochastic Resonance, Brownian Ratchets and the Fokker-Planck Equation

A. Allison* and D. Abbott*

**Centre for Biomedical Engineering (CBME) and*
Department of Electrical and Electronic Engineering, University of Adelaide, SA 5005, Australia.

Abstract. Our immediate aim is to arrive at quantitative realistic estimates of the optimum noise levels for a complete and feasible Brownian ratchet device. A number of difficult "Unsolved Problems of Noise" naturally arise in the course of this work. We pose a general philosophical question: "Is it is possible to relate stochastic resonance and Brownian ratchets in a formal way, through a consideration of the Fokker-Planck equation?" The evidence suggests that some forms of stochastic resonance can be formulated in terms of a Fokker-Planck equation and some cannot. Stochastic resonance may actually be a collection of similar but distinct phenomena.

We present a number of simulations which demonstrate stochastic resonance in a Brownian ratchet. We also show how current is dependent on a number of other parameters.

INTRODUCTION

When engineers design machines at the human scale they tend to regard inertial, elastic and gravitational forces as being the most significant. Even when dissipative forces are considered, these are usually regarded as small additions perturbations to the original theory. At the molecular end of the scale; if we try to design nano-machines then dissipative forces are the most significant forces and the effects of thermal motion of the surrounding medium are very evident. This gives rise to a chaotic agitated movement called Brownian motion. At this scale, it makes sense to attempt to *harness* the random thermal motion in the surrounding medium, rather than to try to resist it or to overpower it or to block it out. This idea can be traced at least as far back as Maxwell and on through Smoluchowski [1] and to Feynman [2]. It is a still topic of active research [3, 4, 5, 6]. Brownian ratchets work by directing particles locally using a potential field. They must operate in at least two modes. An example of a flashing ratchet is shown in Figure 1. A very surprising feature of Brownian ratchets is that in each of the modes, the probability current is "downhill" towards regions of lower potential and yet when the ratchet is switched between the modes, the steady state current can be "uphill" against the average potential gradient.

One of the main motivations for studying ratchets is to understand sub-cellular transport processes [7, 8, 9, 10, 11, 12, 13, 14]. There is considerable evidence that molecular motors in living cells use a ratchet mechanism.

Harmer *et al.* [15] have shown that a very simple discrete-time ratchet can demonstrate the same "uphill" current as the continuous time ratchet. This led to the notion that a combination of two "losing" games can be "winning."

CP665, *Unsolved Problems of Noise and Fluctuations: UPoN 2002: Third International Conference,* edited by S. M. Bezrukov

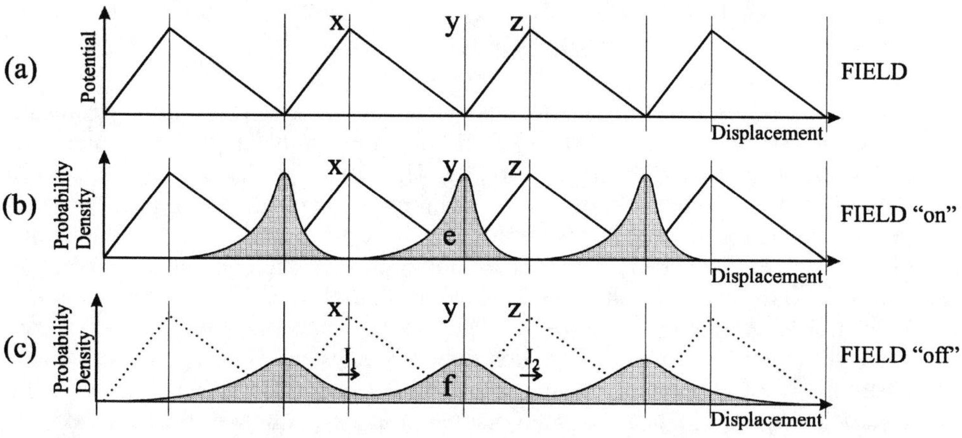

FIGURE 1. The Brownian ratchet requires a potential, Part (a) shows the ratchet shaped potential. The Brownian ratchet requires at least two modes of operation. Part (b) shows the effect when the field is asserted. The charged Brownian particles accumulate near points of lowest electrical potential, such as pont "y." Diffusion prevents the particles from all converging to the same point. This diffusion is the effect of numerous collisions with the particles that make up the surrounding medium. These collisions can also be regarded as noise. Brownian ratchets require the presence of noise. Part (c) shows the effect when the field is turned off and the system "relaxes" as the particles diffuse. The bulk of the distribution is near point "y" and the ratchet is asymmetrical so point "y" is closer to point "z" than it is to point "x." This means that the current density, J_2, past point "z" will be greater than the current density, J_1, past point "x." This inequality is the cause of the steady-state current in the ratchet.

Stochastic Resonance

Stochastic Resonance, SR, [16, 17, 18, 19, 20] is a phenomenon where we can obtain optimal output from a system by adding noise to the system. This idea has been applied to a great number of possible applications in physics, engineering and biology. Much of the early investigation required the use of thresholds or bi-modal potential wells and formulated the problem in terms of a Fokker-Planck equation. See, for example [16].

Stochastic resonance is believed to play a role in the normal operation of the nervous system and it is interesting to note that the Fokker-Planck plays a role in the operation of neural networks [21].

It should be admitted that not *all* forms of SR can be formulated in terms of a Fokker-Planck equation. For example, Bezrukov and Vodyanoy [22] have devised a time dependent Poisson process that has no activation barrier of any kind and is not formulated in terms of a Fokker-Planck equation. It seems that what we notionally call "Stochastic Resonance" is really a collection of several phenomena with similar appearance.

A possible connection between Brownian Ratchets and Stochastic Resonance

The earliest reference that we can find to a connection between Stochastic Resonance and Brownian ratchets is Gang *et al.* [23]. They expand the dynamics into independent subspaces according to the theory of Floquet. They make the equations tractable by only keeping a few important terms from an infinite expansion. Some approximation is involved. All the reasoning is algebraic. There are no experiments or simulations. Doering [24] gives clear analysis of some simple models. He computes the resulting current for both the "fast" flashing and "slow" flashing limits. The flashing is "slow" or "adiabatic" if the temporal period of the potential function in the ratchet is long compared with the characteristic relaxation time of the system. Doering includes graphs demonstrating a non-monotonic change in probability current density $\mathbf{J}$ with increasing mean particle energy, $k_B T$, but the final results are not quantitative. Berdichevsky and Gitterman [25] give a very complex and detailed analysis of SR and ratcheting in a Josephson junction with noise and also conclude that SR and ratchet effects exist in diffusive systems with oscillating barriers [26].

We have shown that Parrondo's discrete-time ratchet exhibits a Stochastic Resonance effect [27]. We have also shown that, with some modification, Parrondo's games can be used to model realistic Fokker-Planck equations and can be used to numerically simulate the dynamics of Brownian ratchets [28]. Our aim here is to simulate realistic Brownian ratchets using Parrondo's games as a numerical method. The aim is to optimize designs for possible ratchets.

Our interest in SR is that we must ensure that the optimum level of noise applies, in order to achieve optimum current in the ratchet device.

OUTLINE OF APPROACH

It is possible to derive a Partial Differential Equation for the probability density of finding a particle at a certain place and time [29, 6]. This is called the "master" equation or "Fokker-Planck" equation:

$$\frac{\partial^2}{\partial z^2}\left(D(z,t)\,p(z,t)\right) - \frac{\partial}{\partial z}\left(\alpha(z,t)\,p(z,t)\right) - \frac{\partial}{\partial t}p(z,t) = 0\,. \tag{1}$$

In this approach we solve for $p = p(z,t)$ which is the probability density of finding a particle near the point z, in space, at time t. We then apply other operators to p in order to calculate any macroscopic parameters of interest, such as probability current density, J, or mean position $E[z]$. We apply finite-difference techniques to Equation 1 to convert it to an equivalent set of difference equations which have the form of Parrondo's games [28]. These difference equations can be evaluated numerically.

Definition of "Noise"

The natural definition of noise would be to use a standardized model of white noise, dB as a reference. This is summarized in the term $g(z,t)\,dB$, in the Langevin equation: $dz(z,t) = \alpha(z,t)\,dt + g(z,t)\,dB$. In this paper we use $D^{(2)} = D = g^2$ as the operational macroscopic definition of "noise." This definition of noise increases monotonically with increasing mean energy of the molecules in the surrounding medium, $\frac{3}{2}kT$. Reif [30] gives an expression for an ideal gas, D =constant$\times(kT)^{3/2}$. The choice of different definitions of noise, g, $D = g^2$ or kT would not alter the basic phenomena; it would only re-scale the axes.

Definition of "Probability Current"

We must apply a law of total probability in the form of a continuity law $\nabla J + \frac{\partial p}{\partial t} = 0$. We define the probability current density as $J = -\nabla(D\cdot p) + (\alpha \cdot p)$ which is a form of Fick's law and is consistent with Risken [29] . In this context we require that $\alpha(z,t) = u\underline{E} = u(-\nabla V(z,t))$ where u is the mobility of the Brownian particle, V is the applied voltage and $\underline{E}$ is the resulting electric field [1]. In three dimensions, we can combine the continuity equation and Fick's law to get the Fokker-Planck equation: $\nabla^2(D\cdot p) - \nabla(\alpha\cdot p) - \frac{\partial p}{\partial t} = 0$. For the one dimensional case, with constant diffusivity, we get

$$D\frac{\partial^2 p}{\partial z^2} - \frac{\partial(\alpha p)}{\partial z} - \frac{\partial p}{\partial t} = 0\,. \tag{2}$$

Risken [29] and Parrondo [6] affirm that the Fokker-Planck equation does describe the fundamental dynamics of the physical process in a Brownian ratchet. This is the physical model that we use. Our aim is to make it realistic and to calculate the effects of our choices.

Boundary Conditions and Initial Conditions

We require the probability current density at infinity to be zero, i.e. $\lim_{z\to+\infty} J(z) = \lim_{z\to-\infty} J(z) = 0$. If we integrate the Continuity Equation together with the boundary conditions the we get a law of total probability: $\int_{-\infty}^{+\infty} p(z,t)dz = 1$ for all values of t. This

[1] If charge becomes crowded then space-charge effects will be significant and the effective electric field will be altered, $\underline{E} = \underline{E_1} + \underline{E_2}$. The first component, $\underline{E_1} = -\nabla V(z,t)$, is due to the applied voltage. The second term is due to space charge and corresponds to Poisson's equation (with zero boundary conditions), $\nabla \underline{E_2} = \frac{(pq)}{\varepsilon}$, where q is the charge per Brownian particle and consequently (pq) is the expected value of the charge density. The constant, ε, is the permittivity of the surrounding medium. In these simulations we have assumed that the sample of Brownian particles is sparse and that $\underline{E_2}$ is not significant, compared with $\underline{E_1}$.

is an important normalization condition. All solutions and simulations must satisfy this condition.

An important consequence of these boundary conditions is that we can define the mean drift velocity in terms of probability density, p, or current density, J:

$$v = \frac{\partial}{\partial t} E[z] = \frac{\partial}{\partial t} \int_{-\infty}^{+\infty} z p(z,t) dz = \int_{-\infty}^{+\infty} J(z,t) dz . \quad (3)$$

We imagine that the particles are injected into a single slot between the teeth of the ratchet at $t = 0$. A reasonable abstraction is $p(z,0) = \delta(z)$. In the numerical simulations, we spread the initial probability across the width of a single slot of the ratchet. This would correspond to the injection of a small sample into a single slot of the ratchet.

Asymptotic Solution

We are interested in the behavior of the ratchet as the distribution spreads out, for large time $t \to +\infty$. The initial conditions do not have a great effect on the parameters of interest, including rates of increase in variance and probability currents, J, in the asymptotic limit. In the simulations this is achieved by running the simulations for a sufficient time, to eliminate the effect of the initial transients.

FINITE DIFFERENCE SIMULATIONS

We have previously shown that Parrondo's games are a particular way of sampling the Fokker-Planck equation [28]. Parrondo's games were originally conceived as a purely heuristic model or "*gedankenexperiment*" but we have shown that they can be generalized to make them realistic.

We have also previously shown that that Parrondo's original games demonstrated an effect which had the same form as Stochastic resonance [27, 31]. In this paper we use more realistic choices for the parameters (including: sampling time and sampling distance, diffusivity, particle mobility and charge) for our simulations of Brownian ratchets and show that these more realistic ratchets also exhibit an effect which has the same form as stochastic resonance. We made choices of the sampling times and distances for a finite element simulation of sodium ions, Na^+, diffusing in water, H_2O. We chose a feature size for the ratchet teeth of $l = 2\ \mu$m, which is well within the limits of existing MEMS technology. We have assumed that the effect of charge crowding is negligible. A typical instant in the numerical simulation is shown in Figure 2(a). The general shape of the distribution is Gaussian although it is clearly modulated by another function, the impression of the teeth of the ratchet. If the field is asserted for a long time then the profile within each slot of the ratchet closely resembles the steady state solution, which is a two sided exponential function.

The time evolution of the ratchet process consists of a series of these functions and is shown in Figure 3. It is possible to apply operators to the distributions in Figures 2 and 3 to calculate quantities of interest, such as probability current density J or mean

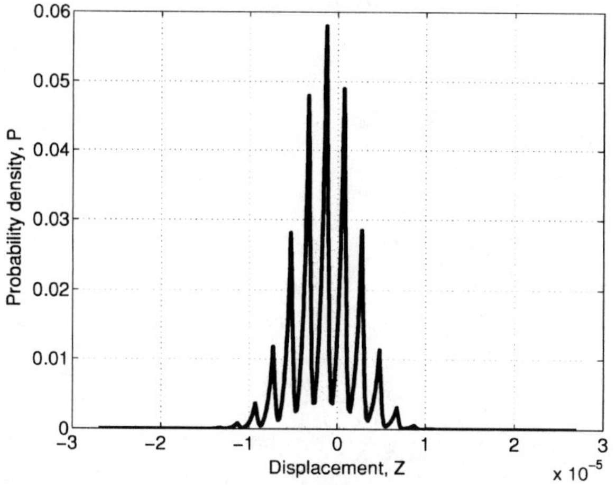

FIGURE 2. A typical time-slice, showing $p(z,t_0)$ for a fixed value of time t_0. The PDF is essentially Gaussian, multiplied by a modulating function, caused by the teeth of the ratchet.

position of the distribution $E[z]$, as shown in Figure 4. We can see how the mean position of the particle shifts in response to the modulation of the electric field of the ratchet. The time interval for which the field is "off" represents noise, since the particle simply undergoes Brownian motion in response to agitation of the surrounding particles. There is no constraint that limits the effect of the noise. This is equivalent to Game "A" in Parrondo's games. The time interval for which the field is "on" represents the action of the ratchet. It is during this time interval that the mean position of the particle moves. If this mode of operation persists for long enough though then the distribution of the particles reaches an equilibrium and detailed balance is achieved and the current of particles will stop. It is necessary to turn off the field and allow some diffusion in order for the current to be maintained.

The presence of noise is needed to maintain the current. More noise produces more current up to a certain point, after which the ratchet is overwhelmed by noise and the current diminishes and eventually stops. This has the form of the stochastic resonance curve, shown in Figure 5.

STOCHASTIC RESONANCE

We can create a parametric graph of the asymptotic rates of probability current in the ratchet as a function of increased diffusivity. This graph is shown in Figure 5 and clearly has the form of a stochastic resonance curve. The rate of probability current in the ratchet can be regarded as a rate of change in the first moment, $\partial E[z]/\partial t$ The operation of the ratchet affects all of the moments. The second central moment or variance, σ^2, contains information about the spread of the distribution. Spread has to be regarded as an undesirable feature of ratchets. Lindler *et al.* have raised the

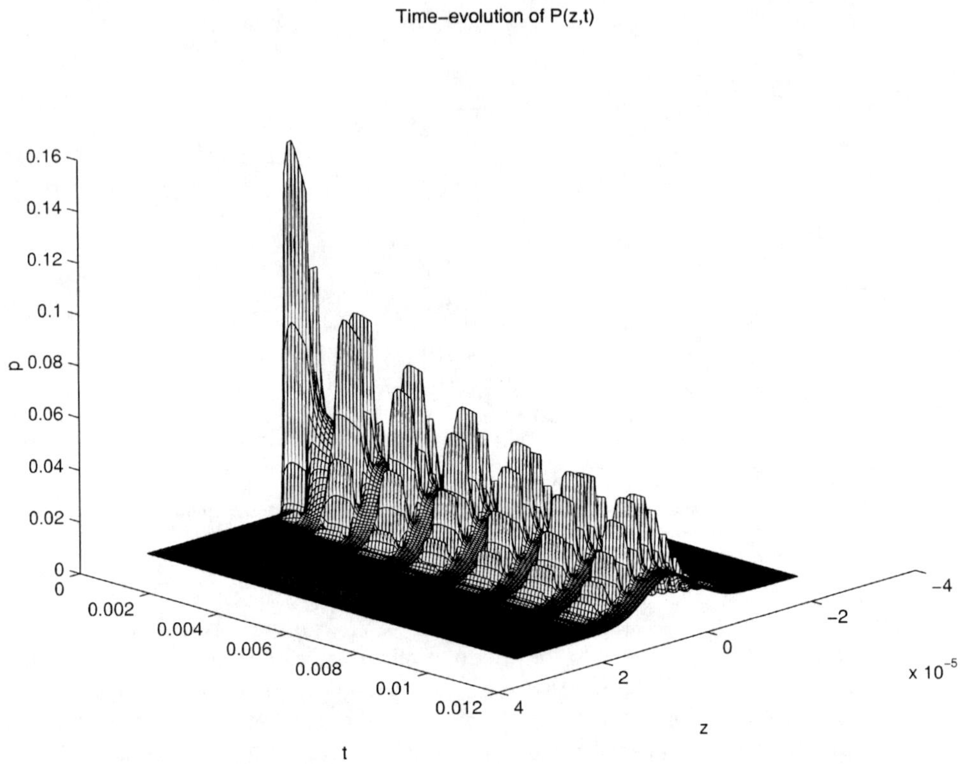

FIGURE 3. A finite-difference simulation of a Brownian ratchet, based on Parrondo's games. z is space, t is time and p is probability density. The spatial period of this ratchet is $L_z = 2\ \mu$m. The temporal period of this ratchet is $T_0 = 1.5$ ms. The physical constants are scaled for the diffusion of hydrated sodium ions in water.

issue of transport coherence [32]. They use a normalized measure of transport and include scaling factors of the spatial period of the ratchet and the effective diffusion coefficient. Einstein's solution to the diffusion equation is a spreading Gaussian curve with variance $\sigma^2 = 2Dt$ and we can use this to define an effective diffusion coefficient for any time varying distribution $p(z,t)$. We can define the effective diffusion coefficient as: $\lim_{t\to\infty}\left(\sigma^2(t)/(2t)\right)$ where $\sigma^2(t)$ is the variance at time t. We can then define the Péclet number [32] as $P_e = vL_z/D_{eff}$ where v is the mean drift velocity of the Brownian particles and L_z is the spatial period of the ratchet. The results for our ratchet are shown in Figure 6 we can see that this ratchet, as constructed is not very coherent. Lindler *et al.* have achieved Péclet number of the order of $P_e = 20$ with some designs. There is a clear Stochastic resonance peak in the graph of the Péclet numbers.

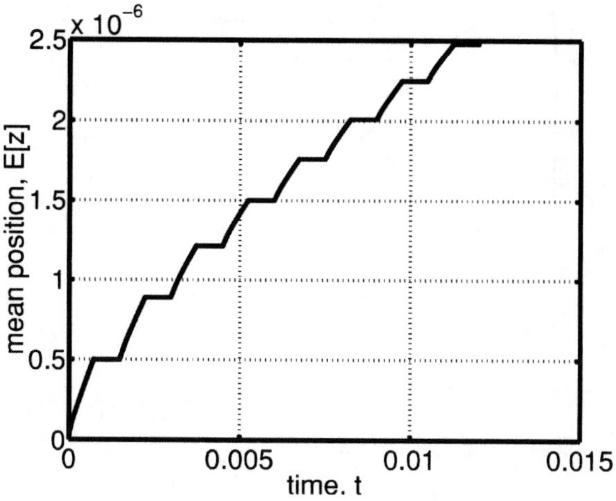

FIGURE 4. Time-evolution of the mean of the distribution $P(z,t)$, called $E[z]$. When the field is asserted, the mean position of the particles moves in a generally "upward" direction. When the field is turned off, the mean remains constant although diffusion causes the field to spread. The total shift in mean position of this ratchet is very modest. Part of the motivation of this work is to optimise the transport effect of the Brownian ratchet, subject to constraints.

CONCLUSIONS AND OPEN QUESTIONS

The conclusions of the present work are: (*i*) that Parrondo's games are not just a heuristic model. They can be made quantitative and can be used as a numerical method and (*ii*) we confirm that the one dimensional Brownian flashing ratchet exhibits stochastic resonance and (*iii*) we can use Parrondo's games as a numerical method which could be used to optimise the performance of a ratchet design.

There are a couple of open questions that naturally arise: (*i*) "Does the Fokker-Planck equation unify some types of Stochastic Resonance?" and (*ii*) "Do other "Fokker-Planck" systems, such as neural-networks have the same *type* of stochastic resonance as Brownian ratchets?"

ACKNOWLEDGEMENTS

We thank Dean Astumian and Lutz Schimansky-Geier for their suggestions that have helped us to formulate some problems more clearly and have suggested fruitful areas for further investigation.

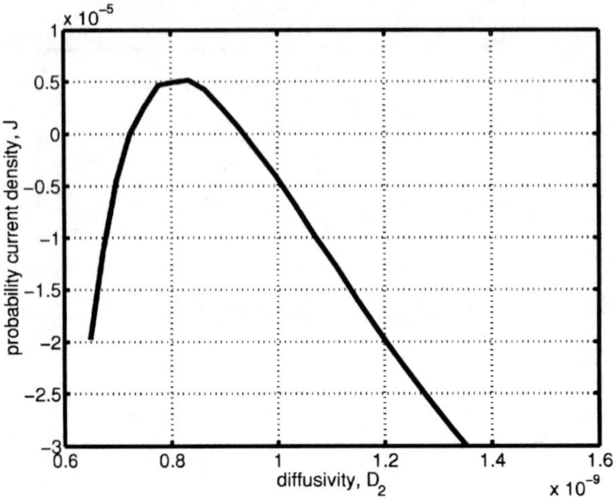

FIGURE 5. Parametric relationship between the Diffusivity, or Fick's law constant, $D = D^{(2)}$ and the probability current density, J. All other parameters are held constant; the feature size of the teeth was $2um$, The temporal period of the ratchet was 1.5 ms, the voltage across each 2 μm tooth was 60 mV. This gives electric field strengths in a similar range to those used in commercial gel electrophoresis. The potential across each tooth is easily small enough to avoid electrolysis since the standard reduction potential of Na is much larger in magnitude, $E^0 = -2.71$ V [33]. The temporal duty cycle was held at a symmetrical 50%:50% and the spatial duty cycle was held at an asymmetrical 80%:20%. This breaking of symmetry is necessary to generate current in the ratchet. The only independent variable here is the diffusivity of the particle, D. The ratchet selects preferentially for particles with a certain range of diffusivity.

REFERENCES

1. von Smoluchowski, M. V., *Phys. Z.*, **13**, 1069–1080 (1912).
2. Feynman, R. P., Leighton, R. B., and Sands, M., *The Feynman Lectures on Physics*, vol. 1, Addison-Wesley, Reading, MA, 1963.
3. Reimann, P., *Physics Reports*, **361**, 57–265 (2002).
4. Reimann, P., and Hänggi, P., *Appl. Phys. A.*, **75**, 169–178 (2002).
5. Wang, H., and Oster, G., *Appl. Phys. A*, **75**, 315–323 (2002).
6. Parrondo, J. M. R., and De Cisneros, B. J., *Appl. Phys A.*, **75**, 179–191 (2002).
7. Peskin, C., Odell, G., and Oster, G., *Biophys. J.*, **65**, 316–324 (1993).
8. Doering, C., B., E., and G., O., *Biophysical Journal*, **69**, 2256–2267 (1995).
9. Bier, M. R., and Astumian, D., *Bioelectrochemistry and Bioenergetics*, **39**, 67–75 (1996).
10. Astumian, R. D., Chock, P. B., Tsong, T. Y., Chen, Y., and Westerhoff, H., *Proc. Natl. Acad. Sci. USA*, **84**, 434–438 (1987).
11. Bier, M., *Contemp. Phys.*, **38**, 371–379 (1997).
12. Bier, M., "A motor protein model and how it relates to stochastic resonance, Feynman's ratchet, and Maxwell's demon," in *Stochastic dynamics*, Springer, Berlin, 1997, vol. 386, pp. 81–87.
13. Astumian, D., *Scientific American*, **285**, 56–64 (2001).
14. Astumian, D., *Appl. Phys. A*, **75**, 193–206 (2002).
15. Harmer, G., Abbott, D., Taylor, P. G., and Parrondo, J. M. R., "Parrondo's Paradoxical Games and the Discrete Brownian Ratchet," in [34], pp. 189–200.
16. Gammaitoni, L., Hänggi, P., Jung, P., and Marchesoni, F., *Reviews-of-Modern-Physics*, **70**, 223–287 (1998).
17. Zozor, S., and Amblard, P. O., *IEEE Transactions on Signal Processing*, **47**, 108–122 (1999).

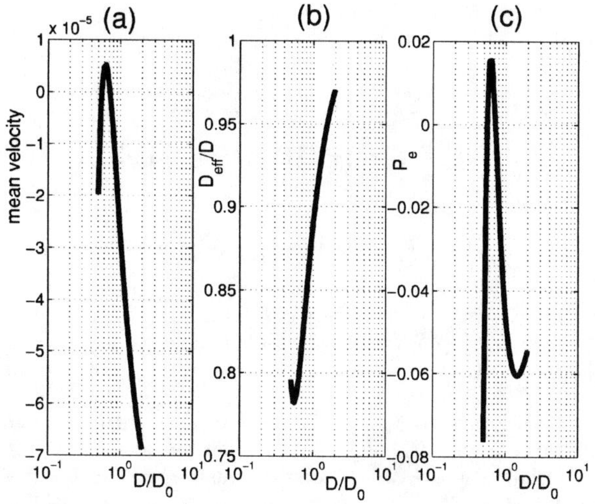

FIGURE 6. In plot (a) we see how the mean drift velocity varies as we vary the diffusion coefficient around a nominal value, $D_0 = 1.3 \times 10^{-9} \mathrm{m}^2\mathrm{s}^{-1}$, in (b) we show the effective diffusion coefficient of the whole device and in (c) we see the normalized drift coefficient or "Péclet" numbers.

18. Luchinsky, D. G., Mannella, R., McClintock, P. V. E., and Stocks, N. G., *IEEE Transactions on Circuits and Systems – II Analog and digital signal processing*, **46**, 1205–1214 (1999).
19. Luchinsky, D. G., Mannella, R., McClintock, P. V. E., and Stocks, N. G., *IEEE Transactions on Circuits and Systems – II Analog and digital signal processing*, **46**, 1215–1224 (1999).
20. Harmer, G. P., Davis, B. R., and Abbott, D., *IEEE Transactions of Instrumentation and Measurement*, **51**, 299–309 (2002).
21. Haykin, S., *Neural Networks*, Macmillan College Publishing Co., New York, 1994.
22. Bezrukov, S. M., and Vodyanoy, I., "In search of a possible statistical basis of stochastic resonance," in [34], pp. 169–178.
23. Gang, H., Nicholis, G., and Nicholis, C., *Phys. Rev. A.*, **42**, 4 (1990).
24. Doering, C. R., *Il Nuovo Cimento*, **17D**, 685–697 (1995).
25. Berdichevsky, V., and Gitterman, M., *Physical Review E.*, **56**, 6340–6354 (1997).
26. Berdichevsky, V., and Gitterman, M., *Phys. A*, **249**, 88–95 (1998,).
27. Allison, A., and Abbott, D., *Fluctuation and Noise Letters*, **1**, L239–L244 (2001).
28. Allison, A., and Abbott, D., *Fluctuation and Noise letters* (2002), in process, preprint: `http://xxx.lanl.gov/ cond_mat/0208470`.
29. Risken, H., *The Fokker-Planck Equation*, Springer, 1989, second edn.
30. Reif, F., *Statistical and Thermal Physics*, McGraw Hill, 1985.
31. Lee, Y., Allison, A., Abbott, D., and Stanley, H. E., *Phys. Rev. Lett.* (2002), submitted, preprint: `http://xxx.lanl.gov/ cond_mat/0205302`.
32. Lindner, B., Kostur, M., and Schimansky-Gier, L., *Fluctuation and Noise Letters*, **1**, R25–R39 (2001).
33. Aylward, G. H., and Findlay, T. J. V., *Chemical Data Book*, John Wiley and sons pty. ltd., Sydney, 1968, fourth edn.
34. Abbott, D., and Kiss, L. B., editors, *Unsolved problems of Noise and Fluctuations (UPoN'99)*, vol. 511, American Institute of Physics (AIP), 2000.

Is Stochastic Resonance Just an Epiphenomenon?

Lawrence M. Ward

Department of Psychology and The Brain Research Centre, University of British Columbia, 2136 West Mall, Vancouver, B.C., V6T 1Z4, Canada

Abstract. Stochastic resonance (SR) has often been demonstrated in dynamic systems and in systems that have thresholds, including animal and human sensory systems. In threshold SR, a discontinuous threshold is not necessary; output signal-to-noise ratio exhibits SR if the noisy input signal is subjected to a "soft" threshold transform. However, signal detection theory, an implementation of statistical decision theory, raises special problems for this conclusion. When recast into this theory, SR appears to be an epiphenomenon of non-optimal placement of a criterion nonlinearity, rather than a genuine increase in sensitivity to weak signals. Here I generalize the above-mentioned results of SR without a discontinuous threshold to the case of a signal detection theory measure of sensitivity, both showing how SR can increase sensitivity to weak noisy signals and indicating limitations of this generalization.

INTRODUCTION

Stochastic resonance (SR), a phenomenon of nonlinear systems in which noise appears to amplify a weak signal, has often been demonstrated in dynamic systems and in simple static systems that have thresholds [1]. Moreover, a discontinuous threshold is not necessary - output signal-to-noise ratio exhibits SR if the noisy input signal is subjected to certain kinds of continuous transforms that implement "soft" thresholds. Signal detection theory, an implementation of statistical decision theory, however, raises special problems for this conclusion. When recast into this theory, "...the phenomenon of stochastic resonance can be considered no more than an epiphenomenon, related to a non-optimal criterion level..." [2, p. 478], rather than a genuine increase in sensitivity of a signal detection system to a weak signal. Since an epiphenomenon is a secondary phenomenon that is a by-product of another, primary, phenomenon, this is a disagreeable conclusion for those who see SR as a fundamental phenomenon illustrating the constructive role noise can play in physical and biological systems. Moreover, it flies in the face of much evidence that SR can increase sensitivity to weak signals in biological sensory systems [3], including those of humans. In what follows I briefly characterize threshold SR, summarize the evidence for SR in sensory systems and define two types of such SR, and generalize the above-mentioned results of SR without a discontinuous threshold to the case of a signal detection theory measure of signal detectability.

CP665, *Unsolved Problems of Noise and Fluctuations: UPoN 2002: Third International Conference*, edited by S. M. Bezrukov

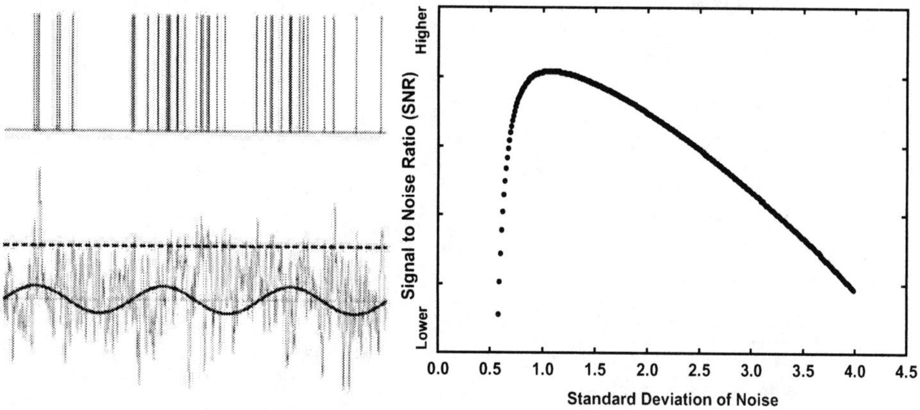

Figure 1. Left: Time density of exceedances of threshold (dotted line) by noisy signal reflects its modulation. Right: Stochastic resonance is indicated by the maximum of the signal-to-noise ratio of spectral power density at the signal frequency occurring at a non-zero noise level.

THRESHOLD STOCHASTIC RESONANCE

In threshold SR, each time a noisy, subthreshold signal exceeds the threshold an exceedance is recorded. Fig. 1 (left panel) shows how the time density of these exceedances reflects the modulations of the signal. The signal-to-noise ratio is defined as the spectral power at the signal frequency when the signal is present divided by that at the signal frequency when the signal is absent. Threshold SR is defined as the case where the SNR is maximal when the amplitude of the noise, measured by the standard deviation the distribution that describes the usually Gaussian noise, is at some non-zero level (Fig. 1, right panel).

SR IN HUMAN SENSORY SYSTEMS

Two types of SR have been demonstrated in human sensory SR experiments. In what I call *Type E* (for *Energy*) experiments, noise is applied throughout a run of experimental trials at the same level, with a sub-threshold stimulus to be detected by the human subject added to that noise on some of those trials or in one of two intervals. In these experiments, the stimulus+noise event has more energy than the noise-alone event. Standard psychophysics predicts that when both events are near the energy threshold the stimulus-plus-noise event will be discriminated from the noise-alone event because of this energy difference, resulting in detection of the stimulus, whereas when noise is well above the energy threshold it will mask the small energy increment provided by the sub-threshold signal; when the noise is well below the threshold, neither stimulus-plus-noise nor noise-alone events will be perceived. This would give rise to the observed optimal detection performance at a

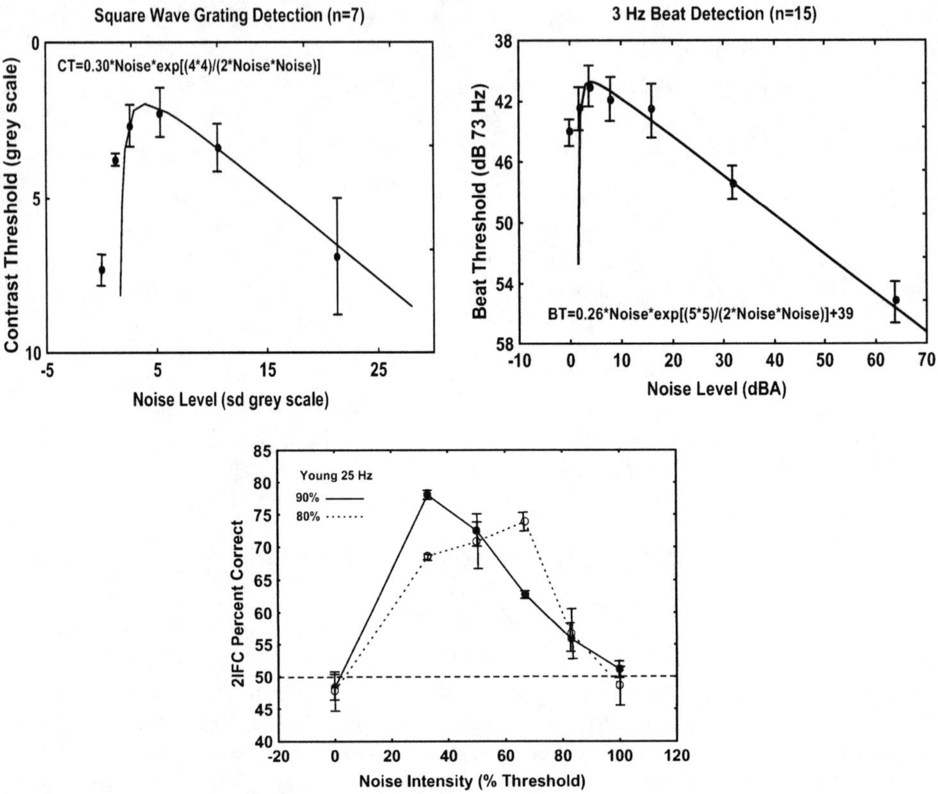

Figure 2. Type I SR. Top left: Human visual system. Top right: Human auditory system. Lower: Human tactile system (chance performance is 50%). All error bars ± 1 s.e. Data from [5].

non-zero noise level. Such effects have been demonstrated in both touch and hearing [4].

In what I call *Type I* (for *Information*) experiments, the energy of stimulus-plus-noise and noise-alone events is equated. Fig. 2 shows representative data for detection of square-wave gratings (vision, left panel), 3-Hz beats in a 70-Hz carrier tone (hearing, center panel), and sinusoidal vibrations applied to the footsole (touch, right panel) [5]. The important aspect of Type I SR is that the information represented by the modulations of the signal is reflected in the threshold exceedances. This is similar to the original scenario for SR, and has been the type of SR demonstrated in all work on sensory and model neurons [1]. Type I SR is not predicted by standard human psychophysics and represents a counterintuitive and truly novel finding.

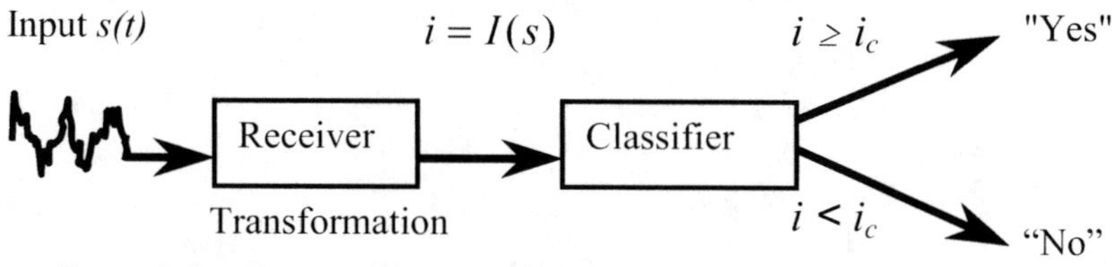

Figure 3. Outline of signal processing events in signal detection theory.

SR AND SIGNAL DETECTION THEORY

All of the human sensory SR results just described were obtained in experimental paradigms that utilized signal detection theory (SDT) to allow the conclusion that sensory sensitivity to the near or sub-threshold stimuli was enhanced by the addition of noise. Arguably SDT provides the most useful account of how physical and biological systems detect signals in noise [6]. Based on the statistical decision theory developed by Neyman and Pearson, SDT separates the sensitivity of a system to signals from the criterion used to classify a given event as arising from either noise alone or from the noise plus a signal. Fig. 3 shows the sequence of events in a single trial of a standard paradigm. The noisy input is transformed into a decision variable, i, typically a function $I(s)$ of the energy in the input, and the decision variable is subjected to the nonlinear classification process. Under typical assumptions, the situation over many trials looks like Fig. 4, with noise-alone and signal-plus-noise trials modeled by two Gaussian probability distributions. The measure of sensitivity is d', the distance between the means of the distributions in units of their (assumed equal) standard deviations. The classification criterion is represented by a vertical line at the level of the evidence variable at which responses change from "no" to "yes" regarding the presence of the signal.

More formally, using the decision rule

$$if\ i \begin{Bmatrix} \geq \\ < \end{Bmatrix} i_c\ then\ say \begin{Bmatrix} yes \\ no \end{Bmatrix} \tag{1}$$

and definitions

$$\begin{aligned} \int_{Z_N = i_c}^{+\infty} f_N(I)\, dI &= p(fa) \\ \int_{Z_{SN} = i_c}^{+\infty} f_{SN}(I)\, dI &= p(hit) \end{aligned} \tag{2}$$

d' is defined as

$$d' = Z_N - Z_{SN} = \frac{M_{SN} - M_N}{\sigma_{N,SN}} \tag{3}$$

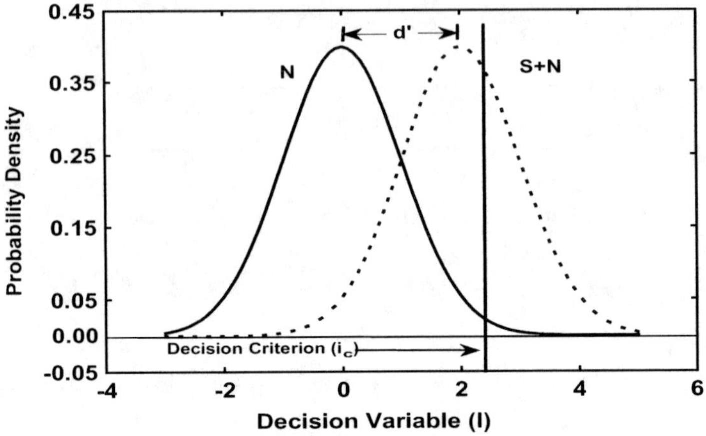

Figure 4. Schematic diagram of signal detection theory.

Since it does not vary with criterion placement, d' is the preferred measure of sensitivity in applications of the theory. In typical applications, either d' is inferred from the various response frequencies, or a modification is used in which some other performance measure can be related to d'. This was the case in the human SR experiments summarized in Fig. 2, in which lower threshold means higher d' and higher percent correct means higher d'.

Recently, Tougaard [2] applied SDT to the SR situation in the context of a simple energy detector. The detector integrates the energy in the squared input (sinusoidal signal plus uncorrelated Gaussian noise or noise alone) over a short interval and uses the output of this process as a decision variable regarding whether the input contains a signal plus noise or only noise. The SDT measure of sensitivity to a noisy sinusoidal signal in this model is

$$d' = \left[\frac{A_s}{A_n}\right]^2 \sqrt{WT} \tag{4}$$

where A_s and A_n are the mean amplitudes of signal and noise, respectively, and W and T are the bandwidth of the noise and the integration time of the detector, respectively. Clearly, d' decreases monotonically with increasing A_n. Thus, SR cannot occur in d' in this formulation. Tougaard [2] was able to show SR only in the measure of percent correct classifications, which varies with criterion placement and is maximum when the criterion is placed exactly where the signal-plus-noise and noise-alone distributions cross. He concluded that SR was merely an epiphenomenon of non-optimal placement of the criterion. This result contrasts sharply both with the several demonstrations of SR in biological systems [3] and human sensory systems [5], and with demonstrations of SR in the output SNR of theoretical systems without a strict threshold [7,8]. The latter demonstrations,

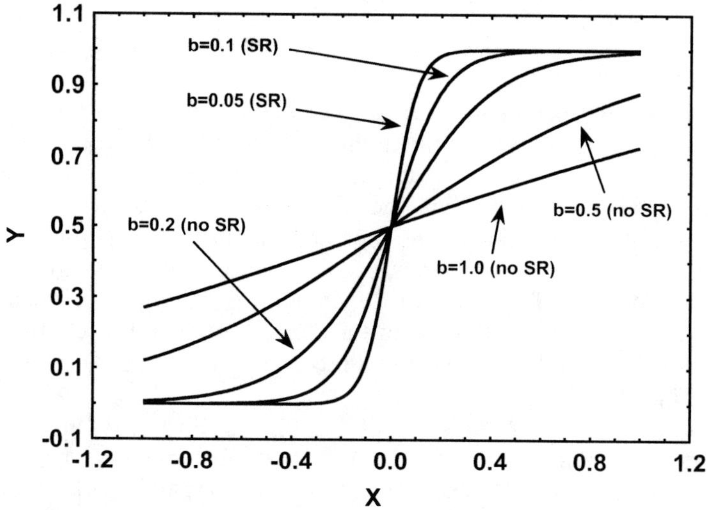

Figure 5. The logistic transform with various values of the parameter b (SR or not in parentheses).

however, are not strictly applicable to this problem since d', although related, is not formally identical to the SNR measures used in those works.

To make progress toward resolving this impasse, I simulated a simple static detector system without a strict threshold but with various continuous input transforms that functioned as a "soft threshold" [9]. This system had no implicit threshold, as possessed by some model systems such as Poisson pulse generators [7], and the output measure calculated was a version of the d' measure of sensitivity from SDT rather than output SNR [8]:

$$d_a' = \frac{(M_{s+n} - M_n)}{\left((\sigma_{s+n}^2 + \sigma_n^2)/2\right)^{1/2}} \tag{5}$$

where M_{s+n} and M_n are the means of the probability distributions of spectral power density at the signal frequency from signal+noise events and noise-alone events, respectively, and σ_{s+n} and σ_n are their respective standard deviations. Spectral power densities in the signal frequency band represented the decision variable values for 1000 simulated events for each distribution. These were obtained either from the power spectra of each of 1000 realizations of 8192-sample epochs of a sinusoidal signal+uncorrelated Gaussian noise or from those of each of 1000 realizations of 8192-sample epochs of the noise alone. Before obtaining the power spectrum by FFT, and in separate runs of simulated events, each sample in the input noisy signal or noise-alone epoch was transformed by one of the following functions: $y = mx + b$, $y = kx^m$, $y = ke^{mx}$, or

$$y = \frac{1}{1 + \exp(-(x-a)/b)} \tag{6}$$

The latter three have been suggested as possible soft threshold transfer functions, and have been shown to result in SR in the SNR when the input-output relationship is "steep" enough [8]. In the case of the power function, an exponent of $m = 3$ seems to be effective [8], whereas the exponential with $m = 1$ yields SR for Poisson pulses [7]. For the logistic transform (6), the a parameter functions as a threshold, whereas the b parameter determines the steepness. As b approaches 0 the logistic approaches a discontinuous threshold (a Heaviside function). Fig. 5 shows the logistic transform over a range of b values. The logistic transform most closely resembles a typical "soft" biological threshold, indeed it is a preferred model in sensory psychophysics [6], and so I have concentrated on it here.

The results of the simulations for the logistic transform for various values of b are summarized in Fig. 5. SR was obtained in d'_a for values of $b \leq 0.1$, which is consistent with the results of Chapeau-Blondeau and Godivier [8]. Fig. 6 shows d'_a as a function of noise level for the logistic with $b = 0.1$ as well as for several different linear transforms. The graph for the logistic is identical in form to that displayed in Fig. 9 of Chapeau-Blondeau and Godivier [8]: the sensitivity index d'_a first decreases rapidly and then rises again before finally decreasing towards an asymptote of 0, showing SR in the secondary peak. Fig. 7 (left panel) shows how this relationship changes as a function of the value of b. Clearly, higher values of b give rise to a similar result to a linear transform, and no SR is observed. As did Chapeau-Blondeau and Godivier [8], I conclude that adding noise can improve

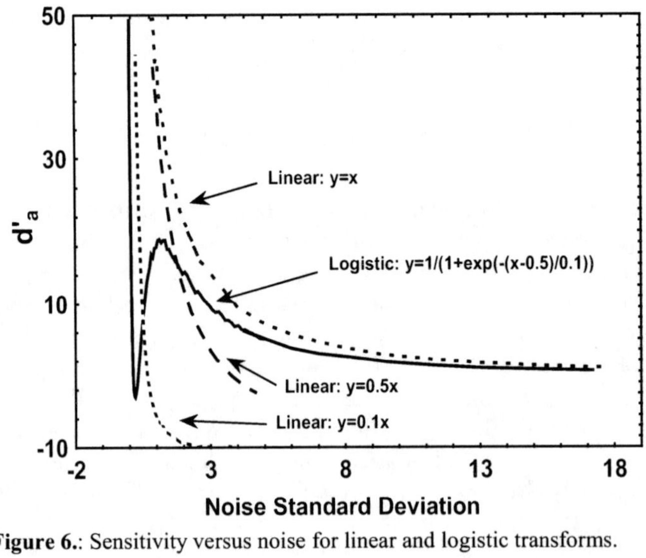

Figure 6.: Sensitivity versus noise for linear and logistic transforms.

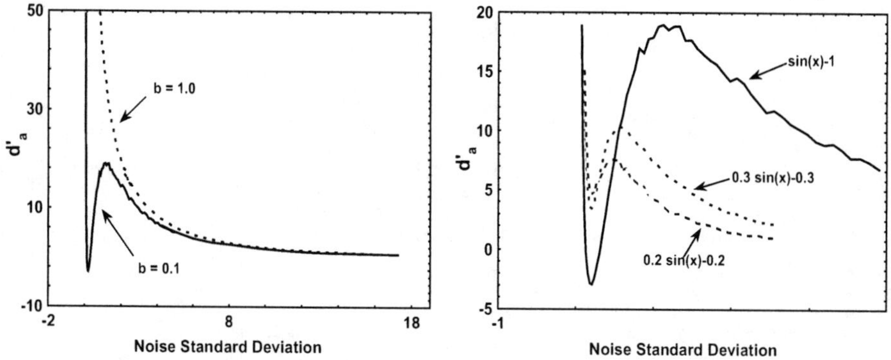

Figure 7. Left: Sensitivity versus noise level for two values of the parameter b in the logistic transform. Right: Dependence of SR on the signal for the logistic transform with $b = 0.1$.

sensitivity to an already somewhat noisy signal without a discontinuous classifier, although sensitivity to a noiseless signal is greater than that for any noisy signal, consistent with linear response theory. Note that here the index of sensitivity, d'_a, was computed directly from the statistics of the relevant probability distributions, rather than by inferring it from classification performance.

As required in SR, the noise must be tuned for optimal performance. Indeed there must be a particular relationship between signal, noise, and input transform in order to observe SR. Fig. 7 (right panel) shows the effects of signal tuning on the relationship between d'_a and noise level for the logistic transform. Clearly, under some conditions the amount of SR might be so small as to be unnoticeable, and be of little practical value, whereas under others, and with an initially somewhat but not too noisy signal, SR could yield large improvements in sensitivity to a weak signal. Clearly, one *open* problem is to obtain a formal understanding of the complex relationship between signal, noise, and the parameters of the logistic input transform so that predictions can be made regarding the presence and magnitude of SR.

It has proven to be difficult to obtain SR in d'_a for the power and exponential input transforms. The signal tuning problem seems to be more severe under my conditions for those functions. Fig. 8 (left panel) shows the effect of signal tuning on the relationship between d'_a and noise level for an exponential input transform. Although the relationship resembles that for the logistic, the required noise values are larger for all of the signals, and the d'_a values are negative for signals whose mean level is too small. Thus, detection of SR would be expected to be more difficult for the exponential input transform. Moreover, the exponential does not have a single parameter that controls steepness; its steepness is always the same but can be manifested differently depending on the values of the parameters. This raises another *open* problem, that of describing the interaction of the mathematical properties of the input transform with those of the signal and the noise. For example, power function transforms with even exponents rectify the noisy signal, whereas

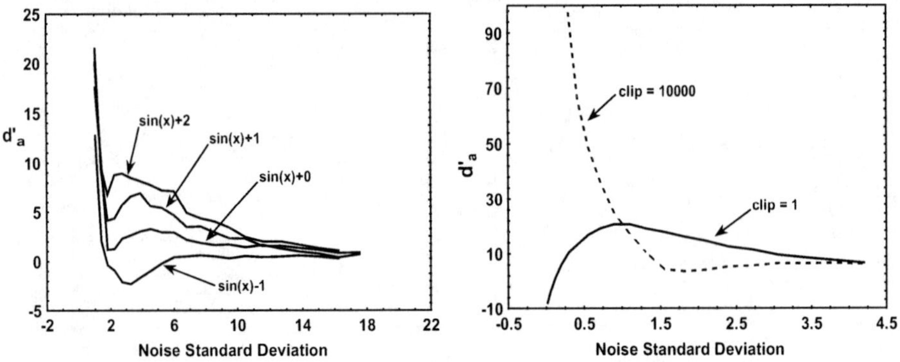

Figure 8. Left: Dependence of SR on the signal for the exponential transform. Right: Effects of clipping the transformed signal on SR for the exponential transform.

those with odd exponents do not. Although this should not affect the relationship between steepness and SR, the mutual tuning required is certainly affected significantly. Moreover, with the "steeper" power functions (larger exponents), many samples of the noisy signal and noise-alone might be magnified to the point where they distort the time series so much that the noise will dominate at far lower levels than for the less steep functions, making it difficult to discern a region of SR.

One way of controlling this magnification by steep input transforms is to clip the output of the transform. Saturation resulting in clipping is a common result of overloading biological and some non-living systems, and it occurs naturally for functions such as the logistic, which squashes the output into the interval between 0 and 1. Fig. 8 (right panel) shows the effect of clipping the output (before doing the spectral analysis) of the exponential transform. The clip value refers to the maximum value of the output; all values exceeding the clip value were set equal to it. Clearly it would be difficult to discern SR in the result of the clip = 1000 case, which represents very little clipping. When clip = 1, however, making the exponential more similar to the logistic with a small b value, SR is large and clear. Thus, the combination of a steep (but continuous) input transform and saturation (or clipping) readily yields SR in d'_a without the need for a classifier. However, it is to be admitted, that this combination, or the logistic alone, does resemble a classifier in its function, as would be expected since the manifestation of the SR phenomenon does require that a system possess a sufficiently steep nonlinearity of some sort. This suggests a remaining *open* problem, that of formally characterizing the steepness of input transforms with diverse mathematical properties. I believe the results of this study help to clarify the circumstances under which SR can be said to improve sensitivity to a (noisy) signal, as well as pointing to some important unsolved problems of noise.

ACKNOWLEDGEMENTS

Preparation of this paper was supported by a grant from the Natural Sciences and Engineering Research Council (NSERC) of Canada. I am also grateful to Frank Moss for valuable discussions.

REFERENCES

1. L. Gammaitoni, P. Hängii, P. Jung, F. Marchesoni, *Rev. Mod. Phys.* **70**, 223-287 (1998).
2. J. Tougaard, *Biol. Cyber.* **83**, 471-480 (2000).
3. L.M. Ward, A. Neiman, F. Moss, *Biol. Cyber.* **87**, 91-101 (2002).
4. J.J. Collins, T.T. Imhoff, P. Grigg, *Nature* **383,** 770 (1996); F.-G. Zeng, Q.-J. Fu, R. Morse, *Brain Res. Interactive,* **869,** 251-255 (2000).
5. L.M. Ward, S. Desai, D. Rootman, M.S. Tata, F. Moss, *Bull. Am. Phys. Soc.,* **46,** N23.002 (2001). Full paper at: http://www.aps.org/meet/MAR01/baps/vpr/layn23-002.html. ; C. Wells, L.M. Ward, J.J. Collins, T. Inglis, *Abstr. Psychon. Soc.*, **6**, 38 (2001).
6. D.M. Green, J.A. Swets, *Signal Detection Theory and Psychophysics.* (Robert E. Krieger Publishing Co., Huntington, NY, 1966); N.A. Macmillan, C.D. Creelman, *Detection Theory: A User's Guide.* (New York: Cambridge University Press, 1991).
7. S.M. Bezrukov, I. Vodyanoy, *Nature* **385,** 319-321 (1997).
8. F. Chapeau-Blondeau, X. Godivier, *Phys.Rev. E*, **55**, 1478-1495 (1997); J.M.G. Vilar, G. Gomila, J.M. Rubí, *Phys. Rev. Lett,.* **81,** 14-17 (1998).
9. L.M. Ward, *Stochastic resonance and signal detection theory in static detectors*. Unpublished ms, University of British Columbia (2002).

Stochastic Synchronization: Analogy with Systems Undergoing Phase Transition

G. Balázsi[*], A. Ordemann[†] and F. Moss[†]

*Northwestern University School of Medicine, Ward Bldg. 6-118, 303 E Chicago Ave, Chicago, IL 60611, USA. Fax: (312) 503-8240. Email: g-balazsi@northwestern.edu
†Center for Neurodynamics, University of Missouri – St. Louis, 8001 Natural Bridge Road, St. Louis, MO 63121, USA. Fax: (314) 516-6152. Email: mossf@umsl.edu
Web: http://neurodyn.umsl.edu/

Abstract. A wide variety of natural oscillator systems undergo the process of synchronization, from lasers to fireflies and clapping in concert halls. The system we study is a two-dimensional network of FitzHugh-Nagumo oscillators with nearest-neighbor coupling in the presence of noise. We consider the case of both positive and negative coupling constants. As the coupling strength or noise intensity is varied, we connect neighboring sites in our system with a bond if the corresponding two oscillators lock their phases. As a result, we expect to establish an analogy between the synchronization in our system and percolation. We also study the analogy between the synchronization of our system and the phase transition towards the ferromagnetic and antiferromagnetic states, for positive and negative coupling constants, respectively. The unsolved problem is the understanding of the route towards synchronization undertaken by extended systems of oscillators, as the coupling strength or noise intensity is varied.

INTRODUCTION

The phenomenon of synchronization was discovered by Christiaan Huygens in 1673 [1], who noticed that two pendulum clocks hung on the same wall always oscillated in antiphase, independent of their initial starting phase. Huygens correctly inferred that the cause of this phenomenon was the coupling between the clocks realized through the wall, because if he hung them on different walls, the motions of the pendulums stopped influencing each other.

Synchronization can thus be defined [2] as the interaction of two oscillators that results in phase locking, e.g. the difference between the phases of the oscillators stays constant. Another signature of synchronization is frequency locking, meaning that the oscillators will have the same frequency after they have been brought into contact.

The notion of synchronization was generalized in 1967 by Stratonovich to include stochastic systems [3]. In the presence of noise, the phases remain locked only for a limited time, after which synchronization is temporarily lost, to be soon regained again. The short events of synchronization loss are called phase slips. The behavior of coupled noisy oscillators is determined by the competition between noise and coupling: the noise tends to destroy synchronization, while the coupling tends to reestablish it. Oscillators that exhibit this behavior are in imperfect synchronization, whereas oscillators that remain phase locked over infinitely long time scales are in

CP665, *Unsolved Problems of Noise and Fluctuations: UPoN 2002: Third International Conference*, edited by S. M. Bezrukov

perfect synchronization. Note that as long as the oscillators maintain a constant phase difference, they are synchronized (and they do not need to oscillate in phase with each other).

Synchronization has been observed in a wide variety of natural and artificial systems [2], such as respiration, circadian rhythm, lasers, neuronal systems, etc. It is important to distinguish between spontaneous synchronization (when the oscillators mutually influence each other), and forced synchronization (when one oscillator drives the other). In the following, we study spontaneous synchronization.

The concept of synchronization can be generalized for extended systems, containing a large number of oscillators. In this case, initially the oscillators have slightly different frequencies, which become equal when coupling is introduced. Typical examples include oscillating chemical media [2], clapping in concert halls [4] and calcium fluctuations in astrocytes [5].

In extended systems of oscillators with no initial coupling, it is interesting to study the modalities that can lead to complete synchronization of all oscillator pairs. First, it is possible to increase the number of coupled oscillator pairs, by coupling oscillators located at arbitrary distances until synchronization is achieved. Second, it is possible to keep the number of coupled oscillators constant (e.g., couple only nearest neighbors), but increase the strength of coupling instead. How noise influences the route to synchronization is an unsolved question to this date. Some studies suggest, however, that the effect of increasing noise could be equivalent to that of decreasing coupling [6]. Particularly, in a two-dimensional system of oscillators with weak coupling, phase slips similar to those of imperfect, noisy synchronization can be observed [5].

METHODS

In this contribution, a square lattice of FitzHugh-Nagumo oscillators with nearest-neighbor coupling is considered:

$$\begin{cases} 0.005\dfrac{dv_{i,j}}{dt} = v_{i,j}(0.5-v_{i,j})(v_{i,j}-1)-w_{i,j}+c_{i,j}+B\xi(i,j,t)+D\nabla^2 v_{i,j} \\ \dfrac{dw_{i,j}}{dt} = v_{i,j}-w_{i,j}-0.2 \end{cases}, \tag{1}$$

where $v_{i,j}$ and $w_{i,j}$ are the variables at site (i,j) in the lattice, $c_{i,j}$ is the excitability parameter of the oscillator at site (i,j) (it determines the frequency of the oscillator), $\xi(i,j,t)$ is a spatiotemporal white noise with uniform distribution between the boundaries $-B/2$ and $B/2$, and D is the coupling constant. It has been shown that strong positive and negative coupling leads to complete synchronization in this system [5]. Nevertheless, the route toward the synchronized state is unclear.

In the following, we characterize the synchronization of any two oscillators by the synchronization index [2]. After the time-dependent phases $\varphi_1(t)$ and $\varphi_2(t)$ of two oscillators have been properly introduced, their wrapped phase difference is defined as $\Phi_{1,2}(t)=[\varphi_1(t)-\varphi_2(t)]_{\mathrm{mod}(2\pi)}$. A prominent peak in the probability distribution $P(\Phi_{1,2})$ of $\Phi_{1,2}(t)$ is a signature of strong coupling. In particular, a delta-peak in $P(\Phi_{1,2})$ is the sign of perfect synchronization. The measure of peak prominence is the synchronization index $\rho_{1,2}$, defined as

$$\rho_{1,2} = \frac{1 - S_{1,2}}{S_{1,2}}, \tag{2}$$

where S is the Shannon-entropy associated with the probability distribution $P(\Phi_{1,2})$:

$$S = -\sum_k P_k \log_2(P_k). \tag{3}$$

To quantify the synchronization of our two-dimensional oscillator system, we use the quantities defined in [5]:

1. The average synchronization index $\langle\rho\rangle$ taken between all possible pairs of a sample set of oscillators;
2. The standard deviation of the oscillation frequencies σ_F, taken for all oscillators in the system;
3. The standard deviation σ_G of the global output $G(t)$, defined as the sum of all variables $v_{i,j}(t)$ at each time step t of the simulation.

RESULTS AND DISCUSSION

Synchronization of a Two-Dimensional Network of Oscillators

As the coupling strength D=0 is increased, neighboring oscillators with nearby frequencies will synchronize first, forming larger and larger clusters of synchronized oscillators. For sufficiently strong coupling, all oscillators in a finite-size system become synchronized. However, if spatiotemporal, uncorrelated noise is introduced, the oscillator pairs that were phase-locked in the absence of noise can slip out of phase (see Eq. 1). The behavior of the system at various values of D and B is illustrated in Fig. 1.

The average synchronization index $\langle\rho\rangle$ has a global minimum at D=0 for every noise intensity. In the absence of noise (B=0), as the coupling strength D increases, the average synchronization index quickly reaches a plateau, which corresponds to the complete synchronization of the system of oscillators. As the noise intensity increases, stronger coupling is needed in order to reach synchronization throughout the system. The contour lines corresponding to average synchronization indices from 0 to 1 can be seen on the inset in Fig. 1.

The standard deviation of the frequencies σ_F reaches a global maximum for D=0 and drops quickly as the coupling becomes stronger, irrespective if D is positive or negative. This measure is therefore a robust indicator of frequency entraining among the oscillators in the presence of noise with various amplitudes, but it does not necessarily indicate phase locking.

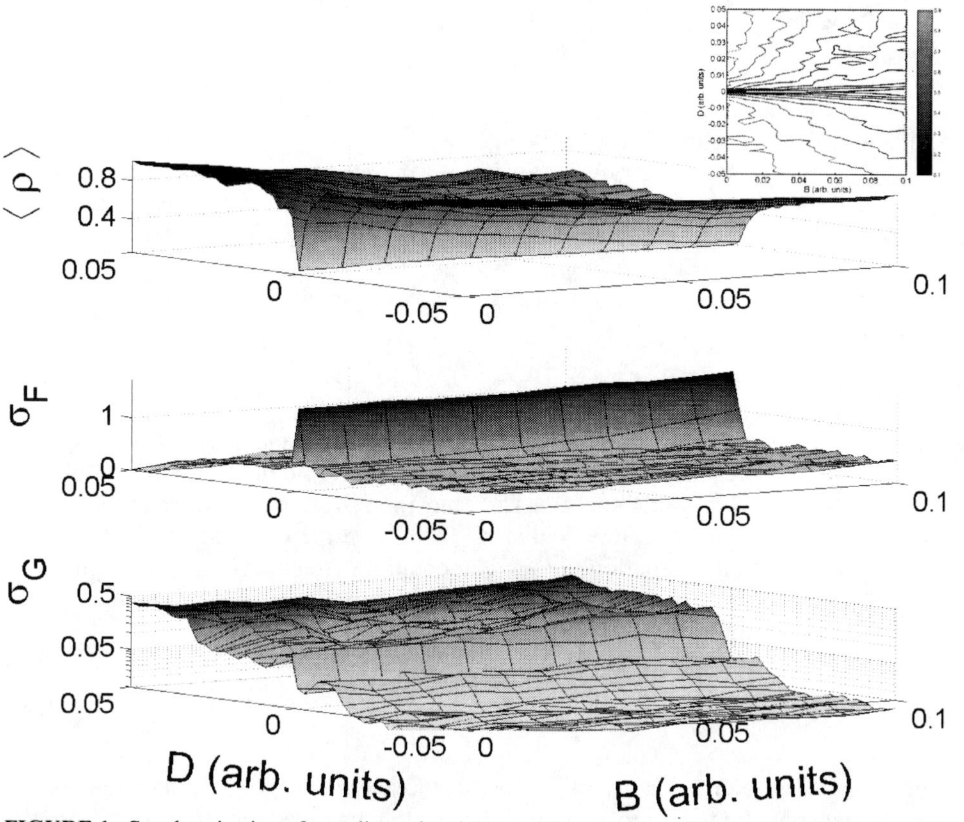

FIGURE 1. Synchronization of two-dimensional FN oscillator systems with nearest-neighbor coupling in the presence of noise. The plots illustrate, from top to bottom: the dependence of the average synchronization index $\langle\rho\rangle$, the standard deviation of the frequencies σ_F and the standard deviation of the global output σ_G upon the coupling strength D and the noise amplitude B. The inset shows the contour lines of $\langle\rho\rangle$, taken at steps of 0.1 between 0 and 1. The system contained 20×20 FitzHugh-Nagumo oscillators, with the following parameters: $\langle c\rangle$=0.2, $-0.025<c-\langle c\rangle<0.025$, N=400. Free boundary conditions were used in these simulations.

The behavior of the standard deviation of the global output σ_G can be elucidated by comparison with a reference value σ_R taken by this quantity at D=0. For D>0, σ_G is in average larger than σ_R; for D<0, σ_G is smaller than σ_R with the exception of a restricted range of D values. If in an experimental system of oscillators the observed value of σ_G is consistently smaller than σ_R, however, one can conclude that the coupling in the system is phase-repulsive (D<0). This conclusion remains valid in the presence of noise with various intensities.

ANALOGY WITH SYSTEMS UNDERGOING PHASE TRANSITION

Our two-dimensional network of coupled oscillators can be compared with two systems in which phase transition has been observed:

1. Bond-percolation. If we progressively join the synchronized (ρ>0.9, see Eq. 2) neighboring oscillators with a bond, the situation resembles that arising in bond-percolation systems [7]. It is known that percolation systems undergo a phase transition for a critical density of bonds.

2. Magnetic systems. The Hamiltonian **H** of a two-dimensional Ising system can be written as

$$\mathbf{H} = -\frac{J}{2}\sum_{<i,j>} S_i S_j - H\sum_{i=1}^{N} S_i - NE_0, \qquad (4)$$

where S_i denotes the i-th spin, J is the coupling constant, the brackets denote summation over nearest neighbors, H is the external magnetic field, N is the number of spins and E_0 is a zero-point energy. The transition from the paramagnetic to the ferromagnetic state can be realized either by increasing the coupling strength J or by increasing the external magnetic field H. The coupling strength J thus corresponds to the coupling constant D in our model. The temperature T enters the dynamics of the model by determining the probability of flipping P_{flip}:

$$P_{flip} = \frac{e^{-\frac{\Delta E}{k_B T}}}{1+e^{-\frac{\Delta E}{k_B T}}}, \qquad (5)$$

where k_B is the Boltzmann-constant and ΔE is the energy change in the system due to the proposed spin-flip [8]. Increasing the temperature can lead to a transition from the ferromagnetic to the paramagnetic state, just as increasing noise intensity can cause the loss of synchronization.

CONCLUSIONS

Despite the apparent analogy between our network of FN oscillators and the two systems mentioned above, the differences between them are also evident. Several studies [2,9,10] point to the fact that a system of oscillators with nearest-neighbor coupling cannot develop a unique phase due to kinetic roughening if the dimensionality of the system $d\leq 2$. Furthermore, [9] points out that even for $2<d\leq 5$, only "frequency order", and not "phase order" can be defined in the system of oscillators. Therefore, the analogy between our system and bond-percolation fails because a spanning cluster cannot develop as the size of our system is extended to infinity. On the other hand, the analogy with the two-dimensional Ising system also fails, because two spins can only have two possible alignments (they are either parallel or anti-parallel), whereas the phase difference of our oscillators can take on continuous values between 0 and 2π.

Even if a two-dimensional system of FN-oscillators cannot undergo a phase transition and develop a unique phase as the system size increases to infinity, it remains an interesting question to understand how the synchronized state is reached as the noise intensity and coupling strength is varied in finite-size oscillator systems. The analogy with the Ising model suggests that the application of a periodic forcing onto the oscillator network can also result in synchronization, provided that the frequency of the forcing is not too far from the average frequency of the oscillators in the system. In analogy with the two-dimensional Ising system, the effect of the variation of three parameters needs to be studied to achieve the synchronization of our system: the coupling strength, the noise intensity, and the strength of the external forcing.

ACKNOWLEDGMENTS

This work was supported by the Graduate School at the University of Missouri – St. Louis, the U.S. Office of Naval Research – Physics Division and the Alexander von Humboldt Foundation. We are grateful to A. Pikovsky, Z. Néda, S. Bezrukov and A. Neiman for valuable comments and discussion.

REFERENCES

1. Huygens, Ch. (Hugenii), *Horologium Oscillatorium*, Apud F. Muguet, Regis et Archiepiscopi typographum, Paris, France (1673).
2. Pikovsky, A., Rosenblum, M., and Kurths, J., *Synchronization (A Universal Concept in Nonlinear Sciences)*, Cambridge University Press (2001).
3. Stratonovich, R. *Topics in the Theory of Random Noise*, Gordon and Breach, New York, Vol. 2. (1967).
4. Néda, Z. et al., *The sound of many hands clapping.* Nature **403**, 849-850 (2000).
5. Balázsi, G. et al., *Synchronization of Hyperexcitable Systems with Phase-Repulsive Coupling.* Phys. Rev. E **64**, 041912 (2001).
6. Neiman, A. et al., *Noise-enhanced phase synchronization in excitable media.* Phys. Rev. Lett. **83**(23), 4896-4899 (1999).
7. Bunde, A. and Havlin, S., editors, *Fractals and disordered systems* (2nd edition). Springer (1996).
8. Leung, K.-t. and Néda, Z., *Nontrivial stochastic resonance temperature for the kinetic Ising model.* Phys. Rev. E **59**(3), 2730-2735 (1999).
9. Sakaguchi et al., *Local and Global Self-Entrainments in Oscillator Lattices.* Prog. Theor. Phys. **77**, 1005-1010 (1987).
10. Grinstein, G., et al., *Temporally Periodic Phases and Kinetic Roughening.* Phys. Rev. Lett. **70**, 3607-3610 (1993).

A Dynamical System Exhibits High Signal-to-noise Ratio Gain by Stochastic Resonance

Peter Makra and Zoltan Gingl

Department of Experimental Physics, University of Szeged
Dóm tér 9., Szeged, H-6720 Hungary
E-mail: phil@neptun.physx.u-szeged.hu; gingl@physx.u-szeged.hu

Abstract. On the basis of mixed-signal simulations, we demonstrate that signal-to-noise ratio (SNR) gains much greater than unity can be obtained in the double-well potential through stochastic resonance (SR) with a symmetric periodic pulse train as deterministic and Gaussian white noise as random excitation. We also show that significant SNR improvement is possible in this system even for a sub-threshold sinusoid input if, instead of the commonly used narrow-band SNR, we apply an equally simple but much more realistic wide-band SNR definition. Using the latter result as an argument, we draw attention to the fact that the choice of the measure to reflect signal quality is critical with regard to the extent of signal improvement observed, and urge reconsideration of the practice prevalent in SR studies that most often the narrow-band SNR is used to characterise SR. Finally, we pose some questions concerning the possibilities of applying SNR improvement in practical set-ups.

1. POSSIBILITIES OF STOCHASTIC SIGNAL IMPROVEMENT

Stochastic resonance (SR) is a rather counter-intuitive phenomenon wherein signal transfer in certain non-linear systems, such as threshold devices, SQUIDs, neuron models, etc, can be improved through increasing the amplitude of the noise at the input. To be more exact, we can speak of stochastic resonance when the quantity reflecting the quality of the transfer (for example, the signal-to-noise ratio — SNR — at the output) shows a maximum at non-zero input noise amplitude. While a certain amount of noise can optimise the quality of the output signal, usually the output signal still remains noisier than the input. Yet the question arose whether one can get, in certain systems, output signals better than the input— in other words, whether stochastic signal improvement is possible. The first attempts to find SR-based signal improvement were largely unsuccessful, and the exceptions soon turned out to be artefacts, due to, for example, aliasing effects. The search for SNR gain took a significant turn with the appearance of linear response theory (LRT), which contained proof that for small-amplitude signals (in the linear-response range), the SNR at the output does not exceed that at the input [1]. As a result, subsequent studies aimed at finding stochastic signal improvement usually assumed the condition that the system is working in the non-linear response (NLR) range. The first theoretical results that showed significant SNR improvement by SR concerned a level-crossing detector (LCD) driven by a model neural signal, a spike train with random initial times of spikes [2]. Later, the same system produced considerable SNR gains also with a

CP665, *Unsolved Problems of Noise and Fluctuations: UPoN 2002: Third International Conference,* edited by S. M. Bezrukov

periodic pulse train [3], and SNR gains greater than unity were also found in a wide range of other systems, including the Schmitt trigger [4] and the Hodgkin-Huxley neuronal model [5].

In this work, we shall revisit our recent study [6], wherein we investigated the possibilities of stochastic signal improvement — namely, SNR gains greater than one — in a dynamical SR model, the double-well potential. We shall show that in such a system driven by a symmetric periodic pulse train to which a Gaussian white noise is added, SNR gains well over unity can be obtained. We shall also examine how the choice of the measure to reflect signal quality may affect the degree of signal improvement observed. In connexion to that, we shall prove that the double-well system may exhibit significant SNR gains even for a sine input if we apply a wide-band definition of SNR instead of the widely used narrow-band one. Finally, we shall draw attention to some open questions concerning the practical applicability of stochastic signal improvement.

2. HIGH SNR GAIN IN A DOUBLE WELL

We considered an archetypal SR model wherein an over-damped particle is moving in a double-well potential, excited by a periodic pulse train and a Gaussian white noise. The dynamics of this system are given by the following Langevin equation:

$$\frac{dx}{dt} = x(t) - x^3(t) + p(t) + w(t), \tag{1}$$

where $p(t)$ is the periodic excitation and $w(t)$ represents a Gaussian white noise. The shape of the periodic signal we used is shown in Figure 1.

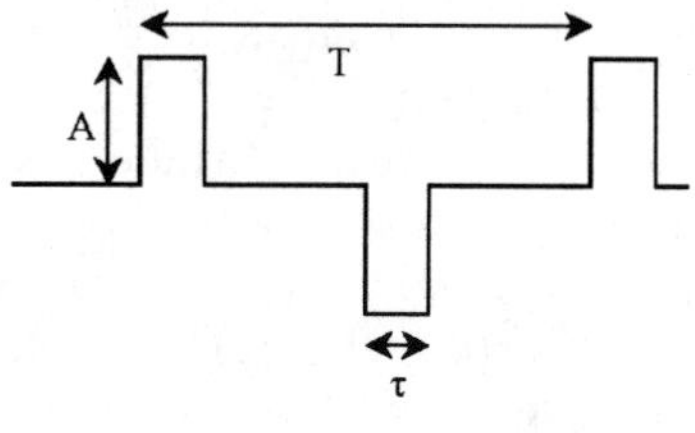

FIGURE 1. The shape of the periodic input signal $p(t)$: A is the amplitude, T stands for the period, τ denotes the width of the pulses and the duty cycle is defined as $2\tau/T$.

In order to solve Equation (1) with analogue methods, we transformed it into an integral equation first (t_0 is an integration constant introduced by the analogue circuitry):

$$x(t) = \frac{1}{t_0} \cdot \int_0^t [x(t') - x^3(t') + p(t') + w(t')]dt'. \tag{2}$$

Then we applied a mixed-signal (ie, both analogue and digital) simulation system wherein analogue circuits realised the operations necessary to solve Equation (2), while the signal and the noise were digitally generated and then converted into the analogue domain. The whole system was driven by a high-performance digital signal processor (DSP) and the data were acquired by high-resolution A/D converters with fast roll-off anti-aliasing filters. We used the same data acquisition equipment (including filters) for input and output signals, guaranteeing the same measurement method and thus avoiding artefacts due to different treatment. A host personal computer running LabVIEW controlled the DSP and performed additional evaluation tasks. The block diagram of the system is shown in Figure 2.

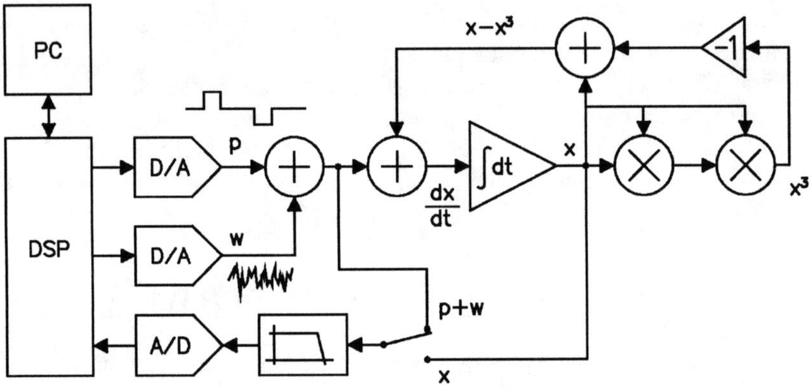

FIGURE 2. Block diagram of the mixed-signal simulation system. The periodic and noisy signals are generated digitally and fed into an analogue simulations system. The solution x(t) is digitised by an A/D converter and the data is processed by a DSP and the host PC.

It is important to note that the analogue realisation introduces a $1/t_0$ factor in the right side of Equation (2), which is equivalent to a rescaling of all frequencies in Equation (1) with t_0 ($1.2\cdot10^{-4}$ s in our case).

We experimentally determined the threshold value of input signal amplitude, that is, the smallest value at which transition between wells occurs without noise, and related all other amplitudes to this level. For every parameter combination, we recorded one thousand samples of $x(t)$, each consisting of 2048 data points, and from these records we calculated the averaged power spectral density (PSD) of the signal using FFT techniques. The sampling frequency was 8 kHz, the frequency of the periodic input signal was set to 31.25 Hz and the bandwidth of the white noise was 50 kHz (these parameters pertain to Equation (2) and correspond to $3.75\cdot10^{-3}$ Hz and 6 Hz in Equation (1)).

We chose two different definitions of the SNR as measures to reflect signal quality: the 'classical' narrow-band definition widely used in the field of SR [7], and a wide-band SNR given by the ratio of the total power of the deterministic part of the signal P_S to the total power of the noise part P_N:

$$SNR := \frac{\lim\limits_{\Delta f \to 0} \int_{f_0-\Delta f}^{f_0+\Delta f} S(f)\,df}{S_N(f_0)}, \tag{3}$$

and

$$SNR_w := \frac{P_S}{P_N} = \frac{\sum_{k=1}^{\infty} \lim_{\Delta f \to 0} \int_{kf_0-\Delta f}^{kf_0+\Delta f} S(f)df}{\int_0^{\infty} S_N(f)df}. \tag{4}$$

In the definitions above, f_0 denotes the frequency of the deterministic signal, $S(f)$ stands for the PSD of the signal and $S_N(f)$ signifies the PSD of the noise component in the signal. The narrow-band and wide-band gains, G and G_w, were calculated as the ratio of the output and input SNRs (with $p(t)+w(t)$ regarded as input and $x(t)$ as output):

$$G := \frac{SNR_{out}}{SNR_{in}}, \tag{5a}$$

and

$$G_w := \frac{SNR_{out,w}}{SNR_{in,w}}. \tag{5b}$$

These gains are plotted as functions of the input noise amplitude in Figure 3. We can clearly see that in the arrangement described above, both kinds of gains can exceed unity, depending on the amplitude of the input signal. The difference between the narrow-band and the wide-band descriptions lies in the fact that the narrow-band gain is smaller than one for low signal amplitudes (70% of threshold), while its wide-band counterpart shows signal improvement for all the three amplitude values we examined. This finding supports the view that the signal improvement observed does depend on the measure we choose to reflect signal quality.

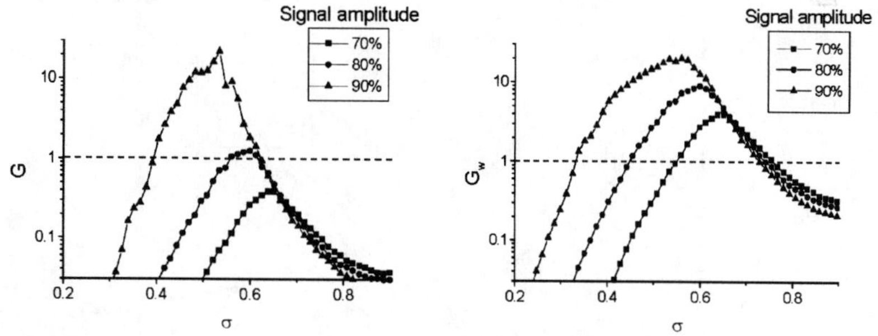

FIGURE 3. Narrow-band (G) and wide-band (G_w) gains as functions of input noise amplitude σ. The amplitude of the signal is 70%, 80% and 90% of the threshold value; the duty cycle is 10%. The amplitude of the noise is denoted by σ, its value is expressed in units normalised by the threshold.

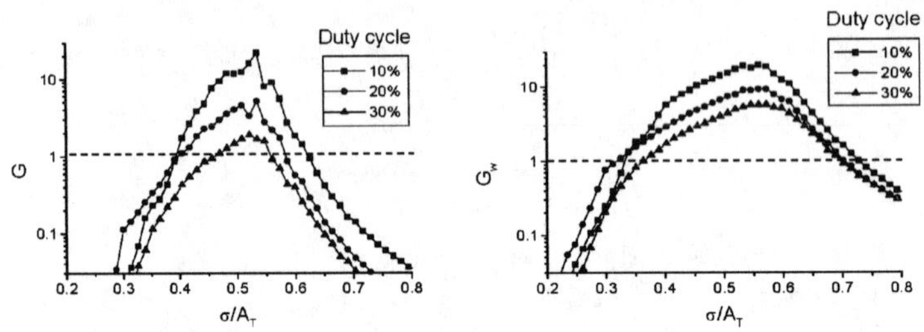

FIGURE 4. Narrow-band (G) and wide-band (G_w) gains as functions of input noise amplitude σ, for three different duty cycles. The amplitude of the noise is scaled by the threshold value A_T; the amplitude of the periodic signal is 90% of the threshold value.

As Figure 4 reveals, smaller duty cycles result in greater gains. To understand this fact, one should consider the noise manifests itself in the output signal mainly in the form of jitter, that is, switch-time fluctuations. Consequently, SNR_{out} does not depend significantly on the duty cycle if the correlation time of the noise is smaller than the width of the pulses in the periodic signal, and the amplitude of the periodic signal is close to the threshold (these requirements are met in our case). On the other hand, the input SNR is considerably higher for greater pulse width: the power of the deterministic component (the numerators in Equations (3) and (4)) increases, while the noise level (the denominators in Equations (3) and (4)) remains constant. The SNR gain, therefore, will be greater if we decrease the value of the duty cycle. This mechanism is illustrated in Figure 5.

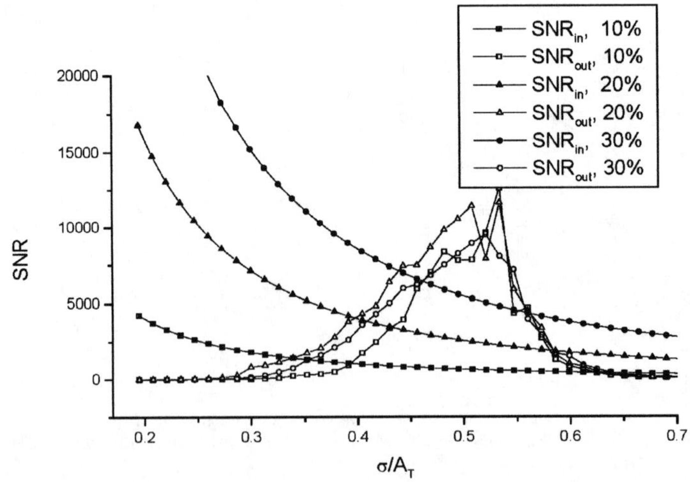

FIGURE 5. Input and output SNRs as functions of input noise amplitude σ, for three different duty cycles. The amplitude of the noise is scaled by the threshold value A_T; the amplitude of the periodic signal is 90% of the threshold value.

3. IMPLICATIONS OF CHOOSING THE MEASURE TO REFLECT SIGNAL QUALITY

In the literature of SR, there have emerged arguments that different predictions for signal improvement stem from the choice of quantity characterising system response; indeed, the narrow-band SNR might not be the optimal measure of signal quality for most signals [8, 9] Here we would like to include a dramatic illustration of this view, in addition to the one already presented in the previous section.

We considered again the double-well system given by (1), only this time $p(t)$ was a simple sine with an angular frequency of 0.1 rad/s. The study we referred to in the above [8] ruled out SNR gains greater than one for sub-threshold signals in such an arrangement, and indeed our numerical simulations with similar parameters yielded a narrow-band gain of 0.82. The wide-band gain, however, was 37.52. Comparing the input and output signals in Fig 6, the improvement of signal quality is visible even to the naked eye. This observation supports the claim that the wide-band SNR characterises the signal-transfer behaviour of the double-well system in a much more realistic way than its narrow-band counterpart. We also find it important to note that while the wide-band SNR offers a more faithful account of the noise content of the signal, it is more flexible in the sense that non-sinusoid signals pose no problem to it, and consequently it is preferred in practical measurements, its concept remains fairly simple and differs only slightly from the conventional definition.

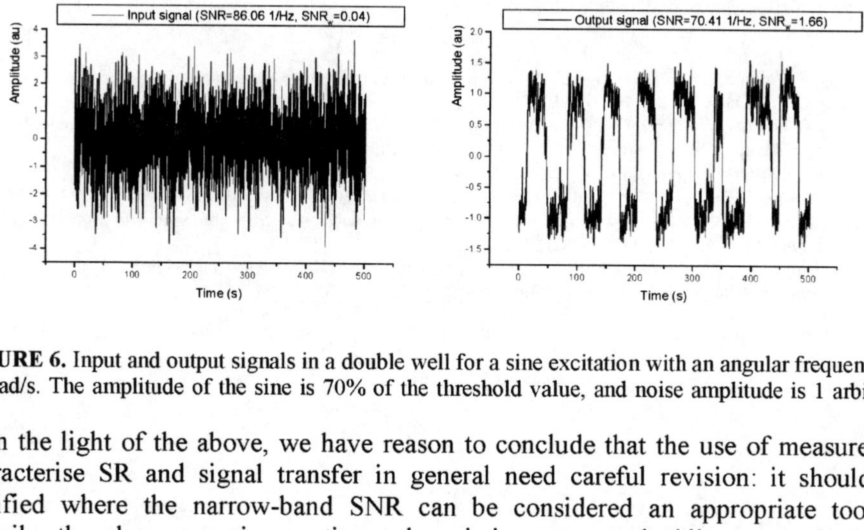

FIGURE 6. Input and output signals in a double well for a sine excitation with an angular frequency of 0.1 rad/s. The amplitude of the sine is 70% of the threshold value, and noise amplitude is 1 arbitrary unit.

In the light of the above, we have reason to conclude that the use of measures to characterise SR and signal transfer in general need careful revision: it should be clarified where the narrow-band SNR can be considered an appropriate tool to describe the phenomena in question, where it is more worthwhile to use the more sophisticated wide-band SNR definition, and what arrangements call for alternative measures such as those based on entropy, channel capacity or general information-theoretic distance measures [9]. In addition, reconsideration of the predictions of the linear response theory concerning SNR gain might well prove necessary as these predictions heavily rely on a narrow-band conception of SNR.

4. QUESTIONS STILL OPEN

While SNR gain has been observed in a number of real-world set-ups such as level-crossing detectors and the Schmitt trigger, the main direction for its application seems to lie in biological systems, where it can account for the detection of sub-threshold neural stimuli [5]. The precise mechanism of this kind of sensory enhancement, however, still remains to be explored.

The possibility of such an impressive signal improvement as that depicted in Figure 6 raises the question whether arrangements based on SR can function as filters. Although the technical applicability of these potential filters is somewhat limited, it is still an open question if there may exist real-world physical and biological systems, with conditions given and defying intervention by us, wherein SR-based set-ups represent the only way of signal optimisation. And, if such systems are to be found, it needs to be clarified whether the requirement of operating in the non-linear limit imposes too strong restrictions on SR-based arrangements.

REFERENCES

1. M I Dykman, D G Luchinsky, R Mannella, P V E McClintock, N D Stein and N G Stocks, *Stochastic resonance in perspective, Il Nuovo Cimento D* **17D** (1995) 661–683.
2. L B Kiss, *Possible breakthrough: significant improvement of signal to noise ratio by stochastic resonance,* in Chaotic, Fractal, and Nonlinear Signal Processing, Proc. Am. Institute Phys., ed R Katz, Mystic, Connecticut, USA (1996) 382–396.
3. K Loerincz, Z Gingl and L B Kiss, *A stochastic resonator is able to greatly improve signal-to-noise ratio, Phys Lett A* **224** (1996) 63–67.
4. Z Gingl, R Vajtai and L B Kiss, *Signal-to-noise ratio gain by stochastic resonance in a bistable system, Chaos, Solitons and Fractals* **11** (2000) 1929–1932.
5. F Liu, Y Yu and W Wang, *Signal-to-noise ratio gain in neuronal systems, Phys Rev E* **63** (2001) 05912.
6. Z Gingl, P Makra and R Vajtai, *High signal-to-noise ratio gain by stochastic resonance in a double well, Fluctuation and Noise Letters* **1** (2001) L181–L188.
7. L Gammaitoni, P Hänggi, P Jung and F Marchesoni, *Stochastic Resonance, Rev Mod Phys* **70** (1998) 223–287.
8. P Hänggi, M E Inchiosa, D Fogliatti and A R Bulsara, *Nonlinear stochastic resonance: The saga of anomalous output-input gain, Phys Rev E* **62** (2000) 6155–6163.
9. J W C Robinson, D E Asraf, A R Bulsara and M E Inchiosa, *Information-theoretic distance measures and a generalization of stochastic resonance, Phys Rev Lett* **81** (2001) 2850-2853.

SENSORY SYSTEMS

Active Amplification by Spontaneous Hair Bundle Oscillations

Frank Jülicher

Max Planck Institut für Physik komplexer Systeme, Nöthnitzerstr. 38, 01187 Dresden, Germany

Abstract. In a quiet environment, the ears of many vertebrates emit so-called spontaneous otoacoustic emissions as manifestations of active processes in the inner ear. It has been suggested that the ear employs dynamical systems which are self-tuned to the critical point of an oscillating instability or Hopf bifurcation. It has recently been demonstrated that hair bundles, which are the mechanosensitive elements of sensory hair cells in the bull frog's sacculum, have the ability to generate noisy oscillations. In the presence of noise, the Hopf bifurcation of an oscillator is concealed by the noise, however the principal signatures remain, such as frequency selectivity and nonlinear response. Using the observed autocorrelation function and the linear response function, the activity of the noisy oscillator can be characterized by its violation of the fluctuation-dissipation theorem.

INTRODUCTION

The task to detect sounds of the outside world imposes stringent demands on the design of the inner ear, where acoustic stimuli are transduced to electrical signals [1]. The cochlea contains highly specialized cells called hair cells, which act as mechanosensors [2]. Each of these cells is responsive to a particular frequency component of the auditory input. Moreover, these sensors need extraordinary sensitivity, since the weakest audible sounds impart an energy, per cycle of oscillation, which is no greater than that of thermal noise [3]. At the same time, they must operate over a wide range of volumes, responding and adapting to intensities which vary by many orders of magnitude. In order to achieve this goal, non-linear amplification is necessary. The familiar resonant gain of a passive elastic system is far from sufficient for the required demands, because of the heavy viscous damping at microscopic scales [4]. The ear relies on active systems to achieve exquisite sensitivity and sharp frequency selectivity [4, 5, 6]. The most striking evidence for active behaviors in the ear are so-called otoacoustic emissions which are sounds emitted from the ears of mammals, birds and amphibians [7]. It has recently been proposed that the cochlea contains active dynamical systems which are close to the critical point of a Hopf-bifurcation [8, 9, 10]. Proximity to the bifurcation can in general be achieved by a self-tuning mechanism [9]. This concept can explain many of the observed features of the ear, in particular, the nonlinearities that are generally observed at resonance conditions [11], the interference effects of multiple frequencies [12] as well as the occurrence of spontaneous emissions. While this concept seems to apply essentially to all vertebrate ears, the question of what are the underlying specific mechanisms at the origin of spontaneous oscillations is a subject of active research.

CP665, *Unsolved Problems of Noise and Fluctuations: UPoN 2002: Third International Conference,* edited by S. M. Bezrukov

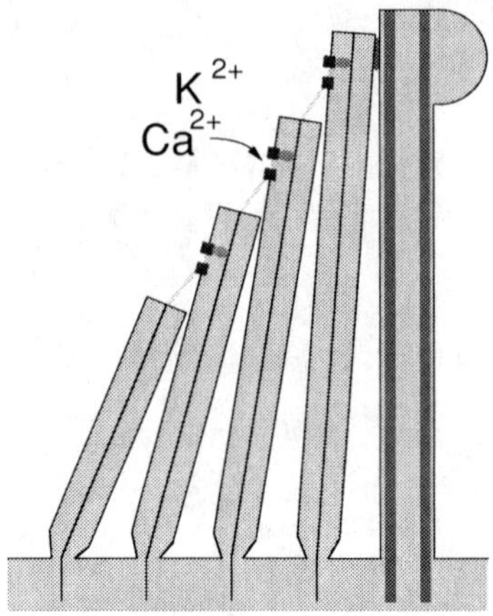

FIGURE 1. Schematic representation of the hair bundle, the sensitive element of mechanosensory hair cells. It consists of 10-50 rod-like stereocilia which have a length of 1-10 μm and are connected by fine filaments, called tip-links. The stereocilia contain ion channels which open as a result of mechanical stimulation.

While in mammals, it is widely thought that outer hair cells are the active elements [5] and thus might contain mechanical oscillators, no spontaneous oscillations have so far been observed. However, in the case of frogs it was discovered that hair bundles of the hair cells themselves could oscillate [13, 14]. The observed noisy oscillations exhibit the signature of a system near a Hopf bifurcation: The response to periodic stimuli was nonlinear for sufficiently large amplitudes near the oscillation frequency [15]. For small amplitudes there was a linear response regime which exhibited stable behaviors. Comparing the linear response to the autocorrelation function it was shown that the fluctuation dissipation theorem was violated, indicative for an active system [16].

The hair bundles of vertebrate hair cells consist of about 50 stereocilia which are stiff, rod-like extensions of the cell with a length of several micron and a diameter of about 300nm, see Fig. 1 [6]. The stereocilia merge at the tip and are grouped in a bundle. Fine filaments, so-called top-links form bridges between neighboring stereocilia [17]. The micromechanical properties of hair bundles in living hair cells can be rich and range from adaptation (slow relaxation involving active elements) [18], twitches (rapid active motion) to spontaneous oscillations [6, 13, 15, 19].

GENERIC BEHAVIORS NEAR A HOPF BIFURCATION

What are the properties that make an oscillatory instability suitable for detection? A dynamical system which is tuned to the proximity of a Hopf bifurcation has generic nonlinear resonance properties and exhibits a compressive nonlinearity. We discuss the general mechanical response properties of an oscillator in presence of a periodic stimulus $f_{\text{ext}} = f_1 e^{i\omega t} + f_{-1} e^{-i\omega t}$. In the periodic limit cycle, we can write the response as a

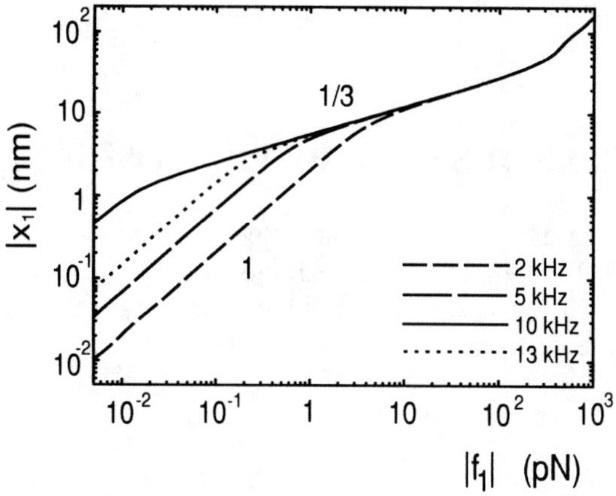

FIGURE 2. Fourier amplitude $|x_1|$ of the response of a dynamical system at a Hopf bifurcation to a stimulus force of amplitude $|f_1|$ at different frequencies. The data is obtained by a numerical solution of a model for oscillations generated by molecular motors, see [9]. The power law of the response is indicated.

Fourier sum

$$x(t) = \sum_n x_n e^{in\omega t} \tag{1}$$

with oscillation frequency ω. Close to the bifurcation point, the first Fourier mode x_1 dominates and obeys to lowest order

$$f_1 = A(C,\omega)x_1 + B|x_1|^2 x_1 \tag{2}$$

where the linear coefficient $A(\omega,C)$ is a function of frequency and a control parameter C. In the absence of a stimulus, spontaneous oscillations can occur with amplitude

$$|x_1|^2 = -\frac{A}{B} \tag{3}$$

and a frequency for which A/B is real. The bifurcation point is characterized by the condition that $A = 0$ at the critical frequency $\omega = \omega_0$. We can therefore write a general expansion

$$A(\omega,C) \simeq \alpha(\omega - \omega_c) + \beta(C - C_c) \tag{4}$$

In order to illustrate the generic mechanical response we consider $C = C_c$ i.e. the system is exactly at the bifurcation. Two different regimes of the response to an oscillating force can be distinguished, see Fig. 2. If the applied frequency is close to the critical frequency, the linear term can be ignored and

$$|x_1| \simeq |B|^{-1/3}|f_1|^{1/3} \tag{5}$$

If the frequency mismatch is larger, $|\omega - \omega_c| \gg |f_1^{2/3}|\ |B|^{1/3}/\alpha$, the cubic term is unimportant and the response becomes linear

$$|x_1| \simeq |A|^{-1}|f_1| \tag{6}$$

This approach can be generalized to situations where more than one frequency are present [12].

SPONTANEOUS HAIR BUNDLE OSCILLATIONS

Spontaneous hair bundle oscillations have been observed experimentally in the sacculus of the frog [13, 14, 19]. The observed noisy oscillations can be characterized for small amplitudes by a linear description. For small amplitudes we write a dynamic equation for the hair bundle deflection x [16]

$$\lambda \frac{dx}{dt} = -kx + F_a + f_{\mathrm{ext}} \tag{7}$$

where, λ and k are a friction and stiffness, respectively, and where we have neglected nonlinearities. Note that we no not include a mass term since at the length scales and frequencies of interest, inertial terms can be ignored. The force F_a is an internally generated force due to active phenomena and active elements such as motor proteins or ion channels. The system is stimulated by a periodic external force f_{ext}. It can oscillate spontaneously, if the active force F_a itself relaxes and provides a positive feedback to x. We therefore write

$$r^{-1} \frac{F_a}{dt} = -F_a - \bar{k}x \tag{8}$$

Here, r is a relaxation rate of the active process and the coupling coefficient $\bar{k}$ has dimensions of a spring constant. Note, that the combined equations for x and F_a represent an active system and cannot be derived from a potential.

In order to simplify our discussion, we focus on the illustrative case where $r = k/\lambda$. Indeed, fitting Eqns. (7) and (8) to the observed linear response function of a hair bundle lead to the observation $r \simeq k/\lambda$ for this case [16]. Writing $F_a = (k\bar{k})^{1/2}y$, where y describes an internal displacement linked to an active process, we can represent the two degrees of freedom by a single complex variable $z = x + iy$. We furthermore add a nonlinearity with complex coefficient B to stabilize the system when it starts to oscillate. We thus arrive at the equation

$$\frac{dz}{dt} = -(r + i\omega_0)z - B|z^2|z + \lambda^{-1} f_{\mathrm{ext}} + \lambda^{-1}\eta \tag{9}$$

which is a standard representation of nonlinear oscillators. Here, the frequency of oscillations is given by $\omega_0 = (k\bar{k})^{1/2}/\lambda$ and forces have been rescaled by the friction coefficient λ. The complex stochastic force η is a noise of zero average describing fluctuations both of thermal origin and due to active processes [20, 21]. If the parameter r changes sign and becomes negative, the system undergoes a Hopf bifurcation and has limit-cycle solutions. This parameter thus pays the role of the control parameter C.

AUTO-CORRELATION AND LINEAR RESPONSE FUNCTION

The linear response function χ is defined as

$$x(t) = \int_{-\infty}^{t} dt' \chi(t-t') f_{\text{ext}}(t') \quad , \tag{10}$$

In the absence of noise $\eta = 0$ it reads in Fourier representation

$$\tilde{\chi}(\omega) \simeq \frac{1/2}{i\lambda(\omega_0 - \omega) + k} + \frac{1/2}{-i\lambda(\omega_0 + \omega) + k} \tag{11}$$

In the presence of noise, the bifurcation is concealed by the noise. The system generates noisy oscillations which can be characterized by their autocorrelation function $C(t) =< x(t)x(0) >$. The noise together with the nonlinear terms lead to a renormalization of the parameters. We can still define the linear response function which is in good approximation of the form given by Eq. (11) if the parameters are replaced by their renormalized values. Assuming Gaussian noise with $< \eta(t)\eta^*(0) >= 4D(t)$, the auto-correlation function in Fourier representation is of the form

$$\tilde{C}(\omega) \simeq \frac{\tilde{D}(\omega)}{k^2 + \lambda^2(\omega - \omega_0^2)} + \frac{\tilde{D}(-\omega)}{k^2 + \lambda^2(\omega + \omega_0)} \tag{12}$$

where k and λ are effective parameters. This is a correlation function characteristic for noisy oscillators which exhibits peaks in the spectrum at the oscillation frequency.

VIOLATION OF THE FLUCTUATION DISSIPATION THEOREM

From the observation of the spectrum alone, one cannot distinguish if the system is a noisy spontaneous oscillator or a damped passive oscillator subject to thermal fluctuations. In the former case of a passive system near thermal equilibrium, however, the linear response function is related to the autocorrelation function via a fluctuation-dissipation theorem (FTD) [22]: $\tilde{C}(\omega) = 2k_B T \tilde{\chi}(\omega)''/\omega$, where $\chi = \chi' + \chi''$, i.e. χ'' is the imaginary part of the linear response function. If the system is an active oscillator, this FDT is broken. We characterize the degree to which the system deviates from equilibrium by defining the effective temperature for every frequency

$$\frac{T_{\text{eff}}(\omega)}{T} = \frac{\omega \tilde{C}(\omega)}{2k_B T \tilde{\chi}(\omega)''} \tag{13}$$

In our oscillator described above, we expect for $\omega \simeq \omega_0$

$$\frac{T_{\text{eff}}}{T} \simeq \frac{\tilde{D}(\omega)}{\lambda k_B T} \left(\frac{\omega}{\omega - \omega_0} \right) \tag{14}$$

The linear response function and the auto-correlation function have been measured for spontaneous hair-bundle oscillations [16]. A fit to the experimental data leads to

$\omega_0 \simeq 8\text{s}^{-1}$, $k \simeq 1.04 \cdot 10^{-4}$N/m, $\lambda \simeq 6.5 \cdot 10^{-6}$Ns/m. The auto-correlation function could be well described by using Gaussian white noise with $\tilde{D}(\omega) = D_0$, with $D_0 \simeq 0.14\text{pN}^2\text{s}$.

DISCUSSION

The concept of self-tuning to a Hopf bifurcation can explain seemingly disconnected but well-known properties of hearing. Sound amplification due to critical oscillations could apply to hearing in all vertebrates. Oto-acoustic emissions, which are a signature for active processes in hearing have been observed in mammals, birds, reptiles and amphibians [23]. These classes of animals however have different ears and different types of hair cells and they might thus use different physical systems to generate oscillations. In the case of mammals, there is some evidence that so-called outer-hair cells are able to generate active motion by contracting the whole cell body [5]. Recently, a protein which could play an active role in these contractions has been identified [24]. Non-mammalian vertebrates do not possess outer hair-cells. Active oscillators are therefore expected to exist within the hair bundle itself. Spontaneous hair-bundle oscillations of amphibian hair-cells have shown that the hair bundle itself is an important candidate for the active system [13, 14, 23]. The fact that the structure of hair bundles in mammals and non-mammalian vertebrates is similar, suggests that hair bundle oscillations could also exist in mammals and that they may represent an important part of the active amplifier in all vertebrates.

Spontaneous oscillations generated by cells are noisy and this noise plays an important role for the detection of weak stimuli by hair cells. The importance of noise has been discussed in the context of stochastic resonance [25, 26]. Near a Hopf bifurcation, there are other aspects of noise which play a role. In the presence of noise, the Hopf-bifurcation is concealed, and noisy oscillations cannot strictly be distinguished from noise. The difference between noisy spontaneous motion and a periodic stimulus lies mainly in the phase coherence of the stimulus. Phase coherence appears in the response of a noisy oscillator. The detection of phase coherence in the nervous signal generated by the hair cells can thus allow the nervous system to detect weak signals.

ACKNOWLEDGMENTS

I thank Sebastien Camalet, Thomas Duke, Pascal Martin, Jim Hudspeth, Björn Nadrowski, Jaques Prost and Thomas Risler for stimulating collaborations.

REFERENCES

1. Hudspeth, A.J. *Nature (London)* **341**, 397-404 (1989) .
2. Dallos, P., Popper A.N. and Fay R.R., *The cochlea* (Springer, New York) (1996).
3. de Vries, H.L. *Physica* **14**, 48-60 (1949).
4. Gold, T.: Proc. R. Soc. London Ser B **135** (1948).
5. Dallos, P., J. Neurosci. **12,** 4575-4585 (1992).

6. Hudspeth A.J., Curr. Opin. Neurobiol. **7**, 480-486 (1997).
7. Probst, R.: Adv. Otorhinolaryngol. **44**, 1-91 (1990).
8. Choe, Y., Magnasco, M., and Hudspeth A.J.: Proc. Natl. acad. Sci. **95**, 15321-15326 (1998).
9. Camalet, S., Duke, T., Jülicher F. and Prost J.: Proc. Natl. Acad. Sci. **97**, 3183-3188 (2000).
10. Eguiluz, V.M., Ospeck, M., Choe Y., Hudspeth. A.J. and Magnasco, M.: Phys. Rev. Lett **84**, 5232-5235 (2000).
11. Ruggero, M.A., Rich, N.C., Recio, A., Narayan, S.S., and Robles, L.: J. Acoust. Soc. Am. **101**, 2151-2163 (1997).
12. Jülicher, F., Andor, D., Duke, T.: Proc. Natl. Acad. Sci. **98**, 9080-9085 (2001).
13. Martin, P., and Hudspeth, A.J.: Proc. Natl. Acad. Sci. **96**, 14306-14311 (1999).
14. Martin, P., Mehta, A., and Hudspeth, A.J.: Proc. Natl. acad. Sci. **97**, 12026-12031 (2000).
15. Martin, P., and Hudspeth A.J.: Proc. Natl. Acad. Sci. **98**, 14386-14391 (2001).
16. Martin, P., and Hudspeth A.J. and F. Jülicher: Proc. Natl. Acad. Sci. **98**, 14386-14391 (2001).
17. Kachar, B., Parakkal, M., Kurc, M., Zhao, Y., and Gillespie, P.G. : Proc. Natl. Acad. Sci. **97**, 13336-13341 (2000).
18. Howard, J., and Hudspeth, A.J.: Neuron **1**, 189-199 (1988).
19. Fettiplace., R., Ricci, A.J., and Hackney, C.M., Trends in Neurosci. **24**, 169-175 (2001).
20. Stratonovich R.S, *Topics in the theory of random noise*
21. Hänggi, P., and Riseborough P.: Am. J. Phys. **51**, 347-352 (1983).
22. Forster, D., *Hydrodynamic fluctuations, broken symmetry, and correlation functions* (Addison-Wesley 1990).
23. Manley, G.A., Kirk, D.L., Köppl, C., and Yates, G.K.: Proc. Natl. Acad. Sci. **98**, 2826-2831 (2001).
24. Zheng J. et al. (2000): Nature **405**, 155.
25. Gammaitoni, L., Hängi, P., Jung, P., and Marchesoni, F., Rev. Mod. Phys. **70**, 223 (1998).
26. Wiesenfeld, K., and Moss, F., Nature **373**, 33 (1995).

Noise in the Outer Hair Cell and Gain Control in the Ear

Mark Ospeck* and Kuni H. Iwasa*

*Biophysics Section, NIDCD, NIH
50 South Drive MSC 8027
Bethesda, Maryland 20892-8027

Abstract We present a minimal model of a local resonance inside the mammalian cochlea in which a feedback loop containing an outer hair cell is poised on a dynamical instability. This model results in an amplifier essentially equivalent to a van der Pol oscillator in its weakly nonlinear limit. The feedback loop automatically seeks the instability where gain and gain compression are optimized. The outer hair cell's average membrane potential is the control parameter which sets amplifier gain. We briefly investigate noise sources inside the outer hair cell which contribute to the feedback loop oscillating in a noise-driven limit-cycle. Experimentally this limit-cycle is increased by 3 to 5 times in high external noise environments. A half millivolt shift due to efferent stimulation can account for the observed 50 percent reduction in both the size of the noisy limit-cycle oscillation and the feedback loop gain.

INTRODUCTION

It has been shown that local resonance is an essential element of the mechanics of the cochlea [1, 2]. Wave propagation starts at the cochlea's high-frequency end: waves traveling along the basilar membrane slow down while growing in amplitude as they approach their respective resonance points and then diminish rapidly as the wave's energy is absorbed at the resonance (Fig. 1A,B) [3, 4]. Here we use our model of local resonance in order to investigate noise and gain control.

The outer hair cell (OHC) is the key element that determines the amplitude gain occuring at the local resonance [5, 6, 7] (Fig. 1B,C). It forms a feedback loop and produces force that amplifies weak signals (Fig. 1D,E).

Previously we have constructed a minimal model of an OHC feedback loop in order to determine the frequency limit at which such an amplifier can function [8]. This model is expected to give a reasonable description of local resonance near the frequency limit. First we restate the model. Assumptions:

1. Bending amplitude of the hair bundles of outer hair cells is proportional to the displacement of the basilar membrane, and is of a similar magnitude.

2. The cell body of outer hair cells behave as piezoelectric elements. Their force production is controlled by their membrane potential.

3. The multiple degrees of freedom in the vibration of the cochlear partition optimizes

CP665, *Unsolved Problems of Noise and Fluctuations: UPoN 2002: Third International Conference*, edited by S. M. Bezrukov
2003 American Institute of Physics 0-7354-0127-6

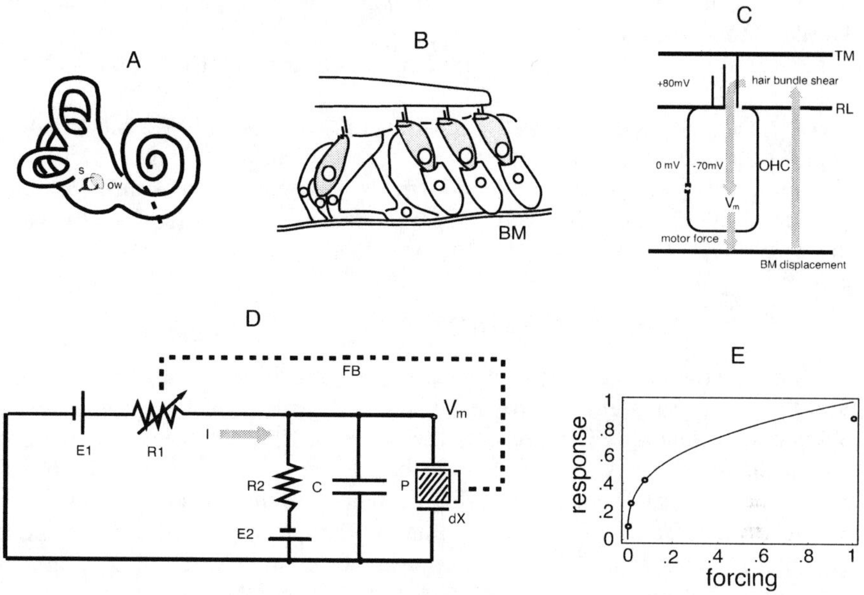

FIGURE 1. A. Snail-shaped mammalian cochlea where sound energy enters the oval window *ow* by the stirrup bone *s* at the cochlea's high-frequency end. B. Cross-section along dashed line in *A* reveals four hair cells tuned to the same frequency. A single inner hair cell at left has afferent innervation and transforms the mechanical oscillation of its hair bundle into a nerve impulse going to the brain. Three cylinder-shaped outer hair cells amplify waves on the cochlea's basilar membrane *BM* and are principally innervated by efferent fibers. C. Outer hair cell feedback loop. A sound wave displaces the *BM* which results in a shear between the reticular lamina *RL* and tectorial membrane *TM*. When the hair bundle is sheared towards its tall end transduction channels admit a positive current depolarizing the membrane potential. The cell wall contains approximately 10^7 voltage-dependent membrane motor proteins which subsequently force the *BM*, completing the feedback loop. D. Equivalent circuit of an OHC feedback loop, an interplay between a variable conductance (hair bundle transduction channels) and a piezoelectric capacitance (membrane motor proteins). $V_m \approx -70$ mV, the d.c. operating point of the OHC. Consider a voltage change to -71 mV, causing an increased E field across the piezoelectric, making it elongate. If the feedback connection is made so that an elongation increases the hair bundle conductance, then this results in a subsequent depolarization and contraction of the piezoelectric. If the feedback connection is made strong enough then the circuit will spontaneously oscillate. When parameter settings poise the circuit at the brink of spontaneous oscillation then the feedback loop is most sensitive to faint signals at its natural frequency. E. Cube-root shaped response versus forcing at the natural frequency for a Hopf bifurcation. We compare data points of the cochlea's basilar membrane response to forcing at the natural frequency at a local resonance [9]. The fit is shown across four orders of magnitude in forcing.

the phase of OHC feedback force so as to counter viscous drag. We call this the "optimized phase" approximation and it allows for a description of the feedback loop while using only a few degrees of freedom.

4.The dominant viscous drag in the OHC feedback loop occurs between the tectorial membrane and the reticular lamina (gap drag) and this drag force is directly coupled to basilar membrane oscillation.

These assumptions lead to Eqns. 1 and 2 below describing an outer hair cell feedback loop *in vivo*.

An outer hair cell membrane with capacitance C_m is charged by a membrane current consisting of a mechano-transducer current at the apical membrane, which is exposed to potassium-rich endolymph at the endocochlear potential e_{ec}, potassium currents at its basolateral membrane with the reversal potential e_K, and current due to transferring motor charge. The membrane potential V_m satisfies

$$C_m \frac{dV_m}{dt} = (g_{hb} P_{hb}[\alpha_1 X])(e_{ec} - V_m) + (e_K - V_m) P_K g_K - \alpha_2 \phi \frac{dX}{dt}. \quad (1)$$

Hair bundle conductance is the product of the maximum hair bundle conductance g_{hb} and the open probability $P_{hb}[\alpha_1 X]$ of the transducer channels in the hair bundle. Here this open probability is represented by a 3rd order Boltzmann function of basilar membrane displacement X (Fig. 2A). g_K is the maximum basolateral potassium conductance and $P_K \approx .75$ is the open probability of the potassium channel, a slow and weakly increasing Boltzmann function of membrane potential. ϕ is a piezoelectric coefficient for the outer hair cell motor. Adjustable phase factors α_n with $n = 1, 2, ...$ are complex numbers with an absolute value of unity.

The displacement of the basilar membrane follows the equation,

$$m \frac{d^2X}{dt^2} = \alpha_3 (V_m - V_0)\phi - m\omega_0^2 X - \gamma \frac{dX}{dt} + F_{ext}, \quad (2)$$

where m is the mass that moves together with a local section of the basilar membrane, zero motor force $V_0 = -68$ mV, ω_0 the natural resonance frequency of the basilar membrane, γ the effective drag coefficient, and F_{ext} the force applied to the basilar membrane by acoustic stimulation.

Eqn. 2 will behave like an undamped harmonic oscillator if drag force can be cancelled by a properly phased OHC feedback force of equal magnitude:

$$\gamma \frac{dX}{dt} = \alpha_3 (V_m - V_0)\phi \quad (3)$$

We have called this "Gold's condition" because he proposed the idea in his pioneering work [10]. Meeting Gold's condition is equivalent to poising the feedback loop on a Hopf bifurcation [11] [12] [13].

In order to find the frequency limit where the system is able to meet Gold's condition, we assume parameter values appropriate to a short outer hair cell and a small acoustic stimulation at frequency ω. When we obtain a perturbation solution Gold's condition becomes:

$$\omega^2 \approx \frac{dP_{hb}}{dX} \frac{\phi g_{hb} (e_{ec} - V_0)}{\gamma C_m}, \quad (4)$$

which gives a frequency limit of approximately 10 kHz, above which the efficiency of the outer hair cell as a cochlear amplifier sharply declines. This frequency limit is within

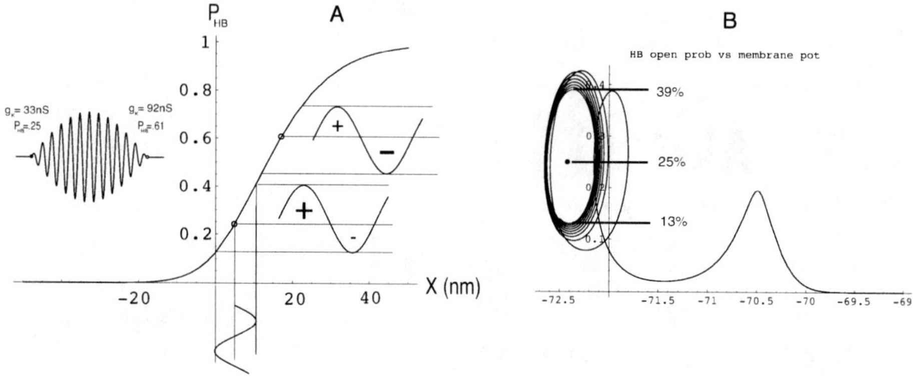

FIGURE 2. A. The feedback loop has two Hopf bifurcations where its hair bundle open probability slope is just steep enought to support spontaneous oscillation (Eqn. 4). The inset shows a transition from a fixed point to a limit-cycle oscillation at hair bundle open probabilities .25 and .61, referred to as the low and high current operating points respectively (g_{HB} is fixed at 11.6 nS for these 5 kHz simulations). At the low current operating point a large symmetric hair bundle oscillation results in an asymmetric bundle conductance oscillation so that each cycle the cell is depolarized, contracted and average P_{HB} decreased. Moreover this coupling between the dynamics and the d.c. operating point results in a self-tuning: increasing P_{HB} above the instability makes an asymmetric limit-cycle opposing the increase. Note that large oscillation at the high current operating point would result in reduced gain via a hyperpolarization increasing average P_{HB}. B. A large depolarizing on-transient at the low current operating point shows the asymmetric bundle conductance oscillation which is depolarizing the cell, turning down the gain and decreasing the size of the limit-cycle oscillation.

the mammalian auditory frequency range. Thus an additional mechanism is required for frequencies above the limit and our model should accurately describe the cochlear feedback loop near the limit [8].

FEEDBACK LOOP SEEKS THE INSTABILITY

In Eqn. 4, for a given piezoelectric motor coefficient, total hair bundle conductance, endocochlear potential, drag, capacitance, and Boltzmann-shaped hair bundle open probability, it is then the hair bundle slope sensitivity *at the operating point* which sets the gain of the system. Keeping this in mind we investigate gain control by simulating the feedback loop at 5 kHz (Eqns. 1 and 2; Fig. 2).

Large gain for small signals and gain compression occur only when parameter settings make the feedback loop poise very near to the Hopf instability. Parameter settings that make large spontaneous oscillations result in lower sensitivity for small signals. Fortunately Hopf type instabilities have the property of resisting large limit-cycle oscillations, in effect actively seeking to be close to the oscillatory side of the instability. For them a robust choice of parameter values results in a "self-tuned critical oscillation" [12]. While our simulations will refer to the low current operating point the results will also hold for the high current operating point (see discussion section).

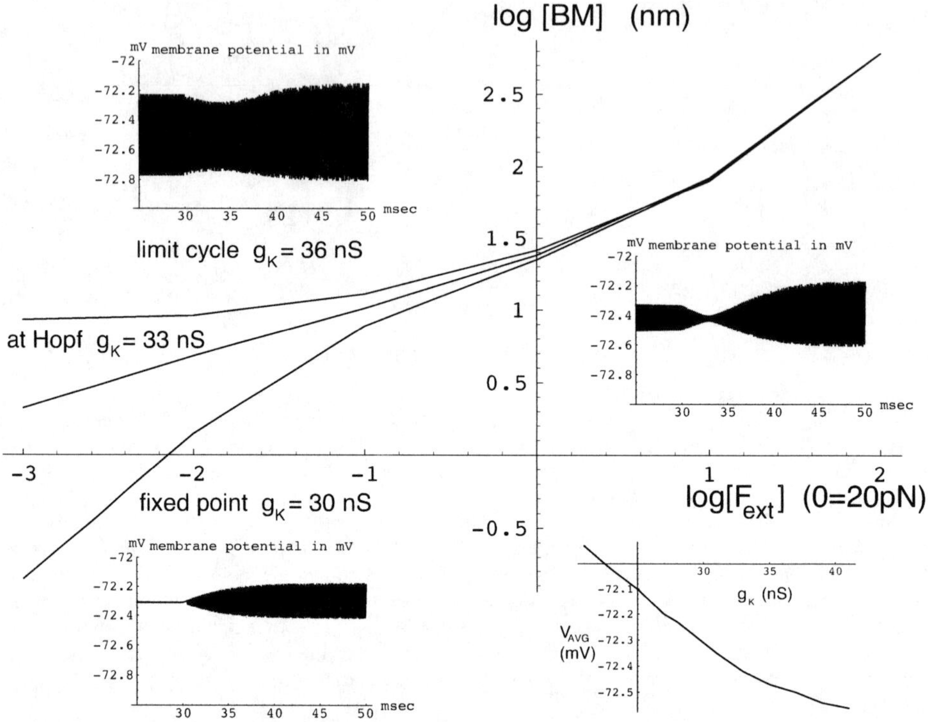

FIGURE 3. Small changes in average membrane potential control the gain of the feedback loop. First the bottom right inset shows that cutting potassium conductance in half (from 40 to 20 nS) near the low current operating point depolarizes the cell by only 0.5 mV. This because as the cell depolarizes it contracts and turns off hair bundle conductance. The main figure is a log-log plot of external forcing versus *BM* displacement. There decreasing the potassium conductance by 20% pushes the feedback loop through the Hopf bifurcation, making a fixed point, even though average membrane potential depolarizes by only 200 μV. The insets show the membrane potential response when the feedback loop is forced at 5.05 kHz starting at 30 msec. Note that lowering gain (making a fixed point) at the high current operating point would require *turning on* potassium conductance.

AVERAGE MEMBRANE POTENTIAL SETS GAIN

The transduction channels in the hair bundle are mechanically gated, however when connected into the feedback loop they effectively become voltage-sensitive:

$$\frac{dP_{hb}}{dV_m} = \frac{dP_{hb}}{dX}\frac{dX}{dF}\frac{dF}{dV_m} = \frac{dP_{hb}}{dX}\frac{\phi}{k} = 1/(5\text{mV}), \tag{5}$$

where ϕ (=100pN/mV) is motor force per volt [14], k (=20pN/nm) is the local stiffness at 5 kHz and dP_{hb}/dX= 1/(25nm) [15].

An additional effect is that a change in the cell's potassium conductance generates a similar response in hair bundle conductance which makes it very difficult to displace the average membrane potential (Fig. 3). This effectively doubles the voltage sensitivity of the hair bundle conductance so that full closed to full open requires only a 2.3 mV hyperpolarization (s for e-fold conductance change ≈ 0.7 mV).

In the vicinity of the low current operating point a 20% decrease in potassium conductance will depolarize the average membrane potential by only about 200 μV. Otherwise it will cause a substantial change in the system's operating point, crossing through the bifurcation and replacing a 0.5 mV limit cycle oscillation with a fixed point (Fig. 3). This makes membrane potential a natural choice for being the system's control parameter which sets gain.

INTERNAL NOISE, EXTERNAL NOISE

We'll compare several of the prominent noise sources in the OHC feedback loop to Johnson noise. At 5 kHz the cell corresponds approximately to a 10 MΩ resistor with a current and voltage noise of approximately 4 pA, 40 μV RMS across 10 kHz bandwidth, because the Johnson noise current spectral density is about $10^{-27}\text{A}^2/\text{Hz}$ and voltage spectral density $10^{-13}\text{V}^2/\text{Hz}$.

Thermal fluctuations of the hair bundle should be about 0.5 nm RMS for a 5 kHz OHC. Drag and compliance low-pass filter these fluctuations across a 15 kHz bandwidth. Given the hair bundle conductance and its sensitivity this results in about 35 pA RMS, significantly larger than Johnson current noise.

The motor proteins will produce a high-pass current noise having a spectral density about ten times larger than Johnson ($\approx 10^{-26}\text{A}^2/\text{Hz}$). However this noise should be less relevant to a 5 kHz feedback loop because it has a high-pass corner frequency near 35 kHz [16].

If at the low current operating point at 5 kHz the cell should contain approximately 2000 M-current channels having a voltage spectral density four orders of magnitude larger than Johnson ($\approx 10^{-9}\text{V}^2/\text{Hz}$). However because these channels are very slow-activating this noise will be rolled off with a corner frequency around 10 Hz [17].

Other possible noise sources are limit-cycle noise if poised above the bifurcation, acetylcholine noise since it is the efferent's neurotransmitter, "battery" noise from perhaps millions of cells sourcing the endocochlear potential, etc. However, among the four which we have considered, current noise due to hair bundle fluctuations would appear to be the largest. Hence we will represent high external noise environments as an increase in the background RMS hair bundle fluctuation.

GAIN CONTROL

Single afferent nerve discharge rates of cats were recorded with and without acoustic stimuli, external broadband noise, and efferent feedback [18]. One recording from a 7.1 kHz afferent nerve fiber showed a discharge rate of 30 spikes/sec in quiet that ramped

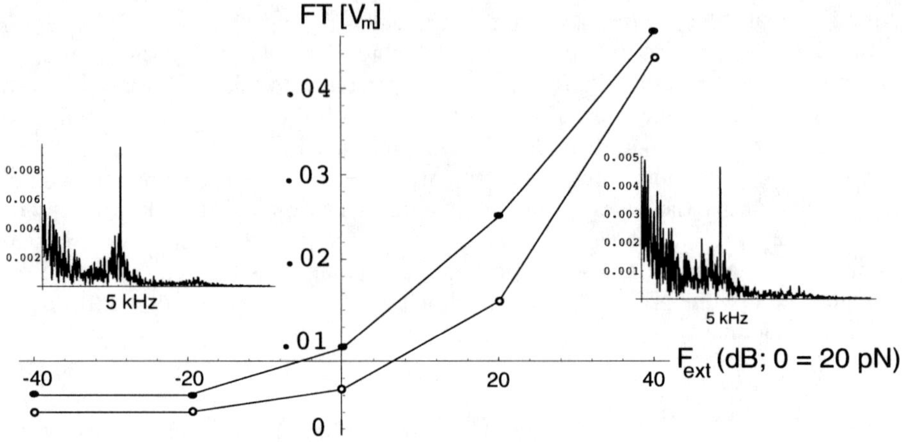

FIGURE 4. Height of the Fourier Transform peak versus external forcing at the natural frequency 5.05 kHz in a high external noise environment (hair bundle fluctuations increased from 0.5 to 5 nm RMS across a 15 kHz band width). Solid points are with g_K= 27 nS. Open circles are as if efferent feedback cuts potassium conductance in half (g_K= 14 nS), which results in only a 0.5 mV depolarization, but cuts the FT peak height in half for 0 dB forcing (insets). Also note that the FT curve is shifted rightward about 15 dB.

to 250/sec as the forcing tone level was increased from 14 to 50 dB SPL. Next the white noise background was increased and the nerve's spontaneous rate became 90/sec. This spontaneous discharge rate then dropped to 40/sec when the efferents to the outer hair cells were stimulated. Also the discharge rate versus tone level curve curve shifted rightward by 16 dB.

Our model can explain the afferent nerve discharge rate experiment, because neural output is proportional to the basilar membrane oscillation [19]. Increasing the RMS hair bundle fluctuation by an order of magnitude, triples the noisy fourier transform peak height (Fig. 4). Then we depolarize the OHC by only 0.5 mV. The FT peak height is cut in half and the FT versus forcing curve is shifted to the right by about 15 dB.

DISCUSSION, OPEN QUESTIONS

The OHC feedback loop contains two Hopf bifurcation points (Fig. 2 A). Since operating in the vicinity of either one will give excellent amplifier performance, which should be chosen? Large oscillations at the low current operating point result in a d.c. depolarization, while at the high current operating point, they would cause a d.c. hyperpolarization. The experimental record has evidence for both, and also for no d.c. offset, and this last would require operating at a Hopf bifurcation very close to steepest slope of the hair bundle Boltzmann [20]. However, it is known that efferent feedback to the OHC, while initially turning on a Ca^{++}/Na^{+} current—that subsequently the Ca^{++} turns on K^{+} conductance [21]. If this is universally true, requiring that gain be turned off would seem to go in favor the high current operating point [22].

Monkeys with intact efferent feedback are able to discriminate between the vowel sounds 'i' and 'u' when presented at 70 dB SPL and imbedded in 95 dB of broadband white noise (2.4 kHz bandwidth). However the same monkey with his efferent feedback cut was only able to discriminate between the two in 80 dB of white noise [23]. So while it is clear that efferent feedback will be able to substantially turn down cochlear gain, it remains unclear how this improves the signal to noise ratio (Fig. 4 insets) as is implied by the vowel discrimination experiment.

CONCLUSIONS

In our model, which should be valid in the vicinity of 5-10 kHz, the outer hair cell feedback loop has two Hopf bifurcations, each occurring where the slope of the Boltzmann-shaped hair bundle conductance becomes just sufficient to suport a spontaneous oscillation. Each of these operating points is able to provide large small-signal gain and gain compression. For each there is a range of average membrane potentials where the feedback loop will automatically seek to poise itself close to the oscillatory side of its respective Hopf bifurcation. The outer hair cell's average membrane potential seems to be a natural choice for the control parameter which both sets amplifier gain and also chooses the high or low current operating point.

RMS hair bundle fluctuations are likely the dominant internal noise source, and noisy external environments are likely fairly represented as an increase in the background of such fluctuations. At the low current operating point if efferent feedback were to depolarize the cell by only half a millivolt then it could account for the observed 50% reduction of the noise-driven limit cycle. Conversely, if at the high current operating point then a half millivolt hyperpolarization would also cut the gain in half, curiously moving away from the dynamical instability by increasing the d.c. current through the cell.

REFERENCES

1. A. F. Huxley. *Nature* **221** 935 (1969).
2. R. Patuzzi. In P. Dallos, A. N. Popper and R. R. Fay, editors *The Cochlea* p.186 (Springer, New York, NY, 1996).
3. G. Zweig. *Cold Harbor Symp. Quantit. Biol.***40**619 (1976).
4. J. Lighthill. *J. Fluid Mech.***100**149 (1981).
5. M. C. Liberman and L. W. Dodds, *Hear. Res.***16** 55 (1984).
6. E. de Boer. *Phys. Rep.***203** 125 (1991).
7. A. E. Hubbard and D. C. Mountain. In H. L. Hawkins, T. A. McMullen, A. N. Popper and R. R. Fay, editors *Auditory Computation* p.62 (Springer, New York, NY, 1995).
8. M. Ospeck, X. -x. Dong and K. H. Iwasa. in Press *Biophys. J.*
9. M. A. Ruggero. *Curr. Opin. Neurobiol.* **2** 449 (1992).
10. T. Gold. *Proc. Roy. Soc. B* **135** 492 (1948).

11. V. M. Eguíluz, M. Ospeck, Y. Choe, A. J. Hudspeth, and M. O. Magnasco. *Phys. Rev. Lett.* **84** 5232 (2000).
12. S. Camalet, T. Duke, F. Julicher, and J. Prost. *PNAS* **97** 3183 (2000).
13. S. H. Strogatz, *Nonlinear dynamics and chaos*,(Perseus Books, Reading, MA, 1994).
14. K. H. Iwasa and M. Adachi, *Biophys. J.* **73** 546 (1997).
15. I. J. Russel, G. P. Richardson and A. R. Cody, *Nature* **321** 517 (1986).
16. X. -x. Dong, D. Ehrenstein and K. H. Iwasa. *Biophys. J.* **79** 1876 (2000).
17. L. J. DeFelice *Introduction to Membrane Noise*,(Plenum Press, New York, NY, 1981).
18. R. L. Winslow and M. B. Sachs. *J. Neurophysiol.* **57**,1002 (1987).
19. S. S. Narayan, A. N. Temchin, A. Recio and M. A. Ruggero. *Science* **282**1882 (1998).
20. J. O. Pickles *An Introduction to the Physiology of Hearing*,(Academic Press, San Diego, CA, 1992).
21. D. J. Jagger, C. B. Greiesinger, M. N. Rivolta, M. C. Holley and J. F. Ashmore. *J. Physiol.* **527.1** 49 (2000).
22. D. O. Kim, X. M. Yang and S. T. Neely. in A. W. Gummer, ed. *Biophysics of the Cochlea: from Molecule to Model* (World Scientific, Singapore, 2002) in press.
23. J. H. Dewson. *J. Neurophysiol.* **31** 122 (1968).

An Information-Theoretic Approach to Cochlear Implant Coding

Robert P. Morse*, David Allingham†, Nigel G. Stocks†

*MacKay Institute of Communication and Neuroscience, Keele University, Keele, Staffordshire, ST5 5BG, UK.

†School of Engineering, University of Warwick, Coventry, CV4 7AL, UK

Abstract. Cochlear implants have become a standard method of providing functional hearing to people who are otherwise profoundly deaf. We present a rigorous design methodology for cochlear implants, based on the premise that neural coding in the normal ear is almost optimal in terms of information transmission. We propose that a cochlear implant should be designed to maximize the information conveyed by the artificially stimulated cochlear nerve. The deliberate addition of noise to cochlear implant signals is an essential part of this strategy.

INTRODUCTION

In the normal acoustic system, sounds received at the outer ear are converted into electrical signals by hair cells within the inner ear, or cochlea. The electrical signals are subsequently conveyed by the cochlear nerve (the nerve of hearing) to the higher auditory centres of the brain. The hair cells within the cochlea are physiologically vulnerable and can be damaged or completely destroyed by some congenital conditions, some diseases (such as meningitis), the side-effects of some life-saving drugs (such as aminoglycoside antibiotics), or by other causes. The severe loss of hair cells results in profound deafness. Fortunately, the cochlear nerve is largely unaffected by the conditions that damage the hair cells, and the hearing of profoundly deaf people can often be partially restored by direct electrical stimulation of the cochlear nerve by a cochlear implant [1]. A standard cochlear implant contains three parts: an array of typically 16 to 22 electrodes surgically implanted within the cochlea, a speech processor outside the body for converting the signal at a microphone into the appropriate electrode signals, and a radio-frequency link for transmitting the speech processor signals across the skull to the electrodes.

Cochlear implants are now made commercially and provide great benefit to a large number of people: in the UK, over 900 adults and 800 children have a cochlear implant; worldwide there are over 30000 implantees. Typically, without the use of lipreading, a cochlear implant user can understand about 80% of the words in a sentence in a good listening environment (no reverberation or competing sounds) [2]. A cochlear implant can therefore be an effective method of providing speech information to people who would otherwise have to rely almost entirely on lipreading. There is, however great variability in the ability of cochlear implant users to

CP665, *Unsolved Problems of Noise and Fluctuations: UPoN 2002: Third International Conference*, edited by S. M. Bezrukov

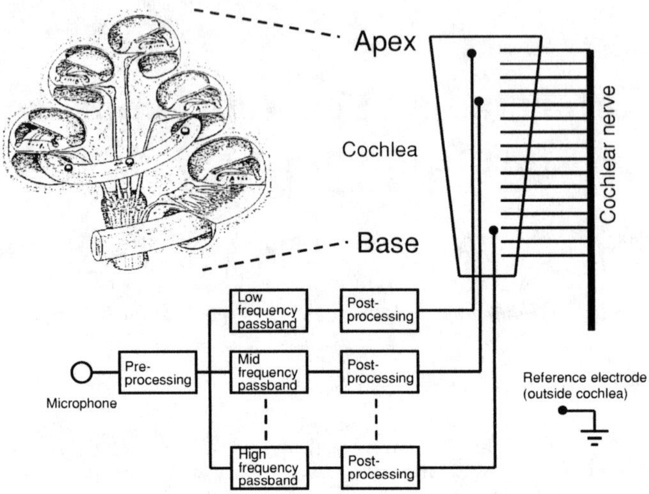

FIGURE 1. Schematic of a cochlear implant showing the electrodes in a cross-section of the cochlea and the filtering used to mimic the place coding of frequency in the normal ear.

comprehend speech: some are able to use a telephone; others get little benefit from the implant [3]. These differences appear to be associated with the physiological state of the surviving cochlear nerve fibres; cochlear implant users who have a shorter duration of complete deafness before implantation tend to have better speech comprehension. There is also a large learning effect after implantation and users often take over a year to show the full benefit of implantation. A learning effect is also apparent if the strategy for converting the microphone input into the appropriate electrode signals (the cochlear implant coding) is changed.

The purpose of the multiple electrodes inside the cochlea is to mimic, albeit crudely, the coding of frequency that occurs with normal acoustic stimulation. In the normal ear, the frequency of an acoustic signal can potentially be coded both in terms of which fibres are most active (place coding) and the temporal pattern of activity of each fibre (time coding). The place coding of information is mimicked by filtering the input signal by a bank of bandpass filters, each with a different passband. In accord with the normal tonotopic arrangement of the cochlea, bandpass filters with low-frequency passbands are used to stimulate electrodes in the apex of the cochlea and those with a high-frequency passband are used to stimulate electrodes in the base (Figure 1).

Ideally, a speech coding strategy would evoke the same patterns of nerve activity with electrical stimulation as those that are evoked in the normal ear with acoustic stimulation. This can never be achieved, however, largely because of the reduced number of healthy nerve fibres present in a damaged cochlear nerve and the limited number of electrodes that can be surgically implanted into the cochlea. Typically, the 20 or so electrodes in the cochlea will stimulate about 15,000 nerve fibres [4]. Furthermore, the electrodes are surrounded by the normal conductive fluids of the cochlea [5] and the currents from the individual electrodes therefore tend to spread throughout the conductive medium. Consequently, nerve fibres are not under the control of single electrodes as desired [6]. Given that the cochlear nerve is the only neural pathway between the ear and higher auditory centres, there is therefore an "information bottleneck" at the electrode-nerve interface [7,8]. Information transfer to higher auditory centres is also limited by the channel capacity of the cochlear nerve.

An early design strategy for cochlear implant coding was to accept unquestioningly the more limited channel capacity in the deafened ear and to therefore reduce the stimulus information in a controlled fashion. The argument was that given a limited capacity, only the features of speech deemed essential for speech perception should be coded. Based on psychophysical experiments on people with normal hearing, the most essential cues were taken to be amplitude, the fundamental frequency (F0) of voiced sounds, and the formant frequencies [e.g. 9,10]; the formant frequencies reflect the resonances of the vocal tract and are used to distinguish vowels and some consonants. In the early Nucleus cochlear implant (Cochlear Corporation), only amplitude, F0 and the second formant frequency (F2) were coded. In later designs, the first formant (F1) and other frequency information was coded. Contrary to the expectations for an information bottleneck, each development of the Nucleus implant led to a significant improvement in speech comprehension for *most* implantees [11-14]. The improvement was particularly noticeable when speech was presented in a background of competing sounds, i.e. noise in the auditory sense of the word. These results lead us to two speculative conclusions. First, that the information content of the pre-processed signals was often less than the channel capacity. Second, that the channel capacity of cochlear implantees differed. These conclusions are speculative because the channel capacity of cochlear implantees has never been measured.

In more modern cochlear implants, speech features are not preselected and the outputs of the bandpass filters are used almost directly to stimulate the appropriate parts of the cochlea. These strategies, which are not speech specific enable better discrimination of environmental sounds [15,16]. Moreover, for many cochlear implantees, non-speech specific strategies also enable better comprehension of speech, especially in noisy environments [13,15]. But even with non-speech specific strategies, the cochlear implant designer must decide which aspects of the normal spatio-temporal pattern of nerve activity are most important for good speech comprehension. Because designers give more consideration to either place or time coding, the coding strategies used by the various manufacturers differ. To-date, the choice of strategy and the settings for a particular patient has been rather *ad hoc*.

PROPOSED STRATEGY

We propose a new design methodology that we believe will circumvent many of the *ad hoc* aspects of contemporary coding strategies and may lead to improved speech comprehension for cochlear implant patients. We are trying to quantify the amount of information that is coded by the cochlear nerve in response to electrical stimulation from a cochlear implant. Furthermore, rather than taking the channel capacity of the deafened ear to be fixed, we are investigating methods to increase it.

The expectation, which will be tested, is that increasing the amount of information transmitted will increase the speech comprehension of cochlear implant users. This would be consistent with findings from the early Nucleus implant. There is other evidence to suggest that coding by the normal nervous system (the neural code) has evolved to maximize the transfer of information. Physiological experiments and theoretical analyses have established that neurones encode information in a highly efficient manner, achieving information rates that are within a factor of two of the theoretical maximum [e.g. 17]. Furthermore, physiological experiments on the vestibular system (the organ of balance) have shown that the level of noise that maximizes the signal-to-noise ratio (SNR)[18] corresponds to the Brownian motion experienced by unrestrained hair bundles. In other words, the vestibular system seems to have evolved to take advantage of the naturally occurring noise.

Measurement of the information transmitted (the transinformation) could be used, in the first instance, to compare proposed cochlear implant designs without the cost and time required for clinical trials of all variations; only those variations that led to a substantial change in the transinformation would be tested clinically. This approach would be particularly useful for studies that would otherwise require a different population of subjects to evaluate each design, for example the investigation of different electrode arrays. Because of patient variability, a large population of implantees is required to establish differences among designs. Our method is also appropriate as a precursor to studies with different coding strategies for a group of subjects that have identical implants. Such experiments are hindered by the effects of learning or familiarity with a standard strategy. With an information-theoretic approach, we aim to predict which strategy is most likely to give the best speech comprehension *after a period of learning* – we hypothesize that this will be the strategy that maximizes the transmission of information.

As introduced above, we are investigating methods for increasing the channel capacity of the deafened ear. In the normal ear, the cochlear nerve spontaneously produces action potentials when there is no acoustic stimulus, i.e. in complete silence; the spontaneous discharge reflects the noisy mechanotransduction of sound by the hair cells [19]. It is often thought that noise (in its generic sense) imposes a fundamental limit on the ability of a sensory system to encode information, but recent research from the field of stochastic resonance (SR) has shown noise can actually enhance neural coding [e.g. 18,20,21]. Noise can be beneficial because a nerve fibre produces action potentials only at the instants when the inherent threshold is exceeded. Without noise, the stimulus between threshold crossing times cannot be deduced from the deterministic series of action potentials. But if noise is added to the stimulus, the instantaneous stimulus amplitude determines the graded probability (i.e. not necessarily 0 or 1) that the threshold will be crossed. The fine time-structure of the discharges then contains information about the stimulus at all points in time, and the amount of information transmitted is potentially greater. Noise may therefore be an essential part of neural coding rather than an obstacle. Effectively, the internal neural noise acts to increase the channel capacity. This is relevant to cochlear implant coding because in the deafened ear, where the hair cells have been destroyed, the inherent noise is much reduced. The reduction, which would be expected to reduce the channel capacity of the deafened ear, is shown by the absence of a spontaneous discharge when the cochlear nerve in the deafened ear is not stimulated [19]. We have therefore advocated the deliberate addition of noise to cochlear implant signals and have shown that noise can enhance the coding of speech signals [22].

Based on what was known about SR at the time, we originally proposed that the information-bearing stimulus without noise should be subthreshold. This was contrary to standard cochlear implant coding where, because noise was not used, the stimuli were always suprathreshold. A new form, however, of SR has been demonstrated. This new form, termed suprathreshold stochastic resonance (SSR)[23,24], differs from conventional SR in that it not only occurs for suprathreshold signals but is actually optimized when the signal is suprathreshold. Also in contrast to conventional SR, SSR occurs only in arrays of parallel threshold devices; it cannot occur in a single device. The main advantage of SSR is that it leads to significantly greater information transmission than can be achieved using conventional SR precisely because the signals are suprathreshold. Therefore, not only are we no longer restricted to subthreshold stimulation but we expect the benefits of noise with conventional suprathreshold cochlear implant signals will be substantially greater than with subthreshold signals.

Ideally, we would like to be able measure the information transmission by the cochlear nerve of a cochlear implant user for a particular coding strategy and to simultaneously measure speech comprehension. Recording from the cochlear nerve of a cochlear implantee, however, would require invasive and ethically unacceptable surgery. We are therefore investigating the correlation between the speech comprehension of cochlear implantees and information transmission in computational models of the electrically stimulated ear. A first step is to demonstrate that we can measure the information transmitted by one channel of a simulated cochlear implant to an array of simulated nerve fibres and that we can change the transinformation by changing the amount of added noise. We have previously made these measurements for slowly varying stimuli where the timescale of the stimuli was much greater than the timescales of the model. In this paper, we demonstrate that we can also make the measurements when the effective stimulus bandwidth is 500 Hz, equivalent to the long-term bandwidth of speech [25].

METHODS

We used an array of leaky integrate-and-fire models to simulate the nerve fibres in the electrically stimulated ear; details of the nerve model are given in [26]. One of the parameters is the intensity of the internal noise.

The global transmission through the array was characterized by the transinformation, I, which for a continuous channel is standardly defined [27] by:

$$\begin{aligned} I &= H(y) - H(y \mid s) \\ &= -\int_{-\infty}^{\infty} P_y(y)\log_2 P(y)dy - \left(-\int_{-\infty}^{\infty} P_s(s)ds\int_{-\infty}^{\infty} P(y \mid s)\log_2(P(y \mid s))dy\right) \end{aligned} \tag{1}$$

where $H(y)$ is the entropy of $y(t)$ and $H(y|s)$ can be interpreted as the amount of encoded information lost in the transmission of the signal. $P_y(y)$ and $P_s(s)$ are the probability density functions (pdfs) of $y(t)$ and $s(t)$ respectively and $P(y|s)$ is the conditional pdf. As described previously [26], $P_y(y)$ can be obtained from a reconstruction of the input signal based on the times of action potentials from the neural array. In the absence of noise $H(y|s) = 0$ and hence $I = H(y)$, and for sufficiently large noise $H(y|s) = H(y)$ and hence $I = 0$. Therefore, I can increase in the presence of noise only if $H(y)$ increases faster with increasing noise than $H(y|s)$. This is precisely what happens in SSR over a range of noise levels.

For generality, the cochlear implant signal from a single channel had a Gaussian amplitude distribution; the autocorrelation of the signal was exponential with a correlation time-constant of 1.6 ms (100 Hz cutoff). The dynamics of the nerve model were 5 times slower than the dynamics of the cochlear nerve [28] and the stimulus was therefore equivalent to that of speech with a long-term bandwidth of 500 Hz. The stimulus amplitudes were defined in dB relative to a threshold stimulus level. This was defined to be twice the standard deviation of the simulated membrane voltage for any one of the neurones in the absence of noise. With this definition, the membrane voltage was below threshold about 99% of the time for a 0 dB signal.

RESULTS

Figure 2 shows the effect of noise on transinformation for a stimulus 6 dB above threshold; results are shown for simulations with 1, 4, 8 or 16 neurones. In the figure, transinformation is plotted against scaled noise, σ, such that $\sigma = \sigma_n^2 / \sigma_s^2$ where σ_n^2 is the variance of the simulated membrane voltage of a neurone in the absence of a signal (i.e. due to noise alone) and σ_s^2 is the variance of the membrane voltage in the absence of noise (i.e. due to signal alone). Consequently, σ is the inverse of the SNR of the membrane voltage. Except for the single neurone condition, there was a non-monotonic dependence of the transinformation on the noise level. For multiple neurones, the transinformation was maximized by a non-zero noise level; this is the signature of the SSR effect [23,24]. With a greater number of neurones, the maximum transinformation was greater. The maximum transinformation also increased with stimulus amplitude (Figure 3), i.e. more suprathreshold signals resulted in higher transinformation. Again this is indicative of the SSR effect.

CONCLUSIONS

In this paper, we have described a rigorous design methodology for cochlear implants: a cochlear implant should be designed to maximize the information conveyed by the electrically stimulated cochlear nerve. To implement this strategy, we need to be able to measure the information conveyed by an array of nerve fibres and to find methods to maximize it. These preliminary results show that these objectives are achievable. Furthermore, SSR, and hence improved information transmission, has now been demonstrated for signal bandwidths comparable to that of speech. The effect of noise was substantial in that the transinformation was doubled by the optimal addition of noise compared with the no noise condition. The results also indicate that a large number of neurones are not required for the SSR effect to be useful.

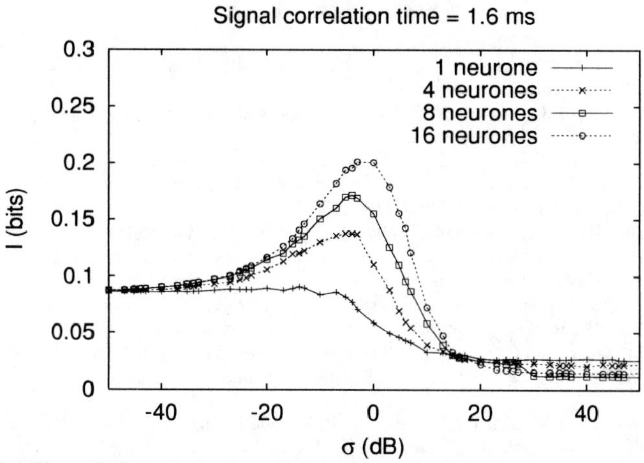

FIGURE 2. The effect of scaled noise intensity, σ, on transinformation, I, for a stimulus 6 dB above threshold. The different curves show the transinformation for either 1, 4, 8 or 16 neurones in the simulated array.

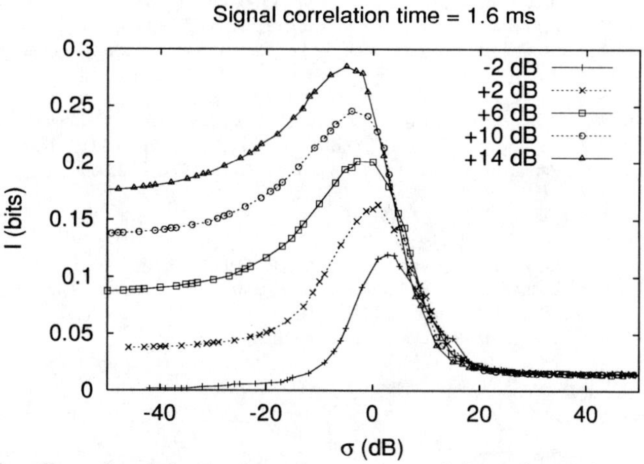

FIGURE 3. The effect of scaled noise intensity, σ, on transinformation, I, for various signal levels from 2 dB below the signal threshold (i.e. the defined threshold in the absence of noise) to 14 dB above threshold.

ACKNOWLEDGMENTS

This work was supported by the Medical Research Council (grant G0001114) and the Engineering and Physical Sciences Research Council (grant GR/35650/01).

REFERENCES

1 Rosen, S., "Cochlear implants," in *Scott-Brown's Otolarngology (6th edition) Volume 2*, edited by D. Stephens, Oxford: Butterworth Heinemann, 1996.
2 Osberger, M. J., and Fisher, L., *Ann. Otol. Rhinol. Laryngol.* **108**, 74-79 (1999).
3 Dorman, M. F., "Speech perception by adults," in *Cochlear implants: audiological foundations*, edited by R. S. Tyler, London: Singular Publishing Group, 1993, pp. 145-190.
4 Otte, J., Schuknecht, H. F., and Kerr, A. G., *Laryngoscope* **8**, 1231-1246 (1978).
5 von Békesy, G., *J. Acoust. Soc. Am.* **23**, 18-28 (1951).
6 Simmons, F. B., and Glattke, T. J., *Ann. Otol. Rhinol. Laryngol.* **81**, 731-738 (1972).
7 Fourcin, A. J., Rosen, S. M., Moore, B. C. J., Douek, E. E., Clarke, G. P., Dodson, H., and Bannister, L. H., *Brit. J. Audiol.* **13**, 85-107 (1979).
8 Clark, G. M., "Cochlear implants: historical perspectives," in *Profound Deafness and Speech Communication*, edited by G. Plant and K.-E. Spens, London: Whurr, 1995, pp. 165-218.
9 Potter, R. K., Kopp, G. A., and Green, H. C., *Visible speech*. New York, Van Nostrand, 1947.
10 Cooper, F. S., Delattre, P. C., Liberman, A. M., Borst, J. M., and Gerstman, L. J., *J. Acoust. Soc. Am.* **24**, 597-606 (1952).
11 Franz, B. K., Dowell, R. C., Clark, G. M., Seligman, P. M., and Patrick, J. F., *Am. J. Otol.* **8**, 516-518 (1987).
12 Skinner, M., Holden, L., and Holden, T., *Ear Hearing* **12**, 3-22 (1991).
13 Skinner, M., Clark, G. M., Whitford, L. A., Seligman, P. M., Staller, S. J., Shipp, D. B., Shallop, J. K., Menapace, C. M., Arndt, P. L., Antogelli, T., Brimacombe, J. A., Pijl, S., Daniels, P., George, C. R., McDermott, H., and Beiter, A. L., *Am. J. Otol.* **15**, 15-27 (1994).
14 Cohen, N. L., and Waltzman, S. B., *Ann. Otol. Rhinol. Laryngol.* **104**, 9-14 (1995).
15 Gantz, B. J., Tyler, R. S., Knutson, J. F., Woodworth, G., Abbas, P., McCabe, B. F., Hinrichs, J., Tye-Murray, N., Lansing, C., Kuk, F., and Brown, C., *Laryngoscope* **98**, 1100-1106 (1988).
16 Tyler, R. S., Moore, B. C. J., and Kuk, F., *J. Speech Hear. Res.* **32**, 887-911 (1989).
17 Rieke, F., Warland, D., Stevenick, R. R., and Bialek, W., *Spikes:exploring the neural code*. Boston, MIT Press, 1997.
18 Jaramillo, F., and Wiesenfeld, K., *Nat. Neurosci.* **1**, 384-388 (1998).
19 Kiang, N. Y. S., *Discharge patterns of single fibres in the cat's auditory cortex*. Cambridge, MIT Press, 1965.
20 Bulsara, A. R., Jacobs, E. W., Zhou, T., Moss, F., and Kiss, L., *J. Theor. Biol.* **152**, 531-555 (1991).
21 Douglass, J., Wilkens, K., Pantazelou, E., and Moss, F., *Nature* **365**, 337-340 (1993).
22 Morse, R. P., and Evans, E. F., *Nat. Med.* **2**, 928-932 (1996).
23 Stocks, N. G., *Phys. Rev. Lett.* **84**, 2310-2313 (2000).
24 Stocks, N. G., *Physical Review E* **63**, 1-11 (2001).
25 French, N. R., and Steinberg, J. C., *J. Acoust. Soc. Am.* **19**, 90-119 (1947).
26 Stocks, N. G., Allingham, D., and Morse, R. P., *Noise and Fluctuation Letters* (2002).
27 Shannon, R. V., *Ann. Otol. Rhinol. Laryngol.* **90**, 13-14 (1981).
28 Morse, R. P., and Evans, E. F., *Hear. Res.* accepted for publication.

Graded Positive Feedback in Elasmobranch Ampullae of Lorenzini

Ad. J. Kalmijn

The Faraday Laboratory, Scripps Institution of Oceanography
University of California, San Diego, La Jolla, CA 92093-0220, USA

Abstract. The acute electrical sensitivity of marine sharks and rays is the greatest known in the Animal Kingdom. I investigate the possibility that the underlying biophysical principles are the very same as those encountered in the central nervous system of animal and man. The elasmobranch ampullae of Lorenzini detect the weak electric fields originating from the oceanic environment, whereas the nerve cells of the brain detect the electric fields arising, well, from the central nervous system. In responding to electrical signals, the cell membranes of excitable cells behave in different regions of the cell as negative or positive conductors. The negative and positive conductances in series, loaded by the cell's electrolytic environment, constitute a positive feedback circuit. The result may be of an all-or-none nature, as in peripheral nerve conduction, or of a graded nature, as in central processing. In this respect, the operation of the elasmobranch ampullae of Lorenzini is more akin to the graded, integrative processes of higher brain centers than to the conduction of nerve action potentials. Hence, the positive-feedback ampullary circuit promises to help elucidate the functioning of the central nervous system as profoundly as the squid giant axon has served to reveal the process of nervous conduction.

ELECTRIC AND MAGNETIC SENSORY WORLD

Although the elasmobranch ampullae of Lorenzini were known anatomically since the 17th century [1], their function remained an enigma until it was ascertained that (i) sharks and rays are highly sensitive to weak, uniform and dipolar, electric fields [2], and (ii) the ampullae of Lorenzini constitute the sense organs mediating the observed behavioral responses [3]. Nonetheless, the weak electric fields in natural waters and the sensory information they provide remained to be identified before sharks and rays could justifiably be accredited with a true electric sense [4].

Early behavioral observations provided compelling evidence that marine sharks and rays (i) zero in on the bioelectric fields of prey, (ii) orient in the weak electric fields of ocean streams [5], and (iii) electrically establish their magnetic compass headings [6]. Electric fields of 5 nV/cm suffice to guide the animals (Fig. 1). Actually, for a foraging shark to rush to its prey at 5 nV/cm, it must have detected the field at 1–2 nV/cm, for the predator needs sufficient time to establish a significant change in the electric field it receives along its path of swimming to initiate its well-directed attack [7].

Sharks and rays have demonstrated the ability to orient in ambient DC electric fields in marine facilities and at sea [8]. Their magnetic compass sense has been ascertained

CP665, *Unsolved Problems of Noise and Fluctuations: UPoN 2002: Third International Conference,*
edited by S. M. Bezrukov

under various earth's magnetic field conditions [9]. Experiments designed to verify the use of their *electric* sense in establishing their *magnetic* headings are in progress. The wealth of behavioral data leaves no doubt as to the animals' astounding sensitivity to electric fields, which present them with a secret sensory world of their own—a world accessed by the ampullae of Lorenzini and governed by Maxwell's equations.

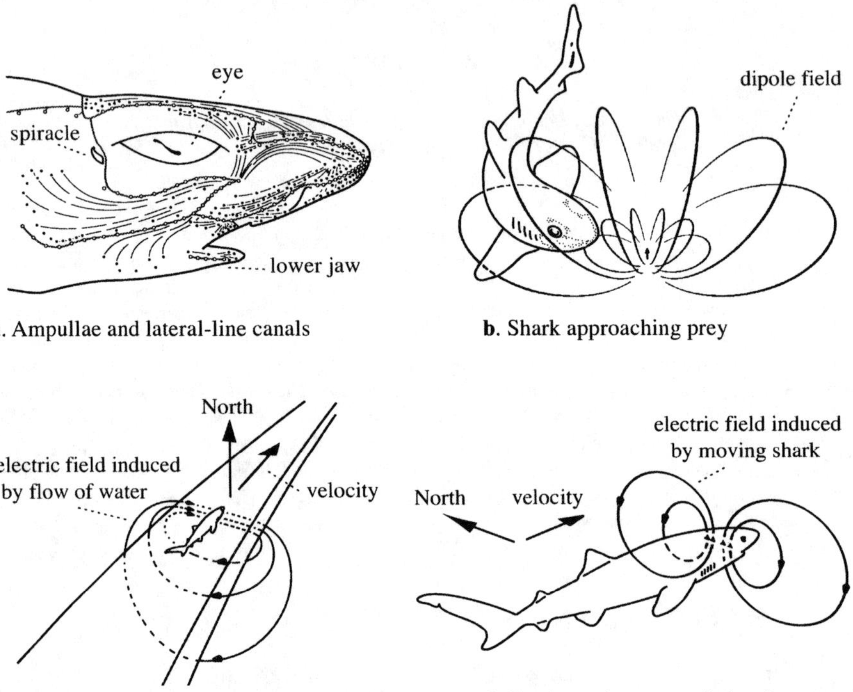

a. Ampullae and lateral-line canals

b. Shark approaching prey

c. Shark sensing drift with water

d. Shark sensing magnetic heading

Figure 1. Morphology and function of marine elasmobranch ampullae of Lorenzini. **a.** Head region of shark. Small solid dots: electroreceptive ampullary pores; small circles: mechanoreceptive lateral-line pores; broken lines: subcutaneous ampullary canals; solid lines: subcutaneous lateral-line canals. **b.** Shark guided by electric field of DC dipole source simulating prey. **c.** Shark electrically sensing its drift with ocean current. **d.** Shark electrically probing its magnetic compass heading.

THE ELASMOBRANCH AMPULLAE OF LORENZINI

The electroreceptive system of elasmobranch fishes operates differentially (Fig 2a). The individual ampullae of Lorenzini detect the electrical potentials at their skin pores single-endedly with respect to the capsular stromata in which the ampullae proper are embedded. The common-mode signals of the capsules are suppressed centrally [10]. Actually, the ampullae are not DC, but very-low-frequency AC devices, featuring an adaptation time constant of several seconds, thereby alleviating the adverse effects of polarization. Thus, in DC fields the animals must move to transpose the signal into the very-low-frequency range of the sense organs. In the small sharks and rays observed to exhibit behavioral responses to fields of ≤ 5 nV/cm, the largest distance between the skin pores of the differentially arranged ampullae was about 10 cm. Thus, the canals connecting the ampullae proper to the farthest apart skin pores averaged about 5 cm in length. All other ampullae had smaller pore separations and shorter canals.

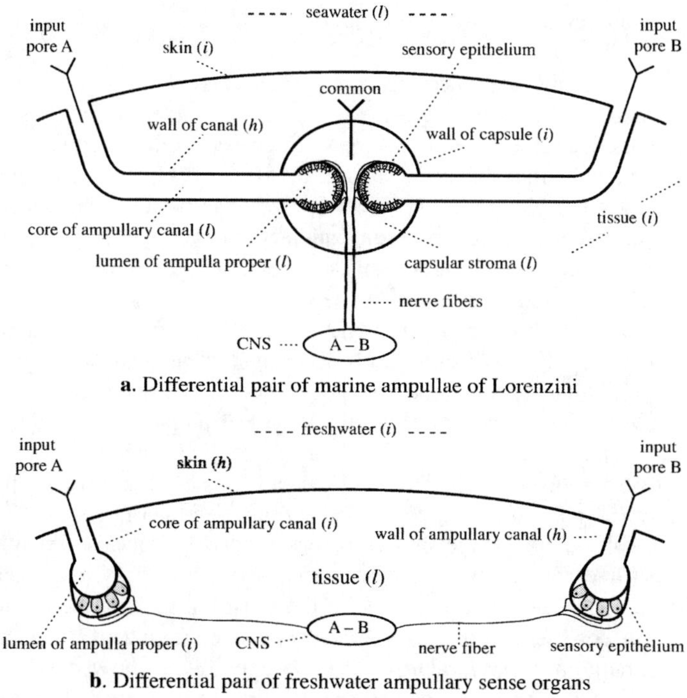

Figure 2. Differentially arranged populations of electroreceptors, one pair each shown. **a.** Ampullae of Lorenzini, marine elasmobranchs, sensing uniform fields of 5 nV/cm. **b.** Ampullae of freshwater catfish, sensing uniform fields of 1 μV/cm. Distance between skin pores, not length of canals *per se*, determines signal across sensory epithelia (see text). (*l*): low, (*i*): intermediate, (*h*): high resistivity.

The high electrical sensitivity results, in first instance, from a great increase in field strength as the potential differences between widely separated skin pores are focused on the thin excitable receptor-cell membranes—in marine elasmobranchs by a factor of ten million, from centimeters to nanometers. Thus, uniform electric fields of 5 nV/cm, producing potential differences of slightly more than 50 nV between skin pores 10 cm apart, give rise to imposed potential differences on the order of 25 nV across each of the oppositely facing sensory epithelia of the differentially arranged ampullae proper.

If the strongest field that ampullae with 5-cm long canals can handle is 500 nV/cm, or perhaps 5,000 nV/cm, the strongest imposed ampullary potentials would be 2.5 μV to 25 μV. Thus, a threshold voltage of 25 nV, imposed across the sensory epithelium, would drive the process of synaptic transmission only through 1/100th to 1/1000th of its dynamic range. If we may reasonably assume that the receptor potentials required to cover the full dynamic range of the synapse measure about 2.5 mV, we must seek an increase in the strength of the input signal (the potential difference between the skin pore and the ampullary capsule) to the resulting output signal (the receptor potentials across the basal membranes of the receptor cells) by a factor of about 100 to 1000.

MARINE VERSUS FRESHWATER HABITAT

Marine sharks and rays are inherently more sensitive to DC and very-low-frequency electric fields than the electrosensitive freshwater catfishes, electric fishes, and lower bony fishes. This does not come as a surprise, because in the freshwater habitat, the electric fields tend to be much stronger than in the ocean [11]. Owing to the high skin resistance of freshwater electrosensitive fishes, electric currents in the aquatic medium hardly penetrate the body tissues and, consequently, the voltage drops develop almost entirely across the skin, leaving the low-ohmic subcutaneous fluids nearly isopotential (Fig. 2b). Hence, the ampullary canals can be very short, just long enough to traverse the skin. In marine elasmobranchs, though, the skin resistance is substantially lower (Fig. 2a), so that in uniform electric fields about half the signal develops across the skin, while the other half appears as a voltage gradient in the body fluids [10].

As a result, in marine sharks and rays, the length of ampullary canals enhances the elasmobranchs' sensitivity by, at most, a factor of two over the situation of canals just long enough to cross the skin. The length of the canals is more important, however, in view of the physiological noise inside the body being no longer negligible compared to the weak electrical signals that the animals detect, rendering it necessary to enclose the ampullae proper in capsules isolating them from the animal's body tissues. Hence, the length of the canals *per se* contributes little to the sensitivity of marine elasmobranchs. Instead, the skin-pore separation largely determines the strength of the signals across the sensory epithelia. In behavioral tests, freshwater catfishes showed to be 200 × less sensitive to uniform electric fields than marine stingrays with similar pore separations, so that we must conclude that the receptor cells in marine elasmobranch fishes are an estimated two orders of magnitude more sensitive than in freshwater catfishes [12].

ELECTRICALLY EXCITABLE ION CHANNELS AS TWO-PORT VOLTAGE-TO-CURRENT CONVERTERS

Electrically sensitive ion channels act as negative (or positive) conductors when, in response to an increase in the membrane potential, the decrease in the current due to a diminishing open probability does (or does not) exceed the increase in current due to the increase in driving voltage. Excitable ion channels thus act as two-port voltage-to-current converters of negative or positive sign. Note that I am speaking of a change in voltage, the signal, with respect to an operating point defined by the *bias* voltage and current, the ratio between which is positive. I further assume that—for small *signal* voltages and currents—the positive and negative conductances are constants. We can make the conductance of a membrane positive or negative by a judicious choice of the kind of ion channel, the pertinent Nernst potential, the gating charge, and the operating point on the *I-V* curve. Truly, negative conductance is a mark of life [13].

It is revealing, perhaps even surprising, how, in an electrical circuit, positive and negative conductances combine. Depending on the ratio between their absolute values, the parallel arrangement of a positive and a negative conductance varies from high to low positive, through zero, to low and high negative, whereas the series arrangement varies from low to high positive, through infinite, to high and low negative. Thus, one can make a high-ohmic combination of low-ohmic positive and negative resistances in parallel, and a low-ohmic combination of high-ohmic positive and negative resistances in series. Ordinary network analysis applies, but it is imperative that between any two points of a circuit containing positive and negative elements, the Thévenin equivalent conductance is positive. When this criterion is not met, the system will be unstable and incapable of processing signals in a *graded* fashion. Where desirable, Nature makes good use of this instability to produce single-minded, all-or-none action potentials.

GRADED POSITIVE-FEEDBACK DRIVEN BY NEGATIVE CONDUCTANCE ION CHANNELS

The graded positive-feedback input circuit advanced for the elasmobranch ampullae of Lorenzini [13] is based on the electrophysiological observations that, at the onset of a uniform, pore-positive field (i) in a live, fully submerged elasmobranch, the *change* in canal current (the signal input current of the ampullary organ) is negative, as it exits through the skin pore *against* the direction of the electric field in the seawater (Fig. 3), (ii) the process of sensory transduction greatly enhances the signal voltage across the basal receptor-cell membranes driving the synapses (Fig. 4), and (iii) the impulse rate of the afferent-nerve action potentials decreases. The graded, positive-feedback model driven by negative conductance elegantly explains the transduction process (Fig. 5).

Imagine that the negative-conductance ion channels prevail in the apical membranes and the positive-conductance channels prevail in the basal membranes of the receptor cells (Fig. 6a,b) and that the series combination of the apical and basal membranes has negative conductance (Fig. 6b,c). Further, imagine that the negative conductance of the ≈ 10,000 receptor cells is larger in absolute value than the positive conductance of the shunt formed by the supporting cells of the sensory epithelium and the nonsensory part of the ampullary wall (Fig. 6c,d). Finally, imagine that the negative conductance of the ampulla proper is smaller in absolute value than the positive conductance of the ampullary canal (Fig. 6d,e), and that the positive conductance of the seawater medium is larger than the absolute value of the negative conductance of the whole sense organ (Fig. 6e). Voilà, one thereby arrives at an extremely powerful positive and negative conductance circuit that, by the daring use of positive feedback, produces an inverted, graded output signal across the basal receptor-cell membranes any desired times larger than the input signal between the skin pore and the ampullary capsule. At ≈8 Hz, the positive feedback turns negative due to phase shift, forming a steep frequency filter.

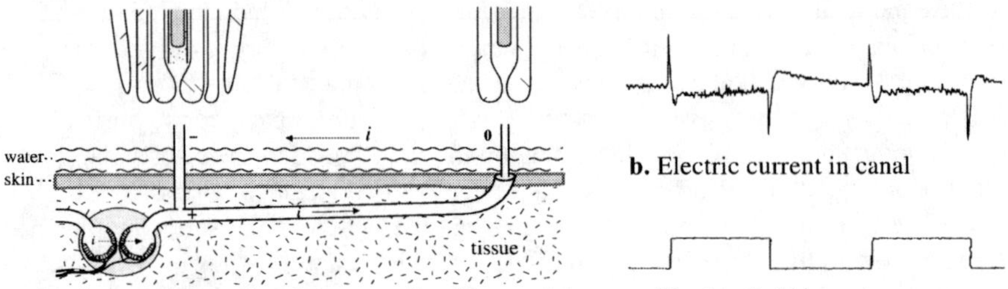

a. In vivo canal recording from ampulla of Lorenzini

b. Electric current in canal

c. Electric field in water

Figure 3. a. Recording of canal current in a live, fully submerged elasmobranch in response to a uniform electric field. Electrode to the right measures voltage in seawater over skin pore. Electrode to the left measures voltage inside ampullary canal near ampulla proper. Notice opposite directions of electric field and electric current outside animal and inside ampullary canal. The left electrode assembly is sucked onto the ampullary canal over a small hole in the canal wall. The central suction electrode is provided with a concentric, measuring electrode and a second concentric, current electrode, employed to render the seal very high-ohmic by negative feedback. Canal recordings often last for several days. The + and – are measured with respect to voltage at skin pore. Negative canal current is driven by sensory epithelia. **b.** Registration of current in ampullary canal. **c.** Uniform stimulus field in seawater.

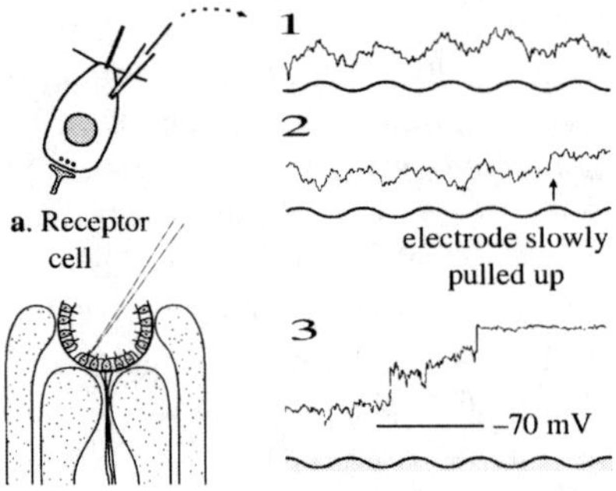

a. Receptor cell

b. Ampulla seat

c. Intracellular recording

Figure 4. Intracellular recording from ampullary epithelium. **a.** Receptor cell, kinocilium, synapse, nerve ending, and apical membranes of adjacent supporting cells. **b.** Suctional ampulla seat. Central electrode recording receptor current and nerve action potentials. Concentric electrode to apply voltage across sensory epithelium relative to seawater bath. **c.** Recording, upper trace; 1-Hz sinewave stimulus, lower trace; line indicates level of –70 mV resting potential. 1: Electrode tip inside receptor cell. 2: At arrow, electrode tip pulled back into supporting cell. 3: Electrode tip pulled back into lumen of ampulla proper. Chance recording, in need of confirmation.

TWO TOUGH QUESTIONS

The positive feedback model shows that sensitivity is not an issue. The real questions are: (i) how does the system deal with the noise of the ion channels, and (ii) how does it assure stability despite the treacherous positive feedback. I will briefly contemplate these issues in the light of present knowledge. An expression for the frightening noise of ion channels will be derived in the next chapter [14]. For definite answers, though, more animal studies will be needed. We could, of course, call on the highly structured kinocilium protruding from the apical receptor-cell membrane to transduce the signal in "a molecular low-noise cascade process" [15] or suggest yet-unknown, low-noise ion channels, or seek the resolve in a refined process of synaptic transmission, but I will not do so. Instead, I want to ascertain to what extent we can explain electroreception based on familiar biophysical principles and well-established experimental results, in particular my own. I expect that the present model is, at least, part of the solution.

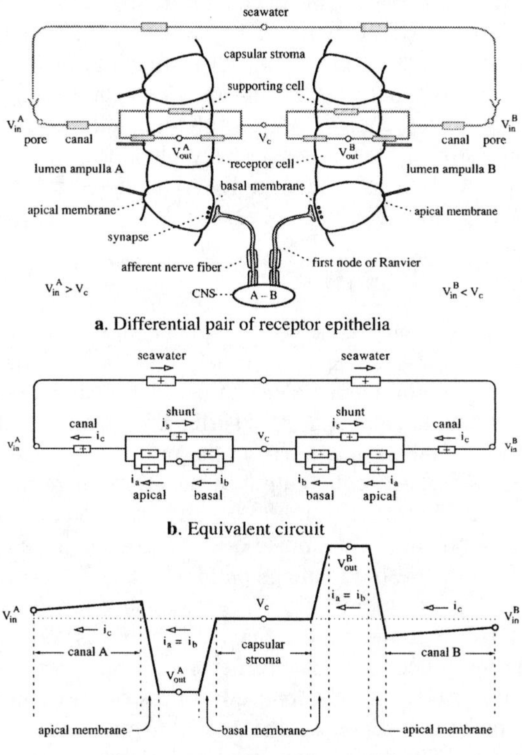

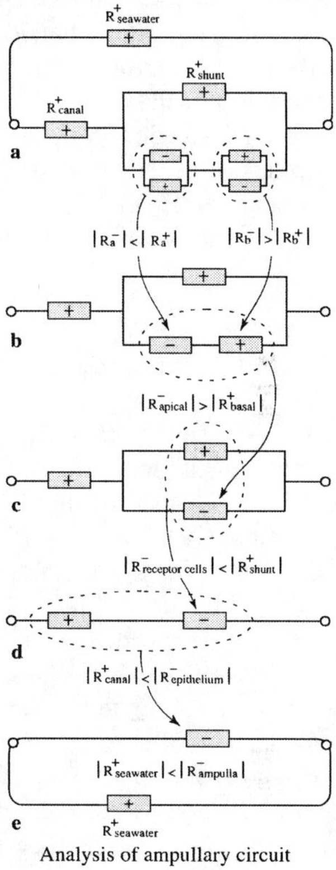

Figure 5. a. Relationship between differential pair of ampullae and **b.** equivalent circuit. **c.** Voltage and current distribution. Signal voltage increases 1.5 × in ampullary canal. Amplification by receptor cell is $R_b/(R_a+R_b)$, i.e., positive basal membrane resistance over *algebraic* sum of negative apical and positive basal membrane resistances.

Figure 6. Analysis of negative- and positive-resistance equivalent circuit. Small capacitance of apical, large capacitance of basal receptor-cell membranes expected to govern frequency response (not shown).

THE STABILITY ISSUE

As for the stability, since the ratio of the output voltage across the basal receptor-cell membranes to the input voltage across the whole sensory epithelium is determined by the resistances of the basal receptor-cell membranes over the difference between the absolute values of the resistances of the apical and basal membranes, the desired high gain requires the two resistances to be nearly the same in absolute value, within one percent or less (Fig. 5). When this difference vanishes, the response turns impulsive, rather than graded. Thus, a change in temperature of a few degrees Celsius renders the system hopelessly unstable and deprives the animal temporarily of its electric sense.

My implementation of the ampullary model, by the use of operational amplifiers, is extremely sensitive to uneven temperature changes, indeed. But so are the ampullae of Lorenzini, to the extent that, in some older physiology texts, they were regarded as the most sensitive thermoreceptors known! Hence, a powerful, all-embracing, negative-feedback loop will be needed to tame the potentially wild behavior of my daring, high-gain, positive-feedback model. Or may the ion-channel noise of the ampullae keep the conductances of the apical and basal cell membranes dangerously close together, yet safely apart, and keep the operating point centered in the negative *I-V* range? May it be true that the ampullae make a virtue of the noise to conquer instability?

THE NOISE ISSUE

The rms thermal noise in the frequency range of the elasmobranchs' electric sense, from near-DC to ≈ 8 Hz, of a 5-cm long, 22-kΩ ampullary canal is roughly the same as the 25-nV signal across the sensory epithelium in a 5 nV/cm field [10]. The number of ampullae with long canals, though rather small, apparently suffices to overcome the canal noise. Ideally, the noise of the ion channels taking part in the process of sensory transduction shall not exceed the canal noise. Since the output of the ≈10,000 receptor cells per ampulla is averaged, after synaptic transmission, by the unmyelinated nerve endings producing the spike-generating potentials, the noise of the individual receptor cells shall not exceed $\sqrt{10{,}000} = 100 \times$ the threshold voltage of 25 nV, that is, 2.5 μV before, and 0.25–2.5 mV after 100–1000 × amplification. The minimal noise would thus cover from 1/10th to the full 2.5-mV dynamic range of synaptic transmission.

In the chain of events constituting the process of sensory detection, each step adds its own intrinsic noise to the signal it receives and is supposed to pass on as clean as possible. If the process of nerve-impulse generation or synaptic transmission adds a certain amount of noise, then the output of the preceding step need not be much less noisy in the low-frequency range of interest. Similarly, if the ampullary canal adds a certain amount of noise, then the input noise of the transduction process need not be much less than the canal noise. Such considerations offer at least some guidance in ascertaining whether the ion-channel noise is compatible with the elasmobranchs' high electrical sensitivity, or whether we must call upon less familiar biophysical principles.

MAY THE SHUNTS COME TO THE RESCUE?

Intuitively, one may expect the system noise to be low when the conductances are high. Experiment shows that the canal current in response to a 25-nV signal across the sensory epithelium measures –1 pA. Given the desired amplification, we can employ the positive-negative-conductance network analysis to calculate the required apical and basal membrane conductances as a function of the positive shunt conductance offered by the supporting cells and other, nonsensory regions of the ampullary proper wall. From the step-by-step analysis illustrated in Figure 6a-e, it is apparent that a positive high-conductance shunt would allow the system to use high-conductance, low-noise, apical and basal receptor-cell membranes. For the same reason, the intrinsic positive-conductance shunts in parallel with the negative conductances of the apical membranes and the negative-conductance shunts in parallel with the positive conductances of the basal membranes may play similar roles. Hence, the opposite-sign conductance shunts may function as effective noise-reducing elements in the ampullary circuit. Whether sharks and rays indeed apply the shunts to alleviate the problem of ion-channel noise and are capable of dealing with the stability issue, only the animals can tell.

ACKNOWLEDGEMENTS

I thank Dr. Sergey M. Bezrukov, Michael C. McClune, and I. Fernando Gonzalez for the lively discussions and Vera Kalmijn for the skillful preparation of the artwork.

REFERENCES

1. Lorenzini, S. Osservazioni intorno alle torpedini. Firenze, 1678, pp 24–25.
2. Kalmijn, A. J. Electro-perception in sharks and rays. *Nature* 1966, 212:1232-1233.
3. Dijkgraaf, S. and Kalmijn, A. J. Untersuchungen über die Funktion der Lorenzinischen Ampullen an Haifischen. *Z. vergl. Physiol.* 1963, 47:438-456.
4. Kalmijn, A. J. The electric sense of sharks and rays. *J. Exp. Biol.* 1971, 55:371–383.
5. Kalmijn, A. J. Electric and magnetic field detection in elasmobranch fishes. *Science* 1982, 218: 916–918.
6. Kalmijn, A. J. Electric and magnetic sensory world of sharks, skates, and rays. In: *Sensory Biology of Sharks, Skates, and Rays*, edited by Hodgson, E. S. and Mathewson, R. F. Washington D.C.: U.S. Government Printing Office, 1978, pp. 507–528.)
7. Kalmijn, A. J. Electric and near-field acoustic detection, a comparative study. *Acta Physiol. Scand.* 1997, 161, suppl. 638, pp. 25–38.
8. Kalmijn, A. J. Detection of weak electric fields. In: *Sensory Biology of Aquatic Animals*, edited by Atema, J. et al. New York: Springer Verlag, 1988, pp. 154–189.
9. Kalmijn, A. J. Biophysics of geomagnetic field detection. *IEEE Trans. Mag.* 1981, 17: 1113–1124.
10. Kalmijn, A. J. Theory of electromagnetic orientation: a further analysis. In: *Comparative Physiology of Sensory Systems*, edited by Bolis, L. et al. Cambridge Univ. Press, 1984, pp. 525–560.
11. Kalmijn, A. J. The detection of electric fields from inanimate and animate sources other than electric organs. In *Handbook of Sensory Physiology, Vol. III/3*, edited by Fessard A. et al., Berlin: Springer Verlag, 1974, pp. 147–200.
12. Kalmijn, A. J., Kolba, C. A. and Kalmijn, V. Orientation of catfish (Ictalurus nebulosus) in strictly uniform electric fields: I. Sensitivity of response. *Biol. Bull.* 1976, 151:415.
13. Kalmijn, A. J. Detection and processing of electromagnetic and near-field acoustic signals in elasmobranch fishes. *Phil. Trans. R. Soc. Lond. B 2000*, 355: 1135–1141.
14. Bezrukov, S. M. Sensing Nature's electric fields: Ion channels as active elements of linear signal transduction. This volume.
15. Dr. I. Vodyanoy, personal communication.

Sensing Nature's Electric Fields: Ion Channels as Active Elements of Linear Amplification

Sergey M. Bezrukov

Laboratory of Physical and Structural Biology, NICHD, National Institutes of Health, Bethesda, MD 20892-0924

Abstract. Given the parameters of familiar cellular elements – voltage-sensitive ion channels, carriers, pumps, phospholipid insulators, and electrolytic conductors – is it possible to construct an amplifier whose sensitivity matches the 5 nV/cm threshold found in behavioral experiments on elasmobranch fish? Or, in addition to clever circuitry that uses commonly known elements and principles, do we need something else to understand this sensitivity? The resolution of this question is important not only for studies in sensory biophysics seeking to reveal underlying mechanisms and molecular structures. More generally, it deepens our appreciation of the stochastic nature of inter- and intra-cellular control circuits. Here I analyze a simplified circuit involving negative differential resistance of voltage-sensitive ion channels. The analysis establishes an off-equilibrium criterion for amplification, shows that ion channels are the dominant noise sources, and, by minimizing channel noise within the given constraints, demonstrates that generic voltage-sensitive ion channels are likely candidates for the active elements of the linear cellular amplifiers. Finally, I highlight a number of unsolved issues.

INTRODUCTION

The crucial role of voltage-sensitive ion channels in *non-linear* amplification is well established. These channels provide the macromolecular basis of nerve excitation where over-threshold stimuli give rise to full-sized action potentials. However, their possible involvement in *linear* amplification has never been analyzed quantitatively. Here I explore the feasibility of linear amplification by considering a cellular system with graded positive feedback employing 'generic ion channels' with parameters typical of voltage-sensitive ion channels in excitable membranes, such as gating charge, conductance, and dwell time in the open state.

As explained in the preceding chapter [1], the crucial element of a cellular amplifier is a membrane with a negative differential resistance. Since the pioneering work by Ehrenstein, Lecar, and Nossal [2] it has been known that properly biased voltage-sensitive ion channels behave like a negative resistor. That is, for sufficiently slow and small changes in voltage, the increase in voltage bias decreases the average current through the channels.

Contrary to ion carriers, which transfer ions one by one, channels regulate ion flow by gating large fluxes on a millisecond time scale – fluxes of about 10^7 ions per second at 'physiological voltages' of about 0.1 V. This leads to a quantization of charge transfer at approximately 10^4 electronic charges per single event of channel opening and, consequently, to a proportionally larger mean-square noise level.

CP665, *Unsolved Problems of Noise and Fluctuations: UPoN 2002: Third International Conference,* edited by S. M. Bezrukov
2003 American Institute of Physics 0-7354-0127-6

Therefore, in view of the astounding sensitivity of the ampulla's receptor cells [1], relying on ion channels as active elements raises the issue of amplifier noise.

RESULTS AND DISCUSSION

Ion Channels as Dominant Noise Sources

Shot noise reveals the random character of charge transfer and is intimately related to the discreteness of charge. When charge is (unidirectionally) transported by independent univalent ions, its low-frequency spectral density is described by the Schottky formula [3]:

$$S_{sh} = 2e\langle I\rangle, \tag{1}$$

where e is the electron charge and $\langle I\rangle$ is the average current.

On the other hand, the low-frequency noise of ion channels that follow single-exponential statistics at low open channel probability can be expressed as:

$$S_{ch} = 4Q\langle I\rangle, \tag{2}$$

where Q is the average charge transported through a channel during its transient open state:

$$Q = h\tau_o(V - E_N). \tag{3}$$

Here, h is the channel conductance, E_N is the Nernst potential pertinent to the particular channel, V is the actual voltage across the membrane, and τ_o is the mean dwell time in the open conformation. Although they give similar dependences on average current, Equations (1) and (2) differ greatly in their numerical factors, which state that the noise introduced by ion channels is $2Q/e$ times larger than the noise expected from ion carriers. The factor 2 reflects the difference in statistical properties of the two processes. In single-electron (or single-ion) shot process elementary events are exactly the same; in noise generation by channels elementary events themselves are random.

To obtain expression (2) we start with the original Machlup formula [4] for the power spectral density, $S_i(f)$, of current fluctuations in a two-state Markov process

$$S_i(f) = \frac{4i^2\tau^2}{\tau_o + \tau_c}\frac{1}{1+(2\pi\tau f)^2}, \tag{4}$$

where f is frequency, i is the current in the open channel state, τ_o is the mean dwell time in the open state, and τ_c is the mean dwell time in the closed state. The 'relaxation time' τ is defined by $\tau = \tau_o\tau_c/(\tau_o+\tau_c)$. The Lorentzian spectrum described by this equation gives a 'white' (that is, frequency-independent) spectral density for frequencies that are smaller than the inverse of $2\pi\tau$. We are interested in the frequency range below 10 Hz [1], while the typical relaxation times for voltage-sensitive ion channels are on the order of milliseconds. Therefore, we can neglect the last multiplier. The probability p_o to find the channel in the open state is $p_o = \tau_o/(\tau_o+\tau_c)$ and the current through the open channel is $i = h(V\text{-}E_N)$. Substituting these values into Equation (4) and multiplying by the total number of channels, N, for the low-frequency spectral density of the channel current noise, $S_I(0)$, we have

$$S_I(0) \quad = \quad 4N\left(h(V - E_N)\right)^2 \tau_o p_o (1 - p_o)^2 . \tag{5}$$

We can leave out the last multiplier when the open channel probability is much smaller that one. Then, by taking into account that the average current through the channels is given by

$$\langle I \rangle \quad = \quad Nhp_o(V - E_N), \tag{6}$$

we arrive to Equation (2). Although the open channel probability does not have to be small to exhibit negative resistance, our analysis of noise minimization (see below) shows that at the optimal noise performance this probability is always smaller than 0.5, so that Equation (2) gives a good order-of-magnitude estimate for the channel noise.

Off-Equilibrium Criterion for Amplification

From Equation (2) it follows that the mean-square noise level is proportional to the total charge passing through the channel during each single transient opening. Therefore, a simple solution of the noise problem would be to reduce the number of ions per transient channel opening by reducing the voltage bias. To be specific, if it were possible to reduce the normal voltage bias corresponding to 10^4 ions per single

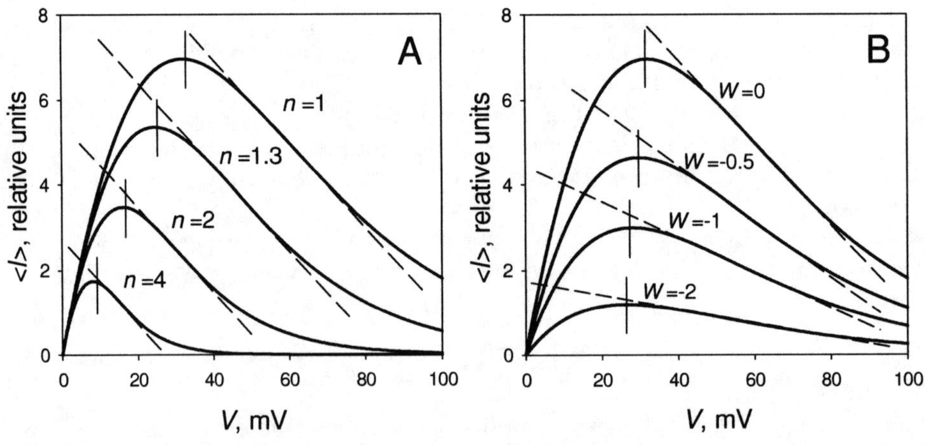

FIGURE 1. Average current through voltage-sensitive channels as a function of applied voltage at $E_N = 0$. **A**: Changing the gating charge, n, at zero free energy difference, $W = 0$, and zero Nernst potential, $E_N = 0$, changes the crossover voltage (vertical solid lines) but does not influence the maximal negative slope (interrupted lines). **B**: Changing the free energy difference, W, at the constant gating charge, $n = 1$, and $E_N = 0$ changes both the crossover voltage and the maximal negative slope.

channel opening by ten thousand times, channel noise would be reduced to the mere level of shot noise. The closer the bias is to equilibrium, the smaller is the channel noise. However, to show negative conductance, ion channels must be kept at a certain voltage distance from equilibrium.

Indeed, the average current through a population of voltage-sensitive ion channels, Equation (6), is proportional to the probability to find a channel in its open state [2]

$$p_o \quad = \quad \left[1+\exp\left(\frac{neV - W}{kT}\right)\right]^{-1}, \tag{7}$$

where n is the channel gating charge, W is the difference in free energy between closed and open channel conformations at $V = 0$, and k and T are the Boltzmann constant and the absolute temperature. Figure 1 gives I-V curves for differing gating charge (A) and free energy difference (B) at $E_N = 0$. To account for differing Nernst potentials, it can be shown that the I-V curves are exactly the same as in Figure 1B if a new independent variable, $V - E_N$, and new free energy, $W - enE_N$, are introduced. The voltage V^* of crossover between positive and negative differential conductance regimes ($d\langle I\rangle/dV > 0$ or < 0) and the maximal negative slope depend on all these parameters.

Therefore, the distance from equilibrium, that is, the difference between the crossover voltage V^* and the Nernst potential E_N, is a function of the gating charge and of the combination of the conformational energy difference and the scaled Nernst potential, $W - neE_N$. Further analysis shows that whatever this combination is, in order to exhibit negative resistance the channels should be biased in excess of kT/ne:

$$V^* - E_N \quad > \quad kT/ne. \tag{8}$$

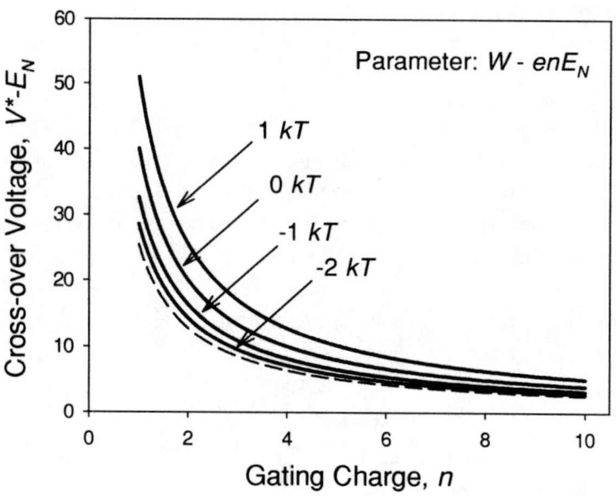

FIGURE 2. Crossover voltage – the voltage bias separating regimes of positive and negative resistance – is a decreasing function of the gating charge. To exhibit negative resistance, ion channels should be driven off equilibrium by at least kT/ne (interrupted line). This limiting value is approached at large negative combinations $W - enE_N$.

Figure 2 shows the dependence of $V^* - E_N$ on the gating charge for different $W - neE_N$ combinations. The maximal negative slope occurs for a bias approximately equal to twice the crossover bias.

These off-equilibrium constraints constitute the criterion of amplification and, according to Equation (2), impose limitations on amplifier noise performance.

Noise Performance of Cellular Amplifier

The minimal equivalent circuit of a cellular amplifier is shown in Figure 3 (see ref. [1] for details). It is a series connection of three elements: canal resistance, R_c, apical membrane resistance, R_a, and basal membrane resistance, R_b. To amplify, the circuit must contain negative resistance. This property is assigned to the channels of the apical membrane. The input voltage, ΔV_{in}, acts on the canal opening; the output voltage, ΔV_{out}, is applied to the basal membrane.

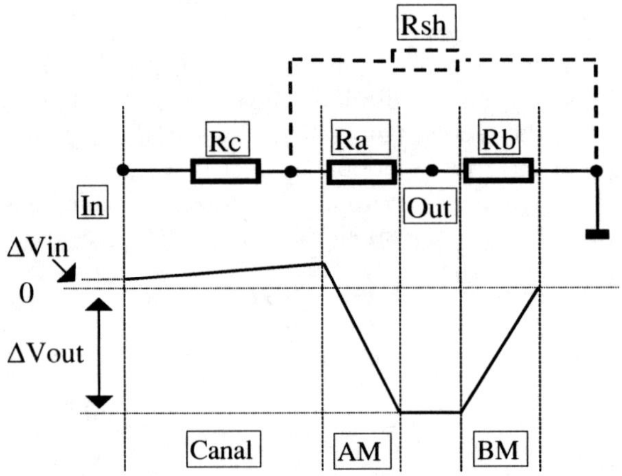

FIGURE 3. The minimal circuit of a voltage amplifier is a series connection of positive resistances of the canal and the basal membrane (BM) and a negative resistance of the apical membrane (AM). The role of the shunt resistance (part of the circuit shown by the interrupted line) is not considered here.

For slow and small signals the system can be linearized and the amplification coefficient can be calculated from Ohm's law as

$$K \equiv \frac{\Delta V_{out}}{\Delta V_{in}} = \frac{R_b}{R_c + R_a^{diff} + R_b}, \qquad (9)$$

where R_a^{diff} is the differential resistance of properly biased voltage-sensitive channels of the apical membrane, which can be calculated from Equations (6) and (7). For the sake of simplicity I assume that both the canal and the basal membrane are linear devices so that their differential resistances coincide with their integral resistances R_c and R_b. Electrophysiological data [1] and the requirement of high amplification suggest that in this circuit canal resistance has to be much smaller than the basal membrane resistance, $R_c << R_b$. Therefore, from Equation (9) it follows that high amplification is achieved when $R_a^{diff} \cong -R_b$. This condition allows us to find the total number of voltage-sensitive channels N as a function of given voltage bias and R_b.

Taking into account that at low frequencies the voltage and current power spectral densities are related by $S_V(0) = S_I(0)R^2$, and using Equation (5) to calculate channel noise, for the equivalent generator of voltage noise of the apical membrane we find

$$S_V(0) \cong 4h\tau_o R_b \frac{(V')^2}{\dfrac{neV'}{kT} - 1 - \exp\left(\dfrac{W'-neV'}{kT}\right)} \frac{1}{1+\exp\left(\dfrac{W'-neV'}{kT}\right)}. \tag{10}$$

Here, V' is the distance from the equilibrium voltage, that is, the difference between the actual voltage at the apical membrane and channel Nernst potential, $V' = V - E_N$, and W' is a modified free energy, $W' = W - enE_N$.

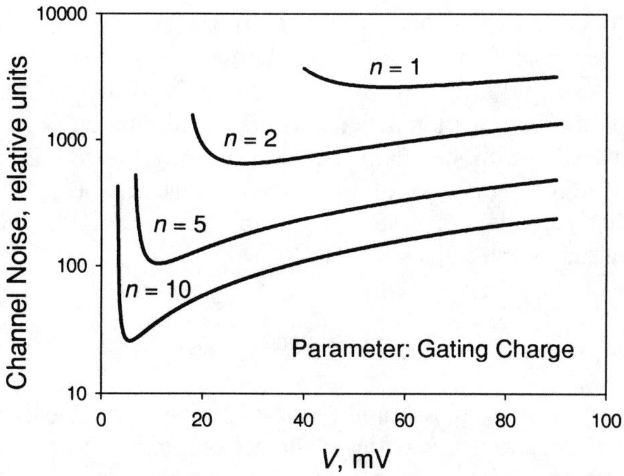

FIGURE 4. Noise of voltage-sensitive ion channels of the apical membrane, Equation (10), as a function of transmembrane voltage at four different gating charge values if $W = E_N = 0$. For simplicity, τ_o is considered to be constant.

For a high-gain cellular amplifier that adjusts the total number of negative-differential-resistance channels to match the resistance of the basal membrane, the channel-induced noise depends strongly and non-monotonicly on voltage (Figure 4). Channel noise at the optimal voltages is reduced proportionally to the square of the gating charge. Thus, the results of the system noise analysis summarized by Equation (10) demonstrate that the channel noise can be minimized by increasing the channel gating charge and by reducing its conductance and dwell time in the open state. Interestingly, varying W' through E_N and W does not reduce noise in the minima. Rather, as our analysis shows, it changes the optimal open probability of the channels within the limits $0 < p_o < 0.5$, consequently changing the 'working point' of the amplifier.

At optimal voltages that give the smallest channel noise, the product of the last two multipliers in Equation (10) is $4(kT/ne)^2$. This gives the following expression for the minimal spectral density of the channel-induced voltage noise of the circuit

$$S_V(0) \cong 16h\tau_o R_b (kT/ne)^2 . \qquad (11)$$

To obtain the root-mean-square noise value one has to account for the effective frequency band

$$Noise_{rms} \equiv \sqrt{S_V(0)\Delta f} \cong 4(kT/ne)\sqrt{h\tau_o R_b \Delta f} . \qquad (12)$$

It is interesting to note that Equation (11) can be presented in a form that expresses the channel-generated noise as a product of the basal membrane thermal noise [5], $4kTR_b$, and a dimensionless "excess noise factor", θ_e

$$\theta_e = 4kTh\tau_o/(ne)^2 . \qquad (13)$$

Plugging in parameters for a generic channel [6]: conductance $h = 10^{-11}$ S, channel dwell-time in the open state $\tau_o = 10^{-3}$ s, gating charge $ne = 10e$, one finds that $\theta_e \cong 60$. For the root-mean-square noise this recalculates into a factor of 8. Therefore, analysis of channel noise of the cellular amplifier permits its minimization to a level that is only an order of magnitude higher than the basic thermal noise. Equation (13) also demonstrates that further reduction of the excess noise factor can be achieved by decreasing the channel conductance and the mean open time and, most significantly, by increasing the gating charge.

CONCLUSIONS

It is widely appreciated that the ability of organisms to detect weak signals is limited by intrinsic fluctuations occurring at the cellular and sub-cellular levels [7-15]. While the idea per se has enjoyed proper attention and has been investigated by many authors, to my knowledge intrinsic noise has never been analyzed for a cellular system capable of *linear signal amplification*.

Here I have considered sensory signal amplification based on negative resistance provided by voltage-sensitive ion channels. The goal was not to analyze any particular type of channels with elaborate gating characteristics and typical voltage-biases [6]. Rather, playing with all these parameters, I was trying to establish the fundamental constraints on the amplifier noise performance.

The main results of this study are: (i) there exists an off-equilibrium criterion for amplification with voltage-sensitive ion channels; to serve as amplifiers, ion-selective channels must be driven away from their equilibrium potential E_N by at least kT/ne, where n is the channel gating charge; (ii) due to the randomness of channel gating, this off-equilibrium condition sets an inescapable restriction on system noise performance; (iii) however, when properly minimized, the root-mean-square excess noise of channels with generic parameters is only one order of magnitude higher than the basic limit imposed by the thermal noise. This makes voltage-sensitive ion channels likely candidates for the active elements of signal amplification by electroreceptor cells.

Nevertheless, the transducer issue remains unsolved. Leaving aside the stability problem, the simplified circuit of Figure 3 appears not to suffice in view of noise considerations. The electrophysiological data [1], and the requirement for significant amplification to prepare the weak electrical input signal for synaptic transduction, would call for prohibitively high values of the negative and positive resistances of the apical and basal receptor-cell membranes. However, by taking into account the accessory-cell shunt resistances of the ampulla proper (part of the circuit shown by interrupted lines in Figure 1, equivalent circuits in ref. [1]), one can explore the possibility of implementing a system with sufficiently low resistances to attain the desired low-noise performance. To get a definite answer, we will have to ask the animals in properly designed experiments.

ACKNOWLEDGMENTS

The author is grateful to Ad. J. Kalmijn and V. Adrian Parsegian for many long enlightening discussions that resulted in this work.

REFERENCES

1. Kalmijn, A.J., Graded positive feedback in elasmobranch ampullae of Lorenzini. *This book*, preceding chapter.
2. Ehrenstein, G., H. Lecar, and R. Nossal, The nature of the negative resistance in bimolecular lipid membranes containing excitability-inducing material. *J. Gen. Physiol.* **55**, 119- 133 (1970).
3. Schottky, W., Uber spontane Stromschwankungen in verschiedenen Elektrizitatsleitern. *Ann. der. Phys.* **57**, 541-567 (1918).
4. Machlup, S., Noise in semicinductors: Spectrum of a two-parameter random signal. *J. Appl. Phys.* **25**, 341-343 (1954).
5. Nyquist, H., Thermal agitation of electric charge in conductors. *Phys. Rev.* **32**, 110-113 (1928).
6. Hille, B., *Ionic Channels of Excitable Membranes*. Sinauer Associates. Sunderland, 2001.
7. Bialek, W., Physical limits to sensation and perception. *Annu. Rev. Biophys. Biophys. Chem.* **16**, 455-478 (1987).
8. Weaver, J.C. and R.D. Astumian, The response of living cells to very weak electric fields: The thermal noise limit. *Science* **247**, 459-462 (1990).
9. Block, S.M., Biophysical principles of sensory transduction. In: *Sensory Transduction*. D.P. Gorey, S.D. Roper, eds., Rockefeller University Press, New York, pp. 1-17, 1992.
10. Lu, J., and H.M. Fishman, Interaction of apical and basal membrane ion channels underlies electroreception in ampullary epithelia of skates. *Biophys. J.* **67**, 1525-1533 (1994).
11. Harvey, J.R., and L.J. Bruner, The spike generation zone of the ampullary electroreceptor. II. Oscillator period noise and the limits of sensitivity. *Biol.Cybern.* **72**, 379-387 (1995).
12. Petracchi, D., S. Lucia, and G. Cercignani, Fundamental sources of noise in sensory transduction. *Nuovo Cimento* D **17**, 883-891 (1995).
13. Galvanovskis, J., and J. Sandblom, Amplification of electromagnetic signals by ion channels. *Biophys. J.* **73**, 3056-3065 (1997).
14. Adair, R.K., R.D. Astumian, and J.C. Weaver, Detection of weak electric fields by sharks, rays, and skates. *Chaos* **8**, 576-587 (1998).
15. Gailey, P.C., Membrane potential and time requirements for detection of weak signals by voltage-gated ion channels. *Bioelectromagnetics* **20**, 102-109 (1999).

Noise-Induced Transition and Synchronization in Paddlefish Electroreceptors

Alexander B. Neiman* and David F. Russell*

*Center for Neurodynamics, University of Missouri at St. Louis, St. Louis, Missouri 63121, USA

Abstract. The electrosensory system of paddlefish contains thousands of electroreceptors, which are spontaneously active systems producing stochastic biperiodic spike trains, with a wide range of natural frequencies (mean firing rates). There is no morphological or physiological evidence for coupling between electroreceptors: they can be considered as a parallel array of uncoupled stochastic oscillators. Using broadband noise stimulation, we show that noise induces parabolic bursts in single electroreceptors, revealing a new slow time scale. Noise stimulation also leads to synchronization of the burst onsets in different electroreceptors, even though their intraburst firing rates may differ.

INTRODUCTION

Perturbations of a nonlinear dynamical system with Gaussian white noise often leads to a global stability in the phase space [1]. That is, if a dynamical system is characterized by several locally stable attractors which can be reached by choosing an appropriate set of initial conditions, then with Gaussian noise perturbations a phase trajectory will eventually visit the whole set of possible states that can be described by a stationary probability density. Thus, noise can be used to highlight dynamical regimes which are available but hidden in a deterministic situation. In neuroscience, stimulation with broadband noise is widely used to estimate general response properties of sensory systems by means of Wiener kernels [2].

In arrays of oscillators, noise can play a positive role in some cases, enhancing [3] or even inducing synchronization, as can occur when a population of *identical* noncoupled oscillators is subjected to a common noisy field [4, 5]. However, in many natural cases the oscillators of an array are not identical, and instead have widely distributed natural frequencies.

In this paper, we show that broadband noise may induce qualitative changes in the firing patterns of sensory neurons. In particular, we demonstrate experimentally a noise-induced transition from a tonic to a bursting mode of firing in paddlefish electroreceptor afferent neurons. We also show experimentally that a population of noncoupled biological oscillators (neurons) with quite different natural frequencies (mean firing rates) can be synchronized by external common broadband noise after undergoing a noise-induced transition to a bursting mode. Bursting and synchronization are both affected by noise, present in any natural system.

Bursting induced by noise has been simulated numerically [6]. Bursts in different neurons can readily be synchronized if the neurons are *coupled*, for instance via synapses,

CP665, *Unsolved Problems of Noise and Fluctuations: UPoN 2002: Third International Conference*, edited by S. M. Bezrukov

as analyzed theoretically [7, 8, 9, 10] and experimentally [11, 12, 13].

Thousands of paddlefish electroreceptors are organized into linear arrays on the elongated rostrum, which acts as an antenna for detecting weak electrical signals from plankton prey [14, 15]. A remarkable property of electroreceptors is that they are spontaneously active [16] and undergo stochastic biperiodic self-sustained oscillations [17]. It is natural then to hypothesize that synchronization among electroreceptors due to the (external) electrical signals from planktonic prey may be an important sensory mechanism for the detection and location of prey. An electroreceptor systems consists of a cluster of skin pores, each leading into a short canal [18], which ends in a sensory epithelium containing $\approx$400 hair cells. All the hair cells of a cluster synaptically excite the terminals of a few primary afferents (sensory neurons), whose axons project to the brain.

A given afferent shows a nearly constant mean frequency of spontaneous firing, but the mean rate in different afferents can vary over a wide range of 30-65 Hz (at 22 ^{o}C). There is no morphological or physiological evidence for coupling between receptors [16, 18], which gives us a unique possibility to study synchronization between different sensory neurons due solely to external stimuli.

NOISE-INDUCED BURSTING

We simultaneously recorded the single-unit spikes from pairs of electroreceptor afferents *in vivo*, using metal microelectrodes. One receptive field was on the left side of the rostrum, the other on the right side. Their locations on opposite sides of the rostrum, which are innervated by different nerves, guaranteed that the pairs of afferent neurons were not coupled [18]. We used uniform-field stimulation of all the electroreceptors: stimulus currents were passed between 15 x 5 cm chlorided silver plate electrodes at the ends of the experimental chamber (see [14, 17, 19] for experimental details). A computer-generated zero-mean Ornstein-Uhlenbeck (OU) noise process was used. The correlation time was set to be 0.002 sec, corresponding to a 500 Hz bandwidth. We generated a sequence of 30 segments of OU noise with incrementing intensities. Each noise segment was 180 sec long, and segments were separated by 5 sec of no stimulus. For comparison, we also used computer generated white noise, high-pass filtered OU noise, or noise from a General Radio model 1390B generator which was lowpass filtered by an 8-pole Bessel filter set to 50 Hz. The latter noise was used in a previous study on electrosensory stochastic resonance [15].

Stimulation with weak noise (<2.5 μV/cm r.m.s.) did not change the firing mode of an electroreceptor afferent, but rather led to the well-known effect of widening the peak in the probability density of interspike intervals and, correspondingly, the power spectrum peak at the mean frequency of the afferent [17] (see Fig.1). When noise of a certain intensity (>2.5 μV/cm r.m.s.) was applied, the firing patterns of the afferents changed drastically such that afferents produced bursts (Fig.1A): spikes were concentrated in clusters of bursts which were separated by quiescent epochs. Bursts were identified using Spike2 software, by imposing the following conditions: (i) a burst had to include at least 3 spikes, and (ii) the maximum time between two spikes in a burst was set to 0.01 sec. The interspike intervals within a burst decreased towards the center of the burst [19],

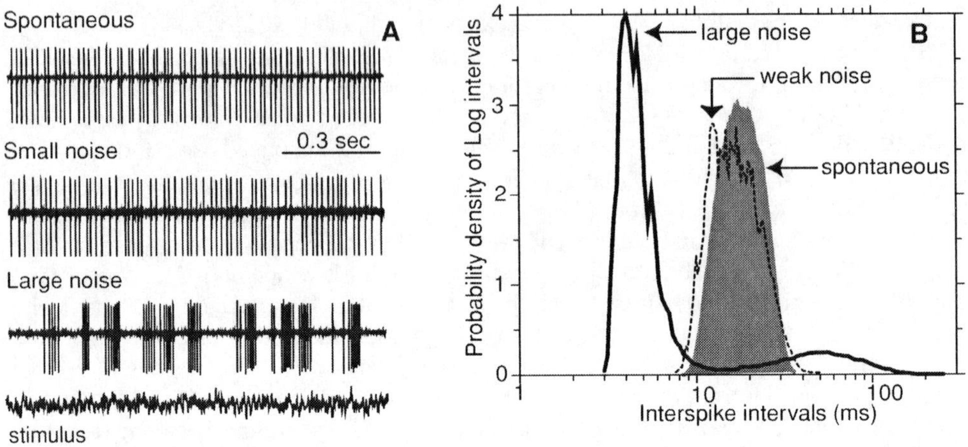

FIGURE 1. Noise-induced bursting in an electroreceptor. A: Afferent firing without stimulation (upper trace), with weak OU noise of 1 μV/cm rms, and with 17 μV/cm rms OU noise. B: Corresponding probability densities of interspike intervals (logarithmic scale). Spontaneous activity corresponds to the shaded curve, weak 1 μV/cm rms OU noise is shown by the dashed line, and a bursting regime at 17 μV/cm rms noise is shown by the solid line.

indicating a parabolic type of bursting [20].

The 500 Hz bandwidth of the OU noise was significantly larger than any of characteristic frequencies of the electroreceptor system, and thus burst initiation cannot be explained simply by slow changes in the stimulus. Similar noise-induced bursting was obtained using computer generated white noise, or OU noise highpass filtered at 20 Hz. Bursts could also be induced by other types of stimuli, for example sine waves, or impulse-like stimuli. The transition to a bursting mode was reflected in the structure of the probability density of interspike-intervals (Fig.1B). With increasing noise intensity, the probability density of interspike intervals changed smoothly from unimodal (for spontaneous firing) to a bimodal distribution during bursting. The two peaks of the bimodal distribution correspond to the slow bursting and to the fast intraburst spiking dynamics, resembling results from other noise-induced transitions [21].

The power spectra of afferent firing (Fig.2) also showed corresponding qualitative changes. The spontaneous power spectrum (left panel in Fig.2) has a usual structure of stochastic biperiodic oscillations [17], with the afferent peak f_a, the epithelial oscillator peak f_e, and a sideband. Weak noise (middle panel in Fig.2) did not alter the epithelial peak f_e [17]. Weak noise did lead to widening of the afferent peak f_a, as expected for noisy self-sustained oscillators [22]. The bursting regime was expressed in drastic changes in the power spectrum (right panel in Fig.2) with appearance in a wide low-frequency peak due to slow interburst dynamics, and a high-frequency peak due to fast spiking inside bursts. A striking feature was observed at the frequency of epithelial oscillations f_e, where a local minimum (instead of a peak) was observed. In experiments with simultaneous recording from an afferent and from a corresponding canal on the rostrum [17], we have verified that noise did not alter the f_e peak in the power spectrum

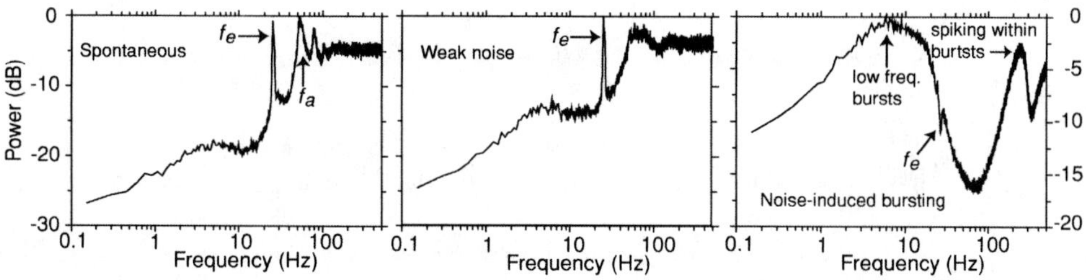

FIGURE 2. Power spectra of the afferent firing corresponding to the probability distributions of interspike intervals in Fig.2.

of canal oscillations. Thus, bursts are generated in the afferent terminals, and not in epithelial cells.

A new slow time scale is introduced by the noise, and can be expressed as the mean interburst interval $\langle \tau_b \rangle$. The mean interburst interval declined exponentially with increasing noise intensity, which was well fitted by the Arrhenius law, $\langle \tau_b \rangle = \nu \exp(\Delta^2/\sigma^2)$ [19]. This implies that burst generation is excitable, and has a well-defined threshold, Δ [23], estimated as $\approx 3\ \mu V/cm$, which is only 3-fold higher than the limit of electroreceptor sensitivity [19].

Thus, the noise-induced transition from tonic to bursting modes of afferent firing involves a new characteristic time scale, reflected in the quiescent epochs between bursts. This slow time scale is distinct and well-separated from the dynamics of fast spiking during bursts.

SYNCHRONIZATION OF NOISE-INDUCED BURSTS

One of the functional implications of bursting regimes is synchronization [8, 9]. Indeed, the existence of bursts implies a slow time scale which makes synchronization of burst onsets easier in comparison with synchronization of individual spikes. In paddlefish electroreceptors, the individual afferents usually possess different mean frequencies (mean firing rates) and different degrees of frequency variability [17]. Moreover, they are noncoupled. It appears, however, that all the electroreceptors have similar slow dynamics, which was revealed by synchronization of noise-induced bursts [19] in different receptors. The transition to burst synchronization occurs in two steps: (i) common noise induces bursts in neurons, but because of differences in their burst thresholds the mean interburst intervals are also different, and thus the burst onsets are not synchronized; (ii) further increase of common noise leads to entrainment of the slow time scale in different neurons, which leads to synchronization of their burst onsets [19].

A representative example of data from two different afferents, recorded simultaneously, is shown in Fig.3. In the absence of stimulation, the individual spikes in these neurons were not synchronized or correlated, since their mean firing rates were different: afferent #1 fired faster (45.1 Hz) than afferent #2 (33.4 Hz). With noise stimulation switched on, each burst started almost simultaneously in the two neurons, even though

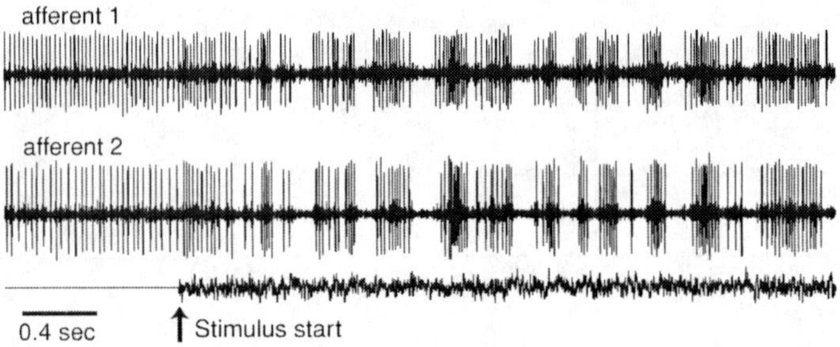

FIGURE 3. Example of simultaneous recordings of spike trains from a pair of electroreceptor afferents. The onset of stimulation with computer-generated OU noise of 16.8 μV/cm rms amplitude is marked by the arrow.

the number of spikes inside a burst was different for the two neurons.

We characterized the coincidence of bursts in pairs of neurons in terms of stochastic synchronization [19], measuring the phases $\phi_{1,2}(t)$ of burst onsets in each neuron, which increases by 2π every time a burst occurs, and interpolates linearly between two sequential burst onsets:

$$\begin{aligned}\phi_1(t) &= 2\pi\frac{t-\tau_k^{(1)}}{\tau_{k+1}^{(1)}-\tau_k^{(1)}}+2\pi k,\\ \phi_2(t) &= 2\pi\frac{t-\tau_m^{(2)}}{\tau_{m+1}^{(2)}-\tau_m^{(2)}}+2\pi m,\end{aligned}$$

where $\tau_k^{(1)}$ and $\tau_m^{(2)}$ are bursts onsets in the first and the second neurons, respectively, and $\tau_k^{(1)} < t < \tau_{k+1}^{(1)}$, $\tau_m^{(2)} < t < \tau_{m+1}^{(2)}$. When stochastic synchronization occurs [22], constant segments of the phase difference $\Delta\phi(t) = \phi_1(t) - \phi_2(t)$ (phase locking) are interrupted by abrupt 2π phase slips. This is illustrated in Fig.4 (left panel) for noise-induced bursting, where the phase differences for three different noise intensities are presented. For a large noise intensity (curve 3), the burst onsets are synchronized, which is expressed in the existence of horizontal epochs of phase locking lasting several seconds. The probability density of the phase difference, $P(\Delta\phi)$, characterizes the degree of synchronization: a well-expressed peak in $P(\Delta\phi)$ indicates synchronization, while a uniform distribution indicates its absence. The probability density is nearly uniform for weak noise, when bursts in the two neurons are not synchronized. With increased noise intensity, the probability density of the phase difference developed a well-defined peak, indicating strong synchronization between the bursting neurons.

CONCLUSIONS

Stimulation of paddlefish electroreceptors with broadband noise was used to study their general response properties. We showed that noise reveals hidden slow excitable

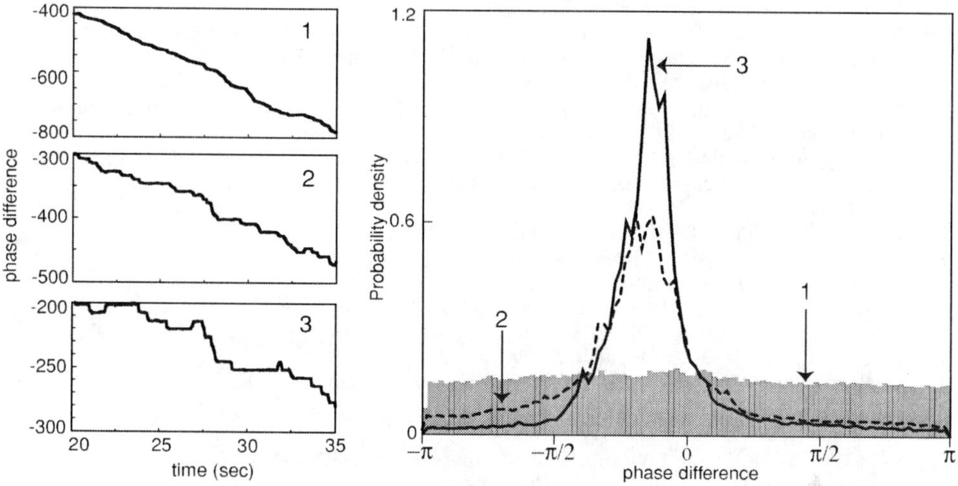

FIGURE 4. Phase differences of burst onsets in a pair of afferents (left panel), and their probability densities (right panel) for the following noise rms amplitude: (1) 2 μV/ cm, (2) 8 μV/ cm, (3) 20 μV/ cm.

dynamics, which have similar properties in different electroreceptors, even though their spontaneous dynamics may be quite different. We also showed that common noise stimulation leads to a specific type of synchronization of the slow dynamics in pairs of non-coupled receptors, when the onsets of their bursts coincide, while there is no coincidence at the level of individual spikes. Synchronous burst responses of a population of sensory neurons may be a neural mechanism for coincidence detection, and may substantially simplify the neural operations that a fish's brain must perform to detect prey and to calculate their position and velocity [24]. The impulse-like electrical signal emitted by an individual plankton prey (e.g. *Daphnia*) moving along the rostrum [25], or the exponentially correlated Gaussian electrical noise generated by swarms of *Daphnia* [26], may be adequate stimuli for evoking synchronized bursting of different electroreceptors during feeding behavior.

ACKNOWLEDGMENTS

This work was supported by the Office of Naval Research - Physics Division, the Fetzer Institute (A.N.), the University of Missouri Research Board, and the National Institutes of Health (R01DC004922-01).

REFERENCES

1. M. Mackey, A. Longtin, and A. Lasota, *J. Stat. Phys.* **60**, 735 (1990).
2. P.Z. Marmarelis, V.Z. Marmarelis, *Analysis of Physiological Systems. The White-Noise Approach* (Plenum Press, New York, 1978).

3. A. Neiman et al. *Phys. Rev. E* **58**, 7118 (1998).
4. A.S. Pikovsky, Radiophys. Quantum Electron. **27**, 576 (1984); A.S. Pikovsky, in: *Nonlinear and Turbulent Processes in Physics* ed. by R.Z. Sagdeev (Gordon and Breach, 1984), pp.1601-1604.
5. C. Zhou and J. Kurths, *Phys. Rev. Lett.* **88** 230602 (2002).
6. A. Longtin, K. Hinzer, *Neural Computation* **8**, 215 (1996); H.A. Braun, et al, *Int. J. Bifurcation Chaos Appl.Sci.Eng.* **8**, 881 (1998).
7. F.C. Hoppensteadt, E.M. Izhikevich, *Weakly Connected Neural Networks* (Springer, NY, 1997).
8. E.M. Izhikevich, Int. J. Bifurcation Chaos Appl.Sci.Eng. **10**, 1171 (2000).
9. E.M. Izhikevich, SIAM Rev. **43** 315 (2001).
10. B. Hu, C. Zhou, Phys. Rev. E **63**, 026201 (2001).
11. E.A. Stern, D. Jaeger, C.J. Wilson, Nature **394**, 475 (1998).
12. R.C. Elson, et al. Phys. Rev. Lett. **81**, 5692 (1998).
13. R. Segev, Y. Shapira, M. Beneveniste, E. Ben-Jacob, Phys. Rev. E **64**, 011920 (2001).
14. L.A. Wilkens, D.F. Russell, X. Pei, C. Gurgens, Proc. Roy. Soc. Lond. B **264**, 1722 (1997).
15. D.F. Russell, L.A. Wilkens, F. Moss, *Nature* **402**, 291-294 (1999).
16. A. Neiman, et al, *Phys. Rev. Lett.* **82**, 660 (1999); *Int. J. Bifurcation Chaos Appl.Sci.Eng.* **10**, 2499 (2000).
17. A. Neiman, D.F. Russell, *Phys. Rev. Lett.* **15**, 3443 (2001).
18. J.M. Jřrgensen, Ê. Flock, J. Wersäll, *Z. Zellforsch.* **130**, 362-377 (1972).
19. A. Neiman and D.F. Russell, *Phys. Rev. Lett.* **88**, 138103 (2002).
20. G.B. Ermentrout, N. Kopell, *SIAM J. Appl. Math.* **46**, 233 (1986).
21. W. Horsthemke, R. Lefever, *Noise-induced Transitions* (Springer, Berlin, 1984).
22. R.L. Stratonovich, *Topics in the Theory of Random Noise*, Vol.2 (Gordon and Breach, New York, 1967).
23. P. Hänggi, P. Talkner, M. Borkovec, *Rev. Mod. Phys.* **62**, 251 (1990).
24. B. Naundorf and J.A. Freund, *Phys. Rev. E* **66**, 040901(R) (2002).
25. W. Wojtenek, M.H. Hofmann, and L.A. Wilkens, *Neurocomputing* **38**, 451 (2001).
26. J.A. Freund, et al., *Phys. Rev.E* **63** 0319100 (2001).

Oscillations and Noise in Paddlefish Electroreceptors

David F. Russell* and Alexander B. Neiman*

*Center for Neurodynamics, University of Missouri at St. Louis, St. Louis, Missouri 63121, USA

Abstract. The electroreceptors of paddlefish (*Polyodon spathula*) contain two types of internal oscillators, including a population of stochastic ≈25 Hz oscillators in the sensory epithelia, and a stochastic 45-65 Hz oscillator in each afferent terminal. The function of the epithelial oscillators is unclear. In other hair cell sensory receptors, oscillations in hair cells have been implicated in frequency tuning, such that the frequency of spontaneous oscillations matches well the best response frequency for external stimuli. However, paddlefish electroreceptors appear to not conform to this concept, since they respond best at ≈5 Hz, while their ≈25 Hz epithelial oscillations are 2-3 octaves higher in frequency. Alternate possible functions are discussed, including inducing extended anticorrelations of spike timing, or encoding the water temperature, or acting as internal harmonic noise.

INTRODUCTION

Electroreceptors in paddlefish (*Polyodon spathula*) [1] offer an accessible and experimentally advantageous biological model for studying nonlinear phenomena such as oscillators, synchronization, bursting, and noise. They are surprisingly complex. Sensory receptors are often portrayed as simple threshold-dependent transducers. However, in paddlefish electroreceptors, we find two distinct types of internal oscillators [2]. There is a population of probably thousands of ≈25 Hz oscillators in the sensory epithelia, which include the electrosensitive hair cells. We do not yet know what is the basis for the observed stochastic oscillatory field potentials that we can record near the sensory epithelia. The electrosensitive hair cells make excitatory chemical synapses onto the terminals of 2nd-order neurons ("afferents") [3], which produce spikes traveling to the brain. Afferents are autoactive, spiking continuously but irregularly at 30-65 Hz, which we attribute to another type of oscillator, located in the afferent terminal.

Thus, the electroreceptors of paddlefish are complex nonlinear systems containing two different types of internal stochastic oscillators. Both types influence the pattern of afferent firing, which, as a result, is stochastically biperiodic. Given the 65×10^6 year evolutionary history of this species, it is likely that this complex organization represents an optimization for carrying out the main function of electroreceptors, namely prey detection during feeding behavior.

In this article, we develop hypotheses about the function of the epithelial oscillations, and point out differences and similarities of the electroreceptor oscillators to oscillators in other types of hair cell – afferent sensory receptors, including those for the senses of hearing and balance. The sensory epithelia of paddlefish electroreceptors are remarkably

CP665, *Unsolved Problems of Noise and Fluctuations: UPoN 2002: Third International Conference,* edited by S. M. Bezrukov

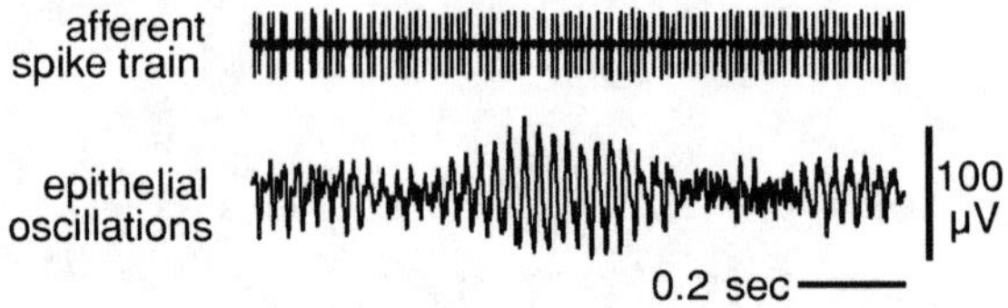

FIGURE 1. Example of spontaneous firing of an afferent, along with epithelial oscillations in the afferent's receptive field.

similar to the cristae and maculae (sensory epithelia) of the vestibular system, and are similar to the auditory organs of lower vertebrates. All have the same two cell types, hair cells and support cells, in a 2-dimensional sheet which is one cell thick. Hence our research may lead to new insights into the operation of auditory and vestibular sensory receptors.

OSCILLATORS IN ELECTRORECEPTORS

Thousands of electroreceptors are organized into linear arrays on the "rostrum", a long flattened paddle-like appendage projecting anterior to the head, for which paddlefish are named. Electrosense in this fish is passive, and is used to detect weak electrical signals from planktonic prey such as *Daphnia* [1, 4]. The anatomical structure of paddlefish electroreceptors is described in [3]. An "electroreceptor" consists of a cluster of 1-35 skin pores, each leading into a short canal, $\approx 200\ \mu$ deep and $\approx 100\ \mu$ in internal diameter. Each canal ends in a sensory epithelium. A cluster of canals is innervated by 2-4 afferents.

We recorded single-unit spikes of electroreceptor afferents having a receptive field on the rostrum, *in vivo*, using a metal microelectrode (see [1, 2] for experimental details). Simultaneous recordings were made of voltage signals in one or two canals of the afferent's receptive field, using a glass pipet electrode inserted into a skin pore. The water conductivity ($\approx 600\ \mu$S/cm) was normal for this species. The water temperature ($\approx 22^o$C) was actually *lower* than the fish were acclimated to in the colony aquaria. The rostrum was always submerged under 5-10 mm of water.

An example of raw data is presented in Fig.1, showing continuous firing of an afferent, along with epithelial oscillations. A given afferent shows a nearly constant mean frequency of background (spontaneous) firing, but the mean firing rate in different afferents can vary over a wide range, 30-65 Hz (at 22^oC). Probability distributions of interspike intervals from spontaneously firing afferents had a well-expressed unimodal peak, and resembled a gamma distribution (i.e. short intervals were more probable). The coefficient of variation (CV) of interspike intervals ranged from 0.1 to 0.3 [5]. The amplitude of epithelial oscillations recorded from a skin pore (canal) was typically $\pm 70\ \mu$V peak-to-peak. The probability distribution of the signal sample values was nearly Gaussian, suggesting that the canal signal may represent the collective outputs from many individual cells. Stochastic oscillations in different canals, even within the same cluster, were statistically independent. Since $\approx$ 10,000 hair cells converge synaptically onto

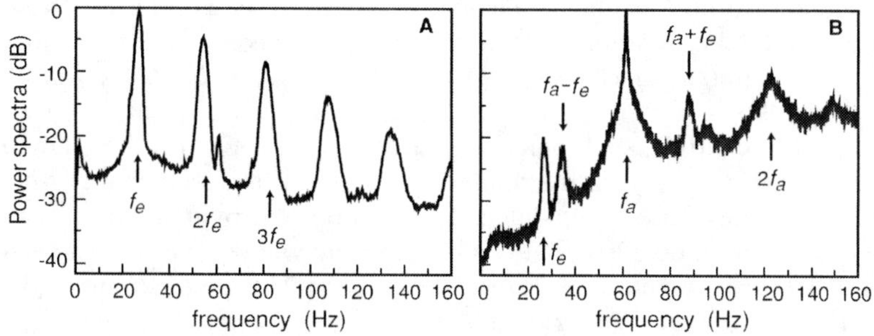

FIGURE 2. Power spectra of canal oscillations (A) and of an afferent spike train (B), calculated from simultaneous recordings of spontaneous canal signals and spontaneous afferent firing, both from the same receptive field.

an afferent, the summed field potentials we can record from canals probably accurately reflect the stochastic periodic forcing of an afferent by epithelial cells.

A typical power spectrum of canal signals is shown in Fig.2A. It shows a well-defined peak at $f_e \approx 27$ Hz and at higher harmonics thereof. The presence of a narrow spectral peak indicates an oscillatory process. Using *in vitro* preparations, and recording directly from epithelial cells, we have shown that these oscillation arise in the sensory epithelium [6]. However, it is not clear whether these oscillations arise in single hair cells, or in supporting cells, or may be due to possible coupling between epithelial cells [7].

Power spectra of spontaneous afferent firing (i.e. of the spike time series) typically showed a biperiodic structure, with the following features: (i) The largest peak at f_a corresponds to the mean firing rate of the afferent spike train. This peak indicates an oscillatory process taking place in the afferent terminal. Higher order harmonics of f_a could sometimes be observed, as in Fig. 2B. (ii) Another peak at f_e corresponds to the frequency of canal oscillations, and is independent of f_a. (iii) There are always two additional smaller peaks at the frequencies $f_{1,2} = f_a \pm f_e$. Thus the power spectrum of afferent firing has the typical structure expected for a periodically driven nonlinear oscillator, including a fundamental peak at the natural frequency of the oscillator (f_a), a peak at the driving frequency (f_e), and sidebands (or combination frequencies) at $f_a \pm f_e$ due to nonlinear mixing [2]. This scheme corresponds well to the morphological organization of electroreceptors, in which epithelial hair cells make excitatory synapses onto the afferent terminals, as unidirectional epithelia-to-afferent coupling.

We did several types of control experiments. (i) Biperiodic afferent power spectra, with all the peaks illustrated in Fig. 2B, were observed without putting any glass pipet electrodes into canals, that is, without changing the resistive loads on the canals. (ii) Attenuated canal oscillations could be recorded with a pipet lifted $\approx$2 mm above a skin pore, such that the canal loading was normal. (iii) Biperiodic afferent spectra persisted when the rostrum was enclosed in a snug-fitting shield made of mu metal, and (iv) also while the flow of ventilatory water into the mouth was stopped briefly.

The existence of two separate unidirectionally coupled oscillators embedded into a single electroreceptor was proven by means of computational [2, 8] and experimental [2, 6] analyses: (i) The coherence between afferent spiking and epithelial oscillations had a well-expressed peak at the frequency f_e. (ii) There is a preferred phase relation between the epithelial oscillations and the afferent pacemaker, consistent with coupling between two oscillators. (iii) The topology of an attractor, reconstructed from the phase angles of interspike intervals, showed a torus-like structure, typical for quasiperiodic oscillations [8]. In contrast, such biperiodic structure disapeared if the epithelial oscillations ceased. (iv) By applying thermal gradients to electroreceptors, we showed that the two oscillators are spatially separated [2, 6]. (v) By destroying all but one canal, and then altering the amplitude of oscillations in the single viable canal, we showed that epithelial oscillations act to entrain the afferent oscillator. This gives strong evidence for a *serial* organization in which the epithelial oscillators are coupled to (synaptically modulate) the afferent oscillator, and excludes alternate models proposing that the f_a and f_e oscillators might be parallel and independent inputs to a third nonlinear element such as a "spike initiation zone".

FUNCTIONAL ROLE OF EPITHELIAL OSCILLATIONS

Electroreceptors respond best near 5 Hz [1, 9]. This characteristic frequency does not match the $\approx$ 25 Hz (at 22°C) natural frequency of the epithelial oscillations (Fig. 3). Phrased differently, there is a 2 to 3 octave mismatch between the best frequency for electroreception and the frequency of canal oscillations. Nevertheless, 25 Hz does correspond to a precipitous decline in the frequency response, so perhaps the epithelial oscillations may be involved in low-pass filtering. Hence the epithelial oscillations in electroreceptors appear to differ from the ones observed in auditory and vestibular hair cells [10, 11, 12, 13], in which the natural frequency of spontaneous oscillations correlates well with the resonant frequency of a given hair cell, thus contributing to frequency selectivity [10, 12]. It was shown recently that auditory hair cell oscillators operate near an Andronov-Hopf bifurcation, providing enhanced sensitivity to weak stimuli [13, 14, 15]. In contrast, external stimuli (weak electric fields) did not affect the presence of epithelial oscillations in electroreceptors, which were always continuous (self-sustained).

External stimuli could partly alter the amplitude of epithelial oscillations, as recorded from canals, but had little effect on their frequency. External stimuli did affect the afferent firing of these electroreceptors. For example, weak sinewave stimuli caused the f_a peak in afferent power spectra to become split, developing sidebands at $f_a \pm f_s$, where f_s was the stimulus frequency. This reveals an unsolved problem, namely how external stimuli can be transduced by hair cells and affect afferent firing via the synapses from hair cells, yet external stimuli have little effect on the epithelial oscillations, which also modulate the afferent.

Thus, self-sustained epithelial oscillations in electroreceptors appear to have a different function than in auditory receptors, despite other shared characteristics of hair cell - primary afferent receptors. In the following, we consider some alternate possible functions.

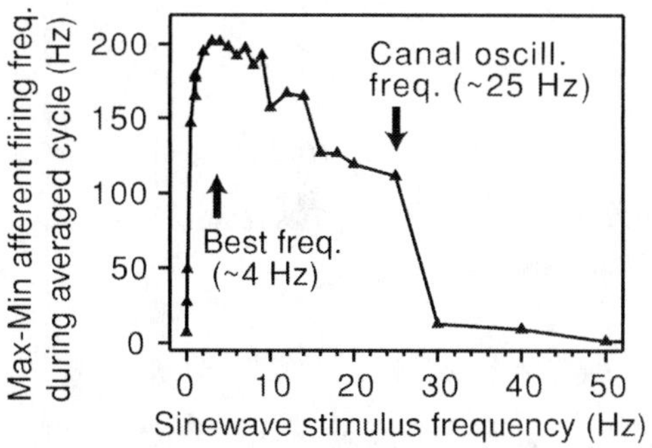

FIGURE 3. Comparison of the frequency response of an electroreceptor to the frequency of canal oscillations. The frequency response of one afferent was measured by presenting constant-amplitude sinewave external electric fields at different frequencies, and subtracting the minimum (Min) from the maximum (Max) of cycle-triggered averages of the instantaneous firing rate.

Epithelial oscillations as noise source

We have demonstrated experimentally that the stochastic epithelial oscillations are responsible for a large part of the variability (irregularity) of afferent neuron firing [2]. For example, we showed that the coefficient of variation (CV) of interspike intervals is a monotonically increasing function of the amplitude of canal oscillations [2]. Another approach is illustrated in Fig. 4, for an electroreceptor in which the epithelial oscillations abruptly declined in amplitude (for unknown reasons). The presence of strong noisy epithelial oscillations correlated with high variability of afferent firing (CV=0.168), seen as a wide distribution of interspikeintervals (left histogram in Fig.4A) and as a wide f_a peak in the power spectrum of spontaneous afferent firing (upper solid trace in Fig.4B). Inversely, after the epithelial oscillations became small, there was nearly periodic firing of the afferent (CV=0.0153), expressed as a narrower distribution of interspike intervals (right histogram in Fig.4A), and as a narrow f_a peak in the power spectrum (lower dashed trace in Fig.4B). Therefore, the epithelial oscillations can be considered as a natural internal source of noise, causing irregularity of an afferent's firing frequency.

One possible role of epithelial noise is *desynchronization of different afferents*. Since epithelial oscillations are statistically independent in different canals, the noise introduced to afferent spike trains may act to prevent inadvertent synchronization between different electroreceptor afferents, which could lead to false alarm incorrect behavioral responses during feeding.

Epithelial oscillatons extend interspike correlations

Although a canal signal is noisy, it possess a high degree of periodicity, expressed as peaks in its power spectrum (Fig.2A). The influence of such narrowband noise on the afferent oscillator results in stochastic quasiperiodicity (biperiodicity), when the instantaneous frequency of the afferent oscillator is stochastically modulated by the

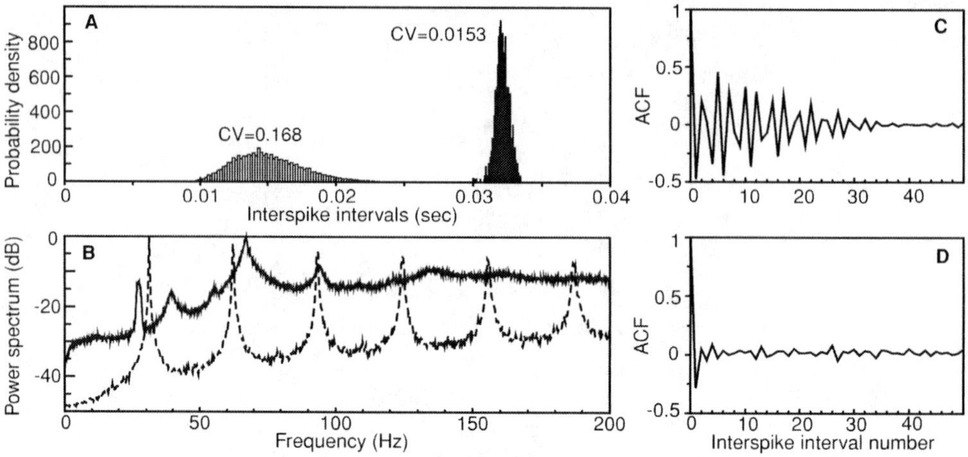

FIGURE 4. Effect of epithelial oscillations on the variability of spontaneous afferent firing, from a receptor in which epithelial oscillations abruptly ceased while we recorded. (A) Probability density of interspike intervals, during the presence (left wide peak) or absence (right narrow peak) of epithelial oscillations.(B) Power spectra of afferent firing from the same data: solid and dashed lines correspond to strong and weak epithelial oscillations, respectively. (C, D) Autocorrelation functions of sequential interspike intervals for strong (C) or weak (D) epithelial oscillations.

epithelial oscillators. Stochastic biperiodicity results in extended correlations between sequential interspike intervals [2]. Such correlations are measured by the autocorrelation function (ACF) (also often called the serial correlation coefficient) [16], defined as $ACF(m) = \langle (T_k - \bar{T})(T_{k+m} - \bar{T}) \rangle / var^2(T)$, where averaging occurs over sequential interspike intervals $T_k, \quad k = 1, ..., N-m$. Examples of ACFs calculated from an afferent's interspike intervals are shown in Fig.4 for strong (C) and weak (D) epithelial oscillations. First, we note that both cases represent nonrenewal point stochastic processes, as the ACFs possess finite correlation times [17]. Second, in both cases the ACFs demonstrate *anticorrelations* which reflect an inverse relation between sequential interspike intervals: short interspike intervals are followed by long intervals, and *vice versa*. This has been attributed to the existence of slow ionic currents at the terminal membrane of afferents, which contribute to driving firing [18, 19], or to short-term changes in spike threshold [20, 21]. Extended oscillatory patterns of the ACF (Fig.4C) are expected in the quasiperiodic regimes of coupled self-sustained oscillators. The sequence of interspike intervals can be considered as a sequence of time intervals between sequential crossings of a Poincaré surface by a phase trajectory of a dynamical system. If a dynamical system is in the regime of a simple limit cycle, then the corresponding map of these return times (also known as a return map) will possess a fixed point, which corresponds to the period of the limit cycle. If noise is taken into account, then this fixed point will be blurred due to amplitude and phase fluctuations. Assuming a broadband character of internal noise (for instance, due to random opening and closing of ion channels in neuronal membranes), the correlations of return time fluctuations will decay very fast, as in Fig.4D. In contrast, if we consider an oscillator which is driven by another noisy oscillator, then the return times and their fluctuations are modulated by the driving process. This in turn is expressed as extended correlations of return times. If the drive is noisy and periodic, there will be decaying oscillatory patterns in the autocorrelation function of return times, as observed in our experimental data (Fig.4C). Thus, noisy epithelial oscillations influence an afferent's firing in two ways. On the one hand, they increase variability, expressed by the CV. On the other hand, however, stochastic epithelial oscillations introduce extended anticorrelations between interspike intervals, and in this sense they *reduce* the randomness of spike time series. It was recently shown that interspike correlations act to improve information transmission about time-varying stimuli [21]. Thus, the noise generated by epithelial oscillator may lead to *increased sensitivity* of paddlefish electroreceptors.

Epithelial noise and stochastic resonance

Stochastic resonance in paddlefish was demonstrated both in feeding behavioral experiments [4] as well as at the level of single electroreceptors [22], for presentations of external noise. We propose that another possible role of epithelial oscillations, acting as internal noise, may be to enhance receptor sensitivity via stochastic resonance, for weak near-threshold stimuli. Theoretical studies have shown that the effect of stochastic resonance can be significantly improved if narrow-bandlimited (harmonic) internal noise is used, similar to epithelial noise, rather than broadband noise [23, 24].

Other possible roles

The preceding proposals are about possible roles of epithelial oscillations in electroreception. However, alternate possible functions can be cited. They may drive secretion of the jelly which fills the canal, and is probably secreted from support cells of the epithelium. That is, epithelial oscillations may play a role like the autoactivity observed in other secretory cells. This hypothesis would require that the support cells undergo membrane potential oscillations similar to the observed canal signals. Also, the epithelial oscillations may transduce the water temperature. Their frequency increases with temperature by 1.2 Hz/^{o}C [6]. Epithelial oscillations could act as a carrier wave, whose frequency encodes the water temperature, such that afferent firing would be frequency modulated, and would simultaneously send to the brain information about two sensory modalities. This hypothesis would require that the brain be able to extract the epithelial frequency from afferent spike trains. This proposal is distinct from previous reports on the thermal sensitivity of electroreceptors [25].

CONCLUSIONS

We have demonstrated the first example of a hair cell sensory receptor that incorporates two distinct types of oscillators, and whose output is biperiodic as a result. The function of the stochastic epithelial oscillations is still unclear. They act as internal harmonic noise for afferent firing, and they induce extended anticorrelations of afferent firing, but the significance of internal noise and spike correlations for receptor operation require further study. They probably do not determine the characteristic frequency responsiveness of paddlefish electroreceptors.

ACKNOWLEDGMENTS

The work was supported by the Office of Naval Research - Physics Divison, the Fetzer Institute (A.N.), the University of Missouri Research Board, and the National Institutes of Health (R01DC004922-01).

REFERENCES

1. L.A. Wilkens, D.F. Russell, X. Pei, C. Gurgens, *Proc. Roy. Soc. Lond. B* **264**, 1722 (1997).
2. A. Neiman and D.F. Russell, *Phys. Rev. Lett.* **86**, 3443 (2001).
3. J.M. Jrgensen, . Flock, J. Wersäll, *Z. Zellforsch.* **130**, 362-377 (1972).
4. D.F. Russell, L.A. Wilkens, F. Moss, *Nature* **402**, 291-294 (1999).
5. The coefficient of variation is defined as $CV = \sqrt{var(T)}/\bar{T}$, where $var(T)$ and $\bar{T}$ are the variance and the mean of interspike intervals, respectively.
6. D.F. Russell and A. Neiman, in preparation (2003).
7. W.T. Clusin, M.V.L. Bennett, *J. Gen. Physiol.* **73**, 685-702 and 703-723 (1979); J. Lu, H.M. Fishman, *Biophysical J.* **69**, 2458-2466 (1995).
8. A. Neiman and D.F. Russell, *J. Neurophysiol.*, submitted (2002).
9. X. Pei, D.F. Russell, L.A. Wilkens, and F. Moss, in *Computational Neuroscience: Trends in Research* eidted by J. Bower, Plenum, New York, pp.245-249 (1998).
10. A.C. Crawford and R. Fettiplace, *J. Physiol. (Lond)***306**, 79 (1980).
11. A. Ruesch and U. Thurm, *Hearing Res.* **48**, 247 (1990).
12. P. Martin, A.J. Hudspeth, and F. Jülicher, *PNAS* **98**, 14380 (2001).
13. M. Ospeck, V.M. Eguíluz, and M.O. Magnasco, *Biophysical J.* **80**, 2597 (2001).
14. S. Camalet, T. Duke, F. Jülicher, and J. Prost, *PNAS* **97**, 3183 (2000).
15. V.M. Eguíluz, M. Ospeck, Y. Choe, A.J. Hudspeth, and M.O. Magnasco, *Phys. Rev. Lett.* **84**, 5232 (2000).
16. H.C. Tuckwell, *Intoroduction to theoretical neurobiology*, Vol.2, (Cambridge University Press, Cambridge, 1988).
17. A renewal stochastic point process is delta-correlated: $ACF(m) = \delta m$. See D.R. Cox and P.A. Lewis, *The statistical analysis of series of events* (Methuen, London, 1966).
18. H.A. Braun, et al, *Int. J. Bifurcation Chaos* **8**, 881 (1998).
19. K. Schäfer, et al, *Pflügers Archiv - European J. Physiology* **429**, 378 (1995); H.A. Braun, et al, *Nature* **367**, 270-273 (1994).
20. R. Ratnam and M.E. Nelson, *J. Neuroscience* **20**, 6672 (2000).
21. M.J. Chacron, A. Longtin, and L. Maler, *J. Neuroscience* **21**, 5328 (2001).
22. D.F. Russell, et. al. *Fluctuation and Noise Letters* **1**, L71 (2001).
23. P. Hänggi, et al. *J. Stat. Phys.* **70**, 25 (1993).
24. R. Li, et al. *Phys. Rev. E* **51**, 3964 (1995).
25. H. Hensel, *Pflügers Archiv* **347**, 89 (1974).

Numerical Simulation of Diffusion of Second Messengers cGMP and Ca^{2+} in Rod Photoreceptor Outer Segment of Vertebrates

H. Khanal*, V. Alexiades†, E. DiBenedetto** and H. Hamm‡

*Department of Mathematics, University of Tennessee, Knoxville, TN 37996-1300
†Department of Mathematics, University of Tennessee, Knoxville TN 37996-1300
and Oak Ridge National Laborotory, Oak Ridge TN 37831
**Department of Mathematics, Vanderbilt University, Nashville TN 37240
‡Department of Pharmacology, Medical Center, Vanderbilt University, Nashville TN 37232

Abstract. A crucial step in the process of phototransduction, whereby light is converted into an electrical response in retinal rod and cone photoreceptors, involves interaction and diffusion of cytoplasmic signaling molecules, termed **second messengers**, in the cytosol. A computational model for the interaction and diffusion of the second messengers (cyclic Guanosine Monophosphate and Calcium) during the activation phase of phototransduction in retinal rod cells is described and results of numerical simulations are presented.

INTRODUCTION

The rod outer segment of vertebrates comprises a stack of equispaced *disc* membranes of thickness about 10 nm and at mutual distance of about 14 nm (for the Salamander).

Each disc is made up of two functionally independent layers of lipidic membrane where proteins are embedded, such as rhodopsin (Rh), the light receptor, G protein (G), also called transducin, and cGMP phosphodiesterase (PDE), the effector. These membrane associated proteins can diffuse on the face of the disc where they are located, but cannot abandon the disc. The lateral membrane of the rod contains cGMP–gated channels of small radius. In absence of light these channels are open and allow a positive influx of sodium and calcium (Ca^{2+}) ions. The space within the rod, and not occupied by the discs, is filled with fluid cytosol, in which cyclic-guanosine monophosphate (cGMP) and Ca^{2+} diffuse.

When a photon is absorbed by a molecule of rhodopsin, located on one of the discs, the rhodopsin becomes activated, and in turn activates any G protein it interacts with. Each of the activated G proteins, is capable of activating one and only one catalytic subunit of PDE on the activated disc, by binding to it upon contact. The bound pair so generated is denoted by PDE*. This cascade takes place only on the activated disc. The next cascade, involving cGMP and Ca^{2+}, takes place in the cytosol.

Active PDE* hydrolyzes cGMP in the cytoplasm, thereby lowering its concentration. The decrease of concentration of the cGMP causes closure of some of the cGMP–gated channels of the plasma membrane, resulting in a lowering of the influx of positive ions, and thus a lowering of the local current J across the outer membrane. Because of the

CP665, *Unsolved Problems of Noise and Fluctuations: UPoN 2002: Third International Conference*, edited by S. M. Bezrukov

$Na^+/K^+/Ca^{2+}$ exchanger which continues to remove Ca^{2+} from the cytoplasm, there is a decrease in the calcium concentration, which in turn results in an increase in cGMP production by stimulation of Ca^{2+}–inhibited guanyl cyclase, and thus a consequent reopening of the channels. The same decrease of calcium closes the cycle by causing deactivation of rhodopsin through stimulation of rhodopsin kinase. The latter ceases activating new G protein. Thus PDE* decays to basal, ending depletion of cGMP.

This cascade is well known and it is supported by a sizable amount of published experimental data [4, 14, 22, 21]. However, despite being one of the best understood *signal transduction* processes, its formal mathematical description, is less developed. The classical methods used to describe signal transduction processes assume a well-stirred aqueous environment, and ordinary differential equations provide solutions which average concentrations within the volume of the cell. Recent investigations [3, 13] suggest that these methods are not adequate to describe the precisely regulated signal transduction processes emanating from these highly localized structures, sometimes called "signalsomes" [3].

We present numerical simulations of the diffusion of the second messengers cGMP and Ca^{2+} in the cytoplasm of the rod outer segment of vertebrates during the activation phase. The simulations build on a mathematical model presented in [1] whose main feature is the inclusion of spatio-temporal variations of the concentration of cGMP and Ca^{2+}. The second messengers cGMP and Ca^{2+}, far from being bulk quantities, are regarded as pointwise functions of space and time. In [1], the existing model of [16], which assumes bulk/lumped quantities of cGMP and Ca^{2+} and based on ordinary differential equations, is extended by incorporating diffusion effects and therefore by giving a pointwise description of the phenomenon by means of evolution partial differential equations. The bulk/lumped source terms are correctly modeled as surface–volume reactions occurring on the faces of the discs where cGMP is hydrolized.

The numerical simulations we have performed trace the space–time dependence of cGMP. Since the latter is linked to the current J through the cGMP–gated channels, the same simulation provide the distribution of current as a space–time function defined on the lateral boundary of the rod outer segment. Such a dependence permits us to verify numerically some facts put forth in the biological literature about the spread of a single photon response [5].

MATHEMATICAL MODEL

Geometry of the Domain: The outer segment of a rod receptor can be considered as a right circular cylinder $\tilde{\Omega}$ of height H and radius R_{rod}, housing a vertical stack of N equispaced parallel discs C_i, $i = 1, 2, \cdots N_{\text{disc}}$, each of radius R_{disc}, thickness ε, and mutually separated by a distance δ. Each of the discs C_i, $i = 1, 2, \cdots N_{\text{disc}}$, carrying the rhodopsin, is assumed to be a short cylinder, coaxial with the rod $\tilde{\Omega}$, of radius R_{disc} and height $\varepsilon << H$. The discs C_i are equally spaced, i.e., the upper face of C_i has distance δ from the lower face of C_{i+1}. The first C_1 has distance $\delta/2$ from the lower face of the rod $\tilde{\Omega}$ and the last $C_{N_{\text{disc}}}$ has distance $\delta/2$ from the upper face of the rod. The cytosol, where diffusion of second messengers takes place, is the subset $\Omega \subset \tilde{\Omega}$ not occupied by the cylinders C_i, i.e., $\Omega = \tilde{\Omega} - \bigcup \bar{C}_i$, $i = 1, \cdots, N_{\text{disc}}$. Cylindrical coordinates (r, z, θ) will be employed. We define $F_i^{\pm}$ = upper/lower disc faces, L_i = lateral boundary of disc C_i (at $r = R_{\text{disc}}$) and $\partial_o\Omega$ = lateral outer boundary (plasma membrane). We also define σ_i to denote the area of disc C_i, σ_{disc} the total area of the N_{disc} discs, V_{cyto} the volume of the cytosol. We also introduce a geometric parameter d, the volume to area ratio, which is approximately $\delta/2$, thus $\mathrm{d} = V_{\text{cyto}}/\sigma_{\text{disc}} \approx \delta/2$.

Basic Equations of the Model: Derivation of the general diffusion equations for the concentrations [cG] of cGMP and [Ca] of Ca^{2+} in the fluid cytosol is based on mass balance in an elementary volume. Since there are no volume-sources for either cGMP or Ca^{2+} in the interior of cytosol, the governing equations for diffusion of the second messengers cGMP and Ca^{2+} in Ω can be expressed as

$$\frac{\partial[\text{cG}]}{\partial t} - \nabla \cdot (D_{\text{cG}} \nabla [\text{cG}]) = 0, \qquad \frac{\partial[\text{Ca}]}{\partial t} - \nabla \cdot (D_{\text{Ca}} \nabla [\text{Ca}]) = 0 \tag{1}$$

in Ω, for $t > 0$, where D_{cG} and D_{Ca} are the respective diffusion coefficients (in $\mu\text{m}^2\text{s}^{-1}$).

Initial Conditions: Initially, the concentrations of free [cG] and [Ca] in the cytosol can be taken as uniform at their steady-state (dark) values.

Boundary Conditions for [cG]: We assume that a small beam of photons hits a disc C_{i_0} on one of its faces, say for example the lower one. Generation and removal of free cGMP concentration in the cytoplasm occurs through binding phenomena on the lower and upper faces of each of the disc. The axial boundary fluxes for [cG] are zero on L_i (lateral boundary of disc C_i) and on $\partial_o\Omega$ (lateral outer boundary). On the upper and lower faces of disc C_i the flux is given by

$$-D_{\text{cG}} \frac{\partial[\text{cG}]}{\partial z}\bigg|_{F_i^{\pm}} = \pm \frac{\alpha_{\max}\,\mathrm{d}}{1 + ([\text{Ca}]/K_{\text{cyc}})^{m_{\text{Ca}}}} \mp k_{\text{hyd}}\,[\text{PDE}]_s[\text{cG}] + \delta_{i_0}\, k^*_{\text{hyd}}[\text{PDE}^*]_s\,[\text{cG}]. \tag{2}$$

The quantity $\alpha_{\max}$ is the maximum rate of synthesis of cGMP by guanylyl cyclase [16, pp. 821]. K_{cyc} is the Ca^{2+} concentration that achieves half of the maximum rate, m_{Ca} is a Hill's constant. Multiplying $\alpha_{\max}$ by $\mathrm{d}(= V_{\text{cyto}}/\sigma_{\text{disc}})$ converts the volumetric flow rate $\alpha_{\max}$ to flux across the disc face. The term $k_{\text{hyd}}\,[\text{PDE}]_s[\text{cG}]$ describes hydrolysis of cGMP by dark-activated PDE, with rate k_{hyd}. Similarly, the last term in (2) describes hydrolysis by light-activated PDE* on activated disc(s) i_0 ($\delta_{i_0} = 1$ only when C_i is an activated disc). $[\text{PDE}]_s$ and $[\text{PDE}^*]_s$ are surface densities of the dark and light activated Phosphodiesterase, related to (volumetric) concentrations via $[\text{PDE}] \cdot \mathrm{d} = [\text{PDE}]_s$ and $[\text{PDE}^*] \cdot \mathrm{d} = [\text{PDE}^*]_s$.

Boundary Conditions for [Ca]: Calcium does not penetrate the disc C_i. We assume that the only calcium fluxes into or out of the cytoplasm are by the cGMP-gated channels and the electrogenic exchanger. The boundary fluxes for [Ca] are zero on L_i (lateral boundary of disc C_i) whereas on $\partial_o\Omega$ (lateral outer boundary)

$$-D_{\text{Ca}} \frac{\partial[\text{Ca}]}{\partial r} = \frac{1}{\Sigma_{\text{rod}} B_{\text{Ca}}\, \mathcal{F}} \left(J_{\text{ex}} - \frac{1}{2} f_{\text{Ca}} J_{\text{cG}} \right). \tag{3}$$

Here J_{ex} is the current across the boundary of the rod due to electrogenic exchange, J_{cG} is the current carried by the cGMP-gated channels (at a fixed membrane voltage), B_{Ca}

is buffering power of the cytoplasm for calcium, f_{Ca} is the fraction of cGMP-activated current carried by Ca^{2+}, and $\mathcal{F}$ is Faraday's constant. The circulating currents J_{ex} and J_{cG} are described respectively by the the following Michaelis-Menten and Hill type relations [16, pp. 820-821]

$$J_{ex} = \frac{j_{ex}^{sat}}{1 + K_{ex}/[Ca]}, \quad J_{cG} = \frac{j_{cG}^{max}}{1 + (K_{cG}/[cG])^{m_{cG}}}, \tag{4}$$

where j_{ex}^{sat} is saturation exchange current, j_{cG}^{max} is maximal cG-current and Σ_{rod} is the lateral surface area of the Rod Outer Segment. Dividing the flow rate in (3) by Σ_{rod}, converts it to flux. Thus, the local flow rate across a patch of area A of membrane (with local concentrations [cG], [Ca]) is given by A times the right-hand side of (3). The total circulating current (J) is the sum $J_{cG} + J_{ex}$.

Activation of PDE: In lieu of a detailed model of the rhodopsin to PDE* cascade taking place on activated discs, we account for the activated PDE by expressing the surface density $[PDE^*]_s$ as a gaussian

$$[PDE^*]_s = \frac{P_0}{4\pi D_a(t - t_{on})} \exp\left(-\frac{r^2}{4D_a(t - t_{on})}\right) \tag{5}$$

with standard deviation $\sqrt{4D_a(t - t_{on})}$, where D_a is a diffusivity of activation of PDE by transducin on the disc surface, P_0 is the number of activated PDE* molecules on a disc, and t_{on} is the time at which light strikes the disc. Note that $[PDE^*]_s$ in (5) is also a solution of 2-dimensional diffusion on the disc face, the response to a Dirac delta source of strength P_0 at $r = 0$ and $t = t_{on}$.

NUMERICAL SIMULATIONS

A computational model of the mathematical problem has been developed based on Finite Volume discretization of the partial differential equations and boundary conditions for the axially symmetric case, and implemented in a Fortran program. Due to the intricate geometry of the cytosol, even the axisymmetric problem involves very intensive computations. We parallelized the scheme for distributed memory clusters of multi-processors or heterogeneous networked computers, employing the MPI (Message Passing Interface) library. The numerical solution of the partial differential equations (1) - (3) gives the evolution of the spatial distribution of concentration of cGMP and Ca^{2+} in the cytosol, from which the circulating currents in equations (4) will be computed.

Parameters: For the numerical experiment, representative parameter values are chosen for the outer segment of Salamander rod cells, and are mostly taken from [16, 20, 11]. Outer segment of the Salamander rod was of radius R_{rod} = 5.515 μm containing a stack of N_{disc} = 800 discs of height ε = 14 nm and radius R_{disc} = 5.5 μm equally spaced at a distance δ = 14 nm apart. Height of the rod outer segment H = 22.4 μm. Typical parameters for the whole cell electrical properties for the dark resting state are taken to be $j_{cG}^{max} = 7000$pA and $j_{ex}^{sat} = 0.17$pA respectively. The values for the rate constants $k_{hyd} = 7 \times 10^{-5}$ molecules$^{-1}\,\mu$m^3 s^{-1} and $k_{hyd}^* = 0.183$ molecules$^{-1}\,\mu$m^3 s^{-1} are obtained by matching the parameters with those in [16]. The diffusion coefficient for cGMP is estimated to be $D_{cG} \approx 50\,\mu m^2 s^{-1}$ by comparing literature values [6, 7, 5, 18] of the experimentally measured effective longitudinal diffusion in cytosol. The diffusion coefficient of calcium in cytosol, $D_{Ca} = 15\,\mu m^2 s^{-1}$, is taken from [15], which accounts for buffering effects. The surface density of PDE molecules present in the disc membrane is $[PDE]_s = 100$ moleculesμm^{-2} [20]. The remaining parameters used in our simulations are as follows: $\alpha_{max} = 50\,\mu M s^{-1}$, $B_{Ca} = 45$, $f_{Ca} = 0.17$, $\mathcal{F} = 96500\,\mathcal{C} mol^{-1}$, $K_{cyc} = 0.15\,\mu M$, $K_{cG} = K_{cG,\,max} = 32\,\mu M$, $K_{ex} = 1.5\,\mu M$, $m_{Ca} = 2$ and $m_{cG} = 2$. In order to study the kinetic response of the concentration of various enzymes, proteins and calcium in the cytoplasmic membrane, we have to activate the receptor protein rhodopsin by the flash of light. Our model allows control of the boundary source terms for [cG] in equation (2) in various ways: activation of PDE molecules, with specific strength P_0 and spread rate D_a, over an arbitrary surface area of either of the faces of any desired number of discs, for any specified duration of time, constantly, or on and off.

Activated PDE: We want the entire disc to be activated within a certain time interval $[0, T]$. Since three standard deviations (s.d.) contains 99% of the gaussian in equation (5), we want 3 s.d. $\approx R_{disc}$ at time T, hence $D_a = R_{disc}^2/(36T)\ \mu m^2 s^{-1}$. Thus, for $T = 1$ sec, $D_a \approx 1\,\mu m^2 s^{-1}$, for $T = 0.1$ sec, $D_a \approx 8\,\mu m^2 s^{-1}$.

One photon is expected to activate about 1000 molecules of PDE* (1 photon $\longrightarrow$ $1R^* \longrightarrow 10^2$ - $10^3 T^* \longrightarrow 10^2$ - 10^3 PDE*, see [9]). Thus, a flash delivering Φ photoisomerizations to the rod is expected to activate about 1000Φ PDE molecules. Each activated disc contains $[PDE]_s \cdot \sigma_{i_0}$ PDE molecules, therefore the number of activated discs $N_{active} = 1 + 1000\Phi/([PDE]_s \sigma_{i_0})$. Distributing the 1000Φ equally among the N_{active} discs, we have $P_0 = 1000\Phi/N_{active}$ = number of PDE* molecules on each activated disc.

Steady-state: With the input of the above mentioned parameter values, we obtain the dark steady state solution as $[cG]_{dark} = 2.91\,\mu M$ and $[Ca]_{dark} = 0.60\,\mu M$. This is slightly less than the Nikhonov, Lamb, and Pugh [16] dark values of 3 and 0.64. By taking $K_{cyc} = 0.165$, we find the more agreeable values of 2.99 and 0.65. We used these values as the initial conditions for concentrations. The dark standing currents are computed using these values in (4) and are used as normalizing factors for the total circulating currents.

Histories and Profiles: At desired intervals, concentrations and current histories at selected locations in the cytosol and concentration/current profiles along the axial and radial direction were outputed. The plots are drawn in terms of relative and relative responses normalized by their dark values. Some of the results of numerical simulations (with a very coarse mesh of 8 nodes radially and 8 axially, per disc unit) are presented in Fig. **1.** and Fig. **2.**. The cGMP concentration behaves like the current and it is not presented.

Computations were performed on Compaq AlphaServer SC clusters at Oak Ridge National Laboratory. A typical 1000 msec simulation with 800 discs takes about 1 hr : 20 min using 21 processors.

Conclusion: The form of cGMP and Ca^{2+} concentrations and the light-induced reduction in standing current in the outer segment of a rod predicted from the numerical simulation show a good match to the general trend of the available experimental data. Thus, the simulations will be helpful for interpretation of available biological data,

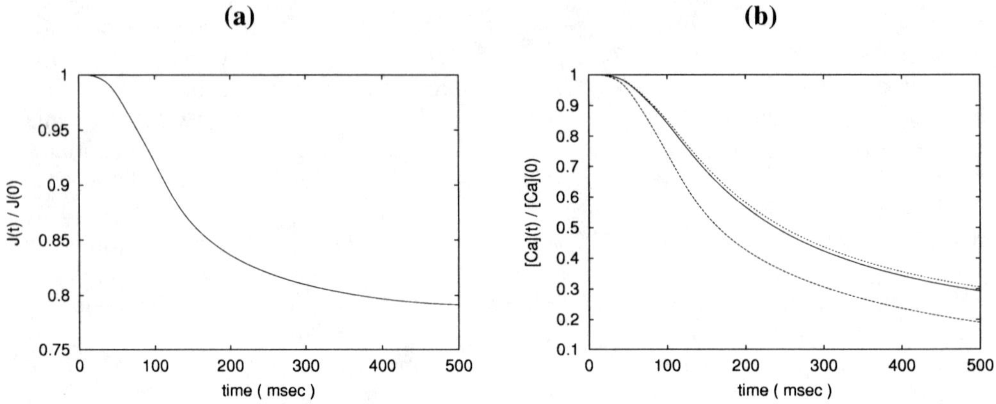

FIGURE 1. Histories of total relative response $J(t)/J(0)$ and calcium relative concentration $[\mathrm{Ca}](t)/[\mathrm{Ca}](0)$ at some fixed location on the cytosol in the vicinity of the activated disc illustrating the light-induced reduction in standing current and concentration. This simulation was performed for a "truncated" rod with 400 discs (i.e., half of the usual size of a typical Salamander photoreceptor rod) activating 51 discs (# 175 - 225) with about 9285 PDE* molecules per disc) via gaussian source of equation (5). The normalizing factors $J(0)$, the total circulating current in the dark steady-state is 65.88 pA and $[\mathrm{Ca}](0)$, the concentration of calcium in the dark is $0.65\,\mu$M. **(a)** Relative response $J(t)/J(0)$ versus time. The current reduces 20% by 400 msec. **(b)** Relative calcium concentration $[\mathrm{Ca}](t)/[\mathrm{Ca}](0)$ versus time at three specified locations in the cytosol is shown. The upper two curves are from disc units # 174 and # 226. The lower curve is from disc # 200. The plot shows that the calcium concentration is depleted by more than 50% in the first 200 msec.

comparision of model predictions with measurements, determinination of sensitivity of output on various model parameters, and to design further biological experiments.

ACKNOWLEDGMENTS

Access to the Compaq AlphaServerSC computer was provided by the Center for Computational Sciences at Oak Ridge National Laboratory and the Evaluation of Early Systems research project, sponsored by the Office of Mathematical, Information, and Computational Sciences, Office of Science, U.S. Department of Energy under Contract No. DE-AC05-00OR22725 with UT-Battelle, LLC.[1] Oak Ridge National Laboratory is managed by UT-Battelle, LLC for the United States Department of Energy under Contract No. DE-AC05-00OR22725.

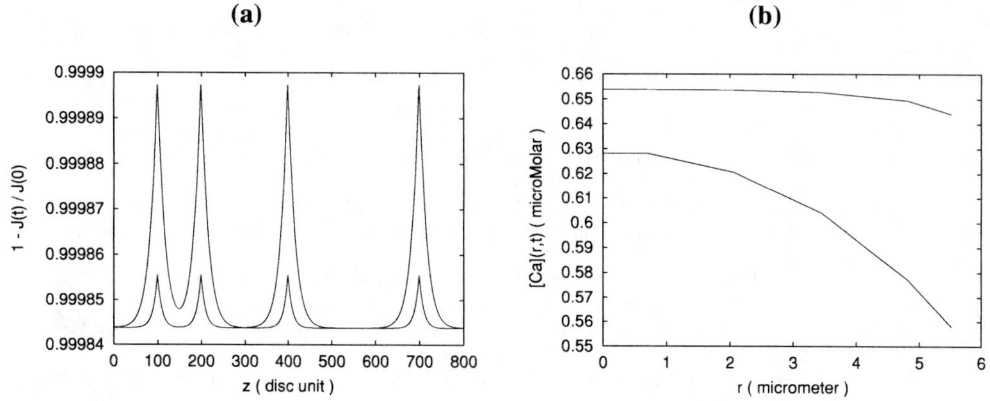

FIGURE 2. Axial (z-direction) and radial (r-direction) profiles at times 150 msec and 600 msec. This simulatation was performed with total number of 800 discs (i.e., full size of a typical Salamander photoreceptor rod) activating only four discs (# 100, 200, 400 and 700) with 5000 PDE* molecules on each disc, via gaussian source of equation (5). **(a)** This describes the spatial variation along the z-direction of the relative response of the total current $1 - J(z,t)/J(z,0)$, where $J = J_{cG} + J_{ex}$. This plot clearly depicts interacting and noninteracting discs. There is separation of the diffusion effect between discs # 400 and # 700 but when the activated discs are closer the responses merge. The spread of the response is about 160 disc units, which amounts to $\approx 4.5\mu$m. **(b)** This describes the spatial variation of the concentration of calcium along the radial direction at a fixed z-location (just below the activated disc units) in the cytosol.

REFERENCES

1. Andreucci, D., Bisegna, P., Dibenedetto, P., Hamm, H.E., "Mathematical Model of the Dynamics of Second Messengers in Visual Transduction: Homogenization and Concentrated Capacity", submitted to *Biophysical J.*, (2002).
2. Baylor, D., "How photons start vision", *Proc. Natl. Acad. Sci. USA*, **93**, 560-565 (1996).
3. Bhalla, U.S., Iyengar, R., "Emergent Properties of Networks of Biological Signaling Pathways", *Science*, **283**, 381-387 (1999).
4. Burns, M.C., Baylor, D.A., "Activation, deactivation, and adaptation in vertebrate photoreceptor cells", *Annual Rev. Neurosci.*, **24**, 779-805 (2001).
5. Gray-Keller, M., Denk, W., Shraiman, B., Detwiler, P.B., "Longitudinal Spread of Second Messenger Signals in Isolated Rod Outer Segments of Lizards", *J. Physiol.*, **519.3**, 679-692 (1999).
6. Koutalos, Y., Nakatani, K., Yau, K.-W., "Cyclic GMP Diffusion Coefficienst in Rod Photoreceptors Outer Segments", *Biophysical J.*, **68**, 373-382 (1995).
7. Koutalos, Y., Nakatani, K., "Calcium Diffusion Coefficients in Rod Photoreceptors Outer Segments", *Biophysical J.*, **76**, A242 (1999).
8. Koutalos, Y., K., Yau, K.-W., "Regulation of Sensitivity in Vertrebrate Rod Photoreceptors by Calcium", *Trends in Neuroscience*, **19**, 73-81 (1996).
9. Kurahashi, T., Gold, G.H., "Ionic Channels Mediating Sensory Transduction", in *Introduction to Cellular Signal transduction*, *, Birkhäuser, Boston, 1990, pp 215-234.
10. Lamb, T.D., "Gain and Kinetics of activation in the G-Protein cascade of phototransduction", *Proc. Natl. Acad. Sci. USA*, **93**, 566-570 (1996).
11. Lamb, T.D., Pugh, E.N., "Amplification and Kinetics of the Activation Steps in Phototransduction", *Biochimica et Biophysica Acta*, **1141**, 111-149 (1993).
12. Lamb, T.D., McNaughton, P. A., Yau, K.-W. , "Spatial Spread of Activation and Background Desensitization in Toad Rod Outer Segment", *J. Physiol.*, **318**, 463-496 (1981).

13. Leskov, I.B., Klenchin, V.A., Handy, J.W., Whitelock, G.G., Govardovskii, V.I., Bownds, M.D., Lamb, T.D., Pugh, E.N., Arshavsky, V.Y., "The Gain of Rod Phototransduction: Reconciliation of Biochemical and Electrophysical Measurements", *Neuron*, **27**, 525-537 (2000).
14. Liebman, P.A., Parker, K.R., Dratz, E.A., "The molecular mechanism of visual excitation and its relation to the structure and composition of the rod outer segment", *Annual Rev. Physiol.*, **49**, 765–791 (1987).
15. Nakatani, K., Chen, C., Koutalos, Y., "Calcium Diffusion Coefficient in Rod Photoreceptor Outer segments", *Biophysical J.*, *82*, 728-739 (2002).
16. Nikonov, S., Lamb, T.D., Pugh, E.N., "The Role of Steady Phosphodiestarase Activity in the Kinetics and Sensitivity of the Light-Adapted Salamander Rod Photoresponse", *J. Gen. Physiol.*, **116**, 795-824 (2000).
17. Nikonov, S., Lamb, T.D., Pugh, E.N., "Photo receptor Sensitivity and Kinetics in Light Adaptation", *J. Gen. Physiol.*, **117**, 365-366 (2001).
18. Olson, A., Pugh, E.N., "Diffusion coefficient of cyclic GMP in Salamander Rod outer segments estimated with two floroscent probes", *Biophysical J.*, **65**, 1335-1352 (1993).
19. Pawson, T., Scott, J.D., "Signaling through scaffold, anchoring, and adaptor proteins", *Science* **278**(5346), 2075-80, (1997).
20. Pugh, E.N., Lamb, T.D., "Phototransduction in Vertebrate Rods and Cones: Molecular Mechanisms of Amplification, Recovery and Light adaptation", in *Molecular Mechanism in Visual Transduction*, edited by D.G. Stavenga, W.J. Degrip, and E.N. Pugh Jr, Elsevier, Amsterdam, 2000, pp. 183-255.
21. Schnapf, J.L., Baylor, D.A., "How Photoreceptor Cells Respond to Light", *Scientific Amer.*, **256**, 40-47 (1987).
22. Wald, G., "The Molecular Basis of Visual Excitation", *Nature*, **219**, 200-207 (1968).

Stochastic Synchronization and Signal Rectification in the Crayfish Caudal Photoreceptor

Sonya Bahar [1] and Frank Moss[2]

[1] *Department of Neurological Surgery*
Weill-Cornell Medical College New York Presbyterian Hospital
525 East 68th Street, Box #99 New York NY 10021
Tel (212) 746-5535 Fax (212) 746-5592 ssb2001@med.cornell.edu

[2] *Center for Neurodynamics*
8001 Natural Bridge Road
University of Missouri at St Louis St Louis MO 63121
Tel (314) 516-6150 Fax (314) 516-6152 mossf@umsl.edu

Abstract. The crayfish possesses two light-sensitive neurons in its abdominal 6th ganglion ("caudal photoreceptors", or CPRs). In addition to being light sensors, the CPRs are secondary neurons in a mechanosensory pathway. These sensory mechanisms, mechanosensitivity and light sensitivity, interact with one another. Pei et al. (1996) demonstrated that light enhances the transduction of weak, periodic hydrodynamic stimuli (measured as an increase in the signal-to-noise ratio at the stimulus frequency in the power spectrum of the recorded neural spikes). This has been interpreted as a stochastic resonance effect, in which added light increases the noise intensity of the input to the photoreceptor (possibly through fluctuations in membrane potential), leading to an enhancement of the signal to noise ratio (SNR). We have demonstrated in the crayfish the correlation between a stochastic-resonance like effect and an increase in stochastic phase synchronization between the neural response and a periodic mechanical stimulus (Bahar et al. 2002). Here, we discuss a novel effect in which light increases the signal-to-noise ratio of the *second higher harmonic* of a periodic input mechanical stimulus, effectively rectifying the input signal. This "second harmonic effect" can be interpreted in terms of stochastic phase synchronization. We hypothesize that the input signal is effectively *full-wave rectified via the summation of two half-wave rectified mechanosensory signals.* This hypothesis can be described by a simple mathematical model which agrees well with our experimental results.

INTRODUCTION

The crayfish has two light-sensitive neurons in its 6th abdominal ganglion (Welsh 1934; Kennedy 1963; Wilkens and Larimer 1972; Edwards 1984; Wilkens 1988). These caudal photoreceptors, or CPRs, increase their firing rate when illuminated by light in the visible range (Bruno and Kennedy 1962). The CPRs are also mechanosensory interneurons: mechanosensory hairs on the crayfish tailfan are mechanically coupled to sensory neurons whose axons enter the 6th abdominal ganglion and synapse onto the CPRs (Flood and Wilkens 1978; Wilkens and Douglass 1994; Douglass and Wilkens 1998). The mechanoreceptor neurons which synapse

CP665, *Unsolved Problems of Noise and Fluctuations: UPoN 2002: Third International Conference,* edited by S. M. Bezrukov

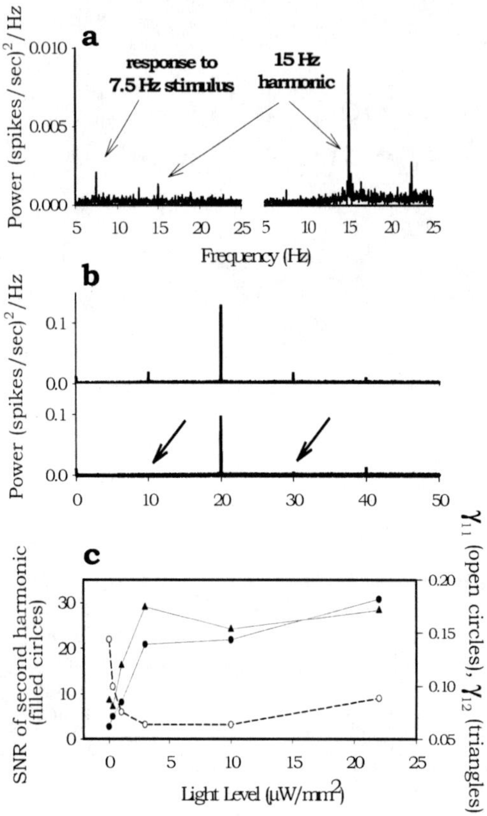

Figure 1 (a) CPR power spectra in dark (left panel, 5 nW/mm^2) and light (right panel, 22 μW/mm^2). Stimulus frequency is 7.5 Hz, amplitude 6 μm. **(b)** CPR power spectra in dark (top panel, 5 nW/mm^2) and light (bottom panel, 22 μW/mm^2). Stimulus frequency is 10 Hz, amplitude 9 μm. Note the light-induced suppression of the odd harmonics at 10 Hz and 30 Hz, indicated by the bold arrows in the bottom panel, and the increase in the 4th harmonic. Note that the broad peak due to faster firing at higher light levels, which is so prominent in the right hand panel of Fig. 1a, is much less dramatic in the right panel of Fig. 1a and bottom panel of 1b. **(c)** SNR_2 (filled circles) vs. light level (stimulus 10 Hz, 7 μm). Note decrease in the 1:1 synchronization index γ_{11} (open circles) as the 1:2 index γ_{12} (triangles) increases. 120 seconds of data were recorded at each light level, after discarding 20 seconds of transients. For details of the experimental setup, see Bahar et al. (2002).

onto the photoreceptors are sensitive to hydrodynamic motions in the frequency range 2-20 Hz, which overlaps with the fin-beat and tail-beat frequencies of fish which typically prey on the crayfish (Bainbridge 1958; Bleckmann et al. 1991; Drucker et al. 1996a,b; Ellerby and Altringham 2001; Webb 2002).

In addition to sharing a common pathway (the photoreceptor axon), these two sensory modalities, mechanosensitivity and light sensitivity, interact with one another (Simon and Edwards 1990). In a particularly striking demonstration of this interaction, Pei et al. (1996) showed that light *enhances the signal-to-noise ratio (SNR)* of weak, periodic hydrodynamic stimuli. This has been interpreted as a stochastic resonance (SR) effect, in which added light increases the noise intensity in the input signal to the photoreceptor, leading to an enhancement of the SNR (Pei et al. 1996).

RESULTS AND DISCUSSION

Recent theoretical studies suggest that an increase in synchronization measures, paralleling an increase in the SNR, should be observed in stochastic resonance effects (Neiman et al. 1998; Neiman et al. 1999). We have shown that light-induced increases in the SNR of the fundamental peak *correlate with an increase in stochastic phase*

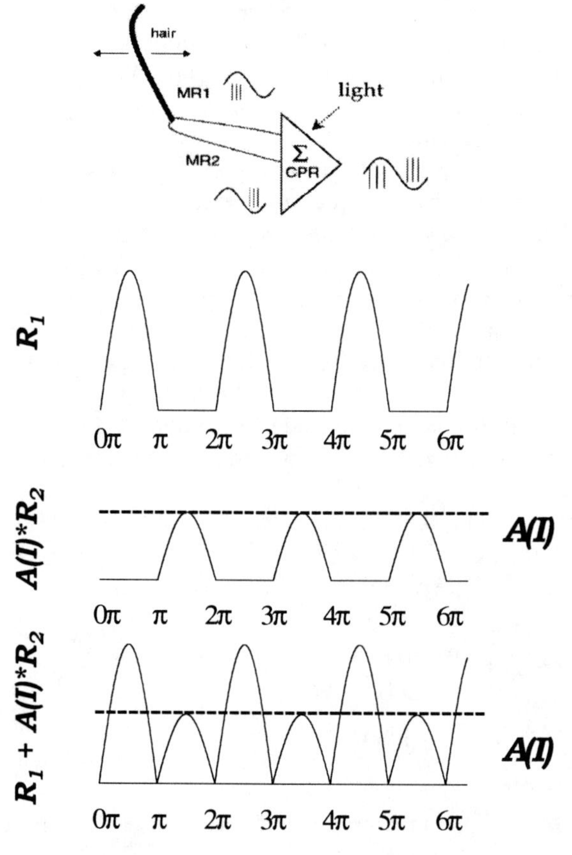

Figure 2 Top Panel: Possible mechanism of full-wave rectification by the CPR. Each mechanosensory hair is innervated by two mechanoreceptor (MR) neurons, whose directional sensitivities differ by 180 degrees (Wiese 1976; Wiese et al. 1976). Summation of the MR inputs at the CPR is enhanced by light, accomplishing full-wave rectification of the input signal and doubling the effective stimulus frequency. **Bottom Panels:** Short time mean firing rate of sensory afferents R_1 (top panel) and R_2 (middle panel). The afferents respond π out of phase. The response of R_1 is normalized to unit amplitude; the response of R_2 is scaled by a factor A(I), which is a function of the light intensity. The bottom panel shows the result of the hypothetical summation $R_1 + A(I)R_2$, in the CPR.

synchronization (Bahar et al. 2002), as predicted by Neiman et al. (1999). To the knowledge of the authors, this is the first demonstration of the correspondence between an SR-like effect and stochastic synchronization in a biological experiment. However, while our results are *consistent* with a stochastic resonance interpretation of Pei's result, an unsolved problem remains. The biochemical effects of light on the photoreceptors remain poorly understood (Kruszewska and Larimer, 1993), and we cannot yet say with any certainty whether light really does increase input "noise" to these neurons.

The mystery deepens with another novel effect of light on signal transduction in the crayfish photoreceptor-mechanoreceptor system. Here, we demonstrate a light-induced increase in the SNR of the *second higher harmonic* (SNR_2) of a periodic input signal, illustrated in Figure 1. Indeed, at higher light levels, *all* odd harmonics are typically suppressed, while the even ones (up to the 4th) are enhanced (see Figure 1b). We also find an increase in the 1:2 synchronization index (Pikovsky et al. 2001; Rosenblum et al. 2001) corresponding to the increase in SNR_2, illustrated in Figure 1c.

We propose that, to a first approximation, the "second harmonic effect" can be accounted for with a simple model based on previous observations of the dual innervation of each mechanosensory hair on the tailfan by two neurons (Wiese, 1976; Wiese et al. 1976). Each neuron responds (by an increased firing rate) to the opposite half of the sinusoidal hydrodynamic displacement cycle (Wiese, 1976, Wiese et al. 1976). At the top of Figure 2, we show a plausible diagrammatic scheme through which *directional rectification of the mechanical stimulus* by the two afferents, MR1 and MR2, *and subsequent light-intensity-mediated summation* in the CPR could account for our observations. The short-time mean firing rates of the two sensory afferents, R_1 and R_2, respond π out of phase with each other as diagrammed in the two middle panels of Fig. 2. Our hypothesis is that, within the CPR, these two responses are summed, and the strength of the summation, $A(I)$, is dependent on the light intensity, I. Here we have normalized the response R_1 of MR1 to unit amplitude. The response R_2 of MR2, after summation, has relative amplitude $A(I)$ with $0 \leq A \leq 1$. The result of the hypothetical summation in the CPR is shown by the response, $R_1 + A(I)R_2$, in the bottom panel of Figure 2.

We propose a simple mathematical model that is consistent with our observations and the rectification hypothesis. The two responses, R_1 and R_2, are given by the functions:

$$R_1(x) = \frac{1}{2}\left[\sin(x) + |\sin(x)|\right]$$
$$A(I)R_2(x) = \frac{A(I)}{2}\left[|\sin(x)| - \sin(x)\right] \quad , \qquad (1)$$

but it is more convenient to represent them, and their sum, as Fourier series. This gives us the three waveforms shown in Fig. 2:

$$R_1(x) = \frac{1}{\pi} + \frac{1}{2}\sin x - \frac{2}{\pi}\left[\frac{\cos(2x)}{3} + \frac{\cos(4x)}{15} + ...\right] \qquad (2)$$

$$AR_2(x) = \frac{A}{\pi} - \frac{A}{2}\sin(x) - \frac{2A}{\pi}\left[\frac{\cos(2x)}{3} + \frac{\cos 4x}{15} + ...\right] \qquad (3)$$

$$R_1(x) + AR_2(x) = \frac{1}{\pi}(1+A) + \frac{1}{2}(1-A)\sin(x) - \frac{2}{3\pi}(1+A)\cos(2x) - \frac{2}{15\pi}(1+A)\cos(4x) - ... \qquad (4)$$

where the frequency of the stimulus is given by $x = 2\pi f$.

The ratios of the peak amplitudes in the power spectrum are given by the square root of the ratio of the SNRs, assuming the background noise maintains an approximately constant level. For $A = 0$, the summed response predicts a power spectrum for which the fundamental peak has the largest amplitude (1/2) followed by decreasing amplitudes for the peaks of the higher harmonics. For $A \rightarrow 1$, the fundamental peak is suppressed; the 2nd harmonic grows in amplitude (4/3π) and is followed by only the even harmonics. In all cases, the 3rd harmonic is absent in the model. Using this model, we can extract the dependence of A on light intensity by comparing the calculations with the amplitudes of the peaks in the measured power spectra.

The SNRs from the experimental data can be interpreted with the corresponding ratios of the coefficients in Eq. (4), thus,

$$\sqrt{\frac{SNR_2}{SNR_1}} = \frac{4}{3\pi}\frac{(1+A_{2:1})}{(1-A_{2:1})} \tag{5}$$

where SNR_1 is the signal to noise ratio of the fundamental peak and SNR_2 is the signal to noise ratio of the second harmonic. Now, inverting to solve for $A_{2,1}$, we obtain

$$A_{2:1} = \frac{\sqrt{\frac{SNR_2}{SNR_1}} - \frac{4}{3\pi}}{\sqrt{\frac{SNR_2}{SNR_1}} + \frac{4}{3\pi}}. \tag{6}$$

Here $A_{2,1}$ denotes the determination of $A(I)$ from the experimental measurements of the ratio of the 2nd harmonic peak to the fundamental. One can obtain similar formulae for the other ratios, for example,

$$A_{4:1} = \frac{\sqrt{\frac{SNR_4}{SNR_1}} - \frac{4}{15\pi}}{\sqrt{\frac{SNR_4}{SNR_1}} + \frac{4}{15\pi}}. \tag{7}$$

Interestingly, the model makes no prediction for $A_{4:2}$, since the corresponding ratio, $\sqrt{\frac{SNR_4}{SNR_2}} = 1/5$, is a constant, and thus *independent of* $A_{4:2}$. As we show below (inset, Figure 3), the prediction of this light-independent, constant value can be used as a consistency check of our model. We expect $A(I)$ to follow a course similar those of SNR_2 and γ_{12} in Figure 1c. The top panel of Figure 3 shows our determinations of $A(I)$ from the measured power spectra using Eq. (6).

We emphasize that this model is *linear in the terms* $R_1(x)$ *and* $R_2(x)$, while the transduction processes that carry the hydrodynamic mechanical stimulus to a firing rate in the CPR are almost certainly nonlinear. For example, quadratic (or higher-order) nonlinearities, or a small offset in the π phase difference between the waveforms of Figure 3, could account for the appearance of the small 3rd harmonic peak at 22.5 Hz in Fig. 1a, though the 3rd and higher odd harmonics are absent from Eq. (4). Moreover, real data can produce a negative value for A (see error bars at lowest light level in the top panel of Figure 3), because it is experimentally possible to obtain $\sqrt{\frac{SNR_2}{SNR_1}} < \frac{4}{3\pi}$. Nevertheless, the linear model captures the gross features evident in our experimental observations.

An estimate of these inaccuracies (and a measure of the degree to which the model is not self-consistent) can be obtained by comparing the two values $A_{2:1}$, $A_{4:1}$ (calculated by applying Eqs. (6) and (7), respectively, to the same data sets) and checking the hypothetically

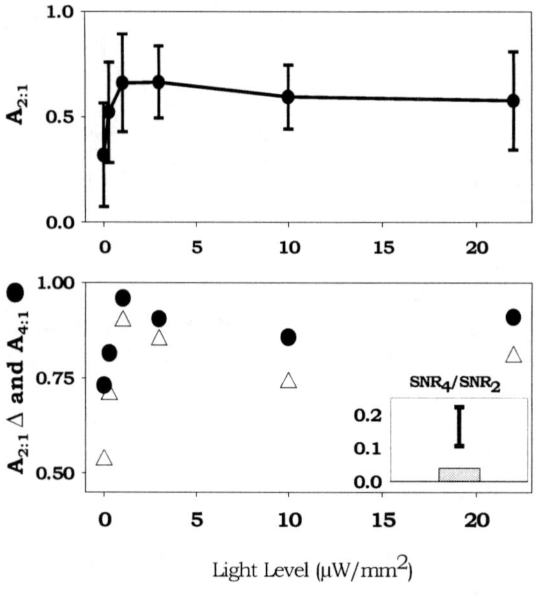

Figure 3 (a) Calculations of $A_{2:1}(I)$ from experimental data, using Eq. (6) (filled circles). N=3; error bars show standard deviation. Measurements were made in three experiments on two different crayfish, with stimulus amplitudes of 7 μm, 6 μm and 2 μm. Hydrodynamic stimuli were delivered at a frequency of 10 Hz. **(b)** Comparison of $A_{2:1}$ and $A_{4:1}$ calculated from a single data set at each light level (stimulation frequency 10 Hz, amplitude 6 μm, recording time 120 seconds). Open triangles show $A_{2:1}$, closed circles show $A_{4:1}$. Curve is drawn through $A_{2:1}$. **Inset** shows the ratio SNR_4/SNR_2, averaged over all six light levels. Error bar shows standard deviation. Our model predicts $SNR_4/SNR_2 \approx 1/25$.

constant value of $\sqrt{\frac{SNR_4}{SNR_2}} = 1/5$ (or, equivalently, $\frac{SNR_4}{SNR_2} = 1/25$) at each light level. Calculations of $A_{2:1}$ and $A_{4:1}$ agree well, as shown in the lower panel of Figure 3. However, our measured ratio of $\frac{SNR_4}{SNR_2}$ is significantly larger than 1/25 (inset, Fig. 3), indicating the model's limitations. The presence of a third harmonic in the experimental data (as for example, in Figure 3a), which is absent in our model, may contribute to this discrepancy. Note that even a small offset in the π phase difference between the waveforms in Eq. (1) could account for the appearance of odd harmonics. An additional complication arises from the fact that the real CPR, in contrast to our simplified cartoon version, receives inputs from ~70 hairs and therefore 70 afferent pairs!

CONCLUSIONS

Both full- and half-wave rectification have been identified in mammalian (e.g. Rowe and Palmer 1995; Chubb and Nam 2000) and invertebrate (e.g. Kondoh et al. 1996; Okuna and Kondoh 1996) nervous systems. *The crayfish system, however, appears to be the first identified neural system in which full-wave rectification of one type of sensory signal is accomplished by stimulation with a different type of sensory input.*

Speculations on the "use" of this effect by the crayfish in its daily routine remain open, though. Light-enhanced mechanical sensitivity may have evolved as a warning mechanism of periodic water motions caused by an oncoming predator when the crayfish is exposed outside its burrow (Pei et al. 1996). Rectifying this signal might relate to the sensitivity range of neurons in the higher nervous system upstream of the CPRs; a higher-frequency signal might be easier for some upstream neurons to extract from a 20-30 Hz spike train, while a lower frequency input might be more easily extractable from a spike train with a lower average frequency.

ACKNOWLEDGMENTS

This work was supported in part by the United States Office of Naval Research – Physics Division. In addition, SB was supported by a National Research Service Award from NIH (NINDS), and FM by the Alexander von Humboldt Foundation. We thank Alexander Neiman, Kevin Dolan, David Russell and Lon Wilkens for useful and stimulating discussions.

REFERENCES

1. Bahar S, Neiman A, Wilkens LA and Moss F. Synchronization and Stochastic Resonance Effects in the Crayfish Caudal Photoreceptor. *Phys Rev E (Rapid Commun).* **65**:050901(R), 2002.
2. Bainbridge R. The speed of swimming of fish as related to size and to the frequency and amplitude of the tail beat. *J Exp Biol* **35**:109-133, 1958.
3. Bleckmann H, Breithaupt T, Blickhan R and Tautz J. The time course and frequency content of hydrodynamic events caused by moving fish, frogs and crustaceans. *J Comp Physiol A* **168:**749-757, 1991.
4. Bruno MS and Kennedy D. Spectral sensitivity of photoreceptor neurons in the sixth ganglion of the crayfish. *Comp Biochem Physiol* **6**:41-46, 1962.
5. Chubb C and Nam JN. Variance of high contrast textures is sensed using negative half-wave rectification. *Vision Res* **40**(13):1677-1694, 2000.
6. Douglass JK and Wilkens LA. Directional selectivities of near-field filiform hair mechanoreceptors on the crayfish tailfan (Crustacea: Decapoda). *J Comp Physiol A* **183**:23-34, 1998.
7. Drucker EG and Jensen JS. Pectoral fin locomotion in the striped surfperch I. Kinematic effects of swimming speed and body size. *J Exp Biol* **199**:2235-2242, 1996a.

8. Drucker EG and Jensen JS. Pectoral fin locomotion in the striped surfperch II. Scaling swimming kinematics and performance at a gait transition. *J Exp Biol* **199**:2243-2252, 1996b.
9. Flood PM and Wilkens LA. Directional sensitivity in a crayfish mechanoreceptive interneurone: analysis by root ablation. *J Exp Biol* **77**:89-106, 1978.
10. Edwards DH. Crayfish extraretinal photoreception. I. Behavioral and motoneural responses to abdominal illumination. *J Exp Biol* **109**:291-306, 1984.
11. Ellerby DJ and Altringham JD. Spatial variation in fast muscle function of the rainbow trout Oncorhynchus mykiss during fast-starts and sprinting. *J Exp Biol* **204**:2239-2250, 2001.
12. Kennedy D. Physiology of photoreceptor neurons in the abdominal nerve cord of the crayfish. *J Gen Physiol* **46**:551-572, 1963.
13. Kondoh Y, Arima T, Okuma J, and Hasegawa Y. Response dynamics and directional properties of nonspiking local interneurons in the cockroach cercal system. *J Neurosci* **13**(6):2287-2305, 1993.
14. Kruszewska B and Larimer JL. Specific second messengers activate the caudal photoreceptor of the crayfish. *Brain Res* **618**:32-40, 1993.
15. Neiman A., Silchenko A, Anishchenko V, and Schimansky-Geier L. Stochastic resonance: noise-induced phase coherence. *Phys Rev E* **58**(6):7118-7125, 1998.
16. Neiman A, Schimansky-Geier L, Moss F, Shulgin B, and Collins JJ. Synchronization of noisy systems by stochastic signals. *Phys Rev E* **60**(1):284-292, 1999.
17. Okuma J, Kondoh Y. Neural circuitry underlying linear representation of wind information in a nonspiking local interneuron of the cockroach. J Comp Physiol [A]. 1996 Dec;179(6):725-40.
18. Pei X, Wilkens LA, and Moss F. Light enhances hydrodynamic signaling in the multimodal caudal photoreceptor interneurons of the crayfish. *J Neurophysiol* **76**(5):3002-3011, 1996.
19. Pikovsky A, Rosenblum M, and Kurths J. *Synchronization: A Universal Concept in Nonlinear Sciences*, Cambridge University Press, 2001.
20. Rosenblum MG, Pikovsky AS, Schäfer C, Tass P, and Kurths J. Phase Synchronization: From Theory to Data Analysis, In: Handbook of Biological Physics, Elsevier Science, Series Editor A.J. Hoff, Vol. 4, Neuro-informatics, Editors: F. Moss and S. Gielen, Chapter 9, pp. 279-321, 2001.
21. Rowe MH and Palmer LA. Spatio-temporal receptive-field structure of phasic W cells in the cat retina. *Vis Neurosci* **12**(1):117-139, 1995.
22. Simon TW and Edwards DH. Light-evoked walking in crayfish: behavioral and neuronal responses triggered by the caudal photoreceptor. *J Com Physiol A Sens Neural Behav Physiol* **166**:745-755, 1990.
23. Webb PW. Kinematics of plaice, Pleuronectes platessa, and cod, Gadus porhua, swimming near the bottom. *J Exp Biol* **205**(14):2125-2134, 2002.
24. Welsh JH. The caudal photoreceptor and responses of the crayfish to light. *J Cell Comp Physiol* **4**:379-388, 1934.
25. Wilkens LA. The crayfish caudal photoreceptor: advances and questions after the first half century. *Comp Biochem Physiol* **91**C(1):61-68, 1988.
26. Wilkens LA and Larimer JL. The CNS photoreceptor of the crayfish: morphology and synaptic activity. *J Comp Phyisol* **80**:389-407, 1972.
27. Wilkens LA and Douglass JK. A stimulus paradigm for analysis of near-field hydrodynamics sensitivity in crustaceans. *J Exp Biol* **189**:263-272, 1994.
28. Wiese K. Mechanoreceptors for near-field water displacements in crayfish. *J Neurophysiol* **39**(4):816-33, 1976.
29. Wiese K, Calabrese RL, and Kennedy D. Integration of directional mechanosensory input by crayfish interneurons. *J Neurophysiol* **39**(4):834-43, 1976.

NEURONS AND NEURONAL NETWORKS

Correlated Noise and Memory Effects in Neural Firing Statistics

J.W. Middleton, M.J. Chacron, B. Lindner and A. Longtin

Department of Physics, University of Ottawa, 150 Louis Pasteur, Ottawa, Ont. Canada K1N 6N5

Abstract. This paper discusses two problems at the forefront of neurobiology and of noise research. They arise from non-renewal firing processes in nerve cells, due to various forms of memory. The combination of short and long-term correlations between firing intervals has been shown to enhance information transfer, namely by causing a minimal variability in the spike count distribution at a specific counting time. The first problem concerns first passage time calculations in a model that combines these two forms of correlations. It is a two-dimensional leaky integrate-and-fire (LIF) model in which the threshold is also a dynamical variable. The second problem concerns the effect of long-range correlations on neuron firing statistics. We show new results on the interspike interval densities as well as the spike count Fano factor for the perfect integrate-and-fire (PIF) model forced by a slow (long-correlation time) Ornstein-Uhlenbeck process, which is a simplification of the previous model. These theoretical results are obtained using a quasi-static noise approximation. There remain, however, many exciting challenges in relating correlations with signal detection in neurobiological systems, some of which are highlighted in our paper.

INTRODUCTION

The long time statistics of neural output is important when studying topics such as signal detection and neural encoding. In order to assess how much information a neuron can transmit in response to certain stimuli we must know the variability of its output spike train under different conditions. Long range correlations are present in many natural stimuli such as natural images [1] as well as music [2], so it is of interest to study neuron models with implicit long range correlation.

It has also recently been shown [3] that the resting discharges of a certain class of electroreceptors have long range correlations. In particular, the ratio of the spike count variance to the spike count mean, also known as the Fano factor, goes through a minimum as a function of the duration of the counting window. This is useful from the point of view of signal detection, since this minimal Fano factor will produce a high signal-to-noise ratio for determining the presence of small amplitude low-frequency stimuli (such as Daphnia on which the weakly electric fish feed). The origin of this effect in a dynamical model of an electroreceptor has been explained using a combination of theory and numerics [4]. The effect relies on a combination of positive and negative correlations in firing time intervals. We have further found that correlations produce a counter-intuitive increase in mutual information for higher frequency signals. A general theory for this effect is presently lacking, but we are making progress, as we will show below.

In fact, there are few (but a growing number of) theoretical results on the statistics of

CP665, *Unsolved Problems of Noise and Fluctuations: UPoN 2002: Third International Conference,* edited by S. M. Bezrukov

correlated firing. Most formalisms, such as first passage time to threshold calculations, have been developed for renewal processes in which there are no correlations between successive first passage times. Also, there is little known about first passage time problems in non-Markovian systems, such as those involving delayed feedback (i.e. where the dynamical equations are formulated as stochastic delay-differential equations); this is true even for white noise forcing, i.e. regardless of whether there are correlations between successive passage times or not. In this paper we report recent progress on our analysis of correlated firing in simple neuron models without delays.

CORRELATED FIRING IN THE LIF MODEL

The model in which we have discovered that correlated firing can improve signal detection is the standard leaky integrate-and-fire model (LIF), with two sources of noise and a memory that persists beyond the firing of a spike [4, 5]. One noise is quasi-white and of moderate strength, and accounts for the noisy phase locking properties of the receptors' firings to the carrier wave emitted by these fish ($\approx$ 1000 Hz); it can be seen as a mixture of conductance noise and synaptic noise. The memory factor is accounted for by a dynamic threshold, which is raised every time the cell fires, and decays between firings. This dynamic threshold produces negative correlations between successive firing intervals, and is responsible for decreasing the Fano factor below its value of one in the Poisson limit. So what accounts for the minimum of the Fano factor? It has to be a mechanism that counteracts this decrease, i.e. which increases the Fano factor at longer counting times.

In our earlier work, we have found that an increase in the Fano factor can occur via the introduction of a small, slow Ornstein-Uhlenbeck process in the model; this is the second noise source in the model, which can be associated with e.g. slower fluctuations in vesicular release rates at synapses. For spike counting times less than the correlation time of this OU noise, the intervals are positively correlated on average. It is as though the noise appears as non-stationary over the counting time. Positive correlations increase the Fano factor. The combination of the decrease (due to memory) and increase (due to weak slow noise) sets the position of the minimum of the Fano factor, and thus the time scale of integration of information over which signal detection is optimal.

Such increases in Fano factors have been seen in various data from different neurons, and in statistical models of neural firing in the form of modulated Poisson processes. Such behavior had not been shown in dynamical models of excitability until our recent study [5]. As mentioned earlier, first passage time to threshold calculations usually assume the "renewal" property, which the interspike interval (ISI) data we are modeling clearly do not have. From the point of view of our model, the "non-renewability" is due to the slow OU process with respect to the counting window, and also to the memory effects carried by the threshold (this is a form of adaptation).

The threshold memory is also difficult to deal with, because the escape time process is then in a two-dimensional space. We have made some progress on calculating the mean first passage time for the case of a simple leaky integrate-and-fire model with Gaussian white noise forcing and dynamical threshold. The idea is to solve the problem self-consistently. The firings act as delta-function forcing on the first-order threshold dynamics. In the asymptotic state, the mean firing rate produces a mean threshold value. This value can be put in the expression for the mean first passage time of the leaky integrate and fire dynamics - this is basically the classic Ornstein-Uhlenbeck process, and one can use the standard formula for its mean first passage time(MFPT). Since this mean rate (the inverse of the MFPT) affects the threshold, and vice-versa, one can find the steady state value for both the mean rate and threshold. The result (not shown) agrees surprisingly well with numerical simulations for a range of integration time constants. This is so even though, in this description, the correlations between firing intervals are not taken into account. It is not clear however how to extend this formalism in a way that will allow the calculation of the correlations.

CORRELATIONS IN THE PIF MODEL

Model

The first passage time with slow OU forcing in the previous model is currently under investigation using a phase representation of the receptors' voltage. The receptor is forced by a 1000 Hz carrier wave; the system is designed to detect modulations of this carrier. Our model predicts that without noise, the receptor simply fires periodically at a high rate, with leakiness playing a secondary role. Thus, its noisy periodic firing dynamics can be approximated by such a phase (rotations of 2π count as firings). The model then amounts to the perfect integrate and fire (PIF) neuron forced by the slow OU process. We will see that many features of the Fano factor and other firing statistics can be understood from this simple model with only one noise source (the slow one) and no dynamical threshold.

The type of input noise we are using is an Ornstein-Uhlenbeck (OU) process; the statistics we are calculating are the interspike interval probability density, its first and second moments, and the Fano-factor which is a measure of spike count variability.

The dynamical equations defining our system are

$$\begin{aligned} \frac{dv(t)}{dt} &= \mu + \eta(t) \\ \frac{d\eta(t)}{dt} &= -\frac{\eta(t)}{\tau} + \sqrt{\frac{D}{\tau}}\xi(t) \end{aligned} \tag{1}$$

where $v(t)$ is the membrane voltage, μ is the average rate of increase of the voltage, and τ and D are, respectively, the time constant and variance of the OU process. $\xi(t)$ is a Gaussian white noise process with zero mean and unity variance.

The voltage increases due to the mean rate μ and the OU process and is reset to zero when it reaches a threshold level, v_{th}; the noise is not reset. In this model the times at which the voltage is reset represent the spike times of the neuron, or the times at which an action potential is initiated.

Quasi-Static Approximation

Since we are looking at the effect of long time correlated noise, we will use a quasi-static approximation of the noise. If τ is known to be much larger than the average ISI, then on that time scale we can approximate η as being constant. In this way we can associate each ISI with a corresponding, unique value of the OU process

$$I_k = \frac{v_{th}}{\mu + \eta_k} \tag{2}$$

where k denotes the index in a sequence of ISIs.

This approximation not only allows us to write down a conditional probability density function (PDF) between I and η, but it also allows us to reduce this conditional PDF to a delta function due to the unique one-to-one correspondence.

$$P(I_k|\eta_k) = \delta\left(I_k - \frac{v_{th}}{\mu + \eta_k}\right) \tag{3}$$

Here the process η_k is a sequence of the values of the OU at times when the voltage process reaches threshold. As will be shown the statistics between $\eta(t)$ and η_k differ slightly due to biased sampling.

Stationary Probability Density Function

The first quantity of interest is the stationary PDF of ISIs. In order to obtain the stationary ISI PDF we can average the conditional PDF between I and η over all values of η

$$P_{st}(I) = \int_{-\infty}^{\infty} d\eta P(I|\eta_k) P_{st}(\eta_k) \tag{4}$$

The statistics of the values η_k is not the same as for the continuous OU process. Imagine we measure the noise value at the beginning of each interval of a long spike train. Then a positive value leads to a shorter interval and hence to more intervals than a negative value of η. This problem is known as biased sampling of a stochastic variable [6] and is resolved by a corrective factor given by the inverse interspike interval (see also [7].) Normalization yields

$$P_{st}(\eta_k) = \frac{e^{-\eta_k^2/2D}}{\sqrt{2\pi D}} \left(1 + \frac{\eta_k}{\mu}\right). \tag{5}$$

The normalization as well as any integration in the remainder of the paper is performed with respect ot the full range of noise values (including $\eta < -\mu$) for simplicity. Inserting

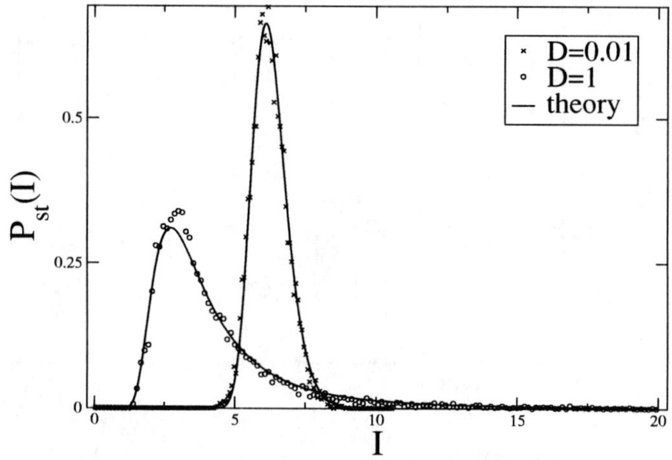

FIGURE 1. Stationary ISI probability density functions. Numerical simulations results for fixed τ and different values of variance, D. The solid lines show the theoretical densities.

equation (5) into (4).yields the PDF for the interspike interval density.

$$
\begin{aligned}
P_{st}(I_k) &= \int_{-\infty}^{\infty} d\eta_k\, \delta\left(I_k - \frac{v_{th}}{\mu+\eta_k}\right) \frac{e^{-\eta_k^2/2D}}{\sqrt{2\pi D}} \left(1+\frac{\eta_k}{\mu}\right) \\
&= \int_{-\infty}^{\infty} d\eta_k\, \delta\left(\eta_k - \frac{v_{th}}{I_k} + \mu\right) \frac{e^{-\eta_k^2/2D}}{\sqrt{2\pi D} v_{th}\mu} (\mu+\eta_k)^3 \\
&= \frac{v_{th}^2}{\sqrt{2\pi D}\mu} \frac{e^{-(\frac{v_{th}}{I_k}-\mu)^2/2D}}{I_k^3}
\end{aligned} \tag{6}
$$

Figure 1 shows the stationary PDF for fixed τ and several values of D from both numerical simulation and the corresponding theoretical curves (in black) using (6). The mean of the sampled OU process is given by

$$
\langle \eta_k \rangle = \int_{-\infty}^{\infty} d\eta_k \frac{e^{-\eta_k^2/2D}}{\sqrt{2\pi D}} \left(\eta_k + \frac{\eta_k^2}{\mu}\right) = \frac{D}{\mu}. \tag{7}
$$

Fano Factor

The Fano factor which is the variance to mean ratio of a counting process, $n(t)$, for a given counting window is useful for determining on which time scales the process is most regular. A simple expression for the Fano-factor for our model neuron can

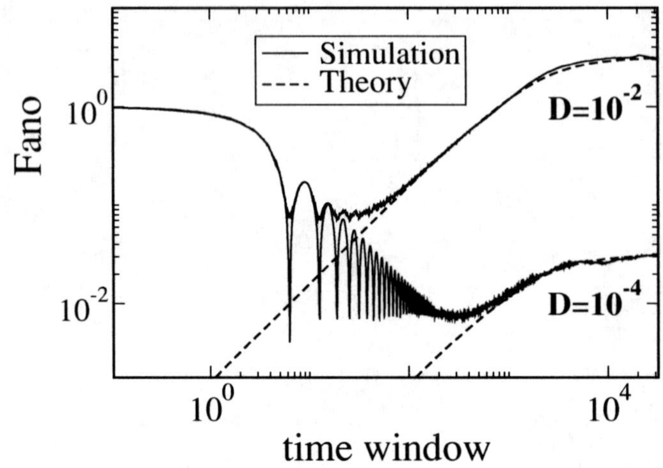

FIGURE 2. The numerical results (symbols) of the Fano-factor for different noise intensities with $\tau = 1000$. The dashed lines are the theoretical curves which are valid in the large counting time limit.

be obtained considering that on a scale much larger than v_{th} fluctuations in the actual membrane voltage, $v(t)$ and fluctuations in the discrete process $n(t)$ times v_{th} become indistinguishable. This is valid provided that there is some nonnegative bias on the dynamics of the voltage. With this in mind we can see that the dynamical equations of our system can be mapped to those of biased Brownian motion. The Fano factor at time t is then equivalent to the variance of the corresponding Brownian motion [8] divided by the mean drift of the Brownian motion

$$F(t) = \frac{2D\tau}{v_{th}\mu}\left(1 - \frac{\tau}{t}\left(1 - e^{-t/\tau}\right)\right) \tag{8}$$

The Fano factor can also be obtained with the SCC and using the relation [6]

$$Var(I_k) = Var(I_1)\left[k + 2\sum_{l=1}^{k-1}(k-l)\rho_l\right] \tag{9}$$

and the equality

$$\lim_{t\to\infty} F(t) = \lim_{k\to\infty} \frac{Var(I_k)}{\langle I\rangle^2 k} \tag{10}$$

This relation holds very well even when the limits are taken to finite values. The approximate equality holds for the times $t = \langle I\rangle k$.

Figure 2 shows $F(t)$ for different variances of numerically generated OU processes with $\tau = 1000$ (solid line). Note that the various simulation values converge with the theoretical Fano curves (8) (dashed lines) at different times depending on the noise variance.

CONCLUSIONS

In this paper we have discussed outstanding unsolved problems on correlated firing in neurons. We have obtained analytic expressions for certain statistical measures for the perfect integrate-and-fire model with additive correlated noise. This model can be seen as a limiting case of the standard leaky integrate-and-fire neuron with vanishing leakiness, or with a strong drive to threshold (i.e. when it is periodically firing at a moderate rate even in the absence of stochastic input). We have derived approximate expressions for the stationary ISI probability density function and for the mid-to-long time behavior of the Fano-factor of the spike count over a given time period.

One interesting consequence from the numerical simulations shown in Figure 2 is that, for a range of noise variances, there is an apparent minimum in the Fano factor. Previously it was thought that a minimum in the Fano factor was only possible in a leaky integrate and fire neuron (LIF) with dynamic threshold [4, 5]. This minimum sets a time scale for the optimal discriminability of two distinct stimuli. This implies that the perfect integrate and fire neuron could also make a good signal detector for a selected range of noise variances if the counting time is chosen to correspond to a minimum of the Fano factor. Short-lived negative correlations will further influence the position of the minimum and are likely to deepen the minimum, leading to higher detectability.

Future work will consider expressions for the full Fano factor curve which will allow us to relate the minimum to specific biophysical factors. We will also consider problems in which there are dynamical memory effects in the form of delayed feedback, in which the spikes propagate back to the cell after a delay time and act e.g. as a simple input current. The presence of this delay clearly renders the whole problem non-Markovian. We have made some progress on understanding escape times in potentials with delayed dynamics [9] for small delays. For larger delays, a new formalism is needed. Here again we can hope that a self-consistent approach can work to some extent, but that might neglect the effects that the delay can induce, such as oscillations in the firing rate. We have made some recent progress here by assuming that the "neuron" is a simple threshold crossing detector (forthcoming work from A. Longtin and R. Morse). It is our hope that such simplified dynamical models with memory will yield some insight into the dynamics of real neural loops with delays. Such circuits are ubiquitous in neurobiology, in particular in the pyramidal cell circuits of the weakly electric fish which receive input from the aforementioned electroreceptors.

ACKNOWLEDGMENTS

This work was supported by NSERC Canada and an Ontario PREA award to AL.

REFERENCES

1. Simoncelli, E. P., and Olshausen, B. A., *Annu. Rev. Neurosci.*, **24**, 1193 (2001).
2. R. F. Voss, J. C., *J. Acoust. Soc. Am.*, **63**, 258 (1978).
3. R. Ratnam, M. E. N., *J. Neurosci*, **20**, 6672 (2000).

4. M. J. Chacron, A. L., and Maler, L., *J. Neurosci.*, **21**, 5328 (2001).
5. M. J. Chacron, M. S.-H., A. Longtin, and Maler, L., *Phys. Rev. Lett*, **85**, 1576 (2000).
6. Cox, D. R., and Lewis, P. A. W., *The statistical analysis of series of events*, Methun, London, 1996.
7. Racicot, D. M., and A., L., *Physica D*, **104**, 184 (1997).
8. Risken, H., *The Fokker-Planck Equation*, Springer, Berlin, 1996.
9. S. Guillouzic, I. L., and Longtin, A., *Phys. Rev. E*, **61**, 4906 (2000).

Noise-induced Synchronized Oscillations in an Inhibitory Neural Network

Jorge V. José* and Paul Tiesinga†

*Center for the Interdisciplinary Research on Complex Systems and Physics Department, Northeastern University, Boston MA
†Department of Physics and Astronomy, The University of North Carolina at Chapel Hill, North Carolina

Abstract. We study the properties of synchronized states in a neuronal network model that represents circuitry of the thalamus or the locust antennal lobe. We find noise-driven and noise-sustained synchronized neural cluster states. We determine the theoretical maximum information output of these networks in terms of the Shannon entropy. Synchronized states are usually characterized by more regular spike trains and reduced information output. We find, however, that even a single neuron can still transmit a significant amount of information. The information content is contained in the changing participation of neurons in neural assemblies. Similar neural assemblies were recently observed in the locust olfactory system.

INTRODUCTION

The brain receives an enormous amount of information transduced by peripheral sense organs. This massive informational influx is coded and decoded in ways that are not yet fully understood in cognitive neuroscience. Over the past decade it has become clear that the synchronization between neurons is an important component of the neural code [17, 3, 13]. The neural mechanisms that underlie synchronization in different parts of the brain are only partly understood [22]. Recent experiments have shown that inhibitory interneurons in the hippocampus [25] and the thalamic reticular nucleus [18] can indeed synchronize neuronal discharges. Subsequent theoretical analysis of networks of interneurons has shown, however, that strong synchronization by mutual inhibition is only moderately robust against neuronal heterogeneities [23] and synaptic noise [21]. In strong synchronization all the neurons fire with a short time-interval of each other.

In most experiments to date one measures the activity of one neuron, or a small population of neurons. Periodic oscillations measured in these experiments are consistent with strong as well as weak synchronization. In weak synchronization the *average* neuronal activity is periodic, without each individual neuron having to fire at each period. Often theoretical analyses, however, have focussed on strong synchronization. Here we conjecture that weak synchronization is robust against neuronal heterogeneities and synaptic noise, and consequently it is much more likely to occur in neurobiological systems. We present numerical results of weak synchronization in a simple model network that supports our conjecture. This type of network is quite common in neurobiology and our results are relevant to the circuitry in thalamus [5] and the locust antennal lobe [2].

CP665, *Unsolved Problems of Noise and Fluctuations: UPoN 2002: Third International Conference,*
edited by S. M. Bezrukov

The hallmark of weak synchronization is multimodal interspike interval (ISI) histograms (ISIH). The ISI occur only near multiples of a particular time-scale, e.g. the period of the population activity T (50-100 ms). Multimodality of the ISIH has been observed in the LGN [4], and was attributed to the action of inhibitory neurons. Multimodal ISIH have also been found in model simulations of the thalamus in the presence of noise [5] and in systems exhibiting stochastic resonance (SR) in response to a periodic drive. In a recent paper, a theoretical mechanism for autonomous stochastic resonance (ASR) was proposed, where the periodic drive was replaced by a periodic mode in an internal kinetic variable. As a result the spikes ride on top of subthreshold voltage oscillations. Similarly, it was shown that noise induced coherent activity in an excitable element [10, 7]. In our work the periodic neuronal activity in the noise-driven system is self-induced by the network. This synchronization is absent in unconnected single elements, or in a single element with autosynaptic feedback [20]. The noise-induced oscillations presented here resemble those obtained from array-enhanced coherence resonance [9, 6, 24, 15].

It has been suggested that the brain may encode information through an ensemble or cluster of neurons that fire within a short time of each other [1]. A particular neuron may be part of a cluster – neural assembly – for a few cycles, before it joins another cluster. This type of dynamics is very similar to the neuronal clusters that form in our model simulations. An important problem is to quantify the information content of these binding-like cluster states. The Shannon entropy has been used as a measure of information content in investigations of sensory neurons in, for instance, crickets, and flies (see Ref. [11]). It is, nonetheless, not known how the brain processes information, and thus it is not clear whether the Shannon entropy is the correct quantity for this purpose. It does, however, provide an upper bound on a theoretical information content for spiking neurons. It does imply that variability in the discharge of neurons contains information, and that "noisy" neurons are transmitting more information compared to regular noiseless spiking neurons. We emphasize that this statement is still controversial (see Ref. [14] and references therein), for the fact that the entropy measure yields consistent results in sensory systems does not guarantee its relevance to the central nervous systems. With these caveats in mind we still proceed to characterize the information content of our neural networks calculations using the well-defined Shannon entropy.

One would like to calculate both the output entropy of the system and the mutual information. The mutual information quantifies how the ensemble of outputs is related to one of the possible realizations of the input, and it involves additional averagings over a conditional probability distribution and is therefore very hard to calculate. Even the simpler calculation of the Shannon entropy from its definition in terms of the spike times is a difficult calculation. Here we shall focus on the Shannon entropy of the neuronal output of a neuron as part of the complete network. We also use some approximations that allow us to estimate the Shannon entropy using the interspike interval time series.

METHODS

Network model

Our single neuron model contains a low threshold Calcium current I_{Ca}, a general leak current I_L, a synaptic current I_{syn}, and a noise current $C_m\xi$, with equations

$$C_m \frac{dV}{dt} = -I_{Ca} - I_L - I_{syn} - C_m\xi \qquad (1)$$

together with the first order (Hodgkin-Huxley type) kinetic equations for the activation m and inactivation h variables for I_{Ca} and the synaptic variable s. We numerically integrate the resulting neuronal dynamics in terms of four variables, V, m, h, and s. We have used the kinetics for I_{Ca} and I_{syn} as specified in Ref. [12]. Our single neuron model captures some of the important features of the dynamics of thalamic neurons, in particular post inhibitory rebound (PIR).

The neurons in our network are connected all-to-all by inhibitory GABAergic synapses. We also have included two types of noises in our model, either Gaussian current noise, $\langle \xi(t)\xi(0)\rangle = 2D\delta(t)$, or Poisson distributed excitatory and inhibitory post-synaptic potentials. In previous work we have shown that these two types of noises are not fully equivalent [19]. The results presented here are obtained with Gaussian noise for a $N = 1000$ network. The total synaptic conductance is $g_s = 2$ mS/cm^2, and the decay time of the synaptic channel $\tau_s = 16$ ms. The noise strength D is expressed in units of mV^2/ms, time in ms, and voltage in mV. The detailed description of our model and its detailed implementation is presented in Ref.[20].

Calculated quantities

The raw model output are the time-traces for V_i, m_i, h_i, and s_i. Spike times of the ith neuron are determined as the time that V_i crosses 0 mV from below. The interspike interval is the difference between two consecutive spike times of the same neuron. We determined the histograms of interspike intervals. The instantaneous firing rate, or frequency f, is defined as the number of action potentials per second in a bin of 2 ms. Both the ISI histogram and f are averaged over all neurons in the network. The current drive due to the synaptic connections with other neurons is proportional to $v_{syn} = \sum_i s_i$. We determined both the instantaneous as well as the time-averaged distribution of h-values at times when v_{syn} crossed 0.01 from above.

Calculation of the Shannon entropy

Here we calculate the one neuron entropy in the network, which is expressed in terms of the distribution of ISIs, as $S \approx -\langle N\rangle \int d\tau P_{ISIH}(\tau)\log_2 P_{ISIH}(\tau)$, with $\langle N\rangle$ the average number of events, equal to the product of the firing rate, f, and the length of the time-interval, T. We find that the ISIH obtained from our simulations is, to a good approximation, a sum of Gaussians (SOG), $P_{ISIH}(\tau) = \sum_i c_i G(\tau|\mu_i, \sigma_i)$. Here $G(\tau|\mu_i, \sigma_i)$ is a Gaussian with average μ_i, standard deviation σ_i, and with c_i the relative weight for the i-th contribution. We find that there are two contributions to the entropy of such a multimodal ISIH. A contribution S_1 due to the jitter around the average ISI value of a given state i. And a contribution S_2 due to the discrete choices of the average ISI. Using the theory of Markov chains to evaluate the Shannon entropy of the spike trains, we find that the ISI correlations add a third term, S_3, decreasing the entropy per spike. The details of these calculations are presented elsewhere [20].

RESULTS

The model considered here contains an inward low threshold Calcium current, I_T, that initiates the Calcium action potentials. It is inactivated at the resting membrane potential (-65.57 mV), and it is de-inactivated at hyperpolarized voltages. For the neuron to be excitable, h has to be de-inactivated (i.e. $h > 0.305$). The necessary hyperpolarization is supplied by the inhibitory postsynaptic potential (IPSP) generated by the activity of other neurons in the network. The strength of the IPSPs is determined by the value of the synaptic conductance g_s and the decay time τ_s. The critical value for periodic oscillations is defined by $\tau_s = \tau_c(g_s, N)$, and depends on the number of neurons in the network. For $\tau_s < \tau_c$ the oscillation dies out after a finite number of action potentials. Sufficiently strong noise can induce and maintain a spiking network state for $\tau_s < \tau_c$. Here we have to distinguish between single elements and the network. As we will see, the noise-induced network oscillations yield a periodic population rhythm. The noise-induced dynamics of a single neuron with autosynaptic feedback will not yield a periodic spike-trace. Instead, the ISI distribution has a peak for short times due to ISI within spike trains, and an exponential type of distribution for the intervals between the end of one, and the start of another spike train. This makes the dynamics different from Stochastic Resonance, since in that case one would obtain a multimodal distribution, with peaks at multiples of the driving frequency.

The network settles in a state where groups of neurons fire simultaneously. It is easy to understand why such cluster states emerge. Starting from random initial conditions each neuron will have a different phase, and thus a different time to reach the AP threshold. The first neurons to fire will cause an inhibition that blocks neurons further from threshold from firing. They can fire after the decay of the inhibition (a few τ_s). The waves of inhibition produced by de-inactivated neurons that fired before, can sustain a periodic oscillation in the network.

We studied the dynamics of the cluster states in a network with $N = 1000$, using the distribution of h values shown in Fig. 1. We focused on four values for $D = 0$, 0.0024, 0.008, and 0.8. An important variable in our analysis is h, the inactivation variable of I_T. The model neuron is only excitable when h is large enough. The distribution of h values in the network will tell us which neurons are excitable, and which ones need to be de-inactivated by further inhibition cycles. The regularity of the network dynamics is further reflected in the periodicity of f and v_{syn} and their autocorrelation function (not shown). Note that v_{syn} itself acts as a drive on the neurons. The larger the distance

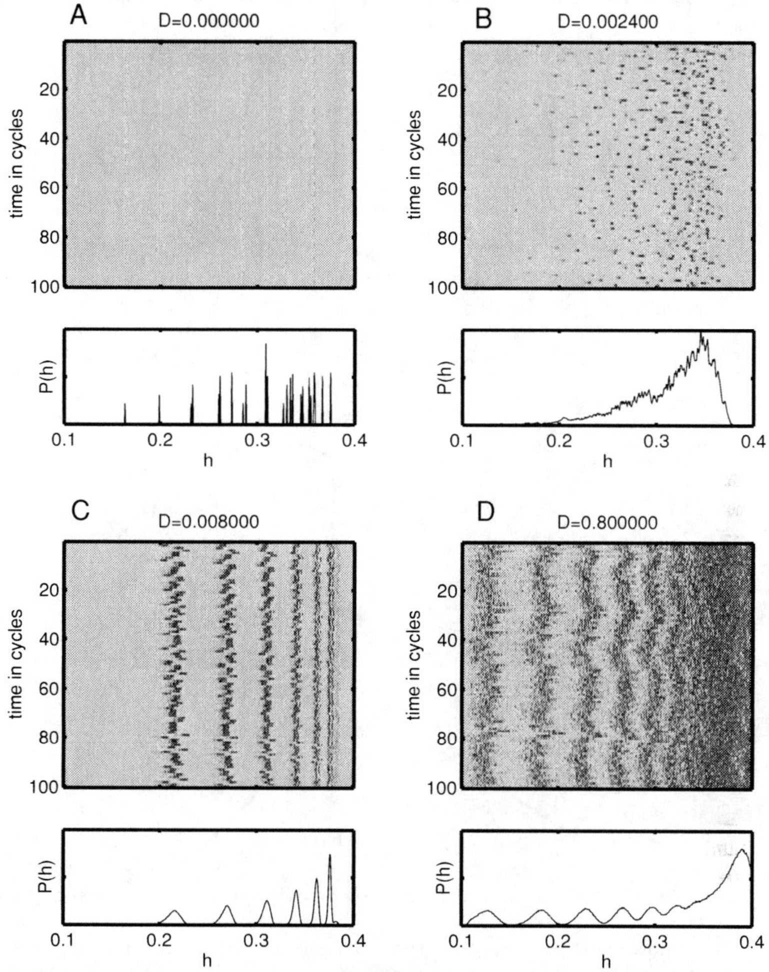

FIGURE 1. Each panel consists of a color coded instantaneous h value distribution for consecutive cycles (top), and the time-averaged h distribution (over at least 500 cycles) (bottom). From left to right, top to bottom the noise-values are (A) $D = 0$, (B) 0.024, (C) 0.08, (D) 0.8. In the upper left panel the delta function peaks in the h-distribution have been broadened to enhance visibility.

between the peak and trough, the stronger the synchronizing force.

We can identify four different types of dynamics. For zero noise (D=0) the system is in a state with five clusters of unequal size. The distribution of cluster sizes is determined by the initial condition. At each instance a neuron can only have one of five h-values. This set of h-values goes through a modulation with a period of five cycles. The derived quantities V, h, v_{syn}, and f go through the same modulation. Immediately after firing, the neuron is de-inactivated partially with each subsequent cycle until it is excitable again. The neuron then has to wait its turn to become disinhibited and fire before other

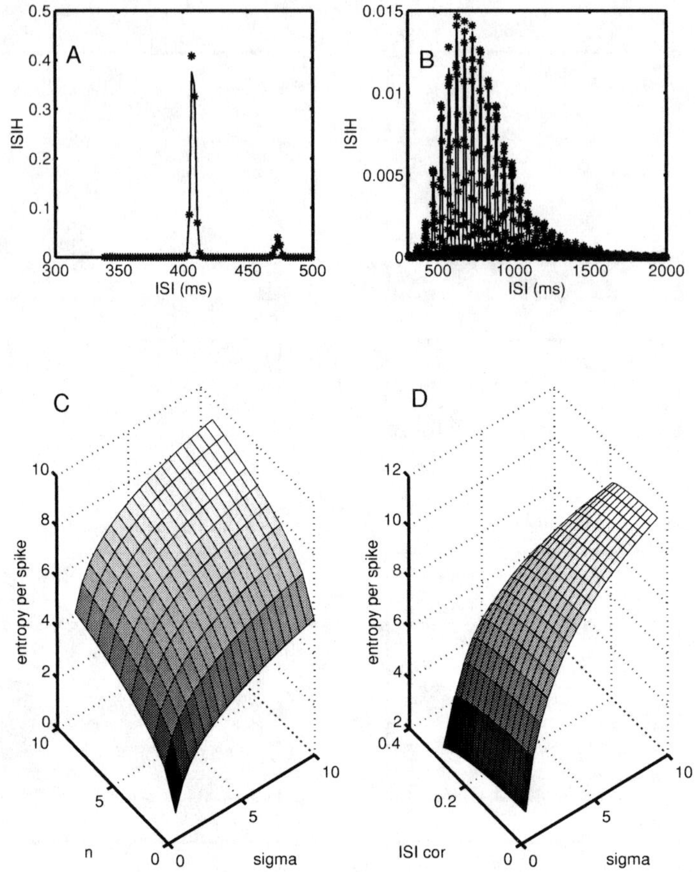

FIGURE 2. (top) We plot the ISIH ($*$) and the corresponding SOG (continuous line) for (a) $D = 0.008$ and (b) $D = 0.8$, both for $N = 1000$, and $\tau_s = 16$. (bottom) We evaluated the entropy per spike (c) as a function of the number of states n and (d) the jitter sigma and also as a function of the amount of correlation between consecutive ISI and sigma (right).

clusters do. When there is noise, there is dispersion in the spike firing times. In the absence of time delay there is only a short interval during which neurons can fire before the inhibition blocks all other firing for one cycle. As a result, clusters lose neurons whose AP has been noise delayed, and other clusters gain those neurons as members. In addition, noise can cause neurons to fire before the rest in their cluster.

Weak noise (D=0.0024) disorders the system. The network starts out with unequal cluster sizes, and large clusters lose more neuron members than smaller ones. Weak noise is not strong enough to equalize their numbers on the time-scale considered (5×10^4 ms). Instead the cluster sizes start to vary in a stochastic fashion, leading to an erratic firing rate and a fluctuating period.

For stronger noise (D=0.008) the average cluster size is stationary. The *h*-values that the neurons of different clusters cycle through are, on average, the same. As a result there are six smooth peaks in the h-histogram. The peaks become sharper for more de-inactivated values, and the h-distribution is stationary, that is on the average the same for each cycle. The actual neurons that fire in each cluster changes with time. This state is stable up to a noise strength of approximately D=1 (for N=1000).

The cycle-to-cycle fluctuations in cluster sizes increases with increasing D. Therefore the inhibition each cluster receives (proportional to v_{syn}) varies, and as a result the width of the peaks in the h histogram increases. The stability of a cluster is defined as the fraction of neurons that are still part of the cluster the next time it fires. This is related to the number and height of the peaks in the ISIH (see 2), and it decreases with D. Finally, for larger noise strengths, $D > 1$, the firing is no longer organized in clusters.

In the top graphs of Fig. 2 we have plotted the ISIH for $D = 0.008$ and $D = 0.8$ obtained from our simulations. In the same figure we plot the corresponding sum of Gaussians (SOG) (see Methods). We also have determined the theoretical value of the Shannon entropy of the SOG, for a given number n of Gaussians in the ISIH, and a width σ (`sigma`). One can increase the entropy by either introducing more jitter σ, or increase the number of discrete states. In 2(d) we determine the effect of correlation (`ISI cor`) between consecutive ISIs on the entropy. As expected, the entropy decreases with increasing correlations.

CONCLUSIONS

A significant amount of interest and controversy has been devoted to try to understand the nature of the information encoding in the brain. At the root of the controversy lies the inconsistency between the regular spiking neurons associated with synchronization in connection to binding, and the high information content of noisy neurons. Previous work indicates that strong synchronization is quenched by high noise levels [21]. Here we find that synaptic noise drives the network to, and maintains it in, a state of weak synchronization. This state is robust against heterogeneities and partial connectivity, and with a Shannon entropy that can attain high values. This means that ordered population rhythms are not inconsistent with high information content. An important issue that needs to be addressed in the future is the calculation of mutual information in the cluster state and its extension to include spatial connectivity. Autonomous stochastic resonance has also recently been found in a single neuron model[8]. Here we found an autonomous or self-induced type of intrinsic stochastic resonance generated by a neural network, that is absent in the individual unconnected neurons of a network.

Clusters of neurons simultaneously discharge, but the participating neurons in a given cluster changes with time. This type of collective state is often associated with a binding-type phenomena [1] or odor-activated neural assemblies in the locust antennal lobe [2]. The instantaneous firing rate may have a well determined period, but the underlying states of the system may be qualitatively very different. The number of clusters and their stability changes with the noise strength and the system size. Individual neurons have a varying information coding capacity as measured by the Shannon entropy of the ISIH. We have found that increasing the noise, increases the encoding capacity while maintaining a periodic population activity.

The cluster states can be quantitatively described by the average cycle length (period), and the periodicity (number of cycles) of the response. We have systematically studied these two quantities as a function of D for different values of τ_s and N [20].

ACKNOWLEDGMENTS

This work has been partially supported by the Center for Interdisciplinary Research on Complex systems (CIRCS) at Northeastern University, the Sloan-Swartz Center for Theoretical Neurobiology at the Salk Institute and the University of North Carolina at Chapel Hill.

REFERENCES

1. BARINAGA M, "Listening in on the Brain", *Science* **280** (1998), 376–378.
2. BAZHENOV M,STOPHER M,RABINOVICH M, HUERTA R, ABABARBANEL HD, SEJNOWSKI TJ, and LAURENT G, "Model of transient oscillatory synchronization in the locust antennal lobe", *Neuron* **30** (2001), 553–567.
3. FRIES P, REYNOLDS JH, RORIE AE,and DESIMONE R, "Modulation of oscillatory neuronal synchronization by selective visual attention", *Science* **291** (2001), 1560–1564.
4. FUNKE K, NELLE E, LI B, and WÖRGÖTTER F, "Corticofugal feedback improves the timing of retino-geniculate signal transmission", *Neuroreport* **7** (1996), 2130–2134.
5. GOLOMB D, and RINZEL J, "Clustering in globally coupled inhibitory neurons", *Physica D* **72** (1994), 259–282.
6. HU B, and ZHOU C, "Phase synchronization in coupled nonidentical excitable systems and array-enhanced coherence resonance", *Phys. Rev. E* **61** (2000), 1001–1004.
7. LEE SG, NEIMAN A, and KIM S, "Coherence Resonance in a Hodgkin-Huxley neuron", *Phys. Rev. E* **57** (1998), 3292–3297.
8. LONGTIN A, "Autonomous stochastic resonance in bursting neurons", *Phys. Rev. E* **55** (1997), 868–876.
9. PHAM J, PAKDAMAN K, and VIBERT JF, "Noise-induced coherent oscillations in randomly connected neural networks", *Phys. Rev. E* **58** (1998), 3610–3622.
10. PIKOVSKY A, and KURTHS J, "Coherence Resonance in a Noise-Driven Excitable System", *Phys. Rev. Letts.* **78** (1997), 775-778.
11. RIEKE F, WARLAND D, DE RUYTER VAN STEVENINCK RR, and BIALEK W, *Spikes: exploring the neural code*, MIT press, Cambridge (1997).
12. RINZEL J, TERMAN D, WANG XJ, and ERMENTROUT B, "Propagating Activity Patterns in Large-Scale Inhibitory Neuronal Networks", *Science* **279** (1998), 1351–1355.
13. SALINAS E, and SEJNOWSKI TJ, "Correlated neuronal activity and the flow of neural information", *Nat Rev Neurosci* **2** (2001), 539–550.
14. SHADLEN MN, and NEWSOME WT, "The variable discharge of cortical neurons: implications for connectivity, computation, and information coding", *J. Neurosci.* **18** (1998), 3870–3896.
15. SHINOHARA Y, KANAMARU T, SUZUKI H, HORITA T, and AIHARA K, "Array-enhanced coherence resonance and forced dynamics in coupled FitzHugh-Nagumo neurons with noise", *Phys. Rev. E* **65** (2002), 051906.
16. SINGER W, and GRAY CM, "Visual feature integration and the temporal correlation hypothesis", *Annu. Rev. Neurosci.* (1995), 555–586.
17. STEINMETZ PN, ROY A, FITZGERALD PJ, HSIAOI SS, JOHNSON KO, and NIEBUR E, "Attention modulates synchronized neuronal firing in primate somatosensory cortex", *Nature* **404** (2000), 187–190.

18. STERIADE M, CORMICK DA, and SEJNOWSKI TJ, "Thalamocortical oscillations in the sleeping and aroused brain", *Science* **262** (1993), 679–685.
19. TIESINGA PHE, and JOSÉ, JV, "Spiking Statistics in Noisy Hippocampal Interneurons.", *Neurocomputing* **26-27** (1999), 299–304.
20. TIESINGA PHE, and JOSÉ, JV, "Synchronous clusters in a noisy inhibitory neural network", *J. Comput. Neurosci.* (2000), 49–65.
21. TIESINGA PHE, and JOSE, JV, "Robust gamma oscillations in networks of inhibitory hippocampal interneurons", *Network* **11** *(2000), 1–23.*
22. TIESINGA PHE, FELLOUS JM, JOSE JV, and SEJNOWSKI TJ, *"Computational model of carbachol-induced delta, theta, and gamma oscillations in the hippocampus"*, Hippocampus **11** *(2001), 251–274.*
23. WANG XJ, and BUZSÁKI G, *"Gamma oscillation by synaptic inhibition in a hippocampal interneuronal network model"*, J. Neurosci. **16** *(1996), 6402–6413.*
24. WANG Y, CHIK D, and WANG ZD, *"Coherence resonance and noise-induced synchronization in globally coupled Hodgkin-Huxley neurons"*, Phys. Rev. E **61** *(2000), 740–746.*
25. WHITTINGTON MA, TRAUB RD, and JEFFREYS JGR, *"Synchronized oscillations in interneuron networks driven by metabotropic glutamate receptor activation"*, Nature **373** *(1995), 612–615.*

Spatiotemporal Organization of Stochastic Synchronization

P. Jung

Department of Physics and Astronomy and Quantitative Biology Institute, Ohio University, Athens, OH 45701

Abstract. We consider a two dimensional array of coupled excitable integrate-and-fire (IF) neurons in the presence of noise and sub-threshold periodic stimulus. Although the single IF-neurons do not phase-lock to the external stimulus, the spatially averaged frequency of the neuronal spikes $< r_s >$ is locked to the frequency of the stimulus Ω at multiple frequency ratios, i.e. $n < r_s >= m\Omega$. It is further shown that the neurons form spatial domains of equal instantaneous interspike intervals and that these domains are separated by phase line-defects.

INTRODUCTION

Excitable dynamics serves as an important paradigm for the mathematical modeling of numerous phenomena in Biology, including neuronal action potentials, cardiac action potentials and intracellular Ca^{2+} release (see e.g. in [1]). On the other hand, these systems are known to be very noisy. The cat's retinal ganglion cells, for example, are high-gain neurons and correspondingly fire approximately 20 action potentials per second in the absence of stimulus [2], and yet, the visual system operates very reliably. Excitable Ca^{2+} release from internal stores, facilitated by small clusters of Ca^{2+} release channels, is subject to substantial noise due to conductance fluctuations.

Synchronization of action potentials in the presence of stimulation across *populations* of neurons is thought to play an important role for the encoding of information in sensory systems. The importance of noise to reproduce observed synchronization patterns in electro-sensitive cells of the paddlefish has been recently established by Neiman et al.[3] and in electroreceptors of weakly electric fish by Chacron et al. [4]. Various aspects of the role of noise for synchronization of excitable systems have been studied recently by Chialvo et al. [5] and Longtin[6] for single neuron models. Neimann et al. [7] have studied phase-synchronization in a spatially extended Fitzhugh-Nagumo model driven only by random noise. In this paper, we show that noise can actually *aid in locking* the firing frequency of a neuronal network to a sub-threshold periodic stimulus. We further show that noise-facilitated *frequency locking is spatiotemporally organized* into domain-structures bounded by closed phase line-singularities.

CP665, *Unsolved Problems of Noise and Fluctuations: UPoN 2002: Third International Conference,* edited by S. M. Bezrukov

THE MODEL

We consider a square array of excitatorily coupled, leaky integrate-and-fire neurons e_{ij} at the cartesian coordinates (ia, ja) with $i, j = 1, \ldots N$ and lattice constant a. Each integrate-and-fire neuron e_{ij} is coupled to a thermal environment, modeled by an Ornstein-Uhlenbeck process for its membrane potential v_{ij}

$$\begin{aligned} \dot{v}_{ij} &= -\frac{1}{\tau} v_{ij} + \sqrt{\frac{\sigma^2}{\tau}} \xi_{ij}(t) + A\sin(\Omega t) \\ &< \xi_{ij}(t)\xi_{kl}(s) >= 2\delta_{ij,kl}\delta(t-s) \\ &< \xi_{ij}(t) >= 0. \end{aligned} \tag{1}$$

When the membrane potential v_{ij} exceeds the threshold-voltage of excitation v_0, neuron e_{ij} fires. After a time delay Δt (synaptic delay), the excitation pulse has travelled to other neurons e_{kl} in its neighborhood and their membrane voltages v_{kl} are increased by $K\exp(-\lambda r^2_{ij,kl}/a^2)$, where r_{ij} denotes the distance between neurons (ij) and (kl). All neurons have an absolute refractory period of τ_r. The variance of the noise σ^2, the signal amplitude A and frequency Ω are uniform across the array. In all simulations in this paper we use a refractory time of $\tau_R = 4\Delta t$. Given a typical synaptic time delay of $\Delta t = 2ms$ this amounts to a refractory period of $8ms$ and a complete neuronal excitation cycle of $10ms$.

Using the dimensionless variables $t \to t/\tau, v_{ij} \to v_{ij}/v_0, K \to K/b, \Omega \to \Omega\tau, A \to A\tau/v_0, \xi_{ij}(t) \to (1/\sqrt{\tau})\xi_{ij}(t)$, and $\sigma^2 \to \sigma^2/a^2$, the system is governed by the discrete stochastic map for the voltages $v_{ij}(t_n)$, where $t_n = n\Delta t$.

$$\begin{aligned} v_{ij}(t_{n+1}) &= \exp(-\Delta t)v_{ij}(t_n) + \sqrt{\sigma^2(1-\exp(-2\Delta t)}G_{ij} \\ &+ K\sum_{kl}\exp\left(-\lambda\frac{r^2_{ij,kl}}{a^2}\right)\Theta(v_{kl}(t_n)-b) \\ &+ \frac{A}{\sqrt{1+\Omega^2}}(\sin(\Omega t_{n+1}+\phi)-\exp(-\Delta t)\sin(\Omega t_n+\phi)). \end{aligned} \tag{2}$$

with the Heaviside function $\Theta(t)$, $\tan(\phi) = -\Omega$, Gaussian random numbers G_{ij} with unit variance and $t_{n+1} = t_n + \Delta t$. In this paper we use $\Delta t = 0.05$.

SINGLE IF-NEURON

We first discuss the single neuron, i.e. $K = 0$. In Fig.1, the average time interval $< T_{ISI} >$ between successive action potentials (averaged over $\approx 10^6$ interspike intervals), is shown as a function of the variance of the noise σ^2 at the stimulus frequency $\Omega = 2\pi$ and several amplitudes lower than the excitation threshold $A_c(\Omega = 2\pi) \approx 6.36$. There is a slight indication of a turning point in the curve for $A = 6$ where $< T_{ISI} >$ matches twice the period of the stimulus T_Ω (see arrow), indicating a possible precursor of a period-2

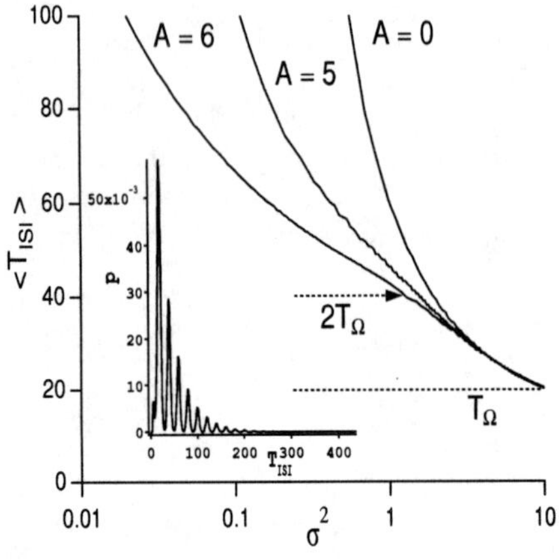

FIGURE 1. The average interspike interval $< T_{ISI} >$ is shown in units of Δt as a function of the variance of the noise for three stimulus amplitudes A and stimulus period $T_\Omega = 2\pi/\Omega = 1 = 20\Delta t$. The inset shows the interspike interval histogram for $A = 6$ and $\sigma^2 = 1$. The time on the horizontal axes is measured in multiples of $\Delta t = 0.05$. The peaks of the distribution are located at multiples of 20, i.e. at multiples of the driving frequency.

locking plateau. (see also [10]). For amplitudes above excitation threshold the neuron-model is oscillatory and phase locking of the spike train to signal has been demonstrated [10] in the absence of noise. The broad interspike interval histogram (inset Fig.(1)) with peaks at multiples of the period $T_\Omega = 2\pi/\Omega$ of the driving at this turning point, however, shows that the range of interspike intervals is wide and thus the matching of the two time scales $2T_\Omega$ and T_{ISH} suffers from large fluctuations. We did not observe indication of a precursor of frequency locking where $< T_{ISI} >= T_\Omega$ since the noise intensity necessary for such a match is so large that the spiking is practically unaffected by the stimulus (note that all curves in Fig.1 are indistinguishable where $< T_{ISI} >= T_\Omega$). For the same reason the step-like features disappear for decreasing T_Ω. A Similar phenomenon has been observed also by Longtin and Chialvo [5] in a Fitzhugh-Nagumo model. In a periodically driven bistable system Shulgin et al. [8], and Neimann et al.[9] found 1 : 1 frequency locking in a bistable system (no other plateaus have been found).

ARRAY OF IF-NEURONS

Next we consider the entire array of pulse coupled integrate-and-fire neurons ($K = 1$). The average time interval between two action potentials of each neuron is shorter than that in the case of a single integrate-and-fire neuron due to wave-mediated spreading of noise-induced excitation [11] across the array. In the absence of a periodic stimulus, the average interspike interval of a single neuron embedded in the array decays monotonously as a function of the noise intensity (see Fig.2).

In the presence of a subthreshold stimulus with period $T_\Omega = 0.3 = 6\Delta t$ and amplitudes

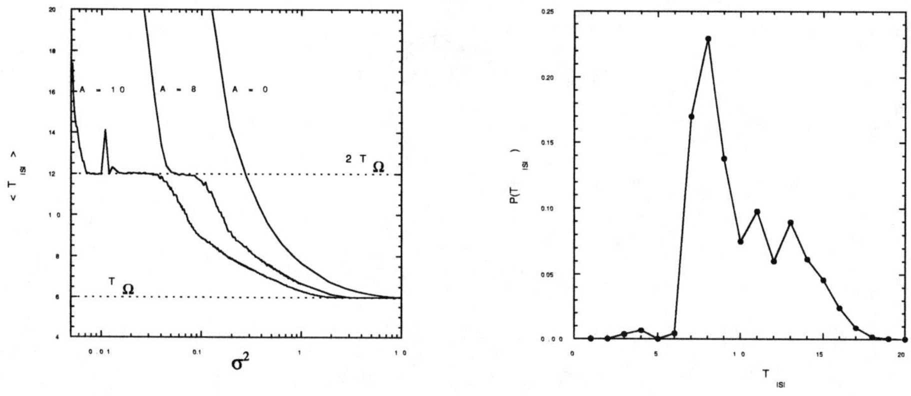

FIGURE 2. In the left panel we show the average interspike interval $< T_{ISI} >$ of a neuron embedded in the array as a function of the variance of the noise without periodic stimulus (A=0) and in the presence of two sub-threshold stimuli $A = 8 < A_c$ and $A = 10 < A_c$. The period of the driving is $T_\Omega = 0.3 = 6$ time steps Δt. In the right panel we show the corresponding interspike interval histogram for $\sigma^2 = 0.02$. The time T_{ISI} is measured in multiples of Δt.

$A < A_c \approx 12$, we observe plateaus and indications of some smaller plateaus in between. In the upper plateau, the average interspike interval $< T_{ISI} >$ matches twice the period T_Ω of the periodic stimulus while in the lower plateau (at larger noise strength) $< T_{ISI} >$ matches the period. The period of the stimulus in Fig.2($T = 0.3$), e.g., is 50% above the refractory period that presents a lower bound for the firing interval of a single neuron. Since the corresponding average interspike interval of isolated integrate-and-fire neuron *at the same parameter values* does not exhibit step or a step-like features, we conclude that the formation of the plateaus is related to network activity. In the right panel of Fig.2, we show the interspike interval histogram (ISIH) of the firing patterns recorded from a single embedded neuron at $\sigma^2 = 0.02$ (in the upper plateau in Fig.2). The intervals are distributed significantly more narrow around $2T_\Omega$ than those of the isolated IF-neuron shown in Fig.1, indicating less interspike interval variability and thus stronger synchronization. This conclusion is supported by Fig.3a, where we show a temporal sequence of interspike intervals of a single embedded neuron. The trace is composed of chained segments. We have marked in Fig.3a a segment with $2:4$ synchronization in which a short interval (shorter than $2T_\Omega$) is followed by a long interval. The sum of two successive intervals, however, equals four times the period of the stimulation almost all the time. Such temporal patterns have been observed in periodically stimulated models of cardiac cells and have been termed *alternans*. The novelty here is that a *proper dose of noise selects* the alternan pattern as the predominant type of excitation pattern. The $(2:4)$-segments are interrupted by segments of periodic spikes with intervals of exactly $2T_\Omega$, i.e. $1:2$ frequency locking and other more irregular patterns. In the lower plateau (Fig.3c), the traces are dominated by segments of periodic spikes that are $1:1$ frequency

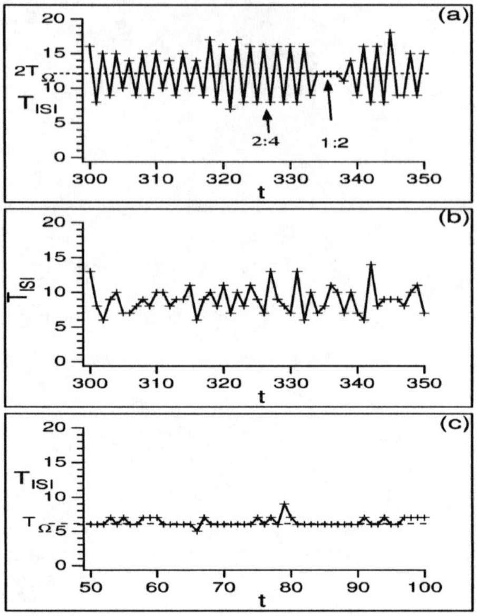

FIGURE 3. Temporal sequence of interspike intervals of an embedded neuron are shown at $\sigma^2 = 0.02$(a), $\sigma^2 = 0.1$(b) and $\sigma^2 = 1$(c). The other parameters are the same as in Fig.2. Times are measured in units of Δt.

locked. The segments of periodic spikes are interrupted by small irregular segments.

Recording simultaneously action potentials from all neurons in the array reveals a remarkable spatiotemporal organization of these segments of frequency locking. To visualize this organization we code neurons with a recent firing interval close to $2T_\Omega$ by light gray and those with a smaller or larger interval (symmetrically) with darker gray. Three snapshots of such interval profiles are shown in Fig.4 for $\sigma^2 = 0.02$ (upper plateau in Fig.2) each 2 periods of stimulus apart. The light lines, indicating neurons locked at that instance of time to $2T\Omega$ (i.e. $1:2$ subharmonic response), are slowly moving. They represent the spatial organization of the temporal segments with $1:2$ frequency locking shown in Fig.3a for a single neuron. When a line is drifting across a group of neurons, they exhibit intermittent 1:2 frequency locking.

Enclosed by the (moving) lines are populations of neurons that fire with an alternan response (one short interval, one long interval) (see also Fig.3a). Neurons in neighbored domains are out of phase by a factor of π. The separating lines represent phase line-defects. Phase line-defects have been observed in oscillatory media at period 2 in chaotic media [12] and recently also experimentally in a BZ-type reaction[13]. Since the line-defects separate populations of neurons that oscillate *at different phase* they are either closed (rings) or end at the boundaries of the array; they cannot end within the array. We have observed spontaneous generation and elimination of closed line defects (see the sequence of snapshots in Fig.4). If the noise strength σ^2 is increased, so that the average interval time lies between the two dominant plateaus in Fig.2, the almost stationary organization into the domains breaks down and is replaced by irregular spatial and temporal dynamics. This can be observed from the local recording in Fig.3b. The synchronization profiles between the plateaus are characterized by rapidly moving phase line singularities (not shown here). Increasing the noise further until the system operates in the lower plateau, the sequence of interspike intervals of an embedded neuron shown in Fig.3c yields a narrow interspike-interval histogram (not shown here) since almost all intervals are of length T_Ω.

DISTRIBUTED PACEMAKERS

In the following, we propose a mechanism responsible for noise-induced and network-facilitated frequency locking. Consider the array of integrate-and-fire neurons in the absence of the periodic stimulus. Assume that at time $t = 0$ a connected cluster of neurons fires spontaneously and elicits a target wave. At the value of the coupling $K = 1$ we have used in the simulations presented here, two adjacent neurons firing simultaneously are sufficient to elicit a target wave. When the refractory period τ_r has passed, the two-cluster is excitable again. The probability of re-excitation of a single neuron Δt after the end of the refractory period can be approximated by the time dependent solution of the Ornstein Uhlenbeck process of the membrane potential $v_{ij}(t)$ (1) as $\rho(t) = (1/2)\mathrm{erfc}[1/\sqrt{2\sigma^2(t)}]$, with $\sigma^2(t) = \sigma^2[1-\exp(-2t/\tau)]$ and the complementary error function $\mathrm{erfc}(x)$[14]. The IF-neurons of the two-cluster thus fire simultaneously at time Δt after recovery with the probability $P_0 = \rho^2(\Delta t)$. This probability is small in the parameter regime we are interested in. For example for $\sigma^2 = 1$, and $\Delta t = 0.05$ it is of the order 10^{-4}. Given the smallness of the probability for re-excitation of a single neuron, the probability for nucleation of a firing two-cluster (which is sufficient to elicit a wave) is proportional to the area enclosed by the circular refractory layer. Given the speed c of the target wave this area increases with time as $\pi c^2(\Delta t)^2$. The probability of re-excitation after one time step Δt is then given by $P_{1\Delta t} = \pi c^2 \Delta t^2 \rho^2(\Delta t)$. At subsequent time-steps, the initiated target wave propagates out radially and an increasing number of neurons enclosed by the refractory layer become excitable. The probability of re-excitation at $t = n\Delta t$is then given by

$$P_{n\Delta t} = \pi c^2 \Delta t^2 \sum_{m=1}^{n} (2m-1)\left(\mathrm{erfc}\left[\frac{b}{\sqrt{2\sigma^2(n\Delta t)}}\right]\right)^2. \tag{3}$$

Since the re-excitation probability is not exactly proportional to the available quiescent area, $P_{n\Delta t}$ exceeds unity for some $n > n_c$. For $n > n_c$, we set P_n to unity. The probabilities that a firing two-cluster forms in the time interval $[n\Delta t, (n+1)\Delta t]$ are then obtained as

$$Q_{n\Delta t} = (1-P_{\Delta t})(1-P_{2\Delta t})\ldots(1-P_{(n-1)\Delta t})P_{n\Delta t}. \tag{4}$$

The probabilities $Q_{n\Delta t}$ are small for small n, increase with increasing n reach a maximum and decrease again. If the window of non-zero probabilities $Q_{n\Delta t}$ is small, the re-excitation is almost periodic. Thus target waves are produced almost periodically, mimicking a pacemaker [15]. Comparison between the average interspike interval T_{ISI}

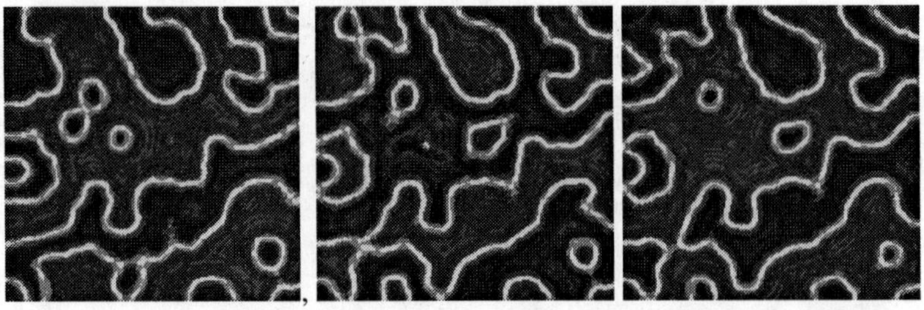

FIGURE 4. Three snapshots of the spatial distribution (twice the period of the stimulus apart) of the interspike intervals at $\sigma^2 = 0.02$ (upper plateau) with time progressing from left to right. The light lines indicate IF-neurons that are 1:2 phase locked with the external signal. They separate regions of 2:4 phase locking that are out of phase by a factor of π. The Parameters are the same as in Fig.2.

obtained from Eq.4 and the average firing interval obtained from the numerical solution shows reasonable agreement. For example, the theory predicts for $\sigma^2 = 1,2,5$ periods of $9.02, 7.1, 5.7$ units of time while numerical results yield approximately $8,7,5$ units of time. The array of coupled neurons develops *robust* noise induced rhythmic activity (distributed pacemakers) with almost perfect periodicity (for a related phenomenon see also [17]). Such a phenomenon cannot be observed at isolated neurons that - driven by noise - elicit random spikes[16].

The periodic stimulus modulates the process of noise induced re-excitation of waves and - if strong enough - entrains the distributed robust pacemakers similar than periodic forcing can entrain limit cycle oscillators yielding the plateaus observed in Fig.2

CONCLUSIONS

In conclusion, stimulating an array of integrate-and-fire neurons with a uniform, *sub-threshold* periodic signal we have observed the following novel phenomena:

1.) The average time interval between action potentials $< T_{ISI} >$ as a function of the noise intensity σ^2 exhibits plateaus (centered around $\sigma^2 = \sigma^2_{n,m}$), where $n < T_{ISI} >= mT_\Omega$ with some selected positive integers n, m (see Fig.2). Such a behavior is not observed at isolated IF-neurons in the same parameter range. These plateaus are fingerprints of the entrainment of self-organized, distributed pacemakers by weak external forcing.

2.) Frequency locking is non-uniform in space and time. Synchronized domains of IF-neurons, firing in phase-shifted alternan-type patterns, are separated by phase line-defects. Phase defect lines can drift across the array. They can end at the boundaries of the array or they can form closed rings. Closed rings can be generated and destroyed spontaneously.

We believe that these observations are of importance for the understanding of sensory neuronal arrays that operate at high gain and are thus noisy. Sensory arrays maybe internally tuned such that their typical noise level aids them to respond in concert.

ACKNOWLEDGEMENTS

This material is based upon work supported by the National Science Foundation under Grant No. IBN-0078055. I further acknowledge Andre Longtin and Alexander Neiman for helpful discussions.

REFERENCES

1. J. Sneyd and J.Keener, *Mathematical Physiolology*, Springer (1999).
2. M. Rowe, and Q. Fischer, Visual Neuroscience **18**, 219 (2001).
3. A. Neiman, X.Pei, D.Russel, W. Wojtanek, L. Wilkins, F. Moss, M.T. Huber, and K. Voigt, Phys. Rev. Lett. **82**, 660 (1999).
4. M.J. Chacron, A. Longtin, M. St-Hilaire and L. Maler, Phys. Rev. Lett. **85**, 1576 (2000).
5. A. Longtin and D.R. Chialvo, Phys. Rev. Lett., **81**, 4012 (1998).
6. A. Longtin, Chaos **5**, 209 (1995);A. Longtin, Chaos, Solitons and Fractals, **11**, 1835 (2000).
7. A. Neimann, L. Schimansky-Geier, Ann Cornell-Bell and F. Moss, Phys. Rev. Lett., **83**, 4896 (1999).
8. B. Shulgin, A. Neimann, and V. Anishenko, Phys. Rev. Lett., **75**, 4157 (1995).
9. A. Neiman, A Silchenko, V. Anishenko, and Lutz Schimansky-Geier, Phys. Rev. **E58**, 7118 (1998).
10. L. Glass et al., J. Theor. Biol. *86*, 455 (1980).
11. P. Jung and G. Mayer-Kress, Phys. Rev. Lett. **74**, 2130 (1995).
12. A. Goryachev, H. Chate and R. Kapral, Phys. Rev. Lett. **80**, 873 (1998); Phys. Rev. Lett. **83**, 1878 (1999).
13. J.S. Park and K.J. Lee, Phys. Rev. Lett. **83**, 5393 (1999).
14. M. Abramowitz and I.A. Stegun,*Handbook of Mathematical Functions*, Dover Publications (1970).
15. P. Jung and P.C. Gailey, Annalen der Physik, **9**, 697 (2000)
16. The pheonomenon of *Coherence resonance* discovered by Pikowsky and Kurths ([18]) describes the generation of periodicty of spike trains in a Fitzhugh Nagumo model for single neurons (coherence resonance). The periodicity due to coherence resonance is weak (wide peak in the power spectrum) and less pronounced than the distributed pacemaker effect we are describing here.
17. H. Hempel, L. Schimansky-Geier, J. Garcia-Ojalvo, Phys. Rev. Lett. **82**, 3713 (1999).
18. A. Pikovsky and J. Kurths, Phys. Rev. Lett. *78*, 775 (1997).

Correlated Neuronal Computation

Jianfeng Feng

COGS, Sussex University, Brighton BN1 9QH, UK
http://www.cogs.susx.ac.uk/users/jianfeng

Abstract. Correlation between neuron activity, a form of noise, is of vital importance for the nervous system to carry out computation. Here we present a concrete example to demonstrate the advantages of correlated neuronal computation. Suppose that a neuron receives two signals masked by random noise. After neuronal transformation, will the masked signals become more separable or more mixed? We prove that with positive correlations between input singles, the output becomes more separable. The critical value at which the total probability of misclassification is zero is analytically obtained and is independent of model parameters.

INTRODUCTION

It is a long standing issue in neuroscience to understand why a neuron in the cortex fires randomly. We all believe that a neuron performs certain computation tasks, but its exact computation objective is not clear. Hence it is difficult to unveil the functional meaning of noise in neuron computation. Many hypotheses have been put forward in the literature: sparse coding, optimization entropy etc. The shortcoming of these approaches is that, for example, even if a neuron does optimize its entropy when it computes, its functional (computational) meaning is still not clear. We all know optimization its entropy implies the neuron will operate in the most random state. What is the computational advantage of the neuron at such a state? Another approach to explore the functional meaning of random firing is the stochastic resonance [8]. As we have pointed out before [4], unfortunately in the stochastic resonance literature, most, if not all, publications treat the dynamics of a neuron as a perturbation of a deterministic system, rather than a genuine stochastic system where the noise strength is proportional to the signal. Therefore, the results of stochastic resonance are illuminating and relevant to some biological systems, but certainly not to neurons in the cortex.
Suppose that a neuron receives p synaptic inputs, its input can be written as

$$\mu dt + \sigma dB_t$$

where μ, σ are constant and B_t is the Brownian motion (see next section). In general the noise magnitude is

$$\sigma \propto \sqrt{p(1+(p-1)c)}$$

where c is the correlation coefficient of afferent neuron activity. When $c = 1$, all input neurons are synchronous and $c = 0$ corresponds to the case of independent inputs. Hence with $c \neq 0$, we have an additional term, i.e. $(p-1)pc$, in the noise term. It is easily seen that even with a small correlation coefficient, the correlated term will play a dominant

CP665, *Unsolved Problems of Noise and Fluctuations: UPoN 2002: Third International Conference,* edited by S. M. Bezrukov

role in noise term since p is usually large. We confine ourselves to the case of $c > 0$ in the current paper. It is interesting to point out that the functional role of negative correlation has been explored in the literature as well. For example, in [7] we have shown that by tuning $c < 0$, a neuron even with Poisson inputs can exhibit stochastic resonance. From statistics point of view, random variables with a mutual negative correlation will be more 'ergodic' than random variables with a positive or zero correlation. For example, denote $N^i(t), i = 1, \cdots, p$ as the counting of spikes of p spike trains (identically distributed) within time $[0,t]$, we have

$$\frac{\sum_i N^i(t)}{\sqrt{pt}} \sim N(a,b)$$

(renewal theorem and central limit theorem) where $N(a,b)$ is the normal distribution withe mean a and standard deviation b. By adjusting $c < 0$, we could have $b \sim 0$. In other words, by adjusting negative correlation, the fluctuation of estimating mean firing rate could be very small, and so an accurate neuron decoding scheme can be achieved. Experimental data show that the correlation between neuron activity could be negative in one time window and positive in the next time window.

For non-negative correlations, H.Barlow has postulated one functional role of correlated neuronal computation–the redundancy reduction principle. In the present paper, we propose a novel approach: both on the setup of the problem and on theory, to attack the issue. Suppose that a neuron receives two set of signals. Both of them are contaminated by noise, as shown in Fig. 1. After neuron transformations, we intend to know whether the signals become more mixed or more separated. This is a typical problem in decision theory. To discriminate between two input signals is of crucial importance for a nervous system.

Due to the space limit we only present theoretical results here and refer the reader to our full paper for numerical results and detailed proofs. We confine ourselves to the integrate-and-fire model [6]. More specifically we consider neuron models with a combination of (coherent) signal inputs and masking 'noise' inputs. We find that with a small fraction of signal inputs, the efferent spike trains of the model contain enough information to discriminate between different inputs (see below for more details).

We also explore the possible function role of inhibitory inputs, another term in noise, on discrimination tasks. A neuron extensively receives both excitatory and inhibitory inputs. It is clear that the excitatory input codes the input information: the stronger the stimuli are, the faster the neuron fires. Less is known about the inhibitory input, much as different, theoretical hypotheses have been put forward in the literature ranging from actually synchronizing the firing of a group of neurons [19], linearizing input-output relationship [5] and increasing neuron firing rates [4] etc. We find that adding certain amount of inhibitory inputs considerably enhances the neuronal discrimination capability if signal inputs are correlated.

The conclusion above seems quite counter intuitive. We all know that increasing inhibitory and correlated inputs to a single neuron model will result in an increase on the variability of its efferent spike trains [4]. The histogram of firing rates will thus become more spread out and, as a consequence, the discrimination of different inputs becomes more difficult. However, this is not the case. To understand the mechanism underpinning the observed phenomena, we then go a step further to theoretically explore the model behaviour. Based upon the IF model, a theory on discrimination tasks is developed. We find that two key mechanisms for achieving a better separation of output signals, in comparison with input signals, are

1. input signals are positively correlated and
2. excitatory inputs and inhibitory inputs are exactly balanced.

Without correlations, no matter how strong the inhibitory inputs are, the separability of the output signals and the input signals is identical: if the input signals are separable, so are the output signals and vise versa. With correlations, the stronger the inhibitory inputs are, the better the separation.

Theoretically the critical value of the coherent inputs at which the output histograms are separable is exactly obtained for the case of correlated and exactly balanced inputs (the most interesting case). The results enable us to assess the dependence of our conclusions on different model parameters and input signals. It is illuminating to see that the critical value is independent of model parameters including the threshold, the decay time and the EPSP and IPSP magnitude.

All the aforementioned results are obtained for the IF without reversal potentials, we further examine our conclusions for the IF model with reversal potentials. Since adding reversal potentials to a model is equivalent to increasing its decay rate (depending on input signals), we would naturally expect that the model with reversal potentials will become more effectively to distinguish different inputs. The conclusion is numerically confirmed (see our full paper).

During the past few years, inhibitory inputs (see for example [10, 11]) and correlated inputs (see for example [16, 17]) are two topics widely investigated in neuroscience. It seems it is generally accepted that they play important roles in information processing in the brain. Our results here provide a convincing and direct evidence to show that they do improve the performance of a single neuron. Such results would also be valuable on practical applications of spiking neural networks [9].

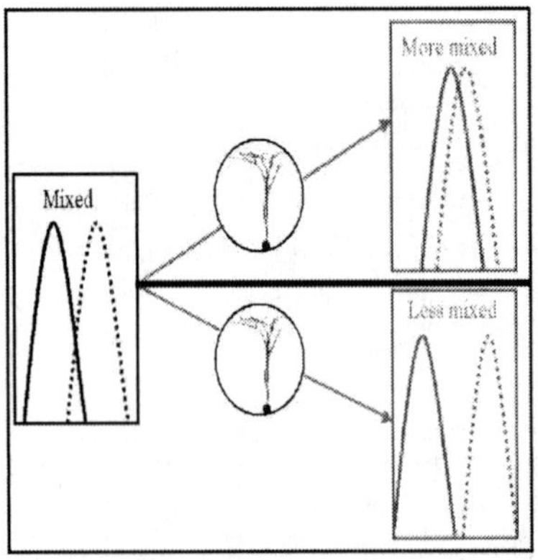

FIGURE 1. Schematic plot of the issue we discussed. Right: after neuronal transformation mixed (masked) signals will become more separable or mixed? Left: input signals to a neuron model.

THE MODELS

We begin with by modelling the inputs. The first neuron model we use here is the classical integrate-and-fire model [2, 3, 18]. When the membrane potential V_t is below the threshold V_{thre}, it is given by

$$dV_t = -L(V_t - V_{rest})dt + dI_{syn}(t) \tag{1}$$

where L is the decay coefficient and the synaptic input is

$$I_{syn}(t) = a\sum_{i=1}^{p} E_i(t) - b\sum_{j=1}^{q} I_j(t)$$

with $E_i(t), I_i(t)$ as Poisson processes with rate $\lambda_{i,E}$ and $\lambda_{i,I}$ respectively, $a > 0, b > 0$ are magnitude of each EPSP and IPSP, p and q are the total number of active excitatory and inhibitory synapses. Once V_t crosses V_{thre} from below a spike is generated and V_t is reset to V_{rest}, the resting potential. This model is termed as the IF model. The interspike interval of efferent spikes is

$$T = \inf\{t : V_t \geq V_{thre}\}$$

We then turn to the situation that a small amount of correlations are added to the synaptic inputs which code coherently moving dots. For the simplicity of notation we assume that the correlation coefficient between ith excitatory (inhibitory) synapse and jth excitatory (inhibitory) synapse is $c > 0$. The correlation considered here reflects the correlation of activity of different synapses, as discussed and explored in [4, 20].
More specifically, synaptic inputs take the following form ($p = q, a = b$)

$$\begin{aligned} I_{syn}(t) &= a\sum_{i=1}^{p} E_i(t) - b\sum_{j=1}^{p} I_j(t) \\ &= a\sum_{i=1}^{p_c} E_i(t) + a\sum_{i=p_c+1}^{p} E_i(t) - a\sum_{i=1}^{p_c} I_i(t) - a\sum_{i=p_c+1}^{p} I_i(t) \end{aligned}$$

where $E_i(t), i = 1, \cdots, p_c$ are correlated Poisson processes with an identical rate λ_1, $E_i(t)$ is Poisson processes with a firing rate λ_i independently and identically distributed random variables from $[0, 100], i = p_c + 1, \cdots, p, I_i(t), i = 1, \cdots, p$ have the same property as $E_i(t)$, but with a firing rate of $r\lambda_i$ for $r \in [0, 1]$ representing the ratio between inhibitory and excitatory inputs.
In numerical simulations [6], we further use diffusion approximations to approximate synaptic inputs [18].

$$\begin{aligned} i_{syn}(t) &= ap_c\lambda_1 t + a\sum_{i=p_c+1}^{p} \lambda_i t - ap_c r\lambda_1 t - a\sum_{i=p_c+1}^{p} r\lambda_i t \\ &+ \sqrt{(a^2 + a^2 r)\lambda_1 p_c(1 + c(p_c - 1)) + (a^2 + a^2 r)\sum_{i=p_c+1}^{p} \lambda_i} \cdot B_t \end{aligned}$$

where B_t is the standard Brownian motion, $\lambda_1 = 25$ Hz or 75 Hz.
In the experiment of Newsome and his colleagues, the input image is a collection of moving dots, say p dots. p_c out of p dots move with an identical (coherent) direction, either upward or downward and $p - p_c$ dots move with random directions. In the literature [1], p_c/p is called coherence level. We assume that an MT neuron exclusively receives information of motion directions of an input image (moving dots), i.e. the cell receives p synaptic inputs, each represents the moving direction of a dot. Denote that $N_i(t), i = 1, \cdots, p$ as a Poisson process with a rate λ_i, where λ_i takes value from $[0, 100]$Hz, i.e. $(\lambda_i/100)2\pi$ is the motion direction of the i-th random moving dot (see [1, 13]) and the motion of each dot is coded by the firing rate of a single synaptic (Poisson) input. Hence $\lambda_i = 25$ Hz represents the input of upward motion, or the ith dot moves upward; $\lambda_i = 75$ Hz represents the downward motion. We refer the reader to our full paper for numerical examples and other experimental results [14, 13, 15].

THEORETICAL RESULTS

In this section we concentrate on theoretical results. Let Λ be the set of input frequency of the model. It will become obvious that all theoretical results are independent of this choice. For a fixed $(\lambda_1 \in \Lambda, \lambda_2 \in \Lambda)$ with $\lambda_1 < \lambda_2$ we have corresponding two histograms $p_1(\lambda)$ and $p_2(\lambda)$ of output firing rates as shown in Fig. 2. Let

$$R_{\min}(\lambda_2) = \min\{\lambda : p_2(\lambda) > 0\}$$

and

$$R_{\max}(\lambda_1) = \max\{\lambda : p_1(\lambda) > 0\}$$

and denote

$$\alpha(\lambda_1, \lambda_2, c, r) = \{p_c : R_{\min}(\lambda_2) = R_{\max}(\lambda_1)\} \tag{2}$$

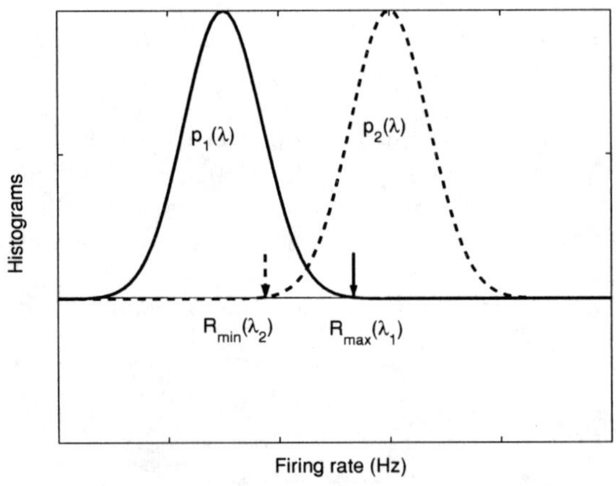

FIGURE 2. A schematic plot of two output histograms, $R_{\min}(\lambda_2)$ and $R_{\max}(\lambda_1)$.

Hence for fixed (λ_1,λ_2), $\alpha(\lambda_1,\lambda_2,c,r)$ gives us the critical value of p_c: when $p_c > \alpha(\lambda_1,\lambda_2)$ the input patterns are perfectly separable in the sense that the the output firing rate histograms are not mixed with TPM=0; when $p_c < \alpha(\lambda_1,\lambda_2)$ the input patterns might not be separable with TPM> 0. Note that we consider the worst case here and in practical applications, the critical value of p_c at which the input patterns are perfectly separable, as found in the previous section, is in general lower than $\alpha(\lambda_1,\lambda_2,c,r)$. From now on, all figures are generated using the same parameters as in the previous section, if not specified otherwise.

Here is the basic idea of our approach. As pointed out before, it is not easy to directly calculate the distribution of $\langle T\rangle$. Nevertheless, the discrimination task is only involved in the most left point of $p_2(\lambda)$, i.e. $R_{\min}(\lambda_2)$, and the most right point of $p_1(\lambda)$, i.e. $R_{\max}(\lambda_1)$, provided that both p_2 and p_1 are positive only in a finite region. This is exactly the case for the model we considered here since neurons fire within a finite region.

Let $\lambda_{max} = max\{\lambda \in \Lambda\}$ Hz, we have

$$\begin{aligned}\alpha(\lambda_1,\lambda_2,c,r) &= \Bigg\{p_c : \int_0^{V_{threL}} g\left(\frac{y-a[p_c\lambda_1+(p-p_c)\lambda_{\max}](1-r)}{a\sqrt{[\lambda_1 p_c(1+c(p_c-1))+(p-p_c)\lambda_{\max}](1+r)}}\right)dy \\ &= \frac{\sqrt{[\lambda_1 p_c(1+c(p_c-1))+(p-p_c)\lambda_{\max}]}}{\sqrt{[\lambda_2 p_c(1+c(p_c-1))]}} \\ &\quad \cdot \int_0^{V_{threL}} g\left(\frac{y-a(p_c\lambda_2)(1-r)}{a\sqrt{[\lambda_2 p_c(1+c(p_c-1))](1+r)}}\right)dy\Bigg\}\end{aligned} \tag{3}$$

When $c = 0$ we have

$$\alpha(\lambda_1,\lambda_2,0,r) = \frac{p\lambda_{\max}}{\lambda_2-\lambda_1+\lambda_{\max}} \tag{4}$$

independent of r. When $c > 0$ we have

$$\alpha(\lambda_1,\lambda_2,c,r_2) < \alpha(\lambda_1,\lambda_2,c,r_1) < \alpha(\lambda_1,\lambda_2,0,r) \tag{5}$$

where $1 \geq r_2 > r_1 > 0$ and furthermore

$$\begin{aligned}\alpha(\lambda_1,\lambda_2,c,1) &= \frac{\sqrt{[(\lambda_2-\lambda_1)(1-c)+\lambda_{max}]^2+4p\lambda_{\max}c(\lambda_2-\lambda_1)}}{2c(\lambda_2-\lambda_1)} \\ &\quad -\frac{(\lambda_2-\lambda_1)(1-c)-\lambda_{\max}}{2c(\lambda_2-\lambda_1)}\end{aligned} \tag{6}$$

From Eq. (4) and Eq. (6), we conclude that with $c > 0$ and $r = 1$, the discrimination capacity is considerably improved. Note that the obtained discrimination capacity is independent of model parameters: the threshold, the decay rate, the magnitude of EPSP and IPSP.

It is certainly interesting to compare the input histograms with output histograms. As before, let Λ be the set of input frequency of the model. For a fixed $(\lambda_1 \in \Lambda, \lambda_2 \in \Lambda)$ with $\lambda_1 < \lambda_2$ we have corresponding two histograms $p_1^i(\lambda)$ and $p_2^i(\lambda)$ of input firing rates, i.e. $p_1^i(\lambda)$ is the histogram of $p_c\lambda_1 + \sum_{i=1}^{p-p_c}\xi_i$ and $p_2^i(\lambda)$ is the histogram of $p_c\lambda_2 + \sum_{i=1}^{p-p_c}\xi_i$.

Define

$$R^i_{\min}(\lambda_2) = \min\{\lambda, p^i_2(\lambda) > 0\}$$

and

$$R^i_{\max}(\lambda_1) = \max\{\lambda, p^i_1(\lambda) > 0\}$$

Then the relationship between $R^i_{\min}(\lambda_2) - R^i_{\max}(\lambda_1)$ and $R_{\min}(\lambda_2) - R_{\max}(\lambda_1)$ characterizes the input-output relationship of neuron signal transformations.
We first want to assess that whether $R_{\min}(\lambda_2) - R_{\max}(\lambda_1) > 0$ even when $R^i_{\min}(\lambda_2) - R^i_{\max}(\lambda_1) < 0$, i.e. the input signal is mixed, but the output signal is separated. It is proved that if $c > 0$ we have

$$R_{\min}(\lambda_2) - R_{\max}(\lambda_1) > 0 \qquad \text{when} \qquad R^i_{\min}(\lambda_2) - R^i_{\max}(\lambda_1) = 0$$

The conclusions above reveal one of the interesting properties of neuronal transformation. Under the assumption that input signals are correlated, the output signals will be separated even the input signals are mixed. As aforementioned, we believe that the fundamental requirement for a nervous system is to tell one signal from the other. The conclusions above tell us that after the transformation of the IF neuron, the input signals could be more easily separable.

CONCLUSIONS

Correlated neuronal computation is considered. We have argued that a tiny correlation, either positive or negative, can play a dominant role in processing information. We have presented theoretical results to demonstrate that with positive correlations between signals and with inhibitory inputs, the discrimination capacity of a neuron is considerably improved.

ACKNOWLEDGEMENT

Partially supported by grants from EPSRC(GR/R54569) and EPSRC (GR/16813), a grant of the Royal Society and an exchange grant between UK and China of the Royal Society.

REFERENCES

1. Britten K.H., Shadlen M.N., Newsome, W.T., Celebrini S., and Movshon J.A. (1992) The analysis of visual motion: a comparison of neuronal and psychophysical performance. *J. Neurosci.* **12** 4745-4765.
2. Brown D., Feng J., and Feerick, S. (1999) Variability of firing of Hodgkin-Huxley and FitzHugh-Nagumo neurons with stochastic synaptic input. *Phys. Rev. Letts.* **82**, 4731-4734.

3. Feng, J.(1997) Behaviours of spike output jitter in the integrate-and-fire model. *Phys. Rev. Lett.* **79** 4505-4508.
4. Feng J. (2001) Is the integrate-and-fire model good enough? –a review. *Neural Networks* **14** 955-975.
5. Leng G. et al. (2001) Responses of magnocellular neurons to osmotic stimulation involves co-activation of excitatory and inhibitory input: an experimental and theoretical analysis. *J. Neurosci.* **21** 6967-6977.
6. Feng J.F., and Liu F.(2002) Discriminating between input signals via single neuron activity (submitted).
7. Feng, J., and Tirozzi B. (2000) Stochastic resonance tuned by correlations in neuronal models. *Phys. Rev. E.* **61**, 4207-4211
8. Gammaitoni L., Hänggi P., Jung P. and Marchesoni F. (1998) Stochastic resonance. *Reviews of Modern Physics* **70** 224-287.
9. Grossberg, S, Maass W., and Markram H. (2001) Neural Networks (Special Issue), vol. **14**.
10. Hopfield J.J., and Brody C.D. (2000) What is a moment? 'Cortical' sensory integration over a brief interval. *Proc. Natl. Acad. Sci. USA* **97** 13919-13924.
11. Hopfield J.J., and Brody C.D. (2001) What is a moment? Transient synchrony as a collective mechanism for spatiotemporal integration. *Proc. Natl. Acad. Sci. USA* **98** 1282-1287.
12. Kast B. (2001) Decisions, decisions ... *Nature* **411** 126-128.
13. Parker A.J., and Newsome W.T. (1998) Sense and the single neuron: probing the physiology of perception. *Annu. Rev. Neurosci.* **21** 227-277.
14. Romo R., and Salinas E.(2001) Touch and go: decision mechanisms in somatosensation. *Annu Rev. Neurosci.* **24** 107-137.
15. Shadlen M.N, and Newsome W.T. (1996) Motion perception: seeing and deciding. *PNAS* **93** 628-633.
16. Salinas E., and Sejnowski T. (2000) Impact of correlated synaptic input on ouput firing rate and variability in simploe neuron mdoels. *Journal of Neuroscience*, **20**, 6196-6209.
17. Salinas E., and Sejnowski T. (2001) Correlated neuronal activity and the flow of neural information. *Nature Reviews Neuroscience* **2**, 539-550.
18. Tuckwell H. C. (1988) *Introduction to Theoretical Neurobiology* Vol 2, Cambridge University Press.
19. van Vereeswijk C., Abbott L.F., and Ermentrout G.B. (1994) When inhibition not excitation synchronizes neural firing. *Jour. Computat. Neurosci.* **1** 313-321.
20. Zohary E., Shadlen M.N., and Newsome W.T. (1994) Correlated neuronal discharge rate and its implications for psychophysical performance. *Nature* **370** 140-143.

Can Intrinsic Fluctuations Increase Efficiency in Neural Information Processing?

Hans Liljenström

Dept. of Biometry and Informatics, SLU, P.O. Box 7013, SE-750 07 Uppsala, Sweden, and Agora for Biosystems, P.O. Box 57, SE-193 22 Sigtuna, Sweden
Email: hans.liljenstrom@bi.slu.se Web:http://www.bi.slu.se/~hanslil

Abstract. All natural processes are accompanied by fluctuations, characterized as noise or chaos. Biological systems, which have evolved during billions of years, are likely to have adapted, not only to cope with such fluctuations, but also to make use of them. We investigate how the complex dynamics of the brain, including oscillations, chaos and noise, can affect the efficiency of neural information processing. In particular, we consider the amplification and functional role of internal fluctuations. Using computer simulations of a neural network model of the olfactory cortex and hippocampus, we demonstrate how microscopic fluctuations can result in global effects at the network level. We show that the rate of information processing in associative memory tasks can be maximized for optimal noise levels, analogous to stochastic resonance phenomena. Noise can also induce transitions between different dynamical states, which could be of significance for learning and memory. A chaotic-like behavior, induced by noise or by an increase in neuronal excitability, can enhance system performance if it is transient and converges to a limit cycle memory state. We speculate whether this dynamical behavior perhaps could be related to (creative) thinking.

INTRODUCTION

Brain structures, such as the olfactory system or the hippocampus, exhibit complex and irregular neurodynamics, including oscillations and chaotic-like behavior, as revealed by EEG and multiunit recordings [1-3]. In addition, there is randomness in structure and function of the neural networks of the brain. Spontaneous firing of neurons seems to be important for maintaining an adequate level of activity (see [4]), but does this "neuronal noise" have any other significance? The functional significance of such behavior is not yet ascertained, but possibly the dynamics of these cortical structures, and perhaps of the brain as a whole, reflect an evolutionary aim for an efficient neural information processing [4-6]. What are the effects of microscopic and mesoscopic fluctuations in the information processing of the brain? Often, one cannot easily determine whether an apparently random process is due to noise, governed by uncontrolled degrees of freedom, or if it is a result of high dimensional "deterministic chaos", but this distinction may not be important for the function of the brain. Stochastic and deterministic processes may co-exist, resulting in a dynamic balance between order and disorder, which with an appropriate regulation may provide the foundation for cognitive functions, such as creative thinking?

CP665, *Unsolved Problems of Noise and Fluctuations: UPoN 2002: Third International Conference,* edited by S. M. Bezrukov

The focus of attention in this work is on the relation between the microscopic fluctuations of ion channels and the dynamics at higher organizational levels, in particular at the neural network level. Both experimental [7,8] and theoretical [9] investigations suggest that spike patterns are directly affected by the stochasticity of single channel openings under certain conditions. In a few instances, specific neurons that apparently function as random generators of impulses have been found. Impulses in high resistance olfactory receptor cells [7] and cultured hippocampal interneurons [8] have been shown to be caused by single channel openings (see Fig. 1). Due to the central role of hippocampus for brain information processing, these results should be highly relevant also to higher cognitive functions such as associative memory and thinking.

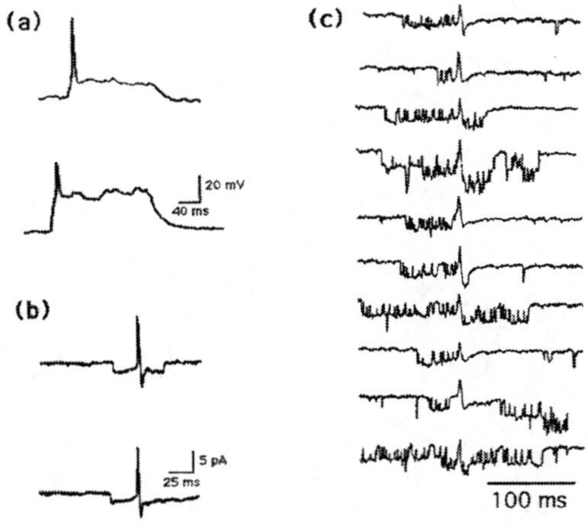

FIGURE 1. Spontaneous activity in small hippocampal interneurons, induced by single channel openings. **(a)** Action potentials associated with plateau potentials, caused by single channel openings. Whole-cell recordings in primary cultures. **(b)** Spontaneous action currents associated with single channel openings, Cell-attached recordings in primary cultures. **(c)** Action currents and single channel openings in "intact" neurons. Cell-attached recordings in hippocampal slice. (Details in Ref. [8]).

METHODS

We use computational methods to investigate relations between the dynamics at different organizational levels of the nervous system, in particular between the microscopic fluctuations at cellular and subcellular level with the complex dynamics at the network level. Our neural network model is mimicking the architecture and dynamics of the olfactory cortex and hippocampus, two structures with a similar, relatively simple structure and with well-studied dynamics and functions, primarily related to learning and memory [4,5]. The network units are modeled with a continuous input-output function, corresponding to the average firing frequency of neural populations. Three different types of network units (cell populations) are organized in three layers, (see Fig. 2). The top layer consists of inhibitory "feedforward interneurons", which receive inputs from an external source (the "olfactory bulb") and from the excitatory "pyramidal cells" in the middle layer. The bottom layer consists of inhibitory "feedback interneurons", receiving inputs only from the excitatory units and projecting back to those. The two sets of inhibitory units are characterized by their different time constants. In addition to the feedback from inhibitory units, the excitatory units receive extensive inputs from each other and from the "olfactory bulb". All connections are modeled with time delays for signal propagation, corresponding to the geometry and fiber characteristics of the real cortex.

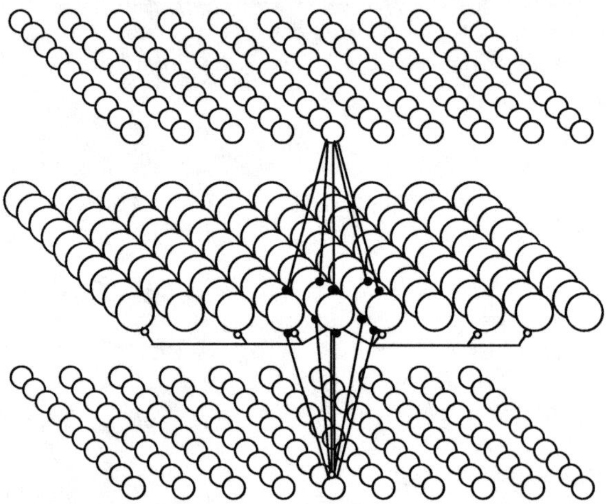

FIGURE 2. The three-layered cortical neural network model used. The top layer consists of inhibitory "feedforward interneurons", the middle layer of excitatory "pyramidal cells", and the bottom layer consists of inhibitory "feedback interneurons". The two sets of inhibitory units are characterized by two different time constants and somewhat different connections to the excitatory units. There are only intra-layer connections for the excitatory units. All connections are modeled with time delays for signal propagation, corresponding to the geometry and fiber characteristics of rat cortex. Spontaneously active units are indicated.

The time evolution for a network of N neural units is given by a set of coupled nonlinear first order differential delay equations for all the N internal states, u. With external input, $I(t)$, characteristic time constant, τ_i, and connection weight w_{ij} between units i and j, separated with a time delay δ_{ij}, we have for each unit activity, u_i,

$$\frac{du_i}{dt} = -\frac{u_i}{\tau_i} + \sum_{j \neq i}^{N} w_{ij} g_j[u_j(t-\delta_{ij})] + I_i(t) + \xi(t) \tag{1}$$

The input-output function, $g_i(u_i)$, is a continuous sigmoid function, experimentally determined by Freeman [10]:

$$g_i = C \cdot Q_i \left\{ 1 - exp\left[-\frac{exp(u_i) - 1}{Q_i} \right] \right\} \tag{2}$$

The gain parameter Q_i determines the slope, threshold and amplitude of the curve for unit i. This gain parameter is associated with the level of arousal, or alternatively, the level of any particular neuromodulator. C is a normalization constant.

The connection weights w_{ij} are initially set and constrained by the general connectivity principles that have evolved for the olfactory cortex. To allow for learning, the weights are incrementally changed according to a learning rule of Hebbian type:

$$\Delta w_{ij} = \eta g_i[u_i(t)]g_j[u_j(t-\delta_{ij})](w_{max}-w_{ij}), \tag{3}$$

where the learning rate is denoted by η, and w_{max} is the maximum strength of an intrinsic synaptic connection.

Noise, or spontaneous neural activity, has been implemented by a Gaussian noise function, $\xi(t)$, such that $<\xi(t)> = 0$ and $<\xi(t)\xi(s)> = 2A\delta(t-s)$, in the differential equation of network node activity. Noise effects are studied by continuously increasing the level A of the additive (and in some cases, multiplicative) noise. In some of the simulations, the noise level is changed equally for all network units, whereas in other simulations, the change is only taking place in a few excitatory or inhibitory units.

RESULTS

Using the cortical network model described above, we address issues on how the neural dynamics at different levels are related. The model has previously been shown to reproduce the essential characteristics of olfactory cortex and hippocampus neurodynamics [4-6,11]. It describes intrinsic oscillatory properties of these structures, and reproduces response patterns associated with a continuous random input signal and with a shock pulse given to the cortex. In the latter case, waves of activity move across the model cortex, consistent with corresponding global dynamic behavior of the functioning cortex. For a constant random input, the network is able to oscillate with two separate frequencies simultaneously, around 5 Hz (theta rhythm) and 40 Hz (gamma rhythm), purely as a result of its intrinsic network properties. A balance between inhibition and excitation, in terms of connection strength and timing of events, is necessary for coherent frequency and phase of the oscillating neural units. Under certain conditions, the system can also display chaotic-like behavior, similar to that seen in EEG traces [1-3]. All of these phenomena have been shown to depend critically upon the network structure, in particular feedforward and feedback inhibitory loops and long-range excitatory connections modeled with distance dependent time delays. Neuromodulation, implemented as a change in the Q control parameter (see Eqn. 2) can result in changes between different dynamical states [11].

We also use our cortical model to study effects of fluctuations at different levels. External or intrinsic fluctuations are usually damped and "ignored" by the system, but in some cases, they may be amplified and have an effect at a macroscopic scale. Simulation results show that noise, as well as neuromodulatory effects, can induce global network oscillations and chaotic-like behavior, depending on the noise level

[12]. An increased noise level in all network units can result in a transition from a stationary to an oscillatory state, or from an oscillatory to a chaotic state, or alternatively, in a shift between two different oscillatory states (see Fig. 3).

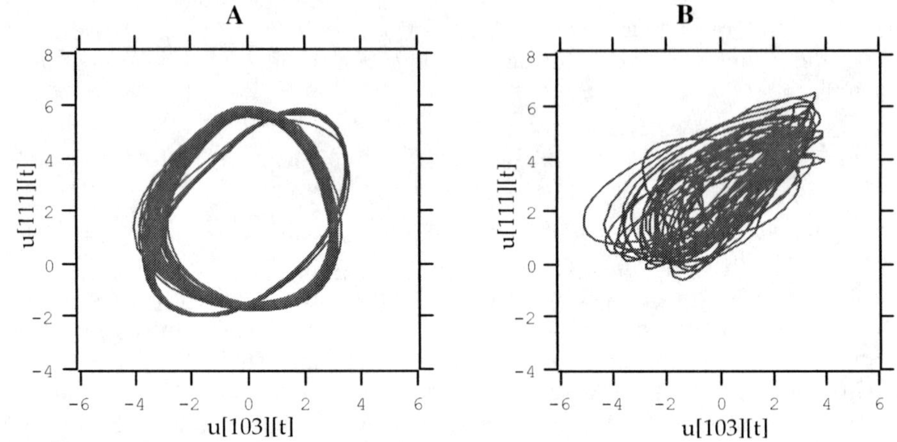

FIGURE 3. **A**) Noise-induced state transition. The "thinner" limit cycle is obtained when the noise level is 0.0. When the noise level is increased to 0.03 the system switches to another ("thicker") limit cycle state. The activity of two excitatory units is plotted against time for 1 simulated second, with the initial 100 ms removed. **B)** Noise-induced "chaos". When the noise level is increased even further, the system does no longer converge to a limit cycle state. Instead, the dynamic behavior becomes more chaos-like. All parameters are the same as in A, except for the noise level, which here is 0.15

Under certain circumstances, a small number of "noisy" network units (with a high intrinsic random activity) is sufficient to induce global coherent oscillatory activity in a network with primarily "silent" units (see Fig. 4). The onset of global oscillatory activity depends on, for example, the noise level, the number and density of the noisy units, the duration of the noise activity, as well as on the network connection strengths. The location and spatial distribution of these units in the network is also important for the onset of oscillations. If the noisy units are separated beyond a certain distance, or if the noise level is too low, no oscillations occur. Likewise, no transition to global oscillations occurs if the noise "frequency" is too low. This is a clear example of how microscopic events, namely the irregular activity of individual network units, can be amplified and give rise to macroscopic and coherent oscillations in the entire network.

In addition to the direct effect on the dynamics, neuronal noise can reduce recall time (convergence time) if noise amplitudes within certain ranges are applied during learning and recall. Consonant with stochastic resonance theory [13-15] we obtain an optimal value of the noise amplitude, A, for which the recall time reaches a minimum. Stated differently, the rate of information processing, which in this case is the rate of convergence to a (near) limit cycle memory state, can be maximized for optimal noise levels. This value seems to be independent on the number of stored patterns in the network. In Fig. 5 we show that a minimum is reached for both additive (triangles) and multiplicative (squares) noise.

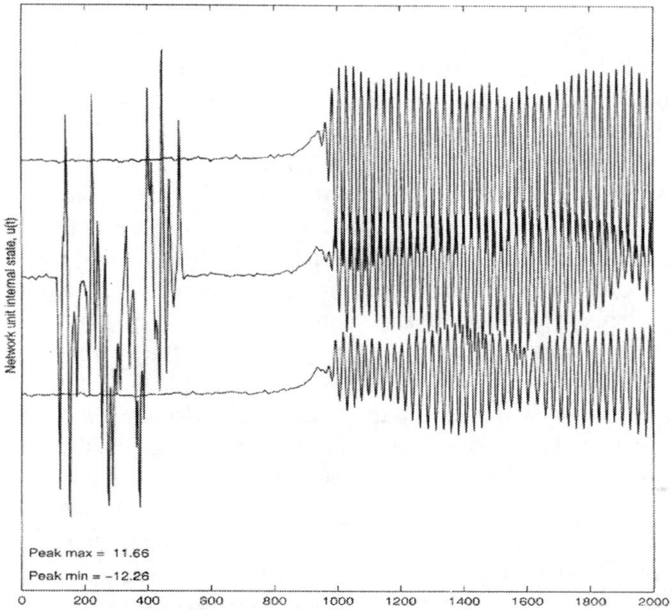

FIGURE 4. Global activity can be induced in the network, even when only a few noisy network units ahave a short burst of activity, lasting for 400 ms. In this case, there is a long period of spatio-temporal chaos-like behavior, before convergence to synchronized global oscillations. The time evolution is given for three excitatory network units for 2000 ms. The middle trace shows the activity of one of five noisy units, where the noise is turned on at t = 100 ms, and turned off at t = 500 ms, and the two other traces are from initially "silent" units.

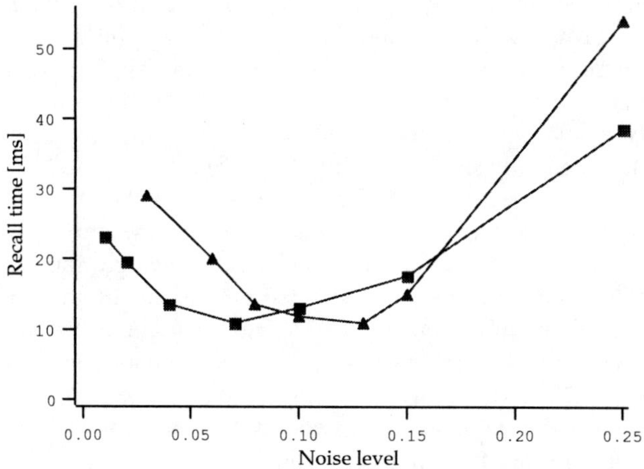

FIGURE 5. Recall time reaches a minimum for an optimal noise amplitude during learning and recall of a single input pattern. For additive noise (triangles) the optimal noise level is higher than for multiplicative noise (squares). The control parameter Q = 10 in both cases.

DISCUSSION

As discussed above, the complex dynamics of our cortical model result in a fast and accurate information processing in associative memory tasks. Neuromodulatory gain control can be used in regulating the accuracy and rate of the recognition process, depending on current demands. Simulation results show that noise, as well as neuromodulation, can reduce recognition and recall time. In addition, state transitions, also to chaos-like behavior, can be induced by noise and neuromodulation. Such regulatory mechanisms can automatically make the system more or less sensitive to external input, and determine the speed and accuracy in the recognition of unknown noisy input patterns, which may change with time.

Oscillatory or complex dynamics can provide a means for fast response to an external input, such as a sensory signal. If sensitivity to small changes in the input is desired, a chaotic–like dynamics should be desired, but a too high sensitivity should be avoided. Oscillations can also be used for enhancing weak signals, and by "resonance" large populations of neurons can be activated for any input. In addition, such "recruitment" of neurons in oscillatory activity can eliminate the negative effects of noise in the input, by canceling out the fluctuations of individual neurons. Finally, from an energy point of view, oscillations in the neuronal activity should be much more efficient than if a static neuronal output (from large populations of neurons) was required. In engineering, great efforts are made to eliminate oscillations in the system, but if the system can perform as well (or better) with oscillations, energy can be saved.

In contrast to chaotic dynamics, where the dynamics can be controlled and easily shift into an oscillatory or other state, noise is not equally controllable. Nevertheless, a certain degree of randomness in the neural activity seems inevitable at every level of organization, and organisms have presumably evolved to cope with it. The question is if it is also of an advantage to the system, and in that case, what could this advantage be? The simplest answer would be that the noise, or (high-dimensional) chaos, could be used for maintaining a baseline activity, even in the absence of external stimuli, keeping the system close to threshold. Indeed, it could be advantageous to have such irregular activity as a "resting activity", in order to avoid unwanted regular activity patterns to interfere with any signal entering the system.

However, the computer simulations we have described above show that noise could induce global synchronous oscillations and shift the system dynamics from one dynamical state to another. This, in turn, could change the efficiency in the information processing of the system. We also demonstrated that system performance could be maximized at an optimal noise level, analogous to the phenomenon of stochastic resonance. Thus, in addition to the (pseudo-)chaotic network dynamics, the noise produced by a few (or many) neurons, could be used for making the system flexible, increasing the responsiveness of the system and for avoiding the system to get stuck in any undesired oscillatory mode. Such activity could also facilitate transitions between different dynamical states, and all of these properties should be crucial for associative memory, and other cognitive processes.

ACKNOWLEDGMENT

Financial support from the Swedish Research Council is gratefully acknowledged.

REFERENCES

1 Bressler S. L. and Freeman, W. J. *Frequency analysis of olfactory system EEG in cat, rabbit, and rat,* Electroenceph. Clin. Neurophysiol., **50:**19-24 (1980).

2 Freeman, W. J. and Skarda, C.A. Spatial EEG Patterns, *Non-linear Dynamics and Perception: the Neo-Sherringtonian View*, Brain Res. Rev., **10**: 147–175 (1985).

3 Skarda, C.A. and Freeman, W. J. *How Brains Make Chaos in Order to Make Sense of the World*, Brain and Behavioral Sciences, **10**: 161–195 (1987).

4 Århem, P. and Liljenström, H. *Fluctuations in Neural Systems: From Subcellular to Network Levels*, In: Handbook of Biological Physics (Eds. F. Moss & S. Gielen) Vol. 4, pp. 83-129 (2001).

5 Liljenström, H. *Autonomous Learning with Complex Dynamics*, Intl. J. Intelligent Systems **10:**119–153 (1995).

6 Liljenström, H. *Cognition and the Efficiency of Neural Processes*, In: Matter Matters? On the Material Basis of the Cognitive Activity of Mind, (Eds. P. Århem, H. Liljenström & U. Svedin), pp. 177-213, Springer (1997).

7 Lynch, J. and Barry, P. *Action potentials initiated by single channels opening in a small neuron (rat ofactory receptor)* Biophys. J. **55**, 755-768 (1989).

8 Johansson, S. and Århem, P. *Single-channel currents trigger action potentials in small cultured hippocampal neurons* Proc. Natl. Acad. Sci. USA. **91**, 1761-1765 (1994).

9 Strassberg, A. and DeFelice, L. *Limitations of the Hodgkin-Huxley formalism: Effects of single channel kinetics on transmembrane voltage dynamics,* Neural Computation **5**, 843-855 (1993).

10 Freeman, W. J. *Nonlinear gain mediating cortical stimulus-response relations*, Biol. Cybern., **33**, 237–247 (1979).

11 Liljenström, H. and Hasselmo, M. E. *Cholinergic Modulation of Cortical Oscillatory Dynamics*, J. Neurophysiol, **74**:288–297 (1995).

12 Liljenström, H and Wu, X. *Noise-enhanced Performance In A Cortical Associative Memory Model*, Intl. J. Neural Systems. **6**,19–29 (1995).

13 A. Bulsara, E. W. Jacobs, T. Zhou, F. Moss, & L. Kiss, "Stochastic resonance in a single neuron model: theory and analog simulation", *J. theor. Biol.*, **152**:531-555 (1991).

14 Anishchenko, V. S., Neiman, A. B., and Safanova, M. A. *Stochastic Resonance in Chaotic Systems*, J. Stat. Phys. **70**, 183-196 (1993).

15 Wiesenfeld, K. and Moss, F., *Stochastic resonance and the benefits of noise: from ice ages to crayfish and SQIDs,* Nature **373**, 33-36 (1995).

CENTRAL NERVOUS SYSTEM

The Human Brain Uses Noise

Toshio Mori, Shoichi Kai

Department of Applied Physics, Graduate School of Engineering,
Kyushu University, Fukuoka 812-8581, Japan
E-mail: toshi-mo@athena.ap.kyushu-u.ac.jp

Abstract. We present the first observation of stochastic resonance (SR) in the human brain's visual processing area. The novel experimental protocol is to stimulate the right eye with a sub-threshold periodic optical signal and the left eye with a noisy one. The stimuli bypass sensory organs and are mixed in the visual cortex. With many noise sources present in the brain, higher brain functions, e.g. perception and cognition, may exploit SR.

Noise plays a surprising role in biological systems [1][2]. Stochastic resonance (SR), which improves signal detection by superimposing noise, has been well demonstrated in the peripheral nervous system and in the cells of sensory organs[3]-[6]. Very recently, SR phenomena in more complex systems such as human tactile sensing, neural networks of mammalian brains and the blood pressure control system in the human brain stem, have been well established [7]-[14]. So far, however, there has been no study of SR for information processing systems in the human brain. That is, it is difficult to confirm this SR because of the challenge of distinguishing between possible SR in the brain and possible SR in the sensory organs through which the signal and added noise must pass in order to reach the brain. Here, for the first time, we exploit the way in which the eyes are "cross-wired" to the visual cortex to bypass sensory organs and observed SR in the human visual cortex itself. While brain wave entrainment only by periodic stimuli has been well established as a phase-locking dynamics [15]-[17], we demonstrate SR and global noise-enhanced entrainment of the brain waves in an information processing area of the central nervous system.

In the human brain, there is a large variety of chaotic firing of neuron networks and columns that function as complex oscillators, as well as spontaneous electrochemical noises[18]. Such noisy signals provide internal noise sources to trigger SR phenomena in the brain. In addition, recently it has been known that spatio-temporal dynamics of brain wave entrainment is closely related to higher order brain functions [19]-[22]. Consequently, we anticipate SR could improve information processing related to cognition and perception.

Brain waves result from a summation of the postsynaptic potential (EPSP). Usually they are detected on the scalp through various tissue layers between it and the brain e.g. the skull, meninges and cerebrospinal fluid. Nevertheless, brain waves appear to indicate rather local information of brain activity [23]. They are classified into five different waves based on their frequency range, typically, 8-13Hz, 14-30Hz, 30-60Hz,

CP665, *Unsolved Problems of Noise and Fluctuations: UPoN 2002: Third International Conference,*
edited by S. M. Bezrukov

0.5-3Hz and 4-7Hz respectively called α, β,γ ,δ and θ waves [24]. Of these, the α wave is considered to be a fundamental brain wave that appears when a human is in a rest state with closed eyes. The other brain waves, on the other hand, appear in active situation of the brain, and are thought to be sensitive to mental, psychological and physical diseases. Applying large amplitude periodic photic stimuli to the eyes can induce a collective firing of neurons synchronizing with the stimuli in many parts of the brain, i.e. entrainment of brain waves [15]-[17][25][26].

The conventional photic activation of the brain wave, in which strong light signals are applied to both eyes of subjects, is frequently used in clinical diagnoses[24]. By strong signals to both eyes, brain waves are immediately synchronized after about 0.2 s from application through phase entrainment. An example is shown in FIGURE 1. Spontaneous brain waves of the subject in the absence of photic stimulus show multiple and broad peaks spectra for any location. In the presence of periodic stimuli of fs=9Hz, however, a single and sharp peak at 9 Hz in the power spectra is observed indicating entrainment which is more remarkable at O_2.

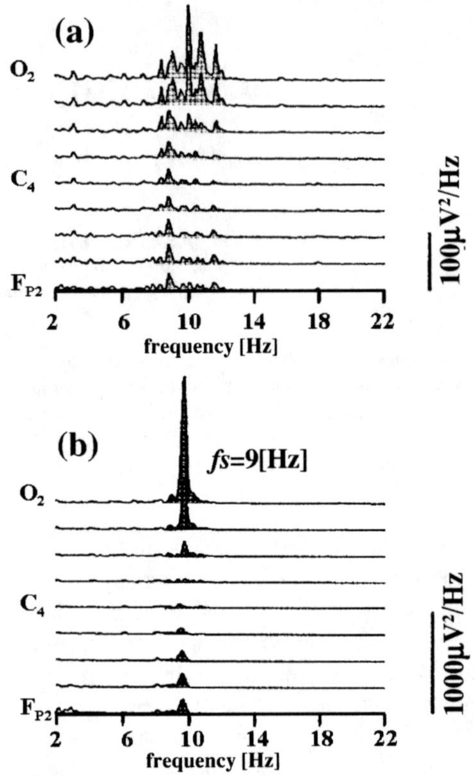

FIGURE 1. The spectrum of the brain wave on each electrode (a) no stimulus (b)application of the light stimulus fs=9Hz (entrainment)

During higher functional activities, the spatio-temporal dynamics of brain waves may change [22]. For example, in a cat study, synchronization of action potentials between areas of the visual and parietal cortex has been reported [21]. In the human perception process, the γ wave is globally synchronized in space and time [22]. This suggests that phase synchronization of brain waves may be evidence of neurons collectively firing in many different parts of the brain. Therefore, spatio-temporal entrainment and transient properties of brain waves (α, β and γ waves) measured on the scalp are associated with cognition, language processing and working memories [19][20][22]. As it is well known that, at an optimum noise intensity, SR enhances synchronization among many coupled oscillators [27]-[32], the possibility emerges that a similar effect could occur in the synchronization of brain waves.

The measurements were performed on five healthy male students(average:22 years old). During measurements of their brain waves, the subjects were in a rest state with their eyes closed. The visual path for photic stimuli from eyes to the brain is shown in

FIGURE 2. A white color light panel was used as a source of external light stimulus to the eyes. Two panels were prepared as periodic and noisy light sources and placed 15 cm from the subject's eyes. The periodic stimulus was applied to the subject's right eyelid and the noisy stimulus to his left eyelid. The characteristics of the applied noise, Gaussian quasi-white noise ranging from 15 to 60 Hz, are shown in FIGURE 2(a)(b). Electroencephalograms (EEGs) were recorded with a 10-channel electroencephalograph where each channel was assigned one of the 10 electrode positions (see FIGURE 2) of the international 10-20 electrode method. The data were simultaneously stored to the hard-disk of a personal computer after analog to digital (AD) conversion. The sampling rate of 500Hz was used with 12 bit resolution using an AD board. Then we calculated the power spectrum density of the brain waves. The detail of the data analysis and experimental setup has been already described elsewhere[15][25][26][33].

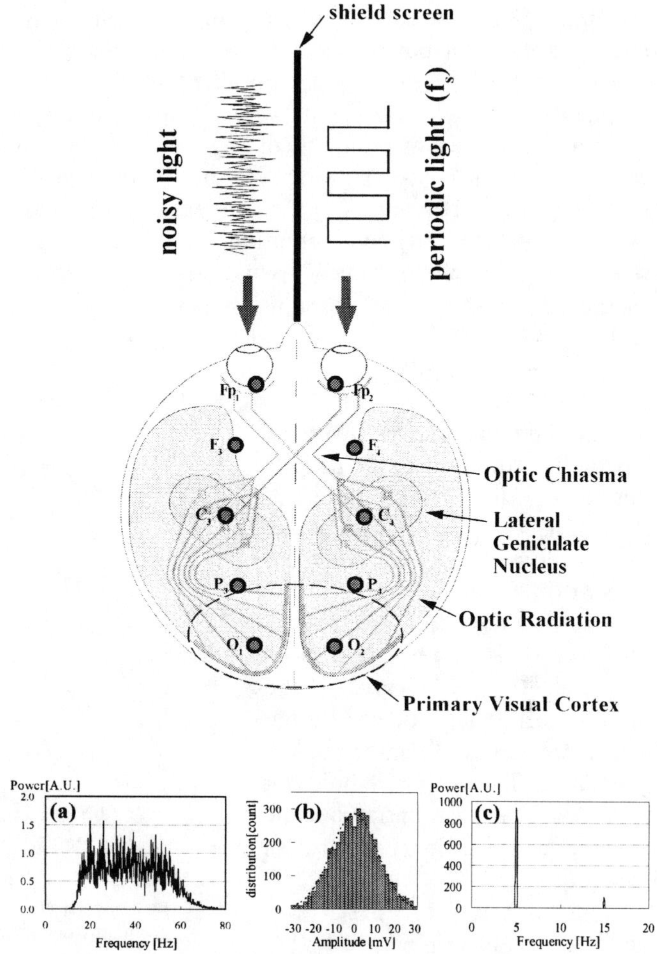

FIGURE 2. The experimental protocol and the trajectory of the photic stimulus from the eyes to the primary visual cortex. And the electrode positions are overwritten on the brain image. (a)Power spectrum of the noisy stimulus; (b)amplitude distribution of the noisy stimulus, the dotted line indicates a Gaussian distribution; and (c)power spectrum of the periodic stimulus.

We can study SR phenomena in the central nervous systems by applying the noisy signal only to the left eye and the periodic one only to the right eye. This is because, with this protocol, the signals are mixed in the visual cortex after the optic chiasma as shown in FIGURE 2. Furthermore, it avoids stimulating SR in the eyes or retina which, like many primary sensory organs

[3]-[6][9][11], may also exhibit SR when the signals are mixed before application to the eyes. In addition, we avoid crosstalk of photic stimuli by inserting a shield screen between the right and left eyes (see FIGURE 2).

First only a periodic stimulus was applied to determine the threshold stimulus for entrainment of brain waves in the α frequency range. The result shows that the stronger the light intensity, the larger the normalized spectrum density, $P_e^N=P_e/P_0$, where P_0 is the total power intensity in the spectrum for each stimulus and P_e, the power intensity of the brain wave entrained at the stimulus frequency. That is, more α oscillators are synchronized with stronger light intensity at the stimulus frequency.

There are two types of entrainment depending on the subject: one at a fundamental frequency (we call it just "entrainment" or "fundamental entrainment") and a second one at a harmonic frequency[24]. Fundamental entrainment is often influenced by the subject's mental and physical condition (e.g. stress) at the moment of testing, while harmonic entrainment is rather insensitive to such uncontrollable factors[24]. The entrainment threshold for stimulus frequency, f_s ($f_s=f(\alpha)$) is determined to be I_{pc}=33.4μW/cm^2. Details are in ref.[33]. It was confirmed that there was no entrainment at I_p=30mW/cm^2 i.e. $I_p < I_{pc}$.

We checked that the intensity of the periodic stimulus, I_p, for the SR experiment was below threshold, I_{pc}, for all subjects. That is, no peaks were observed in the spectrum neither at the fundamental nor at the harmonic frequency for all subjects because of sub-threshold. We note, in general, that the greater the absolute difference between the stimulus frequency, f_s, and $f(\alpha)$, the higher is the intensity threshold for entrainment. Thus, in particular, the threshold at $f_s=1/2f(\alpha)$ is always higher than that for fundamental entrainment, i.e. when $f_s=f(\alpha)$.

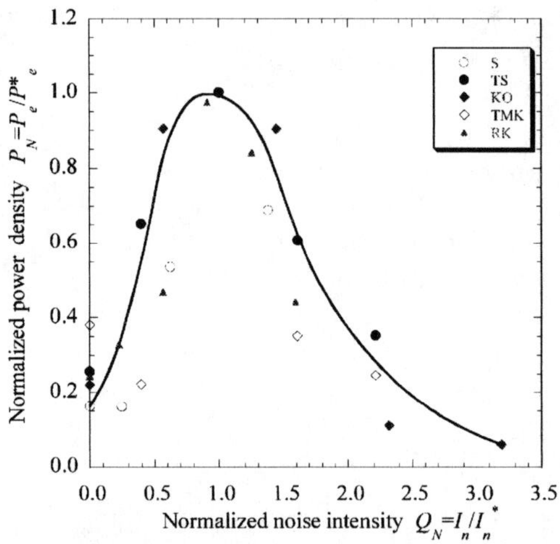

FIGURE3. Noise-stimulus-intensity dependence of the spectrum amplitude at the harmonic entrainment frequency for 5 subjects.

For the SR experiment, with the application of a periodic light stimulus at fixed frequency 5Hz and fixed amplitude I_p=30mW/cm^2, the light intensity, I_n, of the noisy light stimulus was varied. When I_n=0, there is no entrainment peak, P_e, neither at the fundamental, 5Hz, nor at the harmonic, 10Hz.

As the noise intensity, I_n, is increased, a clear entrainment peak is observed in the spectrum at f=10Hz. P_e is maximum at I_n^*=54.5μW/cm^2. When the noise intensity is further increased, P_e decreases. That is, there is an optimum noise intensity for entrainment with the bell-shaped dependence on noise intensity that is the signature of SR[33].

FIGURE 3 shows that the effect is universal i.e. independent of subjects. Here a normalized P_e is plotted against a normalized I_n for all five subjects. As the maximum power spectrum density, P_e^*, and the optimum noise intensity, I_n^*, depend on subjects, the normalized spectrum density, $P_N= P_e /P_e^*$ and noise intensity, $Q_N=I_n/I_n^*$ are used to show the bell-shaped noise-dependence of SR for all subjects. That is, the optimum intensity of the noisy stimulus synchronizes more α oscillators in the human brain. FIGURE 4 shows the resonance curve at reexaminations from the subject RK on subsequent days. It also indicates the bell-shaped dependence i.e. the SR curve. It is suggested that this phenomenon can be reproduced independent of days.

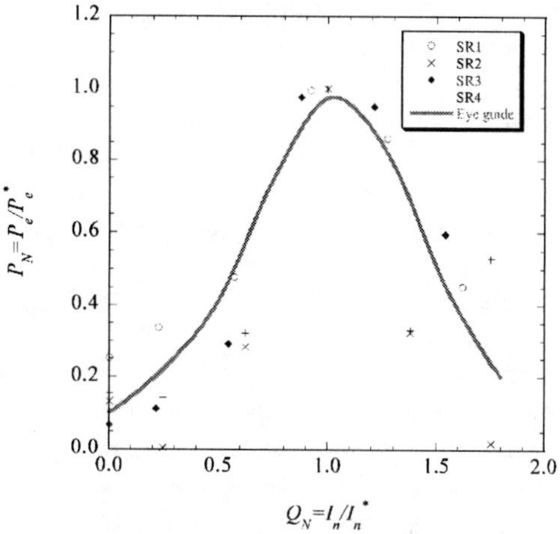

FIGURE 4. Resonance curve at 4 times reexaminations in subsequent days for RK.

We have thus clear evidence for SR at occipital (O_1). The measurement has been done simultaneously for 10 electrode positions. FIGURE 5 shows noise intensity dependence of power peaks at the other electrode positions, Fp_1 and C_3. There is obviously no optimum noise intensity and no clear SR. That is, the SR only occurs at O_1. As is well known, the brain waves observed at Fp_1, at C_3 and at O_1 reflect the activities of the frontal lobe, of somatic or sensory area, and of visual area, respectively. Therefore our result clearly indicates that stochastic synchronization of the brain waves at O_1 is due to SR at visual information area.

We have presented here the first observation of SR in an information processing area

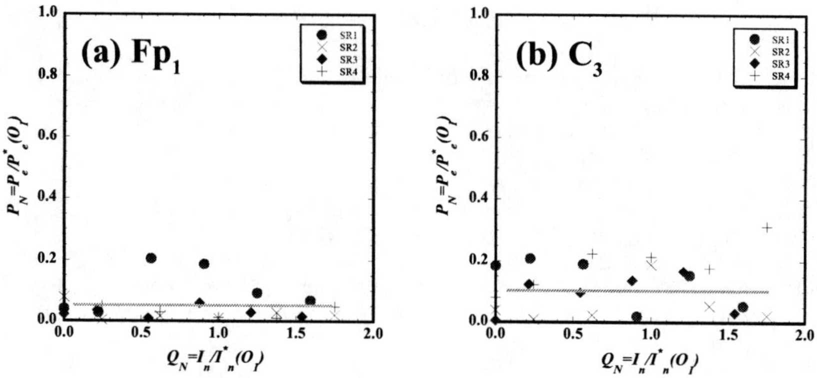

FIGURE 5. Noise intensity dependence of power peaks at electrode position (a)Fp_1 and (b)C_3 for subject RK.

of the central nervous system. By studying noise induced entrainment in the α wave, we found clear evidence of SR which is universal independent of subjects. Here we do not observe SR at input organs such as eyes and retinas because, for each subject, both noisy and sub-threshold periodic photic stimuli are independently applied respectively to the left and right eyes. In this protocol, as shown in FIGURE 2, since both left fields of view are wired to the right cerebral hemisphere and both right fields of view are wired to the left cerebral hemisphere, the signal and noise are mixed not in the eyes but in the visual processing areas of the brain. Thus we bypass sensory organs and observe the SR in the visual cortex itself.

There have been many previous reports of SR in sensory organs and some reports of SR in brain tissue in vitro. However, no one has exploited the present protocol. The SR reported in FIGURE 3, FIGURE 4 and FIGURE 5, observed in multiple subjects and multiple times, must occur in the visual cortex. Since there are many possible internal electrochemical noise sources in the brain, this suggests that higher brain functions, such as perception and cognition, may exploit SR.

Finally we would like to mention some advantages of our protocol and results to clinical applications. Comparing to traditional photic activation method, our method requires very small intensity of light for entrainment of brain waves, typically 1/10 of conventional method. Therefore it is less loading for a patient and may avoid from either mental shock or epilepsy. After every experiments, no subject feels bad and even feels better than before. Therefore they might feel more relaxed after measurements. Combining the result using our protocol with one by the conventional activation method, the location of visual impediment can be determined more preciously in the route between the retina and the visual cortex. Thus, the stochastic synchronization phenomena in the human brain can be expected for various applications in medical and bioengineering fields such as improvement of Alzheimer disease, enhancement of visual information processing of the brain , image processing in biomimetics and so on.

REFERENCES

[1] F.Moss, D.Pierson and D.O'gorman, *Int.J.Bif.Chaos*, **4**, 1383(1994).
[2] L.Glass, *Nature* ,**410**,277(2001).
[3] J.Douglass, L.Wilkens, E.Pantazelou and F.Moss, *Nature*, **365**, 337-340 (1993)
[4] J.E.Levin and J.P.Miller, *Nature* ,**373**, 33(1995).
[5] K. Wiesenfeld and F.Moss, *Nature* , **380**, 165(1996).
[6] P.E. Greenwood, L.M.Ward, D.F. Russell, A. Neiman and F. Moss, *Phys. Rev. Lett.* ,**84**, 4773(2000).
[7] J.J. Collins, T.T. Imhoff and P.Grigg, *Phys. Rev. E*,**56**, 923(1997).
[8] E.Simonotto, M.Riani, C.Seife, M.Roberts, J.Twitty and F.Moss, *Phys. Rev. Lett.*, **78**, 1186(1997).
[9] K.A.Richardson, T.T.Imhoff, P.Grigg and J.J.Collins, *Chaos*, **8**, 599(1998).
[10] B.J. Gluckman, T.I. Netoff, E.J. Neel, W.L. Ditto, M.L. Spano and S.J. Schiff, *Phys. Rev. Lett.*,**77**, 4098(1996).
[11] D.Nozaki, D.J.Mar, P.Grigg and J.J.Collins, *Phys. Rev.Lett.*, **82**, 2402(1999).
[12] W.C.Stacey and D.M.Durand , *J. Neurophysiol.* **83**, 1394(2000).
[13] I. Hidaka, D. Nozaki and Y. Yamamoto, *Phys. Rev. Lett.*, **85**, 17, 3740(2000).
[14] M. Usher and M. Feingold, *Biol.Cybern.* **83**, L11(2000).
[15] K.Harada, S.Nishifuji, S.Kai and K.Hirakawa, *Trans. Inst. Elect. Infom. Commun. Engineer.* **E74**, 1486(1991).
[16] P. Tass M.G.Rosenblum, J.Weule, J.Kurths, A.Pikovsky, J.Volkmann, A.Scnitler and H.-J.Freund, *Phys.Rev.Lett.*, **81**, 3291(1998).

[17] M.G. Rosenblum, A.S. Pikovsky, J. Kurths, C. Schafer, and P. Tass,in *Handbook of Biological Physics*, Vol.4. Neuro-informatics, Chapter 9, pp. 279-321 (Elsevier Science, Amsterdam, 2001) , Eds. F. Moss and S. Gielen.
[18] W.R. Adey , *Int.J.Neurosci.*,**3** , 271(1972).
[19] W. Klimesch, H. Schmike and G. Pfurtschellar, *Brain Topogr.*, **5**, 241(1993).
[20] W. Klimesch, H. Schmike and J. Schwaiger, *Electroenceph clin Neurophysiol.*, **91**, 428(1994).
[21] P.R. Roelfsema, A.K. Engel, P. Konig and W. Singer, *Nature*, **385**, 157(1997).
[22] E.Rodriguez, N.George, J.P. Lachaux, J. Martinerie, B. Renault and F.J. Varela, *Nature*, **397**, 430(1999).
[23] P.Jayakar, M.Duchowby, T.J.Resnick and L.A.Alvarez, *J.Clinic.Neurophysiol.*, **8**, 414(1991).
[24] T.Okuma "Clinical Electroencephalography" (Igaku-Shoin Ltd., Tokyo) 1999.
[25] T.Mori, H.Yamazaki and S.Kai, Jpn. *J. Med. Elect. Biolog. Engineer.* **39**, 284(2001) (in Japanese).
[26] T.Mori and S.Kai, *Trans. Inst. Elect. Info. Commun. Engineer.* **J85** DII,1093-1100 (2002).
[27] J.F.Lindber, B.K.Meadows, W.L.Ditto, M.E.Inchiosa and A.R.Bulsara, *Phys.Rev.Lett.*, **75**, 3(1995).
[28] B.Shulgin,A.Neiman and V.Anishchenko, *Phys.Rev.Lett.*, **75**, 4157(1995).
[29] A.Neiman, A.Silchenko, V.Anishchenko and L.Schimansky-Geier, *Phys.Rev.E*, **58**, 7118(1998),
[30] H.D.I.Abarbanel, M.I.Rabinovich, A.Selverston, M.V.Bazhenov, R.Huerta, M.M.Sushchik and L.L.Rubchinskii, *Physics-Usp*, **39**, 337 (1996).
[31] H.Yamazaki, T.Yamada and S.Kai, *Phys. Rev. Lett.* **80**, 19, 4112(1998).
[32] A. Silchenko T.Kaptitaniak, and V.Anischchenko, *Phys.Rev.E*, **59**, 1593(1999).
[33] T.Mori,S.Kai, *Phys.Rev.Lett.*, **88**, 21,218101(2002)

Noise-Induced Sensitization of Human Brain: Toward the Neurological Application of Stochastic Resonance

Yoshiharu Yamamoto*†, Rika Soma*†, Ichiro Hidaka**, Daichi Nozaki‡, Noriko Iso-o§ and Shin Kwak§

*Educational Physiology Laboratory, Graduate School of Education, The University of Tokyo, 7–3–1 Hongo, Bunkyo-ku, Tokyo 113–0033, Japan
†PRESTO, Japan Science and Technology Corporation, 4–1–8 Honcho, Kawaguchi, Saitama 332–0012, Japan
**National Cardiovascular Center Research Institute, 5–7–1 Fujishirodai, Suita, Osaka 565–8565, Japan
‡Research Institute of National Rehabilitation Center for the Disabled, 4–1 Namiki, Tokorozawa, Saitama 359–8555, Japan
§Department of Neurology, Graduate School of Medicine, The University of Tokyo, 7–3–1 Hongo, Bunkyo-ku, Tokyo 113–0033, Japan

Abstract. In the past decade, it has been recognized that noise can enhance the response of nonlinear systems to weak signals, via a mechanism known as stochastic resonance (SR). Particularly, the concept of SR has generated considerable interest in sensory biology, because it has been shown in several experimental studies that noise can assist neural systems in detecting weak signals which could not be detected in its absence. Recently, we have shown a similar type of noise-induced sensitization of human brain; externally added noise to the brain stem baroreflex centers sensitized their responses in maintaining adequate blood perfusion to the brain itself. Furthermore, the addition of noise has also shown to be useful in compensating for dysfunctions of the baroreflex centers in certain neurological diseases. It is concluded that the statistical physics concept of SR could be useful in sensitizing human brain in health and disease.

INTRODUCTION

It is now well known that noise can enhance the response of nonlinear system to weak signals, via a mechanism known as stochastic resonance (SR) [1, 2]. This concept has proposed a general mechanism for weak signal transmission in a variety of neuronal systems possibly including the brain. In fact, previous experimental results have reported that the sensitivity of sensory neuron to weak signals can be optimized by adding noise [3–7]. Furthermore, several studies have shown that the higher central nervous system can actually utilize the noise enhanced sensory information; it enhances the human tactile sensation [8, 9], the human visual perception [10], or the animal feeding behavior [11].

Whether these functional improvements would be caused by the enhanced sensory afferents at the receptor level or by the effects of noise in the brain was, however, unknown. With regards to this, we recently provided concrete evidence for functional SR in human brain, in which the brain enhances a body function by adding together signal

CP665, *Unsolved Problems of Noise and Fluctuations: UPoN 2002: Third International Conference*, edited by S. M. Bezrukov

and noise from two distinct sources at the periphery of the nervous system [12, 13]. In the current paper, we first summarize such experimental observations in the next section. This will be followed by an introduction to our recent finding showing that the noise-enhanced responses of the brain can be used to "treat" neurological patients with autonomic failure. Finally, we summarize our results and discuss about possible applications of the statistical physics concept of SR to clinical neurology, together with some unsolved problems in this field.

SR IN HUMAN BAROREFLEX

In our recent papers [12, 13], we studied the human "baroreflex" system, in which an increase (or a decrease) in blood pressure is automatically compensated by decreases (or increases) in heart rate and vascular resistance via the autonomic nervous system. This negative feedback system plays a crucial role in maintaining adequate blood perfusion to our brain and, as shown in the next section, the baroreflex failure can easily lead to fainting, annoying and potentially dangerous symptom of patients with autonomic failure.

The baroreflex system has two kinds of pressure-sensitive nerve-cell receptors. Tilting a subject back and forth on a horizontal table, in Ref. [12], we moved blood to the lower part of the body. The draining of blood from the chest area stimulated the "cardiopulmonary baroreceptors" to fire a weak repeating signal which the brain interpreted as a drop in central venous blood pressure (Figs. 1a and 1c). To create neural noise, we randomly added and removed mechanical pressure from the neck. This caused "arterial baroreceptors" to fire randomly, as the artery wall pressure, which normally indicates blood pressure of the arteries, was increasing and decreasing randomly (Figs. 1a and 1b). We hypothesized that injection of noise into the arterial baroreceptor would enhance the response of the baroreflex system to a small signal added to the cardiopulmonary baroreceptors, and thus SR could occur in the baroreflex system via the interaction between signal and noise in the brain stem.

We first measured, for 6 healthy subjects, an individual response threshold for heart interbeat intervals, as measured by an electrocardiographic RR intervals (RRIs), by gradually increasing the local maxima of oscillatory tilt angle with the frequency of oscillation fixed at 0.026 Hz. From the frequency content at 0.026 Hz calculated by a time-frequency analysis [15, 16], the response curves of the magnitude of both RRI and systolic blood pressure (SBP) oscillations at the signal frequency were obtained. Typically, they initially increased slowly, followed by abrupt increases (Fig. 2a). We adopted the maxima of oscillatory tilt angle at this transition point as a fixed signal amplitude. Thereafter, a single 40 minute session where the noise level was gradually increased (Fig. 2b) was conducted to observe transient compensatory responses in RRI oscillations at the signal frequency.

The results of this experiment was very clear in that the magnitude of RRI oscillations by the sub-threshold oscillatory tilt increased initially, reached a maximum, and then gradually decreased as the noise level increased (Fig. 2d). The transient changes in the RRI oscillation magnitudes, as measured by the Wigner-Ville distribution [15] at

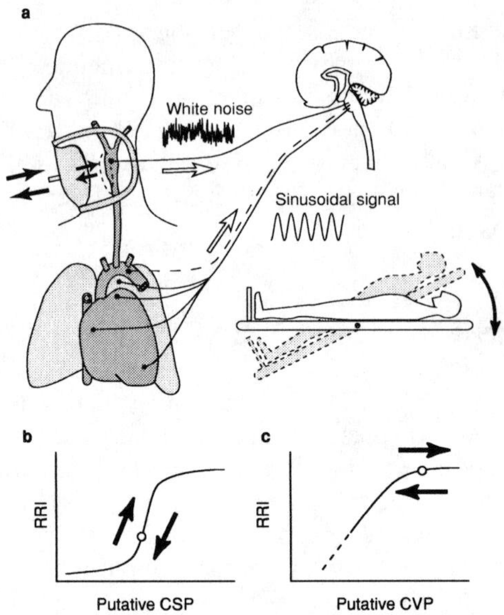

FIGURE 1. Experimental setup and schematic representations of the human baroreflex system. (a) The human baroreflex system has two stretch receptors: the arterial and the cardiopulmonary baroreceptors. The afferent signals from these receptors are independently transmitted to the nucleus tractus solitarius in the brain stem, through which the integrated outputs to peripheral organs are sent via common efferent pathways [14]. In this study, the cardiopulmonary baroreceptor was periodically stimulated by an oscillating tilt angle. This weak periodic signal merged in the brain with noise added to the carotid arterial baroreceptors by randomly changing a pneumatic neck chamber pressure. (b) A typical response curve of heart interbeat intervals, as measured by an electrocardiographic RR intervals (RRIs), to steady-state changes in the carotid sinus pressure (CSP) sensed by the carotid arterial baroreceptors; (c) A typical response curve of RRI to steady-state changes in the central venous pressure (CVP) sensed by the cardiopulmonary baroreceptors. Note that the latter response curve has threshold-like nonlinearity wherein a small change in CVP does not result in appreciable changes in RRI. From Ref. [12].

the signal frequency (WVD_{RRI}), exhibited bell-shaped curves, typically observed in SR phenomenon [1, 2], for all of the 6 subjects (Fig. 2e). This suggested the functional SR in the human baroreflex system and also that the site of interaction between signal and noise was indeed *within* the brain.

In the subsequent study [13], we further conducted a different set of experiments where we used an oscillatory lower body negative pressure to inject a signal into the central venous pressure with physically negligible effects on the arterial system, as there still was a possibility that the interaction of noise with signal could occur not in the brain stem but at the peripheral arterial receptors. Also, we measured muscle sympathetic nerve activity (MSNA) directly controlling vascular resistance. Consequently, the SR in RRI was also observed for all the subjects in such experimental setting and the MSNA responses were enhanced by adding noise to the arterial baroreceptors to overcome given hypotensive stimuli.

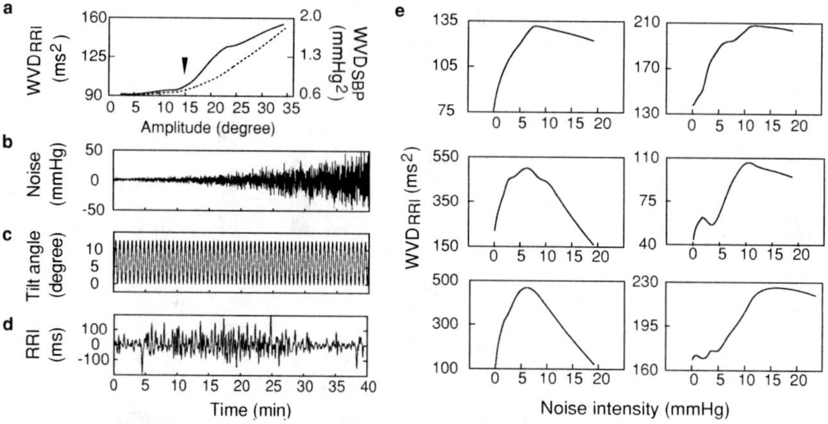

FIGURE 2. Effects of time-varying baroreceptor noise intensity on the magnitude of RRI oscillations. (a) Responses of Wigner-Ville distribution of RRI (WVD_{RRI}; a solid line) and SBP (WVD_{SBP}; a dotted line) at the signal frequency during a graded increase in the local maxima of the oscillatory tilt angle (0.026 Hz). The test signal amplitude was set as the transition point indicated by an arrow. (b) Noise; (c) Signal; (d) A band-pass filtered RRI time series from one subject. Note that the RRI responses were markedly enhanced with an intermediate level of noise. (e) Relationships between noise SD and WVD_{RRI} at the signal frequency for each subject. The bottom left panel is for the subject shown in panel (b)–(d). Note that typical bell-shaped SR curves were observed for all of the 6 subjects. From Ref. [12].

NOISE-BASED TREATMENT FOR NEUROLOGICAL PATIENTS

A natural extension of this line of research is inquiry into whether a similar mechanism would facilitate impaired neural transmission associated with neurological disease. But although several noise-based strategies for treating neurological diseases had been proposed previously [8, 18], none had been successfully applied. This led us to hypothesize, recently in Ref. [17], that adding noise to the arterial baroreceptors might facilitate cardiovascular responses to even strongly hypotensive stimuli in neurological patients with primary autonomic failure (PAF) [19].

PAF is a disorder in which symptoms of autonomic dysfunction, including orthostatic hypotension, impotence, bladder and bowel dysfunction, and sweating defects, are primarily the result of progressive neuronal degeneration of unknown cause. The main pathological finding related to autonomic dysfunction in PAF is severe loss of preganglionic sympathetic neurons in the intermediolateral column of the spinal cord [20]. Consequently, the patients are unable to increase heart rate and vascular resistance against hypotensive stimuli and are susceptible to fainting, annoying and potentially dangerous symptom of the ailment. In contrast to the severe degeneration of the efferent autonomic nervous system, the nucleus solitarius, a relay station receiving baroreceptor afferents [14], remains relatively intact in PAF [21]. Thus, we could use an experimental setting similar to that used in our previous baroreflex studies [12, 13] and tested whether addition of external noise to baroreceptor signaling could improve the blunted autonomic efferent responses to a postural challenge typically seen in PAF patients.

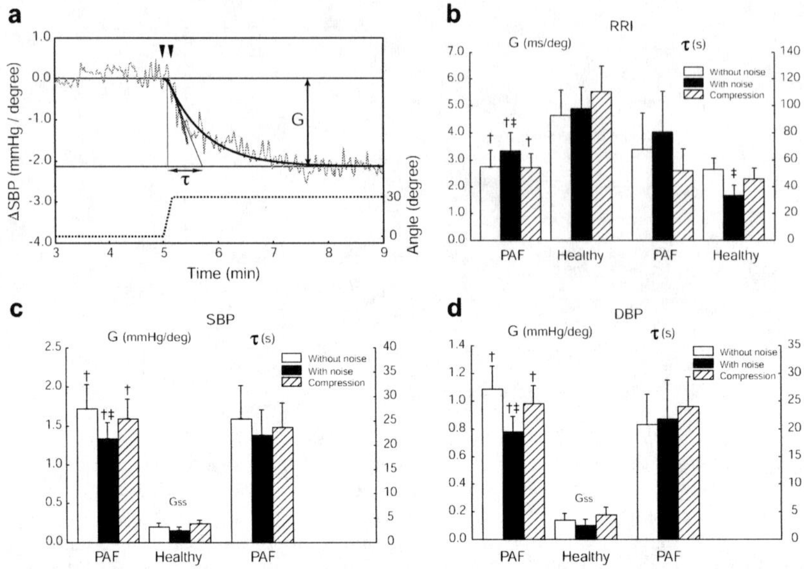

FIGURE 3. Improved heart rate (RRIs) and systolic (SBP) and diastolic (DBP) blood pressure responses in patients with primary autonomic failure (PAF) elicited by baroreceptor noise. (a) Representative trace from one patient showing parameters of the first order model, i.e., the time constant (τ) and the steady state gain (G). The tilt angle (dotted line) was increased linearly for 10 s until it reached 30° for three patients and 60° otherwise; the responses were fit to the model output evoked by the change in tilt angle using nonlinear least squares regression. (b)–(d), Effect of baroreceptor noise and compression, i.e., the application of a continuous positive pressure to the arterial baroreceptors that might also have potential hypertensive effects, on means±SEM of τ and G for RRI, SBP, and DBP in PAF patients (n=5) and healthy subjects (n=8). For healthy subjects, the steady state gain (G_{ss}) was obtained from the mean SBP and DBP during tilt without calculating τ. †$P<0.05$ vs. healthy controls; ‡$P<0.05$ vs. values without noise and with compression. Reprinted from [17], copyright 2002, with permission from Elsevier Science.

Five PAF patients were tested, one in duplicate and another triplicate, for their transient responses of RRI and systolic (SBP) and diastolic (DBP) blood pressures to either 30° or 60° head-up tilt, with and without continuous application of beat-to-beat Gaussian white noise to the arterial baroreceptors. The data normalized by the head-up tilt angle were fit to a first order model to evaluate the speed (by the time constant; τ) and the magnitudes (by the steady state gains; Gs) of RRI and blood pressure responses (Fig. 3a).

Transition of a tilt table from a supine to a head-up position markedly increased the G of RRI in healthy young subjects (Fig. 3b); this autonomic response helped maintain both SBP and DBP at values only slightly lower than baseline, as indicated by smaller G_{ss} (Figs. 3c and 3d, respectively). However, as a consequence of their attenuated compensatory autonomic responses, the Gs of the patients' SBP and DBP responses (Figs. 3c and 3d, respectively) were significantly greater than those of healthy subjects. These convert to average drops in steady state SBP and DBP at a 60° head-up position of >100 mmHg (from 134.6 to 31.1 mmHg) and >60 mmHg (from 77.1 to 12.1 mmHg), respectively, without noise.

In the PAF patients, addition of noise to the arterial baroreceptors significantly improved the G in RRI (Fig. 3b) and diminished the Gs in SBP (Fig. 3c) and DBP (Fig. 3d). The noise-induced decreases in postural hypotensive responses resulted in an average steady state SBP/DBP at a 60° head-up position of 53.4/30.1 mmHg, as compared to 31.1/12.1 mmHg in the absence of noise; the levels of blood pressures with noise were indeed reported to be sufficient to maintain cerebral blood flow and keep their patients asymptomatic when standing [22].

SUMMARY, IMPLICATIONS, AND UNSOLVED PROBLEMS

In the current paper, we introduced our experimental observations for functional SR in baroreflex centers of healthy humans and further showed that the similar noise-based technique could be used to "treat" neurological patients with PAF. We belive these findings to be important for a number of reasons.

First, current therapeutic strategies for presyncope or frank syncope in patients with PAF include increasing peripheral resistance by vasoconstrictors, increasing blood volume by mineral corticoids or high salt diet, and decreasing the capacitance volume by wearing elastic stockings. These "countermeasure" treatments are of limited effects and sometimes rather malicious for PAF patients by generating, e.g., recumbent hypertension. On the other hand, the noise-induced decreases in postural hypotensive responses in the patients could provide an alternative therapeutic strategy to prevent syncope, annoying and potentially dangerous symptom of the ailment. Also, as the effects of noise are considered to be present at the central nervous system level, this treatment strategy may be effective for various symptoms of autonomic dysfunction in PAF other than the orthostatic hypotension. This is an important but unsolved problem when considering the clinical applications.

Second, while most previous studies on the sensory SR have dealt with the noise-induced sensitization in the responses of *normal* sensory neurons to weak input signals [3], we hypothesized that, in neurological diseases with the blunted neural responses even for non-weak signals, noise might ameliorate the impaired neural transmission just like the noise-enhanced detection of small signals in normal sensory systems, and did confirm such scenario. In fact, noise was indeed necessary to induce the improvement of postural hypotension in our PAF patients, since this beneficial effect was not obtained when a continuous positive pressure (compression) equivalent to the root mean square of noise was applied to the arterial baroreceptors (Fig. 3) [17]. This suggests a possibility that *only* noise, not an equivalent facilitatory input, can do such a job: an important perspective in SR researches not only when they are applied to clinical neurology, but also because the problem of "why noise?" has neither been solved experimentally nor theoretically.

Finally, although noise-based strategies utilizing SR in the treatment of neurological diseases have been proposed previously [8, 18], there have been very few convincing reports of its successful clinical application, not withstanding two cases in which electrical noise applied to an auditory brain stem implant enhanced human hearing by about

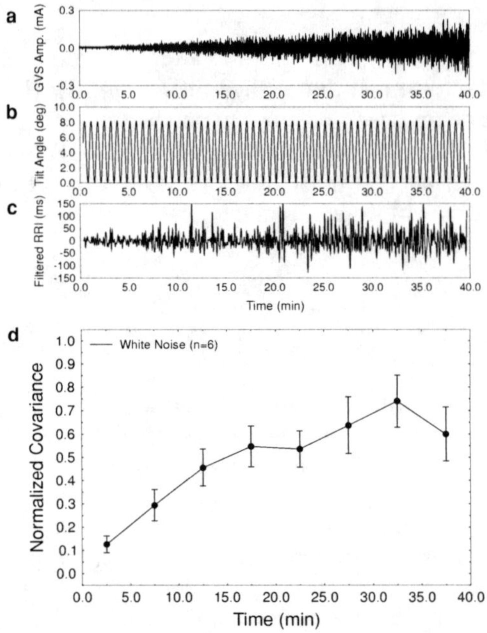

FIGURE 4. Effects of linearly increased vestibular noise intensity on RRI responses during an oscillatory tilt. The subthreshold tilt angle determined by the same procedure as in Ref. [12] was used as a fixed sinusoidal signal amplitude at the frequency of 0.025 Hz. Zero-mean Gaussian white (to 4 Hz) noise was added to a double monaural galvanic vestibular stimulator (GVS). (a) Noise; (b) Signal; (c) A band-pass filtered RRI time series from one subject. Note that the RRI responses were markedly enhanced with the incremental level of noise. (d) Normalized (to $[0,1]$ for each individual) lag-zero covariance between the signal and the filtered RRI averaged over 5 minute bins. The values are means±SEM. The noise-induced increases in the input-output covariance were observed for all of the 6 subjects.

30–60 % [23] and mechanical noise applied to haptic sensory receptors at the periphery decreased vibrotactile detection threshold in patients with stroke and diabetic neuropathy [24]. Thus, these and our findings may lead to a novel noise-based treatment strategy for various neurological diseases, particularly those unresponsive to pharmacological or other currently available therapies. As for PAF, although the neck chamber system we used [12, 13] is not practical for this purpose in terms of its large size and the immobility, other sensory inputs to the brain stem [25], e.g., the vestibular afferents, by portable stimulators may substitute the baroreceptor noise in the near future.

In Fig. 4, we show a preliminary result for such application of the cutaneous galvanic vestibular stimulation (GVS). In this experiment, 6 healthy subjects were tested with exactly the same protocol as in Ref. [12] (Figs. 1 and 2) except that noise was injected by randomly stimulating the bilateral vestibular afferents monaurally [26]. That is, instead of injecting noise through the baroreceptor afferents, the weak (below nociceptive threshold) stochastic GVS was introduced to "activate" the vestibulo-sympathetic pathway in the brain stem [25]. The result was very promising in that, for healthy individuals, lag-zero covariance between the input signal (subthreshold tilt angle) and RRI oscillations was markedly increased by increasing the intensity of stochastic GVS (Fig. 4d). The development of a portable vestibular stimulator is rather easy. A study examining whether the stochastic GVS would be useful for PAF patients in ameliorating their autonomic dysfunctions, a critical as well as unsolved problem, is currently underway.

ACKNOWLEDGMENTS

This study was in part supported by MECSST Grant-in-Aid for Scientific Research and Japan Science and Technology Corporation.

REFERENCES

1. Wiesenfeld, K., and Moss, F., *Nature*, **373**, 33–36 (1995).
2. Gammaitoni, L., Hänggi, P., Jung, P., and Marchesoni, F., *Rev. Mod. Phys.*, **70**, 223–287 (1998).
3. Douglass, J. K., Wilkens, L., Pantazelou, E., and Moss, F., *Nature*, **365**, 337–340 (1993).
4. Levin, J. E., and Miller, J. P., *Nature*, **380**, 165–168 (1996).
5. Cordo, P., Inglis, J. T., Verschueren, S., Collins, J. J., Merfeld, D. M., Rosenblum, S., Buckley, S., and Moss, F., *Nature*, **383**, 769–770 (1996).
6. Collins, J. J., Imhoff, T. T., and Grigg, P., *J. Neurophysiol.*, **76**, 642–645, (1996).
7. Nozaki, D., Mar, D. J., Grigg, P., and Collins, J. J., *Phys. Rev. Lett.*, **82**, 2402–2405 (1999).
8. Collins, J. J., Imhoff, T. T., and Grigg, P., *Nature*, **383**, 770 (1996).
9. Richardson, K. A., Imhoff, T. T., Grigg, P., and Collins, J. J., *Chaos*, **8**, 599–603 (1998).
10. Simonotto, E., Riani, M., Seife, C., Roberts, M., Twitty, J., and Moss, F., *Phys. Rev. Lett.*, **78**, 1186–1189 (1997).
11. Russell, D. F., Wilkens, L. A., and Moss, F., *Nature*, **402**, 291–294 (1999).
12. Hidaka, I., Nozaki, D., and Yamamoto, Y., *Phys. Rev. Lett.*, **85**, 3740–3743 (2000).
13. Hidaka, I., Ando, S., Shigematsu, H., Sakai, K., Setoguchi, S., Seto, T., Hirooka, Y., Takeshita, A., and Yamamoto, Y., *J. Neurophysiol.*, **86**, 559–564 (2001).
14. Andresen, M. C. and Kunzem, D. L., *Annu. Rev. Physiol.*, **56**, 93–116 (1994).
15. Cohen, L., *Proc. IEEE*, **77**, 941–981 (1989).
16. Novak, P., and Novak, V., *Med. Biol. Eng. Comput.*, **31**, 103–110 (1993).
17. Yamamoto, Y., Hidaka, I., Iso-o, N., Komai, A., Soma, R., Kwak, S., *Brain Res.*, **945**, 71–78 (2002).
18. Glass, L., *Nature*, **410**, 277–284 (2001).
19. Bannister, R., *A Textbook of Clinical Disorders of the Autonomic Nervous System*, Oxford Medical Publications, Oxford, 1988, pp267–288.
20. Oppenheimer, D. R., *J. Neurol. Sci.*, **46**, 393–404 (1980).
21. Mizutani, T., *No To Shinkei*, **37**, 675–685 (1985).
22. Harms, H. P. M., Colier, W. N. J. M, Wieling, W., Lenders, J. W. M., Secher, N. H., and van Lieshout, J. J., *Stroke*, **31**, 1608–1614 (2000).
23. Zeng, F. G., Fu, Q. J., and Morse, R., *Brain Res.*, **869**, 251–255 (2000).
24. Liu, W.,Lipsitz, L. A., Montero-Odasso, M., Bean, J., Kerrigan, D. C., Collins, J.J., *Arch. Phys. Med. Rehab.*, **83**, 171–176 (2002).
25. Yates, B. J., and Stocker, S. D., *Exp. Brain Res.*, **119**, 269–275 (1998).
26. Scinicariello, A. P., Eaton, K., Inglis, J. T., and Collins, J.J., *Biol. Cybern.*, **84**, 475–480 (2001).

Development of Demand-Controlled Deep Brain Stimulation Techniques Based on Stochastic Phase Resetting

Peter A. Tass

Institute of Medicine, Research Center Jülich, Germany
Department of Functional and Stereotaxic Neurosurgery, University of Cologne, Germany
P.Tass@fz-juelich.de

Abstract. Stimulation techniques are discussed here which make it possible to effectively desynchronize a synchronized cluster of globally coupled phase oscillators in the presence of noise. To this end composite stimuli are used which consist of a first, stronger stimulus followed by a second, weaker stimulus after a constant time delay. The first stimulus controls the dynamics of the cluster by resetting it, whereas the second stimulus desynchronizes the cluster by hitting it in a vulnerable state. The first, resetting stimulus can be a strong single pulse, a high-frequency pulse train or a low-frequency pulse train. The cluster's resynchronization can effectively be blocked by repeated administration of a composite stimulus. Demand controlled deep brain stimulation with these desynchronizing stimulation techniques is suggested for the therapy of patients suffering from tremor-dominant Parkinson's disease or essential tremor as a milder and more efficient therapy compared to the standard permanent high-frequency deep brain stimulation.

INTRODUCTION

Synchronization processes abound in physics [1], chemistry [2], biology [3, 4, 5], and medicine [6, 7, 8, 9]. In the central nervous system of both animals and humans synchronization of neural oscillations is of great importance [5]. Single neurons may act as oscillators by rhythmically generating action potentials or bursts. Central pattern generators (CPG) [10] are networks of neurons located in the spinal cord and the brain stem, which are able to produce synchronized oscillatory activity intrinsically, in particular, without external rhythmical stimulation [10, 11]. CPG are essential for repetitive motor behavior like walking, running, flying, swimming, and chewing [10, 11]. The coordination of the phase dynamics between different CPG is a fundamental principle for effective locomotion control: Not the absolute timing, but the phase differences between different CPG determine the type of locomotion such as walk, trot or gallop [10, 11].

In the cortex the well-coordinated action of neural oscillations is indispensable to physiological information processing [5]. Whether in-phase synchronization acts as a central mechanism for integrating information processed within and between different brain areas [12] is still a matter of debate [13].

Furthermore, in several neurological diseases, e.g., Parkinson's disease (PD) or essential tremor, brain function is severely perturbed by pathologically enhanced synchronization. Resting tremor in PD is an involuntary shaking with a frequency around 3 Hz

CP665, *Unsolved Problems of Noise and Fluctuations: UPoN 2002: Third International Conference*, edited by S. M. Bezrukov

to 6 Hz which predominantly affects the distal portion of the upper limb and classically decreases or vanishes during voluntary action [8]. This type of tremor appears to be caused by a cluster of neurons located in the thalamus and the basal ganglia which fire synchronously at a frequency similar to that of the tremor [14, 15]. The no-response cells in the anterior nucleus of the ventrolateral thalamus are neither modulated by somatosensory stimuli nor by active or passive movements [15]. These cells fire rather periodically in an intrinsic manner, regardless of any feedback signals. In contrast, under physiological conditions the neurons in this cluster fire incoherently [16]. In patients with Parkinson's disease this cluster acts like a pacemaker and activates premotor areas and the motor cortex [6, 7, 16, 17], where the latter synchronize their oscillatory activity [9]. Similarly, essential tremor also appears to be caused by a central cluster of synchronously firing neurons, located in different brain areas compared to PD [8].

In patients with advanced PD or with essential tremor who do not respond to drug therapy any more, depth electrodes are chronically implanted in target areas like the thalamic ventralis intermedius nucleus or the subthalamic nucleus with millimeter precision [18]. Up to now, electrical deep brain stimulation (DBS) is performed by administering a permanent high-frequency ($>$ 100 Hz) periodic pulse train via the depth electrodes. DBS suppresses the activity of the pacemaker-like cluster which, in turn, suppresses the peripheral tremor [18]. DBS has been developed empirically, mainly based on observations during stereotaxic neurosurgery. The advent of nonlinear dynamics and statistical physics in the field of clinically oriented neuroscience lead to a model-based development of deep brain stimulation techniques [19].

STOCHASTIC PHASE RESETTING

In single neurons the effect of a stimulus depends on the phase of the neural oscillation at which the stimulus is administered [20]. This feature is passed on to networks of neural populations leading to mechanisms like the reflex reversal which is essential for locomotion: Whether an extensor or a flexor muscle is activated depends on the locomotion phase at which a stimulus hits the affected limb [21]. To study phase dependent effects of pulsatile stimuli on neural populations and, in particular, to analyze desynchronizing effects of such stimuli, the concept of phase resetting [3] was extended to populations of non-interacting [22] and interacting [19] oscillators subject to random forces. For this, single neurons are approximated by phase oscillators [2, 23].

A single pulse of the right intensity and duration desynchronizes a fully synchronized cluster of oscillators if the pulse hits the cluster in a vulnerable phase range which corresponds to only a small fraction (5 % or even less) of a period of the oscillation [19]. However, this is tricky to realize under nosiy experimental conditions typically encountered in biological systems. What makes single pulse stimulation even less practicable is that the correct stimulation parameters also depend on the extent of the synchronization of a cluster: A weaker pulse has to be used to desynchronize a weakly synchronized cluster, whereas a stronger pulse is necessary for the desynchronization of a cluster that is in its fully synchronized state. Moreover, not only the strength (i.e. intensity and duration) but also the critical phase at which a pulse has to be administered, crucially depends on the extent of synchronization of the cluster [19, 24].

Recently stimulation techniques have been developed which effectively desynchronize a cluster of oscillators irrespective of the cluster's dynamic state at the beginning of the stimulation [24, 25, 26, 27]. These methods share one particular feature: The stimulus contains two qualitatively different stimuli. The first stimulus is stronger and resets (i.e. restarts) the cluster, whereas the second, weaker stimulus is a single pulse that is administered after a constant time delay and desynchronizes by hitting the cluster in a vulnerable state. The goal of the reset is to control the dynamics of the cluster by restarting the cluster in a stereotypical way. The reset may be achieved by means of (i) a strong single pulse [24], (ii) a high-frequency pulse train (with a frequency that exceeds the cluster's natural frequency by a factor of 20 or more) [25] or (iii) a low-frequency pulse train (with a frequency that is similar to the cluster's natural frequency) [26]. Variants (i) and (ii) cause a *hard reset* (i.e. abrupt reset) during which the collective oscillation runs through less than one period. In contrast, variant (iii) causes a *soft reset* (i.e. slow reset): The collective oscillation is not quickly abrupted. Rather, in the course of an entrainment the influence of the initial dynamic state at the beginning of the stimulation fades away while the collective oscillation runs through several periods.

In this paper we dwell on variant (ii), i.e. a composite stimulus with a high-frequency pulse train as resetting stimulus, and compare it to the standard permanent high-frequency stimulation.

MODEL

The model describing the impact of a stimulus S on a cluster of N interacting phase oscillators subjected to random forces is given by the *Langevin Equation*

$$\dot{\psi}_j = \Omega + \frac{1}{N}\sum_{k=1}^{N}\Gamma(\psi_j - \psi_k) + X(t)S(\psi_j) + F_j(t), \tag{1}$$

where ψ_j denotes the phase of the jth phase oscillator [19]. For the sake of simplicity all oscillators have the same eigenfrequency: $\omega_j = \Omega$. The global coupling Γ is a 2π-periodic function, *i.e.* $\Gamma(x) = \Gamma(x+2\pi)$, depending on coupling parameters (see below). For vanishing stimulation ($S = 0$), Eq. (1) is a well-known standard model of globally coupled phase oscillators [2]. Since in neurons the impact of an electrical stimulus depends on the neuron's phase [20], the stimulus is given by a 2π-periodic function $S(\psi) = S(\psi + 2\pi)$ which depends on stimulation parameters (see below). Switching on and off the stimulus is taken into account by $X(t)$: During the stimulation $X(t) = 1$, whereas without stimulation $X(t) = 0$. The random forces $F_j(t)$ are modelled by Gaussian white noise fulfilling $\langle F_j(t)\rangle = 0$ and $\langle F_j(t)\, F_k(t')\rangle = D\delta_{jk}\delta(t-t')$, where D is the constant noise amplitude.

To study the dynamics of Eq. (1) we first determine the corresponding *Fokker-Planck equation* which is an evolution equation for the probability density $f(\{\psi_l\},t)$, where $\{\psi_l\}$ denotes the vector $(\psi_1,\ldots,\psi_N)$. To crucially simplify the analysis we follow Kuramoto [2] by choosing a more macrospcopic level of description. With this aim in view we introduce the number density

$\tilde{n}(\{\psi_l\};\psi) = N^{-1}\sum_{k=1}^{N}\delta(\psi-\psi_k)$, and, finally, the *average number density* $n(\psi,t)$ according to $n(\psi,t) = \int_0^{2\pi}\cdots\int_0^{2\pi} \mathrm{d}\psi_1\cdots\mathrm{d}\psi_N\, \tilde{n}(\{\psi_l\};\psi) f(\{\psi_l\};t)$. With a little calculation we obtain

$$\frac{\partial n(\psi,t)}{\partial t} = -\frac{\partial}{\partial \psi}\left[n(\psi,t)\int_0^{2\pi} d\psi' \Gamma(\psi-\psi')n(\psi',t)\right]$$
$$-\frac{\partial}{\partial \psi}n(\psi,t)X(t)S(\psi) - \Omega\frac{\partial}{\partial \psi}n(\psi,t) + \frac{D}{2}\frac{\partial^2 n(\psi,t)}{\partial \psi^2}, \tag{2}$$

which holds for large N [19].

To describe the cluster's collective dynamics and to estimate the extent of in-phase synchronization we use the order parameter Z defined by

$$Z(t) = R(t)e^{i\varphi(t)} = \int_0^{2\pi} n(\psi,t)e^{i\psi}\mathrm{d}\psi, \tag{3}$$

where $R(t)$ and $\varphi(t)$ are the real amplitude and phase of the order parameter, where $0 \le R(t) \le 1$ for all times t [2, 28]. Perfect in-phase synchronization corresponds to $R = 1$, whereas in the incoherent state $R = 0$.

SPONTANEOUS DYNAMICS

Before considering the stimulus' impact let us recall the dynamics spontaneously occurring without stimulation, i.e. for $X = 0$. If the coupling terms are of lowest order, e.g., $\Gamma(x) = -K\sin x$ (with positive K), noisy in-phase synchronization emerges out of the incoherent state $n = 1/(2\pi)$ due to a decrease of the noise amplitude D [2] or, analogously, if the coupling K exceeds its critical value D [19]. In the incoherent state Z vanishes, whereas in-phase synchronization, i.e. a one-cluster state, corresponds to a stable limit cycle of the form $Z = Y\exp[i(\Omega+\bar{\Omega})t]$, where Y is a complex constant, and $\bar{\Omega}$ is a real frequency shift term which depends on noise amplitude and coupling parameters, and vanishes if there are no cosine coupling terms [19]. Z's limit cycle is connected with a travelling wave of n.

EFFECTIVELY DESYNCHRONIZING COMPOSITE STIMULI

We consider a composite stimulus consists of a resetting high-frequency train of short pulses which after a pause is followed by a weaker single pulse. The pulse train resets the cluster while the following single pulse desynchronizes it. The pulse train consists of M identical equidistant pulses. Begin and end of the kth pulse of the pulse train are denoted by $t_{\mathrm{b},k}$ and $t_{\mathrm{e},k}$, respectively. The frequency of the pulse train is $\omega_v = 2\pi/(T_1+T_2)$, where $T_1 = t_{\mathrm{e},k} - t_{\mathrm{b},k}$ and $T_2 = t_{\mathrm{b},k+1} - t_{\mathrm{e},k}$ are pulse duration and pause duration, respectively. The purpose of the pulse train is (i) to force the cluster into a well-defined dynamic state and (ii) to suppress the collective oscillation during the pulse train with pulses shorter than the pauses ($T_1 < T_2$). Therefore (i) the pulses of the train have to be sufficiently strong, and (ii) ω_v has to be large compared to the cluster's spontaneous frequency Ω (e.g., $\omega_v/\Omega = 20$). During the pulse train a 1:1 entrained, high-frequency periodic motion of Z rapidly occurs. The entrained Z is captured within a small portion of the Gaussian plane (containing real and imaginary part of Z), so that Z does no longer circulate around the origin of the Gaussian plane, and the collective oscillation is stopped. In this way a reset is achieved with the high-frequency pulse train.

The impact of a single pulse and a composite stimulus on the cluster in the stable synchronized state is compared in Fig. 1 by varying the phase of Z at the beginning of the stimulation within one cycle. Besides the amplitude R from (3) the firing density

$$p(t) = n(0,t) \tag{4}$$

is displayed. The latter is the number density of all firing neurons, where a single neuron fires whenever its phase vanishes (mod 2π) [19]. The single pulse causes a desynchronization only if it hits Z at or close to a vulnerable phase $\phi_B^{crit} \approx 0.38$ (Fig. 1a,c,e). In contrast, the composite stimulus gives rise to a temporary desynchronization no matter at which initial phase it is administered (Fig. 1b,d,f). For further details concerning the dynamics of Z during stimulation with a single pulse and with a composite stimulus I refer to [19, 25].

DESYNCHRONIZATION VS. SUPPRESSION OF ACTIVITY

Since a composite stimulus causes a desynchronization irrespective of the initial state at which it is administered it can be used to block the resynchronization (Fig. 2a). In contrast, the same pulse train without following single pulse stops the individual neurons' firing during stimulation but does not cause any desynchronization (Fig. 2b). Rather we observe a rebound-like, particularly synchronous firing which occurs directly after the pulse train. To suppress the neurons' firing with such a simple pulse train persistently it, thus, has to be administer permanently.

DEMAND-CONTROLLED DEEP BRAIN STIMULATION

The mechanism by which DBS at high frequencies suppresses pathological rhythmic activity has not yet been clarified in detail. The permanent high-frequency stimulation basically mimics the effect of tissue lesioning by suppressing neuronal firing [18, 29]. DBS is reversible and has a much lower rate of side effects than lesioning with thermocoagulation [30]. However, permanent high-frequency stimulation is an unphysiological input which may cause an adaptation of the stimulated neuronal networks. This may be one of the reasons why in a number of patients the stimulation amplitude has to be increased in the course of the treatment in order to maintain a therapeutic effect. As a consequence of the increased stimulation strength, neighboring areas may be affected due to current spread, which leads to severe side effects like dysarthria, dysesthesia, and cerebellar ataxia.

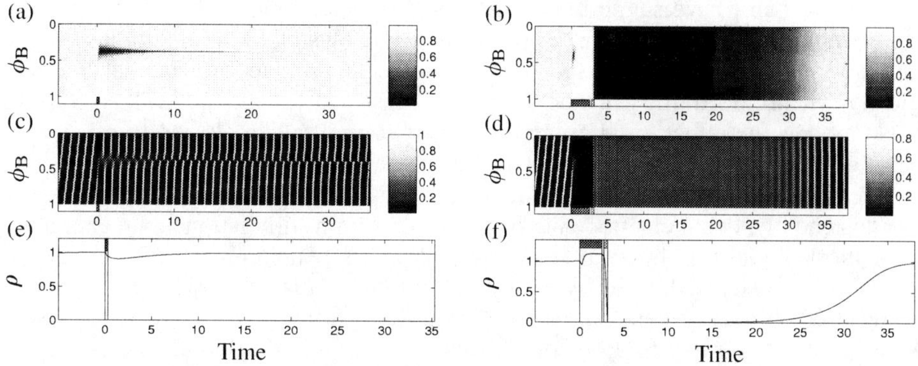

FIGURE 1. Time course of Z's amplitude R from (3) [(a), (b)] and the firing density $p(t) = n(0,t)$ [(c), (d)] before, during and after a single pulse [(a), (c)] and a composite stimulus [(b), (d)], where $\phi_B = \varphi_B/(2\pi)$ mod 1, the phase of Z at the beginning of the stimulation is varied within one cycle. Stimulation starts at $t = 0$. At the bottom of each plot in (a)–(d) the single pulses and the pulse trains are indicated by bars. The desynchronizing effect of the composite stimulus does not depend on the initial phase ϕ_B as additionally visualized in (f) with $\rho(t) = \langle R(t)\rangle / \langle R(0)\rangle$, i.e. the normalized amplitude of the order parameter averaged over the initial phase ϕ_B. $\langle R(t)\rangle = m^{-1}\sum_{j=1}^{m} R^{(j)}(t)$, where $R^{(j)}(t)$ is the order parameter's amplitude $R(t)$ belonging to the jth simulation, where in the $m = 100$ simulations from (b) the initial phase ϕ_B is varied within one cycle $[0,1]$ in equidistant steps. Note that $\langle R(0)\rangle = R(0)$. ρ tells us what happens on average if we stimulate without taking care of the initial phase. In contrast, (e) belongs to (a) and demonstrates how crucially the effect of the single pulse depends on the initial phase. In (e) and (f) begin and end of the single pulses and the pulse train are indicated by vertical lines connected by a shaded region. Model parameters: (a)–(f) $\Gamma(x) = -\sin x$, $D = 0.4$, $\Omega = 2\pi$, $S(\psi) = I\cos\psi$. $I = 7$ in the single pulse in (a) and (c) and in the desynchronizing single pulse (after the high-frequency pulse train) in (b) and (d). $I = 21$ in the high-frequency pulse train in (b) and (d). Single pulse duration $T = 0.4$ in (a) and (c), and $T = 0.41$ in (b) and (d). Pulse train in (b) and (d): pulse duration $T_1 = 0.02$, pause duration $T_2 = 0.03$, number of pulses $M = 50$. (Fig. from [25])

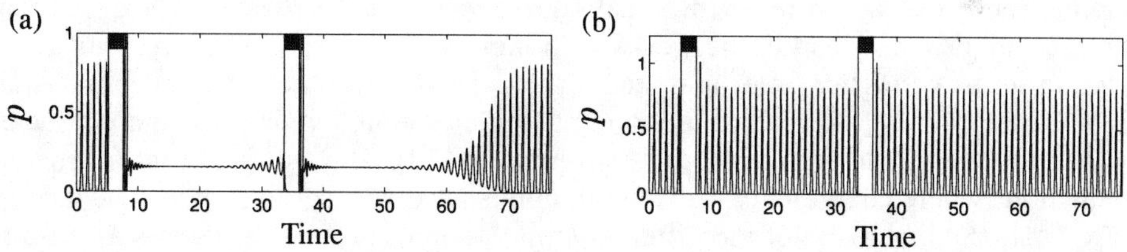

FIGURE 2. Time course of the firing density $p = n(0,t)$ from (4). (a) Two successively administered composite stimuli with identical parameters. The first one desynchronizes the cluster, whereas the second blocks the resynchronization. A composite stimulus consists of a high-frequency pulse train followed by a single pulse. (b) Two successively administered identical pulse trains with parameters as in (a) but without following single pulses only suppress the collective oscillation during the pulse train. After a simple high-frequency pulse train the extent of synchronization is particularly high, and the cluster restarts in a rebound-like manner. Model parameters as in fig. 1. Begin and end of single pulses as well as high-frequency pulse trains are indicated by vertical lines connected by a shaded region. (Fig. from [25])

For this reason I have suggested a different therapeutic approach: Instead of simply suppressing the neuronal firing in the pacemaker-like cluster, the stimulation techniques mentioned in Sec. should be used in order to desynchronize the pacemaker's pathologically synchronized firing in a demand-controlled way [24, 25, 26, 27]. Accordingly, Eq. (1) models the effect of stimulation on the pacemaker-like cluster. In other words, instead of stopping the driving force, I suggest to desynchronize it, so that it is no longer able to entrain other brain areas like premotor areas and the motor cortex.

To this end, the depth electrode has to be used for both stimulation and registration of the feedback signal, i.e. the local field potential (LFP), which in the model corresponds to the firing density p defined by Eq. (4). A desynchronizing bipolar double pulse is administered only and whenever the pacemaker-like cluster becomes synchronized, put otherwise, whenever its LFP exceeds a critical value (Fig. 2a). Both the high-frequency pulse train and the desnychronizing single pulse are delivered to the same site. The goal of this approach is to effectively block the resynchronization and, hence, keep the firing as close to the physiological (i.e. uncorrelated) firing mode as possible. Alternatively, the feedback signal can also be obtained by an epicortical electrode measuring the electrical activity of a cortical area that is strongly coupled to the pacemaker-like cluster (e.g. a premotor area or the primary motor cortex).

CONCLUSIONS

Several stimulation techniques have been developed which make it possible to effectively desynchronize a population of globally coupled phase oscillators [24, 25, 26, 27]. The desynchronizing effect of these stimulation methods does not depend on the coupling pattern in a critical way. Rather desynchronization is achieved, e.g., in networks with global coupling as well as in networks without any coupling [19]. The behavior after desynchronization, however, crucially depends on the coupling pattern. Without coupling there will be no resynchronization. Furthermore, the desynchronizing stimulation techniques also work in networks with more complex couplings containing, e.g., coupling terms of higher order and leading to complex synchronized states like noisy cluster states [19, 24]. Instead of monopolar single or double pulse stimulation one can also use bipolar single or double pulse stimulation [27]. This is significant for medical applications where charge-balanced stimulation is used in order to avoid tissue damage. For a detailed comparison of the different stimulation techniques, for an explanation of their calibration procedure and for details concerning experimental applications I refer to [19, 24, 26, 27].

The clinical effects of the suggested demand-controlled deep brain stimulation are currently being evaluated. Furthermore, the desynchronizing stimulation techniques presented here are just being studied in more microscopic models of networks of electrically stimulated neurons. It is also planned to apply the demand-controlled desynchronizing methods to MPTP monkeys, the animal model of PD [7]. The goal of these coordinated modelling and experimental studies is to establish stimulation techniques that should, in principle, be less aggressive to the brain tissue stimulated and, thus, help to reduce unwanted or adverse side effects.

ACKNOWLEDGMENTS

The work was supported by the German-Israeli Foundation for Scientific Research and Development (grant no. 667/00).

REFERENCES

1. H. Haken, *Advanced Synergetics*, Springer, Berlin, 1983; A. Pikovsky, M.G. Rosenblum, J. Kurths, *Synchronization: a Universal Concept in Nonlinear Sciences*, Cambridge University Press, 2001.
2. Y. Kuramoto, *Chemical Oscillations, Waves, and Turbulence*, Springer, Berlin, 1984.
3. A.T. Winfree, *The Geometry of Biological Time*, Springer, Berlin, 1980; L. Glass and M.C. Mackey, M.C., *From Clocks to Chaos. The Rhythms of Life*, Princeton University Press, 1988.
4. Freund, H.-J., *Physiol. Rev.* **63**, 387 (1983); A. Neiman et al., *Phys. Rev. Lett.* **83**, 4896 (1999); E. Mosekilde, Y. Maistrenko, D. Postov, *Chaotic Synchronization. Applications to Living Systems*, World Scientific Series on Nonlinear Science A, Vol. **42**, 2002.
5. Freeman, W.J., *Mass action in the nervous system*, New York, Academic Press, 1975; H. Steriade, E.G. Jones, R. Llinás, *Thalamic Oscillations and Signaling*, John Wiley & Sons, New York, 1990.; Hari, R., Salmelin, R., *TINS*, 20: 44 (1997).
6. J. Volkmann et al., *Neurology* **46**, 1359 (1996);
7. H. Bergman et al., *Mov. Dis.* **13** Suppl. 3, 29 (1998).
8. R.J. Elble, W.C. Koller, *Tremor*, John Hopkins University Press, Baltimore, 1990.
9. P. Tass et al., *Phys. Rev. Lett.* **81**, 3291 (1998).
10. Grillner, S. and Wallén, P., *Annu. Neurosci.*, *8*, 233 (1985).
11. Brown, T.G., *Proc. R. Soc. Lond. (Biol.)*, 84: 308 (1911); Sherrington, C., *The Integrative Action of the Nervous System.*, New Haven, Yale University Press, 2nd ed., 1947; von Holst, E., *Erg. Physiol.*, 42: 228 (1939); von Holst, E., *Brit. J. Anim. Behav.*, 2: 89 (1954).
12. Eckhorn, R., Bauer, R., Jordan, W., Brosch, M., Kruse, W., Munk, M., and Reitboeck, H.J., *Biol. Cybern.*, **60**, 121 (1988); Gray, C.M., Singer, W., *Proc. Natl. Acad. Sci. USA*, **86**, 1698 (1989).
13. Riesenhuber, M., and Poggio, T., *Neuron*, 24: 87 (1999).
14. R. Llinás, H. Jahnsen, *Nature* **297**, 406 (1982); D. Pare, R. Curro'Dossi, M. Steriade, *Neuroscience* **35**, 217 (1990).
15. F.A. Lenz et al., *Brain* **117**, 531 (1994).
16. A. Nini A. et al., *J. Neurophysiol.* **74**, 1800 (1995).
17. W.W. Alberts, E.J. Wright, B. Feinstein, *Nature* **221**, 670 (1969); Y. Lamarre et al., *Brain Res.* **32**, 246 (1971).
18. A.L. Benabid et al., *The Lancet* **337**, 403 (1991); S. Blond et al., *J. Neurosurg.* **77**, 62 (1992).
19. P.A. Tass, *Phase Resetting in Medicine and Biology – Stochastic Modelling and Data Analysis*, Springer, Berlin, 1999; P.A. Tass, *Prog. Theor. Phys. Suppl. (Kyoto)* **139**, 301 (2000).
20. E.N. Best, *Biophys. J.* **27**, 87 (1979); R. Guttman, S. Lewis, and J. Rinzel, *J. Physiol. (London)* **305**, 377 (1980).
21. Forssberg, H., Grillner, S. and Rossignol, S.,*Brain Res.*, **85**, 103 (1975).
22. P. Tass, *J. Biol. Phys.* **22**, 27 (1996); P. Tass, *J. Biol. Phys.* **22**, 125 (1996).
23. G.B. Ermentrout, N. Kopell, *J. Math. Biol.* **29**, 195 (1991); Grannan et al., *Neural Comp.* **4**, 550 (1992); D. Hansel, G. Mato, C. Meunier, *Europhys. Lett.* **23**, 367 (1993); Hansel, D., Mato, G., Meunier, C., *Concepts in Neuroscience*, 4: 193–210 (1993).
24. P.A. Tass, *Europhys. Lett.* **53**(1), 15 (2001); P.A. Tass, *Biol. Cybern.* **85**, 343 (2001).
25. P.A. Tass, *Europhys. Lett.* **55**(2), 171 (2001).
26. P.A. Tass, *Europhys. Lett.* **57**(2), 164 (2002); P.A. Tass, *Biol. Cybern.* **87**, 102 (2002).
27. P.A. Tass, *Phys. Rev. E* **66**, 036226 (2002).
28. Y. Aizawa, *Prog. Theor. Phys.* **56**, 703 (1976).
29. J.P. Wielepp et al., *Clin. Neurol. Neurosurg.* **103**, 228 (2001).
30. P.R. Schuurman et al., *N. Engl. J. Med.* **342**, 461 (2000).

Delays, Scaling and the Acquisition of Motor Skill

Juan Luis Cabrera* and John Milton†

*Laboratorio de Física Estadística, Centro de Física, IVIC, Apartado 21827, Caracas 1020A, Venezuela
†Department of Neurology, MC 2030, Committees on Neurobiology and Computational Neuroscience, The University of Chicago, 5841 South Maryland, Chicago, IL 60637, USA

Abstract. Motion analysis in three dimensions reveals a number of surprising features of the neural control of stick balancing at the fingertip, namely, 1) on-off intermittency in the controlled variable, and 2) controlling motor forces that exhibit self-similarity. The growing evidence in support of scaling and critical behaviors in neural motor control necessitates a re-thinking of how the nervous systems works.

INTRODUCTION

Motor skill acquisition takes time and practice [1]. The performance of motor tasks that at one time seemed so difficult, such as learning to walk, ride a bicycle, or ice skate, become effortless. Indeed, so much so, that soon, additional complex tasks can be performed concurrently, e.g. playing a musical instrument in a marching band. What is the nature of the changes in the organization of the nervous system that makes all of this possible? An important aspect of this problem is the observation that many skilled movements occur on time scales that are short compared to the latencies, or time delays, in the motor control pathways [2, 3]. How can fast and precise control be achieved in the face of noisy perturbations using time-delayed control mechanisms?

There are two intrinsic properties of the neural control mechanisms that maintain balance: 1) time delays and 2) noise [4, 5]. The fundamental nature of the interplay between noise and time-delayed feedback for controlling balance is captured in the task of stick balancing at the fingertip (Figure 1a). Longer sticks are much easier to balance than short ones: once the stick is sufficiently long its rate of movement become slow relative to the time required by the nervous system to make corrective movements. In appropriate mathematical models of inverted pendulums with time-delayed feedback, the range of corrective forces for which the upright position is stabilized is small and in parameter space the basin of attraction is surrounded by instability [6, 7]. Thus noisy fluctuations in the parameters can result in a crossing of the stability boundary. As a result the balanced stick has a finite probability of falling.

Here we use high speed, high precision motion analysis in three dimensions (Fig. 1a) to obtain a number of unexpected insights into how the nervous systems controls stick balancing at the fingertip and learns to increase its skill [7, 8].

CP665, *Unsolved Problems of Noise and Fluctuations: UPoN 2002: Third International Conference*, edited by S. M. Bezrukov

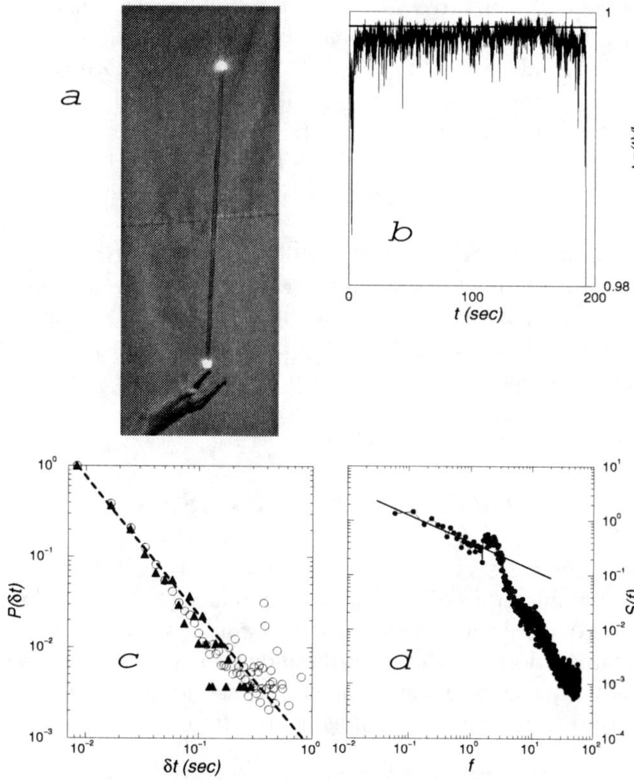

FIGURE 1. a) Stick balancing at the fingertip. Construction materials (e.g. wood, metal, synthetics) were chosen so that the mass remained the same (35 gr) as the length varied. The balls attached to the end of the stick (mass ~ 0.25 gr) are made of reflective material. Infrared light reflected from these markers is detected by specialized motion capture cameras (Qualisys, Inc.) and used to calculate the 3-D coordinates. Spatial resolution was 4-50 μm. b) Fluctuations $\frac{\Delta z}{l}$ as a function of time for a 62 cm stick balanced at the fingertip for a sampling frequency of 120 Hz where Δz is the difference in the vertical coordinate and l is the length of the stick. The horizontal line depicts the threshold position at 1.0005 times the mean value. c) Normalized laminar phase probability distribution, $P(\delta t)$. The black triangles are calculated for the data shown in a) and the open circles represent the mean distribution obtained from 48 realizations. The dashed line represents a power law with exponent $-\frac{3}{2}$. d) Log-log plot of the power spectrum for the fluctuations in $\frac{\Delta z}{l}$. It is characterized by two power law regimes with exponents close to $\frac{1}{2}$ (upper line) and -2.4 (lower line). Data reproduced from [7] with permission.

ON-OFF INTERMITTENCY

The primary sensory input for stick balancing is the visual estimation of the vertical displacement angle, θ [9]. Figure 1b shows a plot of the fluctuations in θ as a function of time. Periods in which small fluctuations in θ occur alternate with shorter periods characterized by larger ones. In order to investigate the nature of these fluctuations in θ we choose a threshold (horizontal line in Fig. 1b) and measured the time intervals between the occurence of corrective movements, i.e. fluctuations that cross the threshold in the upward direction. Figure 1c shows a plot of the logarithm of the probability of measuring a time interval of length δt, $P(\delta t)$, versus the logarithm of δt. The linear relationship over nearly three decades implies that the fluctuations obey the $-\frac{3}{2}$ scaling law characteristic of on-off intermittency [10], i.e.

$$P(\delta t) \sim \delta t^{-3/2}.$$

Further evidence in support of on-off intermittency is the observation that the power spectrum in Fig. 1d of the fluctuations in θ contains a region that obeys a $-\frac{1}{2}$ scaling law [11]. It should be noted that the time delay for stick balancing was estimated to be $\sim 100-200$ msec [7]. Thus the vast majority ($> 90\%$) of the threshold crossings occur at times short compared to the latency of the neural control mechanisms for stick balancing.

CRITICAL PHENOMENA

In stochastic dynamical systems, $-\frac{3}{2}$ scaling arises from the random forcing of an important control parameter across a stability boundary [10] Accordingly the observations in Figure 1 provide strong evidence that the dynamical system controlling stick balancing is operating close to a stability boundary. This prediction has been verified in a mathematical model for an inverted pendulum stabilized by time-delayed feedback with a noisy feedback gain (parametric noise), $r(t)$, for example,

$$\ddot{\theta}(t) = -p\dot{\theta}(t) + q\sin\theta(t) + r(t)\theta(t-\tau) \quad (1)$$

$$r(t) = r_o + \xi(t)$$

where τ is the time delay, p, q, r_o are constants, and $\xi(t)$ is Gaussian distributed white noise [7]. These observations suggest that the dynamical system controlling stick balancing has been tuned be very close (perhaps on) to a stability boundary. The implication is that critical behaviors (e.g. intermittency, self-organized criticality) may be playing a role in stabilizing the upright position of the balanced stick [7, 12].

SELF-SIMILARITY

Figure 2 shows the probability distribution of the change in speed, Δv, made by the finger during stick balancing at the fingertip as a function of the skill level of the subject. The change in speed is proportional to the corrective force applied by the finger to the stick in response to deviations of the balanced stick from vertical. At both skill levels the distributions are non-Gaussian and in particular the shape of the distributions resembles that observed for a truncated Lévy flight (TLF). A Lévy flight is a random walk whose step length comes from a power law distribution [13, 14]. A TLF arises in real physical systems because the probability of taking a step of a given size, $\overline{P}(\Delta v, \Delta t)$, also depends

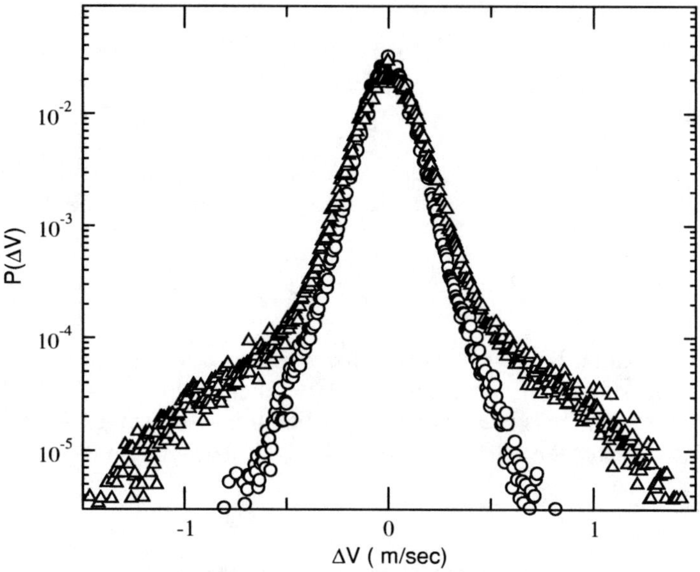

FIGURE 2. Probability distribution of Δv of hand movements while a 62 cm stick is balanced at the fingertip. The reflector closest to the fingertip was used to track the hand movements. Prior to practice this subject could balance 33 % of trials longer than 20s (circles); 100 % after 14 hours practice distributed over 14 days (triangles). The data was pooled from repeated trials and the plots show mean distributions for the pooled trials (variance less than the thickness of the symbols). Sampling frequency was 1000 Hz.

on the limitations of the system, i.e.

$$\overline{P}(\Delta v, \Delta t) = \begin{cases} c_1 L_\alpha(\Delta v, \Delta t) f(\Delta v, \Delta t) & \text{if } |\Delta v| > \Upsilon(\Delta t) \\ c_2 L_\alpha(\Delta v, \Delta t) & \text{otherwise} \end{cases}$$

where c_1, c_2 are normalization constants, Δt is the sampling time, $f(\Delta v, \Delta t)$ is a decreasing function of Δv, $\gamma > 0$ is a scale factor, $\Upsilon(\Delta t)$ is the critical value of Δt at which the distribution begins to deviate from $L_\alpha(\Delta v, \Delta t)$ for a given Δt, and

$$L_\alpha(\Delta v, \Delta t) = \int_0^{+\infty} e^{-\gamma \Delta t q^{-\alpha}} \cos(q \Delta v) dq.$$

Evidence that the nervous system can generate scale-invariant Lévy distributions has been obtained recently in *in vitro* neuronal networks [15].

For both skill levels the probability that the change in speed is 0, $\overline{P}(0, \Delta t)$, obeys the scaling law

$$\overline{P}(0, \Delta t) \sim (\Delta t)^{1/\alpha}$$

where $\alpha \sim 1.3$. Thus the change is skill level is reflected primarily by a change in the truncation of the TLF.

It has been demonstrated that when $f(\Delta v, \Delta t)$ is an exponentially decreasing function, $\overline{P}(0, \Delta t)$ depends on both the decay rate and $\Upsilon(\Delta t)$ [14]. For example, take f to be an exponentially decreasing function of $|\Delta v|$,

$$f(|\Delta v|, \Delta t) = \exp\left[-\left(\frac{|\Delta v| - \Upsilon(\Delta t)}{k(\Delta t)}\right)^{\beta}\right]$$

where $\beta = 2 - \alpha$ and $k(\Delta t)$ is a constant that controls the rapidity by which the Lévy flight is truncated. Then it can be shown that provided that Δt is sufficiently small

$$\overline{P}(0, \Delta t)|_{Gaussian} = \frac{\sqrt{\pi}\alpha\gamma^{1/\alpha}}{\sqrt{2}\Gamma(1/\alpha} \frac{1}{\sigma} \overline{P}(0, \Delta t)|_{TLF}$$

where Γ is the gamma function, and σ is the standard deviation of the distribution of the temporal series sampled at the smallest experimentally realizable Δt. When $\Upsilon(\Delta t)$ increases, σ decreases. This means that the probability of a change in speed of zero for a TLF becomes greater than that for a Gaussian distribution having the same standard deviation.

UNSOLVED PROBLEMS

Our observations present a host of unsolved problems to theoreticians interested in critical phenomena and to neuroscientists interested in motor control.

1. What is the role of feedback control in stick balancing? Traditionally feedback systems are though of in terms of a strategy to minimize an appropriately designed error function. It is difficult to understand the role of feedback for stick balancing from this perspective. Feedback certainly is necessary for a stability boundary to exist (there is no attractor when $\tau = 0$) and likely plays an important role in producing the changes in the distributions shown in Fig. 2. Measurements of the cross-correlation between the movements of the hand and the tip of the stick shows that the role of feedback decreases with increased skill [8]. It is not difficult to imagine that this would be reflected by a change in the truncation of the Lévy flight.
2. What is the nature of the attractor for stick balancing? It has been shown experimentally and mathematically that when the position of an inverted pendulum is controlled by time-delayed negative feedback, the attractor is a low-amplitude limit cycle [16]. However, in stick balancing the pendulum's suspension point changes. This is how parametric noise enters into the control problem [7]. We observed that the agreement between (1) and experimental observation was poor when $\sin\theta$ was approximated by θ.
3. What are the advantages of control strategies that involve critical phenomena over more traditional feedback control paradigms? In our example the advantages are likely related to the necessity of maintaining balance control on time scales shorter than the time delay in the presence of noise.

4. How is it possible to train a neural network to position a dynamical system close to a stability boundary and to alter the properties of a distribution of self-similar forces?
5. How are the results to be reconciled with Fitt's law that describes the trade-off between movement speed and accuracy [1, 17]? Indeed the time delay for the stick balancing task actually appears to increase with increased skill [8]. This observation implies that the beneficial effects on stick balancing due to the changes in the truncation of the Lévy flight must outweigh the detrimental effects of the increased delay. However, it is possible that concerns related to Fitt's law limits the strategies available to the nervous system to increase its balancing skill level.

CONCLUDING REMARKS

Our observations give compelling and indisputable evidence that critical behaviors arise in the control of stick balancing at the fingertip: the corrective forces exhibit self-similarity; the controlled variable exhibits on-off intermittency; numerical simulations of an appropriate mathematical model position the critical parameters to be very, very close (perhaps on) a critical stability boundary. This complexity is contrary to what would be expected if, let us say, an engineer were to design a dynamical system that would be stable in the presence of parametric noise. The "no brainer" choice of parameters would be to place them well inside a stability region so that noisy fluctuations would not lead to crosses of the stability boundary. However, time delays are an intrinsic problem for this control strategy. The downside of parameter choices inside the basin of attraction for a stable attractor is that corrections can not be made faster than the delay. How does the nervous system maintain balance on time scales shorter than the time delay?

Our results suggest an elegant solution to this problem: increase maneuverability by tuning the control system close to instability and then make use of parametric noise (on-off intermittency) to obtain corrective movements on all time scales including those shorter then the delay. The benefit of on-off intermittency is that the fluctuations in θ resemble a random walk with mean value of $\theta \sim 0$, i.e. the balanced position is statistically stabilized [7]. Skill at balance is further increased by tuning a distribution of self-similiar forces (TLF). The importance of noise for motor control has been recognized previously (e.g. [18, 19]). Our observations stand in stark contrast to modern theories in kinesiology that emphasize the properties of the attractors of Newtonian mechanical systems [20].

It is possible that there are advantages to stabilizing unstable equilibria by using critical behavior in the same way that the nervous system learns to balance a stick at the fingertip. Such insights could lead, for example, towards the development of better two-legged robots towards the design of earthquake proof buildings to novel strategies to prevent falling in the elderly and improve neuro-rehabilitation after stroke.

ACKNOWLEDGMENTS

We thank R. Knox and M. MacClellan for assistance in performing these experiments and Qualisys, Inc (East Windsor, CT) and Skilled Technologies (Phoenix, AZ) for technical assistance with the motion analysis. We thank S. A. Campbell, M. Ding, C. W. Eurich, P. Jung, A. Longtin, T. Ohira, M. C. Mackey and S. C. Venkataramani for useful comments. This work was supported by grants from the Brain Research Foundation (BRF) and the National Institutes of Mental Health. J. L. C. acknowledges support from MECD (Spain) and BRF.

REFERENCES

1. Fitts, P. M., and Posner, M. I., *Human Performance*, Prentice/Hall International, Inc., London, 1973.
2. Cochran, A., and Stobbs, J., *Search for the Perfect Swing*, Triumph Books, Chicago, 1986.
3. Jindrich, D. L., and Full, R. J., *J. Exp. Biol.*, **205**, 2803–2823 (2002).
4. Ohira, T., and Milton, J., *Phys. Rev. E*, **52**, 3277–3280 (1995).
5. Eurich, C. W., and Milton, J. G., *Phys. Rev. E*, **54**, 6681–6684 (1996).
6. Stepan, G., *Retarded Dynamical Systems: Stability and Characteristic Functions*, Pittman Research Notes in Mathematical Series 210, Longman Scientific and Technical, Harlow, New York, 1989.
7. Cabrera, J. L., and Milton, J. G., *Phys. Rev. Lett.*, **89**, 158702–1–4 (2002).
8. Cabrera, J. L., and Milton, J. G., *J. Neuroscience* (2002), submitted.
9. Engel, K. C., and Soechting, J. F., *J. Neurophysiol.*, **83**, 3483–3496 (2000).
10. Platt, N., Spiegel, E. A., and Tresser, C., *Phys. Rev. Lett.*, **70**, 279–282 (1993).
11. Venkataramani, S. C., Antonsen, T. M., Ott, E., and Sommerer, J. C., *Physica*, **96D**, 66–99 (1996).
12. Kelso, J. A. S., *Dynamic Patterns: The Self-organization of Brain and Behavior*, MIT Press, Cambridge, MA, 1999.
13. Mantegna, R. N., and Stanley, H. E., *Phys. Rev. Lett.*, **73**, 2946–2949 (1994).
14. Gupta, H. M., and Campanha, J. R., *Physica A*, **268**, 231–239 (1999).
15. Segev, R., Benveniste, M., Hulata, E., Cohen, N., Palevski, A., Kapon, E., Shapira, Y., and Ben-Jacob, E., *Phys. Rev. Lett.*, **88**, 118102–1–118102–4 (2002).
16. Landry, M., Campbell, S. A., and Morris, K. L. (2002), URL `www.math.uwaterloo.ca/~sacampbe/invpend.html`.
17. MacKenzie, I. S., and Ware, C., "Lag as a determinant of human performance in interactive systems," in *Proceedings of the ACM Conference on Human Factors in Computing Systems - INTERCHI'93*, ACM, New York, 1993, pp. 488–493.
18. Matthews, P. B. C., *J. Physiol.*, **492.2**, 597–628 (1996).
19. Harris, C. M., and Wolpert, D. M., *Nature (London)*, **394**, 780–784 (1998).
20. Zatsiorsky, V., *Kinetics of Human Motion*, Human Kinetics, Champaign, Il, 2002.

ION CHANNELS AND MOTORS

Turning a Poor Ion Channel into a Good Pump

Dean Astumian

Department of Physics and Astronomy, The University of Maine, Orono, Maine 04469, USA

Abstract. We consider a membrane protein that can exist in two configurations, either one of which acts as a poor ion channel, allowing ions to slowly leak across the membrane from high to low elctrochemical potential. We show that random external fluctuations can provide the energy to turn this poor channel into a good pump that drives ion transport from low to high electrochemical potential. We discuss this result in terms of a gambling analogy, and point to possible implications for fields as far ranging as population biology, economics, and actuarial science.

Ion channels are proteins that allow selective conduction of ions across a membrane. The direction of flow is always from high to low electrochemical potential - i.e., in the thermodynamically spontaneous direction. Ion pumps on the other hand convert chemical energy to drive thermodynamically uphill transport of ions across biological membranes. These protein molecules span the cell membrane and use energy from adenosine triphosphate (ATP) hydrolysis to generate and maintain the ion electrochemical gradients essential for life [1].

We can view pump proteins as small ($5-10\,\text{nm}$) devices that pump ions between two reservoirs, one the inside and the other the outside of the cell. The electrical potential experienced by an ion is constant in either reservoir, but varies rapidly inside the effectively one dimensional protein tunnel that connects the two, with energy barriers (gates) at the two entrances surrounding an energy well (binding site) in the middle. As ATP is bound to the protein and hydrolyzed on its surface, the protein undergoes shape changes that in turn change the relative gate and binding energies for the ion. The resulting cyclic modulation of the local electric potential within the protein can drive ion pumping.

Because the time between the individual chemical steps of ATP hydrolysis is random, it has long been held that strictly regulated coupling between the chemical events of ATP hydrolysis and mechanical events of transport is essential for the function of a molecular pump [2]. A typical mechanism involves feedback between the protein and its ligands to ensure that neither ATP hydrolysis nor a cycle of ion transport can be completed without the other process occurring.

The rigid requirement for such clock-like coupling has been challenged by recent experiments carried out by Tian Tsong and colleagues at the University of Minnesota [3, 4, 5]. Working under conditions where ATP hydrolysis cannot occur, these researchers applied a randomly fluctuating electric field to a suspension of red blood cells. Amazingly, the zero average external fields were able to drive thermodynamically uphill transport via the ion pump sodium, potassium ATPase (Na,K ATPase) even without ATP hydrolysis. The important point is that there is no possibility whatsoever for control of

CP665, *Unsolved Problems of Noise and Fluctuations: UPoN 2002: Third International Conference,* edited by S. M. Bezrukov

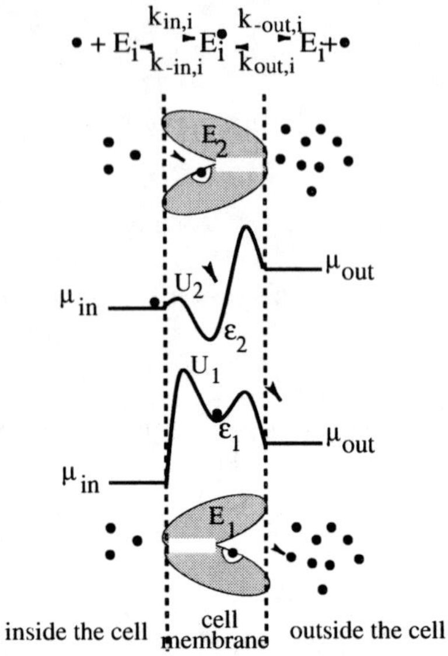

FIGURE 1. Cartoon of a pump protein that can exist in two distinct forms, one, E_1, with the binding site facing the outside of the cell, and the other, E_2, with the binding site facing the inside. In addition to the difference in access, the affinity of the two states for the ion (denoted by *bullet*) is also very different as indicated by the relative energy levels of the binding site (local potential minimum). It is important to note that the cartoon does not indicate that the protein undergoes a full rotation through the membrane, which would be energetically prohibitive. There are many other more subtle structural changes such as local rotation of $\alpha-$ helices that can result in changing the affinity of and access to the binding site with local and relatively small amplitude molecular motions.

the timing of an electric pulse by the occupancy of the ion binding site of an individual protein.

This result can be understood in terms of a model where the field drives the protein between two states - one with a tight ion binding site exposed to the inside, and another with a much weaker ion binding site exposed to the outside (Fig. 1). When the protein is in the inside facing configuration ion binds to the tight site even if the concentration of the ion on the inside is relatively low. Then, when the protein is driven to the outside facing form, the ion is released from the now weak site even if the concentration on the outside is relatively high. This is not a violation of the second law - at equilibrium the energy required for a transition from the tight to weak binding form with an ion at the binding site is large and hence the probability is very small, ensuring that any transport is thermodynamically downhill.

However, because the conformational transition involves a change in dipole moment of the protein, external electric pulses can provide the energy for this transition, albeit at possibly random times, thus driving uphill transport [3, 4, 5, 6] via a “flashing” ratchet [7, 8, 9, 10] mechanism. Either state (configuration) alone is a poor channel - ions are conducted down the electrochemical gradient but only slowly.

We can see this by solving for the current in either state. For simplicity, we consider a model where either 0 or 1 ion can be bound at the binding site of the protein, and treat the reservoirs inside and outside the cell as being each at constant chemical potential of ion, μ_{in} and μ_{out}, respectively. This is an excellent approximation so long as the number of ions in the reservoirs is large compared to the number of pump proteins. When the protein is in the ith state the rate of change of the probability for the binding site to be occupied, P_i is

$$\frac{dP_i}{dt} = (k_{in,i} + k_{out,i})(1 - P_i) - (k_{-in,i} + k_{-out,i})P_i. \tag{1}$$

For notational convenience we take a reference state where the gate height between the inside reservoir and the binding site is U_0 and the well energy is ε_0, and consider states 1 and 2 as perturbations on this. Thus $k_{in,i} = k_{in}^0 e^{\mp\delta u}$, $k_{-in,i} = k_{-in}^0 e^{\pm\delta\varepsilon\mp\delta u}$, $k_{out,i} = k_{out}^0$, and $k_{-out,i} = k_{-out}^0 e^{\pm\delta\varepsilon}$ and we take the upper sign of the "$\pm$" and "$\mp$" symbols for $i = 1$ and the lower sign for $i = 2$. The ratios $k_{in,i}/k_{-in,i} = e^{\mu_{in}-\varepsilon_0\mp\delta\varepsilon}$, $k_{out,i}/k_{-out,i} = e^{\mu_{out}-\varepsilon_0\mp\delta\varepsilon}$, and $k_{in,i}k_{-out,i}/k_{out,i}k_{-in,i} = e^{-\Delta\mu}$ (where $\Delta\mu = \mu_{out} - \mu_{in}$) are fixed by thermodynamics and hence do not depend on the barrier height U_0 or on δu. We measure all energies in units of k_BT, and we have subsumed the constant ion concentrations in the inside and outside reservoir into k_{in}^0 and k_{out}^0, respectively, so all k's are uni- or pseudo uni-molecular rate constants.

Solving equation (1) for the steady state $dP_i/dt = 0$ we find

$$P_{ss,i} = (k_{in,i} + k_{out,i})\tau_i. \tag{2}$$

where $\tau_i = (k_{in,i} + k_{-in,i} + k_{out,i} + k_{-out,i})^{-1}$ is the relaxation time. Note that when $\mu_{in} = \mu_{out} = \mu$ the probability for the binding site to be occupied is given by the Fermi distribution $P_{FD} = (1 + e^{(\varepsilon\pm\delta\varepsilon-\mu)})^{-1}$ for all $i = 1,2$. Now, inserting Eq. (2) into the expression for the rate at which ion is released to the outside

$$J_i = k_{-out,i}P_i - k_{out,i}(1 - P_i), \tag{3}$$

we can express the steady state current as

$$J_{i,ss} = \tau_i k_{in,i}k_{-out,i}(1 - e^{\Delta\mu}), \tag{4}$$

so the current is always in the direction of decreasing electrochemical potential.

The coefficient in Eq. (4) - the conductance - can also be written

$$\tau_i k_{in,i}k_{-out,i} = -\frac{\tau_i^{-1}}{(s+s^{-1})^2}\frac{\partial P_{\mathrm{ss,i}}}{\partial\varepsilon_0}, \tag{5}$$

which is the product of a characteristic frequency τ_i^{-1}, an anisotropy factor $(s+s^{-1})^{-2}$ bounded between 0 and .25 (where $s = \sqrt{k_{-in,i}/k_{-out,i}}$) that reflects the symmetry of the energy profile within the channel, and the derivative of the steady state distribution function with respect to the well energy. This form is reminiscent of the Landauer formula for the conductance of a mesoscopic channel at finite temperature [13], although the Landauer equation is derived based on ballistic conduction where the electrons are not localized within the channel, and equations (1) - (5) here are derived based on the charges being thermally equilibrated within the channel, allowing a Markov description Eq. (1).

When external fluctuations or oscillations drive transitions between the two forms however, the protein can function as a pump. If the fluctuations are fast compared to both τ_i the current in the fluctuating environment is calculated by averaging the rate constants $[\overline{k}_{in} = k^0_{in}\cosh(\delta \mathrm{u}), \overline{k}_{-in} = k^0_{-in}\cosh(\delta \mathrm{u} - \delta\varepsilon), \overline{k}_{out} = k^0_{out}$, and $\overline{k}_{-out} = k^0_{-out}\cosh(\delta\varepsilon)]$ and then solving the Eq. (2) for $\overline{P}_{ss}$ and inserting the result into Eq. (3) to find

$$\overline{J}_{Fluct,ss} = \overline{\tau}\overline{k}_{in}\overline{k}_{-out}[1 - \frac{\cosh(\delta \mathrm{u} - \delta\varepsilon)}{\cosh(\delta \mathrm{u})\cosh(\delta\varepsilon)}e^{\Delta\mu}]. \tag{6}$$

The coefficient of $e^{\Delta\mu}$ in Equation (6) is bounded above by 2 when $\delta \mathrm{u}$ and $\delta\varepsilon$ have opposite sign, is bounded below by 0 when $\delta \mathrm{u}$ and $\delta\varepsilon$ have the same sign, and is 1 when either $\delta \mathrm{u} = 0$ or $\delta\varepsilon = 0$. With $\Delta\mu = 0$ the net pumping to lowest order in δu and $\delta\varepsilon$ is $\tau^0 k^0_{in} k^0_{-out}\delta u\delta\varepsilon$.

The main feature of this model is that modulation of a kinetic barrier height and of the energy of a binding site, both internal parameters governed by the conformation of the protein, can drive net pumping. The interaction between thermal noise, external fluctuations, and the systems kinetic asymmetry for this ion pump model can be partly understood by analogy with a simple gambling game [14, 15] played on a board with two bases, A and B, for the players piece as shown in Fig. 2. The object is to complete a clockwise cycle (corresponding to transport of one ion from inside to outside the cell) before completing a counterclockwise cycle (corresponding to transport of one ion from outside to inside the cell), with the moves between the two bases determined by the roll of a pair of dice. If the game is played according to either of the two sets of rules shown the player loses on average - a counterclockwise cycle is more likely than a clockwise one. However, if we change the game very slightly by allowing a coin flip before each roll of the dice to determine whether the next move is played according to rule set one or rule set two, we surprisingly change the outcome of the game to one where the player on average wins.

In this game, the dice play the role of thermal noise, allowing occasional transitions over the energy barriers, and the net tendency to lose in either game independently (given by the ratio 16/20) represents the overall ion electrochemical gradient. The coin flip plays the role of the random input of energy that allows transition from one state to the other. This simple game is related to what is known in statistics as Simpson's paradox.

In order to connect this picture to the experimental results of Tsong et al. we must realize that the two protein configurations shown in Fig. 1 are in a dynamic equilibrium and the effect of an external perturbation is to cause the equilibrium and rate constants connecting the two to vary. An expanded kinetic scheme illustrating this is shown in Fig. 3a, and a simplified diagram, introduced for notational simplicity in the following mathematical treatment is shown in Fig. 3b.

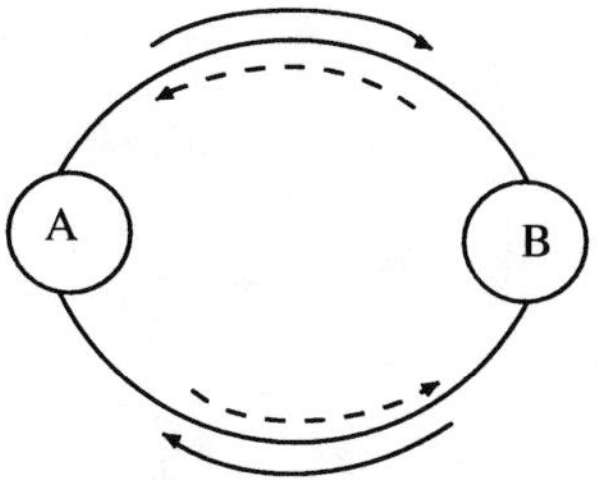

Rule 1

	A	B
cw	7,11 (8)	11 (2)
ccw	2,3,12 (4)	2,4,12 (5)

Rule 2

	A	B
cw	11 (2)	7,11 (8)
ccw	2,4,12 (5)	2,3,12 (4)

FIGURE 2. Gambling analogy for the effect of external fluctuations on a cyclic process. The game can start on either base, and the object is to complete a clockwise cycle around the board before a counterclockwise cycle, where the moves are determined by the roll of a pair of dice according to a table. Two such rule tables are shown below the gameboard. According to rule set one, if the players piece is on base "A", the player makes a clockwise move to "B" if a 7 or 11 is rolled, and a counterclockwise move to "B" if a 2,3, or 12 is rolled, and stays on "A" if any other number is rolled. Play continues in this manner until the player has completed a clockwise cycle for a win, or a counterclockwise cycle, for a loss. Playing according to either rule set 1 or 2, losses are more frequent than wins. The relative probability for a win can be calculated by multiplying the number of ways to make a clockwise move from A by the number of ways to make a clockwise move from B (2x8 = 16), and similarly for a loss (5x4 = 20). Thus playing either game we expect only 80 wins for every 100 losses - not a good bet. However, if after every roll of the dice a coin is flipped to determine whether the next move will be played according to rule set 1 or 2, the player will win more often than lose. Since the coin flip randomizes the rules before each turn, we now calculate the probability to win by taking the product of the average of the number of ways to move clockwise from A and B ((8+2)/2*(8+2)/2 = 25) and similarly for the probability to lose ((4+5)/2*(4+5)/2 = 20.25). Thus we can expect to win 100 times for every 81 losses - a very good bet indeed.

If we denote the probability of a protein to be in state i as E_i, then the E_i's satisfy kinetic equations:

$$\frac{dE_i}{dt} = -\sum_{j=1}^{4} (\omega_{ij} E_i - \omega_{ji} E_j), \tag{7}$$

Since $\sum_{i=1}^{4} E_i = 1$, E_4 can be eliminated from (7),

$$\frac{d\mathbf{E}}{dt} + P \cdot \mathbf{E} = \mathbf{Q}, \tag{8}$$

where

$$\mathbf{E} = \begin{pmatrix} E_1 \\ E_2 \\ E_3 \end{pmatrix}, \quad \text{and} \quad \mathbf{Q} = \begin{pmatrix} \omega_{41} \\ 0 \\ \omega_{43} \end{pmatrix},$$

and

$$P = \begin{bmatrix} \omega_{12} + \omega_{14} + \omega_{41} & \omega_{41} - \omega_{21} & \omega_{41} \\ -\omega_{12} & \omega_{21} + \omega_{23} & -\omega_{32} \\ \omega_{43} & \omega_{43} - \omega_{23} & \omega_{32} + \omega_{34} + \omega_{43} \end{bmatrix}.$$

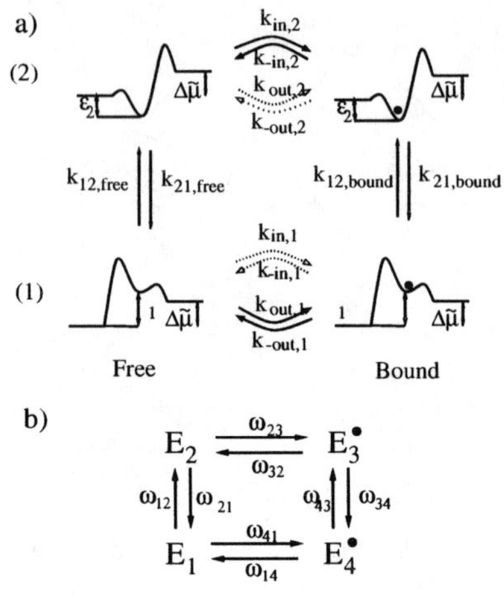

FIGURE 3. a) Expanded kinetic scheme with the two states in Fig. 1 shown in dynamic equilibrium with each other. The dashed lines indicate very slow transitions. b) Simplified diagram introduced for notational convenience. For simplicity we assume that $k_{\pm in,1}$ and $k_{\pm out,2}$ are negligible compared to the other transition coefficients.

When an external electric field is applied, the rate coefficients ω_{ij} are replaced with

$$\omega'_{ij} = \omega_{ij} \exp[z_{ij}\psi], \tag{9}$$

where ψ is the potential of the field, and z_{ij} describe the effects caused by difference of energy in the electric field due to different charge distribution of the states[20].

Due to the exponential functional form of (9), a dichotomic signal will simplify the theoretical calculation greatly and at the same time capture the essential properties of the transport process. We consider a stochastic dichotomic external potential:

$$\psi(t) = \gamma\xi(t), \tag{10}$$

where γ is the amplitude of the signal and $\xi(t) \in \{1,-1\}$, with $\langle\xi(t)\xi(t')\rangle = e^{-2f|t-t'|}$, where f is the frequency of the signal. With this form of external forcing, the expressions of ω'_{ij} are simplified to

$$\begin{aligned} \omega'_{ij} &= \omega_{ij}[\cosh(z_{ij}\gamma) + \sinh(z_{ij}\gamma)\xi(t)] \\ &\equiv c_{ij} + s_{ij}\xi(t). \end{aligned}$$

And now we can get an exact solution of the time-averaged current. As in [17, 18], using the Shapiro-Loginov theorem[19], we can average over ξ and obtain

$$\dot{\mathbf{W}} = -A \cdot \mathbf{W} + \mathbf{y}, \tag{11}$$

TABLE 1. Parameterization used in the numerical calculation. We assumed $z_{12} = z_{21} = z_{34} = z_{43} = 0$, because we do not expect much conformation differences between state 1 and state 2 or between state 3 and state 4.

i,j	1,2	2,3	3,4	4,1
$\omega_{ij}(s^{-1})$	a	$c_1 b$	a	$c_2 b$
h] $\omega_{ji}(s^{-1})$	ab	c_1	ab	c_2
z_{ij}	0	-2	0	4
z_{ji}	0	-3	0	-2

where **W** and **y** are both six-dimensional column vectors with

$$\mathbf{W} = \begin{pmatrix} \langle \mathbf{E} \rangle \\ \langle \xi \mathbf{E} \rangle \end{pmatrix}, \quad \mathbf{y} = \begin{pmatrix} \mathbf{Q_1} \\ \mathbf{Q_2} \end{pmatrix},$$

and A a 6×6 matrix,

$$\mathbf{A} = \begin{bmatrix} \mathbf{P_1} & \mathbf{P_2} \\ \mathbf{P_2} & \mathbf{P_1} + 2f\mathbf{I} \end{bmatrix},$$

where $\mathbf{P_1}$ and $\mathbf{P_2}$ are obtained by replacing ω_{ij} in $\mathbf{P}$ with c_{ij} and s_{ij}, respectively, the same is true for $\mathbf{Q_1}$ and $\mathbf{Q_2}$, and $\mathbf{I}$ is an 3×3 unit matrix. The steady state solution to equation (11) is

$$\mathbf{W}_\infty = A^{-1} \cdot \mathbf{y},$$

and the current is

$$\begin{aligned} J &= \lim_{t \to \infty} \langle \omega'_{23} E_2 - \omega'_{32} E_3 \rangle \\ &= c_{23} W_{\infty 2} + s_{23} W_{\infty 5} - c_{32} W_{\infty 3} - s_{32} W_{\infty 6}. \end{aligned} \tag{12}$$

The values of the parameters used in numerical calculations are listed in Table 1. We also need a scaling factor to convert between the experimental units (amol/cell/hr) and those of the calculation (turnovers/pump/sec). There are about $100 - 1000$ pump proteins per cell so this factor is in the range $0.3 - 6$. We use 3 as the best fit value of this factor.

Fig. 4 shows that the numerical results calculated using this one set of parameters fit to both the frequency and voltage dependence of the experimental data nicely. (The experimental data are from [4].)

In gambling games a fast tongue is all that is often necessary to convince someone to allow the occasional changing of the rules that can turn a losing proposition into a winning one. In natural systems what is required is energy. But in biology this is readilly avaible in the form of chemical reactions kept far from equilibrium by the food we eat, and the random binding of reactants and release of product molecules can for an enzyme play the same role as the flip of a coin in the gambling game described above. Thus a protein can function as a good pump to transport ions uphill even though in any static configuration ions would move downhill as through a poor channel.

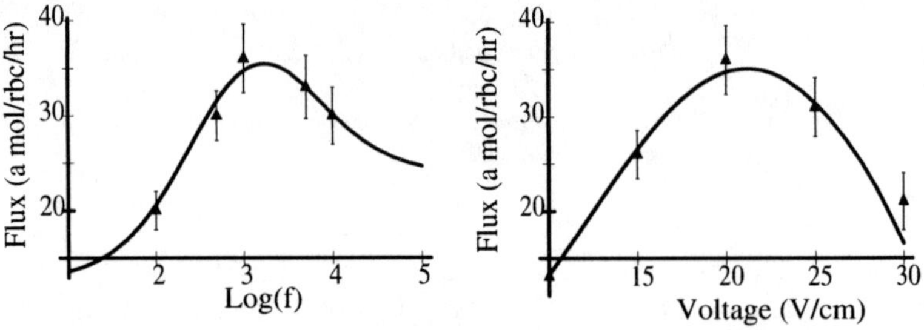

FIGURE 4. The frequency and voltage dependence of the ion current. The points are from experimental data [3]. The Solid line in a) is from the numerical calculation with the voltage set to $20V$ and that in b) is the theoretical curve with the frequency fixed at $1000Hz$. For both curves the parameters used were $b = 3, a = 100, c_1 = 1.2x10^5, and c_2 = 9$.

I thank Drs. Tong Zhou, Martin Bier, and Imre Derenyi, and Prof. T. Y. Tsong for very valuable discussions. This research was supported by NIH grants R29ES06620 and RO3ES08913.

REFERENCES

1. Lauger P, *Electrogenic ion pumps*, (Sinauer, Sunderland, MA, 1991).
2. Jencks, W.P., Meth. Enz.**171**, 145 (1989).
3. D.S. Liu, R.D. Astumian, and T.Y. Tsong, J. Biol. Chem. **265**, 7260 (1990).
4. T.D. Xie, P. Marszalek, Y.D. Chen, and T.Y. Tsong, Biophys. J. **67**, 1247 (1994).
5. T.D. Xie, Y.D. Chen, P. Marszalek, and T.Y. Tsong, Biophys. J. **72**, 2495 (1997).
6. T.Y. Tsong and R.D. Astumian, Bioelectrochem. Bioenerg. **15**, 457 (1986).
7. R.D. Astumian, P.B. Chock, T.Y. Tsong, Y.-D. Chen, and H.V. Westerhoff, Proc. Natl. Acad. Sci. USA **84**, 434 (1987).
8. R.D. Astumian and M. Bier, Phys. Rev. Letts. **72**, 1766 (1994).
9. Prost, J., Chawin, J.-F., Peliti, L., & Ajdari, A. (1994) Phys. Rev. Letts **72**:2652-2655.
10. R. D. Astumian, Science **276**, 917 (1997).
11. Astumian, R. D. and Robertson, B. Journ. Amer. Chem. Soc. **115**, 11063 (1993).
12. V. S. Markin, T.Y. Tsong, R.D. Astumian, and B. Robertson, J. Chem. Phys. **93**, 5062 (1990).
13. R. Landauer, IBM J Res Dev **32** 306 (1988); for review see S. Datta, Electronic Transport in Mesocsopic Systems (Cambridge University Press)(1997).
14. M. Eigen and R. Winkle, *How the Principles of Nature Govern Chance* (Princeton University Press, Princeton, New Jersey, 1981).
15. G.P. Harmer and D. Abbot, Stat. Sci. **14**, 206 (1999).
16. R. D. Astumian, P. B. Chock, T. Y. Tsong and H. V. Westerhof, Phys. Rev. A **39** 6416(1989).
17. A. Fuliński, Phys. Lett. A **193** 267(1994).
18. A. Fuliński, Phys. Rev. Lett. **79** 4926(1997).
19. V. E. Shapiro and V. M. Loginov, Physica (Amsterdam) **91A** 563(1978).
20. R. D. Astumian and B. Robertson, J. Chem. Phys. **91** 4891(1989).

Random Walk Model for Single-File Transport of Water Molecules through Carbon Nanotubes

Alexander Berezhkovskii[1*] and Gerhard Hummer[2]

[1] *Mathematical and Statistical Computing Laboratory, Center for Information Technology, National Institutes of Health, Bethesda, MD 20892*
[2] *Laboratory of Chemical Physics, National Institute of Diabetes and Digestive and Kidney Diseases, National Institutes of Health, Bethesda, MD 20892*

Abstract. A one-dimensional random walk model has been used to describe the single-file transport of water molecules through the interior channel of a carbon nanotube. Here, we derive expressions for the translocation and return probabilities of a molecule that has entered the channel; the distribution of the number of molecules traversing the tube in a unidirectional conduction pulse; the average time spent in the tube by molecules that either translocate through the tube or exit at the entry site; the average duration of unidirectional conduction bursts; and the average reversal time between oppositely directed conduction pulses.

1. INTRODUCTION

Recent molecular dynamics simulations of the transport of water molecules through a short and narrow carbon nanotube [1] showed that molecules densely filled the tube in a single-file arrangement and moved stochastically and concertedly. Most of the molecules entered and escaped from the same side of the tube, but a small fraction traversed the tube. Simulations showed that these conduction events occurred clustered in bursts of unidirectional translocation pulses.

In [2] we developed a continuous-time random walk model that quantitatively reproduced the prominent features of the single-file transport observed in [1], specifically, the translocation and return probabilities of a molecule that had entered the channel; the distribution of the number of molecules traversing the tube during one conduction pulse; the average time spent in the tube by molecules that either translocated or returned, respectively; the average duration of conduction bursts; and the average reversal time between oppositely directed pulses.

In this article, we provide formal derivations of some of the results that are given in [2] as illustrations of our 1D random walk approach. One can find more information about single-file transport of water through carbon nanotubes in [1,2] and references therein.

[*] Permanent address: Karpov Institute of Physical Chemistry, Vorontsovo Pole 10, Moscow, K-64, 103064, Russia.

CP665, *Unsolved Problems of Noise and Fluctuations: UPoN 2002: Third International Conference*, edited by S. M. Bezrukov
2003 American Institute of Physics 0-7354-0127-6

2. TRANSLOCATION AND RETURN PROBABILITIES AND AVERAGE LIFETIMES IN THE TUBE

Consider a continuous-time random walk of a particle that can jump between neighboring sites of a one-dimensional lattice containing M sites. The random walk is symmetric, i.e., the probabilities to jump from site *m* to sites *m-1* and *m+1* are equal to *1/2* when $m \neq 1$ or *M*. A particle on site *1* lands on site *2* or escapes to the left with equal probability *1/2*. Respectively, a particle on site *M* lands on site *M-1* or escapes to the right with equal probability *1/2*. We will assume that the random walk is Markovian and the probability density for the waiting time between successive jumps is *exp(-t)*, i.e., the average lifetime on a site is used as the unit of time.

In this section we derive translocation and return probabilities as probabilities to escape through site 1 on condition that the particle started from site *M* and *1*, respectively. Additionally, we derive average lifetimes for returning and translocating particles as well as the total average lifetimes of particles inside the tube.

Let $f_m(t)$ be the flux escaping through site *1* at time *t* on condition that the random walk started from site *m* at *t=0*. Fluxes $f_m(t)$ ($m = 1, \ldots, M$) satisfy:

$$f_1(t) = \frac{1}{2}\left[e^{-t} + \int_0^t f_2(t-t')e^{-t'}dt'\right]$$

$$f_m(t) = \frac{1}{2}\int_0^t \left[f_{m-1}(t-t') + f_{m+1}(t-t')\right]e^{-t'}dt', \quad m \neq 1, M \tag{2.1}$$

$$f_M(t) = \frac{1}{2}\int_0^t f_{M-1}(t-t')e^{-t'}dt'$$

Laplace transformation results in

$$2(s+1)\hat{f}_1(s) = 1 + \hat{f}_2(s)$$

$$2(s+1)\hat{f}_m(s) = \hat{f}_{m-1}(s) + \hat{f}_{m+1}(s), \qquad m \neq 1, M \tag{2.2}$$

$$2(s+1)\hat{f}_M(s) = \hat{f}_{M-1}(s)$$

As $s \to 0$, $\hat{f}_m(s)$ can be written as

$$\hat{f}_m(s) \cong F_m - s\tau_m \tag{2.3}$$

where

$$F_m = \hat{f}_m(0) = \int_0^\infty f_m(t)dt \tag{2.4}$$

and

$$\tau_m = -\lim_{s\to 0}\frac{d\hat{f}_m(s)}{ds} = \int_0^\infty tf_m(t)dt \tag{2.5}$$

One can derive sets of equations for F_m and τ_m from the set in Eq. (2.2) by passing to the limit $s \to 0$ and using $\hat{f}_m(s)$ in Eq. (2.3). This leads to

$$\begin{aligned} &2F_1 = 1 + F_2 \\ &2F_m = F_{m-1} + F_{m+1}, \quad m \neq 1, M \\ &2F_M = F_{M-1} \end{aligned} \tag{2.6}$$

and

$$\begin{aligned} &2\tau_1 = \tau_2 + 2F_1 \\ &2\tau_m = \tau_{m-1} + \tau_{m+1} + 2F_m, \quad m \neq 1, M \\ &2\tau_M = \tau_{M-1} + 2F_M \end{aligned} \tag{2.7}$$

Solutions to these two sets of equations are

$$F_m = \frac{M+1-m}{M+1} \tag{2.8}$$

and

$$\tau_m = \frac{m}{3(M+1)}\left[2(M+1)^2 - 3(M+1)m + m^2\right] \tag{2.9}$$

We point out that Eq. (2.6) is the standard set of backward equations for the splitting probabilities F_m.

It follows from the definition in Eq. (2.4) that the translocation and return probabilities, respectively, are

$$P_{tr} = F_M = \frac{1}{M+1}, \qquad P_r = F_1 = \frac{M}{M+1} \tag{2.10}$$

One can see that $P_{tr} + P_r = 1$ as it must be. Probability densities for the lifetime of translocating and returning particles are equal to $f_M(t)/F_M$ and $f_1(t)/F_1$. Their average lifetimes respectively are

$$\bar{t}_{tr} = \frac{\tau_m}{F_M} = \frac{1}{3}M(M+2), \qquad \bar{t}_r = \frac{\tau_1}{F_1} = \frac{1}{3}(2M+1) \tag{2.11}$$

One can use these lifetimes to find the total average lifetime of a molecule entering the tube

$$\bar{t} = \bar{t}_r P_r + \bar{t}_{tr} P_{tr} = \tau_1 + \tau_M = M \tag{2.12}$$

The probabilities in Eq. (2.10) and the average lifetimes in Eqs. (2.11), (2.12) are the main results of this section. They will be used in the following derivations of properties of the single-file transport.

3. NUMBER OF PARTICLES TRANSLOCATING IN A SINGLE PULSE

Particles that move in a single-file arrangement cannot pass each other. As a consequence, a "boundary" separates particles that entered through site 1 from those that entered through site *M*. We will call the former "red" particles and the latter "blue" particles. When a red particle traverses the tube and escapes through site *M,* all sites are occupied by red particles. Vice versa, when a blue particle escapes through site 1 all particles in the tube are blue. Transport of molecules through the tube occurs as a sequence of unidirectional translocation pulses of red and blue particles separated by intermittent periods during which the pulse direction is changed. In this section we derive the distribution function for the number of particles translocating in a single pulse, *p(n).*

The probability that the pulse contains only one translocating particle satisfies

$$p(1)=\frac{1}{2}[P_{tr}+P_r p(1)] \tag{3.1}$$

This equation can be derived by straightforward counting of realizations. Solving Eq. (3.1) we find

$$p(1)=\frac{P_{tr}}{1+P_{tr}}=\frac{1}{M+2} \tag{3.2}$$

Note that $p(1)$ is the probability of the pulse reversal. Then $[1-p(1)]$ is the probability that the pulse contains, at least, one more translocating particle. Using these probabilities one can write $p(n)$ as

$$p(n)=[1-p(1)]^{n-1}p(1)=\frac{(M+1)^{n-1}}{(M+2)^n} \tag{3.3}$$

One can check that $p(n)$ is normalized to unity

$$\sum_{n=1}^{\infty}p(n)=1 \tag{3.4}$$

The average number of particles translocating in a single unidirectional pulse is

$$\bar{n}=\sum_{n=1}^{\infty}np(n)=\frac{1}{p(1)}=M+2 \tag{3.5}$$

Using this one can write $p(n)$ as

$$p(n)=\frac{1}{\bar{n}}\left(1-\frac{1}{\bar{n}}\right)^{n-1} \tag{3.6}$$

This is the expression for $p(n)$ given in [2].

4. AVERAGE DURATION OF THE PULSE

In this section we derive an expression for the average duration of unidirectional translocation pulses, $\bar{t}_{up}$. This time can be written as a product of the average duration of the interval between successive translocation events of one pulse, $\bar{t}_{\text{int}}$, and the average number of such events in a pulse that is equal to $\bar{n}-1$,

$$\bar{t}_{up}=(\bar{n}-1)\,\bar{t}_{\text{int}} \tag{4.1}$$

To derive an expression for $\bar{t}_{\text{int}}$, consider, for example, a pulse of red particles. Just after a translocation has occurred the tube is filled with red particles. If a new particle enters through site *1* the rightmost red particle escapes from the tube through site *M* producing the next translocation event. In this case, the interval has the average duration $\bar{\tau}$, i.e., the average time between successive steps of the random walk. We have chosen a time scale with $\bar{\tau} = 1$. The probability of observing this translocation scenario is equal to $(1+\bar{P}_{tr})/2$.

It may happen that after a translocation of a red particle, a blue particle enters the tube through site *M*. For continuation of the red pulse, this blue particle has to escape through site *M*. A translocation event occurs if after that the next particle enters the tube through site *1*. The average duration of the interval in this case is equal to $2\bar{\tau} + \bar{t}_r$ and the probability that it happens is $(1+P_{tr})P_r/4$. In this scenario, a blue particle entered and escaped the tube once.

If the next translocation occurs after blue particles entered and escaped *n* times, the interval duration on average is equal to $(n+1)\bar{\tau} + n\bar{t}_r$. It occurs with the probability equal $(1+P_{tr})P_r^n/2^{n+1}$. The average duration of the interval is given by

$$\bar{t}_{\text{int}} = \frac{1+P_{tr}}{2}\sum_{n=0}^{\infty}[(n+1)\bar{\tau} + n\bar{t}_r]\left(\frac{P_r}{2}\right)^n \tag{4.2}$$

Summing up the infinite series we obtain

$$\bar{t}_{\text{int}} = \frac{1}{1+P_{tr}}\left(2\bar{\tau} + \bar{t}_r P_r\right) = \frac{1}{3}(2M+3) \tag{4.3}$$

where in the second equality we have taken $\bar{\tau} = 1$ and used the expression for $\bar{t}_r$ and P_r derived earlier. Substituting this expression for $\bar{t}_{\text{int}}$ and the expression for $\bar{n}$ given in Eq. (3.5) into Eq. (4.1) we arrive at

$$\bar{t}_{up} = \frac{1}{3}(M+1)(2M+3) \tag{4.4}$$

This coincides with the corresponding expression for $\bar{t}_{up}$ in [2].

5. AVERAGE TIME BETWEEN PULSES

In this section we derive an expression for the average time between oppositely directed translocation pulses. Suppose a red pulse has ended and after a time τ a first blue particle enters the tube. In the simplest scenario this particle traverses the tube and a new blue pulse begins. This happens with probability $(1+P_{tr})/2$ and takes an average time of $\bar{\tau} + \bar{t}_{tr}$.

In a more complex scenario the first blue particle does not traverse the tube. It escapes from the tube through site M. Then a new blue particle enters the tube at time τ after the first blue particle escaped. This particle traverses the tube and initiates the blue pulse. This happens with probability $(1+P_{tr})P_r/2$ and on average takes a time of $2\bar{\tau}+\bar{t}_r+\bar{t}_{tr}$. This scenario describes the initiation of the blue pulse in "the second attempt". When the blue pulse is initiated in the (*n*+1)-th attempt, on average this takes a time of $(n+1)\bar{\tau}+n\bar{t}_r+\bar{t}_{tr}$ and happens with probability $(1+P_{tr})P_r^n/2^{n+1}$.

The average time between successive and oppositely directed pulses can be written as

$$\bar{t}_{bp}=\frac{1}{2}(1+P_{tr})\sum_{n=0}^{\infty}\left[(n+1)\bar{\tau}+n\bar{t}_r+\bar{t}_{tr}\right]\left(\frac{P_r}{2}\right)^n \tag{5.1}$$

Summing up the infinite series we obtain

$$\bar{t}_{bp}=(\bar{\tau}+\bar{t}_{tr})+(\bar{\tau}+\bar{t}_r)\frac{P_r}{1+P_{tr}}=\frac{1}{3}(M+1)(M+3) \tag{5.2}$$

When writing the second equality we have taken $\bar{\tau}=1$ and used the expressions in Eqs. (2.10) and (2.11). The last expression for $\bar{t}_{bp}$ coincides with the corresponding expression in [2].

6. CONCLUDING REMARKS

We have derived some properties of random walks on a 1D lattice between two sites from which the particles can escape. The results of these derivations have been used in [2] to analyze the single-file water motion inside a narrow carbon nanotube, as studied by molecular-dynamics simulation in [1]. We have shown in [2] that this simple model captures the important properties of single-file water transport.

ACKNOWLEDGMENTS

We thank Attila Szabo, Jay Rasaiah, and Robert Zwanzig for many helpful discussions.

REFERENCES

1. Hummer, G., Rasaiah, J. C., and Noworyta, J. P., *Nature (London)* **414**, 188-190 (2001).
2. Berezhkovskii, A., and Hummer, G., *Phys. Rev. Lett.* **89**, 064503 (2002).

$1/f$ Noise in Ion Transport through Nanopores: Origins and Mechanism

Zuzanna Siwy*† and Andrzej Fuliński‡

* *Silesian University of Technology, Strzody 9, 44-100 Gliwice, Poland*
†*Gesellschaft für Schwerionenforschung (GSI), Planckstr.1, D-64291 Darmstadt, Germany*
‡ *Institute of Physics, Jagellonian University, Reymonta 4, 30-059, Krakow, Poland*

Abstract: We have studied the mechanisms generating $1/f$ (flicker) noise (FN). Recently, it has been suggested that FN in biological ion channels is caused by "the equilibrium conductance fluctuations related to the conformational flexibility of the channel pore structural constituents" [1]. It has also been claimed that FN results from the switching of the channels between their different conducting states [2], and is not an "inherent property of ion transport" [1]. To check those conjectures we nanofabricated synthetic pores of similar dimensions to those of biological channels and studied their transport characteristics [3,4]. We have examined the pores produced in two different polymeric materials: polyethylene terephthalate and polyimide. The pore in polyimide has smooth walls, while in PET there remain – sticking from the pore walls – charged, loose ends of cleaved polymer strands, so-called "dangling ends", which move continuously in a more or less random fashion. By comparing the synthetic and biological (BK channel) channels we show that the FN noise indeed originates from the channel's opening-closing process, related in turn to the underlying motions of channel wall constituents. The polyimide pore does not produce FN. Moreover, we have shown the non-Markovianity of the ion transport through the biological and PET nanopores. On the basis of these findings and several recent literature reports we propose the following tentative mechanism for FN: *a necessary but not sufficient condition for appearance of the FN noise in a given process is its coupling to another – deterministic or/and stochastic - background process.*

I. INTRODUCTION

Abundance of $1/f^{\alpha}$ (flicker) noise (FN) is now a well-established empirical fact. FN was observed in many very different phenomena, from electrical conductance (voltage fluctuations in resistors [5] and current fluctuations in biological ionic channels [1,2]), through astrophysics (sunspot activities [6] and light of quasars [6]) and many biological phenomena [2], to neural networks [7], earthquakes [8], music and speech [9]. In spite however of the ubiquity of the FN noise, the mechanisms underlying its origins are still largely unknown, though there are some propositions in literature, best-known being *via* self-organized criticality [10].

Here, we would like to present the conjecture of a partial – at least – solution of this problem, based on analysis of the stochastic properties of ionic currents through nano-

CP665, *Unsolved Problems of Noise and Fluctuations: UPoN 2002: Third International Conference*, edited by S. M. Bezrukov

channels, and on some systems reported in the recent literature. First short report of these findings was published in [1].

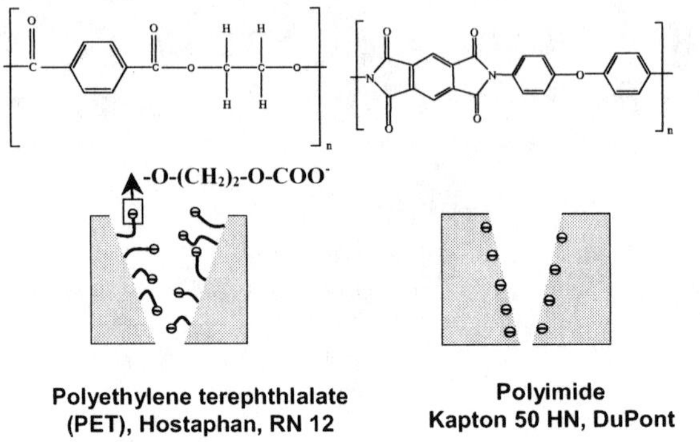

FIGURE 1. Schematic shapes of synthetic nanopores

Ion transport through biological nano-channels is the basis of almost all life processes and it has been studied for many years [12]. Ion current signals however still present a big puzzle to the experimentalists as well as the theoreticians. At constant voltages applied, the ion current switches between discrete levels, often resembling the dichotomous noise. The origins of this behavior and its more detailed statistical properties are not yet fully understood. The statistical analysis including power spectrum has been performed with the aim to get an insight into the physical basis of the channel switching. The FN noise has indeed been found in some biochannels, pointing to the self-similarity in time of the processes underlying the ion transport. Moreover, according to Bezrukov [2], biological membranes constitute "the basic biological level of the generation of FN. Together with ion-transporting protein structures embedded in their lipid matrix, they represent the simplest biological objects still producing this type of noise." They are therefore a wonderful model system for studying the origin of FN. This approach has been used by Bezrukov and Winterhalter in [1]. Their studies on Maltoporin channel provided strong evidence that the presence of the FN is related to the switching of the channel between different conductance levels, which in turn reflect the dynamics of the structural elements of the channel. Those findings and conjectures imply however much more. They suggest that every narrow enough pore, containing parts, which can move, should be able to produce FN. We decided to check this conjecture by studying the transport characteristics of nanopores fabricated in polymer films, of dimensions similar to those of biological channels. The nanopores are much more robust than biochannels and enable the measurements which are impossible or very difficult to perform in biological materials. Specifically, the nanostructuring facilities available at the Gesellschaft fuer Schwerionenforschung (GSI), Darmstadt, Germany, based on irradiating polymer films with single heavy ions followed by chemical etching [13], enable the production of polymeric membranes with one single nano-pore, which makes the examination of individual pore transport characteristics much easier. Moreover, usage of different polymers enables the nanofabrication of channels with walls of various properties.

We have decided to analyze the behavior of two synthetic voltage-gated double cone pores, nano-fabricated in a polyethylene terephthalate (RN12, Hoechst) (PET) [3] and polyimide (DuPont, Kapton 50HN) (KPT) [4] foils, and we have compared their behavior with the properties of a voltage-gated biological channel (BIO) [14]. The PET and KPT channels were produced by irradiation of the polymeric foils with accelerated Au ions of 2.2 GeV kinetic energy (linear accelerator UNILAC, GSI, Darmstadt), and subsequent chemical etching [3,4]. This results in both cases in the formation of carboxylate groups on the pore walls, which renders their negative charge at basic and neutral electrolyte pH. In KPT the carboxylate groups are attached to rigid aromatic rings, which leads to a fairly smooth pore (channel) with relatively stable (unmoving) walls [15]. In the PET pore, on the other hand, due to the presence of the ethylene group in PET chains, the carboxylate groups are the end groups of – sticking from the pore walls – loose ends of cleaved polymer strands, so-called "dangling ends" [16], which move continuously in a more or less random fashion [17, 18] (Fig. 1). The BIO channel (the large conductance locust potassium channel [14]) consists of four protein subunits, which create an inverted cone with charged entrances [19,20].

All three channels have a comparable effective diameter of the narrow part of about 2 nm; the height of the smaller cone in PET and KPT pores is similar to the length of the BIO channel and equal approximately to 20 nm. The magnitudes of measured potassium currents I are also comparable (cf. below), though the synthetic channels are much longer (10-12 μm) than biological ones. We would also like to mention here that the longer cone, which plays here the role of the mechanical support of the short (≈ 20 nm) one, by itself is able to act as a nano-pomp, which transports potassium ions against the concentration gradient, when stimulated with the external field fluctuations [21].

The paper is organized as follows. We start with presenting the ion current time series to be analyzed (section II). Subsequently we examine their stochastic/deterministic character by means of the Smoluchowski-Chapman-Kolmogorov equation, as well as the power spectrum (Section III). Section IV discusses the transport properties of the Kapton pore as well as concludes the presented results and points to unsolved problems.

II. EXPERIMENTAL DATA

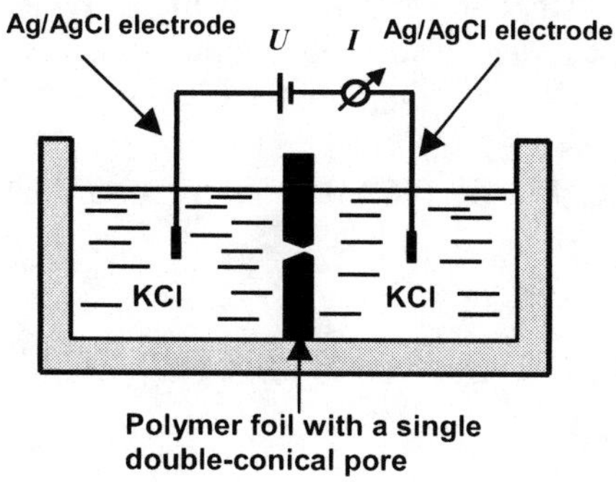

FIGURE 2. Scheme of the experimental set-up with the conductivity cell used for measuring the ion current through a single nanofabricated pore.

BIO channel data are the same as in ref. 14. They were recorded from cell attached patches of adult locust *Schistocerca gregaria* extensor tibiae muscle fibers at 60 mV. The channel examined was a big conductance potassium channel (BK channel). The ion current was low-pass filtered at 10 kHz, and digitized at 22 kHz using a Sony PCM ES-701. The complete data analyzed here consist of one record composed of 250000 values of the channel current, $0 < I < 20.46$ pA, measured at equal intervals Δt = 0.0001 s, the whole duration being 25 s.

PET and KPT synthetic pores currents measurements were performed in a conductivity cell (Fig. 2), connected to the locally built electronic feedback circuit based on a current/voltage converter with pA sensitivity, controlled by software written in LabView 5.0 (National Instruments). The output was digitized at 166 kHz using the National Instruments PCI-MIO-16XE-10 card, and sampled at 1 kHz. The complete raw data consist of several series composed of 500 000 to 1000 000 values of channel currents, excited at 40 – 1000 mV and recorded at 1 ms intervals. In these series we selected the stationary segments, consisting of 130 000 – 300 000 data points, called here *records*. The KPT data were monostable (in the sense described in Sect III A – cf. Figs. 3, 4) in the examined range of voltages. PET data were bistable at intermediate voltages, monostable for low (closed channel, $\langle I \rangle \approx 0$) and for high (open channel, $\langle I \rangle$ proportional to voltage) voltages.

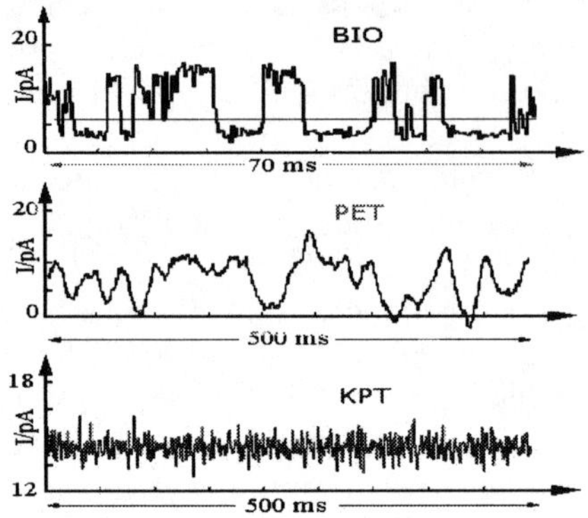

FIGURE 3. Potassium current $I(t)$ through the biological channel (BIO) excited at 60 mV, PET nanopore (PET) and Kapton nanopore (KPT) recorded at 180 mV.

The PET data, $-4.2 < I < 16.5$ pA, recorded at 180 mV (PET 180) and stationary over 130 000 values, is the closest in its behavior to the BIO channel. In both cases the current fluctuates randomly between high (I_o) and low (I_c) values (open and closed channel), with the most probable $I_o \approx 1$ 1pA and $I_c \approx 0$ pA. The only significant difference between BIO and PET 180 data is their time characteristics: the transitions between open and closed states in BIO channel are about 40 times faster than in PET ones (average dwell times $\tau_B \approx 1.7$ ms, $\tau_{PET} \approx 73$ ms). This seems to be the effect of the presence of the long (about 12 μm) supporting cone in synthetic pores, absent in the biological channels (lipid membranes are much thinner than available polymer foils). In the further analysis the following series were also used: monostable PET series recorded at 60 mV ($\langle I \rangle \approx 0$) and 250 mV ($\langle I \rangle \approx 59$ pA), and KPT series at 180 mV ($14 < I < 17.3$ pA, $\langle I \rangle \approx 15.4$ pA) denoted as PET 60, PET 250, and KPT 180, respectively. PET data consisted of 150 000, KPT – of 130 000 stationary current values. Sample data are shown in Fig. 3.

III. STOCHASTIC PROPERTIES OF BIO AND SYNTHETIC CHANNELS

A. Current Distributions

For every analysed record the data were first divided into M groups (bins) of equal width $dI = (I_{max}/M)$ pA. The relative frequencies of visiting these groups were assumed to be equal to the probabilities $P(m)$ of finding a value of the current in the interval $mdI < I < (m+1)dI$, $m=0,\ldots, M-1$.

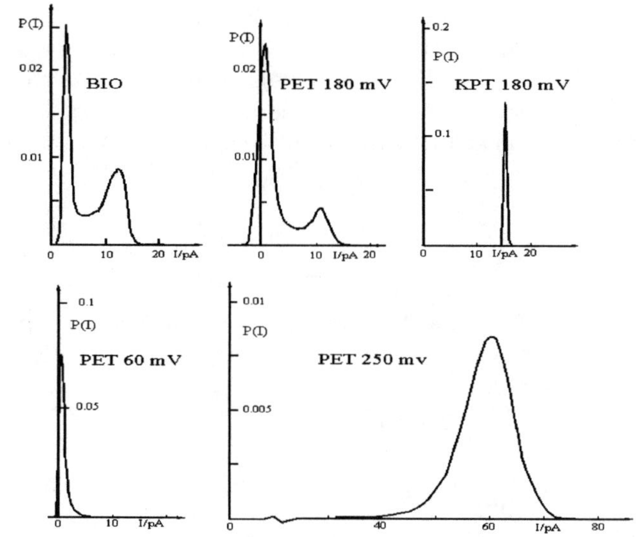

FIGURE 4. One-point current distributions (relative frequencies) for five discussed records of channel currents.

The stationarity of a given record was checked by dividing it into 3-5 smaller sub-records, and by calculating the probabilities for every sub-record separately. It was found that, in every analyzed record, the results were similar (within an experimental error) for every sub-record and for the whole analyzed time series. There was no visible trend in the differences between sub-records either. The probabilities $P(m)$, calculated for M = 500 to 1000, are shown in Fig. 4. As we mentioned above, the current distributions for BIO and PET 180 records are distinctly bistable and similar in shape; the remaining data are monostable.

B. Non-Markovianity

In order to check whether the ion transport through the channels is a Markovian process we applied the procedure described in [14], which here we would like to briefly revisit. One of properties of a Markov process is that it satisfies the Smoluchowski-Chapman-Kolmogorov (SCK) equation, which for a discrete stochastic process reads:

$$P(m,t\,|\,n,0)=\sum_{l}P(m,t\,|\,l,s)P(l,s\,|\,n,0) \qquad 0<s<t \tag{1}$$

where $P(m,t|n,s)$ is the conditional probability (transition frequency) that the system at the time t is in the state m if at the earlier time s was in the state n. Deviations from that equation measure therefore the degree of non-Markovianity.

First we changed the original raw data into a dichotomous record of current values corresponding to open and closed channel states, by procedure identical to that of ref. 14. Again, as in the case of BIO data, the probability distribution for PET 180 is distinctly bimodal (cf. Fig. 4), which enables us dividing the current values into two channel states: closed c ($I < I^*$) and open o ($I > I^*$) ones, with the threshold $I^* = 5.5$

pA. For such two-state series we determined the two-state probabilities $P_2(m,t;n,0) \equiv P_2(m,t_1;n,t_2)$, with $m,n = c,o$, $t=t_1-t_2$. As the time series is stationary, the P_2 were calculated by counting the relative numbers of pairs of the states separated by time interval t, such that $I(t_1)$ belongs to state m and $I(t_2)$ – to state n. The current-current conditional probabilities $P(m,t|n,s)$, $m,n = c,o$ were obtained from the relation $P(m,t|n,s) = P_2(m,t;n,0)/P(n)$. This procedure was applied to both experimental data and to randomly generated Markovian dichotomous series of the same length (control sample). Note that for dichotomous series the following relations are fulfilled:

$$P(c,t|m,s) + P(o,t|m,s) = 1, \qquad P(m,t|c,s)\,P(c) + P(m,t|o,s)\,P(o) = P(c)$$
$$P(c) + P(o) = 1 \tag{2}$$

which implies, therefore, that e.g. $P(c,t|c,0)$ and $P(c)$ represent the whole information on the two-state probabilities of the examined (stationary) series. The calculated conditional probabilities were used for determination of the integral measure (mean square characteristics) of the non-Markovianity G, defined in Ref.14:

$$G = G(\tau,T) = \frac{1}{4T}\left[\sum_{m,n=c,o}\int_{\tau}^{\tau+T} dt D^2(m,n|t,\tau)\right]^{1/2} \tag{3}$$

where D measures the deviation from the SCK equation (2):

$$D(m,n;t,\tau) = P(m,t|n,0) - [P(m,t|c,t-\tau)\,P(c,t-\tau|n,0) + P(m,t|o,t-\tau)\,P(o,t-\tau|n,0)]$$
$$m,n = c,o. \tag{4}$$

The non-Markovian character of the BIO series was found and described in [14]. Here we have focused on the analysis of the PET series in this respect. Maximum values $G_{max} = \sup_\tau G(T,\tau)$ are: $G_{BIO}(5,60) = 0.085$, $G_{PET}(50,60) = 0.090$, whereas $G_{MARK} = 0.0014$ and does not depend on τ. These results point to the non-Markovian character and the similarity of the processes underlying the dichotomous nature of the BIO and PET 180 currents. The probability distributions for PET 60, PET 250, and KPT 180 series are monostable (cf. Fig. 4), and therefore the above bimodal analysis does not apply.

C. Frequency Spectra

The frequency (power) spectra $S(f)$

$$S(f) = \frac{1}{2T}\lim_{T\to\infty}\left|\int_{-T}^{T} dt\, I(t)\,e^{-2\pi ift}\right|^2 \tag{5}$$

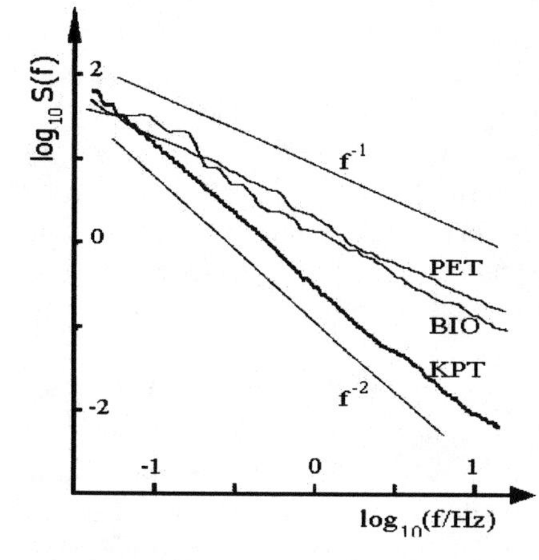

FIGURE 5. Smoothed power spectra for various types of channels: BIO channel, and synthetic: PET 180 and KPT 180 pores. BIO and PET 180 data correspond to bistable (open/closed) region, KPT channel - to monostable (open) region. $S(f)$ in $(pA)^2/Hz$, f in Hz.

were calculated by dividing the experimental data into 5 to 10 shorter segments, computing $S(f)$ for every segment separately, and averaging over the segments. The obtained smoothed power spectra are shown in Fig. 5. It is seen that the spectra for BIO and for PET series are almost identical in the frequency range $0.1 < f < 50$ Hz and show the $1/f^{\alpha}$ behavior with $\alpha = 1.05 \pm 0.05$. The conversion of the current values into a dichotomous (open-closed) series does not change the shape of $S(f)$. This again confirms the close similarity of stochastic properties of the ion current in biological and synthetic channels.

In the above-described random series three presumably different stochastic processes can be singled out: fluctuations of the ion current in the closed state, the current fluctuations in the open state, and random transitions between open and closed states. To find which of these are responsible for the appearance of the flicker noise we have studied power spectra of the time series recorded in the closed state (PET 60), open state (PET 250) and of the dwell-time series. We have found that the open channel spectrum behaves as $f^{-\alpha}$ with $\alpha \approx 1.9$; to fragments of the closed channel and the dwell times spectra one can ascribe the $1/f$ behavior, but such a dependence is not obvious [1].

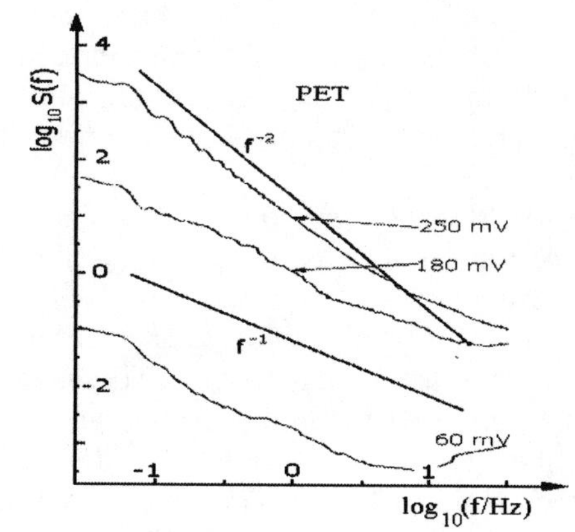

FIGURE 6. Power spectra for different driving voltages: PET 60, PET 180, and PET 250 records.

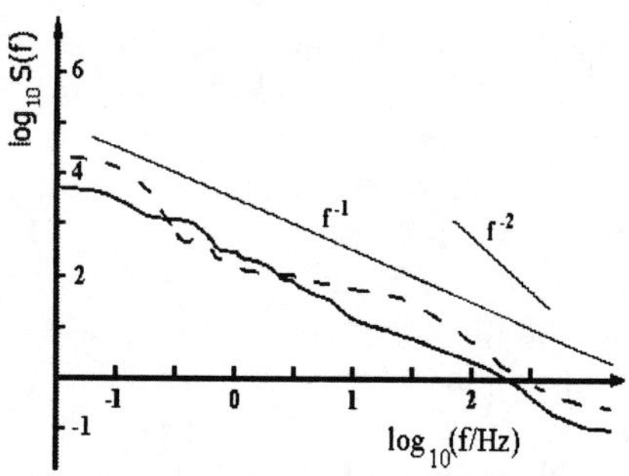

FIGURE 7. Power spectra of data series constructed by changing the subsequent dwell times of PET 180 (continuous line) and Markovian dwell times (dashed line) with 0 and 59 pA, respectively.

As those results did not give any definite answer, still further questions arose (i) are the open-closed transitions responsible for the appearance of FN noise and (ii) do the fluctuations within the states matter? To get a definitive answer, the new "mixed" data series were constructed by taking N_j successive values, N_j being the consecutive dwell times from the PET 180 series, alternately from (i) PET 60 (closed) and PET 250 (open) stationary series, (ii) generated Gaussian-distributed values, peaked at 0 and at 59 pA, with widths adjusted to widths of experimental distributions in (i), and (iii) constant values I_c=0 and I_o=59 pA. The resulting power spectra for all three cases are almost identical, and show distinct $1/f$ characteristics ([1], Fig. 7). Use of dwell times from BIO data gives similar result with $1/f^{\alpha}$ and $\alpha = 1.1 \pm 0.1$.

In order to provide an additional support for those results, the same procedure was repeated for a series of successive dwell times generated as a Markovian process with exponential dwell-time distribution, $\tau_{rnd} = \tau_{PET}$. The character of the power spectrum is therefore Lorentzian. Again all three cases ((i)-(iii) above) are almost identical, without $1/f$ characteristics [1]. Figure 7 shows the power spectra for the dwell times of PET 180 and dwell times of artificially created Markovian process.

These results point to the alternation of the current between low (closed) and high (open) states as the **only** source of the flicker noise in voltage-gated potassium channels. We have also shown that the current fluctuations in these states do not influence in any significant way the shape of the power spectrum. Our findings, therefore, provide a strong evidence for the second Bezrukov's conjecture on the decisive role of the switching of the channels between their different conducting states in appearance of FN.

IV. CONCLUSIONS AND UNSOLVED PROBLEMS

Finally, the question arises what induces the non-Markovian character of the bistable open-closed transitions, which seem to be responsible for the appearance of the flicker noise. According to the mentioned above Bezrukov's conjecture, the "random motion of the channel pore structural constituents" is decisive for the current switchings [1]. To check this, we used the measurements of the potassium current in the Kapton pore.

As it was stated above, in the KPT 180 series the current fluctuates randomly around 15.4 pA ($14 < I < 17.3$ pA). The lack of transitions between open and closed states in all KPT series is the most significant difference between KPT, and BIO and PET 180 channels. This results in the power spectrum (Fig. 5) with totally different characteristics, decaying as $1/f^2$. Identical results were obtained for KPT channel currents measured at 10 and 60 mV - in each case the KPT pore behaves as an open channel (identical as the PET open channel, e.g. at 250 mV), with the only difference of lower current values. This means that the motion of the channels's structural constitutents plays indeed a decisive role in determination of the stochastic properties of ionic currents in nano-channels.

We have examined transport properties of the voltage-gated biochannel (BK channel) and two synthetic nanopores created in PET and KPT polymer films, respectively. The presented close similarity of the stochastic behavior of the BIO and

PET channels seems to be the result of the motions of subunits constituting the channel walls, whereas the KPT channel, lacking of such fluctuating obstacles, is characterized by fundamentally different stochastic behavior. Our results present therefore a strong evidence that FN in the ionic currents through nanopores – both biological and synthetic – is induced by the switching between various conducting states of the channel, which in turn is possible due to the conformational flexibility of the pore's structural constituents. The output recorded in a form of ion current is influenced therefore by the process in the background. When we analyzed carefully the recently reported in the literature examples of $1/f$ noise generation, we found to our surprise that the existence of the process in the background, although not always discussed, indeed has been crucial for the appearance of FN. Let us quote some examples:

(i) The behavior of the walker in lattice gas models (both deterministic and stochastic) is influenced by the presence of other walkers [22].

(ii) The restrictions in the distribution of additional grains along the closed boundaries are considered responsible for FN in the boundary-driven diffusion (ref. [5.b], ref. [22], Sect. 5.2.3).

(iii) In non-equilibrium superconductor-normal conductor phase transition in thin-film bridges, the appearance of FN is ascribed to interactions with another phase transition [23].

(iv) The system of voltage switching with the threshold, which evolved according to a Brownian motion, produces $1/f$ noise [24].

(v) In the model of coupled chemical reactions, the changes of the concentration of the reagent A in time $A(t)$ have a character of the FN noise if there exists a concurrent deterministic process of concentration changes of a reagent B, which drives the process $A(t)$ [25].

(vi) In our present case we have the following situation: input composed of applied voltage (non-fluctuating) and the intrinsic dynamics (presumably more or less random) of "dangling ends" or protein subunits results in the output with FN and non-Markovian characteristics.

We propose therefore the following conjecture: *the basic reason for the omnipresence of the flicker noise in nature lies in the limiting of the dynamics of the examined process by some other process(es) in the background.* Note, that this conjecture needs further more detailed investigations. Moreover, the non-Markovianity of the channel current suggests that the processes behind are not purely random, as non-Markovian stochastic processes contain in most cases a deterministic component [26]. Indeed, it seems that processes exhibiting flicker noise should be also non-Markovian, but – to the authors' best knowledge – there is no theoretical proof or sufficient experimental support for such conjecture.

ACKNOWLEDGMENTS

Z.S. was supported by the Alexander von Humboldt Foundation.

REFERENCES

1. Bezrukov, S.M., and Winterhalter, M. Phys. Rev. Lett. **85**, 202-205 (2000).
2. Bezrukov, S.M. The status of 1/f noise research in biological systems: empirical picture and theories, in *Proc. First Int. Conf. on Unsolved Problems of Noise, Szeged 1996*, edited by Doering, C.R. Kiss, L. B., and Schlesinger, M. F. (World Scientific, Singapore, 1997), pp. 263-274.
3. Apel, P., Korchev, Y.E., Siwy, Z., Spohr, R. and Yoshida, M. a) German and US patent 10044565.9-45 (2000) Verfahren zum Aetzen mindestens einer Ionenspur zu einer Pore in einer Membrane und elektrolytische Zelle zur Praeparierung einer solchen; b) Nucl. Instrum Meth.B **184**, 337-346 (2001).
4. Siwy, Z., Dobrev, D., Neumann, R., Trautmann, C. and Voss, K. a) German and US patent, registration on 26.02.2002, nr 102 08 023.2, Verfahren zur Herstellung von Nanostrukturen in Membranen und Asymmetrische Membran; b) Applied Physics A (2002) – in press.
5. (a) Dutta, P., and Horn, P.M., Rev. Mod. Phys. **53**, 497-516 (1981); (b) Weissman, M.B., Rev. Mod. Phys. **60**, 537-571 (1988).
6. Press, W.H., Commun. Astrophys. **7**, 103-119 (1978).
7. Usher, M., Stemmler, M., and Olami, Z., Phys. Rev. Lett. **74**, 326-329 (1995).
8. Varotsos, P.A., Sarlis, N.V., and Skordas, E. S., Phys. Rev. E **66**, 011902 (2002).
9. Voss, R.F., and Clarke, J., Nature **258**, 317-318 (1975).
10. Bak, P., Tang, C., and Wiesenfeld, K., Phys. Rev. Lett. **59**, 381-384 (1987); Phys. Rev. A **38**, 364-374 (1988).
11. Siwy, Z., and Fuliński, A., Phys. Rev. Lett. **89**, 158101 (2002).
12. Hille, B., *Ionic Channels of Excitable Membranes* (Sinauer Inc., Sunderland, MA, Ed. 2, 1992).
13. R. Spohr, German Patent DE 2951376 C2 (1983), United States Patent 4369370 (1983).
14. Fuliński, A., Grzywna, Z., Mellor, I., Siwy, Z., and Usherwood, P.N.R. Phys. Rev. E **58**, 919-924 (1998).
15. Trautmann, C., Bruechle, W., Spohr, R., Vetter, J., and Angert, N. Nucl. Instr. and Meth. B **111**, 70-74 (1996).
16. Wolf, A., Reber, N., Apel, P.Yu., Fischer, B.E. and Spohr, R., Nucl. Instrum. Meth. B **105**, 291-293 (1995).
17. Doi, M. *Introduction to Polymer Physics*, Oxford University Press (1997).
18. Schmidt-Rohr, K., Hu, W. and Zumdulyadis, N., Science **280**, 714-717 (1998).
19. Doyle, D.A. *et al.* Science **280**, 69-77 (1998).
20. Zhou, Y., Morais-Cabral, J.H., Kaufmann, A. and MacKinnon, R., Nature **414**, 43-48 (2001).
21. Siwy, Z., and Fuliński, A., Phys. Rev. Lett. **89**, 198103 (2002).
22. Jensen, H. J. , *Self-organized criticality* (Cambridge Univ. Press, Cambridge, UK 1988), Sect. 4.4.
23. Koverda, V. P., and Skokov, V. N., Physica A **262**, 376-387 (1999).
24. Davidsen, J., and Schuster, G., Phys. Rev. E **62**, 6111-6115, (2000); **65**, 026120 (2002).
25. Gora, P. F., and Fuliński, A., *1/f noise in chemical reactions* (unpublished).
26. Fuliński, A., Phys. Rev. E **50**, 2668-2681 (1994); P. Hänggi, & P. Jung, *Colored noise in dynamical systems* in *Advances in Chemical Physics*, Vol. LXXXIX. (I. Prigogine, & S. A. Rice, Eds.), Wiley & Sons, Inc., New York 1995, pp. 239-326.

Motor Rumblings: Characterization of Adaptation Motors in Saccular Hair Cells by Noise Analysis

Jonathan E. Frank*, Vladislav Markin†, and Fernán Jaramillo*

*Department of Biology, Carleton College, Northfield, MN 55057, †Department of Anesthesiology, University of Texas Southwestern Medical Center, Dallas, TX 75235.

Abstract. The mechanical sensitivity of hair cells, the sensory receptors of the vestibular and auditory systems, is maintained by *adaptation*, which resets the transducer to cancel the effects of static stimuli. One model of adaptation proposes that myosin motors coupled to transduction channels move along the long axis of the hair bundle's stereocilia, to regulate the tension in the tip links which are thought to gate transduction channels [1, 2]. These motors can be activated by applying a transduction channel blocker to the bundle, causing it to move [3, 4]. We studied the variance in the position of the bundle during these displacements, and found that it increases as the bundle moves to its new position. We can explain both displacement and variance with a simple model in which a single motor acting on the bundle takes ~3.6 nm steps whose frequency declines with the motor's load.

INTRODUCTION

The sacculus provides amphibians with a rugged and sensitive seismic sensor. The sensitivity of this organ presents a problem, namely, how to maintain responsiveness in the presence of large, even saturating stimuli. The answer to this problem is adaptation, which in one of its forms employs motors to regulate the state of mechanically sensitive ion channels in the hair cells which are responsible for mechanoelectrical transduction.

We used the open channel blocker gentamicin to reduce the influx of Ca^{++} through transduction channels open at rest, thus activating the adaptation motors. The resulting increase in motor activity caused the motors to climb, increasing the tension in the tip links, which caused the hair bundle to undergo a slight movement in the negative direction (towards the left in Fig. 1).

METHODS

Preparation. Experiments were performed in the isolated saccular macula of bullfrogs (*Rana catesbeiana*), using techniques described previously [4, 13]. The saccular macula was isolated, and digested for 10 minutes with 75 μg/ml protease XXIV (Sigma) at room temperature (20-25 °C) in a perilymph-like saline ((mM) 110

CP665, *Unsolved Problems of Noise and Fluctuations: UPoN 2002: Third International Conference*, edited by S. M. Bezrukov

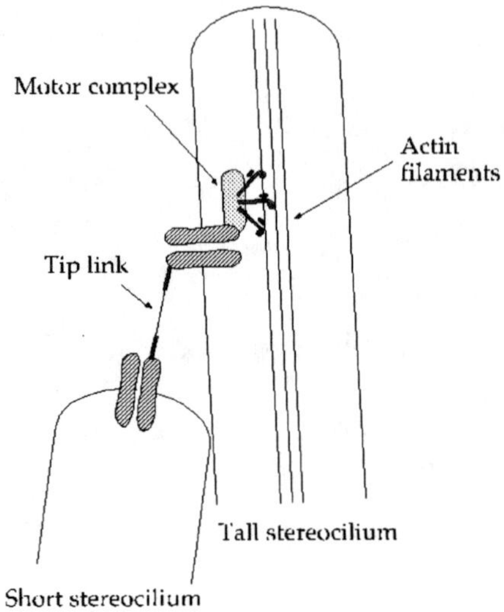

Figure 1. A hypothetical mechanism of *slow* adaptation, which in the *R. catesbeiana* hair cell resets the transduction apparatus within a few tens of milliseconds following a stimulus. The transduction channel closer to the bundle's tip is coupled to a cytoskeletal complex, containing a multimolecular motor which is thought to comprise about 100 myosins (3 shown), whose activity is regulated by calmodulin. For a review see [1].

NaCl, 2 KCl, 0.1 $CaCl_2$, 3 glucose, 5 HEPES, pH 7.25 with 1N NaOH). Following digestion the macula was attached to the glass bottom of a recording chamber using biological glue (Cell-Tak, BD Biosciences Bedford, MA), and the otolithic membrane carefully removed using fine forceps. An upright microscope (Nikon, E600 FN) was used to observe the preparation. Imaging was performed under DIC optics, using a 60x, water immersion objective (N.A. 1.0).

Iontophoresis. Iontophoretic electrodes were used to apply gentamicin to the hair bundle as described previously [4, 15]. Typically about 50 pulses were applied to a given cell, at a frequency of about 2 Hz.

Photodiode detector. To monitor the position of the translucent hair bundles a gold-coated flexible glass probe was attached to the bundle's kinociliary bulb [4, 13]. Probes (~ 0.5-1.0 μm in diameter at their tip) were manufactured from 1 mm borosilicate stock. The image of the probe's distal 2.5 μm was magnified 1000X and projected onto a dual photodiode (EG-G Optoelectronics, Vaudreuil, Quebec), and the photocurrents converted to voltages by current-to-voltage converters with an effective bandwidth of 5.3 kHz. The photocurrents difference was amplified, low pass filtered at 1.0 kHz (four pole Bessel), and sampled at 2.5 kHz by a computer using a 12 bit data acquisition board using LabView (National Instruments). The stiffness of the flexible probes (~1 mN m^{-1}, range 0.39-1.1 mN m^{-1}) was estimated from the probe's variance of motion, which in turn, was estimated from the photodiode's output [13].

EXPERIMENTAL RESULTS AND MODEL

Motion of the bundle in the negative direction, and an increase in the flexible probe position's variance, followed the onset of the iontophoretic pulse (96 out of 124 cells examined). In 76 of the 96 active cells the variance increased monotonically with a roughly exponential timecourse which failed to reach a steady state during the pulse (Fig. 2). In the rest the variance appeared to reach a peak, and in 16 the variance declined to a new steady-state level. The model presented here accounts for both types of observed responses, namely, those in which the variance reached a steady state, and those in which the variance declined following a peak. Here we concentrate on the latter.

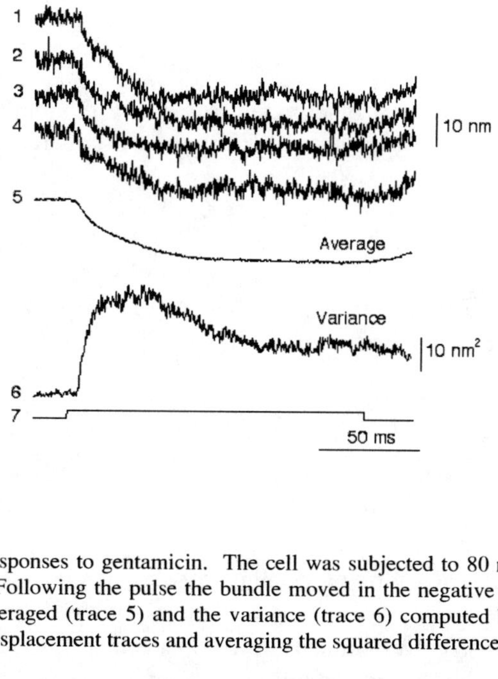

Figure 2. Hair bundle responses to gentamicin. The cell was subjected to 80 nA, 150 ms applications of gentamicin (trace 7). Following the pulse the bundle moved in the negative direction (sample traces 1-4). Responses were averaged (trace 5) and the variance (trace 6) computed by subtracting the mean from each one of the 82 displacement traces and averaging the squared differences.

To model the behavior of an individual motor let x be the displacement of the motor from its resting position along the stereocilium. Let the velocity of the motor's slipping v_{slip} be proportional to the extension of the gating spring which is $x + x_s$, where $-x_s$ is the position when the spring goes slack,

$$v_{slip} = -\alpha_0 (x + x_s) \qquad (1)$$

Let us visualize the motor's climbing as consecutive steps of length h_0 each with frequency f_0. Let us further assume that the stepping frequency length can depend on the tension in the gating spring, i.e. $f = f_0\ (1 - \beta\, x\ )$ (a model without this assumption predicts a monotonically increasing variance). Then

$$v_{c\,limb} = f_0 h_0 (1 - \beta x) \qquad (2)$$

and

$$\frac{dx}{dt} = v_{slip} + v_{c\,lim\,b} \qquad (3)$$

Let us suppose that after the gentamicin pulse the parameters of the system changed correspondingly to f, h, and α. To simplify the equations, let us neglect x_s. Then eq. (3) will transform to

$$\frac{dx}{dt} = -\alpha\, x + hf_0(1 - \beta\, x) \qquad (4)$$

We consider the frequency f of climbing steps as a random parameter. Then the displacement x is a random function and eq. (5) gives the average displacement.

$$< x >= \frac{f_0\, h_0}{\lambda}\left(1 - e^{-\lambda t}\right) \qquad (5)$$

Here f_0 and h_0 are average values of frequency f and step size h, which are different from the corresponding values before the gentamicin application. The variance is

$$\sigma^2 = \frac{f_0 h_0^2}{\lambda}\left[\frac{1}{2}(1-\xi)^2 - \frac{1}{2}(1-\xi)(1+3\xi)e^{-2\lambda t} + \xi^2 \lambda t e^{-2\lambda t} + 2\xi(1-\xi)e^{-\lambda t}\right] \qquad (6)$$

where

$$\lambda = \alpha + \beta\, f_0 h_0 \quad \text{and} \quad \xi = \frac{\beta\, f_0 h_0}{\lambda} \qquad (7)$$

Displacement traces were fitted according to eq.5, yielding estimates for f_0h_0, and λ. Variance traces were fitted to equation 6 with two free parameters (λ and h_0), and using the values of f_0h_0, and λ obtained from displacement to constrain the fits (Fig. 3). The motor's step length (h_0) and stepping frequency (f_0) were estimated at 4.15 ± 2.17 nm, and 297 ± 240 s^{-1}, respectively (s.d., n=9).

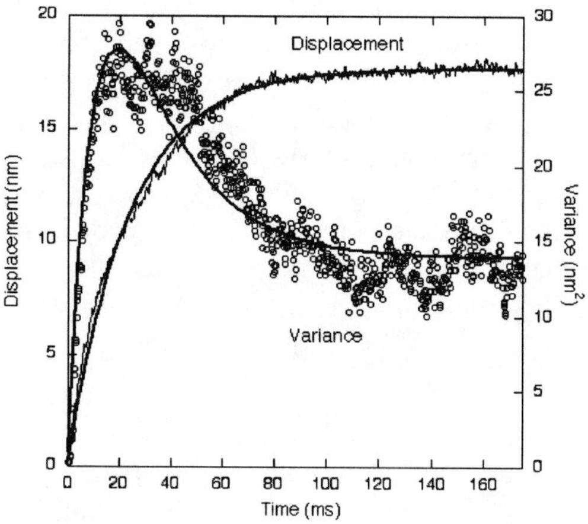

Figure 3. Average hair bundle displacement and displacement variance following repeated gentamicin applications. The displacement and variance baseline were subtracted from their corresponding traces. Displacement and variance traces are fitted to eqs. 5, and 6, respectively. For this cell $h_0 = 5.9$ nm, $f_0 = 130\ s^{-1}$, $\beta = 0.027\ nm^{-1}$, and $\lambda = 0.042\ ms^{-1}$.

This model explains the timecourse of bundle motion, and the gradual decline in motor activity as tension in the tip link increases (eq. 2). This decrease cannot be inferred from the bundle's timecourse of motion (the model predicts an exponential timecourse regardless of β), but only from the time-dependent decrease in the bundle's variance (cells in which both displacement and variance increase monotonically, which we did not fit, can be simply explained by β~0). We attribute this decline to a reduction in the motor's stepping frequency (*f*), and therefore, ultimately to the ATPase rate of the myosins. This interpretation is consistent with the Fenn effect in muscle, which is thought to reflect a decrease in myosin's ATPase rate with an increasing load [5]. A similar effect has been observed for kinesin, whose ATPase rate declines as the load increases, without an observable decline in the motor's working distance [6-8].

How would the model change if instead we assumed a decrease in the motor's working distance with increasing load? Although there is no precedent for such a reduction in working distance with load, the solution of the model would be identical because the action of $1-\beta x$ on eq. 2 can be alternatively viewed as acting on the working distance (*h*).

If we assume our cells to have about 25 intact tip-links (and the same number of motors), a reasonable number following dissection and digestion [9], then a linear contribution by them would imply a stepping frequency per motor of $f_0/25 \sim 297/25\ s^{-1} \sim 12\ s^{-1}$. The motors' stepping frequency could be equated with its ATPase rate, due to the combined ATPase activity of its constituent myosins. The currently favored myosin is Myo1c, whose ATPase rate *in vitro* is ~0.75 s^{-1} [10]. This value, which puts

Myo1c near the low extreme for myosin, could be larger *in vivo*. Nonetheless, less than 16 Myo1c molecules could account for the observed stepping frequency of a motor. Since it is thought that each motor comprises 100 or so myosins [11, 12], the data presented here suggests that only a small fraction of these are engaged at any one time, consistent with the view of myosin as a low duty ratio motor [5].

The step size for the lumped (h_0) motor was estimated at 4.1 nm. This is an apparent motion (the motion registered by the flexible probe attached to the hair bundle). The relationship between the motion of the motor (δ), and the motion of the bundle and attached probe (h_0) is (derivation not shown):

$$\delta = -\left(\gamma + \frac{\kappa_b + \kappa_p}{N\gamma\,\kappa_g}\right) h_0 \qquad (8)$$

Here γ (~0.14), is the lever ratio of bundle displacement to change in tip-link extension), κ_b is the stiffness of the bundle, κ_p is the stiffness of the glass probe, κ_g is the stiffness of a single tip link (~0.5 mN m-1), and N is the number of stereocilia [13]. With the assumed value of 25 for N our estimate of δ would be 3.6 ± 2.17 nm (s.d., n=9). Thus, the apparent working distance of a single motor is comparable to that of myosin (~5 nm) [14]. A working distance for adaptation motors somewhat below that observed for individual myosins is reasonable since a myosin's step does not fully translate into motion of the entire motor because of the stiffness contributed by attached neighbors.

ACKNOWLEDGMENTS

We thank R. A. Eatock, M. A. Vollrath, P. Middleton, and A. Indresano for their comments on this manuscript, and P. G. Gillespie for helpful discussions. This work was supported by NIDCD grant DC02235-02.

REFERENCES

1. Eatock, R.A., Adaptation in hair cells. Annu. Rev. Neurosci., 2000. 23: p. 285-314.
2. Holt, J.R. and D.P. Corey, Two mechanisms for transducer adaptation in vertebrate hair cells. Proc. Natl. Acad. Sci. USA, 2000. 97: p. 11730-11735.
3. Assad, J.A., N. Hacohen, and D.P. Corey, Voltage dependence of adaptation and active bundle movement in bullfrog saccular hair cells. Proc. Natl. Acad. Sci. USA, 1989. 86: p. 2918-2922.
4. Jaramillo, F. and A.J. Hudspeth, Displacement-clamp measurement of forces exerted by gating springs in the hair bundle. Proc. Natl. Acad. Sci. USA, 1993. 90: p. 1330-1334.
5. Howard, J., Mechanics of motor proteins and the cytoskeleton. 2000, Sunderland: Sinauer.
6. Crevel, I.N., et al., Coupled chemical and mechanical reaction steps in a processive Neurospora kinesin. EMBO J., 1999. 18: p. 5863-5872.
7. Meyhöfer, E. and J. Howard, The force generated by a single kinesin molecule against an elastic load. Proc. Natl. Acad. Sci. USA, 1995. 92: p. 574-578.
8. Visscher, K. and M.J. Schnitzer, Single kinesin molecules studied with a molecular force clamp. Nature, 1999. 400: p. 184-189.
9. Assad, J.A., M.G. Shepherd, and D.P. Corey, Tip-link integrity and mechanical transduction in vertebrate hair cells. Neuron, 1991. 7: p. 985-994.

10. Gillespie, P.G., et al., Engineering of the myosin-Ib nucleotide-binding pocket to create selective sensitivity to N6-modified ADP analogs. J. Bio. Chem., 1999. 274: p. 31373-31381.
11. García, J.A., et al., Localization of myosin-Ib near both ends of tip links in frog saccular hair cells. J. Neurosci., 1998. 18: p. 8637-8647.
12. Steyger, P.S., P.G. Gillespie, and R.A. Baird, Myosin-Ib is located at tip link anchors in vestibular hair bundles. J. Neurosci., 1998. 18: p. 4603-4615.
13. Howard, J. and A.J. Hudspeth, Gating compliance associated with gating of mechanoelectrical transduction channels in the bullfrog's saccular hair cell. Neuron, 1988. 1: p. 189-199.
14. Veigel, C., et al., The motor protein myosin-I produces its working stroke in two steps. Nature, 1999. 398: p. 530-533.
15. Jaramillo, F. and A.J. Hudspeth, Localization of the hair cell's transduction channels to the hair bundle's tip by iontophoretic application of channel blockers. Neuron, 1991. 7(409-420).

The Noisy Steps of a Motor Protein

Martin Bier

Dept. of Physics, East Carolina University, Greenville, NC 27858, USA

Abstract. There is experimental evidence to show that the motor protein *kinesin* literally "walks" along the biopolymer *microtubule*. It is possible to depict this "stepping" as the movement of a point along a reaction coordinate. The reorientation of the attached head is an energetically downhill "power stroke." No energy is dissipated (i.e. a flat stretch along the reaction coordinate) during the subsequent diffusive route of the detached head to a new forward docking site. A model that is based on these premises can quantitatively account for the experimentally measured load-velocity diagrams and for the diffusion-drift ratios at different loads. Presently there is a divide in the research on molecular motors and directed transport. Theoreticians study simple, ratchet based mechanisms to come to an understanding of the physics of motion in the overdamped, Brownian realm. Many biophysicists seek to clarify the operation of the motor protein through a detailed study of the 3D molecular structure. This paper attempts to bridge the gap by rigorously modeling the stepping process in an overdamped, Brownian environment.

Over the past decade the amino acid sequence and the 3D structure of an overwhelming amount of proteins has been figured out and catalogued (see http://www.pdb.org). Less progress, however, has been made in understanding how protein structure translates into protein function. The motor protein kinesin is a case in point. The entire 3D structure has been known for a few years now, but it is still largely unclear how the protein operates. Kinesin is a dimer, i.e., it consists of two identical units called "heads." Most textbooks present a model in which these heads function as feet that literally step from one unit of the microtubule to the next unit 8 nanometers further [1] (for an animation see http://valelab.ucsf.edu). But the last word on this issue has not been spoken. The "hand-over-hand" stepping model is far from a certainty. Recently a paper appeared in which evidence for a so-called "inchworm" model was presented [2]. In this model the front head always stays up front and the forward step of the front head is followed by an equally long forward "catch up" step of the hind leg.

In this paper I will use the "hand-over-hand" model as a working hypothesis. But the results, or at least the ideas, are equally well applicable to an inchworm model.

Macroscopic human stepping may look similar to the microscopic stepping of the motor protein, but the underlying physics is entirely different. For a walking human the physics consists of mass, gravity and inertia. A walking human uses energy for the repeated lifting of the anterior leg against gravity and for the subsequent forward acceleration of the same leg to a speed larger than the speed of the torso, so he can eventually put the foot down again in front of him. On the nanometer scale in a liquid the only significant force is friction [3]. Of course, it is not gravity that keeps the motor protein connected to the biopolymer. It is chemical bonds. In the course of stepping these bonds are repeatedly broken and then reestablished. This attachment and detachment cannot just take place in a random and equilibrium fashion. If that were the case no

CP665, *Unsolved Problems of Noise and Fluctuations: UPoN 2002: Third International Conference*, edited by S. M. Bezrukov

forward motion would occur and there would be a significant likelihood for both heads to be simultaneously detached after which the entire dimer would diffuse away from the biopolymer. Directionality and coordination are achieved by coupling the stepping cycle to the hydrolysis of ATP. The conformational changes of the protein as it catalyzes the hydrolysis of one ATP molecule also constitute one forward step. The hydrolysis of ATP is thus energetically coupled to the stepping cycle and, effectively, driving the movement.

When the involved energies are of the order of kT, Brownian noise, i.e., random collisions with molecules from the medium, becomes part of the physics. An important formula is Einstein's Fluctuation-Dissipation Theorem, $D = kT/\beta$, which relates the friction to the strength of the diffusion D.

Microtubule, the biopolymer that kinesin steps on, is anisotropic. Each unit of this polymer is an actual protein of almost 900 amino acids. Going in one direction on the microtubule one sees a different profile than when traveling in the opposite direction. Fig. 1a indicates this with the way the binding sites have been drawn. Kinesin can only step in one direction and this direction is determined by the orientation of the microtubule. Transport over microtubule in the other direction is taken care of by motor proteins other than kinesin [1, 4].

In the stepping cycle of kinesin we can distinguish two phases:

1. A power stroke phase (Fig. 1a left side), i.e. the reorientation of the attached head after the anterior head detaches. This is when force is generated, when power is dissipated and when a load (like a vesicle with chemicals or an organelle) is being pulled.
2. A ratcheted diffusion phase (Fig. 1a right side), i.e., the detached head is randomly diffusing around the neck linker until it hits the posterior docking site. After attachment a next step can commence. This phase has been called a "random diffusional search" [5] and has been described as "fluctuational interactions" or "conformational fluctuations" [6]. But from a physics perspective it looks very much like a random walk between a reflecting barrier (the neck linker attachment) and an absorbing barrier (where docking occurs). Starting at the reflecting barrier at $t = 0$ the average escape time to reach the absorbing barrier is $< T_{esc} >= L^2/2D$ if the distance between the reflecting barrier and the absorbing barrier equals L.

Fig. 1b shows how the stepping process translates into a profile along a reaction coordinate. The power stroke occupies a fraction $1-\phi$ and the diffusive stretch occupies the remaining fraction ϕ. The trajectory of an overdamped, Brownian point particle on this profile describes the progress of the motor protein's catalytic cycle. Many authors have identified the position along the reaction coordinate with the position of the center of mass of the motor protein on the biopolymer. But this may be seriously inaccurate. The essence of the reaction coordinate is that the point particle faces a constant, position independent diffusion coefficient D. For an actual motor protein the different parts of the cycle and the corresponding different segments on the biopolymer may involve very different D's. The analysis of the motion on the reaction coordinate can lead us to durations for the power stroke and for the diffusive stretch, but not to their distances.

In Fig. 1b a power stroke driven by a constant force is represented by a downward

Fig.1a

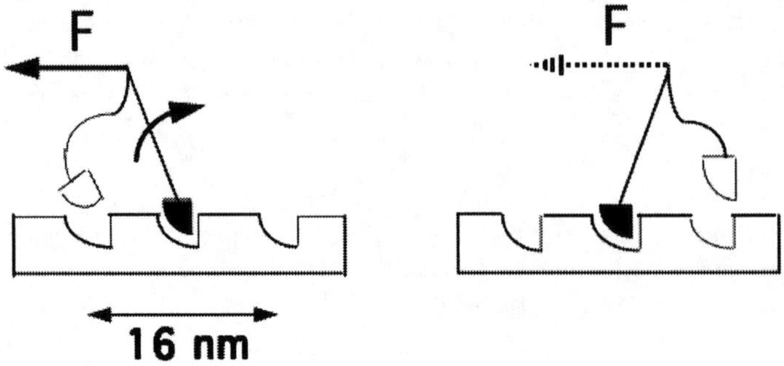

Fig. 1b

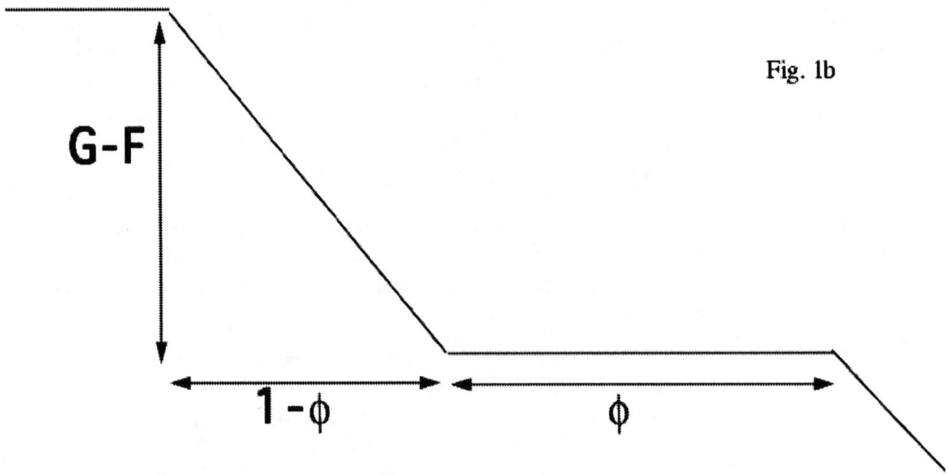

FIGURE 1. The setup for our model. One step of the two headed motor protein (1a) corresponds to traversing one unit in a 1D reaction space (1b). The reorientation of the attached head is the power stroke with energy G that covers a fraction $(1-\phi)$ of the cycle. The subsequent diffusion and docking of the detached head does not dissipate any energy and covers the remaining fraction ϕ. When the motor protein is pulling against a load force F, it is the power stroke that provides the energy. The diffusive segment is unaffected.

slope. We take the stepsize of kinesin, which equals λ =8 nm, to be our unit of length. This implies that if the kinesin is pulling against a load F it releases an energy $G-F$ at every power stroke, where G is the energy of ATP hydrolysis. With energy measured in units of kT, G equals about 22 at physiological conditions. When we also scale the diffusion coefficient D to unity, $\beta = kT/D$ implies a friction coefficient β that equals unity. For $(G-F) > 2$ the diffusive noise on the power stroke constitutes a negligible effect and we take the power stroke to be a deterministic downslide in Fig. 1a. The force driving the power stroke thus equals the power stroke velocity: $F_{ps} = (G-F)/(1-\phi)$. By inverting the total time to traverse one period we obtain a nice and concise formula for the speed of the motor protein:

$$v = \{T_{ps} + T_{diff}\}^{-1} = \{\frac{1}{G-F}(1-\phi)^2 + \frac{1}{2}\phi^2\}^{-1}. \tag{1}$$

It is important to realize that a motor protein is fundamentally different from another energy converter like Na,K-ATPase. Na,K-ATPase and many other proteins convert energy from one storable form to another. Na,K-ATPase takes the chemical energy in ATP and turns it into an electro-osmotic gradient across the cell membrane. Such conversion from one storable form to another can never be accomplished with 100% accuracy if it is to take place within finite time. Some entropy production, i.e. heat loss, must occur. The operation of Na,K-ATPase is, furthermore, reversible. At low ATP concentration Na,K-ATPase can actually let Na ions and K ions flow down the potential and use part of the released energy to produce ATP. A motor protein is fundamentally different. A motor protein employs the energy of ATP hydrolysis to work against friction. The power stroke can actually best be compared to a bullet falling down in a bottle of maple syrup. This process converts potential energy, via friction, into heat and it is obviously irreversible. It is also a process that can achieve a 100% efficiency. It is therefore reasonable to take the full $G = 22$ of ATP hydrolysis as the energy for the power stroke.

When in the overdamped realm without other external forces, the most energy efficient way to move over a distance L in time T is with a constant velocity $v = L/T$ brought about by the work of a constant force. Any variations in speed, i.e. force, will lead to more energy necessary to cover the distance L in time T. In the context of our Fig. 1b this also means that any variation in slope will lead to a larger T_{ps}. It is likely that the motor protein exhibits some variations in the force of the power stroke. After all, the power stroke is constituted by a sequence of discrete conformational changes. This longer time can be construed to derive from a constant slope at a G smaller than 22. Eventually we will see how the best fit is obtained with a constant slope of $G = 16$.

A motor protein is a stochastic stepper. This means that if we do identical experiments starting the motor protein at $x = 0$ at $t = 0$, that we will then measure different arrival times at some $x = L$ in the forward direction down the biopolymer. Formula (1) for the average speed can give only the average arrival time. In 1994 Svoboda et al. were the first to measure standard deviations in arrival times and use the observations to rule out certain models [7]. The standard deviation can be used to formulate an effective diffusion coefficient: $D_{eff} = \frac{1}{2}L^2(\Delta t)^2/ < t >^3$, where $< t >$ is the average time to cover distance L and Δt is the standard deviation. D_{eff} describes how a drifting ensemble of motor proteins spreads and it can be significantly different from D [8]. D_{eff} is commonly absorbed into a dimensionless quantity, $r = 2D_{eff}/(v\lambda)$, which is called the randomness. Here λ represents the spatial period (i.e. the 8 nm steplength) and v represents the average speed. The randomness can be interpreted as a diffusion/drift ratio per period.

It is straightforward to calculate the randomness when drifting down the profile in Fig. 1b. Since we go down the power stroke deterministically, the only source of stochasticity is constituted by the flat segment. When diffusing from a reflecting barrier to an absorbing barrier at a distance ϕ the second moment can be evaluated to be $\frac{5}{12}\phi^4$ [9]. Eventually we derive for the randomness:

$$r = \frac{\frac{1}{6}\phi^4}{\{\frac{1}{G-F}(1-\phi)^2 + \frac{1}{2}\phi^2\}^2}. \tag{2}$$

There is one more complication we have to take care of before we can compare the predictions of our model against actually measured velocity-load and randomness-load graphs. In experimental practice the motion of the motor protein is measured through a silica bead that is connected to the neck linker and visible through a microscope. It appears that between 5% and 10% of kinesin's steps are backward [10]. In the context of the above model the soundest way to explain these backward steps would be as follows. After a full power stroke (i.e. a forward motion of the bead) the detached head accidentally hits and binds at the anterior docking site instead of the posterior one. The anterior binding is then followed by a backward stroke and a detachment of the front head. In the context of Fig. 1b it would be wrong to think of the backstep as an accidental, Brownian upslide up the $G-F$ barrier. With a barrier of the order of 10 kT an upslide in Fig. 1b is many orders of magnitude less likely than the downslide. Incorporating anterior binding would involve adding a dimension to the reaction coordinate. The progress of the motor protein would then be represented by the motion of an overdamped, Brownian point particle in a 2D landscape.

When backward stepping occurs with a probability q (and for the forward stepping probability we thus have $p = 1-q$), the v and r of formulae (1) and (2) need to undergo the following corrections to represent the observed speed and randomness:

$$v_{obs} = (p-q)v, \tag{3}$$

$$r_{obs} = (p-q)r + \frac{4pq}{p-q}. \tag{4}$$

Fig. 2 depicts the power stroke. The power stroke doesn't just drive the orientation of the attached head from θ_2 to θ_1. At the end of the power stroke a Boltzmann equilibrium is established and we have an average orientation $\theta_{eq} = \theta_1 + \Delta\theta$. The power stroke is driven by an energy release $G-F$. Assuming that this energy difference translates into a constant and homogeneous force, we find after integration of θ over the Boltzmann distribution for sufficiently large $G-F$ (larger than a few kT's) that $\Delta\theta = \frac{kT}{(G-F)}(\theta_2 - \theta_1)$. Picturing the detached head as "dangling" from the neck it is realistic to assume that the backstep probability goes up linearly with $\Delta\theta$, i.e. $q(F) = q_0(1-\frac{F}{G})^{-1}$. Obviously q_0 is the backstep probability when $F = 0$.

Next we take the backward stepping probability at $F = 0$ to be equal to $q = 0.05$, which is the lower limit of the 5% to 10% reported in [6, 10]. Schnitzer et al. report no "substantial" increase in the backstep probability for increasing load [6]. Coppin et al. do claim to see an increase in backstepping when the load reaches the stall force [11]. Because a motor protein easily detaches when the load is high, the involved measurements are hard and the margin of error is high. To redimensionalize the F in our model it must be multiplied with $kT/\lambda = 0.53$ pN (where λ represents the 8 nm period

Fig. 2

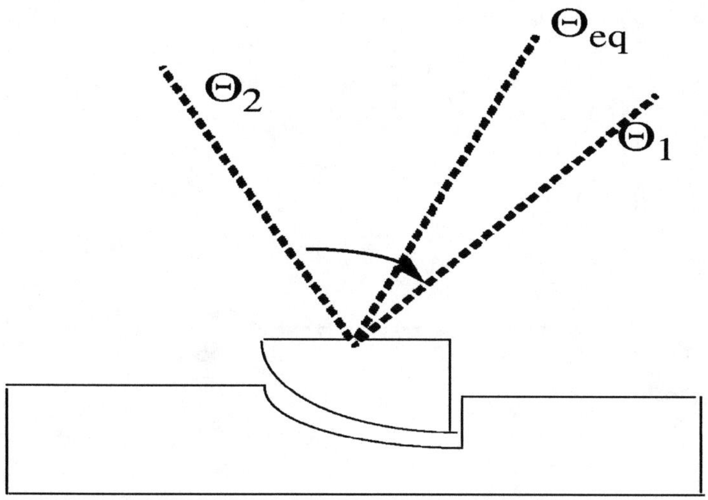

FIGURE 2. A more detailed consideration of the power stroke depicted in Fig. 1a. We assume that going from orientation θ_2 to orientation θ_1 the release of the energy $G-F$ leads to a uniform and homogeneous force. In a Brownian environment a Boltzmann distribution is established at the end of the power stroke and the average position is $\theta_{eq} = \frac{kT}{G-F}(\theta_2 - \theta_1) + \theta_1$. Assuming that the backward stepping probability q increases linearly with θ_{eq} we are led to $q(F) = q_0(1 - \frac{F}{G})^{-1}$, where q_0 is the backward stepping probability at zero load.

of the motion). At $G = 16kT$ we thus get a stall force of about 8 pN. 5 pN corresponds to $F = 9$ and leads to a backstep probability of 11%. These estimates are not inconsistent with experimental observations and constitute a legitimate working assumption.

Fig. 3 shows the experimental data together with the predictions of our model for velocity vs. load and randomness vs. load. The fit was for $G = 16kT$, $\phi = 0.28$ and $D = 5.4\ 10^{-16}\ \mathrm{m}^2/\mathrm{s}$. Our result for the average D is of the same order of magnitude as what other authors have estimated. The average internal friction of the motor protein, i.e. $\beta = kT/D$, does then indeed come out to be an order of magnitude larger than that of the micrometer magnitude bead that is dragged along in the experiments. With $G = 16kT$, $\phi = 0.28$ and $D = 5.4\ 10^{-16}\ \mathrm{m}^2/\mathrm{s}$ we also find that D_{eff} is about 10 times as large as D. This means that the stepping motor protein may provide an example of the recently discussed "enhanced diffusion" [8], i.e., cases where drift can significantly blow up diffusion.

It is remarkable that a simple model, with only very few assumptions and free parameters, can quantitatively account for the experimental data. An independent corroboration of the above results is provided by the work of Yong-Ze Ma and Edwin Taylor. These

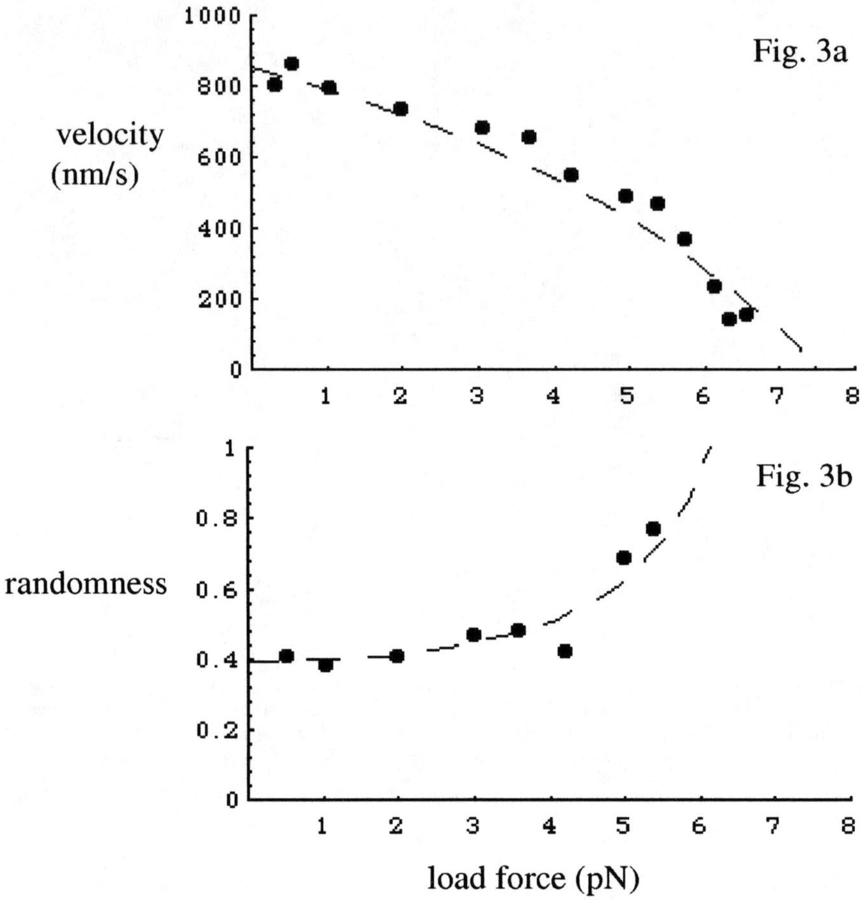

FIGURE 3. A force-velocity characteristic (3a) and a force-randomness characteristic (3b) for kinesin "walking" over microtubule. The dashed curves are the results of the simple stepping model described in the text. The dots represent the experimental data, cf. [10]. In the final analysis these graphs compare theory and experiment for the first and second moment of the stochastic motion.

researchers used a variety of biochemical methods to determine the different conformational states in the stepping cycle of kinesin. At the end of two back-to-back articles in 1997 [12, 13] they present a picture that looks very similar to Fig. 1a. They determined the rates from one state to a next and they found $T_{ps}/T_{diff} \approx 0.75$. At zero load the basic setup of our model (cf. Fig. 1b) leads to $T_{ps}/T_{diff} = \frac{2}{G}(1-\frac{1}{\phi})^2$. Numerically this works out to $T_{ps}/T_{diff} = 0.82$ for the G and ϕ that led to the fit in Fig. 3. All in all, there is excellent agreement between the simple model presented in Fig. 1 and the available data.

REFERENCES

1. D. Bray, *Cell Movements: From Molecules to Motility*, Garland Publishing Inc., New York and London, 2nd ed. 2001
2. W. Hua, J. Chung, and J. Gelles, *Science* **295**, 844 (2002).
3. E.M. Purcell, *Am. J. of Phys.* **45** (1), 3 (1977).
4. B. Alberts, D. Bray, J. Lewis, M. Raff, K. Roberts, and J.D. Watson, *Molecular Biology of the Cell*, Garland Publishing Inc., New York and London, 3rd ed., 1994.
5. R.D. Vale, and R.A. Milligan, *Science* **288**, 88 (2000).
6. M.J. Schnitzer, K. Visscher, and S.M. Block, *Nature Cell Biology* **2**, 718 (2000).
7. K. Svoboda, P.M. Mitra, and S.M. Block, *Proc. Natl. Acad. Sci. USA* **91**, 11782 (1994).
8. P. Reimann, C. Van den Broeck, H. Linke, P. Hänggi, J.M. Rubi, and A. Perez-Madrid, *Nature* **87**, 010602 (2001).
9. C.W. Gardiner, *Handbook of Stochastic Methods*, Springer Verlag, 2nd ed. 1985.
10. K. Visscher, M.J. Schnitzer, and S.M. Block, *Nature* **400**, 184 (1999).
11. C.M. Coppin, D.W. Pierce, L. Hsu, and R.D. Vale, *Proc. Natl. Acad. Sci. USA* **94**, 8539 (1997).
12. Y.Z. Ma, and E.W. Taylor, *J. Biol. Chem.* **272**, 717 (1997).
13. Y.Z. Ma, and E.W. Taylor, *J. Biol. Chem.* **272**, 724 (1997).

Molecular Motions in Ion Channels: a Possible Link to Noise in Single Channels

D. Peter Tieleman

Dept. of Biological Sciences, University of Calgary
2500 University Dr. NW, Calgary, Alberta T2N 1N4, Canada. Fax: (403) 289-9311
Email: tieleman@ucalgary.ca Web:http://moose.bio.ucalgary.ca

Abstract. Molecular dynamics simulations of proteins and lipid bilayers give detailed information on molecular motions on a timescale of up to a microsecond. Collective motions of proteins are thought to play a functionally important role in many water-soluble proteins and simulations of potassium channel structures show that ion transport might be linked to structural fluctuations in key areas of the protein. I describe simulations of two model channels –a channel consisting of parallel alpha-helices formed by the antimicrobial peptide alamethicin, and the bacterial porin OmpF, a large beta-barrel protein that forms three water-filled pores that allow transport of small molecules. Structural fluctuations in alamethicin during ion transport are hypothesized to be a possible source of high-frequency noise observed in single-channel conductance measurements. In a model of the alamethicin channel on a 100 ns time scale almost all the structural fluctuations are in individual helices with no evidence for collective motions of the channel as a whole. In porins, single channel measurements can obtain information on the interaction between permeating molecules and the protein. I present recent simulations that study the interactions between glucose and alanine with OmpF.

INTRODUCTION

Ion channels are proteins designed to conduct ions across the hydrophobic barrier formed by membranes in cells and organelles [1]. They exist in a staggering variety but can be classified in a small number of categories based on the type of ion they conduct or the mechanism that controls their opening and closing. The high-resolution structures of several channels are now available.
(See http://blanco.biomol.uci.edu/Membrane_Proteins_xtal.html for a full list).

Certain types of antimicrobial peptides, viral proteins, and toxins also form channels, and the outer membrane of Gram-negative bacteria and mitochondria contains porins, relatively non-selective channels that passively transport small molecules. Porins and channels formed by antimicrobial peptides such as gramicidin and alamethicin have become popular models for understanding the essential properties of ion channels. From a physical chemistry or physics point of view, these properties are primarily conductance and selectivity. In the selective ion channels such as potassium, sodium and calcium channels, gating (control of opening and closing) is also a key property that is currently not well understood.

Structural fluctuations in ion channels may be linked to their transport function. From an experimental perspective, single channel current measurements make it

CP665, *Unsolved Problems of Noise and Fluctuations: UPoN 2002: Third International Conference,*
edited by S. M. Bezrukov

possible to study fluctuations on a microsecond (and longer) time scale, fast enough to observe slow structural changes and in certain channels interactions between the channel protein and permeating molecules [2, 3] or added polymers [4]. Analysis of current noise in single channels can be used to infer information about the structure of a channel or about the permeation process of molecules through channels. Detailed computer simulations may provide an interpretation in atomic detail of such measurements, with a very high time resolution. Here I consider structural fluctuations in a model of the alamethicin channel and the interactions between permeating glucose and alanine molecules and OmpF, with two questions in mind:

1. What are sources of noise in single channel current measurements?
2. Do protein fluctuations in ion channels play a functional role, analogous to what has been proposed for water-soluble proteins?

In addition, by presenting the current limitations in molecular dynamics simulations I would like to consider what is required to bring simulations and experiments closer together.

METHODS

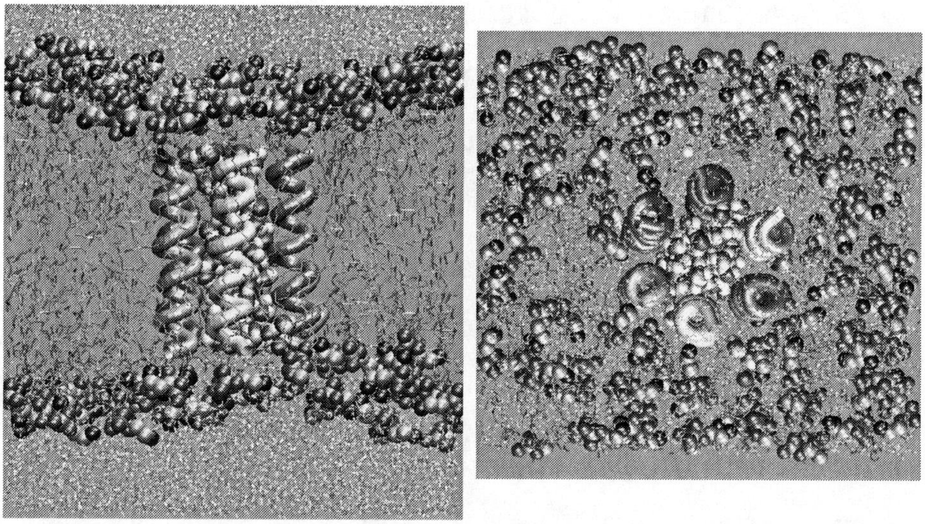

FIGURE 1. Snapshots of a hexameric bundle of alamethicin in a palmitoyl-oleoyl-phosphatidylcholine bilayer. The peptide is shown as schematic tube, water inside the channel and lipid headgroups are shown spacefilling, and lipid chains and bulk water are drawn as thick lines. The dimensions of this system are approximately 6x6x8 nm, with about 20,000 atoms. The simulations assume periodic boundary conditions.

Molecular dynamics simulation is a powerful technique to study lipids and membrane proteins in atomic detail [5]. Interactions between atoms are described by a classical empirical potential function that typically incorporates Lennard-Jones, Coulomb, bond, angle, and dihedral terms. Based on this potential, classical equations of motion are integrated numerically for millions of steps of the order of 2 femtoseconds, for all atoms in the model simultaneously. Typical models in biomolecular simulations include between 10,000 and 100,000 atoms and can be can be studied over time scales of up to a microsecond. In recent years, there have been many applications to ion channels, which conduct ions on a time scale comparable to the time scale that can be reached by molecular dynamics simulations [6]. The 100 ns simulation of the alamethicin N6 bundle (figure 1) in a phospholipid bilayer is an extension of the 20 ns simulation in ref. [7]. The porin simulation is described in more detailed in [8].

Collective motions of a protein can be studied by a covariance analysis, known as quasi-harmonic analysis or essential dynamics [9]. It is found that most of the motion in a protein simulation can be attributed to a small number (of the order of 10) of collective modes. These collective modes may have functional significance and can be used to assess the degree of convergence of the simulations. Hess derived a measure for the overlap between the parts of phase space sampled by two simulations [10], calculated from the covariance matrices of two simulations A and B:

$$overlap = 1 - \sqrt{Tr((\sqrt{A} - \sqrt{B}))^2 / Tr(A+B)} \quad (1)$$

If the covariance matrices A and B describe a completely different space, they are orthogonal and the overlap is 0. If they describe the same space, the overlap is 1. To estimate how much of the major essential modes from the covariance analysis describes functional motion and how much of it resembles random diffusion, its cosine content can be calculated. For random diffusion, without potential, the first few principal components are cosines with the number of periods equal to half the principal component index [11]. The cosine content c_i of principal component i is given by:

$$c_i = \frac{2}{T}\left(\int_0^T \cos(i\pi t)p_i(t)\mathrm{d}t\right)^2\left(\int_0^T p_i^2(t)\mathrm{d}t\right)^{-1} \quad (2)$$

with T the length of the simulation, and $p_i(t)$ the amplitude of the motion along eigenvector i at time t. The cosine content varies between 0 (no similarity to a cosine) to 1 (a perfect cosine).

RESULTS AND DISCUSSION

Alamethicin and OmpF

Figure 2A shows the root mean square deviation (RMSD) matrix of the backbone atoms of the alamethicin bundle. The maximum RMSD is 0.185 nm, which is remarkably low for a 100 ns simulation of a modeled structure (typical values for simulations based on crystal structures are 0.1 – 0.2 nm after 2-10 ns). 'Squares' in the matrix indicate clusters of close related conformations, of which there are many. Figure 2B shows the projection of the trajectory on the first 5 eigenvectors. The first eigenvector looks like a cosine and has a cosine content of 0.9, which is very high. The second and third eigenvector have much lower cosine contents (0.4 and 0.3, respectively). The overlap between the first 50 ns and the last 50 ns as defined above

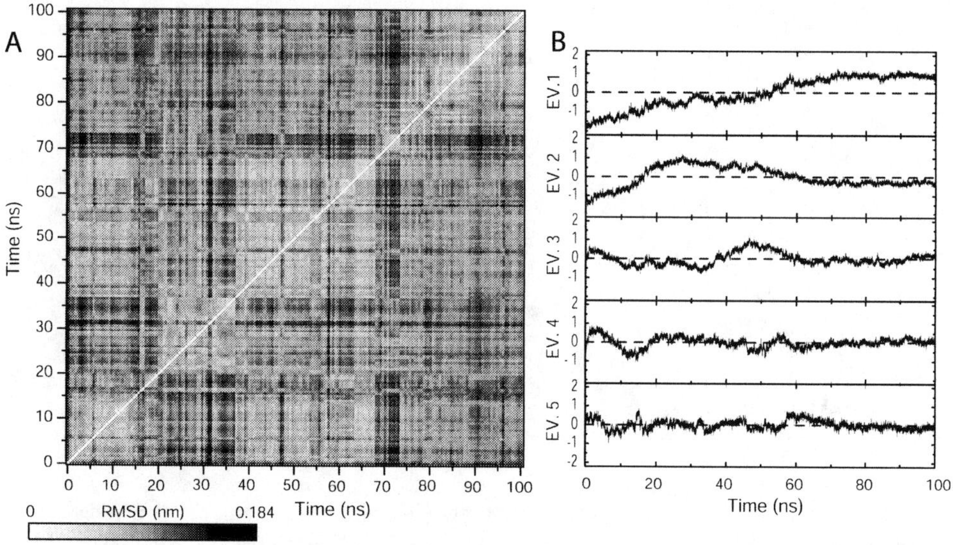

FIGURE 2. A. Root mean square deviation matrix of the backbone atoms of the helix bundle, after fitting on the Cα carbons. B. Projection of the trajectory on the first five eigenvectors.

is only about 0.5, which is quite low. When the trajectory is projected on the first three eigenvectors simultaneously, giving a 3-dimensional cloud of points (data not shown), several large extended clusters are found. Combined, these properties suggest a flat potential landscape with many local minima and low barriers, without sharp boundaries. Thus there is no evidence of well-defined collective motions in this bundle that might cause systematic structural fluctuations on a 100 ns timescale.

In contrast, several simulation studies have suggested that ion transport is coupled to motion of protein and water in gramicidin and KcsA. In gramicidin A, single file motion of water and ion is necessary because of the narrow channel, and protein motions are expected to couple with permeation [12]. In KcsA, the selectivity filter region is a narrow part of the pore lined by carbonyl oxygens from the polypeptide backbone. Several simulations have suggested that motions of the selectivity filter are coupled to ion translocation, e.g. [13]. Thus far, simulations on this protein have been too short to successfully apply essential dynamics analysis, but this will become possible as simulations become longer.

Interactions between permeating molecules and an open channel are also a potential source of noise [2, 3]. Computer simulations can provide a detailed interpretation of such interactions. As example, figure 3 shows the force a permeating molecule exerts on its environment as it is dragged through a pore in a non-equilibrium simulation. During this simulations, hydrogen bonding, salt bridges, energies, and atom-atom contacts were monitored. Although the pulling rate was high, three independent permeating molecules gave very similar results. In all cases, the permeating dipolar molecules are strongly oriented in the local field. The strongly charged eyelet region is likely to present many local minima for permeation of a dipolar molecule, and many different pathways. Extended simulations of large permeating molecules in porins might be able to quantify this idea.

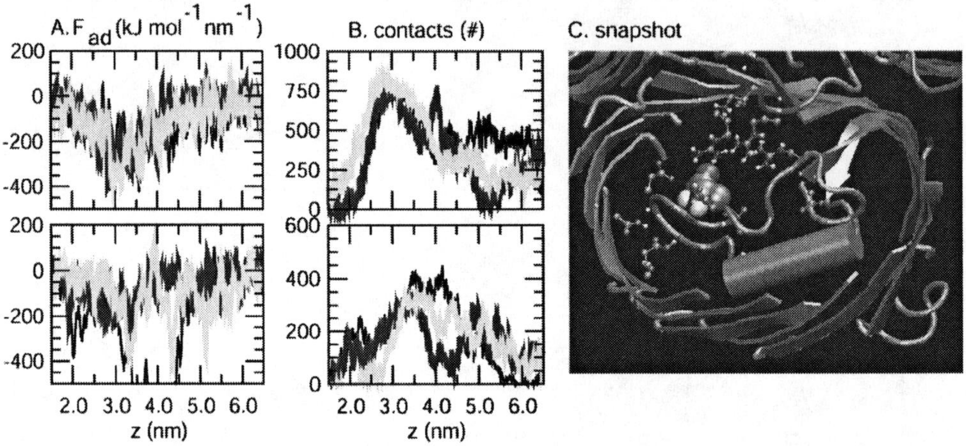

FIGURE 3. A. Adhesion forces of methylglucose (top) and alanine (bottom), for three independent molecules. B. Number of contacts between permeating molecule and porin, defined as a distance of less than 0.6 nm. C. Snapshot of the 'eyelet' region with an alanine molecule in it. The eyelet region has a strong local electric field due to a cluster of positive residues (top) and negative residues (near the loop, middle).

Other Sources of Noise

The two examples given here considered two possible sources: collective motions of the protein and interactions with permeating molecules. Clearly, there are other alternatives. Based on MD simulations, fluctuations on a smaller scale than collective motions of the whole protein, e.g. motions of charged side chains or interactions of permeating ions with the channel, might significantly change the local potential and thereby modulate the current. Another possible source is interactions of the protein with lipids, and the fluctuations on all time scales that occur in lipid bilayers [14]. This has been suggested as a mechanism to explain noise in gramicidin A [15]. Such interactions might change the entrance resistance, or directly change the protein structure. Molecular dynamics simulations have reached a point where collective motions in lipid bilayers can be studied, although these studies are limited to wavelengths that are compatible with the periodic boundary conditions used in MD simulations [16]. In the future, it should also be possible to study the coupling between protein dynamics and lipid bilayer dynamics by direct simulation.

Limitations Of Simulations And Experiments

From a molecular dynamics point of view, any single molecule measurement is exciting because it provides a direct link to the 'single molecule' simulations. In the opposite direction, detailed molecular dynamics simulations may aid in interpreting single molecule experiments. Single-channel measurements are particularly exciting because they allow such an accurate measurement – small changes in concentrations of permeating sugars or single mutations in channel proteins have measurable effects. Unfortunately, the time resolution of single channel measurements is still at least an order of magnitude less than the time scale simulations can reach. Improvements in time resolution would bring measurements closer to molecular dynamics simulations. New techniques for single-molecule studies such as time-resolved fluorescence could also be combined with single channel measurements, which opens up new possibilities for studying fast events in ion channels. Finally, new high-resolution crystal structures of ion channels will also be crucial for simulations of complex channels. A recent structure of the KcsA potassium channel at 2.0 Å significantly improves (relative to the previous 3.2 Å structure, [17]) the accuracy of the structure of the important selectivity filter region [18], which is likely to have a significant effect on simulations starting from these structures.

Molecular dynamics simulations themselves have several important limitations, including the accuracy of pairwise potentials and limits on the time and length scale that can be simulated. Currently, time scales of up to 100 ns for simple systems like alamethicin can be reached on desktop computers, while 300-400 ns would be feasible on a typical Beowulf cluster, but this is still an order of magnitude shorter than the resolution of single-channel measurements. In the foreseeable future, timescales of microseconds can probably be routinely reached. In some cases, smarter algorithms with biased sampling may also be used to alleviate this problem. In spite of these limitations, molecular dynamics simulations and single channel measurements are edging closer and closer, and it will be very exciting when they can shake hands.

CONCLUSIONS

I asked two questions. The first question concerned potential sources of noise in single-channel current measurements. The molecular dynamics simulations presented allow perhaps a bit more informed speculation, but do not provide a single answer. Structural fluctuations might be a source of noise, but the longest simulation of alamethicin (100 ns) does not suggest an obvious candidate for a large-scale type of motion of this particular channel. Interactions with permeating molecules such as observed in alpha-hemolysin [3] and maltoporin [2] are obvious sources of noise, and simulations are beginning to address this in more detail. Coupling of channel motions with lipid motions is still outside the range of what can be studied by molecular dynamics simulations, but there is some experimental justification for it.

The second question was "do protein fluctuations in ion channels play a functional role, analogous to what has been proposed for water-soluble proteins?" In the case of the potassium channel KcsA it seems quite likely that deformations of the selectivity filter are essential for ion permeation, even though such deformations are not seen in the crystal structure [13]. Coupling between ions, water and protein motion has been observed in a molecular dynamics simulation of gramicidin A [12]. In alamethicin channels, no evidence is found for important cooperative motions, but it remains to be seen whether this is due to inaccuracies of the model or due to some intrinsic property of the channel.

ACKNOWLEDGMENTS

This work was supported by the Canadian Institutes of Health Research. DPT is a Scholar of the Alberta Heritage Foundation for Medical Research .

REFERENCES

1. Hille, B., *Ionic Channels of Excitable Membranes*. 2nd ed. 1992, Sunderland, Mass.: Sinauer Associates Inc. 607.
2. Bezrukov, S.M., L. Kullman, and M. Winterhalter, *Probing sugar translocation through maltoporin at the single channel level*. FEBS. Lett. **476**, 224-228 (2000).
3. Bayley, H., and C.R. Martin, *Resistive-pulse sensing - From microbes to molecules*. Chem. Rev. **100**, 2575-2594 (2000).
4. Rostovtseva, T.K., E.M. Nestorovich, and S.M. Bezrukov, *Partitioning of differently sized poly(ethylene glycol)s into OmpF porin*. Biophys. J. **82**, 160-169 (2002)
5. Tieleman, D.P., S.J. Marrink, and H.J.C. Berendsen, *A computer perspective of membranes: molecular dynamics studies of lipid bilayer systems*. Biochim. Biophys. Acta **1331**, 235-270 (1997).
6. Tieleman, D.P., et al., *Simulation approaches to ion channel structure-function relationships*. Quart. Rev. Biophys. **34**, 473-561 (2001)
7. Tieleman, D.P., et al., *Evaluation and analysis of channel models: simulations of alamethicin*. Biophys. J. **83**, 2393-2407 (2002).
8. Robertson, K.M., and D.P. Tieleman, *Orientation and interactions of dipolar molecules during transport through OmpF porin*. FEBS Lett. **529**, 53-57 (2002).
9. Berendsen, H.J.C. and S. Hayward, *Collective protein dynamics in relation to function*. Curr. Opin. Struc. Biol. **10**, 165-169 (2000).
10. Hess, B., *Convergence of sampling in protein simulations*. Phys. Rev. E **65**, art. no. 031910 (2001).
11. Hess, B., *Similarities between principal components of protein dynamics and random diffusion*. Phys. Rev. E, **62**, 8438-8448 (2000).
12. Elber, R., et al., *Sodium in gramicidin: an example of a permion*. Biophys. J. **68**, 906-924 (1995).
13. Shrivastava, I.H., et al., *K+ vs. Na+ ions in a K channel selectivity filter: a simulation study*. Biophys. J. **83**, 633-645 (2002).
14. Bayerl, T.M., *Collective membrane motions*. Curr. Opin. Coll. Inter. Sci. **5**, 232-236 (2000).
15. Armstrong, K.M. and S. Cukierman, *On the origin of closing flickers in gramicidin channels: a new hypothesis*. Biophys. J. **82**, 1329-1337 (2002).
16. Lindahl, E. and O. Edholm, *Mesoscopic undulations and thickness fluctuations in lipid bilayers from molecular dynamics simulations*. Biophys. J. **79**, 426-433 (2000).
17. Doyle, D.A., et al., *The structure of the potassium channel: molecular basis of K^+ conduction and selectivity*. Science **280**, 69-77 (1998).
18. Zhou, Y., et al., *Chemistry of ion coordination and hydration revealed by a K+ channel- Fab complex at 2.0 Å resolution*. Nature, **414**, 43-48 (2001).

Hysteresis in Channel Gating

E. Gudowska-Nowak*, H. Flyvbjerg†, P. Bennekou** and P. Christophersen‡

* *M. Smoluchowski Institute of Physics, Jagellonian University, PL-30059 Kraków, Poland*
†*Materials Research Department, Risø National Laboratory, DK-4000 Roskilde, Denmark*
***August Krogh Institute, Universitetsparken 13, DK-2100, Copenhagen, Denmark*
‡*NeuroSearch, Pederstrupvej 93, DK-2750 Ballerup, Denmark*

Abstract. We present a simple model for the hysteresis observed in the open-state probability of a voltage-activated ion channel. Transitions between the open and closed states are modeled with a reaction coordinate that describes motion of charges. In a simple Kramer's scenario, we assume that the voltage felt by a gate is subject to environmental noise with long correlation times. The stationary states of the reaction coordinate that result in this environment are found as functions of the driving voltage, the fluctuating membrane potential, in an adiabatic approximation. Depending on the intensity of the fluctuations, the open-probability exhibits hysteretic behaviour. The results of this scenario suggest a reanalysis of current traces recorded on the hysteresis cycle, looking for periods of changed characteristics caused by transitions between different branches of stationary states.

INTRODUCTION

Using the patch-clamp technique, *Kaestner et al.* have studied [1] voltage-activated cation channel in the human red blood cell (RBC) membrane, demonstrating that the degree of activation at a given membrane potential depended strongly on the prehistory of the channel. The gating exhibited hysteretic-like behaviour with the quasi steady-state deactivation and activation curves displaced by about 25 mV. The purpose of this contribution is to evaluate a kinetic model for voltage-activated channel gating. This model leads to a delayed response to a cycling external field that is similar to the change of magnetization in ferromagnetic materials under the influence of an external magnetic field.

Hysteresis may occur when a bistable system is driven by an oscillating (or fluctuating) force [2, 3] that varies too fast for the system to equilibrate fully immediately. It is a common phenomenon, occurring in a wide class of equilibrium and non-equilibrium situation, such as ferroelectric transitions, optically bistable systems, bistability in Josephson junctions, and light-induced chemical reactions. Hysteresis is a memory-type phenomenon, and has been also reported for the ligand-gated NMDA receptor [4] and in voltage-gated HERG-like channels [5]. A similar effect has been reported for a vanilloid receptor [6] which is a thermo-sensitive non-selective cation channel that is activated by noxious heat. In order to explain hysteretic behaviour observed in channel response to external stimuli, one has suggested the existence of an ultra-slow inactivation process [5] that effectively introduces a lag in the feedback between voltage and gating.

CP665, *Unsolved Problems of Noise and Fluctuations: UPoN 2002: Third International Conference,* edited by S. M. Bezrukov

Alternatively, one has focused on periodic sensitization of the channel on repeated heat stimulation [6].

Here, we investigate theoretically whether local fluctuations in the membrane potential can cause changes in opening probabilities. We find that in a system that does not display instability by itself, a phenomenon similar to a first-order phase transition can be caused by random variations in the applied voltage.

BISTABILITY AND HYSTERESIS CYCLE

The mechanism of voltage-gated ion channels was introduced in an early work of Hodgkin and Huxley who suggested that Boltzmann statistics describes their probability of being open or closed, including the voltage-dependence of this probability [7, 8]. For the purpose of modeling, we assume the ion channel in question has only two states, open and closed. We associate an internal energy E_{open}, E_{closed} with each of the two states, and assume that Boltzmann statistics describes the probability for a channel to be open or closed, with transitions between the two states being thermally activated barrier crossings in an energy landscape that we do not detail. Then the probability that a channel is open (p_{open}) or closed (p_{closed}) is

$$p_{\text{open},(\text{closed})} = \frac{e^{-E_{\text{open},(\text{closed})}}}{e^{-E_{\text{open}}} + e^{-E_{\text{closed}}}} = \frac{1}{1+e^{\pm\Delta E}} \ , \tag{1}$$

where all energies are given in units of $k_B T$.

In the case of voltage-gated channels, the transition between the open and the closed state involves movement of charge in the trans-membrane field. This causes a change in Gibbs free energy

$$\Delta G = G_{\text{open}} - G_{\text{closed}} + \sum_i z_i F\, \delta_i V \ . \tag{2}$$

Each moving charge i makes a contribution of $z_i \delta_i V$ to ΔG, where z is the valence of the charge, F is Faraday's constant, and $\delta_i V$ the change in membrane potential experienced by the ith charge. ΔG is related to the change of internal energy ΔE through the thermodynamic relation $G = E + pv - TS$, where p denotes pressure, v volume, T temperature, and S entropy. For isothermal, isobaric transitions, $\Delta T = \Delta p = 0$, and consequently $\Delta G = \sum W_e$, where W_e stands for all kinds of works (chemical, elctrochemical) encountered in the biophysical system under study. With ΔG as above, Eq. (1) may be written

$$p_{\text{open}}(V) = \frac{1}{1+e^{-\alpha(V-V_0)}} \equiv f(V) \ , \tag{3}$$

where αV_0 replaces $(G_{\text{closed}} - G_{\text{open}})/k_B T$, V_0 is the voltage at which $p_{\text{open}} = p_{closed} = 1/2$ and αV stands for $(\sum z_i \delta_i V)/k_B T$.

Equation (3) may be derived by assuming Kramer's scenario for the gating kinetics [9]. This approach offers a description in which the configurational change of the gating protein is modeled by a multi-component "reaction coordinate" x. Since the potential energy of the entire system (the gate plus membrane environment) is a function of hundreds of relevant coordinates of the system, x can be understood as an effective variable similar to its definition in the Marcus theory of charge-transfer reactions [10]. The open and closed states of a channel can then be described as $x \approx 1$ and $x \approx 0$, respectively, with a potential (activation) barrier separating these two regions on the x-axis, and x being confined to the regions near 0 and 1 by the potential taking large values elsewhere. Opening and closing of a channel is then modeled by a process similar to an activated uni-molecular chemical reaction pictured by a kinetic scheme

$$\text{Open} \underset{k_o}{\overset{k_c}{\rightleftarrows}} \text{Closed} \quad . \tag{4}$$

For a sufficiently high barrier separating open and closed states, the probability of x taking values in between vanishes and we can work with the integrated probability of x being near 0 or 1, p_{closed} and p_{open}. They satisfy $p_{\text{closed}} + p_{\text{open}} = 1$ and the linear kinetic equation

$$\frac{dp_{\text{open}}}{dt} = k_o p_{\text{closed}}(t) - k_c p_{\text{open}}(t) \quad . \tag{5}$$

Under stationary conditions the ratio of the reaction rates can be related to the difference in the free energy by the equilibrium constant

$$K = \frac{k_c}{k_o} = \frac{p_{\text{closed}}}{p_{\text{open}}} = e^{-\Delta G/k_{\text{B}}T} \quad . \tag{6}$$

Rearrangement of this equation gives a formula for the probability of the channel being open Eq.(3):

$$p_{\text{open}} = \frac{1}{1 + k_c/k_o} = \frac{1}{1 + e^{-\Delta G/k_{\text{B}}T}} \quad . \tag{7}$$

Thus ΔG is related to the ratio of kinetic rates for the "forward" and "backward" reactions. Experimental estimates of the opening and closing rates can be found from the voltage-dependent dwell times determined from experimental recordings as inverse of averages *i.e.* $k_o = \langle T_c \rangle^{-1}$, $k_c = \langle T_o \rangle^{-1}$ where $\langle T_{c,o} \rangle$ are average dwell times in the closed (open) states. By assuming the rates to have a general Arrhenius form, their ratio would be constrained (*cf.* Eq.(3)) by:

$$\frac{k_c}{k_o} = \exp(-\alpha(V - V_0)) \tag{8}$$

We now additionally assume that the membrane potential felt by a channel is not exactly constant, but fluctuates, albeit slowly so compared to the rates of channel opening and closing:

$$V(t) = \langle V \rangle + \xi(t) \quad . \tag{9}$$

Through V, the height of the potential barrier dividing wells of open and closed states is now itself a random variable, so the dynamics of transitions between these states becomes dependent on the statistical properties of the random variable controlling the barrier height [16, 17, 18]. From other cotexts, it is known that this presence of noise (fluctuations) may lead to several nontrivial and previously unexpected effects, such as resonant activation and resonant inhibition phenomena [17], or enhancement of the signal registered from open channels [18].

Here we limit ourselves to fluctuations $\xi(t)$ modeled by a Gaussian stochastic process [11, 12, 14] with a correlation time much longer than the characteristic evolution time of the open state probability $p_{\text{open}}(t)$:

$$d\xi(t) = -\varepsilon^2\xi(t)dt + \varepsilon\sigma dW(t) \ . \tag{10}$$

here $dW(t)$ denotes an increment of the Brownian motion (Wiener process) and $1/\varepsilon^2$ is the correlation time of ξ:

$$\langle \xi(t)\xi(t+s)\rangle = \frac{\sigma^2}{2}\exp(-\varepsilon^2 s) \ . \tag{11}$$

The pair process (p_{open}, ξ) described by a set of differential equations (the deterministic Eq.(5) and the Langevin Eq.(10)) is Markovian and can be shown to obey the Fokker-Planck equation [11, 12]:

$$\begin{aligned}\frac{\partial P(p_{\text{open}},\xi,t)}{\partial t} &= -\frac{\partial}{\partial p_{\text{open}}}[k_o - (k_o + k_c)p_{\text{open}}]P(p_{\text{open}},\xi,t) \\ &\quad +\varepsilon^2\left(\frac{\partial}{\partial \xi}\xi + \frac{\sigma^2}{2}\frac{\partial^2}{\partial \xi^2}\right)P(p_{\text{open}},\xi,t)\end{aligned} \tag{12}$$

where $P(p_{\text{open}},\xi,t)$ represents the probability density for observing values p_{open} and ξ at time t and $P_s(p_{\text{open}},\xi)$ is its time independent analogue in the stationary version of the same Fokker-Planck equation

$$\varepsilon^2\left(\frac{\partial}{\partial \xi}\xi + \frac{\sigma^2}{2}\frac{\partial^2}{\partial \xi^2}\right)P_s(p_{\text{open}},\xi) = \frac{\partial}{\partial p_{\text{open}}}\left[k_o - (k_o + k_c)p_{\text{open}}\right]P_s(p_{\text{open}},\xi) \tag{13}$$

Now consider the quasi-steady states that the system settles in after all fast transient modes in $p_{\text{open}}(t)$ have died out due to k_o and k_c being large compared to slow kinetics of $\xi(t)$ in V. The quasi-stationary opening probability p_{open} is then itself a randomized variable whose (quasi-stationary) probability density $P_s(p_{\text{open}})$ can be obtained. Such an approximation assumes [14, 12] that the system is always in a quasi-stationary state with respect to the value of the slowly fluctuating parameter V. In other words, since the noise in voltage is much slower than the evolution of $p_{\text{open}}(t)$, the latter stays always close to the curve $p_{\text{open}}(V)$ given by Eq.(3). The explicit form of $P_s(p_{\text{open}})$ follows then from the stationary probability density of the noise $p_s(\xi)$ transformed (*cf.* Eq.(3)) by the change of variable $V = f^{-1}(p_{\text{open}})$ [14].

In fact, from the standard probabilistic calculus we recall [12, 13] that if X is a random variable with probability density function $f_X(x)$ and g is a strictly increasing differentiable function of X, then $Y = g(X)$ defines a random variable with a density function $f_Y(y) = f_X(x)[g'(x)]^{-1}$ where $y = g(x)$. For static fluctuations (at the lowest order of ε) this procedure yields

$$P_s(p_{\text{open}}) = \int d\xi\, p_s(\xi)\delta(p_{\text{open}} - f(\xi)) = p_s(f^{-1}(p_{\text{open}}))|\frac{d}{dp_{\text{open}}} f^{-1}(p_{\text{open}})| \quad (14)$$

where fluctuations $\xi = f^{-1}(p_{\text{open}})$

$$\xi = V - \langle V \rangle = V_0 + \frac{1}{\alpha}\log(\frac{p_{\text{open}}}{1 - p_{\text{open}}}) - \langle V \rangle \quad (15)$$

are characterized by a standard probability density $p_s(\xi)$

$$p_s(\xi) = \frac{1}{\sqrt{(2\pi)}\sigma}\exp(-\frac{\xi^2}{2\sigma^2}) \quad (16)$$

Direct evaluation of the formula (14) results in

$$P_s(p_{\text{open}}) = \frac{1}{\sqrt{(2\pi)}\sigma\alpha}\frac{1}{p_{\text{open}}(1 - p_{\text{open}})} \times \exp\left(-\frac{(V_0 + \frac{1}{\alpha}\log(\frac{p_{\text{open}}}{1-p_{\text{open}}}) - \langle V \rangle)^2}{2\sigma^2}\right) \quad (17)$$

This form of of the probability density function predicts the existence of two branches of opening probabilities p_{open} that can be read off the equation for the extrema of $P_s(p_{\text{open}})$:

$$dP_s(p_{\text{open}})/dp_{\text{open}} = \alpha\sigma^2(2p_o^* - 1) - V_0 + \langle V \rangle - \frac{1}{\alpha}\log(\frac{p_o^*}{1 - p_o^*}) = 0 \quad (18)$$

Interpreted as a functional equation of $p_o^* = f(V)$, the above relation leads now to a modified curve $f_{\sigma^2}(V)$ (cf. Figure 1) due to the fluctuations effects in the voltage V. Beyond a critical value of variance $\sigma^2 = 1$, the probability density $P_s(p_{\text{open}})$ becomes bimodal resembling a first order transition in binary mixtures: Hysteresis in this case may be induced by sweeping the voltage from negative to positive values $V_0 - \langle V \rangle$ at a constant variance of the noise, $\sigma^2 > 1$.

OPEN ENDS

The notes above present in brief a problem we have started to study [1, 15]. Hysteretic voltage-dependent gating reported by Kaestner *et al.* is persistent: it has been observed repeatedly for the same patch after a period at a deactivating (negative) potential. It seems that memory effects in channel gating and hysteresis may well be a normal kinetic behaviour with potentially important physiological implications. Hysteretic phenomena in channels are, however, not understood at the molecular level. A formal description of the effect was achieved here by assuming time dependent rate parameters due either to intrinsic environmental fluctuations or/and external time-variating fields [6, 16, 17,

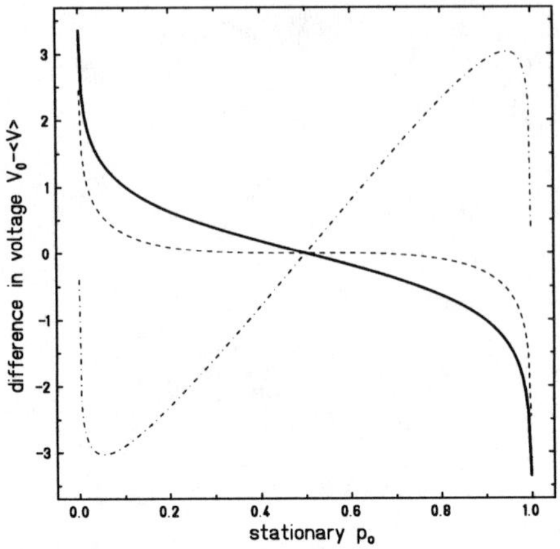

FIGURE 1. Modified switching curve between open and closed state due to the noise effect in voltage. Curves present extrema of the probability density $P_s(p_{\text{open}})$ as a function of the driving potential V for various intensity of fluctuations σ^2 and $\alpha = 1$. Solid line: $\sigma^2 = 0.1$, dashed line: $\sigma^2 = 1.0$, dotted-dashed line: $\sigma^2 = 5.0$.

18]. For slowly varying fields, the effect can be described in terms of an adiabatic elimination of the state variable which effectively pictures situation with many substates [17] and hence, a whole ensemble of (static) activation barriers amongst the states. For comparison, in bistable systems driven by periodic forcing, hysteresis may be a natural consequence of periodicity.

ACKNOWLEDGMENTS

E.G-N expresses her gratitude to the organizers of the UPoN2002 meeting for the hospitality.

REFERENCES

1. L. Kaestner, P. Christophersen, I. Bernhards and P. Bennekou, *The Non-Selective Voltage-Activated Cation Channel in the Human Red Blood Cell Membrane: Reconciliation Between Two Conflicting Reports and Further Characterization*, Bioelectrochem. **52** (2000) 117-125.
2. P. Jung, G. Gray, R. Roy and P. Mandel, *Scaling Law for Dynamical Hysteresis*, Phys. Rev. Lett. **65** (1990) 1873-1878.
3. M.C. Mahato and A.M. Jayannavar, *Some Stochastic Phenomena in a Driven Double-Well System*, Physica A **248** (1998) 138-154.
4. L. M. Nowak and J. M. Wright, *Slow Voltage-Dependent Changes in Channel Open State Probability Underlie Hysteresis of NMDA Responses in Mg^{2+}-Free Solutions*, Neuron **8** (1992) 181-187.
5. P. S. Pennefather, W. Zhou and T. E. DeCoursey, *Idiosyncratic Gating of HERG-like K^+ Channels in Microglia* J. Gen. Physiol. **11** (1998) 795-805.
6. H. Xu *et al.*, *TRPV3 is a Calcium-Permeable Temperature Sensitive Cation Channel* Nature **418** (2002) 181-186.

7. A.L. Hodgin and A.F. Huxley, *A Quantitative Description of Membrane Current and Its Application to Conduction and Excitation in Nerve*, J. Physiol. (Lond.) **117** (1952) 500-544.
8. B. Hille, *Ionic Channels of Excitable Membranes*, Sinauer Associates Inc, Sunderland, MA (1992).
9. D. Sigg, H. Qiau and F. Bezanilla, *Kramers Diffusion Theory Applied to Gating Kinetics of Voltage-Dependent Ion Channels* Biophys. J. **76** (1999) 782-803.
10. A.M Kuznetsov and J. Ulstrup, *Electron Transfer in Chemistry and Biology. An Introduction to the Theory*, Wiley, Chichester, (1999).
11. H. Risken, *The Fokker Planck Equation*, Springer-Verlag (1989).
12. W. Gardiner, *Handbook of Stochastic Methods*, Springer-Verlag (1983)
13. M. Kac, *Probability, Number Theory and Statistical Physics*, MIT Press (1979).
14. L. Arnold, W. Horsthemke, R. Lefever, *White and Coloured External Noises and Transition Phenomena in Nonlinear Systems*, Z. Phys. B. **29** (1978) 367-373.
W. Horsthemke and R. Lefever, *Noise Induced Transitions: Theory and Applications in Physics, Chemistry and Biology*, Springer-Verlag (1984).
15. E. Gudowska-Nowak, P. Bennekou, P. Christophersen, H. Flyvbjerg, *Modelling Hysteresis Observed in a Voltage-Activated Ion Channel*, work in progress.
16. A. Fuliński, *Noisy Chemical Equilibria and Flow Reactor's Yield*, J. Chem. Phys. **114** (2001) 3109-3119.
17. J. Iwaniszewski, I. Kaufman, P.V.E. McClintock and A. J. McKane, *Resonances While Surmounting a Fluctuating Barrier*, Phys. Rev. E **61** (2000) 1170-1175.
18. S.M. Bezrukov and I. Vodyanoy, *Stochastic Resonance in Non-Dynamical Systems Without Response Thresholds*, Nature **385** (1997) 319-321.

A PDE Formulation of Non-Equilibrium Statistical Mechanics for Ionic Permeation

Zeev Schuss*, Boaz Nadler†, Amit Singer* and Robert S. Eisenberg**

*Department of Mathematics, Tel-Aviv University, Ramat-Aviv 69978, Israel
†Department of Mathematics, Yale University, P.O. Box 208283, New Haven, CT, 06520
**Department of Molecular Biophysics and Physiology, Rush Medical Center, 1750 Harrison St., Chicago, IL 60612

Abstract. When there is a steady net flux in a system of interacting particles, the microscopic structure of the system can no longer be determined from the Boltzmann equilibrium distribution (partition function). Nonetheless, the microscopic structure of a finite system of diffusing interacting particles can be described by Poisson-Nernst-Planck-type partial differential equations. These equations, defined in a finite domain, are the non-equilibrium generalization of the BBGKY hierarchy of equilibrium statistical mechanics. Indeed, when no-flux conditions are imposed on the domain boundaries, equilibrium results are recovered. When non-homogeneous boundary conditions are given for these equations, the solutions describe densities and electrostatic potentials of particle systems not in equilibrium. The construction of a pair correlation function under these conditions will be a new result in statistical physics. As in the equilibrium case, a closure relation between a higher and a lower order correlation function has to be assumed. However, since we are considering a finite system, boundary conditions for the higher order correlation functions must also be derived. In applications to the permeation of ions through protein channels of biological membranes the computation of the pair correlation function will lead to a prediction of current through an open channel, given the spatial structure and fixed charge distribution. The pair correlation function contains finite size effects that lead to blocking in a narrow channel and possibly to selectivity.

1. DEFINITIONS

Permeation of ions from one electrolytic solution to another through a protein channel is a biological process of considerable importance [1], [2]. Permeation occurs on a time scale of micro- to milliseconds, far longer than the femtosecond time scales of atomic motion. Direct simulations of atomic dynamics are not yet possible for such long time scales; thus, averaging is unavoidable. The question is what and how to average. Specifically, we consider a solution confined to a finite region Ω, containing N_h ions of type h ($h = \text{Na}^+, \text{Cl}^-, \text{Ca}^{++}, \ldots$), $\sum_h N_h = N$. We denote the coordinates of a point by $x = (x,y,z)$. We number the ions in the solution at time $t = 0$ and denote the vectors of coordinates and velocity of the j-th positive ion at time t by $x_j^h(t)$ and $\dot{x}_j^h(t)$, respectively. The coordinate vector of all ions in configuration space is denoted $\tilde{x} = \left(x_1^{h_1}, \ldots, x_{N_{h_k}}^{h_k}\right)$, while in analogy, the vector of all velocities is denoted as $\dot{\tilde{x}}$ or $\tilde{v}$. For future reference, the vector of coordinates of all $N-1$ ions, excluding the j-th ion of type h is denoted $\tilde{x}_j^h$.

We assume that all diffusing ions interact with each other, so their motion can be described by a system of coupled Langevin equations. Specifically, for

CP665, *Unsolved Problems of Noise and Fluctuations: UPoN 2002: Third International Conference*, edited by S. M. Bezrukov

$h = \mathrm{Na}^+, \mathrm{Cl}^-, \mathrm{Ca}^{++}, \ldots$

$$\ddot{x}_j^h + \gamma^h\left(x_j^h\right)\dot{x}_j^h = \frac{f_j^h(\tilde{x})}{m^h} + \sqrt{\frac{2\gamma^h(x_j^h)k_BT}{m^h}}\,\dot{w}_j^h, \quad (j = 1,2,\ldots,N_h), \tag{1}$$

where a dot on a variable denotes differentiation with respect to time, $f_j^h(\tilde{x})$ denotes the total force acting on the j-th ion of type h, $\gamma^h(x)$ is the location dependent friction coefficient per unit mass of the ionic species of type h, and $\dot{w}_j^h$ are, by assumption, independent standard Gaussian white noises. The parameter k_B is Boltzmann's constant, T is absolute temperature, and m^h is an effective mass of an ion of type h. The force on an ion consists of inter-ionic attractions and repulsions, the imposed external field, and a self force near dielectric interfaces. When Lennard-Jones repulsions are included and no flux boundary conditions are imposed, the equilibrium theory of [6]-[8] applies. However, if fixed, but different concentrations are maintained on different parts of the boundary of the finite domain Ω, the system is no longer in equilibrium, because it carries a current. In this case the trajectories of eq.(1) are reflected at impermeable boundaries, denoted $\partial\Omega_R$, and are created and terminated by a feedback mechanism at a part $\partial\Omega_F$ of the boundary, where concentrations are kept fixed. This is the usual situation in biology. Therefore, a theory is needed for the description of non-equilibrium diffusion of interacting particles in a finite domain.

2. PROBABILITY DENSITIES

We denote by $p_j^h(x)$, $j = 1,\ldots,N^h$, the stationary probability density function (pdf) of the location of the j-th ion of species h. The joint stationary pdf of the locations of all N ions is denoted $p_N(\tilde{x})$. By definition, the pdf of a single particle is simply the marginal density of the pdf of all particles integrated over all particle locations except the j-th ion of species h,

$$p_j^h(x_j^h) = \int_{\Omega^{N-1}} p_N(\tilde{x})\, d\tilde{x}_j^h. \tag{2}$$

Since all ions of a given species are indistinguishable, $p_j^h(x) = p^h(x)$, for all j. Thus, the time averaged steady state concentration of ions of species h, denoted $c^h(x)$ is given by

$$c^h(x) = N^h p^h(x). \tag{3}$$

The concentration $c^h(x)$ is the physical density of ions and it integrates to the total number of ions of species h in the solution.

Similarly, the joint pdf of ions of types h, h' at points x, y is denoted $p^{h,h'}(x,y)$. The conditional pdf of an ion of type h' at y, given an ion of type h at x is denoted $p^{h'|h}(y\,|\,x)$. It is related to the joint densities by

$$p^{h'|h}(y\,|\,x) = \frac{p^{h,h'}(x,y)}{p^h(x)}. \tag{4}$$

Similarly, $p^{h,h',h''}(x,y,z)$ denotes the joint density of ions of types h,h',h'' located at x,y,z, respectively, $p^{h''|h',h}(z\,|\,y,x)$ denotes the conditional density of an ion of type h'' at z, given ions of types h' and h located at y and x, respectively, and so on. We have

$$p^{h''|h',h}(z\,|\,y,x) = \frac{p^{h,h',h''}(x,y,z)}{p^{h,h'}(x,y)}. \tag{5}$$

3. NERNST-PLANCK EQUATIONS FOR ONE AND TWO ION DENSITIES

Our computation of the one and two ion densities in a non-equilibrium steady state system begins with the derivation of partial differential equations for the computation of the one and two ions probability density functions. We note that an equation for the pdf $p^h(x)$ is readily converted into an equation for the single particle density $c^h(x)$ by the relation (3).

In the overdamped limit, the joint pdf of all particles in configuration space, $p_N(\tilde{x})$, satisfies the Fokker-Planck-Smoluchowski equation in Ω^N [13],

$$0 = \sum_h \sum_j \nabla_{x_j^h} J_{x_j^h}(\tilde{x}) \tag{6}$$

where $J_{x_j^h}$ is the three dimensional probability flux density of the j-th ion of species h, given by

$$J_{x_j^h}(\tilde{x}) = -\nabla_{x_j^h} \cdot \left(\frac{f_j^h(\tilde{x})}{m^h \gamma^h(x_j^h)} p_N(\tilde{x}) - \frac{k_B T}{m^h \gamma^h(x_j^h)} \nabla_{x_j^h} p_N(\tilde{x}) \right).$$

By integrating (6) with respect to all variables but one, the following equation for the single particle probability density was derived in [9]-[11],

$$0 = \nabla_x \cdot J^h(x) = -\nabla_x \cdot \left(\frac{\bar{f}^h(x)}{m^h \gamma^h(x)} p^h(x) - \frac{k_B T}{m^h \gamma^h(x)} \nabla_x p^h(x) \right), \tag{7}$$

where $J^h(x)$ is the probability flux density of a single ion of type h and

$$\bar{f}^h(x) = \int_{\Omega^{N-1}} f_j^h(\tilde{x}) p_{N-1}(\tilde{x}_j^h \,|\, x_j^h = x)\, d\tilde{x}_j^h \tag{8}$$

is the average force on an ion of type h.

Equation (7) is a Fokker-Planck type equation for the probability density of a single ion of species h. It is defined in the finite domain Ω, and contains an average force $\bar{f}^h(x)$ whose calculation, according to eq.(8), depends on the conditional probability density of all ions, given the phase space coordinates of the j-th ion of species h. However, this conditional density depends on the solution of the full Fokker-Planck equation (6) for all N particles. Therefore all of these quantities are coupled, and the Fokker-Planck equation for the stationary probability density of a single ion cannot be solved independently of the full Fokker-Planck equation of all other ions. Obviously, equation (7) is not very useful as long as its force term is not known.

Even if the force term is known a boundary condition on $\partial\Omega$ must be specified in order to determine the unique solution $p^h(x)$ of eq.(7) in the finite region Ω. Obviously, $p^h(x)$ satisfies no flux boundary conditions on $\partial\Omega_R$,

$$J^h(x)\cdot\nu\bigg|_{x\in\partial\Omega_R} = 0. \tag{9}$$

According to our assumptions, an external feedback mechanism maintains the concentrations $c^h(x)$ at fixed known values on the remaining boundaries, denoted $\partial\Omega_F$. Therefore, regardless of the exact method by which the feedback mechanism maintains these concentrations,

$$p^h(x)\bigg|_{x\in\partial\Omega_F} = \frac{c^h(x)}{N^h}. \tag{10}$$

3.1. The average force

In the case of charged ions in solution, the ion-ion interaction forces are pair-wise additive, and thus the force on the j-th ion of species h can be written as

$$f_j^h(\tilde{x}) = f_{ed}^h(x_j^h) + \sum_{(i,h')\neq(j,h)} f^{h,h'}(x_i^{h'}, x_j^h), \tag{11}$$

where $f^{h,h'}$ is the ion-ion interaction force that an ion of type h' exerts on an ion of type h. It includes both Coulombic interactions as well as short range interactions, such as excluded volume or Lennard-Jones potentials. Note that this force is not necessarily radial, due to boundary conditions on $\partial\Omega$ for the corresponding Poisson equation. The force f_{ed}^h contains both the effects of an applied external field as well as the dielectric self-force near dielectric boundaries [11], [12].

As shown in [11], with the specific form (11) for the force on a single ion, equation (8) for the average force $\bar{f}^h$ simplifies to

$$\begin{aligned}\bar{f}^h(x) &= f_{ed}^h(x) + \\ &(N^h-1)\int_\Omega f^{h,h}(y,x)p^{h|h}(y|x)\,dy + \sum_{h'} N^{h'}\int_\Omega f^{h,h'}(y,x)p^{h'|h}(y|x)\,dy.\end{aligned} \tag{12}$$

According to equations (12), (4), equation (7) for the single particle density $p^h(x)$ depends on the pair probability density functions $p^{h,h'}(x,y)$.

Thus, we now write an equation for the pair density. Employing similar methods to those of [11], we obtain that $p^{h,h'}(x,y)$ satisfies the equation

$$\nabla_x J_{x^h}^{h,h'}\left(x,y\right) + \nabla_y J_{y^{h'}}^{h,h'}\left(x,y\right) = 0, \tag{13}$$

where $J^{h,h'}_{x^h}(x,y)$ is given by

$$J^{h,h'}_{x^h}(x,y) = -\nabla_x \cdot \left(\frac{\bar{f}^{h,h'}(x,y)}{m^h\gamma^h(x)} p^{h,h'}(x,y) - \frac{k_B T}{m^h\gamma^h(x)} \nabla_x p^{h,h'}(x,y) \right), \tag{14}$$

and $J^{h,h'}_{y^{h'}}$ is given by an analogous expression.

The force $\bar{f}^{h,h'}(x,y)$ appearing in (14) is the average force on an ion of species h, given an ion of species h' located at y. In general, it is given by

$$\bar{f}^{h,h'}(x,y) = \int_{\Omega^{N-2}} f_{x_i^h}(\tilde{x}) p_{N-2}\left(x_{i,j}^{h,h'} | x_i^h = x, x_j^{h'} = y \right) d\tilde{x}_{i,j}^{h,h'}, \tag{15}$$

while for the case of pairwise additive forces, it can be simplified to

$$\begin{aligned} \bar{f}^{h,h'}(x,y) &= f^h_{ed}(x) + f^{h,h'}(x,y) + (N^h - 1) \int_\Omega f^{h,h}(x,z) p^{h|h,h'}(z|x,y)\, dz \\ &\quad + (N^{h'} - 1) \int_\Omega f^{h,h'}(x,z) p^{h'|h,h'}(z|x,y)\, dz \\ &\quad + \sum_{h'' \neq h,h'} N^{h''} \int_\Omega f^{h,h''}(x,z) p^{h''|h,h'}(z|x,y)\, dz \end{aligned} \tag{16}$$

for $h \neq h'$, and a similar expression for $h = h'$.

As seen from (16), the determination of the forces $\bar{f}^{h,h'}(x,y)$ in eq. (13) requires the determination of the triplet densities $p^{h,h',h''}(x,y,z)$. In the study of infinite equilibrium or non-equilibrium systems, closure relations relating the triplet densities to the pair densities are usually employed in order to close the system [6], [10].

However, since in our case we are concerned with a finite system, a closure relation between the triplet and the pair densities is not enough to close the system. Specifically, the Smoluchowski type equation (13) [13] is defined in the finite domain $(x,y) \in \Omega \times \Omega$. Therefore boundary conditions are necessary to determine the solution uniquely, e.g. at $(x,y) \in \partial\Omega \times \Omega$ and $(x,y) \in \Omega \times \partial\Omega$. Only after these boundary conditions are specified, the pair correlation function $p^{h,h'}(x,y)$ is completely determined, provided the forces $\bar{f}^{h,h'}(x,y)$ are known.

As in the case of the single ion densities, the boundary conditions for the pair densities should also be determined by the action of the feedback mechanism. For the case of the single ion densities we assumed that the action of the feedback mechanism is to keep an average fixed concentration on the domain boundaries. This leads to the natural concentration boundary conditions (10). However, the above description of the feedback mechanism is too general and does not determine boundary conditions for higher order densities. In the next section we propose a more detailed description of the feedback mechanism and derive the corresponding boundary conditions for the pair and higher order densities.

4. BOUNDARY CONDITIONS AND THE FEEDBACK MECHANISM

To understand the need for boundary conditions, we consider the full Fokker-Planck equation (6) for the joint pdf of all particles, $p_N(\tilde{x})$. Since all trajectories are confined to the finite domain Ω, this equation is defined in the $3N$-dimensional domain

$$\tilde{\Omega} = \bigotimes_{j=1}^{N} \Omega_j$$

in configuration space, where $\otimes$ denotes Cartesian product, and Ω_j is the three-dimensional physical domain of the j-th ion, identical to Ω.

Boundary conditions have to be specified on the boundary of $\tilde{\Omega}$ for the unique determination of $p_N(\tilde{x})$. These boundary conditions have to represent the feedback mechanism that maintains the fixed average concentrations at the boundary $\partial\Omega_F$. For the derivation of these boundary conditions, we first consider the exactly solvable case of a system of a single species of independent (non-interacting) particles diffusing in an external potential $V(x)$ between two fixed concentrations. Specifically, we assume now that the interaction forces are pairwise additive, so that the total potential of the system, denoted U_N, is given by

$$U_N(\tilde{x}) = \sum_{1 \le i \le N} V(x_i). \tag{17}$$

We also assume, for the sake of simplicity, that the friction is state-independent. Thus the Langevin equations (1) take the form

$$\ddot{x}_i + \gamma \dot{x}_i + \frac{1}{m} \nabla_{x_i} U_N(\tilde{x}) = \sqrt{2\frac{\gamma k_B T}{m}}\, \dot{w}_i, \qquad i = 1, 2, \ldots, N, \tag{18}$$

In an overdamped system the stationary Smoluchowski equation for the joint pdf of all particles takes the form

$$0 = \sum_i \nabla_i \cdot \left[\frac{1}{m\gamma} \nabla_i U_N(\tilde{x}) p_N(\tilde{x}) + \frac{k_B T}{m\gamma} \nabla_i p_N(\tilde{x}) \right] = -\sum_{j=1}^{N} \nabla_{x_j} \cdot J_j(\tilde{x}). \tag{19}$$

Since the particles are independent, the solution of the Smoluchowski equation (19) is given by

$$p_N(\tilde{x}) = \prod_{j=1}^{N} p_1(x_j), \tag{20}$$

where $p_1(x_j)$, the probability density of the j-th ion, is the solution of the three-dimensional Smoluchowski equation

$$0 = \nabla_{x_j} \cdot \frac{1}{m\gamma} \left[\nabla_{x_j} V(x_j) p_1(x_j) + k_B T \nabla_{x_j} p_1(x_j) \right] = -\nabla_{x_j} \cdot \mathscr{J}_j(x_j) \tag{21}$$

with vanishing flux conditions on reflecting boundaries $\partial\Omega_R$, and given concentrations on the other (feedback) boundaries. The boundary conditions determine the boundary fluxes, $\mathscr{J}_j(x_j)\cdot\nu(x_j)=0$ for $x_j\in\partial\Omega_R$ and $\mathscr{J}_j(x_j)\cdot\nu(x_j)\neq 0$ for $x_j\in\partial\Omega_F$, which are the same for all j. We denote the single ion probability flux density on the boundary $\partial\Omega$ by $\mathscr{J}^1(x)$ and assume henceforward that it is known. Inserting the solution (20) for $p_N(\tilde{x})$ into the Smoluchowski equation (19) with $U_N(\tilde{x})$ given in eq.(17) shows that the flux of the j-th particle is given by

$$J_j(\tilde{x})=\mathscr{J}^1(x_j)\prod_{i\neq j}p_1(x_i)=\mathscr{J}^1(x_j)p_{N-1}(\tilde{x}_j)=\mathscr{J}^1(x_j)p_{N-1}(\tilde{x}_j\,|\,x_j). \tag{22}$$

The last equality in eq.(22) expresses the independence of x_j of the remaining $\tilde{x}_j$. This conditioning anticipates the case of dependent ions (see [14]). This means that the boundary condition for the joint pdf $p_N(\tilde{x})$, when $x_j\in\partial\Omega_F$ and $\tilde{x}_j\in\tilde{\Omega}_j$ is eq.(22). Equation (22) with the known $\mathscr{J}^1(x)$ means that the feedback mechanism removes and inserts ions into the system at the boundary, depending on the configuration of ions inside the domain, with a flux that maintains the boundary concentrations. With this interpretation in mind, we adopt the relation (22) as the connection between the j-th component of the $3N$-dimensional flux and the 3-dimensional singlet flux of the j-th ion on the boundary for the case of interacting particles. Thus the boundary condition for the joint pdf $p_N(\tilde{x})$ is

$$J_j(\tilde{x})\cdot\nu(x_j)=\mathscr{J}^1(x_j)\cdot\nu(x_j)p_{N-1}(\tilde{x}_j\,|\,x_j)\quad\text{for } x_j\in\partial\Omega,\ \tilde{x}_j\in\tilde{\Omega}_j, \tag{23}$$

where $\nu(x_j)$ is the unit normal to the boundary at the point $x_j\in\partial\Omega$, and $\mathscr{J}^1(x)$ is the single ion boundary flux density, determined by the single ion density in eq.(21). The boundary condition (23) can be equivalently written as

$$\begin{aligned}-k_BT\frac{\partial\log p_N(\tilde{x})}{\partial\nu(x_i)}-\frac{\partial U(\tilde{x})}{\partial\nu(x_i)} &= m\gamma\mathscr{J}^1(x_i)\cdot\nu(x_i)\frac{p_{N-1}(\tilde{x}_i\,|\,x_i)}{p_N(\tilde{x})} \\ &= m\gamma\frac{\mathscr{J}^1(x_i)\cdot\nu(x_i)}{p(x_i)}\quad\text{for } x_i\in\partial\Omega,\ \tilde{x}_i\in\tilde{\Omega}_i.\end{aligned} \tag{24}$$

Integration of eq.(23) over the domain with respect to all particle coordinates, except those of the first k particles, gives the boundary conditions

$$J_i(x_1,x_2,\ldots,x_k)\cdot\nu(x_i)=\mathscr{J}^1(x_i)\cdot\nu(x_i)p(x_1,x_2,\ldots,x_k\,|\,x_i) \tag{25}$$

for $x_i\in\partial\Omega,\ x_1,\ldots,x_{i-1},x_{i+1},\ldots,x_k\in\Omega,\ (1\leq i\leq k)$. The boundary condition (25) holds for the flux of each species h, where the one-dimensional flux $\mathscr{J}^1(x_i)$ is that of species h. This condition is equivalent to the boundary condition

$$\begin{aligned}-k_BT\frac{\partial\log p_k(x_1,x_2,\ldots,x_k)}{\partial\nu(x_i)} &+ \bar{f}_i(x_1,x_2,\ldots,x_k)\cdot\nu(x_i) \\ &= m\gamma\frac{\mathscr{J}^1(x_i)\cdot\nu(x_i)}{p(x_i)},\end{aligned} \tag{26}$$

The force $\bar{f}_i(x_1,x_2,\ldots,x_k)$ is the average force acting on ion i at the boundary point x_i, given the coordinates of the first k ions, $(x_1,x_2,\ldots,x_k)$. In particular, the boundary conditions for the pair density are

$$J_i^{h,h'}(x_i,x_j)\cdot\nu(x_i) = \mathscr{J}^h(x_i)\cdot\nu(x_i)p^{h'|h}(x_j\,|\,x_i) \quad \text{for } x_i\in\partial\Omega,\ x_j\in\Omega, \tag{27}$$

or equivalently,

$$\begin{aligned}&-k_BT\frac{\partial}{\partial\nu(x_i)}\log p^{h,h'}(x_i,x_j)+\bar{f}_i^{h,h'}(x_i,x_j)\cdot\nu(x_i)\\ &=m^h\gamma^h\mathscr{J}^h(x_i)\cdot\nu(x_i)\frac{p^{h'|h}(x_j\,|\,x_i)}{p^{h,h'}(x_i,x_j)}=m^h\gamma^h\frac{\mathscr{J}^h(x_i)\cdot\nu(x_i)}{p^h(x_i)} \quad \text{for } x_i\in\partial\Omega,\ x_j\in\Omega.\end{aligned} \tag{28}$$

Note that in the general case the single ion flux $\mathscr{J}^h(x)$ is not known on the boundary, due to the unknown forces $\bar{f}_i(x_1,x_2,\ldots,x_k)$ for all combinations of species. Thus the boundary conditions eq.(28) *couple* the equation and boundary conditions for $p^{h,h'}(x_i,x_j)$ to those of $p^h(x)$, and the same is true for all $p^{h_1,\ldots,h_k}(x_1,x_2,\ldots,x_k)$ of all combinations of species. The probability flux density components $J_{x_i}^{h,h'}(x_i,x_j)\cdot\nu(x_i)$ for $x_i\in\partial\Omega,\ x_j\in\Omega$, of the six-dimensional flux of $p^{h,h'}(x_i,x_j)$ are the three-dimensional "normal components" of the boundary flux. With these boundary conditions, the Smoluchowski equations for all $p^h(x)$ and $p^{h,h'}(x,y)$ are coupled both through the single ion fluxes $\mathscr{J}^h(x)$ and through the forces. The concentration and flux boundary conditions determine the single ion density and the joint pdfs, provided the forces are known.

We note, however, that the computation of the forces by a suitable non-equilibrium closure relation that is valid for complex geometries and dielectric interfaces, is still an open problem that requires further analysis.

5. SUMMARY

We have formulated the problem of determining the non-equilibrium pair and higher correlation functions of interacting particles diffusing in a finite domain in terms of boundary value problems for partial differential equations. In this formulation the need for a partition function is avoided. The new contribution of this paper is the formulation of boundary conditions for the equations derived in [9]-[11]. This renders the problem of non-equilibrium diffusion of interacting particles solvable in a finite domain with prescribed boundary concentrations.

The problem is relevant to ionic permeation in protein channels of biological membranes, because the permeation process is inherently non-equilibrium, due to the current flowing through the system. The boundary conditions of given concentrations on the boundaries of the system is relevant here, because these are the conditions in the baths on both sides of the membrane.

The relevance of the pair correlation function to the permeation problem is in that the ion-ion interactions in the confined geometry of a protein channel are believed to be responsible for much of the channel phenomenology, e.g., single filing phenomena, selectivity, and more [1, 15].

Acknowledgment: This research was partially supported by research grants from the Israel Science Foundation, US-Israel Binational Science Foundation, and DARPA.

REFERENCES

1. B. Hille, *Ionic Channels of Excitable Membranes*, 3rd edition, Sinauer Associates, 2001.
2. R.S. Eisenberg, "Ionic channels in biological membranes: electrostatic analysis of a natural nanotube", *Contemp. Phys.*, 39, pp.447-466 (1998).
3. Kurnikova, M. G., Coalson, R. D., Graf, P., Nitzan, A., A lattice relaxation algorithm for three-dimensional Poisson-Nernst-Planck with application to ion transport through the gramicidin A channel, *Biophys. J.*, **76**, pp.642Ű 656 (1999).
4. Cárdenas, A. E., Coalson, R. D., Kurnikova, M. G., "Three-Dimensional Poisson-Nernst-Planck Theory Studies: Influence of Membrane Electrostatics on Gramicidin A Channel Conductance", *Biophys. J.*, **79**, p. 80Ű93 (2000).
5. Hollerbach, U., Chen, D.P., Busath, D., Eisenberg, R.S., "Predicting function from structure using the Poisson-Nernst-Planck equations: Sodium current in the gramicidin A channel", *Langmuir*, **16** (13), pp.5509-5514 (2000).
6. D. Henderson, *Fundamentals of Inhomogeneous Fluids*, Marcel Dekker, NY 1992.
7. R. S. Berry, S. Rice, J. Ross, *Physical Chemistry*, 2nd edition. Oxford University Press, NY 2000.
8. H. T. Davis, *Statistical Mechanics of Phases, Interfaces, and Thin Films*, (Advances in Interfacial Engineering Series), John Wiley & Sons, NY, 1995.
9. W. Ebeling, R. Feistel, G. Kelbg, R. Sandig, *J. Non-Equilibrium Thermodynamics*, **3** p.11 (1978)
10. J.M.G. Barthel,ăH. Baumgartel (Editor), H. Krienke, *Physical Chemistry of Electrolyte Solutions: Modern Aspects*, Steinkopf, Dietrich Pub. 1998.
11. Z. Schuss, B. Nadler, R.S. Eisenberg, "Derivation of PNP Equations in Bath and Channel from a Molecular Model", *Phys. Rev. E*, **64** (2-3) 036116, pp. 1-14, September (2001).
12. R. Allen, J-P Hansen and S. Melchionna, "Electrostatic potential inside ionic solutions confined by dielectrics: a variational approach", *Phys. Chem. Chem. Phys.*, **3**, pp.4177-4186, (2001).
13. Z. Schuss, *Theory and Application of Stochastic Differential Equations*, Wiley, N.Y. 1980.
14. Z. Schuss, B. Nadler, "Langevin trajectories between fixed concentrations", (submitted for publication).
15. W. Nonner, L. Catacuzzeño, R.S.Eisenberg, "Binding and selectivity in L-type calcium channels: A mean spherical approximation", *Biophys. J.* **79** (4) pp.1976-92 (2000).

SINGLE MOLECULES

Single-Molecule Approach to Dispersed Kinetics and Dynamic Disorder: Probing Protein Conformational Fluctuation

X. Sunney Xie

Department of Chemistry and Chemical Biology, Harvard University
12 Oxford Street, Cambridge, MA 02138

Abstract. I give a perspective of the single-molecule approach to protein conformational dynamics. The single-molecule approach is extremely useful in studying dispersed kinetics and dynamic disorder. In particular, fluorescence lifetime experiments of natural fluorophores quenched by electron transfer reactions yielded new dynamic information previously unavailable. Our single-molecule experiments have provided the picture and added quantitative understanding that an enzyme molecule is a dynamic entity with conformational fluctuation spanning a vast range of time scales. Further studies of the conformational and enzymatic dynamic with the single-molecule approach will shed light on how enzyme work at a level of great details[1].

INTRODUCTION

Perhaps the most profound contribution of the single-molecule approach to biophysical chemistry or even chemical biology is the ability to record movies of molecular motions and to follow the chemical activities of individual macromolecules or their complexes. When the first structural models of proteins became available, Jeremy Knowles, a renowned enzymologist at Harvard, observed that "making a model of a horse from photographs does not necessarily tell us how fast it can run." While structural information is essential, the dynamics are important to the understanding of functions. The famous 1878 fast photography of a running horse by Eadweard Muybridge, the father of motion pictures, pointed to the obvious importance of motion pictures in observing dynamics.[2] The advent of ultrafast spectroscopy has allowed monitoring of molecular motions with femtosecond time resolution.[3] However, these experiments have been conducted with large ensembles of molecules and often require coherent motions of the molecules. Such coherence is created by short laser pulses and is only maintained on the subpicosecond time scale, an event comparable to a pack of racehorses leaving the gate. Under this condition, ensemble-averaged dynamics is identical to single-molecule dynamics. At longer time scales, however, the coherence is lost. This corresponds to a situation where different horses have different actions at different times. Most enzymatic reactions occur at the time scales of millisecond to second, at which molecules do not evolve synchronously. Single-molecule experiments circumvent the necessity of synchronization of many

CP665, *Unsolved Problems of Noise and Fluctuations: UPoN 2002: Third International Conference,* edited by S. M. Bezrukov

molecules, and provide detailed dynamical information hidden in ensemble-averaged experiments.

Molecular dynamics (MD) simulations are virtual single-molecule experiments that provide vivid movies of molecular motions. MD has substantially enhanced our knowledge of macromolecular dynamics.[4] Unfortunately, the accessible time scale of MD has been limited to microseconds,[5] while the actual motions of a biomolecule span from femtoseconds to seconds or longer. It has been argued that the ultrafast conformational dynamics probed by MD simulations to date have little functional importance besides a few fast photochemical reactions such as the vision process or primary events of photosynthesis, and only the energetics probed by MD is of biochemical importance.[6] On a single-molecule basis, the event of a chemical reaction occurs on a subpicosecond time scale. However, the waiting time for the reaction to occur is usually much longer because of the low probability for acquiring enough energy to reach the transition state. The subpicoseond conformational dynamics only affects the short-lived transition state and determines the transmission coefficient in transition state theory. [7,8] It has been well established that dynamics beyond the MD simulation time scale, such as conformational gating of chemical reactions and conformational changes triggered by either ligand binding or chemical reactions is crucially important to macromolecule functions. However, for the reason described above, it has been difficult to study the slow conformational changes in ensemble-averaged experiments. Single-molecule experiments allow the study of these motions and their influence on chemical reactions at a greater level of detail.

With the advent of a variety of single-molecule techniques, movies or trajectories of single enzyme molecules at work have been recorded for many enzymatic systems.[9-17] More and more single-molecule assays are being developed. Some of these assays are easier or better than the ensemble-averaged assays, while others have no ensemble-averaged equivalents. It is fair to say that, due to the efforts of many groups, the single-molecule approach has changed the way not only biophysical but also biochemical problems are addressed. New insights derived from the single-molecule approach continue to emerge.

The hallmark of the single-molecule movies or trajectories is the ability to capture transient intermediates, which are difficult to detect in ensemble-averaged experiments because of their low concentrations. In some cases, rather long-lived intermediates, such as RNA folding intermediates,[14] can be observed through their spectroscopic identities; in other cases, the existence of the intermediates in enzymatic reactions, such as the Michaelis-Menten intermediates[12] or those involved in ATP hydrolysis,[11] cannot be detected directly because of the lack of observables, but can be inferred from the statistical analyses of single-molecule trajectories. A short review article on this subject has appeared.[18]

Another emerging contribution of single-molecule research is the study of conformational fluctuations at enzymatically relevant time scales and their influence on biochemical reactions. Observations of conformation fluctuation are made through optical measurements such as spectral diffusion of emission spectra,[12] fluorescence resonance energy transfer,[19] and photo-induced electron transfer.[20] This is a subject related to the long-standing issues of disperse kinetics and dynamic disorder.[21] In this article, we give a survey of the phenomenon of dynamic disorder associated with conformational fluctuation and the new information extractable from single-molecule experiments.

DISPERSED KINETICS AND DYNAMIC DISORDER

Dispersed kinetics exists for many condensed-phase systems, biological systems in particular. A well known example is the ligand-recombination kinetics of myoglobin following photodissociation. [22-25] Dispersed kinetics has also been observed in ion channels [26,27] using the patch clamp technique,[28] the first single-molecule technique. Ensemble-averaged experiments often report multiexponential population decays. However, given more than three exponentials, it is difficult to distinguish them in practice. The stretched exponential, known as Kohlrausch-Williams-Watts function, is a phenomenological fitting function in the form of $\exp[-(t/\tau_0)^\gamma]$. It was first used to describe mechanical creep in glassy fibers,[29] and more than a century later adapted to describe dielectric constants of glassy and polymeric materials.[30] The universality of the stretched exponential in nature has been a subject of extensive inquiry. [31-33] Practically, the stretched exponential function conveniently describes the dynamics spanning a broad range of time scales with only two adjustable parameters instead of numerous exponentials.

Let us consider a population decay, such as that of an excited fluorophore in protein, or that of an enzyme-substrate complex undergoing a catalytic reaction. Concrete examples of both will be given below. The general procedure is to fit the experimentally-determined population decay or survival probability, $P(t)$, with multiple exponential decays with different rate constants, k_i and amplitudes, a_i.

$$P(t) = \sum_i a_i \exp(-k_i t) \quad . \tag{1}$$

The decay rate constant can also have a continuous distribution:

$$P(t) = \int_0^\infty \tilde{P}(k)\exp(-kt)dk \ , \tag{2}$$

with the probability density of decay components, $\tilde{P}(k)$, being the inverse Laplace transform of $P(t)$,

$$\tilde{P}(k) = L^{-1}\{P(t)\} = \frac{1}{2\pi}\int_{b-i\infty}^{b+i\infty} P(t)e^{kt}dt \cdot \tag{3}$$

In principle, $\tilde{P}(k)$ is deducible from $P(t)$. For example, the $\tilde{P}(k)$ for the stretched exponential $P(t)$ is known.[34] In practice, however, $\tilde{P}(k)$ cannot be accurately computed from discrete experimental data of $P(t)$ because of the numerical instability of the inverse Laplace transform,[35] i.e, a small error in $P(t)$ results in a large error in $\tilde{P}(k)$. As an alternative, the maximum entropy method (MEM) has been used to determine the $\tilde{P}(k)$ from ensemble-averaged decay data.[36] Whereas MEM provides better estimates of exponential components and their amplitudes, it does not provide accurate and unique distributions of the rates.

What is implied in Eqs. 1 and 2 is a static distribution or a slow interconversion among the states with different rate constants. For two conformational states with decay rate constants, k_1 and k_2, the total population is a double exponential decay

$$P_0(t) = a_1\exp(-k_1t) + a_2\exp(-k_2t), \tag{4}$$

where a_1 and a_2 are the fraction of the two states. Now let us introduce the interconversion between the two states, with a rate constant v, and discuss its effect on Eq. 4:

$$\begin{array}{ll} & k_1 \\ \text{State 1} & \rightarrow \\ v\uparrow\downarrow v & \\ & k_2 \\ \text{State 2} & \rightarrow \end{array} \tag{5}$$

By solving the master equation for the kinetic scheme in Eq. 5 using matrix algebra[37], the total population decay is still a double exponential decay,

$$P(t) = \chi_+\exp(-\kappa_+t) + \chi_-\exp(-\kappa_-t) \tag{6}$$

with the two rate constants (eigenvalues) being

$$\kappa_\pm = \frac{1}{2}\left(k_1 + k_2 + 2v \pm \sqrt{k_1^2 - 2k_1k_2 + k_2^2 + 4v^2}\right), \tag{7}$$

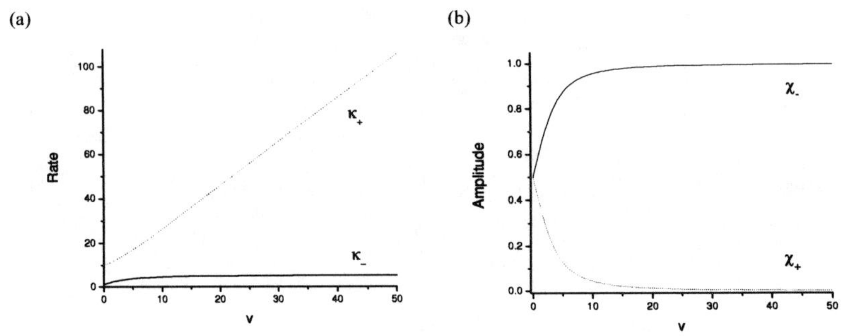

Figure 1. The effect of motional narrowing. For the kinetic scheme of Eq. 5, the total population decay is a biexponential decay, $P(t) = \chi_+\exp(-\kappa_+t) + \chi_-\exp(-\kappa_-t)$. κ_+ and κ_- (a) χ_+ and χ_- (b) are plotted as functions of v, assuming $k_1 = 10$, $k_2 = 1$, $a_1 = a_2 = 0.5$ in Eq. 4 and 5. When v = 0 (static disorder), κ_+ and κ_- are k_1 and k_2, respectively. With increasing v, the amplitude (λ_+) of the faster component ($\kappa_+ \rightarrow 2v$) approaches zero and P(t) becomes a single exponential decay with the rate constant (κ_-) approaching the mean of k_1 and k_2.

and the two amplitudes, χ_+ and χ_-, dependent on a_1 and a_2 at time zero.

Figure 1 shows how κ_+ , κ_-, χ_+ and χ_- change as a function of v, assuming k_1=1, k_2=10, a_1=a_2=0.5. When v >>k_1, k_2, $\kappa_- = (k_1 + k_2)/2$, and $\kappa_+ = 2\nu$. Interestingly, χ_+, the amplitude of the faster component, approaches zero with increasing v. Therefore, P(t) becomes a single exponential, with a decay constant being the mean of k_1 and k_2 at large v. This is similar to the motional narrowing phenomenon of spectral line shapes.[38,39] Such "motional narrowing" exists for fast interconversion not only between two states but also among multiple or continuous distribution of states.

Here we stress several points: First, P(t) can be meaningfully decomposed to sum or distribution of exponential decays of individual states (Eqs. 1, 2) only when interconversion among states is much slower than the individual decay rate constants. Second, it is practically impossible to extract the interconversion dynamics from P(t). For example, one cannot determine if a double exponential decay is due to two static states (Eq. 4) or two interchanging states (Eq. 6). Finally, even an apparent single exponential decay can arise from fast interconversion among states of vastly different decay rate constants.

In summary, there are two difficulties associated with knowing only the population decay P(t). First, it is practically impossible to determine the distribution of rate constants from an experimentally-determined P(t). Second, it is impossible to distinguish static heterogeneity from dynamic fluctuation of the decay rate constant. Single-molecule measurements provide not only the distribution of the rates but also the dynamics of interconverison within the distribution. Below, we give an example in the context of conformational dynamics. The methodology is generally applicable to other processes.

CONFORMATIONAL DYNAMICS

Fluorescence decay is a good probe for the local environment of a fluorophore, and is usually measured with the time-correlated single-photon counting method,[40] by which a train of high-repetition laser pulses excites the sample. For each detected photon, one measures the delay time with respect to its excitation pulse (Figure 2a). Then one can compile a histogram of the delay times (Figure 2b), which is the convolution of P(t) with an instrumental response function. This method works for both ensemble-averaged and single-molecule experiments. Figure 2b depicts such a histogram of a single dye molecule,[41] reported shortly after Eric Betzig's seminal demonstration of single-molecule imaging at room temperature with near-field microscopy.[42] The decay was single exponential, which indicates either no fluctuation of the fluorescence decay rate constant or much faster fluctuation than the fluorescence decay (motionally narrowed). Multiexponential decays of single molecules were subsequently reported [43-45] indicating conformational fluctuation at a time scale longer than the fluorescence decays.

Figure 3a shows the multiexponential decay of a single natural fluorophore, flavin adenine dinucleotide (FAD), in the binding pocket of an enzyme molecule, flavin: NADH oxidoreductase (Fre). The single-molecule decay is similar to the ensemble-

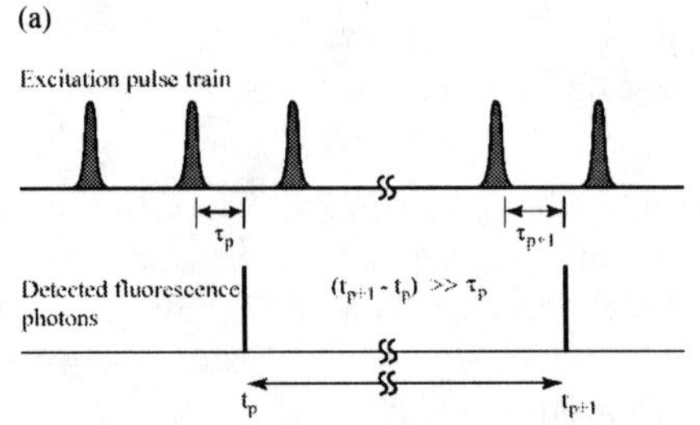

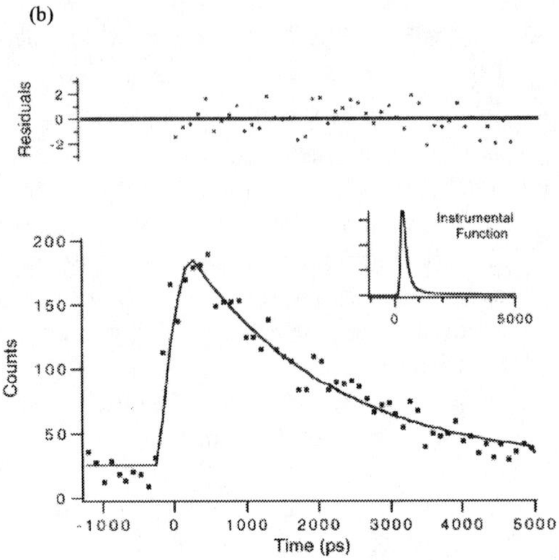

Figure 2. (a) Schematic of a train of laser excitation pulses and detected fluorescence photons. A histogram of photon delay time, τ_p, is compiled in a time-correlated single-photon counting experiment. (b) Fluorescence decay of a single sulforhodamine 101 molecule on a glass surface. The decay is best fit with a single exponential (2.0 ns) convoluted with the instrument response (inset). The weighted residuals are shown above. Adapted from reference 41.

averaged decay, which is best fit by MEM with four components, ~0.03ns(25%), ~0.20ns(52%), ~0.45ns(21%) and ~3.1ns(1.2%). The distribution of decay rate constants, for reasons described above, cannot be deduced from either the ensemble-averaged or the single-molecule histograms. The multiexponential decay of the single

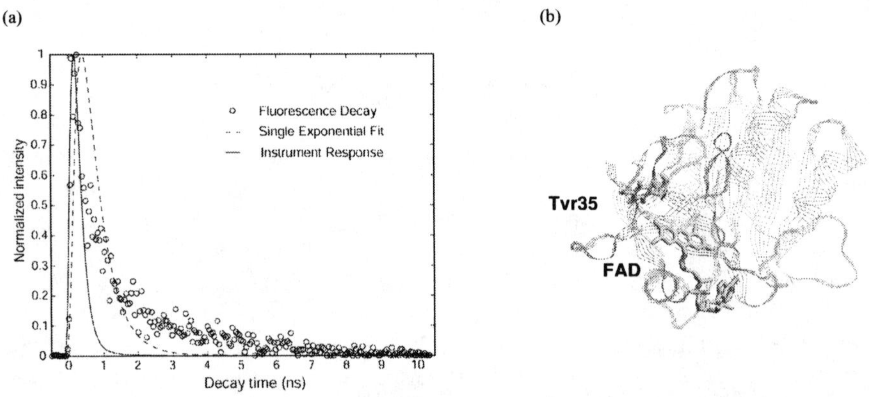

Figure 3. (a) Fluorescence decay (histogram of τ_p) for a single Fre/FAD complex. The decay is multiexponential, as indicated by a poor fit with a single exponential decay (dashed line). Also shown is the instrumental response function. See reference 20 for experimental details. (b) Structure of the Fre/FAD complex based on M. Ingelman, S. Ramaswamy, V. Niviére, M. Fontecave, and H. Eklund, *Biochemistry* **38**, 7040 (1999). Natural fluorescence from the flavin ring of FAD is quenched by the nearby Tyr35 residue through excited-state electron transfer. The ET rate is used to probe fluctuation of the edge-to-edge distance, x, between the donor (tyrosine) and acceptor (flavin ring).

molecule points to the fact that the fluorescence decay rate constant fluctuates during the course of measurements. However, the histogram cannot tell how fast the decay rate constant fluctuates with time, which can be extracted from the single-molecule trajectory. The easiest analysis is the autocorrelation function of k^{-1} fluctuation,

$$C(t) = <\Delta k(0)^{-1} \Delta k(t)^{-1}>, \tag{8}$$

where $\Delta k(t)^{-1} = k(t)^{-1} - <k^{-1}>$, and the bracket indicates time average. C(0) gives the variance. The decay of C(t) gives the time scale of the fluctuation, i.e. the memory time.

The fluorescence decay rate constant of a fluorophore is affected by the presence of a quencher either through fluorescence energy transfer (FRET)[46] or electron transfer (ET)[20] In this way, the lifetime serves as a distance-dependent probe for conformation fluctuation. The FRET rate constant is dependent on the center-to-center distance between the donor (fluorophore) and acceptor (quencher) to the inverse sixth power, and is sensitive to motions on the order of Förster radius, usually a few nanometers. FRET has found a broad range of applications in studies of conformational dynamics. While ET in biological systems has been extensively studied, [48] the use of ET in conformational analyses[20] is yet to be fully explored. ET rate constants have an exponential dependence on the edge-to-edge distance between the donor and acceptor, and are sensitive to motions on the angstrom scale. We thus have

$$k(t) = k_0 e^{-\beta x(t)}, \tag{9}$$

where $\beta \sim 1.0\text{-}1.4 Å^{-1}$ for proteins.[49-51] Although the k fluctuation might be induced by orientation and energetics fluctuations, it is most sensitive to distance fluctuation because of the exponential dependence.[20] For example, the fast fluorescence quenching in Figure 3 is dominated by ET from a nearby tyrosine residue (donor) to the flavin fluorophore (acceptor). The positions of the ET donor and acceptor in the 26 kD protein is shown in Figure 3b.[20]

Before we present the experimentally measured correlation functions, let's consider two simple models. First, a two-state model assumes two conformers with different lifetimes, k_1^{-1} and k_2^{-1}, and interconversion rate constants of v_{12} and v_{21} between them. It can be shown that the correlation function of fluorescence lifetime is:[52]

$$C(t) = \frac{(k_1^{-1} - k_2^{-1})^2 v_{12} v_{21}}{(v_{12} + v_{21})^2} \exp[-(v_{12} + v_{21})t] \cdot \tag{10}$$

This model describes the behavior of dynamic equilibrium between the two states. It corresponds to thermally-activated barrier crossing between the potential wells of two conformational states.

The second model is the Brownian diffusion model, which deals with a continuous distribution of rather than discrete decay rate constants. The fluctuation donor-acceptor distance, x, can be modeled as Brownian diffusion within a harmonic well using the Langevin equation[53]:

$$\mu \frac{d^2x}{dt^2} = -\mu\omega^2(x - x_{eq}) - \eta\frac{dx}{dt} + F(t), \tag{11}$$

where μ is the reduced mass, a harmonic well $V(x) = \frac{1}{2}\mu\omega^2(x - x_{eq})^2$ is located at the equilibrium position x_{eq}, η is the friction coefficient, and F(t) is the random fluctuating force. Under the overdamped condition, the acceleration (left side) is zero, so Eq.11 becomes[21]

$$\frac{dx}{dt} = -\lambda(x - x_{eq}) + f(t), \tag{12}$$

where λ is the drifting coefficient, $\lambda = \mu\omega^2/\eta$, and the random variable $f(t) = F(t)/\eta$ is assumed to be white noise with mean $\langle f(t)\rangle = 0$. The fluctuation dissipation theorem[53] relates the fluctuation of f(t) with λ such that

$$\langle f(t)f(t')\rangle = 2\lambda\theta\delta(t - t'), \tag{13}$$

where $\theta = \langle\Delta x^2\rangle = k_BT/\mu\omega^2$, and k_B is the Boltzman constant.
It follows that the autocorrelation of x is

$$\langle \Delta x(0)\Delta x(t)\rangle = \theta e^{-\lambda t}. \tag{14}$$

In the time-correlated single-photon counting experiment, we do not measure x(t) directly, but stochastic realization of k. The correlation function of k^{-1} is easier to obtain and less noisy than that of x(t).[47] We have shown that the autocorrelation function of k^{-1} associated with ET is

$$C(t) = \left\langle \Delta k(t)^{-1}\Delta k(0)^{-1}\right\rangle = k_0^{-2}\exp\left(2\beta x_{eq} + \beta^2\theta\right)\left[-1 + \exp\left(\beta^2\theta e^{-\lambda t}\right)\right]. \tag{15}$$

Note that Eqs. 14 and 15 are characterized by a single time scale λ^{-1}.

In a single-molecule experiment, we cannot determine the k(t) at a particular moment of time because of the stochastic arrival times of individual photons. The conventional method is to bin detected photons in long time intervals. The drawback is that it sacrifices the time resolution and limits the range of accessible time scales given a finite number of detectable photons before photobleaching of the fluorophore. In a recent paper by Yang et al.,[47] we present a new method analyzing lifetime trajectories on a photon-by-photon basis, which not only improves the time resolution of C(t) but also extends the detectable time scales to a broader range. In order to have high time resolution, we need short bins. The uncertainty in each bin is increased. However, the uncorrelated fluctuation in different time bins will only contribute to the variance, C(0), but not to C(t) beyond zero time. With this method, time resolution of the C(t) can be as high as one over the photon countrate.[47] In principle, submicrosecond time resolution is possible, allowing direct comparison between MD simulations and single-molecule experiments in the future.

Shown in Figure 4a is the C(t) of an FAD/Fre complex in linear logarithm scale, which was obtained with the photon-by-photon method.[20] This is an interesting result in regard to the dynamical properties of proteins. If there were no distance fluctuation, C(t) would be a spike only at time zero. Instead, the decay of C(t) spans four decades of time scales, reflecting conformational changes, fluctuation of x in particular, occurring from 1 millisecond to 10 second. The C(t) in Figure 4a can be fit well with a stretched exponential, $\exp[-(t/\tau_0)^\gamma]$, with γ=0.17, and τ_0=1.4ms.

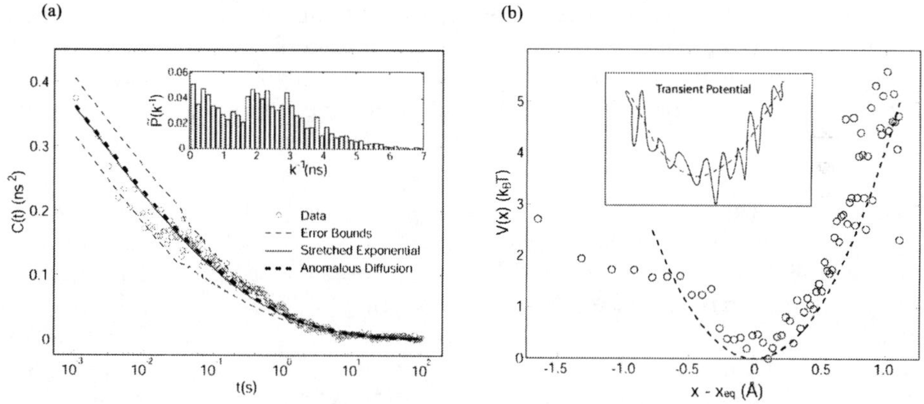

Figure 4. (a) Autocorrelation function of fluorescence lifetime fluctuation, C(t), of the same Fre/FAD complex as in Figure 3, plotted in logarithmic scale in time. The data is fit with a stretched exponential decay, $\exp[-(t/\tau_0)^\gamma]$, with $\gamma = 0.17$, and τ_0=1.4ms. The *C*(*t*) unravels protein conformational fluctuation spanning a broad range of time scales. The C(t) also fits well with the anomalous diffusion model (Eq. 16). Inset: Distribution of fluorescence lifetimes of the same single Fre/FAD complex. (b) Potential of mean force calculated from the lifetime trajectory of the same Fre/FAD complex. The dashed line is a fit to a harmonic potential of mean force, $V_0(x) = k_B T(x - x_{eq})^2/2\theta$. T is 300K and θ =0.13 Å^2. Inset: A sketch of a rugged transient potential that has a distribution of fluctuating barrier heights.

The time scale of fluctuation at room temperature is rather long, so much so that some people were surprised by its existence. In an ensemble-averaged study of conformational changes of myoglobin after photodissociation of a bound ligand, stretched exponential kinetics of conformational relaxation ranging from picosecond to microsecond time scale was observed at room tempertaure.[25] This was probed with a near infrared spectral transition of the heme, which is highly sensitive to the protein environment, However, as discussed earlier, the ensemble-averaged measurement could not distinguish dynamic disorder from static disorder. Our measured C(t) provides a direct measurement of the interconversion among the conformers at an even longer time scale (100μs –10s) for a protein smaller than myoglobin.

An important distinction between our single-molecule experiment and the previous ensemble-averaged photodissociation experiments [22-25] is that our experiment probes equilibrium fluctuation. In contrast, the previous experiments probed nonequilibrium relaxation. Although photoexcitation of the fluorophore in our experiment leads to electron transfer on the excited state, the back transfer completes within a few nanoseconds such that both donor and acceptor return to their original states. The C(t) is independent of excitation intensity, indicating the observed fluctuation originated from the spontaneous fluctuation of the ground electronic state, rather than photoinduced processes.[20] Although nuclear magnetic resonance spectroscopy is capable of probing spontaneous motions from microsecond to millisecond time scale,[54] it is also an ensemble-averaged experiment.

In addition to C(t), we can determine the distribution of k by constructing a histogram of k measured in fixed time intervals (course grained), as shown in the inset of Figure 4a. The potential of mean force, can be experimentally determined by $V(x) = -k_BT\ln p(x)$. p(x) is the probability density of x for the entire trajectory, mapped from $\tilde{p}(k)$, and is shown in Figure 4b. The bottom of the potential of mean force is near harmonic.

The stretched exponential C(t) cannot be explained by the two-state or Brownian diffusion model discussed above. It points to the existence of multiple conformational states at a wide range of time scales, which is consistent with the energy landscape picture popularized by Hans Frauenfelder.[55] According to this picture, conformational states correspond to the minima of multidimensional potential energy surfaces. This intuition resulted from extensive ensemble-averaged experiments on recombination kinetics of photodissociated ligands in heme proteins.[23] At low temperatures, the recombination kinetics is highly dispersed because of static heterogeneity of the conformal states. At elevated temperatures, the kinetics becomes less dispersed (motionally narrowed) because of interconversion among the conformational states. Current computational approaches can calculate the free energies of the conformational states, e.g. the local minima; those within a few k_BT will be populated significantly. However, it is a difficult task to compute the barrier heights connecting the local minima, some promising methodology not withstanding.[56-59] There are small barriers and large barriers. Assuming an Arrhenius process with a prefactor of $10^{12}s^{-1}$, if a barrier height is 0.5 kcal/mol, a conformational transition takes place on the picosecond time scale; if a barrier height is 16 kcal/mol, a conformational transition takes place on the second time scale.

The experimentally accessible one-dimensional coordinate, x, varies as the three dimensional structure of the protein/flavin complex changes. The projection of the multidimensional potential to the x coordinate gives fluctuating 1D potential. Its short time average is the "transient potential" depicted in the inset of Figure 4b, which is a rugged function of x. The barrier height for escaping a particular x has a broad distribution, which is associated with the broad range of survival times for the 3D conformational states with the particular x. The long-time average of the transient potential is the smooth potential of mean force. The single-molecule trajectories provide much richer information regarding the properties of the energy landscape and the dynamics within than previous ensemble experiments.

Our data indicated that the trapping time at a particular x can be rather long, resulting from the metastable 3D structures trapped in deep potential wells. The stretched exponential C(t) cannot be explained by the Brownian diffusion model (Eq. 15). This is because the model implies a distribution of trapping time (τ) at a particular x with a finite mean. For example, the finite mean can be the time constant of an exponential τ distribution associated with equal barrier heights in the 1D potential. In contrast, there might be a power law distribution of the trapping time, $1/\tau^{1+\alpha}$, $0<\alpha<1$, which has no finite first moment. This might arise from a distribution of the fluctuating barrier height in the rugged 1D potential. This is the sub-diffusion regime of anomalous diffusion, in which the mean square displacement is $<\Delta x^2> \propto t^\alpha$, in the absence of bound potential, V(x)=0. Anomalous diffusion has been extensively studied [60] and has been used to address protein motion.[61] Recently, Klafter and coworkers developed a theory to describe sub-diffusive motion within a potential based on a fractional Fokker-Planck equation (FFPE).[62] Taking $V_0(R) = k_B T(x - x_{eq})^2/2\theta$ as the potential (Figure 4b), we solved the FFPE for C(t) [15],

$$C(t) = k_0^{-2}\exp(2\beta\, x_{eq} + \beta^2\theta)\left[\exp(\beta^2\theta\, \exp(-\eta_\alpha t^\alpha/\Gamma(1+\alpha)))-1\right] \tag{16}$$

where η_α is the generalized drift coefficient and has a fractional unit of $s^{-\alpha}$, and $\Gamma(z) = \int_0^\infty y^{z-1}e^{-y}dy$ is the gamma function. As shown in Figure 4a, Eq. 16 fits the experimental data extremely well[20], with α = 0.17 and $\eta_\alpha = 2.9\, s^{-\alpha}$. Although other theoretical models, such as spin glass model[63-66] might also account for the observation, the advantage of the FFPE approach is that it provides a simple physical picture and has only two adjustable parameters.

The microscopic picture of the conformational states is an unsolved problem. The barrier heights among the conformational states are significant, but not unphysically high, and are comparable to breaking and reforming a few hydrogen bonds. The power law distribution of the trapping time arises from an exponential distribution of the barrier heights.[23,67] The existence of large barrier heights results in conformational fluctuation at the time scale of enzymatic reactions. Single-molecule studies of dynamic disorder have been the subject of an increasing amount of theoretical work,[68-76,47,52] Several theoretical investigations have been focused on the cholesterol oxidase experiment[12], with some dealing with discrete conformational states[68,73,75] and others using Brownian diffusion models.[68, 73,74] It is interesting to note that recent theoretical work on single-molecule enzymology emphasizes the effect of energetics on the nonequlibrium nature of enzymatic reactions.[77,78] We expect the dynamic disorder to be a general phenomenon in enzymology, which has been otherwise masked in ensemble-averaged measurements. The availability of conformations with different enzymatic rates opens up possibilities for control. The physiological relevance of the dynamic disorder phenomenon, if any, will be best investigated by single-molecule experiments in cellular environments, which are currently being pursued in many laboratories.

ACKNOWLEDGEMENTS

I thank current and former members of my research group, particularly Haw Yang, Guobin Luo, and Long Cai, for their contributions to the work summarized here. Our work in this area has benefited from collaborations and interactions with many colleagues, in particular, Luying Xun, Martin Karplus, Yossi Klafter, Samuel Kou, Jun Liu, David Reichman, and Ioan Andricioaei. The work is funded by National Institute of Health and Department of Energy.

REFERENCES

1. This article is a condensation of a review by X.S. Xie, to appear in J. Chem. Phys, in Dec. 2002.
2. Eadweard Muybridge, *Animal Locomotion*, 1887.
3. For an overview, see articles in *Femtochemistry and Femtobiology: Ultrafast Reaction Dynamics at Atomic-Scale Resolution, Nobel Symposium 101*, V. Sundstrom, ed. 1997.
4. For an overview, see the special issue on molecular dynamics simulation, edited by M. Karplus, Acc. Chem. Res. **35**, 321, (2002).
5. Y. Duan and P. A. Kollman, Science, **282**, 740 (1998).
6. A. Warshel, Acc. Chem. Res. **35**, 385 (2002).
7. D. Chandler, Chapter 8 in *Introduction to Modern Statistical Mechanics* (Oxford University Press, Oxford, 1987).
8. E. Neria and M. Karplus, Chem. Phys. Lett. **267**, 23 (1997).
9. R. Yasuda, H. Noji, K. Kinosita, and M.Yoshida, Cell, **93**, 1117 (1998); R. Yasuda, H. Noji, M. Yoshida, K. Kinoshita, and H. Itoh, Nature, **410**, 898, (2001).
10. T. Funatsu, Y. Harada, M. Tokunaga, K. Saito, and T. Yanagida, Nature, **374**, 555 (1995);
11. M. J. Schnitzer and S. M. Block, Nature, **388**, 386, (1997); M. Wang, M. S. Schnitzer, H. Yin, R. Landick, J. Gelles, and S. M. Block, Science, **282**, 902 (1998).
12. H. P. Lu, L. Xun, and X. S. Xie, *Science*, **282**, 1877 (1998); X. S. Xie, and H. P. Lu, J. Biol. Chem. **274**, 15967 (1999).
13. G. J. L Wuite, S. B. Smith, M. Young, D. Keller, and C. Bustamante, Nature, **404**, 103 (2000).
14. X. Zhuang, L. E. Bartley, H. P. Babcock, R. Russell, T. Ha, D. Herschlag, and S. Chu, Science, **288**, 2048 (2000).
15. T. Strick, J. F. Allemand, V. Croquette, and D. Bensimon, Physics Today, **54**, 46 (2001).
16. H. Yin, I. Artsimovitch, R. Landick, and J. Gelles, PNAS, **96**, 13124 (1999); K. M. Dohoney and J. Gelles Nature, **409**, 370, (2001).
17. H. Sosa, E. J. G. Peterman, W. E. Moerner, and L. S. B. Goldstein, *Nature Struc. Biology* **8**, 540 (2001).
18. X. S. Xie, Single Molecules, **2**, 229 (2001).
19. S. Weiss, Science, **283**, 1676 (1999).
20. H. Yang, G. Luo, P. Karnchanaphanurach, T.-M. Louie, L. Xun and X.S. Xie, 2002 preprint.
21. R. Zwanzig, Acc. Chem. Res. **23**, 148 (1990).
22. H. Frauenfelder, F. Parak, and R. D. Young, *Annu. Rev. Biophys. Chem.* **17**, 451 (1988).
23. R. Austin, K. W. Beeson, L. Eisentein, and H. Frauenfielder, Biochemistry, **14**, 5355 (1975).
24. J. W. Petrich, J.-C. Lambry, K. Kuczera, and M. Karplus, C. Poyart, J.-L. Martin, Biochemistry, **30**, 3975, (1991); K. Kuczera, J.-C. Lambry, J.-L Martin, and M. Karplus, Proc. Natl. Acad. Sci. USA, **90**, 5805 (1993).
25. M. Lim, T. A. Jackson, and P. A. Anfinrid, Proc. Natl. Acad. Sci. USA, **90**, 5801 (1993); T. A. Jackson, M. Lim, and P. A. Anfinrud, Chem. Phys. **180**, 131 (1994); S. Hagen, and W. A. Eaton, J. Chem. Phys. **104**, 3395, (1996).
26. K. Rubinson, Biophys. J. **61**, 463 (1992).
27. S. M. Bezrukov and M. Winterhalter, Phys. Rev. Lett. **85**, 202 (2000).
28. B. Sakamann and E. Neher, eds. *Single-Channel Recording*, 2nd ed., Plenum Press, New York, 1995.
29. F. Kohlrausch, Pogg. Ann. Phys. **119**, 352 (1863).
30. G. Williams and D. C. Watts, Trans. Faraday Soc. **66**, 80 (1970).
31. H. Scher, M. F. Shlesinger, and J. T. Blendler, Phys. Today, **44**, 26 (1991).
32. J. Ross and M. Vlad, Annu. Rev. Phys. Chem., **50**, 51 (1999).
33. J. Laherrere and D. Sornette, Eur. Phys. J. B **2**, 525 (1998).

34. For γ=1/2, $\widetilde{P}(k) = \frac{1}{2\sqrt{\pi\tau_0}} k^{-3/2} \exp\left(-\frac{1}{4k\tau_0}\right)$; for γ=1/2, $\widetilde{P}(k) = \int_0^\infty \frac{1}{\pi} \exp\left[-kx - \left(\frac{x}{\tau_0}\right)^{\gamma} \cos(\gamma\pi)\right] \sin\left[\left(\frac{x}{\tau_0}\right)\gamma \sin(\gamma\pi)\right] dx$.
see L. S. Liebovitch and T. I. Toth, Bull. Math. Biol. **53**, 443 (1991); and C. P. Lindsey and G. D. Patterson, J. Chem. Phys. **73**, 3348 (1980).
35. J. G. McWhirter and E. R. Pike, J. Phys. A. Math. Gen. **11**, 1729 (1978).
36. A. K. Livesey and J. C. Brochon, Biophys. J. **52**, 693 (1987); J. C. Brochon, Method Enzymol. **240**, 262 (1994).
37. N. G. van Kampen, *Stochastic Processes in Physics and Chemistry*. Elsevier Science Publishers, Amsterdam, (1992).
38. R. Kubo, in: *Fluctuation, Relaxation, and Resonance in Magnetic Systems*, Oliver and Boyd, London, 1961, p.23; R. Kubo, N. Toda, N. Hashitsume, *Statistical Physics II*, Springer-Verlag, Berlin, 1985.
39. P. W. Anderson, J. Phys. Soc. Jpn. **9**, 316 (1954).
40. D. V. O'Connor and D. Phillips, *Time-Correlated Single Photon Counting*, Academic Press, 1984.
41. X. S. Xie and R. C. Dunn, Science **265**, 361 (1994).
42, E. Betzig and R. Chichester, Science, **262**, 1422 (1993).
43. L. Edman, U. Mets, and R. Rigler, Proc. Natl. Acad. Sci. U.S.A. **93**, 6710 (1996).
44. Y.-W. Jia, A. Sytnik, L. Li, S. Vladimirov, S. Cooperman, and R. M. Hochstrasser, Proc. Natl. Acad. Sci. USA **94**, 7932 (1997).
45. C. Eggeling, J. R. Fries, L. Brand, R.Gunter, and C. A. M. Seidel, Proc. Natl. Acad. Sci. USA **95**, 1556 (1998).
46. L. Stryer, Ann. Rev. Biochem. **47**, 819 (1978).
47. H. Yang and X. S. Xie, J. Chem. Phys. to appear in Dec. 2002.
48. For reviews, see R. A. Marcus and N. Sutin, Biochem. Biophys. Acta **811**, 265 (1985); J. Jortner and M. Bixon, Eds., *Electron Transfer-From Isolated Molecules to Biomolecules*, vol. 106 John Wiley & Sons, New York, (1999).
49. C. C. Moser, J. M. Keske, K. Warncke, R. S. Farid, P. L. Dutton, Nature **355**, 796 (1992).
50. H. B. Gray and J. R. Winkler, Ann. Rev. Biochem. **65**, 537 (1996).
51. D. N. Beratan and S. S. Skourtis, Current Opinion Chem. Biol., **2**, 235 (1998).
52. H. Yang and X. S. Xie, Chem. Phys. **284**, 423 (2002).
53. R. Zwanzig, *Nonequilibrium Statistical Mechanics*, Oxford University Press (2001)
54. A. G. Palmer, C. D. Kroenke, and J. P. Loria, *Meth. Enzymol.* **339**, 204 (2001); A. G. Palmer, *Annu. Rev. Biophys. Biomolec. Struct.* **30**, 129 (2001).
55. H. Frauenfelder, S. G. Sligar, and P. G. Wolynes, Science **254**, 1598 (1991).
56. P. G. Bolhius, D. Chandler, C. Dellago, and P. L. Geissler, Ann. Rev. Phys. Chem. **53**, 291, (2002).
57. R. Elber, A. Ghosh, and A. Cardenas, Acc. Chem. Res. **35**, 396 (2002).
58. S. V. Krivov, S. F. Chekmarrev, and M. Karplus, Phys. Rev. Lett. **88**, 038101 (2002).
59. J. Shimada, E. Kussell, and E. I. Shakhnovich, J.Mol.Biol. **308**, 79 (2001).
60. For a review, see J. Klafter, M. F. Shlesinger, and G. Zumofen, Physics Today, **49**, 33 (1996).
61. A. E. Garcia, R. Blumenfeld, G. Hummer, and J. A. Krumhansl, Physca D. **107**, 225 (1997).
62. R. Metzler, E. Barkai, and J. Klafter, Phys. Rev. Lett. **82**, 3563 (1999).
63. J. D. Bryngelson and P. G. Wolynes. Proc. Natl. Acad. Sci. USA, **84**, 7524 (1987);
64. E. I. Shakhnovich and A. M. Gutin, Europhys. Lett. **9**, 569 (1989).
65. C. L. Lee, C. T. Lin, G. Stell, and J. Wang, preprint 2002.
66. D. Reichman, preprint 2002.
67. A. Blumen, J. Klafter, and G. Zumofen, *Fractals in Physics*, eds. L. Pietronero, E. Tosatti, p399, (1986).
68. G. K. Schenter, H. P. Lu, and X. S. Xie, J. Phys. Chem. A **103**, 10477 (1999).
69. J. Wang and P. Wolynes, Phys. Rev. Lett. **74**, 4317 (1995).
70. J. N. Onuchic, J. Wang, and P. G. Wolynes, Chem. Phys. **247**, 175 (1999).
71. E. Geva , J. L. Skinner, Chem. Phys. Lett. **288**, 225 (1998).
72. A. M. Berezhkovski, A. Szabo, and G. H. Weiss, J. Chem. Phys. 110, 9145 (1999); A. M. Berezhkovskii, M. Boguna, and G. H. Weiss, Chem. Phys. Lett. **336**, 321 (2001).
73. J. Cao, Chem. Phys. Lett., **327**, 38 (2000). S. L. Yang and J. S. Cao, J. Phys. Chem. B **105**, 6536 (2001); J. Cao, and S. Yang, J. Chem. Phys. this issue.
74. N. Agmon, J. Phys. Chem. B, **104,** 7830 (2000).
75. G. H. Weiss and J. Masoliver, Physica A (Amsterdam, Neth.) **296**, 75 (2001).
76. V. Chernyak, M. Schulz, and S. Mukamel, J. Chem. Phys. **111**, 7416 (1999); V. Barsegov, V. Chernyak, and S. Mukamel, J. Chem. Phys. **116**, 4240 (2002).
77. D. Astumian, Appl. Phys. A **75**, 1 (2002).
78. H. Qian and E. Elson, Biophys. Chem., in press.

Unzipping DNA: From Pulling to Pores and Back Again

David K. Lubensky

Bell Labs, Lucent Technologies, 700 Mountain Ave., Murray Hill NJ 07974

Abstract. We recapitulate recent theoretical work on the mechanical unzipping of double-stranded DNA (dsDNA), emphasizing the effects of heterogeneity in the base sequence. When the dsDNA is pulled apart with a *constant force*, there is a thermodynamic unzipping transition at a critical force of order 10–15 pN. The behavior in the vicinity of this transition is markedly different for heteropolymers and for homopolymers. Whereas homopolymer unzipping is driven by the *entropy* gained from fluctuations in the number of denatured base pairs and is described by an almost flat energy landscape, the energy of a heteropolymer fluctuates strongly as successive base pairs are opened. When the applied force approaches the critical value, the dsDNA unravels in a series of discrete, sequence-dependent steps that allow it to reach successively deeper *energy* minima. Plots of extension versus force thus take the striking form of a series of plateaus separated by sharp jumps. Similarly dramatic effects are expected in the dynamics of unzipping; anomalous behavior, dominated by the presence of large energy barriers, should persist for several pN above the thermodynamic transition. This prediction has been partially confirmed by recent experiments. Similar qualitative features, driven by the presence of a rugged energy landscape, should appear in micromanipulation experiments on folded RNA molecules and proteins.

INTRODUCTION

The past decade has seen a remarkable shift in the way experimental biophysics research is carried out. Traditionally, the properties of biomolecules could be probed solely in the aggregate, through bulk measurements on macroscopic samples. It was thus only possible to learn about ensemble averages; large fluctuations on the level of individual molecules were smoothed out, and sparsely populated excited states could rarely be detected. Recently, however, a growing arsenal of techniques has been developed that allows experiments to be carried out on single molecules [1]. Particularly remarkable are devices such as optical tweezers [2] and the atomic force microscope (AFM) [3] that make it possible to push and pull on individual molecules, and thus to learn about their structure and mechanical properties. A substantial number of such micromechanical studies have been reported in the past several years, on systems ranging from proteins [4] to ribozyme domains [5] to duplex DNA [6, 7].

Inspired by this wealth of experimental work, we report here on theoretical studies of a relatively simple single molecule experiment, the *unzipping* of double-stranded DNA (dsDNA). Our goal is both to understand recent experiments on dsDNA (and its close relative duplex RNA) and, equally importantly, to emphasize some general features of mechanical denaturation experiments that should also appear in more complicated systems such as proteins and folded RNA molecules. In particular, we show that unzipping

CP665, *Unsolved Problems of Noise and Fluctuations: UPoN 2002: Third International Conference*, edited by S. M. Bezrukov

heteropolymers is very different from unzipping homopolymers. The presence in the heteropolymer of a variable sequence, and thus of variable pairing energies, generates a complex energy landscape with deep wells and high barriers that dominate the unzipping process. Since most biomolecules are heteropolymers, this is expected to be a fairly common situation in micromechanical experiments.

In the remainder of this contribution, we first introduce the constant force ensemble for pulling on dsDNA and give a more precise definition of the energy landscape. We then show how the changes in this landscape as a function of the applied force give rise to a sharp denaturation transition. In the vicinity of this transition, for a *random* DNA sequence, one can give a comprehensive analytic description of the equilibrium and the dynamical behavior of the dsDNA as it is pulled apart. They are determined, respectively, by the locations of the deepest minima and highest maxima of the energy landscape. These ideas are next applied to analyze recent experiments on the unzipping of dsDNA being forced through a narrow pore. We conclude by mentioning some open problems to which the reasoning described here might be applied, notably the inference of RNA secondary structure from micromechanical experiments.

This paper summarizes work with several collaborators over the past few years, some of which has previously been published in another form. A comprehensive account of the work in the next two sections, done with David Nelson, can be found in [8, 9]. The experiments on unzipping dsDNA with pores were done by Alexis Sauer-Budge, working in Dan Branton's lab; both the experiments and their analysis are described in more detail in [10]. Related theoretical work appears in [11, 12].

THE ENERGY LANDSCAPE AND EQUILIBRIUM

Figure 1 shows a sketch of the situation studied in this paper. Just as the thermodynamic properties of a gas can be studied in conjugate ensembles in which either the pressure or the volume are held constant, so one can think of pulling apart dsDNA either by applying a steady force $\mathbf{F}$ or by fixing the distance between the ends of the two single strands (the *extension* $\mathbf{r}$). In single molecule unzipping experiments, there is no large parameter to define a thermodynamic limit; as a result, the relationship between force and extension in the two ensembles is not the same. Here, we study the constant force ensemble. Consider a dsDNA molecule, as in Fig. 1, whose two constituent single strands are being pulled apart by a force $\mathbf{F}$, so that m base pairs are opened. Define $\mathscr{E}_m$ to be the free energy of the system (in the presence of the force) when exactly the first m nucleotides are unzipped.[1] There are two contributions to $\mathscr{E}_m$: the free energy cost $-\sum_{i=1}^{m}\varepsilon_i$ of breaking the m pairs, where $\varepsilon_i < 0$ is the pairing energy of the ith base pair,[2] and the free energy $2m\varepsilon_{\text{U}}(F)$ that the $2m$ unzipped single-stranded nucleotides gain from aligning with the applied force. The free energy $\varepsilon_{\text{U}}(F) < 0$ of a single nucleotide is expected to monotonically decrease as the magnitude F of the pulling force increases; its precise form depends on

---

[1] Throughout, we assume that all bases other than those opened by the force remain paired—an excellent approximation in physiological conditions.

[2] For simplicity, we neglect nearest-neighbor terms.

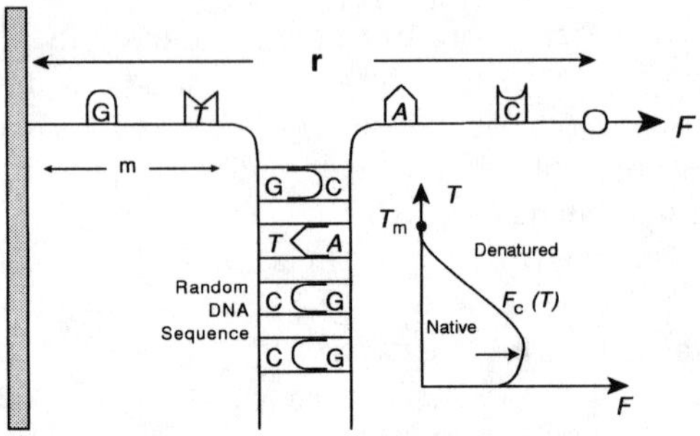

FIGURE 1. Sketch of the DNA unzipping experiment. One of the single strands of a dsDNA molecule is attached by its end to a solid surface, and the other is pulled away from the surface with a constant force **F**. As a result, the double strand partially denatures, separating m base pairs ($m = 2$ in the figure). The distance between the ends of the two single strands, or *extension*, is **r**. *Inset*: Schematic phase diagram in the temperature-pulling force (T-F) plane of a dsDNA molecule in three dimensions. At low enough T and F, the polymer is in the native, double-stranded phase. At the phase transition line $F_c(T)$, the DNA denatures and the two strands separate. Thermally-induced *melting* occurs at zero force at a temperature T_m. As indicated by the arrow, this paper considers instead the *unzipping* transition, in which the phase transition line is crossed at non-zero F.

the detailed elastic properties of the single strand. Combining these two terms, we can write

$$\mathscr{E}_m = 2m\varepsilon_U(F) - \sum_{i=1}^{m} \varepsilon_i \equiv mf + \sum_{i=1}^{m} \eta_i \,. \tag{1}$$

With the second equality, we have rewritten $\mathscr{E}_m$ in terms of the average free energy difference f between zipped and unzipped base pairs and the sequence-dependent fluctuations η_i about this average: With $\overline{\varepsilon} \equiv (1/N)\sum_{i=1}^{N} \varepsilon_i$ for a molecule of length N nucleotides, $f \equiv 2\varepsilon_U(F) - \overline{\varepsilon}$ and $\eta_i \equiv \overline{\varepsilon} - \varepsilon_i$.

One can think of f as the average slope of the one-dimensional energy landscape $\mathscr{E}_m$. When $f > 0$, as will be the case for small enough forces, $\mathscr{E}_m$ tends to grow with m, and the number of unzipped nucleotides m remains finite. When $f < 0$, in contrast, the landscape slopes downwards on average, and it is energetically favorable to completely denature the dsDNA. In the limit $N \to \infty$, there is a sharp first order phase transition between the paired and unpaired states at $f = 0$, which corresponds to a critical force F_c satisfying $\varepsilon_U(F_c) = \overline{\varepsilon}$ [8, 9]. As indicated in the inset to Fig. 1, F_c tends to zero as the temperature is increased, and the unzipping phase transition line ends at the thermal melting transition at $F = 0$ [13]. In physiological conditions, $F_c \sim 15$pN.

In many ways more relevant to micromechanical experiments than the existence of the phase transition is the behavior of the equilibrium average number of unzipped bases $\langle m \rangle$ above the transition. One expects that $\langle m \rangle$ will diverge as $f \to 0$; for a homopolymer ($\eta_i = 0$ for all i), one can easily verify that $\langle m \rangle$ blows up as $1/f$ by taking a derivative of the partition function $Z = \sum_{i=0}^{N} \exp(-mf/k_B T)$ with N large. To gain some insight into how $\langle m \rangle$ depends on F in the case of a heteropolymer, it is useful to consider the idealization of a random DNA sequence with at most short-ranged correlations between

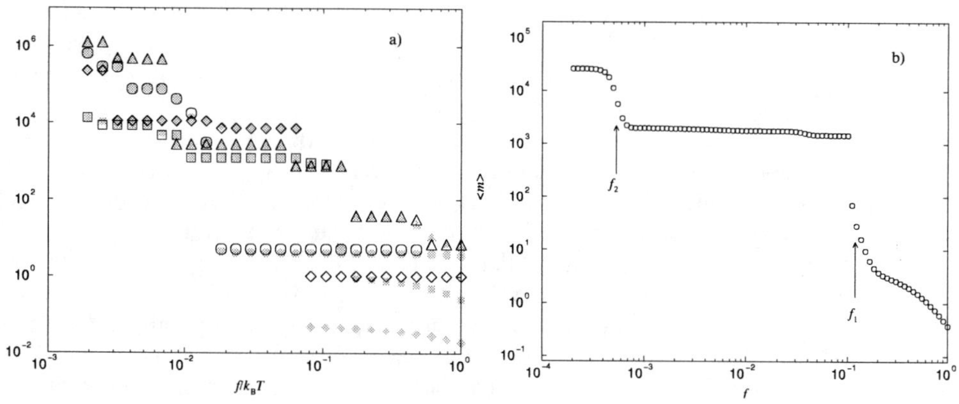

FIGURE 2. Log-log plots of the average number of bases opened $\langle m \rangle$ (closed symbols) and the location of the absolute minimum m_{min} of $\mathscr{E}_m$ (open symbols) as a function of the distance f from the unzipping transition. a) Both variables are plotted for each of four individual polymers, represented by four different symbol shapes, with independently chosen random sequences (variance $\Delta = 9(k_B T)^2$) of length $N = 5 \times 10^6$ bases. Note that, except when $m = \mathscr{O}(1)$, $\langle m \rangle$ and m_{min} coincide very well. b) $\langle m \rangle$ alone for the bacteriophage lambda genome. The jumps between the successive plateaus occur at biases f_1 and f_2 as indicated in the figure.

bases.[3] The random contribution $\sum_{i=1}^{m} \eta_i$ to $\mathscr{E}_m$ will then typically grow like $\sqrt{\Delta m}$, where Δ is the variance of a single pairing energy η_i. One can gain an idea of how large $\langle m \rangle$ will be for a typical random sequence by balancing this random contribution to $\mathscr{E}_m$ with the average contribution mf. They are of the same order at $m* \sim \Delta / f^2$. The crossover scale $m*$ diverges more strongly than the homopolymer average $\langle m \rangle \sim k_B T / f$, suggesting that sequence disorder is *relevant* near the unzipping transition. Indeed, an exact calculation confirms that the average over all random sequences $\overline{\langle m \rangle} \sim \Delta / f^2$ has the same scaling as $m*$ [8, 9].

Much more striking than $\overline{\langle m \rangle}$, however, is the dependence of $\langle m \rangle$ on F for a given dsDNA molecule with a given random sequence. Figure 2a) shows some examples of force-extension curves for individual random heteropolymers. They take the form of long plateaus followed by sharp jumps. Moreover, except when $\langle m \rangle$ is of order 1, $\langle m \rangle$ corresponds remarkably well with the location m_{min} of the absolute minimum of $\mathscr{E}_m$. This suggests that, near the unzipping transition, the system behaves as if it were at zero temperature: It seeks to minimize the energy, with any fluctuations about the minimum playing a secondary role. Indeed, one can show that, for a random sequence, $\langle m \rangle / m_{\text{min}} \to 1$ with probability 1 as $f \to 0$ [9]. This is in marked contrast to the situation with a homopolymer, where $\langle m \rangle$ diverges as $f \to 0$ even as the minimum of $\mathscr{E}_m = mf$ remains at $m = 0$ as long as $f > 0$. The physical reason for this behavior (which is known to exist in other random systems [15]) is that $\mathscr{E}_m$ at m is not determined by the pairing energies in the immediate vicinity of m, but by the *sum* of contributions from all

[3] Although real DNA sequences are not strictly random, they are often closer than one might expect [14].

of the broken base pairs. This typically leads to energy scales that grow with m. At small enough f, and thus large enough m, the characteristic scale of $\mathscr{E}_m$ is much larger than $k_B T$, so that the chance that there will be a second local minimum with an energy within a few $k_B T$ of $\mathscr{E}_{m_{\min}}$ becomes vanishingly small.

Knowing that $\langle m \rangle$ coincides with $m_{\min}$ near the transition, we are in a position to explain the phenomenon of plateaus and jumps seen in Fig. 2. As f decreases, the energy landscape tilts toward the horizontal. As a result, the relative stability of different local minima of $\mathscr{E}_m$ changes, with the energy of minima at larger m decreasing faster than that of minima at smaller m. Over a finite range of f, the location $m_{\min}$ of the absolute minimum will be in one local potential well, giving rise to a plateau. Then, that well will be overtaken by another local minimum at higher m, leading to a sharp jump in $\langle m \rangle$. This process continues until the duplex unzips completely. For our idealized random DNA sequence, one can give a complete statistical description of this process. For example, the distribution over different random sequences of $\langle m \rangle$ at fixed f is a scaling function of $\Delta \langle m \rangle / f^2$, and the size m_{jump} of a jump that occurs at a bias f is similarly a function only of $\Delta m_{\text{jump}}/f^2$; the endpoint f_2 of a plateau that begins at a bias f_1 is uniformly distributed between f_1 and 0 [9]. The qualitative physics behind the presence of plateaus and jumps remains unchanged for more realistic biological sequences. For example, Fig. 2b) shows a computationally generated force-extension curve for the unzipping of a portion of the lambda phage genome that exhibits 3 plateaus separated by jumps.

UNZIPPING DYNAMICS

Just as the deepest wells in the energy landscape $\mathscr{E}_m$ govern the equilibrium properties of unzipping, so the highest energy barriers dominate the dynamics of strand separation. As we noted in the previous section, because $\mathscr{E}_m$ is a sum of a large number of terms, its typical energy scale tends to grow with distance (as $\sqrt{m}$ for the idealized case of a random sequence). Together with the fact that the probability of opening exactly m base pairs in equilibrium is proportional to $\exp(-\mathscr{E}_m k_B T)$, this implies that at long enough length scales, the equilibrium chance of being in the vicinity of $m_{\min}$ may be much higher than the probability of being near some other local minimum. Essentially the same arguments tell us that the time to unzip a duplex will be set by one or a few very large barriers; the only difference is that instead of looking at an equilibrium probability proportional $\exp(-\mathscr{E}_m/k_B T)$, one instead must consider a barrier crossing time proportional to $\exp(\mathscr{E}_m/k_B T)$. Although there will be many small barriers to unzipping, the time required to make it over the rare large barriers is so much longer that it essentially determines the total time required to open the duplex. In fact, one can think of a coarse-grained energy landscape obtained by gradually removing smaller barriers and shallower wells until the full landscape $\mathscr{E}_m$ is replaced by a caricature in which only the locations and energies of the most pronounced extrema are retained, and $\mathscr{E}_m$ is assumed to vary smoothly between them [16, 17]. On long time scales, the dsDNA then unzips by hopping among the few remaining minima, over the large barriers imposed by the few remaining maxima. Precisely such behavior has recently been observed in the mechanical denaturation of lambda phage DNA [17]: The molecule opens in fits and starts, with long pauses at particular sequence-dependent points separated by events in which the polymer rapidly unzips until it reaches a new minimum.

Somewhat more quantitative estimates of the dynamical effect of large energy barriers can be obtained with the assumption of a random DNA sequence. In the simplest case, when the opening and closing timescales of individual base pairs are slower than the equilibration times of other degrees of freedom (e.g. the liberated single strands), the dynamics of opening is essentially a biased random walk in the number of opened base pairs m. The problem then maps directly onto a well-studied model of diffusion in a one-dimensional random medium [18]. Exactly at the unzipping transition, when $\mathcal{E}_m$ is an unbiased random walk, one expects that energy barriers grow as $\sqrt{\Delta m}$. The time required to go a distance m is thus $t(m) \sim \exp(\sqrt{\Delta m}/k_B T)$, whence we conclude that the number of bases $m(t)$ opened at time t typically scales as $m(t) \sim \ln(t)^2$. Just above the transition, one can show that the opening is sublinear in time: $m(t) \sim t^{\mu}$ with μ proportional to $f/\sqrt{\Delta}$. This behavior persists for several pN above F_c. Similar behavior, albeit with different exponents, is predicted when the basic microscopic timescale is m-dependent, as for example when it is set by rotational drag on the dsDNA or by the relaxation of the single-stranded DNA (ssDNA) [9].

UNZIPPING WITH PORES

The central idea of the previous section—that, because each base pair contributes additively to the total energy landscape, the dynamics of dsDNA unzipping is usually dominated by the time to cross a few barriers—can be applied to a variety of experiments, including some that do not involve pulling apart the dsDNA in a strict sense. An example of such a situation is denaturation induced by forcing one of two single strands through a narrow pore (Fig. 3). In recent experiments [10], dsDNA molecules with single-stranded overhangs were driven by an electric field through a nanometer-scale channel large enough to accommodate ssDNA but not dsDNA. At high enough field strengths, the duplex was torn apart so that the single strands could pass through the pore. A computational analysis of the energy landscape for unzipping indicated, in agreement with experiment, that the opening dynamics for one of the constructs used would be determined by two barriers and for the second by a single barrier. From the known locations of the maxima and minima of the energy landscape, we could obtain an estimate of the effective charge q per base (including, for example, the effects of counterions) on the ssDNA in the pore. One expects that if the voltage across the membrane is V, the energy gained from translocating one nucleotide across the membrane will be qV. If the distance between one minimum and the succeeding maximum is $m_{\text{barrier}} \equiv m_{\max} - m_{\min}$, then the rate constant k for barrier crossing should depend on the voltage as $k = k_0 \exp(-m_{\text{barrier}} qV)$. By fitting measured rate constants to this expression, we were able to extract the estimate $q \sim 0.1e$, where e is the elementary electron charge.

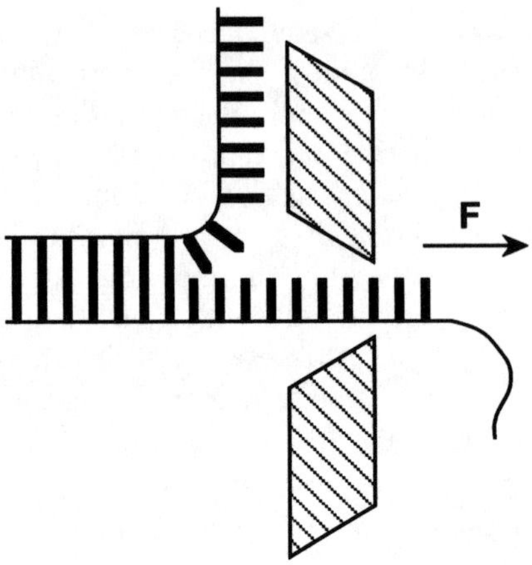

FIGURE 3. Schematic of dsDNA unzipping through a narrow pore [10]. The pore is assumed to be large enough that single-stranded DNA, but not double-stranded DNA, can fit through it. Under the influence of an electric field or comparable force **F**, one single strand inserts into the channel and is gradually pulled through. As the strand is drawn across the pore, it must unzip from its complement.

DISCUSSION AND UNSOLVED PROBLEMS

In this contribution, we have given an overview of recent work on the force-induced denaturation, or unzipping, of double-stranded DNA. The central idea to emerge from these studies is that the energy landscape describing the unzipping process in the constant force ensemble is typically rugged. The energy to open m base pairs is a sum of contributions from each pair. Since almost all DNA molecules are heteropolymers, the contributions from successive nucleotides will be distributed about an average value. Runs of stronger and weaker than average base pairs then give rise, respectively, to local maxima and minima in the energy landscape; in most cases, the scales of these runs grow with m. For the idealized case of a completely random sequence, one can show that the deepest of these minima determine the extent to which the polymer unzips in equilibrium, giving rise to a very non-uniform force-extension curve consisting of long plateaus followed by sharp jumps. Similarly, the highest maxima dominate the dynamics of unzipping, with the result that it is far slower than diffusion in a flat landscape. Qualitatively similar features are predicted in biological sequences that are not truly random and have been observed in simulations [9] and in experiments [17].

Armed with these basic results, we can now consider extensions and practical applications of the approach presented here. For example, in the constant extension ensemble, it has been shown that the basic features of the energy landscape remain unchanged, but that the unzipped single strands (whose end to end distance is fixed) act like a weak spring that confines the unzipping fork to a given section of the sequence [12]. If the fork is in addition stuck in a local minimum of $\mathcal{E}_m$ while the spring is extended by moving apart the ends of the two single strands, then the measured force will grow linearly with distance. Such "sawtooth" runs are observed experimentally [7], suggesting that the idea that minima of the energy landscape dominate equilibrium properties can usefully be applied to the constant extension ensemble. An immediate consequence of this observation is that it should be difficult to detect mutations in a DNA sequence from its force-extension signal in unzipping [19], as has recently been proposed [7]. A more serious challenge is to use results on the unzipping of single duplexes to understand micromechanical experiments on more complicated RNA secondary structures containing multiple stems [5]. A means to infer the secondary structure from force-extension measurements would be particularly interesting. Although many of the basic elements needed to accomplish this seem to be in place, it remains an open challenge to synthesize a workable method from them.

ACKNOWLEDGMENTS

I would like to thank Alexis Sauer-Budge and Dan Branton for teaching me about the behavior of polymers in nanopores and for fruitful joint work and Claudia Danilowicz and Mara Prentiss for discussions on their dsDNA unzipping experiments. I am especially grateful to David Nelson for pointing me toward the topic of DNA unzipping.

REFERENCES

1. See, for example, the contributions by S. Winters-Hilt, J.J. Kasianowicz, and S. Xie in this volume.
2. Ashkin, A. and Dziedzic, J.M., *Science* **235**, 1517–1520 (1987).
3. Binning, G., Quate, C.F., and Gerber, Ch., *Phys. Rev. Lett.* **56**, 930–933 (1986).
4. Reif, M., Gautel, M., Oesterhelt, F., Fernandez, J.M., and Gaub, H.E., *Science* **276**, 1109–1112 (1997); Kellermayer, M.S.Z., Smith, S.B., Granzier, H.L., and Bustamante, C., *ibid.* **276**, 1112–1116 (1997); Tskhovrebova, L., Trinick, J., Sleep, J.A., and Simmons, R.M., *Nature (London)* **387**, 308–312 (1997).
5. Liphardt, J., Onoa, B., Smith, S.B., Tinoco, I. and Bustamante, C., *Science* **292**, 733–737 (2001); Liphardt, J., Dumont, S., Smith, S.B., Tinoco, I., and Bustamante, C., *ibid.* **296**, 1832–1835 (2002).
6. Smith, S.B., Finzi, L., and Bustamante, C., *Science* **258**, 1122–1126 (1992).
7. Essevaz-Roulet, B., Bockelmann, U., and Heslot, F., *Proc. Nat'l. Acad. Sci. USA* **94**, 11935–11940 (1997); Bockelmann, U., Thomen, P., Essevaz-Roulet, B., Viasnoff, V., and Heslot, F., *Biophys. J.* **82**, 1537–1553 (2002).
8. Lubensky, D.K. and Nelson, D.R., *Phys. Rev. Lett.* **85**, 1572–1575 (2000).
9. Lubensky, D.K. and Nelson, D.R., *Phys. Rev. E* **65**, 031917 (2002).
10. Sauer-Budge, A.F, Nyamwanda, J.A., Lubensky, D.K., and Branton, D., eprint cond-mat/0209414.
11. Bhattacharjee, S.M. and Marenduzzo, D., *J. Phys. A* **35**, L349–L356 (2002); Cocco, S., Marko, J.F., and Monasson, R., *Comptes Rendus Phys.* **3**, 569–584 (2002); and references therein.
12. Gerland, U., Bundschuh, R., and Hwa, T., *Biophys. J.* **81**, 1324–1332 (2001).
13. Poland, D. and Scheraga, H.A., *Theory of Helix-Coil Transitions in Biopolymers*, Academic, New York, 1970.
14. Buldyrev, S.V., *et al.*, *Phys. Rev. E* **51**, 5084–5091 (1995), and references therein.
15. Fisher, D.S. and Huse, D.A., *Phys. Rev. B* **38**, 373–386 (1988).
16. le Doussal, P., Monthus, C., and Fisher, D.S. *Phys. Rev. E* **59**, 4795–4840 (1999).
17. Danilowicz, C., Coljee, V., Bouzig, C., Lubensky, D.K., Nelson, D.R., and Prentiss, M., "Multiple metastable intermediate states in the unzipping of DNA under a constant force." Submitted.
18. Bouchaud, J.-P. and Georges, A., *Phys. Rep.* **195**, 127–293 (1990).
19. Lubensky, D.K., in preparation.

Phase Transitions of a Polymer Threading a Membrane: Character of the Transition When the Molecule Can Undergo a Helix-Random Coil or an Equilibrium Polymerization Transition

Edmund A. Di Marzio[†] and John J. Kasianowicz[‡]

[†]*University of Maryland, College Park, MD, 20742-2111*
Polymers Division, NIST, Gaithersburg, MD 20899-8544
[‡]*Biotechnology Division, NIST, Gaithersburg, MD 20899-8313*

Abstract. The polymer threading a membrane transition (PTM) which is a first-order thermodynamic phase transition for an isolated linear polymer in the limit of infinite molecular weight is coupled to the helix to random coil (HRC) transition and the polymerization (P) transition. In each case an exact expression for the partition function of the coupled pair is given as a 1-dimensional summation over products of the individual partition functions. For HRC-HRC there are 8 possible translocation modes. For example, as we raise the temperature we can have H1→H2→R1→R2 in obvious notation. The PTM-P mode is most interesting because polymer can be created in the PTM side even though the P side is in the depolymerization regime! These exact model solutions provide a base for the study of the kinetics of significant technological problems such as the translocation of DNA through pores in membranes. They also throw light on the nature of polymer-membrane-pore interactions in living cells and viruses.

INTRODUCTION

In a previous publication [1] it was shown that polymers undergo at least 10 separate kinds of phase transitions. Five of them occur also in non-polymeric systems. They are 1) two component liquid-liquid phases, 2) liquid-crystals and plastic crystals, 3) crystals, 4) glass and sol-gel transitions, 5) soaps-block copolymers-membranes-vesicles-micelles. But 5 are unique to the isolated polymer molecule. They are 6) the polymer threading a membrane (PTM) transition, 7) helix-random coil (HRC) transition, 8) collapse (C) transition, 9) surface adsorption (SA) transition, 10) equilibrium polymerization (P). Because these 5 transitions occur only in polymers, polymers display by far the most complex and interesting thermodynamic phase transitions. The ten transitions are each interesting in themselves and a basic, albeit incomplete, understanding of each of them has been obtained [1].

However, these transitions can be coupled to each other in various combinations and very little is known about the behavior of such coupled systems. This paucity of knowledge arises partly from the difficulty of problems and partly from the perceived lack of applications. We address both questions.

CP665, *Unsolved Problems of Noise and Fluctuations: UPoN 2002: Third International Conference*, edited by S. M. Bezrukov
2003 American Institute of Physics 0-7354-0127-6

Concerning the difficulty of the problems we note the following. First, there are very many problems to be solved. The number of coupled pairs is 10x9/2=45 and the number of coupled triplets is 10x9x8/2x3=120 and hardly any of these systems have been studied. Further, each class of transitions contains subclasses. For example the class liquid crystals contains nematic, cholesteric and 7 smectic phases so that in coupling them to polymerization as was done in an attempt to understand sickle-cell anemia [2,3] we have 9 a priori possibilities rather than just one. The total number of coupled systems is in the thousands. Each of these systems is more difficult to understand than the component transitions which constitute them and which have yielded to increased understanding only through many years of diligent study. Another difficulty involves the expression of these transitions in time and space. The kinetics and pattern formation aspects of the individual transitions are less well understood than the equilibrium aspects and coupled systems introduce added complexities.

It is clear then that coupled polymer phase transitions is a vast and complex field of study that we are just now beginning to approach. It is equally clear that the field of coupled non-polymer phase transitions is not nearly as vast, complex or rich in its possibilities. For example, the number of coupled pairs is 5x4/2=10 rather than 45 and the number of coupled triplets is 5x4x3/2x3=10 rather than 120. Obviously we better have good reason to study the more difficult coupled systems, especially polymer systems.

An application from biology is the attempt to understand sickle cell anemia as a coupling of the polymerization of haemoglobin into fibrils with liquid crystal formation [2,3], all within the red blood cell. For a constant volume fraction of fibrils the longer the fibrils the easier it is to form the liquid crystal phase [4]. It has been argued previously that an understanding of polymeric phase transitions, especially the coupled phase transitions, is essential to an understanding of biological self-assembly [1]. The formation of the molecules of life, DNA, RNA, proteins, and Polysaccharides; their aggregation into larger structures such as actin, microtubules, the various membranes; processes such as translocation though pores; are intimately related to the occurrence of polymer phase transitions. It is difficult to understand these materials and processes without first understanding polymer phase transitions. It is probably no accident that the molecules of life, DNA, The various RNA's, proteins and polysaccharides are each polymeric. Certainly polymers are sufficient for life since our earth-life is polymeric. But it may well be that life is **necessarily** polymeric. If so it is undoubtedly because of the many additional phase transitions and resultant structures and processes available to polymers that are impossible to non-polymeric materials. These additional structures and processes may well be necessary to the manifold requirements of life. The possibility is intriguing and deserves investigation.

In this paper we will limit ourselves to the polymer threading a membrane transition [5] (PTM) and its coupling to the other 4 transitions of the isolated polymer (see Fig. 1). Several preliminary attempts to derive the translocation kinetics exist [6-13]. This paper hopes to derive the basic thermodynamic underpinning upon which the full kinetics of the translocation process can be built. We first will derive the formulas (partition functions) for each of the pairs. It is shown that for the general case the partition function (PF) for the pair is a one variable sum of terms and each term is a product of two terms, one for each side of the membrane. Since formulas exist in the literature for each of the separate transitions the problem is easily solved on a computer for finite molecular weights. See Sec. 2. In Sec. 3 we describe the general character of the phase behavior that is both understandable and marvelously complex and, undoubtedly provides a material base for some of the complexities of biology. A possible application to the function of the rough endoplasmic reticulum is given in Sec. 3.

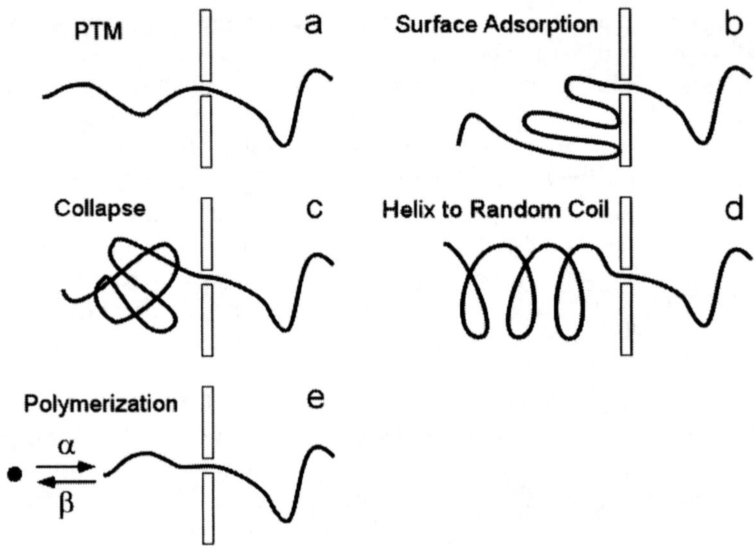

FIGURE 1. Illustration of the complexities involved in the equilibrium aspects of the polymer threading a membrane (PTM) transition. If the solutions on either sides of the membranes allow it we can have a) PTM-PTM transition, b) surface adsorption-PTM (SA-PTM) transition, c) collapse-PTM (C-PTM) transition, d) helix to random coil-PTM (HRC-PTM) transition, e) polymerization-PTM (P-PTM) transition. Not shown are the SA-SA, C-C, HRC-HRC, P-P, HRC-C, HRC-SA, HRC-P, C-SA, C-P, and the SA-P transitions. Higher order couplings such as (HRC+C)-PTM, of which there are many are not discussed. Ten of these 15 coupled systems (all but those involving polymer collapse) are exactly solvable. In this paper we will give an exact treatment of the PTM-PTM and P-PTM transitions and discuss the marvelous complexities of the HRC-HRC transition. An interesting feature of this problem is that it combines complete comprehensibility with extraordinary complexity.

2. THEORY

Denote the partition function (PF) for n segments residing on side 1 of the membrane by $q_{1,k}(n)$ and the PF for the N-n segments residing on side 2 of the membrane by $q_{2,j}(N-n)$. One end of each polymer is at the pore opening. The product of the two q's represents a polymer of N segments that has n segments on side 1 and N-n on side 2. We are assuming the membrane thickness to be zero, but finite thickness is easily accommodated. We do not allow multiple threading. Here k, j = PTM, HRC, C, SA, P represent the PTM transition, helix-random coil transition, collapse transition, surface adsorption transition, polymerization transition. The total PF, not counting escape into the reservoir on either side, is denoted by Q

$$Q(N)=\sum_{0\le n\le N} q_{1,k}(n)q_{2,j}(N-n) \tag{1}$$

where $q_{1,k}$ is the PF for that portion of polymer residing in side 1. This portion of the polymer samples all conformations consistent with its end being at the pore entrance. Fortunately, many of the expressions for the q's have been developed previously and are accessible in the literature. The number of ways to have k and j differ is 5x4/2=10, and the total number of ways for $j = k$ is 5. An enumeration of the total number (15) of distinct transitions is PTM-PTM, SA-SA, HRC-HRC, C-C, P-P; PTM-SA, PTM-HRC, PTM-C, PTM-P, SA-HRC, SA-C, SA-P, HRC-C, HRC-P, C-P. We shall in this paper discuss only PTM-PTM, HRC-HRC, and PTM-P.

2.1 The PTM-PTM Transition

For the pure PTM transition

$$q_{1,ptm}(n)=z_1^n\exp(-n\varepsilon_1/kT)/n^{\theta},\, q_{2,ptm}(N-n)=z_2^{N-n}\exp(-(N-n)\varepsilon_2/kT)/(N-n)^{\theta} \tag{2}$$

where θ depends on the nature of the constraint [1,5]; chain attached to surface only at one end, $\theta = 1/2$; chain attached to surface at both ends but one end free to roam on the surface, $\theta = 3/2$; Chain attached to surface at both ends but neither end free to roam $\theta = 5/2$. The form of Eq. 2 encompasses other geometries as well. If the surface has the shape of a cone or of a wedge and the chain threads through the apex then the only effect is to alter the value of θ to a value determined by the wedge [14] or cone [15] angle. Additionally, one can approximately accommodate [16,17] the effects of excluded volume by adjusting the value of θ. In all cases the transition is first-order (in the limit of infinite molecular weight) and the temperature of the transition is independent of θ. The effect of different θ is only to give transitions variable width for finite N; the larger θ the more narrow the width. The location of the transition does not depend on θ and is determined only by z_1, z_2, ε_1, ε_2. Fig. 2 illustrates the transition behavior for $\theta = 0$. Since it is first-order in the limit of infinite molecular weight and since larger θ give sharper transitions for each molecular weight these $\theta > 0$ transitions are first-order also. This fact will be used below in discussing coupled transitions.

Before leaving the pure PTM case we should point out that the reason for the occurrence of the transition lies in the fact that the monomers are covalently connected in a line which allows the monomers to be distinguishable from each other by virtue of the fact that they occur at different positions along the chain. If the monomers were not connected they would be indistinguishable from each other and there would be no transition. See [1] for a discussion of this important point. The occurrence of each of the 5 phase transitions of the isolated polymer is due to their linear covalent connectivity [1]. In this paper we wish to explore the effects of coupling the PTM transition to the other transitions. We begin with the HRC-HRC transition.

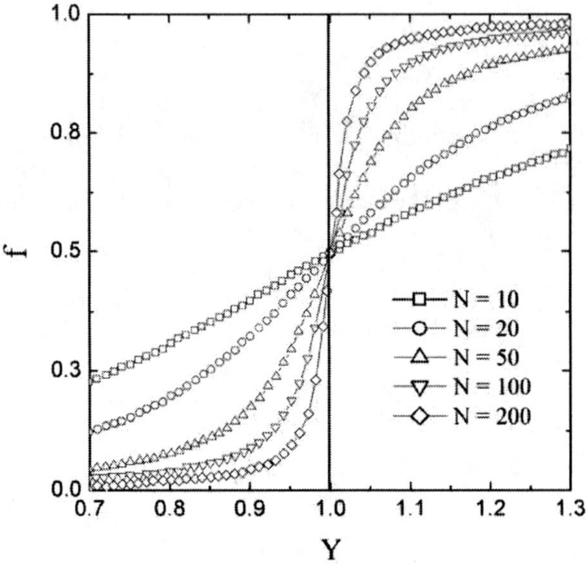

FIGURE 2. The polymer threading a membrane (PTM-PTM) transition. We plot the fraction of segments residing in region 1 versus Y the ratio of partition functions per monomer ($Y = z_1\exp(-\varepsilon_1/kT)/z_2\exp(-\varepsilon_2/kT)$) for chains of different length *N*. A first-order transition occurs in the limit of infinite molecular weight. These curves are for $\theta = 0$ that corresponds to minimal chain constraints. Additional chain constraints result in even sharper transition curves. See text for details.

2.2 The HRC-HRC Transition

There have been several treatments that couple the helix-random coil transition (HRC) to surface adsorption [18,19]. The transitions can be altered dramatically. The second-order transition for two stranded DNA becomes first-order [18] and more surprisingly the diffuse transition in single-stranded polypeptides can become first-order when coupled to surface adsorption (SA) [19]. We really have the case of a triplet coupling when we couple PTM to HRC because of the ever present surface. The triplet coupling PTM-HRC-SA is more complicated than we care to deal with here. The difficulty is that for fixed *n*, that is ignoring the threading, there is a complexity of transition behavior for the HRC-SA case that we have not yet digested. We postpone treating triplet coupling until we have studied aspects of the pair couplings. So, we use for a polymer chain tied to the surface at one end the HRC results for a polymer chain of length *n* located far away from the surface. This has the distinct advantage that such results exist in the literature.

Can we justify this approach? Consider Eq. 2. We already know that the location of the transition is not affected by the value of θ, only the width of the transition. So, as an approximation we shall take the deep space partition function for HRC and multiply it by $n^{-\theta}$. This should be a good approximation for non-attractive surfaces. For very long polymers, since most of the polymer is far from the surface we expect tethered HRC to not differ from the deep space HRC. So, we have

$$q_{1,hrc}(n) = q_{1,h}(n)/n^{\theta}, q_{2,hrc}(N-n) = q_{2,h}(N-n)/(N-n)^{\theta} \tag{3}$$

where $q_{k,h}(n)$ is the partition function for the helix to random coil transition and is taken from the literature. A good source is the book by Poland and Scheraga [20]. For attractive surfaces the above approximation is a poor one and we need to treat the coupled triplet (PTM-HRC-SA) problem.

2.3 The Polymerization-PTM Transition

If k denotes polymerization [21] and j denotes PTM then we have infinite sums on both! By inspection of the following equation for the grand canonical ensemble we see that every term that should appear does appear and appears only once.

$$\Xi(N) = \left[\sum_{n} q_{1,p}(n)\lambda_1^n\right]\left[\sum_{m} q_{2,j}(m)\lambda_1^m\right] \tag{4}$$

Because only side 1 is the feedstock a chain of length $n + m$ must have $n + m$ factors of λ_1. Although a λ_1 term appears in the term representing side 2 in Eq. 4 it is the λ_1 appropriate to side 1. We have obtained a remarkable result because it means that polymerization on one side only controls the number of segments on the other side. For clarity consider the case where both sides 1 and 2 have the PTM value for $q_{1,k}$, $q_{2,j}$.

$$q_{1,k} = q_{1,ptm} = z_1^n \exp(-n\varepsilon_1/kT)/n^{\theta}, q_{2,j} = q_{2,ptm} = z_2^m \exp(-m\varepsilon_2/kT)/m^{\theta} \tag{5}$$

When $z_2 \exp(-\varepsilon_2/kT)\lambda_1 > 1$ or < 1, the polymer is either infinitely long or exponentially short on side 2. It is obvious that there are 4 possibilities. The polymer can be short on both sides, short on side 1 and infinite on side 2, short on side 2 and infinite on side 1, or infinite on both sides of the membrane. For the case when side 1 gives an infinite molecular weight, it is perhaps not surprising that the chain on the other side can be infinitely long. One side acts as feedstock for the other side. But the case when side 1 taken alone gives a finite and small molecular weight (this corresponds to depolymerizing) is surprising. The number of chains of length n on side 1 is exponentially small in n, yet side 1 still acts as feedstock for side 2! Thus, a depolymerization process occurring in side 1 can still result in polymerization in side 2 merely by having a pore connecting the two solutions. By connecting two spatially separated free energies with a polymer, we have changed a depolymerization process in side 1 to a polymerization process in side 2. This idea has to be important to biology. We will return to this point below.

3. RESULTS AND DISCUSSION: APPLICATIONS TO PROBLEMS IN BIOLOGY

Let us consider the case where the individual q's each display one first order transition. In Fig. 3 we schematically illustrate the free energies $F_1/n = -kT \ln(q_1(n))$ and $F_2/m = -kT \ln(q_2(m))$ as a function of temperature. Given that F_1/n lies below F_2/m for small T, we have 8 possibilities as shown in Fig. 3. Suppose that we are considering the helix to random coil transition for the case where the transition is sharp. Then, we have the following 8 possible behaviors as we raise the temperature.

H1→R1; H1 → R2; H1→H2→R2; H1→R1→R2; H1→R2→R1
H1→R1→H2→R2; H1→H2→R2→R1; H1→H2→R1→R2

H1 corresponds to helix, 1 to side 1 and R corresponds to random coil, 2 to side 2. The polymer can translocate as many as 3 times. For example, H1→H2→R1→R2 means helix in side 1 translocating to helix in side 2 translocating to random coil in side 1 translocating to random coil in side 2. In the above, depending on which of the free energy diagrams of Fig. 3 are applicable, the polymer can translocate 0, 1, 2, or 3 times.

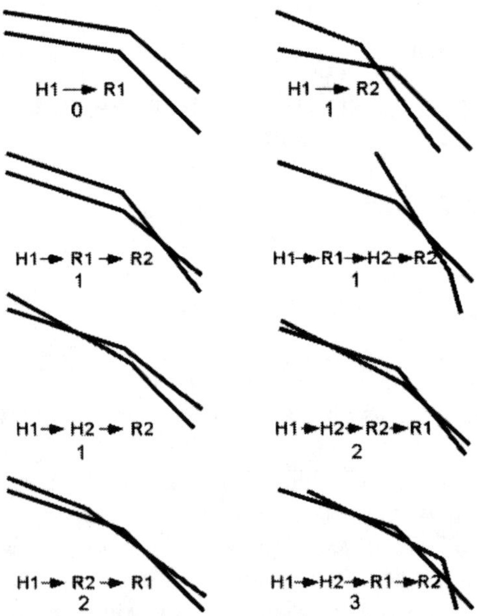

FIGURE 3. The HRC-HRC transition. Free energy per monomer versus T curves for a long polymer in side 1 or 2. There are 8 pairs of curves corresponding to the 8 ways the curves can cross. In each case, the lower curve at small T is the side 1 curve. The breaks in the curves represent first-order transitions. The curves are labeled assuming there is a sharp HRC transition for increasing T. A translocation occurs when the curves intersect. The system free energy is given by the composite curve consisting of the lowest connected line segments. There can be as few as 0, or as many as 3 translocations.

If the polymer can undergo two transitions, that is, if the figures representing free energies in Fig. 3 show two breaks rather than 1 the possibilities enlarge. This may be relevant to the collapse transition, and by implication to globular proteins. It is possible that polymer collapse is a two-stage process. The polymer first collapses to a dense liquid with (mean square) dimensions varying as $n^{2/3}$. It then collapses further to a dense solid with dimensions also varying as $n^{2/3}$. The existence of globular proteins with set structure implies that this is the case. Such a system would have two breaks in the free energy curve.

In this paper we have observed that there are 15 different variations of the polymer threading a membrane (PTM) problem. They arise from coupling the PTM transition to 4 other transitions. These are the surface adsorption transition (SA), the helix to random coil transition (HRC), the collapse transition (C), the equilibrium polymerization transition (P). We allow the solutions to be sufficiently different that one kind of transition can be occurring in solution 1 while another is occurring in solution 2. The 15 variations are PTM-PTM, SA-SA, HRC-HRC, C-C, P-P; PTM-SA, PTM-HRC, PTM-C, PTM-P, SA-HRC, SA-C, SA-P, HRC-C, HRC-P, C-P. An exact solution to the PTM-PTM transition was given previously and was reproduced here. In this paper we solved exactly the PTM-P transition. Of the remaining 13 variations only 5 are not solvable exactly. They are the 5 containing the collapse transition. This is because an exact solution to the excluded volume problem has never been obtained. However, accurate, but non-exact solutions do exist [22,23]. The remaining 8 transitions are each solvable exactly because the PTM, SA, HRC, and P transitions are each solvable exactly and because combining them involves only a one-dimensional sum of analytical expressions. Although the solutions are exact they are of models that may not be very realistic. For example the PTM treatment, which is the least realistic but easiest to solve, assumes that each monomer has a partition function given by $z \exp(-\varepsilon/kT)$ where z allows for the entropic freedom of rotation and ε allows for an energy of monomer-solution interaction.

The transitions of our model systems display interesting complicated behavior that is, nonetheless, completely comprehendible because our solutions are exact. As an example we have examined the HRC-HRC transition and find that for large molecular weight the molecule can translocate back and forth across the membrane zero, one, two, or three times as we lower temperature or change chemical potential depending on the nature of the solutions (see Fig. 3).

The PTM-P case is even more interesting. Depending on the nature of the solutions and the chemical potential for polymerization we can have long-long, short-long, long-short or short-short polymer. The P side in which polymerization is occurring acts as a feedstock for the other side. The case long-short is most interesting because it allows depolymerization in the P side to create long polymer in the PTM side. The explanation is that for equilibrium polymerization, depolymerization results in short polymer whose probability of being of length n is exponentially small in n. But these short chains by virtue of fluctuations can still act as feedstock for the chain on the PTM side and these chains can be long, even infinitely long.

It is always proper to circumscribe what we have done by listing what we have not done. This serves both to humble the authors and to suggest directions for future research. 1) Although the partition functions have been written down we have gone no further in obtaining the thermodynamic quantities, except for the cases PTM-PTM, HRC-HRC, PTM-P. Perhaps it is best to wait for the pressure of (possible) future applications. 2) We have not made numerical calculations for finite molecular weight. The transitions should each be broadened and it would be nice to know how big the effect is. 3) We did not calculate the effects of an applied force on the end or ends of the polymer. Experimentally one could conceivably apply such a force by use of optical tweezers, magnetic particles or the tip of an atomic force microscope. 4) We have not treated realistic models for the polymers. Our models are vanilla models. The use of more realistic polymer models and the proper treatment of copolymers are needed if more meaningful connections to biology are to be obtained. 5) We did not allow for the finite thickness of the membrane or any other details of its structure. These details will be especially important when considering the kinetics of translocation. Nevertheless, our treatment has given us added insight into the polymer translocation process.

There are more than a few biological problems involving translocation. By discussing them briefly applications or directions for future research may be suggested.

For example, the nuclear pore membrane allows and/or assists in the translocation of RNA and proteins as a normal process of cell metabolism/function [24,25]. Also, viruses can infect by translocating their (DNA) genes into the cell nucleus [24,25]. Finally, biotechnology procedures exist that manage to breach the nuclear membrane [26]. These various translocations sometimes use helper molecules as essential elements of the process.

Viruses also need to breach the cell membrane in order to have their genetic content enter the cell. In T2 or T4 viruses the process looks very much like a "translocation through a pore" process, as does also the reverse process of filling the newly created but empty virus capsules. Chaperon or helper molecules are sometimes involved but perhaps not always [27]. One wonders whether during the evolution of T2 and T4 viruses there was a time when translocation did not involve helper molecules, seeing that unadorned translocation is possible and that helper molecules seem to make the process more robust. By examining the more primitive viruses one might be able to determine if translocation with helper molecules evolved from unadorned translocation.

There are many signaling molecules and some are polymers. To what extent is translocation or partial translocation involved, either in the process of signaling or of incorporation of the signaling polymer into the membrane?

A crude "going fishing" model was proposed for transferring low molecular weight materials across a membrane [5] (for another possible practical extension, see [28]). During the process the polymer is also translocated. Is this the way helper molecules evolved? Should helper molecules be monomers or polymers? Obviously our model needs to be extended to address such questions.

The use of surfaces and pores allows the cell to separate the polymerization of proteins and their collapse to globular forms from the subsequent interaction of the globular proteins among themselves. It would be a real mess if the thousands of proteins involved in the various cell functions were not so sequestered and instead had to both polymerize and interact with each other simultaneously. Ribosomes gather near the rough endoplasmic reticulum membrane and helper molecules are involved in the process of translocation of polymerizing protein polymer into the endoplasmic reticulum [29]. Our polymerization insight suggests why ribosomes might be associated with the membrane. Suppose the ribosomes were in the depolymerization range. Then away from the membrane they would not manufacture protein. This is fine for the cytosol. Imagine that the thousands of proteins performing their many cytosolic functions needed also to be polymerized within the cytosol. In that case, perhaps we would have the equivalent of scrambled eggs. How much more reasonable it is to have the polymerization and subsequent preparation of the proteins occur in a sequestered space and then be released from the sequestered space only when they were fully formed and ready for action? Our polymerization-depolymerization insight above shows that the ribosome-membrane system can be in the depolymerization range in side 1 and still grow polymer in side 2. The helper molecules simply serve to make the protein polymer long on side 2. More simple systems, those with fewer proteins in their cytosol, such as yeast cells apparently get by with ribosomes that are in the polymerization range within the cytosol.

REFERENCES

1. Di Marzio, E.A., The ten classes of polymeric phase transitions: their use as models for self-assembly, *Prog. Polym. Sci.*, **24**, 329-377 (1999).
2. Minton, A.P., Thermodynamic analysis of chemical inhibition of sickle-cell hemoglobin gelation, *J. Mol. Bio.*, **95**, 289-307 (1975); Minton, A.P., Relations between oxygen saturation and aggregation of sickle-cell hemoglobin, *J. Mol. Bio.*, **100,** 519-542 (1976).
3. Herzfeld, J., Entropically driven order in crowded solutions: from liquid crystals to cell biology, *Acc. Chem. Res.*, **29**, 31-37 (1996).
4. P.J. Flory derives the relation $xV_x=8$ (where x is the length to width ratio of the rigid polymer and V_x the volume fraction) giving the volume fraction when packing interferences first begin to align the polymer rods. See page 99 in Chapoy, L. L., Recent advances in liquid crystalline polymers, New York, Elsevier Applied Science, 1985.
5. Di Marzio, E.A. and Mandell, A.J., Phase transition behavior of a linear macromolecule threading a membrane, *J. Chem. Phys.*, **107**, 5510-5514 (1997).
6. Sung, W. and Park, P.J., Polymer translocation through a pore in a membrane, *Phys. Rev. Lett.*, **77**, 783-786 (1996).
7. Park, P.J. and Sung, W., Dynamics of a polymer surmounting a potential barrier: the Kramers problem for polymers, *J. Chem. Phys.*, **111**, 5259-5266 (1999).
8. Lubensky, D.K. and Nelson, D.R., Driven polymer translocation through a narrow pore, *Biophys. J.*, **77**, 1824-1838 (1999).
9. deGennes, P.G., Flexible polymers in nanopores, *Adv. Polym. Sci.* , **138**, 91-105 (1999).
10. Muthukumar, M., Polymer translocation through a hole, *J. Chem. Phys.*, **111**, 10371-10374 (1999).
11. Chern, S.S. and Cardenas, A.E., Three-dimensional dynamic Monte Carlo simulations of driven polymer transport through a hole in a wall, *J. Chem. Phys.*, **115**, 1772-1782 (2001).
12. Muthukumar, M., Translocation of a confined polymer through a hole, *Phys. Rev. Lett.*, **86**, 3188-3191 (2001).
13. Ambjörnsson, T., Apell, S.P., Konkoli, Z., Di Marzio, E.A. and Kasianowicz, J.J., Charged polymer membrane translocation, *J. Chem. Phys.*, **117**, 4063-4073 (2002); see also; Kasianowicz,

J.J., Brandin, E., Branton, D., Deamer, D.W., Characterization of individual polynucleotide molecules using a membrane channel. *Proc. Natl. Acad. Sci. (USA)*, **93**, 13770-13773 (1996), Henrickson, S.E, Misakian, M., Robertson, B. and Kasianowicz, J.J., Driven asymmetric DNA transport in a nanometer-scale pore. *Phys. Rev. Lett.*, **85**, 3057-3060 (2000);

14. Lauritzen, J.I. and Di Marzio, E.A., Configurational statistics of a chain confined to a wedge of interior angle α, *J. Research NBS*, **83**, 381-385 (1978).
15. Guttman, C.M. and Di Marzio, E.A., The polymer in a cone, *J. Research NBS*, **85**, 273-282 (1980).
16. Fisher, M.E., Effect of excluded volume on phase transitions in polymers?, *J. Chem. Phys.*, **45**, 1469-1473 (1966).
17. Douglas, J.F., Nemirovski, A.M. and Freed, K.F., Polymer-polymer and polymer-surface excluded volume effects in flexible polymers attached to an interface-comparison of renormalization group calculations with Monte Carlo and direct enumeration data, *Macromolecules*, **19**, 2041-2054 (1986).
18. Di Marzio, E.A. and Bishop, M., Nature of the helix-random coil transition of biological molecules attached to a surface, *Biopolymers*, **13**, 2331-2348 (1974).
19. Carri, G.A. and Muthukumar, M., Coupling between adsorption and the helix-coil transition, *Phys. Rev. Lett.*, **82**, 5405-5408 (1999).
20. Poland, D.C. and Scheraga, H.A., Theory of the helix-coil transition, Academic Press, New York, 1970.
21. We are treating only the case of equilibrium polymerization.
22. deGennes, P.G., Scaling concepts in polymer physics, Cornell Univ. Press, Ithaca NY, 1979.
23. Freed, K.F., Renormalization group theory of macromolecules, John Wiley and Sons, New York, 1987.
24. Lodish, H., Berk, A., Zipursky, S.L., Matsudaira, P., Baltimore, D. and Darnell, J.E., Molecular biology of the cell, 4th edition, W. H. Freeman and Co., New York, 2000.
25. Alberts, B., Bray, D., Lewis, J., Raff, M., Roberts, K. and Watson, J. D., Molecular biology of the cell, 3rd edition, Garland Publishing Co., New York, 1994.
26. Salman, H., Zbaida, Z., Rabin, Y., Chatenay, D. and Elbaum, M., Kinetics and mechanism of DNA uptake into the cell nucleus, *Proc. Nat. Acad Sci.*, **98**, 7247-7252 (2001).
27. Ellis, R.S. and van der Vies, S.M., Molecular Chaperones, *Annu. Rev. Biochem.*, **60**, 321-347 (1991).
28. Kasianowicz, J.J., Henrickson, S.E., Weetall, H.H. and Robertson, B., Simultaneous multianalyte detection with a nanopore. *Analytical Chemistry*, **73**, 2268-2272 (2001)
29. Elston, T.C., Models of post translational protein translocation, *Biophys. J.*, **79**, 2235-2251 (2000).

Highly Accurate Real-Time Classification of Channel-Captured DNA Termini

Stephen Winters-Hilt

Center for Biomolecular Science & Engineering, University of California, Santa Cruz
Computer Science Department, University of California, Santa Cruz

Abstract. A computational method is briefly described for classification of individual DNA molecules measured by an α-hemolysin channel detector. Classification is performed with better than 99% accuracy for DNA hairpin molecules that differ only in their terminal Watson-Crick base pairs. Signal classification was initially done on synthetic data streams, where sampling on real mixtures of hairpins was modeled in order to establish performance metrics (i.e., where train and test data were of known type, via single-species data files). Signal classification was then performed on observations from real mixtures of DNA hairpins. Hidden Markov Models (HMMs) were used with Expectation/Maximization for de-noising and for associating a feature vector with the ionic current blockade of the DNA molecule. Support Vector Machines (SVMs) were used as discriminators, and were the focus of off-line training. A multi-class SVM architecture was designed to place less discriminatory load on weaker discriminators, and novel SVM kernels were used to boost discrimination strength. The tuning on HMMs and SVMs enabled biophysical analysis of the captured molecule states and state-transitions; structure revealed in the biophysical analysis was used for better feature selection. This analysis is presented in more detail, but with less discussion, in Winters-Hilt et al. 2003.

INTRODUCTION

Molecular classification using nanometer-scale channels provides a powerful new tool for efforts in biophysics and biotechnology. Such nanopore detectors use a nanometer-scale channel to relate ionic current blockade measurements to single molecule translocation by the pore (Akeson et al., 1999; Kasianowicz et al., 1996; Meller et al., 2000; Meller et al. 2001), or to single molecule capture by the pore (Vercoutere et al., 2001). Biologically based α-hemolysin channels are elegant in this regard in that they self-assemble in lipid bilayers (Gouaux et al., 1994; Song et al., 1996), thereby providing inexpensive and reproducible nanopores. The size of the α-hemolysin pore is also well suited to DNA measurement in that single-stranded DNA (ssDNA) translocates while double stranded DNA (dsDNA) does not, being held instead in a vestibule of the pore (Vercoutere et al., 2001). For end-capture on dsDNA, extensive characterization of ionic current blockades is possible.

CP665, *Unsolved Problems of Noise and Fluctuations: UPoN 2002: Third International Conference,* edited by S. M. Bezrukov

Modifications to the α-hemolysin channel have been examined (Bayley 2000), and semiconductor nanopores are being developed (Li et al., 2001). In this paper a brief description is given of how a nanopore detector, coupled with pattern recognition, can be used to discriminate between DNA hairpin termini with high accuracy. A more extensive discussion can be found in (Winters-Hilt et al., 2003).

In the nanopore signal analysis a Hidden Markov Model (HMM) is used to extract a feature vector from each blockade example. HMMs (Chung et al., 1990; Chung and Gage, 1998; Colquhoun and Sigworth, 1995) can characterize current blockades by identifying a sequence of sub-blockades as a sequence of state emissions. The parameters of an HMM are usually estimated using a method called Expectation/Maximization (Durbin 1998). Although HMMs can be used to discriminate among several classes of input, multi-class computational scalability tends to favor their use as feature extractors. In particular, HMMs are well suited to extraction of aperiodic information embedded in stochastic sequential data. Support Vector Machines (SVMs) are then used to classify the feature vectors obtained by the HMM (for each individual blockade event). SVMs are fast, easily trained, discriminators (Vapnik 1998; Burges 1998), for which strong discrimination is possible without the over-fitting complications common to neural net discriminators (Vapnik 1998).

METHODS

Nanopore Implementation And DNA Hairpin Design

Each experiment was conducted using one α-hemolysin channel inserted into a diphytanoyl-phosphatidylcholine/hexadecane bilayer, where the bilayer was formed across a 20-micron diameter horizontal Teflon aperture (Vercoutere et al., 2001). The bilayer separates two seventy-microliter chambers containing 1.0 M KCl buffered at pH 8.0 (10 mM HEPES/KOH). The nine base-pair hairpin molecules examined share an eight base-pair hairpin core sequence, with addition of one of the four permutations of Watson-Crick base-pairs that may exist at the blunt end terminus, i.e., 5'-G•C-3', 5'-C•G-3', 5'-T•A-3', and 5'-A•T-3'. Denoted 9GC, 9CG, 9TA, and 9AT, respectively. The full sequence for the 9CG hairpin is 5' <u>CTTCGAACG</u>TTTT<u>CGTTCGAAG</u> 3', where the base-pairing region is underlined. An eight base-pair DNA hairpin with a 5'-G•C-3' terminus was also tested. The prediction that each hairpin would adopt one base-paired structure was tested and confirmed using the DNA mfold server (http://bioinfo.math.rpi.edu/~mfold/dna/form1.cgi), which is based in part on data from (SantaLucia 1998). The nanopore construction and the DNA synthesis tools are described in (Winters-Hilt et al., 2003).

Sampling Protocol And Signal Acquisition

The solution sampling protocol used periodic reversal of the applied potential to accomplish the capture and ejection of single DNA molecules (added to the cis chamber in 20 μM concentrations). The current blockade data was filtered at 10 kHz bandwidth using an analog low pass Bessel filter and recorded at 20 μs intervals using an Axopatch 200B amplifier coupled to an Axon Digidata 1200 digitizer (Axon Instruments, Foster City, CA). A time-domain finite state automaton (FSA; Cormen et al. 1989) with eight states performed the identification and acquisition on the first 100 ms of blockade signal (Acquisition Stage, Figure 1). The effective duty cycle for acquiring the desired 100 ms blockade measurements was one reading every 0.4 seconds. Further details on voltage toggling protocol the time-domain FSA are in (Winters-Hilt et al., 2003).

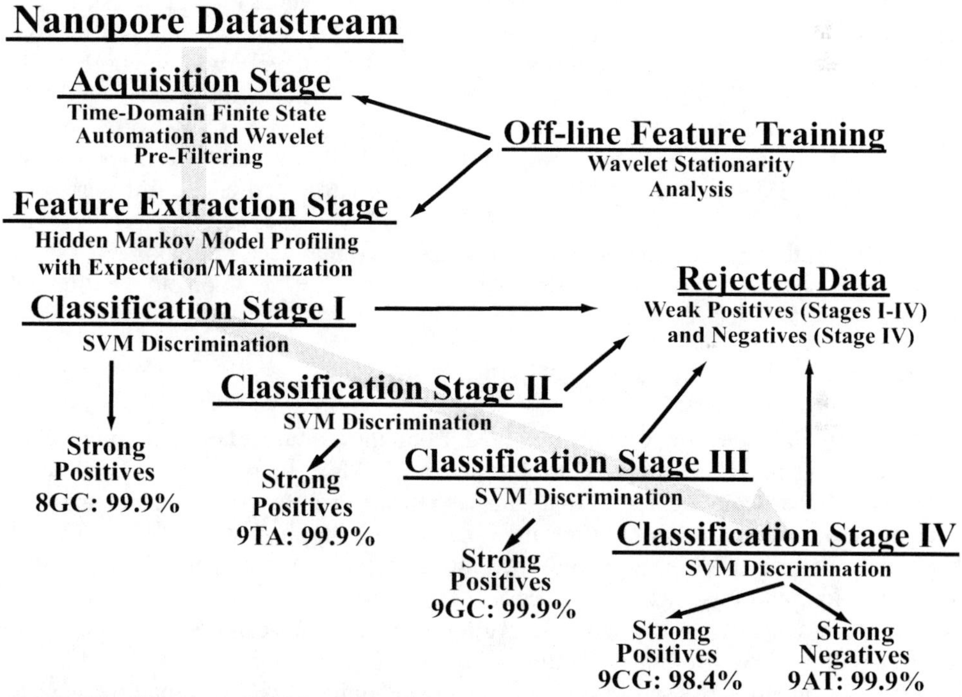

FIGURE 1. Signal acquisition was performed using a time-domain, thresholding, Finite State Automaton. This was followed by adaptive pre-filtering using a wavelet-domain Finite State Automaton. Feature extraction on those acquired channel blockades was done by Hidden Markov Model processing; and classification was done by Support Vector Machine. The optimal SVM architecture is shown for classification of molecules 9CG, 9GC, 9TA, 9AT, and 8GC. The linear tree multi-class SVM architecture benefits from strong signal skimming and weak signal rejection along the line of decision nodes. Scalability to larger multi-class problems is possible since the main on-line computational cost is at the Hidden Markov Model feature extraction stage. The accuracy shown is for single-species mixture identification upon completing the 15th single molecule sampling/classification (in approx. 6 seconds).

Signal Preprocessing And Feature Extraction

Each 100 ms signal acquired by the time-domain FSA consisted of a sequence of 5000 sub-blockade levels (with the 20 μs analog-to-digital sampling). Signal preprocessing was then used for adaptive low-pass filtering. For the data sets examined the preprocessing led to length compression on the sample sequence from 5000 to 625 samples (later HMM processing then only required construction of a dynamic programming table with 625 columns). The signal preprocessing makes use of an off-line wavelet stationarity analysis (Off-line Wavelet Stationarity Analysis, Figure 1, Diserbo et al., 2000). With completion of preprocessing, an HMM (Durbin 1998) was used to remove noise from the acquired signals, and to extract features from them (Feature Extraction Stage, Fig. 1). The HMM was implemented with fifty states, corresponding to current blockades in 1% increments ranging from 20% residual current to 69% residual current. The HMM states, numbered 0 to 49, corresponded to the 50 different current blockade levels in the discrete sequences that it processed. The state emission parameters of the HMM were initially set so that the state j, $0 <= j <= 49$ corresponding to level $L = j+20$, could emit all possible levels, with the probability distribution over emitted levels set to a discretized Gaussian with mean L and unit variance. All transitions between states were possible, and initially were equally likely. Each blockade signature was de-noised by 5 rounds of Expectation-Maximization (EM) training on the parameters of the HMM. After the EM iterations, 150 parameters were extracted from the HMM (further details in Winters-Hilt et al., 2003). The resulting parameter vector, normalized such that vector components sum to unity, was used to represent the acquired signal in discrimination at the Support Vector Machine stages.

Classification Training

The normalized feature vectors obtained from the feature extraction stage were classified using binary Support Vector Machines (SVMs). Binary SVMs are based on a decision-hyperplane heuristic that incorporates structural risk management by attempting to obtain a training-instance void, or "margin," around the decision hyperplane. Binary SVMs can be grouped into a classifier tree to perform multi-class discrimination, and this was done here for the five classes of DNA hairpin (shown in classification stages I-IV in Figure 1). Tuning on the multi-class SVM architecture itself was done for performance optimization, and separate tuning was also done on the polarization strength used in the data cleaning. Tuning was also done on the SVM internals, over families of kernels based on regularized distances (Jaakkola 1998) and regularized information divergences. In the former case, the squared Euclidean distance between feature vectors **x** and **y**, $\mathbf{d^2(x,y) = \Sigma_k(x_k\text{-}y_k)^2}$, also known as the squared $\mathbf{l_2}$-norm on **(x-y)**, $\mathbf{[l_2(x\text{-}y)]^2 = d^2(x,y)}$, is associated with the Gaussian kernel: $\mathbf{K_G(x,y) = exp(-d^2(x,y)/2\sigma^2)}$. In the latter case, the information divergence (relative entropy) between probability vectors **x** and **y**, $\mathbf{D(x||y) = \Sigma_k x_k log(x_k/y_k)}$, can be associated with the "Entropic kernel:" $\mathbf{K_E(x,y) = exp(-[D(x||y)+D(y||x)]/2\sigma^2)}$. The terminating SVM node of the classifier tree (stage IV in Figure 1) used the Entropic kernel. The other nodes of the classifier tree used a regularized-distance type kernel, the "Indicator kernel," based on the square root of the $\mathbf{l_1}$-norm, where $\mathbf{l_1(x\text{-}y) = \Sigma_k|x_k\text{-}y_k|}$, with kernel $\mathbf{K_I(x,y) = exp(-\sqrt{l_1(x\text{-}y)}/2\sigma^2)}$. The kernels considered were not restricted by Mercer's conditions. Instead, attention was focused on exploring kernels based on regularized information divergences as a parallel to the very successful kernels based on regularized distances (such as the Gaussian kernel). The Gaussian kernel (which satisfies Mercer's conditions) was outperformed in all cases studied by the Entropic and Indicator kernels.

Discriminator Implementation

The SVM discriminators were trained by solving their KKT relations using the Sequential Minimal Optimization (SMO) procedure (Platt 1998). A Chunking (Osuna et al., 1997; Joachims, 1998) variant of SMO was employed to manage the large training task at each SVM node. The multi-class SVM training was based on over ten thousand blockade signatures for each DNA hairpin species. The data cleaning needed on the training data was accomplished by an extra SVM training round (further details on data cleaning in Winters-Hilt et al., 2003).

Testing Protocol

The test data consisted of over two thousand blockade signals for each DNA hairpin species and was drawn from experiments that were run on days (and nanopores) different from those used to acquire the training data. Testing on single-species mixture calling was done directly, with classification on observations from single-species solutions in the cis chamber. One goal of the study was to find how many classification attempts were required to classify the single-species solutions with very high confidence. Scoring was possible by tracking the known labels on the test data. For the mixture tests some of the train data was used for an added calibration. An extra calibration was required because true mixtures of hairpins are sensitive to the different (entropic) acceptance rates and (discriminator) rejection rates by the nanopore instrument for the different hairpin species.

RESULTS

We were able to determine which of five species of DNA hairpin had been added to the cis chamber of the nanopore device. This was done in less than six seconds with 99.6% accuracy. The five DNA hairpins consisted of four hairpins that differed only in their terminal base-pairs, and a control hairpin. These results were data drawn from nanopores established on days other than those used to generate the training data. At 75% weak signal rejection, approximately 15 classification attempts were needed to classify the type of single-species solution being sampled; final solution classification was obtained in six seconds on average. If training and testing were done on data drawn from the same set of days of nanopore operation, albeit different samples,

99.9% calling was obtained with 15% rejection, and throughput was about one call every half second.

Identification of two hairpins in mixtures was also attempted. Figure 2 shows the percentage of 9TA classification in a 3:1 mixture of 9TA to 9GC. (Although the mixture preparations are estimated to be ±10% of their stated mixture ratios, calibration and testing of aliquots from the same mixture compensates for such common error.) The assay on 9TA concentration asymptotes to 75% ± 1%, consistent with the 3:1 ratio, and the assay error drops to 1% after approximately 100 individual molecule classification attempts (completed in 40 seconds).

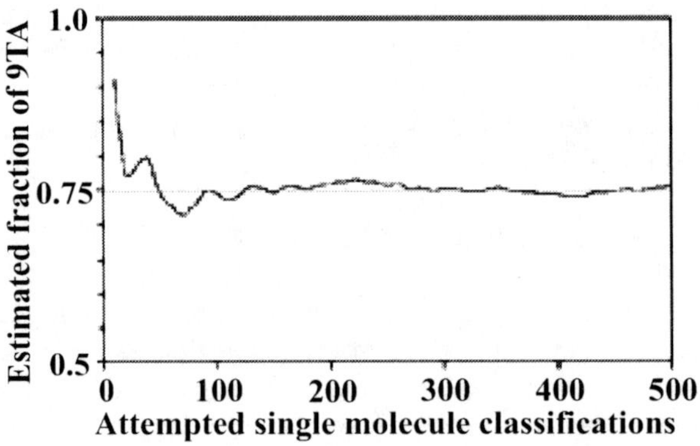

FIGURE 2. Classification on a 3:1 mixture of 9TA and 9GC hairpin molecules as a function of single molecule acquisitions.

HMM/EM characterization on the five classes of hairpin signatures revealed the existence of two major conductance blockade levels, one minor level intermediate between them, and one to three other statistically relevant levels depending on the hairpin. By examining the transition probabilities between the various levels it was found that blockades typically began in the less common intermediate level and from there almost always transitioned to the greater conductance blockade level.

DISCUSSION

Calibration And Feature Extraction By HMM

A single HMM/EM process was used to perform feature extractions in the experiments. If separate HMMs were used to model each species, the HMM/EM processing could also be operated in a discriminative mode. This requires multiple HMM/EM evaluations (one for each species) on each unknown signal as it is observed. Increased computational burden would thus be added at the worst place: the expensive feature extraction stage. For future work, semi-scalable, species-specific processing is being considered for the HMM/EM in an indirect manner, by using prior HMM/EM characterization of the species to identify a reduced set of features relevant to each species. Traditional signal analysis on the data, using power spectral analysis methods, reveals approximately Lorentzian noise, indicating an approximately two-state random process from which rate constants for transitions back and forth (or time constants) can be extracted. In comparison, the HMM/EM reduced feature set is far more detailed, and can provide a complete (statistically optimized) physical characterization of the captured molecule, via blockade states, their time-constants, and allowed state transitions.

Tests with mixtures of hairpins required an added calibration due to the nanopore's different acceptance rates for different hairpins (i.e., there are different free energy barriers to capture). This finding was consistent with a model for hairpin capture in which hairpins are captured by an entropically accessible binding site. It is also in agreement with the brief intermediate level state typically observed at the start of the signal blockades.

Classification By SVM Hierarchy

Novel SVM kernels were used to obtain the results described here. The novel kernels are based on a generalization from regularized square-distances to regularized information divergences. One of the kernels (the Entropic kernel at Classification Stage IV in Figure 1) used the Kullback-Leibler information divergence (Cover and Thomas, 1991). Entropic-type kernels may offer advantages when all or part of the feature vector can be interpreted as a probability vector. SVM confidence for a classification is a function of the distance from the feature vector coordinates (in feature space) to the separating, discriminatory, hyperplane (where greater distance represents higher confidence in discriminating between two signals). Since the kernel defines the notion of distance, tuning on kernels also provides a powerful means to manage rejection behavior. Multi-class SVM discrimination can then obtained by grouping binary SVMs into a decision-tree architecture (Vapnik, 1998; Bredensteiner and Bennett, 1999) using rejection of low confidence data at earlier stages to postpone decisions to more appropriate later stages.

Re-establishing the α-hemolysin channel on a day-to-day basis presents a major complication to the pattern recognition task. The class training data that would normally map to a single cluster is shattered into a cluster of clusters, with greater dispersion and class overlap in the SVM feature vector space. SVM classification in such circumstances faces weaker training convergence and poorer signal calling. For the five classes considered here, a passive stabilization approach was used that optimized the kernels for high rejection. More active (computational) stabilization methods are being studied for larger multiclass problems and improved accuracy overall.

Blockade Mechanism For Nine Base-Pair DNA Hairpins

In a forthcoming manuscript (Winters-Hilt et al., in preparation), analysis will focus on details of the current blockade mechanism, so only a preliminary description is given here (Figure 3). The intermediate level (IL) conductance state initiates most blockades and always transitions to the upper level conductance state (UL). This is explained by binding of the hairpin terminus to the vestibule interior (IL) followed by desorption of the DNA from the protein wall and orientation of the stem along the axis of the electric field (UL). Transitions from the UL state were either back to the IL state or to the lower level conductance state (LL). From the LL state there were brief transitions to nearly full blockade, denoted by S for spike conductance state. The LL and S states are both thought to involve binding between the hairpin's terminal 5′ base and the pore's limiting aperture. The brief S state behavior is explained by a terminus-fraying event that is accompanied by extension by the terminal 3′ base into the limiting aperture. Part of the evidence for this is a strong spike (fraying) frequency correlation with the different terminus binding energies. Asymmetric base addition or phosphorylation (at the terminal 3′ and 5′ positions) is part of the evidence for the asymmetric roles for 5′ binding (LL and S) and 3′ fraying/extension (S).

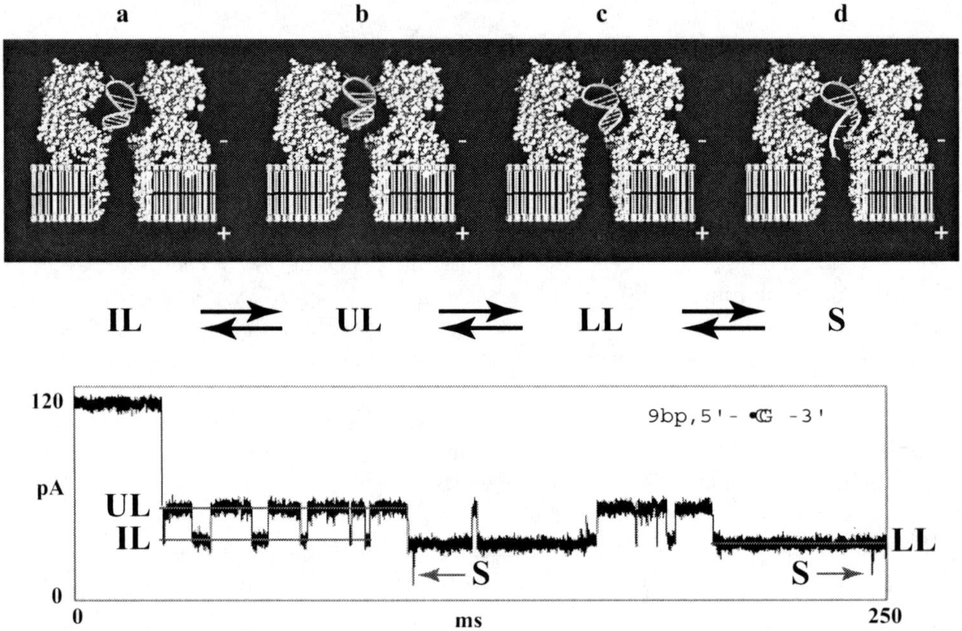

FIGURE 3. Molecular mechanisms underlying the observed current transitions. a) When a 9bp DNA hairpin initially enters the pore, the loop is perched in the vestibule mouth and the stem terminus binds to amino acid residues near the limiting aperture. This results in the IL conductance level. b) When the terminal base pair desorbs from the pore wall, the stem and loop may realign, resulting in a substantial current increase to UL. Interconversion between the IL and UL states may occur numerous times, or UL may convert to the LL state c). The LL state corresponds to binding of the stem terminus to amino acids near the limiting aperture

but in a different manner from IL. d) From the LL bound state, the duplex terminus may fray resulting in extension and capture of one strand in the pore constriction.

Force/Geometry Probing Using DNA Hairpins

In another forthcoming manuscript (DeGuzman et al., in preparation), a variety of DNA hairpins are used as probes of the α-hemolysin protein channel geometry. The same experiments also serve to reveal the forces at various points in the channel. Building on the work of (Vercoutere et al., 2001), a series of blunt-ended DNA hairpins are used to probe the depth of the vestibule (where the dsDNA stems can reach). The blockade signal exhibits a single blockade level for hairpins with stem lengths ranging from three base-pairs (3bps) to seven base-pairs (7bps). For the 8bp hairpin a telegraph signal appears, with the primary blockade level at the greater resistance. For 9bp hairpins, and those with longer stems, there appear to be three main levels (the 9bp case is discussed above). The geometric bottom of the vestibule is reached with a 9bp hairpin, ±1bp. The indeterminacy results from the unknown positions of the binding events thought to correlate with the lower conductance levels and is being studied further. Using the 9bp hairpin as a base, and taking into account the abovementioned 3′-fraying/extension hypothesis, single-stranded DNA overhangs of varying length were added to the base at the 3′ terminus. This permits critical probing through the trans-membrane part of the channel in a very controlled manner, by a single captured molecule event. Preliminary results indicate two significant trans-membrane constrictions, one at the limiting aperture, and one near the trans-opening. The resolving power of the limiting-aperture/trans-opening constrictions is of critical importance in DNA-sequencing and biosensor applications, and is undetermined as of yet.

Sequencing Possibilities

For sequencing, the single molecule basis of nanopore measurements may permit Sanger-type sequencing on DNA molecules separated by capillary electrophoresis. If ssDNA translocation for α-hemolysin can be slowed enough, by use of single-enzyme couplings or servo-electronics, then single-molecule DNA sequencing may prove possible as well. For single-molecule sequencing to be successful, however, the deconvolution problem must be solved for the collection of bases at the main current restrictions (where, presumably, the greatest physical imprint is made on the ionic current). Deconvolution of base content from a single blockade signal may be possible if dominant contributions to resistance span only 20 Å or so (amounting to about three nucleotides length of ssDNA). If the critical nanopore thickness can be made less than 20 Å it may prove advantageous to work with dsDNA. Such an approach would probably require symmetry breaking on the dsDNA strand via base-pairing a native ssDNA strand with a set of Watson-Crick substitute nucleotides. Although dsDNA takes about twice as many bases to cover the same distance, the information imparted in those bases is much richer than that for ssDNA (due to Watson-Crick type bond formation, etc.), and this may prove critical to obtaining a working sequencer, particularly given the much greater information that can be extracted by introducing excitation to the dsDNA bonds during observation.

Other Applications

One of the key strengths of nanopore detectors is that they analyze populations of single molecules. With signal processing and pattern recognition, this information enables a new type of cheminformatics based on channel current measurements. Single molecule observations are also of interest in biophysics; binding/conformational changes on captured dsDNA end regions, for example, might be tracked and understood using the nanopore blockade signal. DNA regions away from the ends may eventually be studied in a similar manner, using pore-translocation that resolves (and is dominated by) the distinctive conductance/binding imprint of those bases threading the limiting aperture constriction. For single nucleotide polymorphism (SNP) identification, small sample volumes can be used, such that PCR amplification may not be needed. Non-PCR expression analysis, in general, may offer a new method for biological experimentation on live cells using patch-clamp methods.

CONCLUSION

Five species of DNA hairpin were examined, four of which differed only in their terminal base-pairs. Classification of a single 100 ms hairpin event, with no rejection, was 77% accurate on average. Accuracy was boosted above 99% if longer event durations were used or if multiple short events were used with nonzero rejection. For purposes of rapid mixture analysis, the latter approach was adopted, with single species identification with 99.6% accuracy in six seconds, and two species mixture identification in 40 seconds with less than 1% error in the majority species percentage. The signal processing architecture that accomplished this used HMMs for feature extraction and SVMs for classification. The HMMs were implemented with Expectation/Maximization and the SVMs were implemented with novel Kernels. The on-line signal processing was designed to be scalable to hundreds of species, or more, while at the same time performing the classification in less time than the duration of the signal acquisition itself (100 ms). This was accomplished on an inexpensive PC. An unconstrained training process, as used here, has scalability complications due to rapid growth in multiclass combinatorics, but for five species was easily automated (on a network of five PCs). If scalability requirements are relaxed, allowing species-specific HMM processing for example, discrimination accuracy (or speed) can be boosted even further. The processing architecture is directly applicable to other channel current analysis situations by simply re-training the machine learning components.

ACKNOWLEDGEMENTS

I thank my collaborators from the larger exposition of this work (Winters-Hilt et al., 2003): W. Vercoutere, V. S. DeGuzman, D.W. Deamer, M. Akeson, and D. Haussler.

I benefited greatly from discussions of biophysics, biochemistry, and nanopore operation with W. Vercoutere, V. S. DeGuzman, D.W. Deamer, and M. Akeson. Discussions with D. Haussler provided insight and clarity to the development of the machine learning solution. A special thanks to Mark Akeson for a careful reading of early drafts. Thanks also to Sam Ridino and Joseph T. Rodgers for help with data acquisition. This work was funded by the National Human Genome Research Institute and by the Howard Hughes Medical Institute.

REFERENCES

Akeson M, D. Branton, J.J. Kasianowicz, E. Brandin, D.W. Deamer. 1999. Microsecond Time-Scale Discrimination Among Polycytidylic Acid, Polyadenylic Acid, and Polyuridylic Acid as Homopolymers or as Segments Within Single RNA Molecules. Biophys. J. 77(6):3227-3233.

Bayley, H. 2000. Pore planning: Functional membrane proteins by design. J. Gen. Physiol. 116. 1a.

Bredensteiner, E.J.; and K.P. Bennett. 1999. Multicategory classification by support vector machines. Comput. Optim. and Appl. 12. 53-79.

Burges, C.J.C. 1998. A tutorial on support vector machines for pattern recognition. Data Min. Knowl. Discov., 2. 121-67.

Chung, S.H., J.B. Moore, L. Xia, L. S. Premkumar, and P. W. Gage. 1990. Characterization of single channel currents using digital signal processing techniques based on Hidden Markov models. Phil. Trans. R. Soc. Lond. B 329. 265-285.

Chung, S-H., and P. W. Gage. 1998. Signal processing techniques for channel current analysis based on hidden Markov models. *In* Methods in Enzymology; Ion channels, Part B. P. M. Conn editior. Academic Press, Inc., San Diego. 420-437.

Colquhoun, D., and F. J. Sigworth. 1995. Fitting and statistical analysis of single-channel products. *In* Single-channel recording. B. Sakmann and E. Neher editors. Second edition. Plenum Publishing Corp., New York. 483-587.

Cormen, T.H., C. E. Leiserson, and R. L. Rivest. 1989. Introduction to Algorithms. MIT-Press, Cambridge, USA.

Cover, T. M. and J. A. Thomas. 1991. Elements of Information Theory. Wiley-Interscience, New York.

Diserbo M, P. Masson, P. Gourmelon, and R. Caterini, 2000. Utility of the wavelet transform to analyze the stationarity of single ionic channel recordings. J. Neurosci. Methods 99(1-2):137-141.

Durbin R. 1998. Biological sequence analysis : probalistic models of proteins and nucleic acids. Cambridge, UK New York: Cambridge University Press. xi, 356 p.

Gouaux J.E., O. Braha, M.R. Hobaugh, L. Song, S. Cheley, C. Shustak, and H. Bayley, 1994. Subunit stoichiometry of staphylococcal alpha-hemolysin in crystals and on membranes: a heptameric transmembrane pore. Proc. Natl. Acad. Sci. USA 91:12828-12831.

Jaakkola, T. S., and D. Haussler. 1998. Exploiting generative models in discriminative classifiers. *In* Advances in Neural Processing Systems 11. Cambridge, MA, 1999. MIT Press.

Joachims, T. 1998. Making large-scale SVM learning practical. *In* Advances in Kernel Methods -- Support Vector Learning. B. Scholkopf, C. J. C. Burges, and A. J. Smola editors. MIT Press, Cambridge, USA. Ch. 11.

Kasianowicz, J.J., E. Brandin, D. Branton, and D.W. Deamer, 1996. Characterization of Individual Polynucleotide Molecules Using a Membrane Channel. Proc. Natl. Acad. Sci. USA 93(24):13770-13773.

Li, J., C. McMullan, D. Stein, D. Branton, and J. Golovckenko. 2001. Solid state nanopores for single molecule detection. Biophys. J. 80. 339a.

Meller A, L. Nivon, E. Brandin, J. Golovchenko, and D. Branton, 2000. Rapid nanopore discrimination between single polynucleotide molecules. Proc. Natl. Acad. Sci. USA 97(3):1079-1084.

Meller A, L. Nivon, and D. Branton, 2001. Voltage-driven DNA translocations through a nanopore. Phys. Rev. Lett. 86(15):3435-3438.

Osuna, E.; R. Freund, and F. Girosi. 1997. An improved training algorithm for support vector machines. *In* Neural Networks for Signal Processing VII. J. Principe, L. Gile, N. Morgan, and E. Wilson editors. IEEE, New York. 276-85.

Platt, J. C. 1998. Fast Training of Support Vector Machines using Sequential Minimal Optimization. *In* Advances in Kernel Methods -- Support Vector Learning. B. Scholkopf, C. J. C. Burges, and A. J. Smola editors. MIT Press, Cambridge, USA. Ch. 12.

SantaLucia J. 1998. A unified view of polymer, dumbbell, and oligonucleotide DNA nearest-neighbor thermodynamics. Proc. Natl. Acad. Sci. USA 95(4):1460-1465.

Song L., M.R. Hobaugh, C. Shustak, S. Cheley, H. Bayley, and J.E. Gouaux, 1996. Structure of Staphylococcal Alpha-Hemolysin, a Heptameric Transmembrane Pore. Science 274 (5294):1859-1866.

Vapnik, V. N. 1999. The Nature of Statistical Learning Theory (2nd ed.). Springer-Verlag, New York.

Vercoutere W., S. Winters-Hilt, H. Olsen, D.W. Deamer, D. Haussler, and M. Akeson, 2001. Rapid discrimination among individual DNA hairpin molecules at single-nucleotide resolution using an ion channel. Nat. Biotechnol. 19(3):248-252

Winters-Hilt, S., W. Vercoutere, V. S. DeGuzman, D.W. Deamer, M. Akeson, and D. Haussler, 2003. Highly Accurate Classification of Watson-Crick Base-Pairs on Termini of Single DNA Molecules. Biophys. J. 84 (2) 1-10.

GENETICS

Stochastic Model of Evolution of Conserved Protein Coding Sequences

Vladimir A. Kuznetsov

The Laboratory of Integrative and Medical Biophysics, National Institute of Child Health and Human Development, NIH, 13 South Drive, Bethesda, MD, 20892, USA; vk28u@nih.gov

Abstract. We study statistical distributions of events appearing for a variety of evolutionary conserved sequences encoded in a variety of genomes. Skewed empirical distributions of the number of non-redundant domain-to-protein links in bacterial, archaeal and eukaryotic proteomes [1,2], and the number of links between conserved order orthologous groups of genes in genomes of prokaryotes [3] are analyzed. The Kolmogorov forward differential-difference equations for a linear birth-death evolution process are used to explain these distributions. These equations provide a limiting solution in form of the terms of hypergeometric Gauss series ${}_2F_1(a,1;b+1;\theta)$. We show that the limiting probability function associated with this series fits both empirical distributions well. For the case of domain-to-protein links distributions, the degenerate ${}_2F_1$-function at $\theta=1$ (the Waring distribution) holds true for eukaryotic proteomes; however, for many archaeal and bacterial proteomes the distribution degenerates to the Yule distribution. These results indicate the critical changes of the conserved protein-coding sequences organization in the course of progressive evolution from prokaryotic to eukaryotic organisms and predict the number of the domains in the eukaryotic proteomes.

INTRODUCTION

In this work, we study statistical properties of the number of links between certain subsets of evolutionary conserved protein-coding sequences contained in a variety of genomes. First, we analyze statistical distributions of the number of the non-redundant protein domain-to-protein links in proteomes. In this work, a proteome is considered as a complete set of "distinct" proteins encoded by protein-coding genes in the entire genome. Proteins may be considered as compositions of evolutionary conserved "building blocks", - protein domains [1,2,4,5]. A protein domain is an amino acid sequence that is essential to the biological function(s) of the protein in which it occurs. By working definition, a protein domain is often considered as a *mutually disjoint class of elements* (i.e. homologous amino acid sequences). As a class, the *distinct protein domain* can be represented by a single polypeptide sequence for a protein domain or by two or more homologous polypeptide patterns. Such equivalence classes constitute a partition of polypeptide sequences set, S. A union of distinct domains is S. The same distinct domain may be repeated in a given protein many times, as well as

CP665, *Unsolved Problems of Noise and Fluctuations: UPoN 2002: Third International Conference*, edited by S. M. Bezrukov
2003 American Institute of Physics 0-7354-0127-6

occurring in many different proteins of the proteome. Let m_i denote the number of proteins containing one or more instances of distinct domain i in the sample proteome of an organism. In this study, we will count the number of non-redundant occurrences of the domain i in all proteins of an organism. By non-redundant, we mean that a distinct domain is counted only once even if it occurs several times in the same protein. We also refer to the number m_i as the number of non-redundant domain-to-protein links of a distinct domain in the proteome of the organism. Currently, only some fraction of the distinct proteins, and some fraction of the protein domains existing in nature are known, and, correspondingly, the list of non-redundant domain-to-protein links seen in any proteome is incomplete. Therefore, for a given organism, we do not generally know the true numbers of distinct domains N_D, and the true number of proteins N_P in the proteome. Let N denote the number of distinct domains in the sample proteome. Note $N \le N_P$. Thus, $M = \sum_{i=1}^{N} m_i$ denote the sum of the numbers of domain-to-protein links in the sampled proteome of an organism. M can characterize the proteome complexity. We call M the *connectivity number* of the domain-to-protein network of a proteome.

Second, we will also analyze the distribution of the number of links in "conserved orthologous groups" of genes in genomes of prokaryotes [3]. Genes that are significant homologs across different species were clustered into the orthologous groups and then any two orthologous groups were merged into the "conserved" group if they co-occur in the same potential operon two or more times in two or more species that are not closely related.

In the both cases, the empirical frequency distributions of the number of non-redundant domain-to-protein links of a distinct domain in the proteome and the number of links per conserved orthologous group show a very similar distribution pattern, where only a few classes of conserved protein-coding sequences (distinct domains or orthologous groups, respectively) combine with many other classes and most of these classes have only one or few link patterns.

In this work we consider a model of the distribution of the number of protein domains containing within proteins of a proteome in terms of a Markov birth-death stochastic process assuming four independent mechanisms: duplication and combination of domains in different ways in proteins (birth) and loss (death) of already existing domains, *spontaneous* emergence (innovation) and *spontaneous* elimination (mutation, gene modification) of domains [1,6]. The same mathematical model we apply to analysis of the distribution of the number of links between conserved orthologous groups of genes in genomes of prokaryotes.

RESULTS AND DISCUSSION

A Hypergeometric Model of the Birth, Death, and Creation Evolution

Let X denote the number of non-redundant domain-to-protein links of a random domain within proteins in a given proteome. Then we can define the domain occurrence probability function (DOPF) $p_m = P(X = m)$ for occurrence values $m = 0,1,2,...$, as a probability that a random distinct domain occurs non-redundantly exactly *m* times within the proteome. A simple model of the DOPF can be described in terms of a Markov stochastic process [7]. Let us briefly describe the model. We will characterize a proteome of an organism by its N distinct domains $d_1, d_2, ..., d_N$. Let m_i denote the number of proteins containing d_i. Note $\sum_{i=1}^{N} m_i = M$. Given an organism, we may imagine the evolutionary path starting from its unknown ancestor species at time zero that leads to the evolution of this organism. Evolutionary progress along this path, in general, results in the appearance of new domains, and the increase/decrease in the number of uses of domains, and even the dropping of domains (i.e. due to lineage-specific gene loss). Many distinct species may occur on this path as precursors of the given organism.

Let the random variable $D_t(d,P)$ be the number of proteins containing the distinct domain d in the organism occurring at time t in the evolutionary path P of some end-point organism. $D_t(d,P)$ is a realization of a continuous-time stochastic process $\{D_t, t \geq 0\}$. We assume that time parameter t is a continuous parameter. This process can be considered as a birth-death random process that has a net change across an infinitesimal time interval which can never be other than -1, 0, or +1 and where protein domains are "born" and "die" during evolution for some evolutionary path. A birth indicates an increase in the number of proteins that contain the random fixed domain d, whereas death indicates a decrease in this number. We assume that the proteome size associated with any evolutionary path is finitely-bounded above (due to selection factors, self-organization etc.) and becomes nearly constant as evolution progresses along that path over long stretches of geological time. Let functions $\lambda_m(t)$ and $\mu_m(t)$ be the intensities of the birth and death processes for a given domain represented by m proteins in the proteome, respectively. $m = 1,2,...$. The intensities $\lambda_m(t)$ and $\mu_m(t)$ are related to the transition probability $P_{i,j}(t,s)$ which is the probability that $D_{t+s} = j$ given that $D_s = i$. For birth-death process, $P_{i,j}(t,s)$ depends only on i, j and time t. Thus we may write $P_{i,j}(t,s) = P_{i,j}(t)$ and, also, $P_{i,j}(t,s) = 0$ for $|i-j| > 1$. An additional fact is that the time, T_m, that process D_t stays in state m before making a transition into a different state is assumed to be exponentially-determined so

$P(T_m \le \alpha) = 1 - \exp(-\alpha / v_m)$, where $v_m = E(T_m)$. We can specify the meaning of $\lambda_m(t) = v_m P_{m,m+1}(t,s)$ and $\mu_m(t) = v_m P_{m,m-1}(t,s)$.

Let $p_m(t) = P(D_t = m)$ denote the probability function associated with the random process $\{D_t, t \ge 0\}$. Then the rate of the probability functions p_m ($m = 0,1,2,...$) can be described by the forward Kolmogorov differential-difference equations [7]:

$$dp_0(t)/dt = -\lambda_0(t)p_0(t) + \mu_1 p_1(t); \tag{1}$$

$$dp_m(t)/dt = -(\lambda_m(t) + \mu_m(t))p_m(t) + \lambda_{m-1}(t)p_{m-1}(t) + \mu_{m+1}(t)p_{m+1}(t), \tag{2}$$

where $m = 1,2,...$.The initial probabilities $p_m(0) \ge 0; (m = 0,1,2,...)$ follow to the condition

$$\sum_{m \ge 0} p_m = 1.$$

Probably in the most evolving near end-point organisms, the random birth and death processes of domains are keeping near equilibrium. This equilibrium solution can be written explicitly by stating $dp_m / dt = 0; m = 0,1,...$ in Eqs. (1)-(2) as

$$\hat{p}_m = p_0 \prod_{i=1}^{m} \frac{\lambda_{i-1}}{\mu_i}, \tag{3}$$

$$p_0 = (1 + \sum_{m=1}^{\infty} \prod_{i=1}^{m} \frac{\lambda_{i-1}}{\mu_i})^{-1} \tag{4}$$

At $m \to \infty$, a necessary and sufficient condition for the existence of the non-trivial stationary solution (3)-(4) is provided by convergence of the series

$$Q = \sum_{m=1}^{\infty} \prod_{i=1}^{m} \eta_i, \tag{5}$$

where $\eta_i = \frac{\lambda_{i-1}}{\mu_i} \le v < 1$. This condition exists when starting from some $i = i_c$ the condition $\eta_i \le v < 1$ takes place for all $i \ge i_c$ (i.e. on the right tail of the DOPF).

In the case of the DOPF, this inequality implies that, starting from some large enough size i_c of multidomain proteins, the intensity of the birth process (leading to the appearance of a more complex protein in the proteome) must be less or equal to the intensity of the death process (leading to disappearing of more complex protein in the proteome) for all consequential multidomain proteins containing i distinct domains for which $i \ge i_c$.

Using Eqs. (3)-(4), we can obtain the non-zero limiting probability function for the random process $D_{t \to \infty}$:

$$P(X = m) = p^*_m = \lim_{t \to \infty} p_m(t); \tag{6}$$

If we assume that the limiting probability distribution p^*_m exists, then all dp_m / dt ($m = 0,1,2,...$) would necessarily converge to 0 as $t \to \infty$ and we can obtain:

$$p^*_m = \hat{p}_m = p_0 \prod_{s=1}^{m} \eta_s. \tag{7}$$

In general, the birth and death intensity rates can be approximated by the polynomial functions of integer argument s. Then Eq. (3) (or (7)) can be considered as the product of the rational functions $\eta_s = \frac{\lambda_{s-1}}{\mu_s} = \theta \frac{(s^i + a_{i-1}s^{i-1} + ... + a_1 s + a_0)}{(s^j + b_j s^{j-1} + ... + b_1 s + b_0)}$ of argument s and constants θ, a_i, b_i (i=0,1...), multiplied by the positive constant p_0. Using a major theorem of algebra, we can present the probability (3) (or (7)) as the *m*-th term of the generalized hypergeometric series ${}_pF_q$ [8] (Kuznetsov, Fluctuation and Noise Letters, in press). For our application purposes, we will consider a Markov birth-death random process such that the intensity rates are the linear functions of s [1,6]:

$$\lambda_s = \lambda_1^* + \lambda_2^* s \tag{8}$$

and

$$\mu_s = \mu_1^* + \mu_2^* s \tag{9}$$

($s = 1,2,...$), where constants $\lambda_1^* > 0, \lambda_2^* > 0, \mu_1^* > 0, \mu_2^* > 0$. Hence, during an interval $(t, t+h)$ where h is small, we assume four independent processes: the spontaneous "birth" and "death" of a domain, with constant intensities λ_1^* and μ_1^*, respectively, and the ``flows" of the domains with the intensities proportional to the number of proteins already containing that domain $\lambda_2^* s$ and $\mu_2^* s$. Note that, probabilistically, the intensities λ_1^* and μ_1^* are the intensities of Poisson processes. During an interval $(t, t+h)$, where h is small, the intensity λ_1^* is proportional to a transitional (birth) probability of sporadic increase of the number of protein-coding genes, which implies gene duplication events, mutations and horizontal gene transfer to an organism from another species. During the same interval $(t, t+h)$, the intensity μ_1^* is proportional to a transitional (death) probability of spontaneous deleting and/or modification of a domain.

For steady state distribution we could estimate three independent parameters of the steady-state random process. Let denote $a = \lambda_1^* / \lambda_2^*$, $\theta = \lambda_2^* / \mu_2^*, b = \mu_1^* / \mu_2^*$. Let also denote factorial power $z^{[m]} = z(z+1)...(z+m-1)$, where $m=0,1,2,...$; $z^{[0]} = 1$. Using Eqs. (3), (4), (7) and Eqs. (8)-(9), we can obtain the unique limiting probability function for the process (1)-(2) with the intensities given by Eqs. (8) and (9):

$$p^*_m = b p_0 \frac{a^{[m]}}{b^{[m+1]}} \theta^m = p_0 \frac{\Gamma(a+m)\Gamma(b+1)}{\Gamma(a)\Gamma(b+m+1)} \theta^m = p_0 \frac{B(b+1,m)}{B(a,m)} \theta^m, \tag{10}$$

$$p_0 = (1 + \sum_{m=1}^{\infty} (\prod_{i=1}^{m} \frac{(a-1+i)}{(b+i)} \theta))^{-1}, \tag{11}$$

$m=0,1,2,...$. $\Gamma(x)$ is the gamma function and the $B(x,y)$ is beta function [8]. In the specific case $b > a > 0$ and $p_0 = (1 - \frac{a}{b})$ the distribution (10)-(11) will be the well known Waring distribution [9]. We call the Eqs.(10)-(11) the Kolmogorov-Waring probability function.

The sum of probabilities (10) can be written in a form of hypergeometric series:

$$1 = p_0 (1 + \frac{a}{b+1}\theta + \frac{a(a+1)}{(b+1)(b+2)}\theta^2 + ... + \frac{a(a+1)...(a+m-1)}{(b+1)(b+2)...(b+m)}\theta^m + ...) \tag{12}$$

or

$$1 = p_0 \cdot {}_2F_1(a,1;b+1;\theta) \tag{13}$$

where ${}_2F_1(a,1;b+1;\theta)$ is the hypergeometric Gauss series [8]

$${}_2F_1(\alpha,\beta;\gamma;\theta) = \sum_{m=0}^{\infty} \frac{\alpha^{[m]}\beta^{[m]}}{\gamma^{[m]}} \frac{\theta^m}{m!} \tag{14}$$

at $\alpha = a, \beta = 1, \gamma = b+1$. The probability function (10) has probability generating function $g_{KW} = {}_2F_1(a,1;b+1;\theta z) / {}_2F_1(a,1;b+1;\theta)$.

Using Eqs. (13) and (14), we can obtain the following results.

1. If ${}_2F_1(a,1;b+1;\theta) < \infty$, then $p_0 = \dfrac{1}{{}_2F_1(a,1;b+1;\theta)} > 0$; (15)

2. If $\theta < 1; a > 0; b > 0$, then $p_0 = \dfrac{1}{b\int_0^1 (1-s)^{b-1}(1-s\theta)^{-a}ds}$; (16)

3. If $b > a > 0$ and $\theta \to 1-0$, then $\lim_{\theta\to 1-0} p_0 = (1-\frac{a}{b})$. (17)

Using Eqs. (10), (11), (15) and integral presentation of the Beta and hypergeometic Gauss functions, we can prove

Theorem 1. If $b+1>a>0$ and $\theta \le 1$, then the limiting probability function is

$$p^*_m = p_0 \frac{B(a+m, b+1-a)}{B(a, b+1-a)} \theta^m = \frac{\int_0^1 s^{b-a}(1-s)^{a+m-1}ds}{{}_2F_1(a,1;b+1;\theta) B(a, b-a+1)} \theta^m \tag{18}$$

and

$$P(X \ge m) = \sum_{s=m}^{\infty} p^*_s = p_m^* \cdot {}_2F_1(a+m,1;b+m+1;\theta) = p^*_m / p_{0,m}, \tag{19}$$

where $p_{0,m} = [(b+m+1)\int_0^1 (1-s)^{b+m-1}(1-\theta s)^{-(a+m)}ds]^{-1}$ and $m=0,1,2,\ldots$.

Existence of the Non-zero Limiting Distribution

In Eqs. (10)-(11), the non-zero limiting probability p_m^* ($m = 1,2,\ldots$) exists when $p_0 > 0$ is at $t \to \infty$. Therefore, it is important to define the conditions which provide a convergence of the series

$$Q = \sum_{m=1}^{\infty} (\prod_{i=1}^{m} \Psi(i)\theta), \tag{20}$$

where $\Psi(i) = (a-1+i)/(b+i)$. Let us define the convergence conditions of this series.

Corollary. The series (14) is converged at $\theta < 1$ or at ($\theta = 1$; $b > a$).

Proof.

Let $\lambda_1^* > 0, \lambda_2^* > 0, \mu_1^* > 0, \mu_2^* > 0$. The formula (20) can be presented as follows:

$$Q = \sum_{m=1}^{\infty} \exp\{\log[\prod_{i=1}^{m} \Psi(i)\theta]\} = \sum_{m=1}^{\infty} \theta^m \exp\{\sum_{i=1}^{m} \log \Psi(i)\}.$$

Let $\lim_{i\to\infty} [\frac{|O(i^{-k})|}{i^{-k}}] = const > 0$. Using MacLaurin decomposition, we obtain $\Psi(i) = 1 + (a-b-1)/i + O(i^{-2})$.

Using $\sum_{i=k}^{m} \frac{1}{i} = \log(m/k) + O(1/k)$ and $\sum_{i=k}^{m} O(\frac{1}{i^2}) < const \sum_{i=k}^{\infty} \frac{1}{i^2} = O(1/k)$ we obtain

$$Q_k := \sum_{m=k}^{\infty} \theta^m (m/k)^{(a-b-1)} \exp(O(1/k)) \tag{21}$$

According to the comparison test of convergence, this series converges at $\theta < 1$ and diverges at $\theta > 1$. In the critical point $\theta = 1$ (or $\lambda_2^* = \mu_2^*$) the power series $\sum_{m=k}^{\infty} (m/k)^{(a-b-1)}$ converges, when $b > a$ (i.e. $\lambda^*_1 < \mu_1^*$) and diverges, when $b \le a$ (i.e. at $\lambda^*_1 \ge \mu_1^*$).

Theorem 2. For the intensity functions defined by Eqs. (8)-(9), the non-zero limiting solution of the Eqs. (1)-(2) exists if

$$\lambda_2^* < \mu_2^* \tag{22}$$

or

$$(\lambda_2^* = \mu_2^* \text{ and } \lambda_1^* < \mu_1^*) \tag{23}$$

Thus, conditions (22) and (23) provide the existence of the non-zero steady state probability function (10)-(11).

Dynamics of Mean Value and Dispersion

Let denote $m_X(t) = E[X(t)] = \sum_{i\ge 0} i p_i(t)$ and $D_X(t) = \sum_{i\ge 0} (i - m_X(t))^2 p_i(t)$). Then, for (1)-(2) we can derive differential equations for mean value $m_X(t)$ and dispersion $D_X(t)$ and the higher moments of random variable $X(t)$. In particular, at (8)-(9) we can obtain

$$\frac{dm_X(t)}{dt} = (\lambda_1^* - \mu_1^*) - (\mu_2^* - \lambda_2^*) m_X(t) + \mu_1^* p_0(t), \tag{24}$$

$$\frac{dD_X(t)}{dt} = (\lambda_1^* + \mu_1^*) + (\mu_2^* + \lambda_2^*) m_X(t) - \mu_1^* p_0(t)(1 + 2m_X(t)) - 2(\mu_2^* - \lambda_2^*) D_X(t) \tag{25}$$

and prove

Theorem 3. If $\lambda_2^* < \mu_2^*$ (or $\theta < 1$) and $\lambda_1^* > \mu_1^*(1 - p_0)$, where p_0 is defined by Eq. (16), then for any initial value $m_X(0) \ge 0$ the random process exponentially approaches a steady state at characteristic time $1/(\mu_2^* - \lambda_2^*)$ giving

$$\lim_{t\to\infty} m_X(t) = (\lambda^*_1 - \mu^*_1(1-p_0))/(\mu^*_2 - \lambda_2^*) \text{ and } \lim_{t\to\infty} D_X(t) < \infty; \quad (26)$$

if $\lambda_2^* = \mu_2^*$ (or $\theta = 1$) and $\lambda_1^* < \mu_1^*$, then $\lim_{t\to\infty} m_X(t) = \lambda^*_1/(\mu^*_1 - \lambda^*_1 - \lambda^*_2)$,

$$E[p_o(t)] = \frac{1}{t}\int_0^t p_0(s)ds = (1 - \lambda_1^*/\mu_1^*) \text{ and if } \mu_1^* > \lambda_1^* + 2\lambda_2^* \text{ then } \lim_{t\to\infty} D_X(t) < \infty. \quad (27)$$

The conditions $\lambda_2^* = \mu_2^*, \lambda_1^* < \mu_1^*$ for Eqs. (26)-(27) imply the Waring distribution. We can see that the Waring distribution corresponding to the critical steady state of a linear Markov birth and death stochastic process, when 1) for a given class (i.e., a given distinct domain) the intensity of transition from *m*-th (counting by the number of occurrences, *m*, of components of such classes in the system) in *m+1*-th state equals the intensity of transition from *m*-th to *m-1*-th state and 2) the rates of the sporadic loss of components of that class are greater than the rate of sporadic external input of the component from the external environment ($\mu_1^* > \lambda_1^*$).

Family of Skewed Frequency Distributions

We can present Eq. (10) with the following recursive formula

$$\eta_m = p^*_{m+1}/p^*_m = \theta\frac{(a+m)}{b+m+1}, \quad (28)$$

where $m = 0,1,...$. Using Eqs. (10), (11), (28) and properties of gamma function, beta function and hypergemetric series [8], we can obtain several important degenerate forms of the KW distribution:

Theorem 4. If $\theta \to 1, b > 0$, then we obtain the zero-truncated Waring distribution [9] and as $m \to \infty$: $p^*_m \sim (b-a)\frac{\Gamma(b)}{\Gamma(a)}\frac{1}{m^{b+1}}\theta^m$, (29)

i.e. the probability distribution (10) approximates the Champernowne probability function (at $\theta \to 1$) [10] and the discrete Pareto probability distribution (at $\theta = 1$) [12] in the tail.

If p_0>0, we can define the zero-truncated limiting probability distribution $P(X = m \mid X = 0) = p_m^{\{0\}^*}$ as

$$p^*_m{}^{\{0\}} = p^*_m/(1-p_0) = \frac{B(b-a+1, a+m)}{B(b-a, a+1)}\theta^m, \quad (30)$$

where *m=1,2,...* and then to prove

Theorem 5.

1. If in (10) $a \to 0+; \theta \to 1; b > 0$, then

$$\lim_{a\to 0+, \theta\to 1} p^{\{0\}*}_m = \lim_{a\to 0+} \frac{(b-a)\Gamma(a+m)\Gamma(b+1)}{\Gamma(a+1)\Gamma(b+m+1)} \lim_{\theta\to 1}\theta^m = bB(b+1, m), \quad (31)$$

where $B(b+1, m)$ is the Beta function [8].

2. If in Eq. (10) $a \to 0; b \to 0$ and $\theta < 1$, then it approaches the Fisher logarithmic series distribution $f_m = f_1\theta^{m-1}/m$, where $f_1 = -\theta/\log_e(1-\theta)$, *m=1,2,...*.

3. If in (10) $a \to \infty; b \to \infty$ but $a\theta / b \to const = q < 1$, then Eq. (10) approximates the geometrical distribution $f_m = q(1-q)^m$ where $m = 0, 1, 2, \ldots$

For case 1 of Theorem 5, the probability function (31) depends on one positive parameter, b $(b = \mu_1^* / \mu_2^*)$, and this function approximates the Yule distribution[1] [10,11] at $\lambda_1^* / \lambda_2^* \to 0; \lambda_2^* / \mu_2^* \to 1$ and, thus, $\lim_{a \to 0+, \theta \to 1} p_0 = 1$. Simon has constructed a birth random process model of words distribution in the text [10]; the limiting

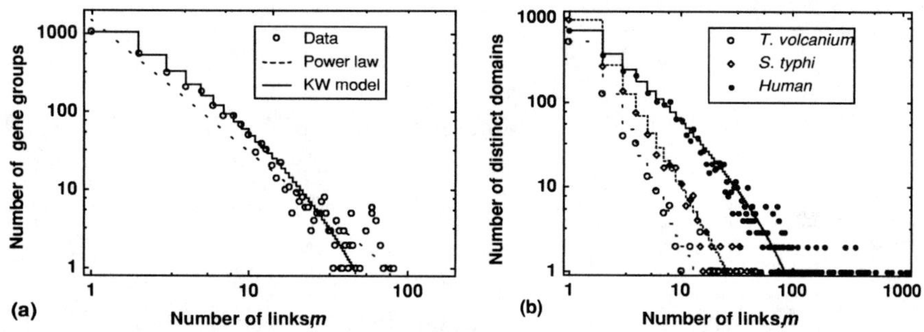

FIGURE 1. Fitting of the Kolmogorov-Waring (KW) model to the empirical frequency distribution of the number of links per conserved orthologous gene group in the 38 prokaryotic genomes [3] and the number of non-redundant domain-to-protein links in three sample proteomes **(a)** ∘: the number of conserved orthologous gene groups; m=1,2,...,110. *Step-function*: best-fit KW model at a=0.879; b=1.476; $\theta = 0.955$ for data ∘; $n(m) = Np^{*T}{}_m$, where N is the total number of conserved orthologous gene groups (N=3033) [3] and $p^{*T}{}_m$ is calculated by the Eq.(32). Dashed line: best-fit power law model ($P(X = m) := f_m = cm^{-(k+1)}$, where $c = \sum_{i=1}^{J} f_i$, $i=1,2,\ldots 110$) at k=0.687. **(b)** ∘,◊,•: the number of distinct domains, $n(m)$, in the *T. volcanium*, *S. typhi* and human sample proteome, respectively; m=1,2,... *Step-function*: best-fit KW model at a=0.017; b=2.412; θ =1.046 for data ∘; best-fit KW model at a=063; b=1.198; θ =1.00 for data ◊; best-fit function $n = n(m)$ for the KW model at a=0.433; b=0.653 and θ =0.98 for data •; $n(m) = Np^{*T}{}_m$, where N is the total number of distinct domains in the sample proteome and $p^{*T}{}_m$ is calculated by the Eq.(32) at the best-fit parameters. Estimates of the parameters obtained using at maximization of the Model Selection Criterion, MSC [12,13]. MSC ranges between excellent (11-8], very good (8-6], satisfactory (6-4], and poor (4-1). The MSC-values are closely correlated to the average absolute error (AAE) and F-test values. MSC for the KW and the power law model is 7.3 and 1.8, respectively; AAE for the KW and the power law model is $7 \cdot 10^{-4}$ and $44 \cdot 10^{-4}$, respectively.

[1] Shubert and Glänzel [11] have reported the same relation between the zero-truncated Waring and the Yule distributions. Irwing [9] has noted that the Yule distribution is specific case of the Waring distribution, however, he made an error claiming that it is true at $a = 1$.

distribution of this model approaches the Yule distribution (31) at $b>1$ and in this case the Kolmogorov-Waring distribution characterizes the generalized discrete Pareto distribution [1,11,13]. Cases 2 and 3 can be simply tested using Eq. (28).

Truncated Distribution

In order to apply the probability function (10) to the data, let us assume that the random variable X is doubly truncated, i.e., random variable X is restricted to the range $1,2,\ldots, J; J<\infty$. Using Eq. (19), the probability function of the resulting truncated distribution is written

$$p_m^{*T} = p^*_m / (\sum_{s=1}^{s=J} p^*_s) = \frac{p^*_m}{1-p_0-p^*_{J+1}/p_{0,J+1}} \tag{32}$$

This probability function corresponds to a typical situation in which the occurrence values 0 and $J+1, J+2,\ldots,\infty$ are not observed. At $\lambda_2^*/\mu_2^* \to 1\text{-}0$, Eq. (32) is transformed in the expression $p_m^{*T} = \dfrac{p^*_m}{a/b-(b+J+1)p^*_{J+1}/(b-a)}$, which at $J\to\infty$ could be simplified to $p_m^{*T} = bp^*_m/a$.

Fitting of the Kolmogorov-Waring Distribution

Recently, based on analysis of some biochemical and protein databases, a claim has been made that statistical distributions in many interconnected biomolecular systems exhibit power-law behavior [4,5,15]. Unfortunately, accurate theoretical analysis of this model and goodness of fit analysis of alternative models [1,2] was omitted. Snel et al. 2002 also have claimed that the distribution in Figure 1(a) is power law distribution [3]. The dotted line on Figure 1(a) shows our best-fit line by the simple power law distribution and the step-function line by the KW distribution. The dotted line on Figure 1(a) shows our best-fit line by the simple power law distribution and the step-function line by the KW distribution. Analysis of the frequency distributions in many large-scale complex systems display strong systematic deviations data from the power law and alternative distribution models have been developed [see for example, 2,9,10,11,12,13,14]. Our goodness of fit analysis and theoretical modeling of distributions of the number of non-redundant domain-to-protein links in proteomes of 12 Archaea, 50 Bacteria and 8 Eukaryotic proteomes (InterPro, Mach 12, 2002; www.ebi.ac.uk/interpro/) showed poor-fit of the power law distribution to the data and, moreover, the empirical skewed distributions exhibit dependence of the slope of the probability distribution on the sample size [1,2]. Here, using Eq. (32), we fitted the doubly truncated KW distribution to the latest InterPro data sets (June 6, 2002) for 13 fully-sequenced genome organisms (Table 1) and to the empirical distribution of the number of links in "conserved orthologous groups" of genes in genomes of prokaryotes [3] (Figure 1a). Figures 1(a) and 1(b) (and goodness of fit analysis by our criteria [1,12]) clearly show that all of the distributions are well fitted to the data. That includes the right size region and the left long tail (not power law regions). It seems there are no theoretical and goodness of fit statistical analysis reasons why the simple power law model should be the case.[2]

Table 1. Parameters of the domain occurrence probability distribution for 13 Archaeal, Bacterial and Eukaryotic organisms. G: number of genes/ORFs; M: total number of non-redundant domain-to-protein links; N: number of distinct domains; J: maximum number of non-redundant domain links per protein in a same proteome; p_1 : fraction of proteins containing one distinct domain; p_0, a, b, θ : parameters of the KW model.

Organism	G	M	N	J	p_1	p_0	b	a	θ
Archaea									
Thermoplasma volcanium	1523	1460	778	35	0.68	0.998	2.412	0.017	1.046
Pyrococcus abyssi	1765	1859	900	44	0.67	0.999	2.431	0.003	1.053
Halobacterium sp. NRC-1	2425	2048	910	39	0.62	0.93	2.263	0.138	1.033
Archaeoglobus fulgidus	2400	2286	1009	53	0.64	0.917	2.581	0.183	1.04
Bacterea									
Ureaplasma parvum	611	622	432	17	0.81	0.999	6.598	0.003	1.167
Aquifex aeolicus	1551	1818	1068	27	0.71	0.999	2.653	0.003	1.033
Deinococcus radiodurans	3084	2909	1224	61	0.63	0.999	2.20	0.001	1.075
Rhizobium meliloti	6149	5275	1357	212	0.56	0.984	1.503	0.023	1.02
Rhizobium loti	7259	6780	1461	269	0.52	0.945	1.198	0.063	1.00
Eukaryota									
Schizosaccharomyces pombe	5047	4911	1484	120	0.49	0.600	3.601	1.232	1.027
Caenorhabditis elegans	18859	17967	1838	620	0.36	0.652	1.213	0.423	1.00
Mus musculus	22444	21121	2543	520	0.32	0.503	1.988	0.993	1.00
Homo sapiens	42743	41573	2758	1146	0.26	0.504	0.653	0.433	0.98

Table 1 shows that the connectivity number, M, the number of distinct domains, N, a sample mean $\bar{m}_X$ ($\bar{m}_X = M/N$), a fraction of proteins containing only single distinct domain and parameters of the KW distribution strongly correlate to the number of protein-coding genes in genomes, G. These correlations allow us to make a number of predictions. For example, for pair G and M we obtained r =0.997 and, using simple regression $G \approx M$, we have estimated 36,600-42,700 protein-coding genes for humans [1,2,6]. Table 1 also shows: 1) that the steady state distributions for all organisms relate to degeneration condition $\theta \cong 1$ (or $\lambda_2^* \cong \mu_2^*$), 2) estimated parameters a ($a = \lambda_1^* / \lambda_2^*$) and p_0 are dramatically

[2] After this paper has been submitted, Koonin et al. published the review article [16], where they refer to their own in the press manuscripts which confirmed our results in [1,2] showing that the generalized Pareto function as well as the steady-state frequency distribution from birth-death and creation process like in [1,6], better fit the empirical DOPFs than the simple power law model.

different for eukaryotic and for prokaryotic organisms; with respect to our classification (Theorems 4 and 5), the distributions of non-redundant domain-to-protein links in eukaryotic proteomes are associated with the two-parameters Waring distribution, but in prokaryotic organisms they follow to the (degenerated) single-parameter Yule distribution (see Eq. (31)). The limiting probability distribution of the random process exhibits a slightly unstable behavior at $a \cong 0; \theta \geq 1$ (approximated by the Yule distribution) or a quasi-steady state behavior at $b > a > 0; \theta \cong 1$ (approximated by the Waring distribution). The estimate $\theta > 1$ implies that a proteome might be in progress of increasing in size (see for example, *Ureaplasma parvum* or *Deinococcus radiodurans* data in Table 1).

Using formula $N_P = N/(1 - p_0)$, where N_P is the total number of distinct domains in a proteome, we can estimate N_P. Hence, for highly eukaryotic organisms *C. elegans, Mus musculus, Homo sapiens* we obtained 5281, 5117, 5560 protein domains, respectively. The numbers are consistent with our estimates obtained by another approach [1].

ACKNOWLEDGMENTS

I thank Oleg Senko for his assistance in goodness of fit calculations and Ted Crump for his corrections of the text. I also thank three anonymous referees for providing helpful comments and recommendations.

REFERENCES

1. Kuznetsov, V.A, Pickalov, V.V., Senko, O.V., Knott, G.D. *J. Biol. Systems*, **10** 381-408 (2002).
2. Kuznetsov, V.A., "Statistics of the numbers of transcripts and protein sequences encoded in the genome", in: *Computational and Statistical Methods to Genomics*, edited by W. Zhang, and I. Shmulevich, Boston: Kluwer, 2002, pp. 125-171.
3. Snel, B., Bork, P., and Huynen, M.A. *Proc. of Natl. Acad. Sci. USA* , **99** 5890-5895 (2002).
4. Apic, G., Gough, J., Teichmann, S.A. *J. Mol. Biol.,* **310** 311-325 (2001).
5. Rzhetsky, A., Gomez, S.M. *Bioinformatics,* **17** 988-996 (2001).
6. Kuznetsov, V.A., Pickalov, V.V. "The number of protein domain sequences and protein coding genes in the evolved proteomes". In: *Proceedings of the third international conference on bioinformatics of genome regulation and structure: July 14-20, 2002*, Novosibirsk: IC&G, Russian Acad. of Sci., Russia, 2002, v. **3**. pp.160-163.
7. Feller, W. *An Introduction to Probability Theory and Its Applications* (3-d Edition), v.1, New York:Willey, 1968, pp. 454-457.
8. Lebedev, N.N., *Special Functions and Their Applications*, Englewood Cliffs, New Jersey: Prentice-Hall, 1965.
9. Irwin, J.O. *J. Royal Stat. Soc. A.,* **126** 1-45 (1963).
10. Simon, H.A. *Biometrika,* **42** 425-440 (1955).
11. Shubert, A., Glänzel, W. *Scientometrics,* **6** 149-167 (1984).
12. Kuznetsov, V.A.. *EURASIP J. on Applied Signal Processing,* **4** 285-296 (2001).
13. Kuznetsov, V.A., Knott, G.D., Bonner, R.F. *Genetics,* **161** 1321-1332 (2002).
14. Reed, W.J., McKelvey, K.S. *Ecological Modeling,* **150** 239-254 (2002)
15. Jeong, H., Tombor, B., Albert, R., Ottval, Z.N., Barabasi, A.-L. *Nature,* **407** 651-654 (2000).
16. Koonin, E., V., Wolf, Y.I., Karev, G.P. *Nature,* **420** 218-223 (2002).

NOISE AND NOISE ANALYSIS FOR MEDICAL APPLICATIONS

Quantifying Heartbeat Dynamics by Magnitude and Sign Correlations

Plamen Ch. Ivanov*, Yosef Ashkenazy†, Jan W. Kantelhardt† and H.Eugene Stanley†

*Center for Polymer Studies and Department of Physics, Boston University, Boston, MA 02215 and Beth Israel Deaconess Medical Center, Harvard Medical School, Boston, Massachusetts 02215, USA

†Center for Polymer Studies and Department of Physics, Boston University, Boston, MA 02215

Abstract. We review a recently developed approach for analyzing time series with long-range correlations by decomposing the signal increment series into magnitude and sign series and analyzing their scaling properties. We show that time series with identical long-range correlations can exhibit different time organization for the magnitude and sign. We apply our approach to series of time intervals between consecutive heartbeats. Using the detrended fluctuation analysis method we find that the magnitude series is long-range correlated, while the sign series is anticorrelated and that both magnitude and sign series may have clinical applications. Further, we study the heartbeat magnitude and sign series during different sleep stages — light sleep, deep sleep, and REM sleep. For the heartbeat sign time series we find short-range anticorrelations, which are strong during deep sleep, weaker during light sleep and even weaker during REM sleep. In contrast, for the heartbeat magnitude time series we find long-range positive correlations, which are strong during REM sleep and weaker during light sleep. Thus, the sign and the magnitude series provide information which is also useful for distinguishing between different sleep stages.

A broad class of physical and biological systems exhibits complex dynamics, associated with the presence of many components interacting over a wide range of time or space scales. These often-competing interactions may generate an output signal with fluctuations that appear "noisy" and "erratic" but reveal scale-invariant structure. One general approach to study these systems is to analyze the ways that such fluctuations obey scaling laws [1, 2, 3].

We consider the time series formed by consecutive cardiac interbeat intervals (Fig. 1a) and focus on the correlations in the time *increments* between consecutive beats. This time series is of general interest, in part because it is the output of a complex integrated control system, including competing stimuli from the neuroautonomic nervous system [4]. These stimuli modulate the rhythmicity of the heart's intrinsic pacemaker, leading to complex fluctuations. Previous reports indicate that these fluctuations exhibit scale-invariant properties [5, 6, 7], and are anticorrelated over a broad range of time scales (i.e., the power spectrum follows a power-law where the amplitudes of the higher frequencies are dominant) [8]. By long-range anticorrelations we also mean that the root mean square fluctuations function of the integrated series is proportional to n^{α} where n is the window scale and the scaling exponent α is smaller than 0.5. In contrast, for uncorrelated behavior $\alpha = 0.5$, while for correlated behavior $\alpha > 0.5$.

The time series of the fluctuations in heartbeat intervals can be "decomposed" into

CP665, *Unsolved Problems of Noise and Fluctuations: UPoN 2002: Third International Conference*, edited by S. M. Bezrukov

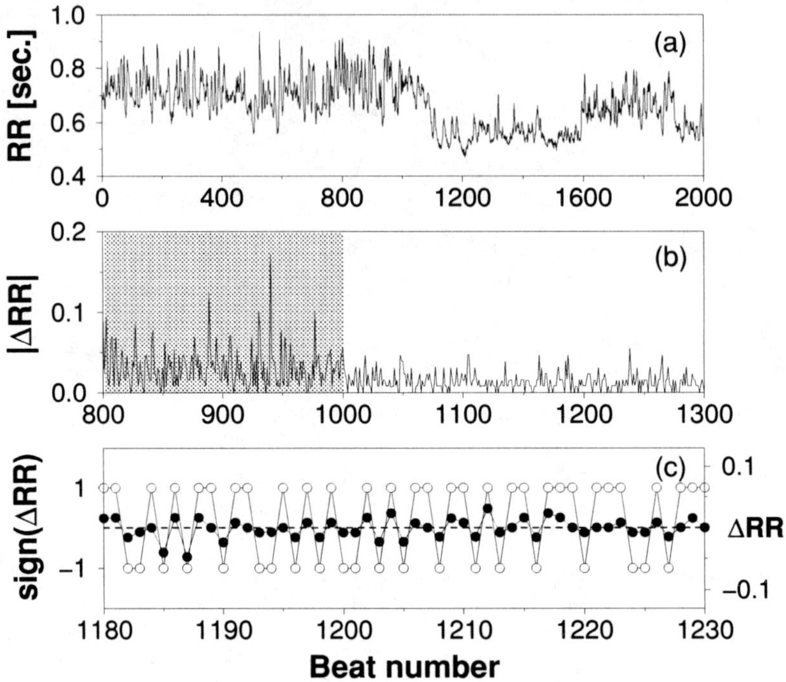

FIGURE 1. (a) An example of 2000 time intervals between successive R peaks (RR intervals) taken from electrocardiogram (ECG) recording of a healthy subject during daytime. (b) The magnitude series of a portion of the RR series (beat numbers 800-1300) shown in (a). Patches of more "volatile" increments with large magnitude (beat numbers 800-1000) are followed by patches of less volatile increments with small magnitude (beat numbers 1000-1300), consistent with our quantitative conclusion that there is correlation in the magnitude time series. (c) The sign series (○), as well as the ΔRR series (•) of a portion of the RR series (beat numbers 1180-1230) shown in (a). The positive sign $(+1)$ represents a positive increment, while the negative sign (-1) represents a negative increment in the RR series of interbeat intervals. The tendency to alternation between $+1$ and -1 is consistent with our quantitative conclusion that there is (multiscale) anticorrelation in the sign time series. Adapted from ref. [14].

two different time series. We analyze separately the time series formed by the magnitude and the sign of the increments in the time intervals between successive heartbeats (Fig. 1b,c). We use 2nd order detrended fluctuation analysis [9, 10, 11, 12] and not the conventional power spectrum, since it has the ability to accurately estimate correlations in the heartbeat fluctuations even when they are masked by linear trends. The 1st order detrended fluctuation analysis (DFA) eliminates constant trends from the original series (or, equivalently, linear trends from the integrated series); the 2nd order DFA removes linear trends, and the kth order DFA eliminates polynomial trends of order $k-1$.

The magnitude/sign decomposition consists of the following steps: (i) given a series of successive interbeat intervals RR_i we create the increment series, $\Delta RR_i = RR_{i+1} - RR_i$; (ii) we decompose the increment series into a magnitude series ($|\Delta RR|$) and a sign series ($\text{sign}(\Delta RR)$); (iii) to avoid artificial trends we subtract from the magnitude and sign series their average; (iv) because of limitations in the accuracy of the detrended

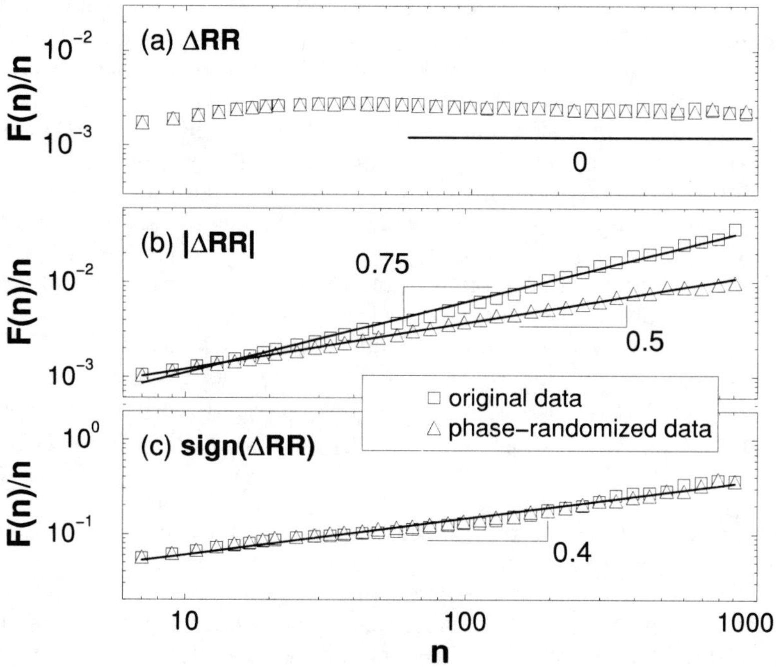

FIGURE 2. (a) Root mean square fluctuation, $F(n)$, for $\approx$ 6 hour record ($\approx 32{,}000$ data points) for the interbeat interval RR_i series (□) of healthy subject. Here, n indicates the time scale (in beat numbers) over which each measure is calculated. The scaling is obtained using 2nd order detrended fluctuation analysis, and indicates long-range anticorrelations in the heartbeat interval increment series ΔRR_i [17]. As expected, the scaling properties of the heartbeat interval increment series remain unchanged after the Fourier phase randomization (△). (b) The root mean square fluctuation of the integrated magnitude series (□) indicates long-range correlations in the magnitude series $|\Delta RR_i|$ (group average exponent of $\alpha - 1 = 0.74 \pm 0.08$ where $F(n)/n \propto n^{\alpha-1}$). After Fourier phase randomization of the interbeat interval increment series we find random behavior with exponent 0.5 (△). This change in the scaling (after removing the nonlinear features in the time series) suggests that the magnitude series carries information about the nonlinear properties of the heartbeat dynamics. (c) The root mean square fluctuation of the integrated sign series (□) indicates anticorrelated behavior in sign(ΔRR_i) (group average exponent of $\alpha - 1 = 0.42 \pm 0.03$ where $F(n)/n \propto n^{\alpha-1}$). The scaling properties of the sign series remain unchanged after the Fourier phase randomization (△), which suggests that the sign series relates to linear properties of the heartbeat interval time series. We note the apparent crossovers at $n \approx 20$ beats and $n \approx 100$ beats. A gradual loss of anticorrelation in the sign series is observed at time scales larger than $n \approx 100$ beats. The numbers in the figure indicate the scaling exponents before and after the phase-randomization procedure. Adapted from ref. [14]

fluctuation analysis method for estimating the scaling exponents of anticorrelated signals ($\alpha < 0.5$), we integrate the magnitude and sign series; (v) we perform a scaling analysis using 2nd order detrended fluctuation analysis on the integrated magnitude and sign series; (vi) to obtain the scaling exponents for the magnitude and sign series we measure the slope of $F(n)/n$ on a log-log plot, where $F(n)$ is the root mean square fluctuation function and n is the scale of analysis (in beat numbers).

We find for each subject in a group of 18 healthy individuals [13], that the time series of the magnitudes exhibits correlated behavior (Fig. 2b) (unlike the original heartbeat increment time series, which is anticorrelated, Fig. 2a). The sign series, however, exhibits anticorrelated behavior (Fig. 2c). Correlation in the magnitude series indicates that an increment with large magnitude is more likely to be followed by an increment with large magnitude. Anticorrelation in the sign series indicates that a positive increment is more likely to be followed by a negative increment. Our result for the temporal organization of heartbeat fluctuations thus suggests that, under healthy conditions, a large increment in the positive direction is more likely to be followed by a large increment in the negative direction [14]. We find that this empirical "rule" holds over a broad range of time scales from several up to hundreds of beats (Fig. 2) [15].

To show that fluctuations following an identical scaling law can exhibit different time ordering for the magnitude and sign, we perform a Fourier transform on a heartbeat interval increment time series, preserving the amplitudes of the Fourier transform but randomizing the Fourier phases. Then we perform an inverse Fourier transform to create a surrogate series. This procedure eliminates non-linearities, preserving only the linear features (i.e. two-point correlations) of the original time series [16]. The new surrogate series has the *same* power spectrum as the original heartbeat interval increment time series, with a scaling exponent indicating long-range anticorrelations in the interbeat increments (Fig. 2a). Our analysis of the sign time series derived from this surrogate signal shows scaling behavior almost identical to the one for the sign series from the original data (Fig. 2c). However, the magnitude time series derived from the surrogate (linearized) signal exhibits *uncorrelated* behavior — a significant change from the strongly *correlated* behavior observed for the original magnitude series (Fig. 2b). Thus, the increments in the surrogate series do not follow the empirical "rule" observed for the original heartbeat series, although these increments follow a scaling law identical to the original heartbeat increment series. Moreover, our results raise the interesting possibility that the magnitude series carries information about the nonlinear properties of the heartbeat series, while the sign series relates importantly to linear properties.

We test our analysis on a group of 12 subjects with congestive heart failure [13]. Compared to the healthy subjects, the magnitude exhibits weaker correlations with a scaling exponent closer to the exponent of an uncorrelated series. The change in the magnitude exponent for the heart failure subjects is consistent with a previously reported loss of nonlinearity with disease [18, 19, 20, 21]. The sign time series of heart failure subjects shows scaling behavior similar to the one observed in the original time series, but significantly different from the healthy subjects (Table 1).

Next, we investigate how the heart rhythms of healthy subjects change within the different sleep stages. Typically the differences in cardiac dynamics during wake or sleep state, and during different sleep stages are reflected in the average and standard deviation of the interbeat interval time series [22, 23]. Recent studies show that changes in cardiac control due to circadian rhythms or different sleep stages can lead to systematic changes in the correlation (scaling) properties of the heartbeat dynamics. In particular it was found that the long-range correlations in heartbeat dynamics change during wake and sleep periods [24], indicating different regimes of intrinsic neuroautonomic regulation of the cardiac dynamics, which may switch on and off with the circadian rhythms.

Healthy sleep consists of cycles of approximately 1–2 hours duration. Each cycle is

TABLE 1. Comparison of the statistics of the root mean square fluctuation, $F(n)$ (calculated using the 2nd order detrended fluctuation analysis method [9, 11, 12] where n is the time scale in beat numbers over which each measure is calculated), and the scaling exponents for 18 healthy subjects and 12 subjects with heart failure [13] (obtained from 6-hour records during the day). The scaling features of the magnitude and sign change significantly for the subjects with heart failure, raising the possibility of bedside applications. α is the best fit to the range $6 < n < 1024$. $F(n)$ is estimated at the crossover position ($n = 16$) (Fig. 2b) where the largest separation between the two groups is estimated. Since we observe two apparent crossovers in the scaling behavior of the sign series, we calculate the scaling exponents in three different regions : (i) the short range regime for time scales $6 < n < 16$ with scaling exponent α_1, (ii) the intermediate regime for time scales $16 \leq n \leq 64$ with scaling exponent α_2, (iii) and the long range regime for time scales $64 < n \leq 1024$ with scaling exponent α_3. For each measure, the group average $\pm$ 1 standard deviation is presented. The values which show highly significant differences ($p \leq 0.01$ by Student's t-test) between the healthy and heart failure groups are indicated in boldface. We note, surprisingly, that the short range and the intermediate range scaling exponents α_1 and α_2 of the sign series may provide even more robust separation between healthy and heart failure compared to previous reports [17] based on the scaling exponents of the original heartbeat series.

magnitude			
measure	healthy	heart failure	p value
$\log_{10}F(n)$	$\mathbf{-1.49 \pm 0.16}$	$\mathbf{-1.92 \pm 0.17}$	1×10^{-7}
α	$\mathbf{1.74 \pm 0.08}$	$\mathbf{1.66 \pm 0.06}$	0.01
α_1	1.55 ± 0.08	1.6 ± 0.08	0.13
α_2	1.66 ± 0.08	1.61 ± 0.08	0.14
α_3	$\mathbf{1.82 \pm 0.1}$	$\mathbf{1.71 \pm 0.1}$	4×10^{-3}
sign			
measure	healthy	heart failure	p value
$\log_{10}F(n)$	$\mathbf{0.14 \pm 0.05}$	$\mathbf{0.02 \pm 0.06}$	1×10^{-6}
α	1.42 ± 0.03	1.44 ± 0.02	0.08
α_1	$\mathbf{1.43 \pm 0.12}$	$\mathbf{1.15 \pm 0.12}$	7×10^{-7}
α_2	$\mathbf{1.27 \pm 0.07}$	$\mathbf{1.41 \pm 0.07}$	1×10^{-5}
α_3	1.53 ± 0.065	1.49 ± 0.04	0.04

characterized by a sequence of sleep stages usually starting with light sleep, followed by deep sleep, and rapid eye movement (REM) sleep [25]. While the specific functions of the different sleep stages are not yet well understood, many believe that deep sleep is essential for physical rest, while REM sleep is important for memory consolidation [25]. Different sleep stages during nocturnal sleep were found to relate to a specific type of correlations in the heartbeat intervals [26], suggesting a change in the mechanism of cardiac regulation in the process of sleep.

We considered 24 records of interbeat intervals obtained from 12 healthy individuals during sleep [28]. The records have an approximate duration of 7.5 hours. The anno-

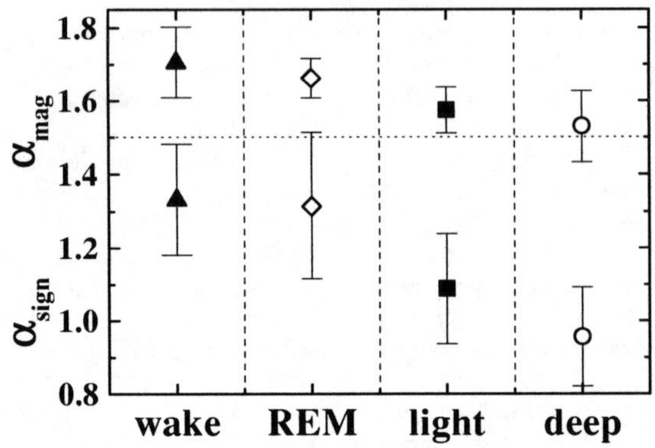

FIGURE 3. The average values of the fluctuation exponents α_{mag} for the magnitude series and α_{sign} for the sign series for the different phases (wake state, REM sleep, light sleep, and deep sleep). For each of the 24 records from 12 healthy subjects the corresponding 2nd order DFA fluctuation functions $F(n)$ have been fit in the range of $8 \leq n \leq 13$ and $11 \leq n \leq 150$ heartbeats for α_{sign} and α_{mag}, respectively, where the most significant differences between the sleep stages occur. After ref. [28]

tation and duration of the sleep stages were determined based on standard procedures [27]. We applied the detrended fluctuation analysis (DFA) method [9, 10, 11, 12] on both the sign and the magnitude time series. We found that the sign series exhibits anti-correlated behavior at short time scales which is characterized by a correlation exponent with smallest value for deep sleep, larger value for light sleep, and largest value for REM sleep. The magnitude series, on the other hand, exhibits uncorrelated behavior for deep sleep, and long-range correlations are found for light and REM sleep, with a larger exponent for REM sleep. The observed increase in the values of both the sign and magnitude correlation exponents from deep through light to REM sleep is systematic and significant. We also found that the values of the sign and magnitude exponents for REM sleep are very close to the values of these exponents for the wake state [28].

The mean values and their standard deviations for the different sleep stages are shown in Fig. 3. We estimate the exponents α from the slopes in the log-log plot of $F(n)$ versus n for all records. Since the most significant differences for the short-range sign correlations occur in the range of $8 \leq n \leq 13$ heartbeats, we use this fitting range for the exponents α_{sign}. For the magnitude exponent α_{mag}, we use the range $11 \leq n \leq 150$, since the long-range correlations occurring in light and REM sleep can be observed best in this region. We find that there is a significant difference in the sign series exponent α_{sign} observed for all three sleep stages (the p-value, obtained by the Student's t-test, is below 0.001). The magnitude correlation exponents for REM sleep and for intermediate wake states are significantly larger than those for the non-REM stages (light and deep sleep). Here also, the p-values are less than 0.001. Note that we do not find a significant difference between the average exponents for REM sleep and for the intermediate wake states. This is not surprising because heartbeat activity during REM sleep is very close to heartbeat activity during the wake state and the heartbeat time series during REM and wake exhibit similar scaling properties [24, 26]

More significant than the differences for the average exponents are the differences

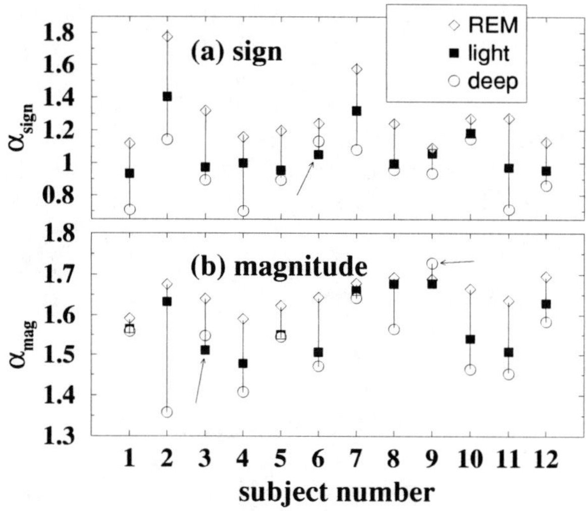

FIGURE 4. The values of the effective fluctuation exponents α for the integrated sign series (a) and the integrated magnitude series (b) are shown for all 12 healthy subjects (second night of recording). While the α values are fluctuating, for REM sleep the α is larger than the α for light sleep, which is larger than the α for deep sleep (the 3 arrows indicate the cases which are not ordered in the same way as the majority). The exponent values have been determined over the fitting ranges as described in the caption of Fig. 3. After ref. [28].

between the exponents for each individual. Figure 4 shows the α values for REM, light, and deep sleep for all 12 healthy subjects (second night only). In almost all cases the exponent of the REM sleep is the largest, the exponent of the light sleep is intermediate, and the exponent of the deep sleep is smallest (there are three exceptions, indicated by arrows). In our group of 24 records from 12 healthy individuals, we find larger exponents in REM sleep than in deep sleep for 100% of the sign series and for 88% of the magnitude series.

Our results suggest that the empirical "rule" mentioned above also applies to REM sleep, while in deep sleep small and large increments seem to appear in a random fashion. On the other hand, the stronger sign anticorrelations in deep sleep indicate that a positive increment is more likely — even more likely than in REM sleep — to be followed by a negative increment. Thus, the correlation behavior of the heartbeat increments and their signs and magnitudes during daytime activity is similar to the behavior we find in REM sleep, but quite different from the behavior we observe in deep sleep. This is consistent with our finding [Fig. 3] of average exponent values for the wake episodes similar to the exponent values for REM sleep.

We conclude that series with identical correlation properties can have completely different time ordering which can be characterized by different scaling exponents for the magnitude and sign series. Moreover, we show that the magnitude series carries information regarding the nonlinear properties of the original series while the sign series carries information regarding the linear properties of the original series. The significant decrease in the short-range scaling exponent for the sign series in heart failure may be related to perturbed vagal control affecting relatively high frequency fluctuations. The decrease of the long-range scaling exponent for the magnitude series of the heart failure patients indicates weaker correlations and loss of nonlinearity which may be related to impaired feedback mechanisms of neurohormonal cardiac regulation. Further we observe short-range anticorrelations in the sign of the interbeat interval increments which are stronger during deep sleep, weaker during light sleep, and even weaker during REM sleep. In contrast, the magnitude of the increments is long-range correlated with a larger exponent during REM sleep suggesting stronger nonlinear contributions to the heartbeat dynamics in this stage compared with weaker nonlinear contributions in the non-REM stages.

ACKNOWLEDGMENTS

We thank A. Bunde, A.L. Goldberger, S. Havlin, C.-K. Peng, and T. Penzel for discussions and major contributions to the results reviewed here which represent collaborative research efforts. We thank the NIH/National Center for Research Resources (P41 RR13622), the Israel-U.S.A Binational Science Foundation and the Deutscher Akademischer Austauschdienst (DAAD) for support. The healthy volunteers were recorded as part of the SIESTA project funded by the European Union grant no. Biomed-2-BMH4-CT97-2040.

REFERENCES

1. M.F. Shlesinger, *Ann. NY Acad. Sci.* **504**, 214 (1987).
2. T. Vicsek, *Fractal Growth Phenomenon* 2nd edn (World Scientific, Singapure, 1993).
3. H. Takayasu, *Fractals in the Physical Sciences* (Manchester Univ. Press, Manchester, UK, 1997).
4. R.M. Berne and M.N. Levy, *Cardiovascular Physiology*, 7th edn (Mosby, St. Louis, 1996).
5. M. Kobayashi and T. Musha, *IEEE Trans. Biomed. Eng.* **29**, 456 (1982).
6. P.Ch. Ivanov, M.G. Rosenblum, C-K Peng, J. Mietus, S. Havlin, H.E. Stanley, and A.L. Goldberger, *Nature* **383**, 323 (1996).
7. P.Ch. Ivanov, L.A.N. Amaral, A.L. Goldberger, S. Havlin, M.G. Rosenblum, H.E. Stanley, Z. Struzik, *Chaos* **11**, 641 (2001).
8. C.-K. Peng, J. Mietus, J. M. Hausdorff, S. Havlin, H. E. Stanley, and A. L. Goldberger, *Physical Review Letters* **70**, 1343 (1993).
9. C.-K. Peng, S.V. Buldyrev, S. Havlin, M. Simons, H.E. Stanley, and A.L. Goldberger, *Physical Review E* **49**, 1685 (1994).
10. J.W. Kantelhardt, E. Koscielny-Bunde, H.H.A. Rego, S. Havlin, and A. Bunde, *Physica A* **294**, 441 (2001).
11. K. Hu, P.Ch. Ivanov, Z. Chen, P. Carpena, and H.E. Stanley, *Physical Review E* **64**, 011114(19) (2001).
12. Z. Chen, P.Ch. Ivanov, K. Hu, and H.E. Stanley, *Physical Review E* **65**, 041107(15) (2002).
13. MIT-BIH Normal Sinus Rhythm Database and BIDMC Congestive Heart Failure Database available at http://www.physionet.org/physiobank/database/#ecg.
14. Y. Ashkenazy, P.Ch. Ivanov, S. Havlin, C.-K. Peng, A.L. Goldberger, and H.E. Stanley, *Physical Review Letters* **86**, 1900 (2001).
15. Heartbeat increment series were investigated by A. Babloyantz and P. Maurer, Phys. Lett. A **221**, 43 (1996) and P. Maurer, H.-D. Wang, and A. Babloyantz, Phys. Rev. E **56**, 1188 (1997). These studies

differ from ours because we investigate, quantitatively, normal heartbeats by evaluating the scaling properties of the magnitude and sign series. In addition, our calculations are based on window scales larger than 6 and up to one thousand heartbeats.
16. P.F. Panter, *Modulation, Noise, and Spectral Analysis Applied to Information Transmission* (New York, New York, 1965). We also applied a test for nonlinearity using the phase randomization procedure described in J. Theiler, S. Eubank, A. Longtin, B. Galdrikian, and D.J. Farmer, Physica D **58**, 77 (1992), and find that the magnitude scaling exponent drops to 0.5 while the sign scaling exponent remains unchanged.
17. C.-K. Peng, S. Havlin, H.E. Stanley, and A.L. Goldberger, *Chaos* **5**, 82 (1995).
18. J. Kurths, A. Voss, P. Saparin, A. Witt, H.J. Kleiner, and N. Wessel, *Chaos* **5**, 88 (1995).
19. G. Sugihara, W. Allan, D. Sobel, and K.D. Allan, *Proc. Natl. Acad. Sci. USA* **93**, 2608 (1996).
20. P.Ch. Ivanov, M.G. Rosenblum, L.A.N. Amaral, Z.R. Struzik, S. Havlin, A.L. Goldberger, and H.E. Stanley, *Nature* **399**, 461 (1999).
21. L.A.N. Amaral, P.Ch. Ivanov, N. Aoyagi, I. Hidaka, S. Tomono, A.L. Goldberger, H.E. Stanley, and Y. Yamamoto. *Physical Review Letters* **86**, 6026 (2001).
22. H. Moelgaard, K. E. Soerensen, and P. Bjerregaard, *Am. J. Cardiology* **68**, 77 (1991).
23. H. V. Huikuri, K. M. Kessler, E. Terracall, A. Castellanos, M. K. Linnaluoto, and R. J. Myerburg, *Am. J. Cardiology* **65**, 391 (1990).
24. P. Ch. Ivanov, A. Bunde, L. A. N. Amaral, S. Havlin, J. Fritsch-Yelle, R. M. Baevsky, H. E. Stanley, and A. L. Goldberger, *Europhysics Letters* **48**, 594 (1999).
25. M. A. Carskadon and W. C. Dement, in: *Principles and Practice of Sleep Medicine*, edited by M. H. Kryger, T. Roth, and W. C. Dement (W. B. Saunders, Philadelphia, 1994), pp. 16-25.
26. A. Bunde, S. Havlin, J. W. Kantelhardt, T. Penzel, J.-H. Peter, and K. Voigt, *Physical Review Letters* **85**, 3736 (2000).
27. A. Rechtschaffen and A. Kales, *A Manual of Standardized Terminology, Techniques, and Scoring System for Sleep Stages of Human Subjects* (U.S. Government Printing Office, Washington, 1968).
28. J.W. Kantelhardt, Y. Ashkenazy, P.Ch. Ivanov, A. Bunde, S. Havlin, T. Penzel, J.-H. Peter, and H.E. Stanley, *Physical Review E* **65**, 051908(6) (2002).

Cardiovascular Dynamics – Multiple Time Scales, Oscillations and Noise

Aneta Stefanovska*†, Andriy Bandrivskyy† and Peter V.E. McClintock†

*Group of Nonlinear Dynamics and Synergetics, Faculty of Electrical Engineering, University of Ljubljana, Ljubljana, Slovenia
†Department of Physics, Lancaster University, Lancaster, UK

Abstract. Modelling the cardiovascular system (CVS) presents a challenging and important problem. The CVS is a complex dynamical system that is vital to the function of the human organism, and it reflects numerous different states of health and disease. Its complexity lies in a combination of oscillatory modes spanning a wide frequency scale that can synchronize for short episodes of time, coupled with a strong stochastic contribution. Motivated by these properties, we discuss the problem of characterising dynamics when there is a combination of oscillatory components in the presence of strong noise and, in particular, where the characteristic frequencies and corresponding amplitudes vary in time. We show that, where there are several noisy oscillatory modes, the slower modes are difficult to characterise because the length of the recorded time series is inevitably limited in real measurements. We argue that, in the case of strong noise combined with a limited observation time, such oscillatory dynamics with several modes may appear to manifest as a $1/f$-like behaviour. We also show that methods of time-frequency analysis can provide a basis for characterising noisy oscillations, but that a straightforward characterisation of multi-scale oscillatory dynamics in the presence of strong noise still remains an unsolved problem.

INTRODUCTION

Nature abounds with rhythms of diverse origin. The day and the night rhythmically exchange, influencing the rhythmicity of light- and temperature-dependent physiological processes. But natural clocks seem to be adaptive. As the length and intensity of the daylight change within a year, the basic frequencies of many natural rhythms are modulated and thus vary in time.

One of the most vital rhythmical functions for humans, and for higher animals generally, is the beating of the heart. The heart is a part of the cardiovascular system (CVS) which serves to keep the blood in continuous motion. The CVS rhythmically distributes nutrients and oxygen to every cell and takes away the products of their metabolism. It consists of the heart, lung, and a network of contractile vessels that are actively involved in the regulation of blood pressure and blood flow. The flow of blood is regulated with respect to the needs of each cell of the body, as well as to the spontaneous fluctuations arising from naturally occurring perturbations to the CVS. Thus the heart frequency is not constant, but varies within certain limits, adapting to the current state and needs of the organism.

The CVS is an extremely complex system, and it reflects the different states of health or disease. How best to parameterize CVS dynamics using appropriate cardiovascular signals, measured noninvasively, has been a continuing challenge over the last twenty

CP665, *Unsolved Problems of Noise and Fluctuations: UPoN 2002: Third International Conference*, edited by S. M. Bezrukov

years. In spite of intensive research, an optimal characterization able unambiguously to describe the dynamics of the CVS, and reliably to distinguish between its states, has yet to be found. In particular, we would like to establish a way of detecting changes due to diseases in their early stages, before symptoms become obvious. The characterization involves two major questions –

1. How to define the appropriate level of complexity at which the system can best be characterized and modelled?
2. What is the underlying dynamics that best characterize its properties?

The first problem involves understanding the structure and function of the CVS, i.e. the underlying physiological background. The second problem is usually approached through the analysis of cardiovascular time series. The recent development of noninvasive sensors based on electrical, optical, and auditory detection of cardiovascular functions, of powerful computers, and of methods of nonlinear and stochastic time series analysis, has motivated huge interest in understanding the dynamics of the cardiovascular system. However, the complex fluctuations which we can observe and measure in the blood flow, blood pressure and heart rate, are yet to be understood.

It has been established that signals, derived from the human CVS contain several oscillatory components [1, 2, 3, 4, 5, 6], independent of the measurement site and function. On a time scale of around one minute, the dynamics of the blood distribution system can be characterized by five oscillatory components with slowly varying frequencies and amplitudes [7, 8, 9, 10]. The frequency and amplitude variations are reminiscent of phenomena observed in coupled oscillators. Moreover, another phenomenon characteristic of coupled oscillators was observed – the synchronization of two of the rhythms: the cardiac and respiratory [11, 12, 13, 14, 15, 16]. One possibility, therefore, is to characterize cardiovascular signals in terms of a model consisting of five coupled autonomous noisy oscillators [17, 18].

The other approach comes from analyses of the statistical properties of heart rate variability (HRV) [19, 20, 21, 22, 23], and considers the chaotic behaviour [24, 25], scaling properties [26, 27, 28] and multifractal characteristics [23] detected in the HRV signal. HRV was considered to result from $1/f$ dynamics and it was suggested that the state of the system defines the scaling properties of the power spectrum of the HRV and that the state can be classified in terms of the slope in the log-log representation of its power spectrum.

Where, then, is the origin of the difference in these two approaches to CVS dynamics? Is the dynamics primarily oscillatory – spanning multiple time scales, or is it mainly stochastic, or is it a combination? Studies concerned with the stochastic nature of the system are mainly based on analyses of the HRV signal. Those concerned with its oscillatory nature are based on analyses of blood flow and blood pressure, as well as of the HRV signal. They are motivated by the physiological importance of the local regulatory mechanisms for blood flow and blood pressure, which manifest as low frequency oscillations (<0.05 Hz) apparently reflecting the vascular and endothelial immune and cytotoxic reactions.

In this paper we illustrate problems that arise in the analysis of noisy time-varying oscillations from a finite data set, and we show that they may lead to differences in interpretation. We first analyze two cardiovascular signals, HRV and blood flow, and then apply exactly the *same* analysis to a model signal whose dynamics is completely known. For illustration several methods of time series analysis are applied: discrete Fourier transform (DFT), time averaged DFT, wavelet transform using the Morlet wavelet, time averaged wavelet transform, and detrended fluctuation analysis (DFA).

TIME SERIES AND THEIR CHARACTERIZATION

Cardiovascular signals

In this section we present analyses of cardiovascular time series. A set of cardiovascular signals, i.e. the electrical activity of the heart (ECG), respiratory movements of the lung, blood pressure, and peripheral blood flow, was simultaneously recorded from a healthy, relaxed subject for 30 minutes using the procedure described in [9]. In our discussion we will consider, first, the HRV "signal" derived from the ECG as an estimate of the instantaneous cardiac frequency defined between two successive R-peaks and, secondly, the skin blood flow signal. The signals were sampled at 400 Hz for precise detection of R-peaks. After generation of the HRV, they were resampled to 10 Hz for analysis. The resultant signals are presented in figures 1(a) and 2(a) respectively.

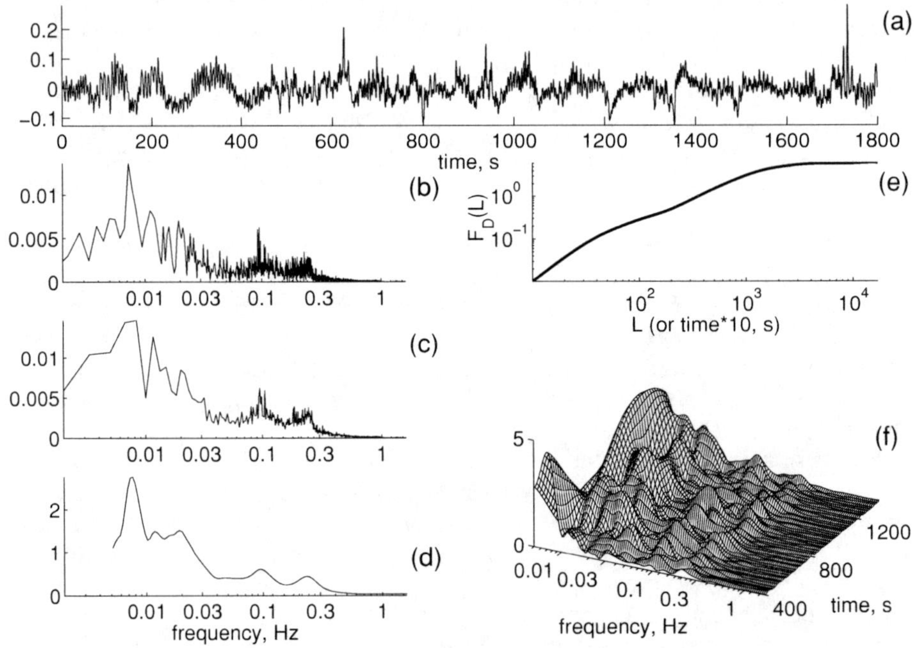

FIGURE 1. (a) The HRV signal and its (b) DFT, (c) time-averaged DFT, (d) time-averaged wavelet transform, (e) detrended fluctuation function and (f) its wavelet transform. The variations in (a) are in seconds around the mean value (1 s) that was subtracted for the analyses shown below.

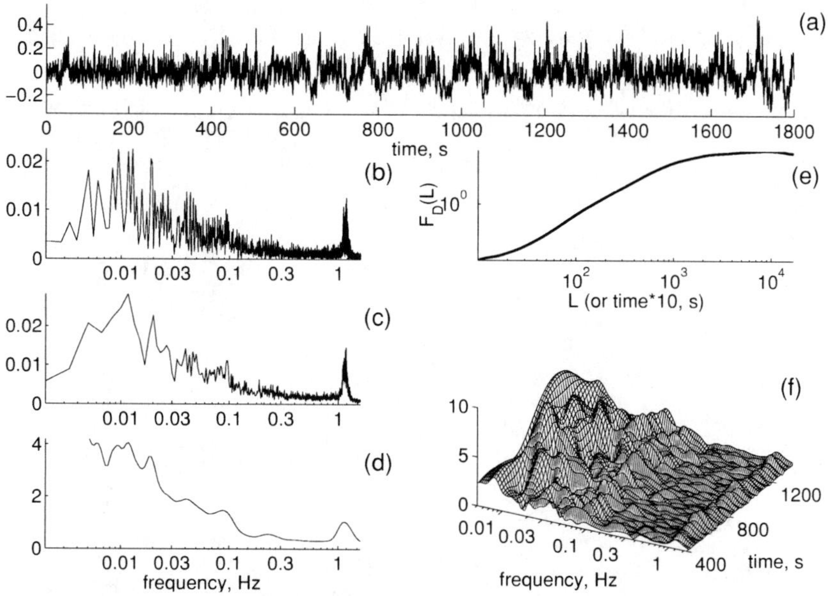

FIGURE 2. (a) The skin blood flow signal and its (b) DFT, (c) time-averaged DFT, (d) time-averaged wavelet transform, (e) detrended fluctuation function and (f) time-frequency projection of the wavelet transform.

Spectral analysis reveals the presence of low frequency components in both signals; note that, by definition, the HRV signal cannot contain the ($\sim$1 Hz) cardiac frequency itself. However, markedly different interpretations of the very low frequency components are possible. One may think of oscillatory components which undergo strong complex modulation by noise and from the rest of the system, or one can study the scaling properties of the power spectra which can be classified by the α coefficient given by detrended fluctuation analysis (DFA) [21, 22]. As demonstrated in figures 1(e) and 2(e) both cardiovascular signals manifest $1/f$-like behavior.

Model time series

Let us now consider a Van-der-Pol oscillator using the same methods of analysis. Figure 3(a) presents a time series obtained from one oscillator of the system

$$\ddot{x}_i + \varepsilon_i(1 - x_i^2)\dot{x}_i + w_i^2 x_i = D_i \xi(t), \tag{1}$$

where x_i is the coordinate of the i-th oscillator and $\xi(t)$ is white Gaussian noise of intensity D_i. The length of the time series, the integration time step and the characteristic frequencies are all chosen to mach the statistical properties of the cardiovascular time series and the conditions under which they are analysed. Results are shown both in the absence of noise (left) and with strong noise (right). In the absence of noise the system

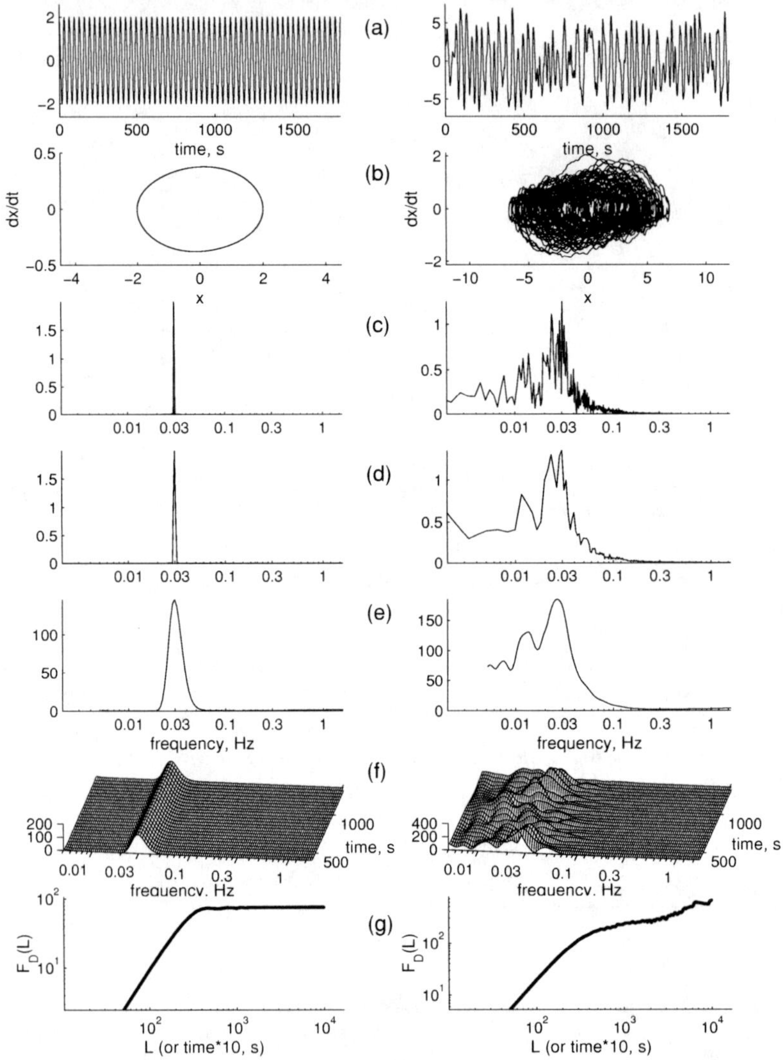

FIGURE 3. (a) A signal of one noise-free Van-der-Pol oscillator with frequency of 0.03 Hz (left), and in presence of strong noise (right). (b) Phase portrait, (c) DFT, (d) time-averaged DFT, (e) time averaged wavelet transform and (f) its wavelet transform and (g) detrended fluctuation function for noise-free (left) and noisy oscillator (right). L in (g) is the window length.

(1) possesses a stable periodic orbit (limit cycle) and produces an almost harmonic signal for small ε_i. Strong noise applied to the oscillator results in significant variations of its instantaneous amplitude and frequency, strikingly similar to the behaviour already observed in cardiovascular signals. In the absence of noise the problem of characterising

the dynamics of a single oscillator from time series is of course trivial. Each of the existing methods, either in phase space, or in the time, frequency, or time-frequency domains, reconstructs its properties unambiguously. The addition of strong noise, however, transforms this trivial problem into one that remains unsolved. Note that the phase portrait of the single noisy oscillator can no longer be characterised by a circle and no obvious origin can be determined. The DFA then results in a curve that may be interpreted as $1/f$-like, although the scaling interval(s) cannot easily be defined. A more precise characterisation may be attempted by spectral methods, and by time-frequency analysis in particular. Here, however, we face difficulties in frequency determination. The higher the frequency is, the better we are able to define it, as there will be a larger number of cycles within the observation period. For the slower oscillations, however, with fewer complete cycles, the effect of noise is inevitably more pronounced. Note that the choice of the window length represents a compromise between time and frequency localisation. The actual window for the wavelet transform in our case was chosen to include 5 periods of the corresponding frequency being estimated.

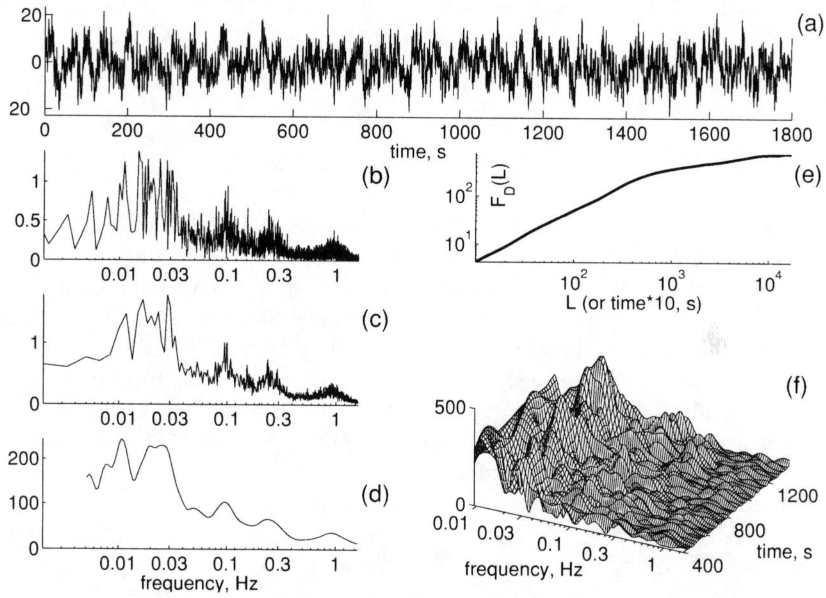

FIGURE 4. (a) The model signal and its (b) DFT, (c) time averaged DFT, (d) time averaged wavelet transform, (e) detrended fluctuation function and (f) its wavelet transform.

Now let us consider a more complex model with five noisy oscillators with characteristic frequencies $f_i = \{1, 0.26, 0.1, 0.03, 0.018\}$ Hz. The choices of the number of oscillators, and their frequencies, were motivated by the idea that CVS can be modelled as a system of five coupled oscillators, each corresponding to a different CVS activity (cardio, respiratory, neurogenic, myogenic and metabolic) [9]. We scale the parameters w_i, ε_i and D_i in order to get exactly the same oscillators but acting on different time scales. So if k is a scaling factor so that $w_i = kw_1$, then $\varepsilon_i = k\varepsilon_1$ $D_i = k^2 D_1$. The time series in figure 4(a) are obtained as $x = \sum_i x_i$ for parameters $\varepsilon_1 = 0.4$ and $D_1 = 1000$. Note that all five oscillators (1) are identical, so that they produce identical time series in a statistical sense, but scaled in time and in frequency. Note also that in reality the cardiovascular oscillators are coupled. In our model we use noise to introduce the effect of couplings. In this way, although the model is complex, e.g. consists of five modes, we are still able to treat it both analytically and numerically.

Spectral analyses using frequency and time-frequency methods are presented in figures 4(b)–4(e). Note that the wavelet transform has better statistical averaging (averaging in time and frequency domain). However, as already noted, optimal resolution in time and in frequency cannot be obtained simultaneously. The Morlet wavelet allows a compromise with good frequency resolution in the low frequency (LF) domain and a good time resolution in the high frequency (HF) domain. It allows the best estimation of the lowest frequency peaks compared to the results obtained by DFT and time averaged DFT. Yet, as can be seen from figure 4, the two spectral peaks lowest in frequency are not well resolved in the spectra. That means that the available time of observation, which is chosen to include about 30 realizations of the slowest oscillation, is insufficient for appropriate estimation of the LF oscillatory components due to poor statistical averaging. Figure 4(e) gives the DFA function of the model signal in 4(a). Again it shows $1/f$-like behavior with two different scaling regions. We note that DFA cannot provide information about the underlying oscillatory dynamics.

CONCLUSION

We have discussed the problem of characterising oscillatory dynamics from time series in the presence of strong noise, introduced in such a way as to produce time-variability of the characteristic frequencies as observed for the cardiovascular oscillations. Using model time series that mimic the main properties of the cardiovascular signals we show that, when the noise is strong and the frequency variability is large, the underlying dynamics cannot be determined unambiguously. Nonetheless, time-frequency methods have the potential advantage over DFA and phase-space techniques that they can in principle separate the different oscillatory components, and can thus provide information specific to particular physiological processes.

The time-variable characteristic oscillations can be taken as indicating of a non-autonomous system. A time-frequency, or time-phase representation is therefore needed for characterisation of the corresponding dynamics. The wavelet transform, as a time-frequency method, enables the instantaneous frequencies to be traced as they evolve in time. Where several oscillatory components are present, however, and because the time of observation in reality is inevitably finite, we face difficulties in estimation of the lower oscillation frequencies. The longer their period in comparison to the total observation time, the more obtrusive the noise becomes. The lower frequencies may thus be misestimated as the result of poor statistical averaging. We note that the problem of identifying characteristic frequencies in the presence of noise is widespread, because biological systems are characterised by inherently noisy oscillatory dynamics.

ACKNOWLEDGMENTS

This work has been supported by the Slovenian Ministry of Education, Science and Sport, Joy Welch Trust (UK), the EPSRC (UK) and INTAS.

REFERENCES

1. Hyndman, B. W., Kitney, R. I., and Sayers, B. M., *Nature*, **233**, 339–341 (1971).
2. Peňáz, J., *Automedica*, **2**, 135–141 (1978).
3. Akselrod, S., Gordon, D., Ubel, F. A., Shannon, D. C., C., B. A., and Cohen, R. J., *Science*, **213**, 220–222 (1981).
4. de Boer, R. W., Karemaker, J. M., and Strackee, J., *Psychophysiology*, **22**, 147–155 (1985).
5. Camm, A. J., Malik, M., Bigger, J. T., and et al., *Circulation*, **93**, 1043–1065 (1996).
6. Di Renzo, M., Parati, G., Pedotti, A., and Zanchetti, A. E., *Frontiers of Blood Pressure and Heart Rate Analysis*, IOS Press, Amsterdam, 1997.
7. Stefanovska, A., and Krošelj, P., *Open Syst. and Inf. Dyn.*, **4**, 457–478 (1997).
8. Kvernmo, H. D., Stefanovska, A., Bračič, M., Kirkeøen, K.-A., and Kvernebo, K., *Microvasc. Res.*, **56**, 173–182 (1998).
9. Stefanovska, A., and Bračič, M., *Contemporary Physics*, **40**, 31–55 (1999).
10. Bračič, M., McClintock, P. V. E., and Stefanovska, A., "Characteristic frequencies of the human blood distribution system," in *Stochastic and Chaotic Dynamics in the Lakes*, edited by D. S. Broomhead, E. A. Luchinskaya, P. V. E. McClintock, and T. Mullin, American Institute of Physics, Melville, New York, 2000, pp. 146–153.
11. Kenner, T., Passenhofer, H., and Schwaberger, G., *Pflügers Archiv.*, **363**, 263–265 (1976).
12. Hildebrandt, G., "The autonomous time structure and its reactive modifications in the human organism," in *Temporal Disorder in Human Oscillatory System*, edited by L. Rensing, U. an der Heiden, and M. C. Mackey, Springer, Berlin, 1987, pp. 160–174.
13. Raschke, F., "Coordination in the circulatory and respiratory systems," in *Temporal Disorder in Human Oscillatory System*, edited by L. Rensing, U. an der Heiden, and M. C. Mackey, Springer, Berlin, 1987, pp. 152–158.
14. Schäfer, C., Rosenblum, M. G., and Kurths, J., *Phys. Rev. E*, **60**, 857–870 (1999).
15. Bračič Lotrič, M., and Stefanovska, A., *Physica A*, **283**, 451–461 (2000).
16. Stefanovska, A., Haken, H., McClintock, P. V. E., Hožič, M., Bajrović, F., and Ribarič, S., *Phys. Rev. Lett.*, **85**, 4831–4834 (2000).
17. Stefanovska, A., Bračič, M., Strle, S., and Haken, H., *Physiol. Meas.*, **22**, 535–550 (2001).
18. Stefanovska, A., Luchinsky, D. G., and McClintock, P. V. E., *Physiol. Meas.*, **22**, 551–564 (2001).
19. Kobayashi, M., and Musha, T., *IEEE Trans. Biomed. Eng.*, **29**, 456–457 (1982).
20. Peng, C. K., Mietus, J., Hausdorff, J. M., Havlin, S., Stanley, H. E., and Goldberger, A. R., *Phys. Rev. Lett.*, **70**, 1343 (1993).
21. Hausdorff, J. M., and Peng, C. K., *Phys. Rev. E*, **54**, 2154 (1996).
22. Stanley, H. E., Amaral, L. A. N., Goldberger, A. L., Havlin, S., Ivanov, P. C., and Peng, C. K., *Physica D*, **270**, 309–324 (1999).
23. Ivanov, P. C., Amaral, L. A. N., Goldberger, A. L., Havlin, S., Rosenblum, M. G., Struzik, Z. R., and Stanley, H. E., *Nature*, **399**, 461–465 (1999).
24. Babloyantz, A., and Destexhe, A., *Biol. Cybern.*, **58**, 203–211 (1988).
25. Poon, C. S., and Merrill, C. K., *Nature*, **389**, 492–495 (1997).
26. Amaral, L. A. N., Goldberger, A. L., Ivanov, P. C., and Stanley, H. E., *Phys. Rev. Lett.*, **81**, 2388–2391 (1998).
27. Ashkenazy, Y., Ivanov, P. C., Havlin, C. K., Peng, A. L., Goldberger, A. L., and Stanley, H. E., *Phys. Rev. Lett.*, **86**, 1900–1903 (2001).
28. Bernaola-Galván, P., Ivanov, P. C., Amaral, L. A. N., and Stanley, H. E., *Phys. Rev. Lett.*, **87**, 168105 (2001).

Noisy Ventilation Improves Lung Function

Béla Suki, Stephen P. Arold, Adriano Alencar, Kenneth R. Lutchen and *Edward P. Ingenito

Dept. of Biomedical Engineering, Boston University and
**Brigham and Women's Hospital, Boston, MA, 02215.*

Abstract. It has been shown that mechanical ventilation in the setting of acute lung injury may propagate additional injury within the lung and numerous studies have been carried out to determine the optimal method of minimizing ventilator induced lung injury while still maintaining life-sustaining gas exchange. We have found that noise added to tidal volume and frequency, called noisy ventilation, during mechanical ventilation improves both lung mechanics and oxygenation in a rodent model of acute lung injury. Additionally, the standard deviation of the noise appears to be directly related to the magnitude of improvements seen with this ventilation modality in a manner similar to stochastic resonance. Furthermore, healthy guinea pigs that underwent with noisy ventilation exhibited increased surfactant content and reduced plasma proteins than their conventionally ventilated counterparts within the alveolar space of the lung. This suggests that not only did noisy ventilation induce endogenous surfactant release, but also served to reduce ventilator induced lung injury in this animal model. In conclusion, noisy ventilation improves blood oxygenation during acute lung injury and also serves to enhance lung function and reduce ventilator induced lung injury in healthy lungs.

INTRODUCTION

Positive pressure ventilation is an essential component of supportive therapy among patients suffering from acute lung injury (ALI). However, it has also been shown to contribute to worsening of pre-existing injury when administered using protocols which impart high levels of stress to the lung tissue (1). It is thought that this ventilator-induced lung injury (VILI) arises from two main sources. First, due to the heterogeneous nature of ALI, significant regions of the lung are often completely collapsed and/or flooded from local edema, shunting significant portions of the tidal volume (VT) into the more compliant regions of the lung resulting in over-distention of the surrounding tissue or volutrauma (2). Second, regions of the lung may collapse at end-expiration only to re-open at the subsequent application of positive pressure. This repetitive collapse and re-opening is thought to impart significant shear forces on the airway and alveolar walls inducing additional injury in the epithelial cells (3). Thus, the currently accepted approach used to manage patients with ALI applies sufficient positive end expiratory pressure (PEEP) to prevent airway collapse, in conjunction with small VT to minimize alveolar over-distention (4). While this approach appears promising, it still has some inherent drawbacks and mortality remains unacceptably high (33%) (4).

CP665, *Unsolved Problems of Noise and Fluctuations: UPoN 2002: Third International Conference,* edited by S. M. Bezrukov

Alternatives to this ventilation modality that seek to limit alveolar over-distention have recently been introduced with promising results. Lefevre et al. (5) observed that randomly varying the delivered breath rate and VT with each breath significantly increased arterial blood oxygen tension (paO2) and lung compliance without causing an increase in mean airway pressures in a porcine model of ALI. Furthermore, the wet-dry weight ratio in the variably ventilated lungs was significantly lower than those ventilated with conventional ventilation techniques suggesting decreased pulmonary edema in their experimental group.

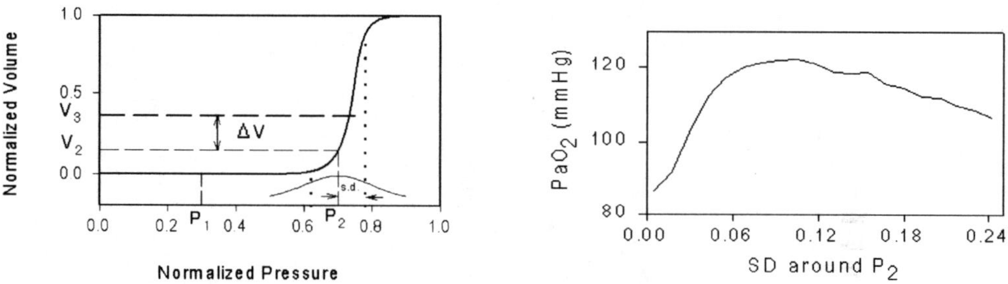

FIGURE 1. Noise added to peak inspiratory pressure improves lung physiology. Left: Solid line is a model of the normalized pressure-volume curve of a collapsed lung region. Ventilation is applied between end-expiratory pressure (P1) and peak inspiratory pressure (P2). V2 corresponds to recruited volume. When noise with standard deviation (SD=0.075) is added to P2 (Gaussian under pressure-volume curve), recruited volume increases from V2=0.15 to V3=0.363. The difference DV is the improvement. Right: The predicted arterial blood oxygen pressure as a function of the SD of the noise exhibits stochastic resonance (reprinted from Ref. 6, with permission from *Nature*).

NOISY VENTILATION: MODEL

An explicit physiological mechanism to explain the apparent improvements in gas exchange seen by the Leferve group was subsequently put forth by Suki et al. (6) They hypothesized that oxygenation may be improved during variable mechanical ventilation by exploiting the non-linear nature of the pressure-volume curve for an atelectic lung during ALI. Figure 1A shows a model of the pressure-volume curve of a lung region in the case of severe, acute lung injury. Suki et al. (6) suggested that due to the highly non-linear nature of this curve, it may be possible to increase lung volume by adding random noise in peak inspiratory pressures (PIP) over the course of mechanical ventilation. By placing the mean PIP near the lower inflection point (P2) in the pressure-volume curve, a small increase in PIP will result in a dramatic increase in lung volume while comparable decreases in PIP will result in a relatively insignificant loss of lung volume. They suggest that this "noisy" ventilation (NV) after many repeated breathing cycles will result in a considerable net increase in lung volume (V), and a resulting increase in surface area for gas exchange without increasing mean airway pressures. Additionally, it was predicted that the amount of noise placed about the PIP can be fine tuned to provide an optimal level of oxygenation in a manner analogous to a phenomenon called stochastic resonance (SR) (7). In SR, adding noise to the input, in this case PIP, results in amplification of the output, paO2. Gradually increasing noise in a system that demonstrates SR results in an increasing output signal until at some level of noise a maximum magnitude of the output is reached. Thereafter, adding additional noise to the input only serves to decrease the magnitude of the output, represented in Figure 1B. Should the injured lung behave in such a manner during mechanical ventilation then this observation may have important clinical consequences.

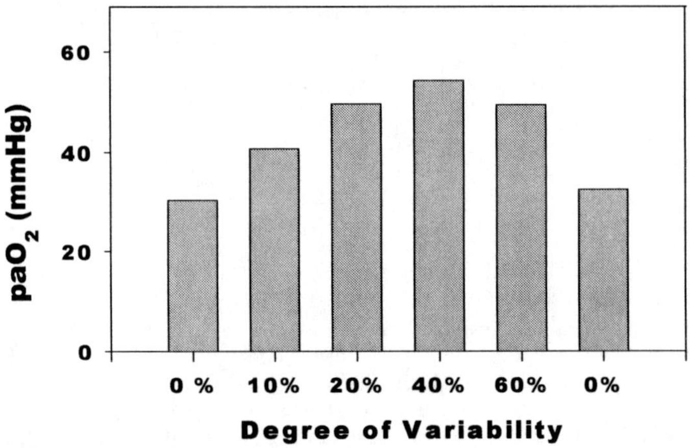

Figure 2. Arterial oxygen tension as a function of the coefficient of variation of the noise added to tidal volume in an example animal. The data are consistent with the possibility that paO2 follows SR.

EFFECTS OF NOISY VENTILATION ON LUNG PHYSIOLOGY

We recently tested the hypothesis that gas exchange can be optimized by the addition of noise to mechanical ventilation in a rodent model of acute lung injury (8). Guinea pigs with endotoxin-induced lung injury were ventilated for 5, 30 minute intervals whereby in each interval VT's were randomly selected from a given probability distribution. These distributions ranged from 0% variability (constant volume ventilation, CVV) to NV of 60% variability (having an equal probability of being delivered a VT between 40% and 160% of the mean VT). Frequency of ventilation was adjusted on a cycle by cycle basis to ensure constant minute ventilation, while lung mechanics and gas exchange were measured at the conclusion of each ventilation period.

Figure 2 shows paO2 for one animal ventilated with this protocol. Notice that as the level of noise in VT is increased oxygenation improves stepwise until 40% variability is reached. At this point, paO2 increased by nearly 100%, which in clinical setting, would be a very significant achievement. Addition of further noise results in a decline in oxygenation, in apparent agreement with the stochastic resonance hypothesis (6). Population averages for normalized lung elastance (defined as the ratio of the change in pressure and change in volume) and normalized paO2 are summarized in Fig. 3A and B as a function of percent variability. Notice that tissue elastance decreases stepwise as variability increases suggesting lung volume recruitment with additional variability. The paO2, on the other hand, demonstrates a slightly different trend whereby it seems to plateau near 40% to 60% variability. An interesting point to note is that the degree of variability associated with the largest improvement in lung elastance did not always correspond to that at which gas exchange was optimized. Every animal demonstrated maximal reductions in lung elastance during 60% variability, whereas many animals demonstrated optimal oxygenation with 40% variability. This may be a consequence several factors that could result in a de-coupling of improvement in lung elastance from paO2. Thus, a maximum paO2 and hence an optimal level of variability in VT exists at approximately 40-60% in agreement with the stochastic resonance model (6).

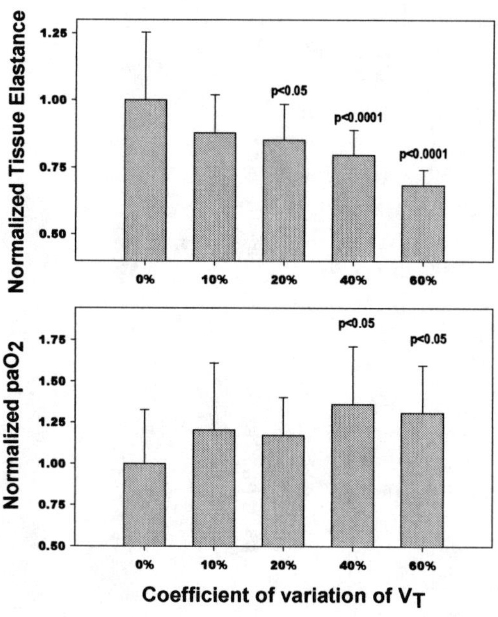

Figure 3. Population averages (mean±SD) for normalized tissue elastance and normalized paO2 as function of the coefficient of variation in VT. Statistical significant differences are between the measured values after ventilation at a given level of variability and after the initial period of CVV.

EFFECTS OF LONG-TERM VENTILATION ON SURFACTANT

In order to further elucidate the effects of variable ventilation on lung biology, we ventilated two groups of healthy guinea pigs with either CVV or a form of NV for a period of 3 hours (9, 10). Again, lung mechanics and gas exchange were measured periodically. At the conclusion of the experiment the animals underwent whole lung lavage and the broncho-alveolar lavage fluid was assayed for total protein and phospholipid content. Phospholipid, the primary constituent of lung surfactant, later underwent thin-layer chromatography to determine the measured lipids were derived from cell membrane associated or surfactant associated phospholipid.

While both groups of animals showed significant increases in tissue elastance, as seen in Fig. 4A, only the CVV group demonstrated a significant decline in PaO2, Fig. 4B, and the NV group demonstrated a significantly higher final paO2 than its counterpart. This demonstrates that the improvement in oxygenation is not solely due to lung recruitment and may also be a result of other factors. When phospholipid and protein contents between the two groups were compared to those of an unventilated control, Table 1, it was seen that the NV group had a nearly two-fold increase in phospholipids in comparison with the CVV group, while CVV group also demonstrated a significant increase in protein content. Meanwhile the lipid composition index, a ratio of cell membrane associated phospholipids to surfactant associated phospholipids, was significantly elevated in the conventionally ventilated group. The observation that animals receiving conventional ventilation demonstrated significantly increased levels of protein and cell-membrane-associated phospholipids in the lavage fluid, suggests the presence of capillary leakage and cellular injury. These findings indicate development of VILI in the CVV group. It is likely that failure to develop capillary leakage in the alveolar compartment following NV is, at least, in part due to the ability of this ventilation strategy to increase surfactant levels, thus reducing interfacial surface tensions and reducing re-opening stresses. Indeed, in a recent study, Boker and coworkers (11) reported that in a porcine model of ALI a similar ventilation mode reduced alveolar cytokine concentrations in comparison to animals ventilated with CVV suggesting that this ventilation mode reduces the risk of VILI in agreement with our results.

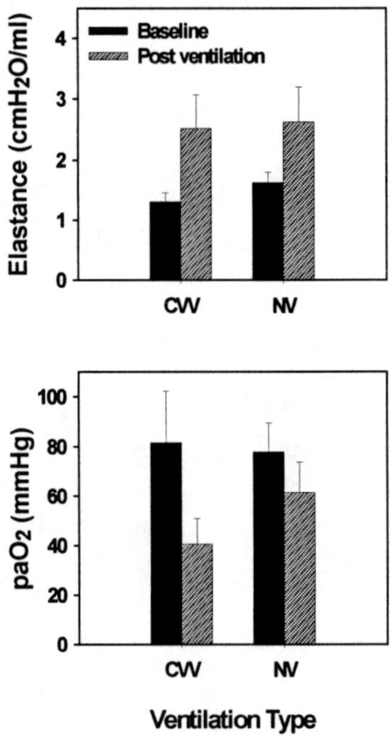

Figure 4. Elastance and paO_2 in two groups of animals in baseline and after 3 hours of ventilation. All differences are statistically significant, except the decrease in paO_2 in the NV group.

TABLE 1. Broncho-alveolar lavage protein and phospholipid content

Ventilation Type	Phopholipid content (mg/kg)	Protein content (mg/kg)	Lipid composition Index
Unventilated	98.0±19.5	10.0±3.4	.13±.04
CVV	84.0±29.3	17.2±3.9*	.36±.05*
NV	160.7±27.5 †	12.0±1.6	.19±.06†

Values represented as mean±SD in terms of mg of return per kg animal body weight. (* p<.05 vs. UNVENT), (†<.05 vs CVV).

The mechanism behind enhanced surfactant secretion following NV is likely due to the effects of the stress pattern that result from its application on the lung epithelium. Wirtz and Dobbs (12) have demonstrated that stretching cultured alveolar epithelial type II cells resulted in enhanced surfactant secretion and that the magnitude of secretion increased nearly exponentially with the change in surface area of the underlying basement membrane. It is conceivable that intermittent larger volume excursions that occur during NV stretch the lung epithelial layer, and promote surfactant secretion in a similar fashion. Tschumperlin and colleagues (13) have demonstrated that, while epithelial cells tolerate moderate degrees of stretch, excessive strain results in cell membrane disruption and death. Therefore, it is likely that a specific threshold exists such that distribution of VT's which generate stretch exceeding this threshold injure the lung, while those that do not exceed the threshold have beneficial effects on surfactant homeostasis. This suggests that there may be some optimal degree of variability in VT that optimizes surfactant release, however applying a range of VT's that surpasses this range would have a detrimental effect on cell viability likely reducing any benefits that may have realized from NV.

SUMMARY AND FUTURE WORK

We have demonstrated that NV improves lung function and gas exchange in a rodent model of acute lung injury via SR. We also found that NV improves lung biology as well as gas exchange during long-term mechanical ventilation in healthy animals. Should these findings also hold true in human patients there are obvious clinical applications. First, NV could serve as a simple and inexpensive therapy for sustainable improvements in gas exchange during life-support for patients suffering from ALI. Second, NV may prove to be a reliable surfactant replacement therapy, whereby endogenous release of surfactant may be stimulated by simple addition of appropriate amount of noise to tidal volume delivered by commercial mechanical ventilators.

Despite the above encouraging results, there are still a number of open questions that need to be addressed in future studies. To date, we only have demonstrated SR in endotoxin exposed guinea pigs. The optimal range or distribution of noise in human patients suffering from ALI may be different and depending on the underlying conditions that led to lung injury and disease progression, may vary significantly from patient to patient. Thus, measurements in larger animals and ultimately in human subjects will be necessary. Another important question is the following: Can similar improvements as seen during NV also be obtained through the application of periodic sighs or deep inflations of the lung? Our preliminary data and other studies suggest that sighs are not as effective as noise, suggesting thereby the existence of a true SR phenomenon.

The mechanism behind the enhanced surfactant secretion seen in the NV animals is also deserving of additional investigation. While we hypothesize that the stretch patterns applied to the surfactant secreting cells of the lung up-regulated surfactant release, this is purely speculation and further work possibly with cell culture systems could elucidate this hypothesis. Is the surfactant release accompanied by gene expression? At the moment, the answer to this question is unclear. Nevertheless, SR-induced gene expression and up-regulation of surfactant production is likely and would be a significant finding. Additionally we observed increased surfactant secretion in healthy guinea pigs. Whether this phenomenon holds true in a system stressed by acute lung injury also merits further work. Finally, it will be necessary to develop models that describe the physics of noisy stretch patterns on cells and the surfactant producing and secreting cellular organelles.

In conclusion, we have demonstrated that noisy ventilation improves lung mechanics and gas exchange in a rodent model of acute lung injury and improves lung biology in healthy guinea pigs. While further basic research is required to study the underlying mechanisms, the clinical application of noisy ventilation could have a significant impact on morbidity and mortality of patients suffering from ALI.

Below is an example equation created with Word 97's Equation Editor. To move this equation, highlight the entire line, then use cut and paste to the new location. To use this as a template, select the entire line, then use copy and paste to place the equation in the new location.

ACKNOWLEDGMENTS

Supported by: NSF BES-0114538 and NIH HL 59215-04.

REFERENCES

1. Dreyfuss, D., and Saumon G. Ventilator-induced lung injury: lessons from experimental studies. *Am J Respir Crit Care Med* **157(1)**:294-323 (1998).
2. Dreyfuss, D., Basset G., Soler P., and Saumon, G. Intermittent positive-pressure hyperventilation with high inflation pressures produces pulmonary microvascular injury in rats. *Am Rev Respir Dis* **132(4)**:880-4 (1985).
3. Muscedere, J.G., Mullen, J.B., Gan, K., and Slutsky, A.S. Tidal ventilation at low airway pressures can augment lung injury. *Am J Respir Crit Care Med* **149(5)**:1327-34 (1994).
4. The Acute Respiratory Distress Syndrome Network Ventilation with lower tidal volumes as compared with traditional tidal volumes for acute lung injury and the acute respiratory distress syndrome. *N Engl J Med* **342(18)**:1301-1308 (2000).
5. Lefevre, G.R., Kowalski, S.E., Girling, L.G., Thiessen, D.B., and Mutch, W.A. Improved arterial oxygenation after oleic acid lung injury in the pig using a computer-controlled mechanical ventilator. *Am J Respir Crit Care Med* **154(5)**:1567-1572 (1996).
6. Suki, B., Alencar, A.M., Sujeer, M.K., Lutchen, K.R., Collins, J.J., Andrade, J.S., Jr., Ingenito, E.P., Zapperi, S., and Stanley, H.E. Life-support system benefits from noise. *Nature* **393(6681)**:127-8 (1998).

7. Wiesenfeld, K., and Moss, J., Stochastic Resonance and the Benefits of Noise-From Ice Ages to Crayfish and Squids. *Nature* **373 (6509):** 33-36 (1995).
8. Arold, S.P., Mora, R., Lutchen, K.R., Ingenito, E.P., and Suki, B. Variable tidal volume ventilation improves lung mechanics and gas exchange in a rodent model of acute lung injury. *Am J Respir Crit Care Med* **165(3):** 366-371 (2002).
9. Ingenito, E.P., Arold, S.P., Lutchen, K., and Suki, B. Effects of noisy ventilation (NV) and open lung ventilation (OLV) on lung mechanics, gas exchange, and surfactant content and properties (abstract). *Am J Respir Crit Care Med* **163:** A483 (2001).
10. Arold, S.P., Suki, B., Alencar, A.M., Lutchen, K.R., and Ingenito, E.P. Variable Ventilation Induces Endogenous Surfactant Release in Normal Guinea Pigs. (*Submitted*)
11. Boker, A., Graham, M.R., Walley, K.R., McManus, M.B., Girling, L.G., Walker, E., Lefevre, G.R., and Mutch, W.A. Improved arterial oxygenation with biologically variable or fractal ventilation using low tidal volumes in a porcine model of acute respiratory distress syndrome. *Am J Respir Crit Care Med* **165(4)**: 456-462 (2002).
12. Wirtz, H.R.W., and Dobbs, L.G. Calcium mobilization and exocytosis after one mechanical stretch of lung epithelial cells. *Science* **250:**1266-1269 (1990).
13. Tschumperlin, D.J., and Margulies, S.S. Alveolar epithelial surface area-volume relationship in isolated rat lungs. *J Appl. Physiol* **86:** 2026-2033 (1999).

Fluctuations and Noise in Stochastic Spread of Respiratory Infection Epidemics in Social Networks

Renat Yulmetyev*, Natalya Emelyanova*, Sergey Demin*, Fail Gafarov*, Peter Hänggi† and Dinara Yulmetyeva (M.D.)**

*Department of Physics, Kazan State Pedagogical University, Mezhlauk Street 1, 420021 Kazan,Russia
†Department of Physics, University of Augsburg, Universitätsstrasse 1,D-86135 Augsburg,Germany
**Division of Therapy, Republican Clinical Hospital, Orenburgskii Trakt 79, 420064 Kazan, Russia

Abstract. For the analysis of epidemic and disease dynamics complexity, it is necessary to understand the basic principles and notions of its spreading in long-time memory media. Here we considering the problem from a theoretical and practical viewpoint, presenting the quantitative evidence confirming the existence of stochastic long-range memory and robust chaos in a real time series of respiratory infections of human upper respiratory track. In this work we present a new statistical method of analyzing the spread of grippe and acute respiratory track infections epidemic process of human upper respiratory track by means of the theory of discrete non-Markov stochastic processes. We use the results of our recent theory (Phys. Rev. E 65, 046107 (2002)) for the study of statistical effects of memory in real data series, describing the epidemic dynamics of human acute respiratory track infections and grippe. The obtained results testify to an opportunity of the strict quantitative description of the regular and stochastic components in epidemic dynamics of social networks with a view to time discreteness and effects of statistical memory.

INTRODUCTION

Now it has been well recognized that the study of epidemic and disease dynamics in social networks is a relevant theoretical issue of spreading viruses and diseases of various nature [1]-[4]. It is connected with understanding of the phase transitions in models of agent spreading in nonsteady systems. In particular, we deal with the development of the models describing epidemic spread, forest fires, the growth of populations, the activity of catalyzers, the formation of stars and galaxies. The study of the epidemic processes related to the increase in the fraction of population with respiratory infections, represents significant interest in modern biomedicine, ecology and economy [5]-[22]. The epidemic means that a large number of people in a certain country develop a disease at the same period of time. In case of "pandemic" the people in several countries develop the same disease at the same period of time. The first cases of respiratory infections with fatal outcome in USA and Europe were described in documents at the beginning of the 20th century. About 20 million people died then of "Spanish fever" within the period of 10 months. Pandemics of "Asian" and "Hongkong" grippe with smaller quantity of fatal outcomes [5]-[8] were observed in middle of the 20th century. The period of pandemic

CP665, *Unsolved Problems of Noise and Fluctuations: UPoN 2002: Third International Conference,* edited by S. M. Bezrukov

processes is about 30-40 years. As a rule, epidemics break out in the autumn or in the winter in Northern Hemisphere and in the spring or in the summer in Southern Hemisphere. The duration of epidemics is about 1-3 months. Seasonality is one of the typical demonstrations of the phase development of epidemic processes. Respiratory infections affect people of any age. However children aged 1-14 develop the disease four times more often than adults. Each epidemic process of respiratory infections is accompanied by the increasing death rate. The parameters of grippe death rate in the world is $0.01 - 0.2\%$. The greatest number of deaths caused by grippe is connected with complications after the infection. Children (up to 2 years old) and elderly people (over 65 years old) are more liable to this infection and die more often.

Expenses caused by epidemics affect the successful development of economy due to the amount of money needed for financing scientific actions to prevent epidemics. These expenses are commensurable only with the sums earmarked to prevent heart diseases [9]-[16]. In many countries of the world great money is spent on the development of vaccines and serums, as well as prevention of epidemic outbreak flashes. However viruses of grippe and some other respiratory infections have great variability. There is a great variety of grippe virus cultures. Therefore the coming grippe variant is difficult to predict. The identification of a virus presents a great problem; consequently the prevention of epidemics outbreaks is complicated. Thus new preventive measures and methods of predicting possible epidemic outbreaks [17]-[22] are of great interest.

Now various methods of description and prediction of epidemic and pandemic processes are developed in medicine [17]-[22]. As a result of these studies hypotheses about risk factors have been formulated. The standard statistics indices [22] are used in traditional epidemic studies. The basic data yield the number of diseases, their intensive parameters, and average sizes. The extensive parameters, the cumulative data, the relative number of presentation, the ratio parameters and standardized parameters are used depending on the specific features of the epidemic process. However the existing methods are of no sufficient efficiency.

In this work we offer a new statistical approach in the research of epidemic processes with the help of the statistical theory of non-Markov stochastic discrete processes [23, 24, 25]. This method has already been used by the authors to study various real objects in cardiology [23, 25], seismology [24] and neuropsychology [26, 27]. With the help of this method we intend to find the full spectrum of the dynamic characteristics of the complex system. Taking in the account the discrete behavior and nonstationarity of the dynamic and kinetic epidemic parameters we shall receive a more detailed representation of the stochastic dynamics of epidemic processes.

This work deals with the study of statistical properties of epidemic dynamics. The key distinctions of grippe and respiratory infections epidemic processes are defined with the help of the long-range memory effects and statistical effects of non-Markovity in these processes. In particular, one of the key problems consists in revealing the role of randomness, regularity and predictability of epidemic processes.

More detailed statement of the theory and the description of quantities used here can be found in the works published earlier [23]-[27]. Below we present the basic equations of our theory necessary for calculation of time correlation function (TCF) and other spectral characteristics only.

For the nonstationary discrete processes the TCF has the form $(t = m\tau, N-1 > m > 1)$

$$a(t) = \frac{1}{(N-m)\sigma(0)\sigma(t)} \sum_{j=0}^{N-1-m} \delta x(T + j\tau)\delta x(T + (j + m\tau)), \tag{1}$$

where $\sigma(0)$ and $\sigma(t)$ is the variances of the initial $(t = 0)$ and final (at moment t) states of a systems, correspondingly. The kinetic equation for the TCF $a(t)$ has the form of a closed set of the finite-difference equations of the non-Markov type

$$\frac{\Delta a(t)}{\Delta t} = \lambda_1 a(t) - \tau\Lambda_1 \sum_{j=0}^{m-1} M_1(j\tau)a(t - j\tau). \tag{2}$$

Introducing the corresponding projection operators we come to the following chain of connected non-Markov finite-difference kinetic equations $(t = m\tau, n = 1, 2, ...)$

$$\frac{\Delta M_n(t)}{\Delta t} = \lambda_{n+1} M_n(t) - \tau\Lambda_{n+1} \sum_{j=0}^{m-1} M_{n+1}(j\tau)M_n(t - j\tau). \tag{3}$$

EXPERIMENTAL DATA AND DATA PROCESSING

The registration and accounting materials, as well as the results of certain studies are the initial data for epidemic research in medicine. We have made use of the accounting materials of Kazan sanitary - epidemiological stations located in industrial districts. The data were given by the sanitary - epidemiological station of the Privolzhskii district of Kazan covering the period from 10.27.1995 to 03.05.2002. The experimental data involve the total population of the Privolzhskii district (about 300 thousand people). The first group of the data represents the six-year dynamics of grippe, the second group describes the six-year dynamics of acute respiratory track infections in this district. The obtained data were processed with the help of the above introduced technique. The set of three memory functions was calculated for each sequence of the data. The power spectra for each of these functions are obtained by the fast Fourier transform (FFT). We will also show the phase portrait in plan projections of the multidimensional space of the dynamic orthogonal variables. For a more detailed diagnostics of the system we will consider the frequency spectrum of the first three points of the statistical spectrum of the non-Markovity parameter. In this study we will use the frequency spectrum of non-Markovity parameter [23]-[25] $\varepsilon_i(\omega) = \{\mu_{i-1}(\omega)/\mu_i(\omega)\}^{\frac{1}{2}}$. Here $i = 1, 2..$ and $\mu_i(\omega)$ is a power spectrum of the ith level. The statistical spectrum of non-Markovity parameter $\varepsilon_i(\omega)$ constitutes an information measure of Markovity and non-Markovity or randomness and regularity in time evolution of underlying systems.

DISCUSSION OF THE RESULTS

In this section the quantitative and comparative analysis of the six-year period of grippe chaotic dynamics (Fig.1a) and of acute respiratory track infection (Fig.1e) will be

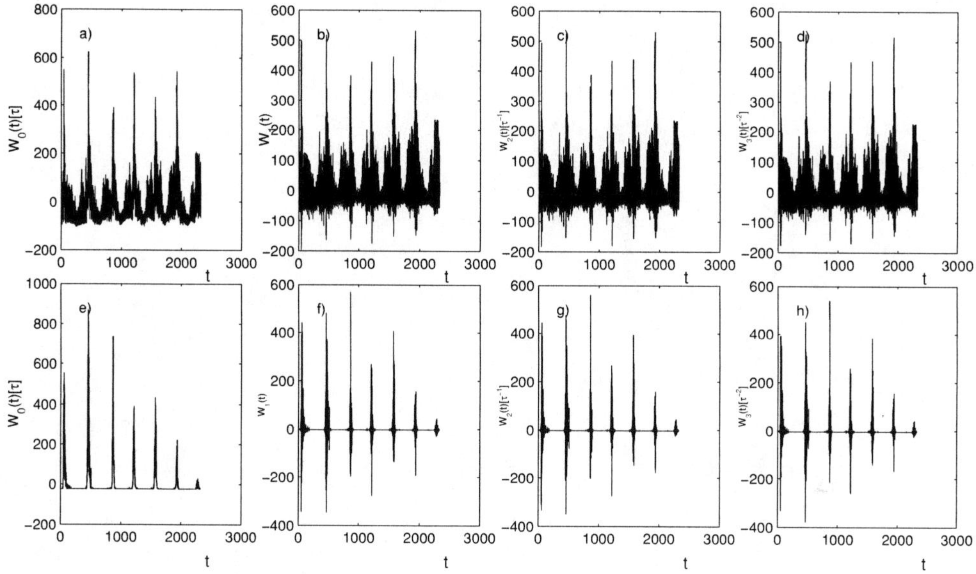

FIGURE 1. Time series of orthogonal variables W_0 (a, e), W_1 (b, f), W_2 (c, g), W_3 (d, h) of chaotic dynamics of grippe and acute respiratory track infections (ARTI)

carried out. The time series of the initial signals, the phase portraits of the dynamic variables, the power spectra of TCF and the junior functions of memory as well as frequency dependence of the first three points of the statistical non-Markov parameter are submitted in Figs. 1-4. The six-year sampling of grippe and acute respiratory track infections (ARTI) dynamics are submitted in Fig. 1 (Figs.1 a-d for grippe , Figs.1 e-h for ARTI). The time series of the orthogonal variable W_0 (Fig. 1a), W_1 (Fig. 1b), W_2 (Fig. 1c), W_3 (Fig. 1d) for the chaotic dynamics of grippe and ARTI have appreciable symmetry relative to straight line $W_i = 0$, where $i = 1.3$. The analysis of the time series of variable W_0 (Fig.1a) for a grippe epidemic shows that the moment of the grippe outburst or epidemic corresponds to the most significant fluctuations of this dynamic variable. Epidemic outbursts of the disease are located at equal distances from each other. This is the evidence of periodicity and interval constancy between grippe epidemics within the time of observation. The time series for three orthogonal variables W_1 (Fig. 1b), W_2 (Fig. 1c), W_3 (Fig. 1d) of grippe epidemics are almost perfectly symmetrical. The time series of dynamic variables W_0 (Fig. 1e), W_1 (Fig. 1f), W_2 (Fig. 1g), W_3 (Fig. 1h) of ARTI epidemiological processes present a different picture. The visible fluctuations of the dynamic variables for ARTI are observed in the autumn-winter period same as in case of grippe epidemics. However dynamic noises can be observed between the outbursts of ARTI (epidemics). They are represented by monotonously inequable parabolas. One of the half-parabola corresponds to one of the epidemics and another corresponds to the subsequent one. The time development of orthogonal variables W_1 (Fig. 1f), W_2 (Fig. 1g), W_3 (Fig. 1h) of the ARTI epidemiological process represents symmetric series about straight line $W_i = 0$. There are seven outbursts in the time behaviour of the dynamic

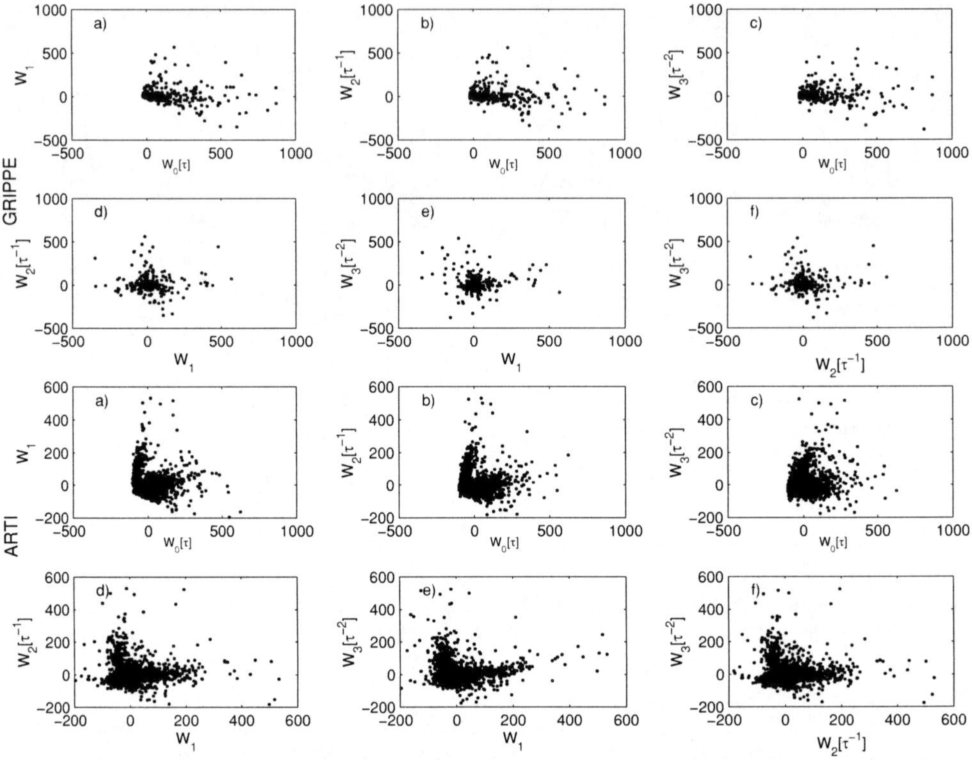

FIGURE 2. The phase clouds for epidemic processes of grippe and acute respiratory track infections (ARTI) in plane projections for the various combinations of orthogonal variables W_i, W_j where $i, j = 0...3$.

variables in the seven autumn - winter periods (from 10.27.1995 till 03.05.02). They corresponds to the seven cycles of grippe and ARTI epidemics. However the time series of orthogonal variables W_i $(i = 1,2,3)$ for ARTI differ from grippe by the strongly expressed asymmetry and the existence of a clearly expressed noise.

In Fig. 2 the phase clouds in the six plane projections of the four first dynamic variables W_i, $i = 0...3$ for epidemic processes of grippe and ARTI are submitted. In case of grippe the phase clouds have a well defined asymmetry about the centre of coordinates. All the phase clouds contain a centralized nucleus and an aggregate of points scattered on perimeter in a fan-shaped way. The interval of the dispersal makes up 900 τ. Owing to such asymmetric stratification the phase clouds get an elongated shape. The phase portraits of the ARTI epidemiological process have smaller symmetry about the centre of coordinates. The nucleus of the portrait gets the shape of a right angle and resembles a "boot". Apical emissions of separate points are appreciable on each side of the "boot". The interval of the dispersal of the points decreases up to 700 τ. The stratification of the phase clouds of a grippe epidemiological process is more essential, than in case of an ARTI epidemic. The analysis of the data in other Kazan districts reveals a prominent phase portrait of a grippe epidemic. It is the elongation of

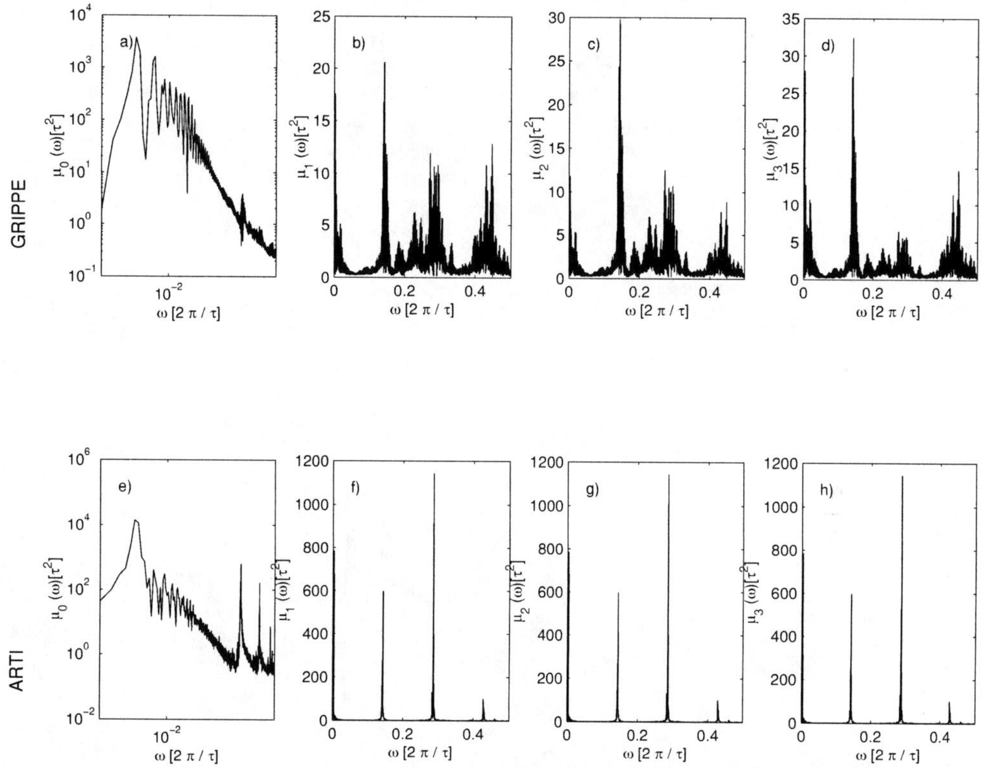

FIGURE 3. The power spectra of time correlation function (TCF) $\mu_0(\omega)(a)$ and of the three junior memory functions $\mu_i(\omega), i = 1(b), 2(c), 3(d)$ for the epidemic processes of grippe and acute respiratory track infections (ARTI). The spectral peaks in the registered spectra are the evidence of of long-range memory in the system.

the phase portrait for an ARTI epidemic namely its "boot-type" shape.

In Fig. 3 the power spectra of TCF $\mu_0(\omega)$ and three junior memory functions $\mu_i(\omega), i = 1,2,3$ for grippe and ARTI epidemiological processes are submitted. The frequency spectra of $\mu_0(\omega)$ are given in double logarithmic scale for a more detailed analysis of the data. In case of grippe the fractal dependence in the medium and high frequencies is found in the power spectrum of the initial TCF $\mu_0(\omega)$. The area of these frequencies is divided by a small outburst of power on frequency $\omega = 3 * 10^{-1} f.u.$ $(1 f.u. = 2\pi/\tau)$. In the domain of the low frequencies sharp breaks of the spectrum are observed. The frequency dependencies of the next three junior memory functions $\mu_i(\omega), i = 1,2,3$ are accompanied by spectral outbursts at equal intervals and by condensation of spectral lines near the outbursts. A spectral noise is present in all diagrams of the frequency dependencies. The spectrum of the initial TCF $\mu_0(\omega)$ for an ARTI epidemic process differs from a grippe epidemic only in the high frequency area. Additional spectral peaks are appreciable on frequency $\omega = 10^{-1} f.u.$ and $\omega = 10^{-1/2} f.u.$ The power spectra of the three junior memory functions $\mu_i(\omega), i = 1,2,3$ for an ARTI epidemic have bright

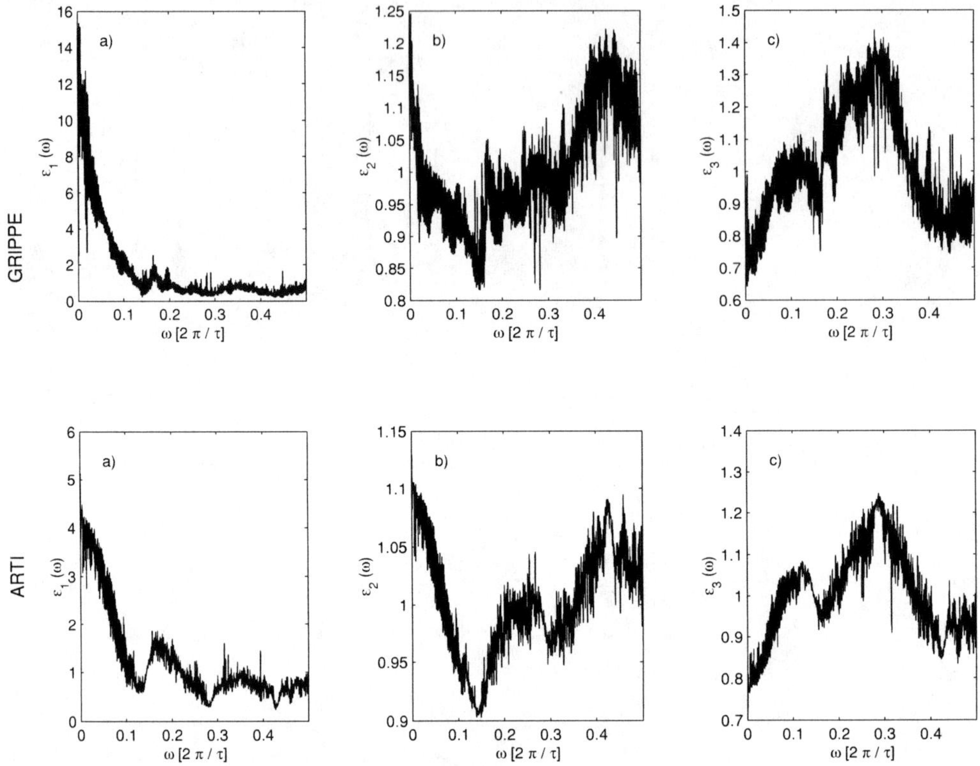

FIGURE 4. The frequency dependence of the first three points of statistical non-Markov parameter $\varepsilon_i(\omega)$ for epidemic processes of grippe and acute respiratory track infections (ARTI). This reflects the existence of long-range memory order in the system.

fractal dependence with several spectral peaks at every $0.15 f.u.$ In Fig. 4 the spectra of the first three points of statistical non-Markov parameter $\varepsilon_i(\omega)$, where $i = 1, 2, 3$ are presented. Fractalness is appreciable only in the behaviour of $\varepsilon_1(\omega)$. With approximation to frequency $\omega = 0$ the monotonous amplification of markovization is observed. The spectra of the non-Markovity parameter for the second point of $\varepsilon_2(\omega)$ for ARTI and grippe epidemics accept value $\varepsilon_2(\omega) = 1$ and are almost identical but have small fluctuations. It testifies to strong non-Markovity. The sinuosity with two frequency crests in area $\omega = 0.1\ f.u.$ and $\omega = 0.3\ f.u.$ is appreciable in frequency spectra of the third point of statistical non-Markovity parameter $\varepsilon_3(\omega)$.

Judging by the behaviour of the parameters the distinction between the epidemic of grippe (I) and that of ARTI (II) comes to the following. The initial values of the intensity $\mu_i(\omega = 0)$ for all $i = 0, 1, 2, 3$ differ for systems I and II by more than one order. It is the evidence of serious distinction in relaxation (correlation) times . For system I these times are almost 20-50 times shorter, than for system II. The intensity scales also differ almost by 50 times. Thus the initial values of $\varepsilon_1(\omega = 0)$ differ almost by 5 times for systems I and II.

On the basis of the aforesaid it is possible to come to certain conclusions about a great randomness of grippe dynamics in comparison with ARTI dynamics. The last one has greater regularity and predictability.

Another feature of the investigated epidemics is the existence of the three additional well - defined spectral splashes except the basic annual. From figs. 3 b-d and 3 f-h one can see that these groups of spectral peaks correspond to the characteristic frequencies equal accordingly to $0.14f.u.$,$0.28f.u.$ and $0.43f.u.$ The intensity of the high-frequency peak is almost an order lower, than the two other peaks of lower frequencies.

The strong distinction in the dynamic behaviour of grippe and ARTI epidemics is investigated in local time dependence of relaxation constants λ_1 , λ_2, λ_3, Λ_1 and Λ_2. It is necessary to remind, that parameters λ_i, $i = 1,2,3...$ remind Luapunov's exponents for various relaxation levels. For system II great regularity is observed. All parameters of λ_1, λ_2, λ_3 are negative, which is the evidence of the stability of the system. For system I (grippe) owing to the general instability the parameter $\lambda_1(t)$ gets small positive values on tail. This corresponds to the slump of epidemic.

CONCLUSIONS

In this paper we have shown that the chaotic dynamics of epidemic processes of grippe and ARTI can be considered as a stochastic discrete non-Markov process. It is more convenient to use the statistical theory of the discrete non-Markov processes for real objects and alive systems [23] - [25] to study the similar processes. The used theory allows to define essential distinctions between the epidemic processes of grippe and acute respiratory track infections by degrees of randomness, regularity and predictability. By processing the experimental data about patient's quantity we obtained the time series of dynamic variables $W_i(t)$ and calculated the memory functions and the first three points of the statistical non-Markov parameter. We used the power spectra, which were obtained with help of the fast Fourier transform to analyze the time functions (correlation and memory functions).

It is possible to come to the following conclusions on the basis of the conceptions about long-range memory and localization of parameters. The epidemic process of grippe causes a great danger to mankind because of its randomness and unpredictability. Each epidemic process of grippe is characterized by a sudden outburst and sudden attenuation. The peak of an epidemic lasts a comparatively short period of time (60 days). This means efforts to liquidate the consequences of a grippe epidemic should be made during this period of time. The epidemic process of ARTI is characterized by a greater regularity and predictability. It is accompanied by a small outburst at the beginning of the epidemic, then the number of patients diminishes. After this, a drastic outburst corresponding to the epidemic peak is observed. The process ends with small short-term splashes which later quickly fade. Unlike grippe, an ARTI epidemic is characterized by a less drastic outburst and attenuation. During the studied interval insignificant outbursts of respiratory infections are observed followed by periodic outbursts of ARTI.

We have revealed and explained essential distinctions in phase portraits, frequency spectra, and quantitative estimations of memory functions and statistical non-Markov parameters between the epidemic processes of grippe and ARTI. Relative distinctions of the quantitative data on epidemics of grippe and ARTI are the evidence of the distinction in the long-range memory, order, and organization of each process in the same system. These distinctions allow us to define the degree of importance of each of the epidemics. In particular, we have shown that the epidemic process of grippe and respiratory infections have essentially different quantitative measure of randomness and regularity.

The obtained results can be of practical value while researching other diseases and comparatively estimating epidemic danger.

In this paper we have demonstrated that the set of relaxation, kinetic, and spectral parameters and characteristics of discrete non-Markov stochastic processes are valuable for the description of the role of randomness, regularity, and predictability of epidemic processes, the spread of grippe and ARTI in the world.

The results obtained here generally relate to non-stationary and non-ergodic processes in stochastic systems. Therefore, they are certainly of value for the physics of the disordered matter in which similar processes can occur on molecular or structural levels.

ACKNOWLEDGMENTS

This work supported by the Graduiertenkolleg 283: Nonlinear Problems in Analysis, Geometry and Physics, of Deutsche Forschungsgemeinschaft (DFG) (P.H. and R.Y.), Russian Humanitarian Science Foundation (Grant N 00-06-00005a) and Russian Foundation of Basic Researches(Grant N 02-02-16146). The authors acknowledge Dr. L.O. Svirina for technical assistance.

REFERENCES

1. M.E. Newman, *Phys. Rev. E* **66**, 016128 (2002).
2. R. Pastor-Satorras, and A. Vespignani, *Phys. Rev. E* **65**, 035108 (2002).
3. P. Grassberger, H. Chate, and G. Rousseau, *Phys. Rev. E* **55**, 2488 (1997); M. Ipsen, and A. Mikhailov, *Rhys. Rev. E* **66**, 046109 (2002).
4. L.A. Amaral, A. Scala, M. Barthelemy, and H.E. Stanley, *Proc. Natl. Acad. Sci. USA* **97**, 11149 (2000); M. Barthelemy, and L.A.N. Amaral, *Phys. Rev. Lett.* **82**, 3180 (1999); S. Mossa, M. Barthelemy, H. E. Stanley, and L.A.N. Amaral, *Phys. Rev. Lett.* **88**, 138701 (2002); J. Camacho, R. Guimera, and L.A.N. Amaral, *Phys. Rev. E* **65**, 030901(R) (2002).
5. R.F. Burk, W. Schaffner, M.G. Koenig, *Arch. Intern. Med.* **127**, 1122 (1971).
6. S.D. Collins, J. Lehman, Excess deaths from influenza and pneumonia and from important chronic diseases during epidemic periods 1918-51, *Public Health monographs* **10**, 1 (1953).
7. S.W. Schwarzmann, J.L. Adler, R.J. Sullivan, W.M. Marine, *Arch. Intern. Med.* **127**, 1037 (1971).
8. C.H. Stuart-Harris, *Nature* **225**, 850 (1970).
9. C.K. Peng, S.V. Buldyrev, A.L. Goldberger, S. Havlin, M. Simons, H.E. Stanley, *Phys.Rev.E* **47**, 3730 (1993).
10. C.K. Peng, S. Havlin, H.E. Stanley, A.L. Goldberger, *Chaos* **6**, 82 (1995).
11. L.A.N. Amaral, S.V. Buldyrev, S. Havlin, M.A. Salinger, H.E. Stanley, *Phys.Rev.Lett.* **80**, 1385 (1998).

12. Y. Ashkenazy, P.Ch. Ivanov, S. Havlin, C.K. Peng, A.L. Goldberger, H.E. Stanley, *Phys.Rev.Lett.* **86**, 1900 (2001).
13. L.A.N. Amaral, P.Ch. Ivanov, N. Aoyagi, I. Hidaka, S. Tomono, A.L. Goldberger, H.E. Stanley, Y. Yamamoto, *Phys.Rev.Lett.* **86**, 6026 (2001).
14. V. Schulte-Frohlinde, Y. Ashkenazy, P.Ch. Ivanov, L. Glass, A.L. Goldberger, H.E. Stanley, *Phys.Rev.Lett.* **87**, 068104 (2001).
15. Z. Chen, P.Ch. Ivanov, K. Hu, H.E. Stanley, *Phys.Rev.E* **65**, 041107 (2002).
16. P.Ch. Ivanov, L.A.N. Amaral, A.L. Goldberger, S. Havlin, M.G. Rosenblum, H.E. Stanley, Z.R. Struzik, *Chaos* **11**, 641 (2001).
17. R.G. Webster, W.J. Bean, T.O. Gorman, *Microbiol.* **56**, 159 (1992).
18. H. Scheiblauer, M. Reinacher, M. Tashiro, R. Rott, *J. Infect Dis* **166**, 783 (1992).
19. F.M. LaForce, K.L. Nichol, N.J. Cox, *Am. J. Prev. Med.* **10**, 31 (1994).
20. P.O. Hokanen, T. Keistinen and S-L. Kivela, *Public Health* **110**, 163 (1996).
21. Y. Ghendon, *Euro J Epidemiology* **10**, 485 (1994).
22. R. Snacken, J.C. Manuguerra , P. Taylor, *Method of Information in Medicine* **37**, 266 (1998).
23. R.M. Yulmetyev, P. Hänggi, F.M. Gafarov, *Phys. Rev. E* **62**, 6178 (2000).
24. R.M. Yulmetyev,F.M. Gafarov, P. Hänggi, R.R. Nigmatullin, Sh. Kayumov, *Phys. Rev. E* **64**, 066132 (2001).
25. R.M. Yulmetyev, P. Hänggi, F. Gafarov, *Phys. Rev. E* **65**, 046107 (2002).
26. R.M. Yulmetyev, F.M. Gafarov, D.G. Yulmetyeva, N.A. Emelyanova, *Physica A* **303**, 427 (2002).
27. R. Yulmetyev, N. Emelyanova, P. Hänggi, F.Gafarov, A. Prokhorov, *Physica A* **316**, 361 (2002).

NONLINEAR DYNAMICS

Drastic Reduction of the Activation Barrier by a Moderately Weak Periodic Driving

S.M. Soskin*†, R. Mannella**†, A.N. Silchenko† and M. Arrayás‡†

*Institute of Semiconductor Physics, Ukrainian National Academy of Sciences, Kiev, Ukraine
†Department of Physics, Lancaster University, Lancaster LA1 4YB, UK
**Dipartimento di Fisica, Università di Pisa and INFM UdR Pisa, 56100 Pisa, Italy
‡Escuela Sup. de Ciencias y Tec., Universidad Rey Juan Carlos, 28933 Mostoles, Madrid, Spain

Abstract. We have found mechanisms by which a moderately weak *non-adiabatic* periodic driving may *significantly* reduce an activation barrier for a noise-induced escape from a potential well in an *underdamped* multi-well system. Even in a conventional case, the reduction may be comparable with the potential barrier while the driving amplitude is still small; such a reduction therefore exceeds by a few times the maximum reduction which would be provided by the adiabatic driving of the same amplitude. In the zero-dispersion case, the reduction comparable with the potential barrier may be achieved at yet smaller scales of the driving amplitude.

In the conventional case, the mechanism of the significant reduction has been understood and described quantitatively in the major part of the relevant frequency range, being also verified in simulations. In the zero-dispersion case, the mechanisms of the reduction have been understood while the reduction and the optimal driving parameters were correctly estimated semi-quantitatively and verified in simulations; but their rigorous quantitative description is a challenging open problem.

INTRODUCTION

The influence of a weak *non-adiabatic* periodic driving on noise-induced escape is a fundamental problem whose solution is far from completion, despite numerous studies (e.g. [1-6]). It is also relevant to many applications, e.g. to the destruction of metastable states in devices based on Josephson junctions[1, 4] or in mechanical electrometers [7], and to directed diffusion [5, 6].

Unlike most of the works on stochastic resonance (see [8, 9] for reviews) and early works on directed diffusion (e.g. [10]), which relate to *adiabatic* driving (when the escape rates are determined by the instantaneous value of the driving force), the escape rate for *non-adiabatic* driving does not manage to follow changes of the driving force. If the system is underdamped while the driving amplitude is much smaller than temperature, the main effect of the driving was shown [3] to be the enhancement of the diffusion over energy, which increases only the *prefactor* in the escape rate. But in the opposite limit, when temperature is smaller than the driving amplitude, the effect of non-adiabatic driving was shown [5] to be much stronger: the mechanism in [5] is based on the positive work by the force pushing the system resonantly with the eigenoscillation at the resonant energy, thus freeing the noise from this work in the range of energies close to the resonant one, which lowers the *activation barrier*.

However, the increase of the escape rate predicted in [3, 5] *diverges* as the damping Γ

CP665, *Unsolved Problems of Noise and Fluctuations: UPoN 2002: Third International Conference*, edited by S. M. Bezrukov

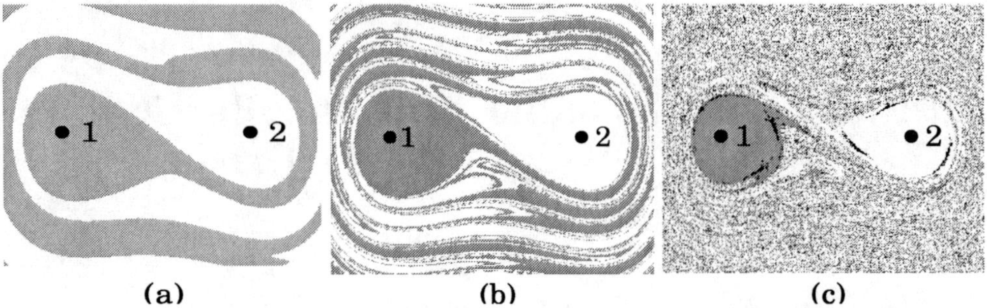

FIGURE 1. Stroboscopic ($\Omega t = 0, 2\pi, 4\pi, ...$) Poincaré section $\dot{q} - q$ of the noise-free ($T = 0$) system (1)-(2) for $A = 0.07$, $\Omega = 1.7$ as Γ decreases: (a) $\Gamma = 0.07$, (b) $\Gamma = 0.025$, (c) $\Gamma = 0.005$. Attractors corresponding to the linear responses in the left and right wells are marked by dots and labels 1 and 2 respectively. Their basins of attraction are shown by different shades of grey (small black areas in (c) are basins of attraction of period-3 orbits). The mixing of basins is: (a) absent, (b) already present, (c) well developed.

goes to zero. On one hand, this suggests that the system should be *underdamped* in order for the increase to be *maximally* possible. On the other hand, neither of these theories can answer what are (i) the maximum increase, (ii) the proper conditions for it, (iii) the underlying mechanism. Thus, it is extremely important to study the problem in the underdamped limit.

CONVENTIONAL CASE

We showed recently [11] that, unlike the case of vanishing small driving amplitudes $A \ll \Gamma$ when the major mechanism of the activation barrier reduction is the resonant mechanism [5], the major mechanism is typically different if A is *moderately small* i.e. small enough for the linear response to exist but large enough for the reduction to be comparable with the activation barrier in the undriven system (i.e. with the potential barrier). Namely, the major mechanism then is the lowering of the exit energy due to the onset of the layer of *transient chaos* arising in the phase space around the separatrix of the undriven system [12] (see Fig. 1): the lower energy boundary of the layer is lower than the top of the potential barrier, and it is enough for the noise to activate the system up to the boundary of the layer (rather than to the very top of the potential barrier, associated with the saddle of the undriven system) while the further escape may continue in a noise-free manner i.e. without any activation barrier.

The transient chaos layer in the *noise-free* system can be readily generated on computer and it is not difficult then to find its lowest (within the relevant potential well) energy, which constitutes the *exit energy* for the noise-driven conventional system provided the periodic driving is moderately small. Comparison of the theory with the results of simulations of the subjected to noise periodically driven Duffing oscillator (1),(2),

$$\Pi(\ddot{q}, \dot{q}, q, t) = \xi(t), \quad \Pi = \ddot{q} + 2\Gamma\dot{q} + U'(q) - A\cos(\Omega t),$$

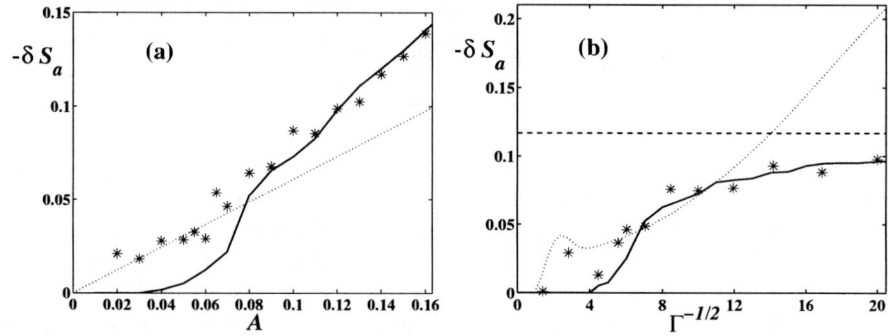

FIGURE 2. $-\delta S_a$ as function of (a) A (for $\Gamma = 0.025$), or (b) $\Gamma^{-1/2}$ (for $A = 0.07$); $\Omega = 1.7$ in both cases. Stars are obtained by means of an adjustment of S_a in (4) such that the flux J measured in the system (1),(2) (simulated on computer) at different temperatures fits Eq. (4); dotted lines are theory, based on the resonant mechanism [5]; solid lines are theory, based on the layer mechanism [11]. The dashed line in (b) shows the non-dissipative asymptote for the layer mechanism [11].

$$\langle \xi(t)\xi(t')\rangle = 4\Gamma T\delta(t-t') \tag{1}$$
$$U(q) = -q^2/2 + q^4/4. \tag{2}$$

is presented in Fig. 2. Stars show the driving amplitude and friction dependences of the reduction of the activation barrier

$$\begin{aligned} -\delta S_a &\equiv \Delta U - S_a, \\ \Delta U &\equiv U_{\max} - U_{\min} = 0.25, \\ U_{\max} &\equiv U(q = q_{\max} = 0), \qquad U_{\min} \equiv U(q = q_{\min} = -1), \end{aligned} \tag{3}$$

where ΔU is the potential barrier ($q_{\max}$ and $q_{\min}$ are positions of respectively the top of the barrier $U_{\max}$ and the bottom of the well $U_{\min}$) while the activation barrier S_a is extracted from data of simulations by means of fitting the activation law for the escape flux

$$J \propto \mathrm{e}^{-\frac{S_a}{T}} \tag{4}$$

The theoretical curves are evaluated using two different approximations of the method of large fluctuations (see [13, 14] for complementary general reviews), namely that based on the resonant mechanism [5] (dotted lines) and that based on the layer mechanism [11] (solid lines). The latter is, in fact, just the lowest energy in the transient chaos layer within the well.

It is worth mentioning that, in the underdamped (i.e. most relevant) limit, $-\delta S_a$ is nearly twice larger than the maximally possible (at the given A) reduction by the adiabatic periodic force of the same amplitude (the latter reduction is $\approx A|q_{\max} - q_{\min}| = A$): see the non-dissipative asymptote in Fig. 2(b).

Fig.2 clearly shows that, in the range $\Gamma \ll A$, $-\delta S_a$ is nicely described by the layer mechanism alone. It was therefore assumed in [11] that, as $\Gamma/A \to 0$, the resonant mechanism [5] vanishes due to the shrinking of the relevant band of energies around the resonant energy. However, more careful analysis [15] (also reviewed in Sec. 5.3 of [16]) has shown that such explanation is irrelevant. Moreover, in the asymptotic limit $A \to 0$ (but still $\Gamma \ll A$), the contribution from the range of resonant energies is of the order of the energy width of the lowest nonlinear resonance i.e. $\propto \sqrt{A}$, while the lowering of the exit energy due to the layer mechanism is much less (roughly, the latter is $\propto A \ll \sqrt{A}$). Thus, the contribution from the resonant energy range dominates and the corresponding mechanism may be called *nonlinear resonance* mechanism. How then to explain that the results of the simulations presented in Fig.2 are nicely described in the assumption that the resonant mechanism does not contribute? This paradox is resolved below.

The above argument concerning the nonlinear resonance is valid only in the limit $A \to 0$, when the layer of transient chaos is distinctly separated from the lowest nonlinear resonance. But, if the latter is much closer to the barrier level $U_{\max}$ than to the bottom $U_{\min}$, then there exists a range of *moderately weak* driving amplitudes which are weak enough for a linear response to exist but large enough to provide a complete absorption of the energy range of the lowest nonlinear resonance by the layer of transient chaos. When such an absorption takes place, then obviously only one mechanism of the decrease of S_a remains relevant – the lowering of the exit energy due to the transient chaos layer.

In the case of a conventional potential well like that of the Duffing potential, $\omega(E)$ decreases *monotonically* from ω_0 to 0 as energy varies from $U_{\min}$ to $U_{\max}$, with the major decrease occurring in a narrow vicinity of the barrier level. For $\omega(E)$ of such type, the absorption described in the previous paragraph is valid in the major part of the relevant driving frequency range $\Omega \lesssim \omega_0$, namely in the whole range $\Omega \lesssim \omega_0$ except the close vicinity of the bottom eigenfrequency ω_0 or of any of its harmonics of moderate order. On the other hand, if Ω lies in the close vicinity of ω_0, the linear response disappears even at quite weak driving and the problem of the escape from the well becomes irrelevant. Thus, if one is interested by such a driving which does not destroy the linear response while reducing the activation barrier for a value comparable with the potential barrier, then, in the major part of the relevant frequency range, the dominating mechanism of the reduction is the lowering of the exit energy due to the onset of the layer of transient chaos.

ZERO-DISPERSION CASE

The situation becomes essentially different in the *zero-dispersion* case [15, 16], when $\omega(E)$ possesses an extremum at some energy E_e,

$$d\omega(E_e)/dE_e = 0 \tag{5}$$

(cf. Fig.3(b)). If the driving frequency Ω is close[1] to the extreme eigenfrequency $\Omega_m \equiv \omega(E_e)$, then the two major nonlinear resonances may be very wide in energy while the linear response is still not destroyed. Besides, the chaotic layers around the nonlinear

[1] In fact, the requirement of the closeness is not very strict: the width of the relevant band of frequencies around Ω_m is typically comparable with Ω_m itself [15, 16].

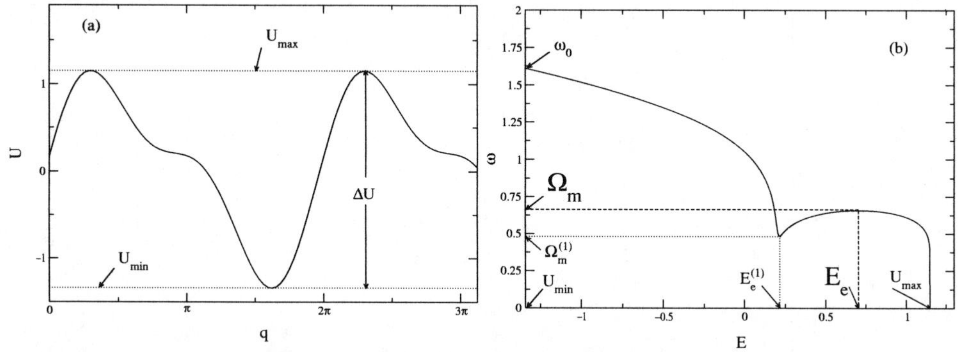

FIGURE 3. (a) The potential (a); and (b) the corresponding dependence of eigenfrequency on energy in between the minimum and maximum values of the potential: the most relevant extremum of $\omega(E)$ is indicated by the dashed lines and labels E_e and Ω_m for E and ω respectively; another extremum, indicated by the dotted lines and labels $E_e^{(1)}$ and $\Omega_m^{(1)}$, is relevant only to a very narrow range of energies around it and therefore its influence on the reduction of activation barrier is negligible.

resonances may overlap each other and the chaotic layer associated with the barrier [17, 16], which drastically lowers the exit energy. The onset of the overlap between the chaotic layers occurs at extremely small values of A [15-17], as can be seen from Fig. 4 which relates to the periodic potential

$$U(q) = \sin(q) + 0.45\sin(2q+0.4), \tag{6}$$

shown in Fig. 3(a) (the overlap is manifested in Fig. 4 as the drastic jump of the lowest energies in the corresponding layers). At the same time, the contributions from both the layer and nonlinear resonance mechanisms are of the same order in this case, namely of the order of the energy width of the nonlinear resonances. Moreover, these mechanisms cannot be resolved in this case. As can be seen from the comparison between the results of simulations and the theoretical calculations for the lowest energies in the layers (Fig.4), the jump-wise increase of the latter does provide correct qualitative estimates both for the lowest A at which $-\delta S_a$ becomes comparable with ΔU and for the corresponding $-\delta S_a$ itself. However, there is a non-small quantitative discrepancy. Even the exit energies do not coincide in this case with the lowest energies in the layers. The rigorous calculation of the reduction of the activation barrier in the zero-dispersion case is an important and challenging *open* problem.

It is worth noting that, as can be seen from the results of simulations shown in Fig. 4, a significant (about 20%) reduction of the activation barrier in the zero-dispersion system comes into play at *extremely small* driving amplitudes $A \sim 0.01$: in order to obtain the same reduction by means of the *adiabatic* periodic driving, one would need to use the amplitude approximately 10(!) times larger.

Finally, we comment that, despite the equality of the activation barriers for the escape to the left and to the right, the prefactors still differ due to the asymmetry of chaotic transport in the transient chaos layer (cf. [18]). This means that the significant reduction of the activation barriers by a weak periodic driving should lead to a drastic enhancement

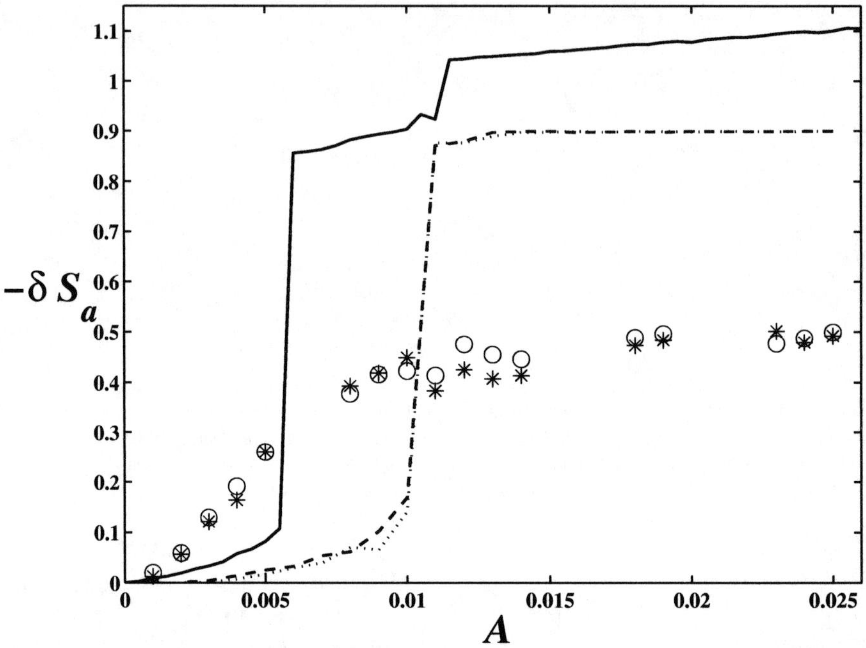

FIGURE 4. The reductions of activation barriers as functions of A, for the system (1),(6), with $\Omega = 0.63$ and $\Gamma = 5 \times 10^{-4}$ (the periodic potential $U(q)$ is shown in Fig. 3(a)). Results of simulations for the escape over the barrier to the left or right of an initial well are shown by stars and circles respectively. The theoretically calculated lowest within the initial well energy in the set of basins of attraction of attractors situated to the left/right from the barrier to the left/right from an initial well is shown by the dotted/dashed line. The non-dissipative approximation of the lowest energies is shown by the solid line.

of the ratchet effect (directed diffusion) in spatially periodic potentials, especially in those possessing the zero-dispersion property (5).

CONCLUSIONS

We have demonstrated that the reduction of the activation barrier by means of *non-adiabatic* periodic driving is most strong in *underdamped* systems, and it may become comparable with the original (i.e. potential) barrier at driving amplitudes smaller than those in case of an adiabatic driving. The difference is especially strong (by an order of magnitude) in *zero-dispersion* systems.

In case of a conventional system, the major mechanism of such a strong reduction of the activation barrier is the lowering of the exit energy due to the onset of the layer of transient chaos, whose lowest energy in the potential well is lower than the top of the potential barrier. The lowering is of the order of the driving amplitude and it can be quantitatively evaluated theoretically by means of a simple numerical procedure.

In case of a zero-dispersion system, there are two major mechanisms of the strong reduction: the lowering of the exit energy due to the transient chaos and the strong change of the most probable escape path in the energy range of nonlinear resonances. Both mechanisms are extremely strong in the zero-dispersion case, which is why the reduction becomes comparable with the potential barrier already at very small driving amplitudes. On the other hand, the quantitative description has not yet been achieved in this case and its derivation is an important and challenging *open* problem.

ACKNOWLEDGMENTS

The work was supported by the INTAS Grant 00-00867.

REFERENCES

1. M.H. Devoret, J. M. Martinis, D.Esteve, and J. Clarke, *Phys. Rev. Lett.* **53**, 1260 (1984); M.H. Devoret, D.Esteve, J. M. Martinis, A. Cleland and J. Clarke, *Phys. Rev. B* **36**, 58 (1987).
2. B. Carmeli and A. Nitzan, *Phys. Rev. A* **32**, 2439 (1985).
3. A.I. Larkin and Yu.N. Ovchinnikov, *J. Low Temp. Phys.* **63**, 317 (1986); B.I. Ivlev and V.I. Mel'nikov, *Phys. Lett. A* **116**, 427 (1986); S. Linkwitz and H. Grabert, *Phys. Rev. B* **44**, 11888 (1991); **44**, 11901 (1991).
4. R.L. Kautz, *Rep. Prog. Phys.* **59**, 935 (1996).
5. M.I. Dykman, H. Rabitz, V. N. Smelyanskiy and B. E. Vugmeister, *Phys. Rev. Lett.* **79**, 1178 (1997).
6. D.G. Luchinsky, M.J. Greenall, P.V.E. McClintock, *Phys. Lett. A* **273**, 316 (2000).
7. A.N. Cleland and M.L. Roukes, *Nature* **392**, 160 (1998).
8. M.I. Dykman, D.G. Luchinsky, R. Mannella, P.V.E. McClintock, N.D. Stein, N.G. Stocks, *Nuovo Cimento D* **17**, 661 (1995).
9. L. Gammaitoni, P. Hanggi, P. Jung, and F. Marchesoni, *Rev. Mod. Phys.* **70**, 223 (1998).
10. M. Magnasco, *Phys. Rev. Lett.* **71**, 1477 (1993).
11. S.M. Soskin, R. Mannella, M. Arrayás, and A.N. Silchenko, *Phys. Rev. E* **63**, 051111 (2001).
12. A.J. Lichtenberg and M.A. Liebermann, *Regular and Stochastic Motion* (Springer, New York, 1992).
13. M.I. Freidlin and A.D. Wentzell, Random Perturbations of Dynamical Systems, 2nd ed. (Springer, New York, 1998).
14. D.G. Luchinsky, P.V.E. McClintock, M.I. Dykman, *Rep. Prog. Phys.* **61**, 889 (1998).
15. S.M. Soskin, R. Mannella, A.N. Silchenko, unpublished.
16. S.M. Soskin, R. Mannella, and P.V.E. McClintock, Zero-Dispersion Phenomena, *Phys. Rep.*, in press.
17. S.M. Soskin, O.M. Yevtushenko, and R. Mannella, submitted to *Phys. Rev. Lett.*
18. S. Flach, O. Yevtushenko, Y. Zolotaryuk, *Phys. Rev. Lett.* **84**, 2358 (2000).

Universality in Escape from a Modulated Potential Well

M.I. Dykman*, B. Golding*, J.R. Kruse*, L.I. McCann† and D. Ryvkine*

*Department of Physics and Astronomy, Michigan State University, East Lansing, MI 48824
†Department of Physics, University of Wisconsin-River Falls, River Falls, WI 54022

Abstract. We show that the rate of activated escape W from a periodically modulated potential displays scaling behavior versus modulation amplitude A. For adiabatic modulation of an optically trapped Brownian particle, measurements yield $\ln W \propto (A_c - A)^{\mu}$ with $\mu = 1.5$. The theory gives $\mu = 3/2$ in the adiabatic limit and predicts a crossover to $\mu = 2$ scaling as A approaches the bifurcation point where the metastable state disappears.

Activated processes, such as escape from a metastable state, are exponentially sensitive to time-dependent external perturbations. Escape in driven systems has been investigated in studies of Josephson junctions [1, 2] and infrared photoemission [3]. It has been also explored recently in the context of thermally activated magnetization reversal in nanomagnets driven by a time-dependent magnetic field [4] or a spin-polarized current [5]. The strong dependence of the rates of activated processes on the parameters of the driving field leads to a variety of stochastic resonance phenomena [6, 7]. It also enables selective control of the rates, which is important for many applications, for example control of rate and direction of diffusion in spatially periodic systems [8].

In this paper we investigate the dependence of the rate of activated escape W on the amplitude A of a periodic driving field. We show that this dependence displays universal behavior for large amplitudes close to a critical value $\mathcal{A}_c$ and investigate the scaling of $\ln W$ with A. Experimental results are obtained for a colloidal particle trapped in a modulated optically-generated potential. This system was used earlier [9] for a quantitative test of the Kramers theory [10] of activated escape over a stationary potential barrier. We observe a power-law behavior of $\ln W$ as A approaches $\mathcal{A}_c$. The critical exponent for adiabatic modulation is $\mu = 1.5$, in agreement with theory. Adiabaticity is always violated close to the bifurcation point, and we predict the emergence of a different scaling, with $\mu = 2$.

Field dependence of the escape rate is well understood for adiabatically slow modulation, where the field frequency ω_F is small compared to the reciprocal relaxation time t_r^{-1}. The driven system remains instantaneously in thermal equilibrium. The adiabatic escape rate $W_{\rm ad} \propto \exp[-R_{\rm ad}/k_BT]$ depends on the field primarily through the instantaneous activation energy $R_{\rm ad}(t)$. Even where the modulation of $R_{\rm ad}$ is small compared to the zero-field value of $R_{\rm ad}$, it may still substantially exceed k_BT, leading to strong modulation of $W_{\rm ad}$.

Finding the escape rate becomes more complicated for nonadiabatic modulation, where $\omega_F \gtrsim t_r^{-1}$, because a driven system is far from thermal equilibrium. The problem

CP665, *Unsolved Problems of Noise and Fluctuations: UPoN 2002: Third International Conference,* edited by S. M. Bezrukov

has been addressed theoretically for different models of fluctuating systems [1, 11, 12, 13, 14, 15]. The underlying idea is that activated escape results from an optimal fluctuation, which brings the system from a metastable to an appropriate saddle-type periodic state. We use it here to draw general conclusions about the scaling of the period-averaged escape rate $\bar{W}$ with the field amplitude A.

The field dependence of $\ln \bar{W}$ for $\bar{W} \ll t_r^{-1}, \omega_F$ is sketched in Fig. 1. For small A, the field is a perturbation. Its major effect is to heat the system, with the change of the effective energy-dependent temperature quadratic in A (the lowest order term after period averaging). As a result, $\ln[\bar{W}/W_0] \propto A^2$, where W_0 is the escape rate in the absence of modulation [1, 2]. The range of A is limited to $\ln[\bar{W}/W_0] \lesssim 1$.

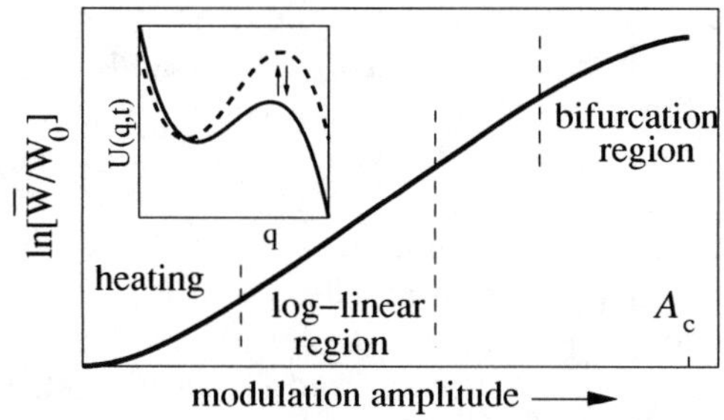

FIGURE 1. Amplitude scaling of the period-averaged rate of escape $\bar{W}$ from the modulated potential well shown in the inset (W_0 is the escape rate in the absence of modulation). The metastable state of forced vibrations disappears once the modulation amplitude reaches the critical value $\mathcal{A}_c$.

For stronger modulating fields, the activation energy of escape $R \propto \ln \bar{W}$ becomes linear in the amplitude A [12]. This behavior occurs because the optimal fluctuation leading to escape is a real-time instanton [16], with duration $\sim t_r$. In the absence of modulation, the probability density for such fluctuation to happen is independent of time. Modulation lifts the time degeneracy and synchronizes escape events. As a result, even for zero-mean modulation, $\ln \bar{W}/W_0 \propto A$. For an overdamped Brownian particle, an explicit solution was obtained [13] throughout the heating and log-linear regions in Fig. 1, and the results have been confirmed by extensive simulations [17].

As we show in this paper [18], there is another region where the dependence of $\bar{W}$ on A is universal and unexpected: the vicinity of the saddle-node bifurcation point $\mathcal{A}_c$ where the metastable state of forced vibrations merges with a saddle-type periodic state [19]. In this range one of the motions of the system becomes slow. The universality is related to the corresponding critical slowing down. However, it turns out that the system can display different types of critical behavior depending on the relationship between ω_F and the relaxation time in the absence of driving t_{r0}.

Experiments on amplitude-dependent escape from a sinusoidally modulated potential were performed using an optically-trapped Brownian particle in water. A configurable optical potential was constructed by focusing two parallel laser beams through a single microscope objective lens. Each beam creates a stable three-dimensional trap as a result of electric field gradient forces exerted on a transparent dielectric spherical silica particle of diameter 0.6 μm. At the focal plane, the beams are typically displaced by 0.25 to 0.45 μm, creating a double-well potential. The depths of the potential wells and the height of the intervening barrier, which is typically 1 to 10 k_BT, are readily controlled by adjusting beam intensities and separations.

The two HeNe lasers (17 mW, 633 nm) that create the traps are stabilized by electro-optic modulators and imaged into a sample cell by a 100x objective. The beams are mutually incoherent and circularly polarized when they enter the microscope. A single trapped colloidal sphere is imaged onto a digital camera operating at 200 frames/s. The x and y coordinates of the sphere's center are computed using a pattern matching routine that yields spatial resolution better than ± 10 nm, whereas the z coordinate parallel to the light wave vector is extracted by analysis of the image as the particle position fluctuates about the focal plane. The overall analysis and storing of the sphere's (x,y,z) coordinates is accomplished in less than a frame duration so that no images need to be recorded.

The stability of the system is sufficient to compute the full 3-dimensional optical potential from long-time measurements of the particle probability density [9]. The absolute transition rates calculated from the Kramers expression using measured curvatures at the stable and unstable points of the potential are in quantitative agreement with the experimental data over 4 orders of magnitude.

The electro-optic modulators that stabilize the optical traps also enable ac modulation with arbitrary waveforms derived from an electronic signal generator. When the two beams overlap at the separations required to form a potential barrier in the range 5 to 8 k_BT, each contributes to the position and shape of both potential minima and the potential in the intermediate range. As a consequence of this nonlocality, a small modulation of one beam causes a nearly antisymmetric change of both barrier heights.

The results on over-barrier transitions from one of the wells for the sinusoidal modulation $\delta U(t) = Ad\cos\omega_F t$ are plotted in Fig. 2. The amplitude A is the effective force amplitude, and d is the scaling factor of order of the interwell distance. In Fig. 2 A is normalized to k_BT, where T = 298 K. The intrawell relaxation rate in the absence of modulation $t_{r0}^{-1} \approx 10^3$ s^{-1} [9] is much larger than ω_F, so the data fall into the adiabatic regime. In this case escape is most likely to occur when the instantaneous potential barrier $\Delta U(t)$ is minimal. For large modulation amplitude and for temperatures used here, the maximal escape rate W_m becomes comparable to the modulation frequency, and therefore it is W_m rather than the period-averaged rate $\bar{W}$, that is the more relevant measure. In obtaining W_m we used the distribution of the dwell time measured for the modulated double-well system as the time interval between the appearance in, and escape from, a well.

The solid line in Fig. 2 represents a least-squares fit to the data for large A using the function $\ln W_m/W_0 = C_1 - C_2(A_c - A)^\mu$, where $\mu = 1.5$, and the critical modulation amplitude A_c and $C_{1,2}$ are fitting parameters [20].

To interpret these results we will consider activated escape in the system for A close to the critical amplitude $\mathcal{A}_c$ where the metastable and unstable periodic states merge. In the adiabatic approximation, the stable and unstable states correspond to the potential well and the barrier top. Once per period, when the driving force $|F(t)|$ is maximal (for example when $t = n\tau_F, n = 0, \pm 1, \ldots$), the stable and unstable states are closest to each other and the barrier height $\Delta U(t)$ is a minimum. We take $F(0)$ positive, $F(0) = A$. The well and barrier top merge for $A = A_c$, which is the critical amplitude $\mathcal{A}_c$ in the adiabatic approximation.

For small $\delta A = A - A_c$ and $|t - n\tau_F|$, the dynamics of the system near the metastable state is slow [19]. It is described by the "soft mode" with coordinate $q(t)$. For negative

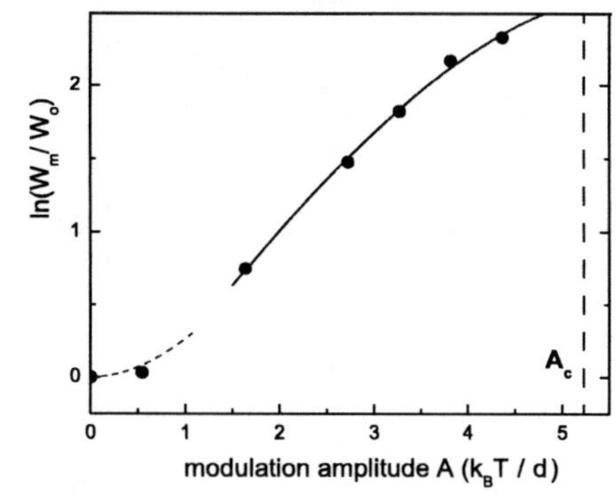

FIGURE 2. Modulation amplitude dependence of the maximal transition rate W_m normalized to W_0, the unmodulated transition rate. The data points represent the transition rates for an unmodulated barrier $\Delta U_0/k_BT = 5.8$ ($W_0 = 0.069\ \mathrm{s}^{-1}$). The solid line is a least-squares fit to the data with an asymptotic form described by the theory (see text). The modulation frequency $\omega_F/2\pi = 0.4$ Hz.

δA the system has a stable and an unstable state, q_a and q_b. We set $q = 0$ at the saddle-node state where they merge in the limit of small ω_F for $A = A_c, t = n\tau_F$. Close to A_c and $n\tau_F$ the equation of motion for q, neglecting fluctuations, takes the form

$$t_{r0}\frac{dq}{dt} = q^2 + F(t) - A_c \tag{1}$$

(t_{r0} is the relaxation time for $F = 0$, $\omega_F t_{r0} \ll 1$). In the adiabatic approximation, $F(t)$ depends parametrically on t. Activated transitions occur over the instantaneous barrier $\Delta U(t) = (4/3)[A_c - F(t)]^{3/2}$.

The relaxation time of the system $t_r = t_{r0}[A_c - F(t)]^{-1/2}$ becomes large close to the bifurcation point. Therefore even a slowly varying field will ultimately become too fast for the system to follow adiabatically. To allow for nonadiabaticity we expand $F(t)$ in the region of small $\omega_F|t - n\tau_F|$. The equation of motion (1) in scaled variables $Q = q/\Omega$, $\tau = \Omega(t - n\tau_F)/t_{r0}$ is then (cf. [21])

$$\dot{Q} = K(Q,\tau),\ K = Q^2 + \zeta - \tau^2, \tag{2}$$

with $\zeta = \Omega^{-2}(A - A_c)$. In Eq. (2) $\dot{Q} \equiv dQ/d\tau$, and $\Omega = (-\frac{1}{2}t_{r0}^2 d^2F/dt^2)^{1/4}$ (the derivative is evaluated at $t = 0$ and $A = A_c$). For sinusoidal modulation $\Omega = (A_c t_{r0}^2 \omega_F^2/2)^{1/4}$.

The dynamics (2) is determined by one dimensionless parameter ζ. For $-\zeta \gg 1$, the adiabatic approximation applies, and the stable and unstable states are given by $Q_{a,b}(\tau) = \mp(-\zeta + \tau^2)^{1/2}$. On the other hand, for $|\zeta| \lesssim 1$ the time dependence of the states becomes distorted and asymmetric, cf. inset in Fig. 3. For $\zeta = 1$, the states given by Eq. (2) merge and $Q_a(\tau) = Q_b(\tau) = \tau$. The corresponding weakly nonadiabatic bifurcational value of the amplitude is $A_b = A_c + \Omega^2$. This value is close to the "true" bifurcational value of the amplitude $\mathcal{A}_c$, see below.

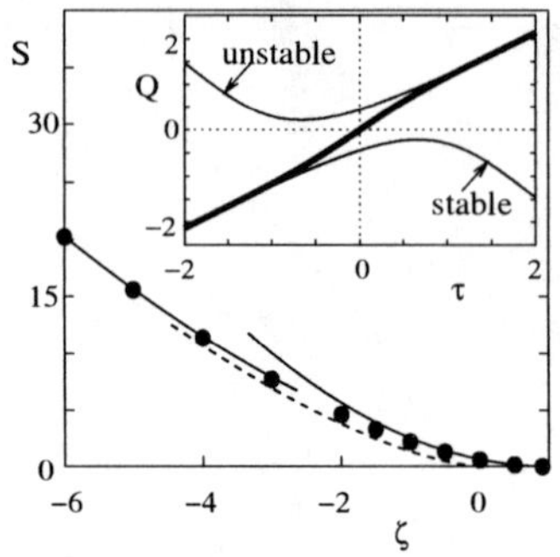

FIGURE 3. The scaled activation energy of escape $S \approx -\tilde{D}\ln W_t$ (dots), obtained numerically from the variational problem (3), vs. $\zeta = \Omega^{-2}(A - A_c)$. In the adiabatic approximation $S = \frac{4}{3}(-\zeta)^{3/2}$ (the dashed line), which explains the results in Fig. 2. The nonadiabatic bifurcation point is $\zeta = 1$. The solid lines show the corrected adiabatic expression for large $|\zeta|$, Eq. (4), and the nonadiabatic behavior $S \propto (1-\zeta)^2$, Eq. (5). Inset: the attracting and repelling states Q_a and Q_b (thin lines) for weakly nonadiabatic driving, Eq. (2) with $\zeta = 0.95$, and the most probable escape path (thick line).

Escape from the attractor can be quite generally described by adding a random Gaussian force to Eq. (1). Because the system dynamics is slow, this force is effectively δ-correlated (white noise), with intensity $D \propto k_BT$ for thermal noise. The corresponding force $\xi(\tau)$ in Eq. (2) for $\dot{Q}$ has a correlator $\langle\xi(\tau)\xi(\tau')\rangle = 2\tilde{D}\delta(\tau-\tau')$, $\tilde{D} = D/\Omega^3 t_{r0}$.

The noise-driven system is most likely to escape during the time when $F(t)$ is close to its maximum. The total probability to escape over this time is $W_t = \bar{W}\tau_F$, where $\bar{W}$ is the period-averaged escape rate. We will calculate W_t assuming that the noise intensity D is the smallest parameter of the theory, in which case $W_t \propto \exp(-R/D)$. The calculation can be done by generalizing the instanton technique [17] to the Langevin equation $\dot{Q} = K(Q,\tau) + \xi(\tau)$. The activation energy of escape $R = (D/\tilde{D})S$ is given by the solution of the variational problem

$$S = \frac{1}{4}\min\int_{-\infty}^{\infty} d\tau(\dot{Q}-K)^2. \tag{3}$$

The boundary conditions are $Q(\tau) \to Q_a(\tau)$ for $\tau \to -\infty$ and $Q(\tau) \to Q_b(\tau)$ for $\tau \to \infty$. They correspond to the picture in which, prior to escape, the system is performing small fluctuations about the metastable state Q_a. Escape occurs as a result of a large fluctuation that brings the system to the saddle state Q_b. The most probable among such fluctuations is the one in which the system moves along a trajectory [the most probable escape path (MPEP)] that minimizes the functional (3). From Eq. (3), the activation energy R is a function of one parameter ζ, shown in Fig. 3.

An explicit expression for R can be obtained in the adiabatic limit $-\zeta \gg 1$. Here, one can ignore the term τ^2 in K (2) for typical $|\tau| \lesssim \zeta^{-1/2}$. Then the MPEP is an instanton $Q_{\rm opt}(\tau,\tau_0) = |\zeta|^{1/2}\tanh\left[|\zeta|^{1/2}(\tau-\tau_0)\right]$ centered at an arbitrary τ_0. The adiabatic activation energy is $R \propto S = (4/3)(-\zeta)^{3/2} \propto (A_c - A)^{3/2}$ [21, 22]. This agrees with the experimental data shown in Fig. 2.

The time-dependent term in K lifts translational invariance of the instanton $Q_{\rm opt}(\tau,\tau_0)$ and synchronizes escape events. The action S is minimal for $\tau_0 = 0$. To lowest order in $1/\zeta$

$$S \approx (4/3)|\zeta|^{3/2} + (\pi^2/6)|\zeta|^{-1/2}. \tag{4}$$

From (4), the nonadiabatic correction to the activation energy diverges as $|\delta A|^{-1/2}$ as A approaches $A_{\rm c}$.

The activation energy may also be found explicitly close to the nonadiabatic bifurcation point $\zeta = 1$. In this case escape is most likely to occur while the coexisting attracting and repelling trajectories $Q_a(\tau)$ and $Q_b(\tau)$ stay close to each other. Because of the special structure of the states $Q_{a,b}(\tau)$, the variational problem (3) for the MPEP can be linearized and solved, giving

$$S = (\pi/8)^{1/2}(1-\zeta)^2,\ 1-\zeta \ll 1. \tag{5}$$

Eq. (5) shows that, near the nonadiabatic bifurcation point $A_{\rm b}$, there emerges another scaling region, where the activation energy $R \propto (A_{\rm b} - A)^2$. The nonadiabatic exponent is thus $\mu = 2$, in contrast to $\mu = 3/2$ in the adiabatic limit. It describes $R(A)$ for the modulation amplitude lying between $A_{\rm c}$ and $A_{\rm b}$, as seen from Fig. 3. The width of this interval is $\propto \omega_F$ provided $\omega_F t_{\rm r0} \ll 1$.

The above theory does not apply in the exponentially narrow range $1 - \zeta \lesssim \exp[-C/\omega_F t_{\rm r0}]$ $(C \sim 1)$. This is because, as A approaches $A_{\rm b}$, the relaxation time of the system $t_{\rm r}$ becomes logarithmically long. Then the expansion of the driving force in t/τ_F that leads to the equation of motion (2) becomes inapplicable. For $\omega_F t_{\rm r} \gg 1$ the duration of the MPEP greatly exceeds the period τ_F. In this case the activation energy again displays the $\mu = 3/2$ power-law dependence on the distance to the true bifurcation value of the amplitude $\mathcal{A}_{\rm c}$. The $(\mathcal{A}_{\rm c} - A)^{3/2}$-law also applies asymptotically if the driving is nonadiabatic for small A.

The experimental data discussed above refer to very small $\omega_F t_{\rm r0} < 10^{-3}$. Therefore the nonadiabatic region of modulating amplitudes is narrow and the activation energy there is less than $k_B T$. Higher modulation frequencies will be required to detect the crossover from the observed $\mu = 3/2$ scaling to the nonadiabatic $\mu = 2$ scaling. It appears that the $\mu = 3/2$ scaling function describes the data in a broad range of A. It thus provides a good interpolation of the activation energy from the linear in A to the critical region.

In conclusion, we have identified several regions where the activation energy R of escape from a metastable potential displays scaling behavior as a function of the amplitude A of an externally applied periodic modulation. We have demonstrated with a system whose potential was directly measured that, near the amplitude $A_{\rm c}$ where the local minimum and maximum of the potential contact one another, $R \propto (A_{\rm c} - A)^{3/2}$. We have also shown that, because of nonadiabaticity near the bifurcation point, there necessarily emerges a region where the scaling exponent changes from $\mu = 3/2$ to $\mu = 2$. We expect that these scalings can be observed in other systems, such as modulated Josephson junctions [2] and nanomagnets [4, 5, 23]. The ideas discussed here can be used to gain additional insight into the physics of these systems and more generally, activated effects in nonequilibrium systems.

ACKNOWLEDGMENTS

This research was supported by the NSF DMR-9971537 and NSF PHY-0071059.

REFERENCES

1. A.I. Larkin and Yu.N. Ovchinnikov, J. Low Temp. Phys. **63**, 317 (1986); B.I. Ivlev and V.I. Mel'nikov, Phys. Lett. A **116**, 427 (1986); S. Linkwitz and H. Grabert, Phys. Rev. B **44** 11888, 11901 (1991).
2. M.H. Devoret, D. Esteve, J.M. Martinis, A. Cleland, and J. Clarke, Phys. Rev. B **36**, 58 (1987); E. Turlot, S. Linkwitz, D. Esteve, C. Urbina, M.H. Devoret, and H. Grabert, Chem. Phys. **235**, 47 (1998).
3. F. Pisani, J. L. Fabre, S. Guizard, P. Palianov, Ph. Martin, F. Glotin and J. M. Ortega, Phys. Rev. Lett. **87**, 187403(4) (2001).
4. W. Wernsdorfer, E.B. Orozco, K. Hasselbach, A. Benoit, B. Barbara, N. Demoncy, A. Loiseau, H. Pascard, and D. Mailly, Phys. Rev. Lett. **78**, 1791 (1997); W. Wernsdorfer, E. B. Orozco, K. Hasselbach, A. Benoit, D. Mailly, O. Kubo, H. Nakano, and B. Barbara, Phys. Rev. Lett. **79**, 4014 (1997).
5. E. B. Myers, F. J. Albert, J. C. Sankey, E. Bonet, R. A. Buhrman, and D. C. Ralph, cond-mat/0203487.
6. M.I. Dykman, D.G. Luchinsky, R. Mannella, P.V.E. McClintock, N.D. Stein and N.G. Stocks Nuovo Cimento D **17**, 661 (1995); L. Gammaitoni, P. Hänggi, P. Jung, and M. Marchesoni, Rev. Mod. Phys. **70**, 223 (1998).
7. C.R. Doering and J. C. Gadoua, Phys. Rev. Lett. **69**, 2318 (1992).
8. F. Jülicher, A. Ajdari, and J. Prost, Rev. Mod. Phys. **69**, 1269 (1997);
9. L.I. McCann, M.I. Dykman, and B. Golding, Nature **402**, 785 (1999).
10. H. Kramers, Physica (Utrecht) **7**, 240 (1940).
11. R. Graham and T. Tél, Phys. Rev. Lett. **52**, 9 (1984); Phys. Rev. A **31**, 1109 (1985).
12. V.N. Smelyanskiy, M.I. Dykman, H. Rabitz, and B.E. Vugmeister, Phys. Rev. Lett. **79**, 3113 (1997); M.I. Dykman and B. Golding, Fluct. Noise Lett. **1**, C1 (2001).
13. V.N. Smelyanskiy, M.I. Dykman, and B. Golding, Phys. Rev. Lett. **82**, 3193 (1999).
14. J. Lehman, P. Riemann, and P. Hänggi, Phys. Rev. Lett. **84**, 1639 (2000); Phys. Rev. E **62**, 6282 (2000).
15. R.S. Maier and D.L. Stein, Phys. Rev. Lett. **86**, 3942 (2001).
16. J.S. Langer, Ann. Phys. **41**, 108 (1967).
17. M.I. Dykman, B. Golding, L.I. McCann, V.N. Smelyanskiy, D.G. Luchinsky, R. Mannella, and P.V.E. McClintock, Chaos **11**, 587 (2001) and references therein.
18. M.I. Dykman, B. Golding, J.R. Kruse, L.I. McCann, and D. Ryvkine, cond-mat/0204621.
19. J. Guckenheimer and P. Holmes, *Nonlinear Oscillators, Dynamical Systems and Bifurcations of Vector Fields* (Springer-Verlag, NY 1987).
20. The results of the least squares fit in Fig. 2 are: $A_c/k_BTd = 5.2 \pm 0.7, C_1 = 2.6 \pm 0.3, C_2(k_BTd)^{-\mu} = 0.27 \pm 0.05$. We also allowed the critical exponent μ to vary with a four-parameter fit, which led to the uncertainty $\delta\mu = \pm 0.1$.
21. N. Berglund and B. Gentz, J. Phys. A **35**, 2057 (2002).
22. For a metastable state near a termination point, $\mu = 3/2$ is the mean-field exponent describing the vanishing of the free-energy barrier as a termination point is approached along the approriate variable. For (quasi)stationary systems, it was discussed in various contexts, such as Josephson junctions [J. Kurkijärvi, Phys. Rev. B **6**, 832 (1972)] and magnets [R.H. Victora, Phys. Rev. Lett. **63**, 457 (1989)], as well as fluctuating dynamical systems [M.I. Dykman and M.A. Krivoglaz, Physica A **104**, 480 (1980); R. Graham and T. Tél, Phys. Rev. A **35**, 1328 (1987)].
23. R.H. Koch,G. Grinstein, G. A. Keefe, Yu Lu, P. L. Trouilloud, W. J. Gallagher, and S. S. P. Parkin, Phys. Rev. Lett. **84**, 5419 (2000).

Fluctuational Escape from Chaotic Attractors

I. A. Khovanov*, D. G. Luchinsky†, R. Mannella**, P. V. E. McClintock†
and A. N. Silchenko†

*Department of Physics, Saratov State University, Astrahanskaya 83, 410026, Saratov, Russia
†Department of Physics, Lancaster University, Lancaster, LA1 4YB, UK
**Dipartimento di Fisica, Università di Pisa and INFM UdR Pisa, 56100 Pisa, Italy

Abstract. Fluctuational transitions between two coexisting attractors are investigated. Two different systems are considered: the periodically driven nonlinear oscillator and the two-dimensional map introduced by Holmes. These two systems have smooth and fractal boundaries, respectively, separating their coexisting attractors. It is shown that, starting from a cycle embedded in the chaotic attractor, the periodically-driven oscillator escapes to a saddle cycle at the boundary of the basin of attraction, and does so through sequential transitions between saddles cycles embedded in the attractor. In the case of discrete dynamics with locally disconnected fractal boundaries, it is shown that escape from an attractor always seems to occur through an accessible boundary orbit and further through the specific homoclinic points forming a fractal structure of the boundary. It is shown that analysis of fluctuational transitions between attractors can be used to solve a problem of the energy-optimal migration of a chaotic system. The deterministic optimal control functions are identified with the corresponding optimal fluctuational forces in the limit of small noise intensity. We discuss possible applications and related unsolved problems of stochastic dynamics.

INTRODUCTION

The way in which fluctuational escape occurs from a chaotic attractor (CA) subject to noise has been a major unsolved problem ever since the first attempt to generalize the classical escape problem to cover this case [1]. We have recently used measurements of the prehistory probability distribution to analyse [2] escape from a chaotic attractor to a coexisting stable limit cycle in a periodically driven nonlinear oscillator – a situation where the basin boundaries are smooth. In the present work, we summarise these results, set them in context, and describe a completely new study in which we have extended the investigation to treat the discrete dynamics of a 2-dimensional map [3] for which the basin boundaries are *fractal*.

As in our earlier study [2], we bring together two seemingly quite different unsolved problems: first, that of fluctuational escape from a chaotic attractor and, secondly, the energy-optimal stimulation of escape in the absence of noise. We show that, remarkably, a solution of the first problem leads naturally to a solution of the second one.

The methods already available [4] for the control of chaos include entraining to a chosen "goal dynamics" that necessarily requires large modifications of the system's dynamics, and a variety of minimal forms of interaction which have hitherto been restricted by the linear approximations adopted. Solution of the energy-optimal control problem constitutes an important extension of the range of model-exploration objectives achievable through minimal control techniques.

CP665, *Unsolved Problems of Noise and Fluctuations: UPoN 2002: Third International Conference*,
edited by S. M. Bezrukov

The main difficulty in tackling these problems stems from the complexity of the system dynamics near a CA and in general near boundaries, which may be fractal. The problems are related, in particular, to delicate questions concerning the uniqueness of the solution and the boundary conditions at a CA with fractal basin boundaries. The approach that we apply is based on the analysis of an oscillator interacting with a thermal bath. In the zero-noise-intensity limit, a consistent theoretical development [5, 6] from the microscopic to the macroscopic equations of motion leads to descriptions of both its deterministic (dissipative) and fluctuational dynamics within the framework of Hamiltonian formalism [7]. It can be shown on physical grounds, and rigorously, that the Wentzel-Freidlin Hamiltonian [7] arising in this approach is equivalent to the Pontryagin Hamiltonian in the control problem [8] with an additive linear unrestricted control; the corresponding optimal control function is equivalent to the optimal fluctuational force. The solution of the energy-optimal control problem can thus be found to an excellent approximation by building the *prehistory probability distribution* [9] of fluctuational escape trajectories.

We illustrate the approach by analysing and comparing two very different cases: escape from the chaotic attractor of a periodically driven nonlinear oscillator where the basin boundaries are smooth [2]; and escape from the chaotic attractor of an invertible 2-dimensional map with fractal basin boundaries [3].

CONTINUOUS-TIME DYNAMICS

We consider a periodically driven nonlinear oscillator of form

$$\begin{aligned} \dot{q}_1 &= q_2, \\ \dot{q}_2 &= -2\Gamma q_2 - \omega_0^2 q_1 - \beta q_1^2 - \gamma q_1^3 - h\cos(\Omega t) + \xi(t), \\ &\quad \langle \xi(t) \rangle = 0, \quad \langle \xi(t)\xi(0) \rangle = D\delta(t). \end{aligned} \tag{1}$$

Here $\xi(t)$ is the Gaussian white noise. Parameters were chosen such that the potential is monostable ($\beta^2 < 4\gamma\omega_0^2$), the dependence of the energy of oscillations on their frequency is nonmonotonic ($\beta^2/\gamma\omega_0^2 > 9/10$), and the motion is underdamped $\Gamma \ll \Omega \approx 2\omega_0$. This model is of interest in number of applications and allows a theoretical analysis to be carried out over a wide range of parameters values. It is a system in which chaos can conveniently be observed at relatively small values $h \approx 0.1$ of the driving force amplitude.

For the given parameters $\omega_0 \approx 0.597$, $\beta = 1$, $\gamma = 1$ and, with damping constant $\Gamma = 0.025$, the amplitude and frequency of the driving force were chosen such that the chaotic attractor coexists with the stable limit cycle. The chaotic state appears as the result of period-doubling bifurcations and thus corresponds to a nonhyperbolic attractor (NHA). The boundary of attraction $\partial\Omega$ of the NHA is non-fractal and is formed by the stable manifold of the saddle cycle of period 1.

Direct application of Hamiltonian formalism [7] consists of consideration of a boundary problem and requires the identification of the generally unknown boundary conditions on the CA and on the boundaries of attraction (for fractal boundaries) and relies

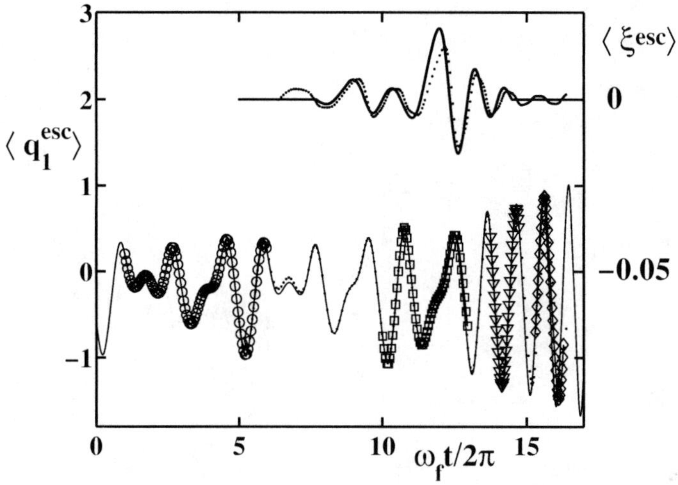

FIGURE 1. The most probable escape path (bottom, full curve) from the NHA to the boundary saddle cycle of period 1, found in numerical simulations with $\Gamma = 0.025$, $h = 0.13$, $\Omega = 0.95$, $D \approx 0.01$. Single periods of the unstable saddle cycles of period 5, 3 and 1 are shown by open circles, squares and triangles respectively; the stable limit cycle is shown by rhombus. The corresponding optimal force (top) after filtration. The optimal path and the optimal fluctuational force found by solution of the boundary value problem for motion equations with Wentzel-Freidlin Hamiltonian, corresponding to the system 1, are shown as dotted lines. After [2].

on the existence of the so-called most probable escape (optimal) path and corresponding optimal fluctuational force, which coincides with the optimal control function in the control problem. The use of the experimental prehistory technique [9] can solve all these problems and gives accurate approximation for the optimal control force.

The idea underlying our experimental technique is that, when the system (1) is driven by a random force $\xi(t)$ it will occasionally fluctuate to $\partial\Omega$. In doing so in the limit where the noise intensity tends to zero, the system will follow very closely the optimal escape path.

The method involves following the dynamics of the system continuously and collecting all successful realizations $(q_1^{esc}(t), q_2^{esc}(t), \xi^{esc}(t))$ that move it from the NHA to $\partial\Omega$. An approximate control function is then found as an ensemble average over the corresponding realizations of the random force $\langle\xi^{esc}(t)\rangle$. The boundary conditions are found from an analysis of how the energy-optimal escape path $(\langle q_1^{esc}(t)\rangle, \langle q_2^{esc}(t)\rangle)$ merges with the NHA.

A typical optimal escape path and the corresponding optimal force, obtained [2] by averaging a few hundred escape trajectories, are shown in Fig. 1. Analysis of the optimal path reveals that the system leaves the NHA along the unstable manifold of the saddle limit cycle of period 5 (with multipliers $\mu_1 = 0.04157 < 1$ and $\mu_2 = 4.60403728 > 1$) embedded in the NHA. At this moment the optimal fluctuational force $\langle\xi^{esc}(t)\rangle$ switches on, driving the system to $\partial\Omega$ via the saddle cycle of period 3 (with multipliers $\mu_1 = 0.04873 < 1$ and $\mu_2 = 7.608312 > 1$). Near the saddle cycle that forms the boundary of

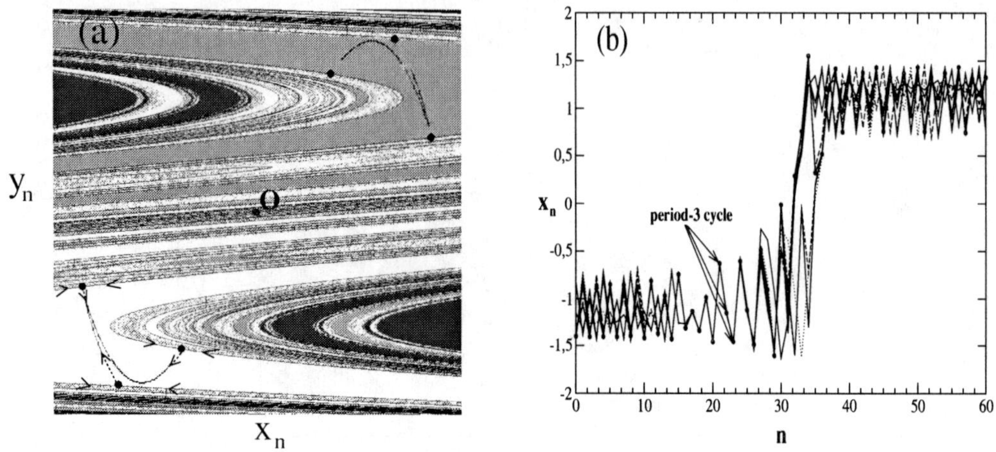

FIGURE 2. (a) Two co-existing chaotic attractors (full lines) in (2) at $a = 2.7, b = 0.2$ and their basins of attraction illustrated in grey and white respectively. The points of the period-3 homoclinic saddles cycle are labeled by black dots, and the point O corresponds to the saddle at $(0,0)$; (b) Some typical escape trajectories collected with $D = 10^{-5}$.

the basin of attraction the optimal force dies out. Note that no action is required to bring the system from S1 to the stable limit cycle.

Note that the saddle cycle of period 3 is not embedded in the NHA. It is probably the nearest saddle cycle to the boundary of the basin of attraction of the NHA in terms of the action variable, and can be considered as the boundary of the NHA itself.

DISCRETE-TIME DYNAMICS

As a second and very different example we consider fluctuational transitions between chaotic attractors separated by a fractal basin boundary in the two-dimensional map introduced my Holmes [3]:

$$\begin{aligned} x_{i+1} &= y_i, \\ y_{i+1} &= -bx_i + ay_i - y_i^3 + \xi_i, \end{aligned} \tag{2}$$

where ξ_i is white Gaussian noise with $\langle \xi_i \rangle = 0$ and $\langle \xi_i, \xi_j \rangle = 2D\,\delta_{ij}$. We choose (2) for our investigation because it exhibits a generic type of locally disconnected fractal basin boundary (see [10]), and because it reproduces most of the important features in the chaotic dynamics of the periodically driven Duffing oscillator [3]. The use of a discrete model brings the additional advantage of allowing us to speed up our numerical calculation for the case of low noise intensity.

We choose values of the control parameters a and b in (2) such that, in the purely deterministic case ($D = 0$), there are two co-existing chaotic attractors in phase space separated by a locally disconnected fractal basin boundary (see Fig. 2(a)). The fractal dimension of the boundary has been determined numerically as $d = 1.8441$ by using the

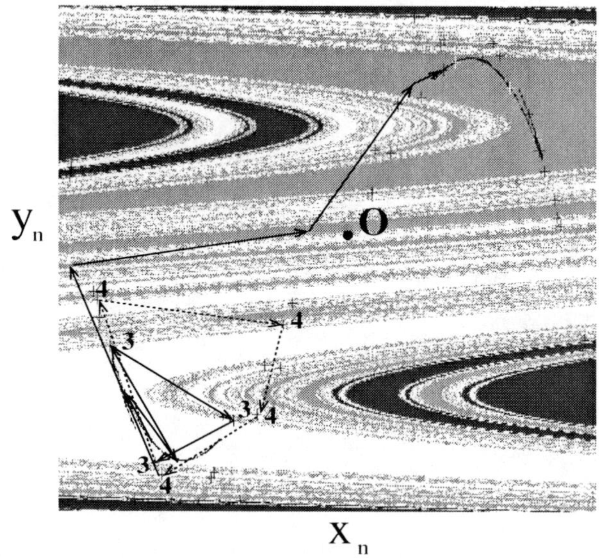

FIGURE 3. The escape trajectory in (2) from one chaotic attractor (full lines) to the other passes through the period-3 (full lines) and period-4 (dashed lines) original saddle cycles. The values of the control parameters and noise intensity are the same as in Fig. 1.

numerical "uncertainty exponent" technique introduced in [12]. We excite our system (2) with weak noise and collect trajectories that include escape paths from one chaotic attractor to the other (see Fig. 2(b)). The noise intensity was chosen in such a way that the mean escape time was essentially large; the characteristic relaxation time of an invariant measure on the corresponding chaotic set was estimated by us as 3×10^7 iterations. As can be clearly seen in Fig. 2(b), all escape trajectories have a part corresponding to the period-3 saddle cycle, apparently implying the presence of a boundary point located near, or directly on, the fractal boundary. Simple calculations have shown that the period-3 saddle cycle does exist for the chosen parameter values and that it lies on the boundary. Moreover, its stable manifold (full line in Fig. 2(a)), lying in the boundary, detaches the open neighborhood including the chaotic attractor from the fractal basin boundary itself. One part of its unstable manifold belongs to the homoclinic structure forming the fractal boundary, whereas the other part (labeled by the dashed line in Fig. 2(a)) approaches the attractor. Thus, we can classify this period-3 saddle as an accessible boundary point [13]. Indeed, by the definition given in [13] a boundary point P is accessible from a given region if there is a curve of finite length connecting P to an attractor in the interior of the region, such that no point of the curve lies in the boundary except for P. In our case, the part of the unstable manifold approaching the chaotic attractor plays the role of such a curve. Thus, in the present case, the period-3 saddle point plays the role of the boundary condition.

It is well known that the global behavior of a chaotic dynamical system is in many respects determined by a homoclinic structure, i.e. by a set consisting of homoclinic saddle cycles resulting from tangencies of the stable and unstable manifolds of a saddle

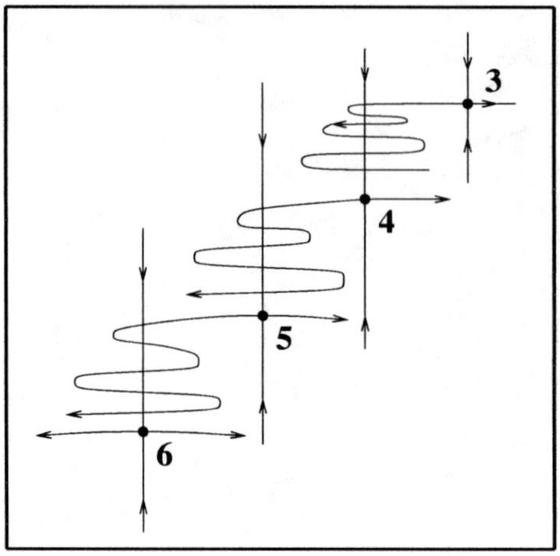

FIGURE 4. Heteroclinic crossings of the stable and unstable manifolds of different original saddles.

point. In our case, we observe an infinite sequence of saddle-node bifurcations of period 3,4,5,6,7..., which occur at parameter values $d_3 < d_4 < d_5 < d_6 < d_7$... and are caused by the sequential tangencies of the stable and unstable manifolds of the saddle point O at $(0,0)$. The homoclinic orbits appearing as the result of these bifurcations were classified earlier as *original saddles* [13]. It was also shown that their stable and unstable manifolds cross each other in a hierarchical sequence: the unstable manifold of the period-3 saddle crosses the stable manifold of the period-4 saddle, the unstable manifold of the period-4 saddle crosses the stable manifold of the period-5 saddle, etc. (see Fig. 4). Our numerical calculations have shown that these original saddles play a key role in the escape through a fractal basin boundary and that their hierarchy defines the structure of the escape paths. Fig. 3 shows a typical escape path obtained in numerical simulations. As clearly seen from this figure, a phase trajectory leaving the chaotic attractor penetrates into the fractal basin boundary through a small neighborhood of the period-3 saddle cycle, makes a few turns, and then approaches another period-4 original saddle point. After that it moves to the basin of the other chaotic attractor, reaching it in the next two or three iterations. In fact, the heteroclinic structure formed by numerous sequential crossings of the stable and unstable manifolds of original homoclinic saddles plays the role of the "staircase" allowing a trajectory to pass over the fractal basin boundary and defining the structure of the escape paths. Moreover, a hierarchical relation between original saddles can be revealed if we characterize them by a parameter μ equal to the ratio $\mid \lambda_s(P) \mid / \lambda_u(P)$ of the stable and unstable eigenvalues of the linearized deterministic flow at a saddle point P. Simple calculations show that, for the original saddles with periods 3,4,5,6,7,8... in (2), the following hierarchical sequence of index μ values occurs: $\mu_3 = 3.339$, $\mu_4 = 3.08$, $\mu_5 = 2.999$, $\mu_6 = 2.339$, $\mu_7 = 1.958$, $\mu_8 = 1.539$. The estimates of probabilities to find a fragment of the corresponding original saddle cycles in the escape trajectories are in a very good agreement with the hierarchical relation obtained above. By using the procedure described above for the continuous system, we

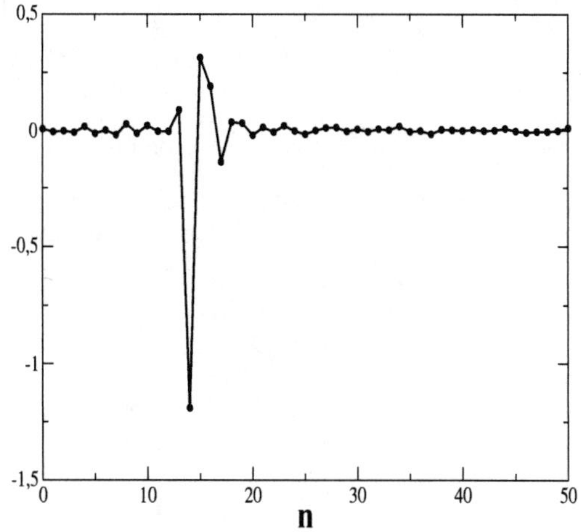

FIGURE 5. The optimal fluctuational force calculated at $a = 2.7, b = 0.2$ and $D = 1.25 \cdot 10^{-5}$.

were able to calculate the optimal fluctuational force migrating the phase trajectory from one attractor to the other (see Fig. 5).

Thus we conclude that the control function and the corresponding boundary conditions can be found using the same technique [2] for both continuous and discrete time chaotic dynamics, and with both smooth and fractal basin boundaries around the chaotic attractor. Moreover the escape problem has in this case been reduced to the analysis of transitions between saddle cycles, in qualitative agreement with the well known statement that unstable cycles provide detailed invariant characterizations for dynamical systems of low intrinsic dimension.

The probability of escape from a chaotic attractor can now be calculated, therefore, as a product of the probabilities of transitions between saddle cycles. To date, the theoretical procedure for calculation of these probabilities remains an unsolved problem. It requires calculation of the invariant measure of the saddle cycles in the presence of noise, and the development of methods for calculation of the transition probability between cycles as a function of noise intensity. Although it is a difficult problem, the present work has brought a solution closer and will be applicable in many areas of physics, including e.g. the stability problem for ac-biased Josephson [1] junctions and the problem of fluctuational transitions in lasers [11].

REFERENCES

1. R. Kautz, Phys. Lett. A **125**, 315-319 (1987).
2. I.A. Khovanov, D.G. Luchinsky, R. Mannella and P.V.E. McClintock, *Phys. Rev. Lett.* **85**, 2100-2103 (2000).
3. P. Holmes, Phil. Trans. R. Soc. **292**, 419-448 (1979).
4. A. L. Fradkov and A. Y. Pogromsky, *Introduction to Control of Oscillations and Chaos*, Vol. 35 of *series on nonlinear science A* (World Scientific, Singapore, 1998).
5. R. Zwanzig, J. Stat. Phys. **9**, 215-220 (1973).

6. M. I. Dykman and M. Krivoglaz, in *Soviet Physics Reviews*, edited by I.M. Khalatnikov (Harwood Academic Publishers, New York, 1984), Vol. 5, pp. 265–442.
7. M. I. Freidlin and A. D. Wentzel, *Random Perturbations in Dynamical Systems* (Springer, New-York, 1984).
8. L.C. Young, *Lectures on the Calculus of Variations and Optimal Control Theory* (Saunders, Philadelphia, 1969); P. Hagedorn, *Non-linear Oscillations* (Clarendon Press, Oxford, 1982).
9. M. I. Dykman et al, Phys. Rev. Lett. **68**, 2718-2721 (1992).
10. S. V. McDonald, C. Grebogi, E. Ott, and J. A. Yorke, Physica **17D**, 125-153 (1985).
11. V. N. Chizhevsky, E. V. Grigorieva, S. A. Kashchenko, Opt. Comm., **133**, 189-195 (1997).
12. C. Grebogi, S. W. McDonald, E. Ott, and J.A. Yorke, Phys. Lett. **99A** 415-418 (1983).
13. C. Grebogi, E. Ott, and J.A. Yorke, Physica **24D**, 243-262 (1987).

Two-State Description of Stochastic Excitability

Benjamin Lindner*, Lars Meinhold† and Lutz Schimansky-Geier**

**Department of Physics, University of Ottawa, 150 Louis Pasteur, Ottawa, Canada KIN 6N5*
†IWR - Biocomputing, Heidelberg University, Im Neuenheimer Feld 368, D-69120 Heidelberg
***Institute of Physics, Humboldt-University of Berlin, Invalidenstr. 110, D-10115 Berlin*

Abstract. We review analytic results of two-state dynamics that results from the approximation of excitable dynamics. By means of power spectrum, coeffient of variation and diffusion coefficient of the number of excitations we demonstrate coherence resonance in the Fitzhugh-Nagumo model and in a cluster of calcium ion channels.

INTRODUCTION

Excitability is encountered in fields as different as neurobiology and physical chemistry. It is commonly described by two-dimensional dynamical systems that respond to stimulation in an all-or-none fashion, i.e. either the input strength suffices to elucidate a strong nonlinear response (spike or pulse seen in one of the variables) or the system remains essentially silent. When noise is present in such an excitable system, unexpected phenomena like coherence resonance (CR) can be observed: the excitation sequence (pulse train) is most regular for a finite (optimal) amount of input noise [1]. This can be easily verified in computer simulations of the activator-inhibitor system; the effect is, however, more complicated to treat analytically at this level.
Here we show analytic results for CR in two different two-dimensional models using a two-state description of stochastic excitability. First, we will study the stochastic FitzHugh-Nagumo (FN) neuron model and explain the main idea of our approximation. Second, we consider a model for a cluster of calcium ion channels. The origin of noise is quite different in these two systems: while in the FN model noise is mainly given by random synaptic input, fluctuations occurring in the calcium dynamics are due to the finite number of interacting channels and hence in this case we deal with internal noise.

STOCHASTIC FITZHUGH-NAGUMO MODEL

This model is frequently used in computational studies of neurons. In terms of the fast voltage (or activator) variable $v(t)$ and the slow recovery (or inhibitor) variable $u(t)$ it reads

$$\tau \dot{v} = F(v) - u, \quad \dot{u} = \gamma v - u + b + \sqrt{2D}\xi(t) \tag{1}$$

where $\xi(t)$ is Gaussian white noise, τ is the (small) time scale ratio of the two variables and b is a constant determining positions of the null clines $u = F(v)$ and $u = \gamma v + b$.

CP665, *Unsolved Problems of Noise and Fluctuations: UPoN 2002: Third International Conference*, edited by S. M. Bezrukov

The nonlinearity in v is either given by a cubic function $F(v) = v - v^3$ or by a piecewise linear approximation of it

$$F_{pwl}(v) = \begin{cases} -1-v & , \quad v \leq -1/2 \\ v & , \quad -1/2 < v < 1/2 \\ +1-v & , \quad v \geq 1/2 \end{cases} \tag{2}$$

Under the influence of the noise the model generates spikes in the voltage variable that are separated by the so called interspike intervals (cf. Fig. 1, l.h.s.). How can the regularity of this spike sequence be quantified?
First, we may consider the statistics of the interspike interval, in particular its relative standard deviation, also known as coefficient of variation (CV)

$$R = \sqrt{\langle T^2 \rangle - \langle T \rangle^2} / \langle T \rangle \tag{3}$$

where $\Delta x^2 = (x - \langle x \rangle)^2$ denotes here and in the following the square displacement of a stochastic quantity. The CV is small for a regular periodic sequence of spikes and large for irregular spiking (a Poisson spike train yields $R = 1$).
Second, we may look at the power spectrum of $v(t)$

$$N(\omega) = \lim_{T\to\infty} \langle | \int_0^\infty dt v(t) e^{i\omega t} |^2 \rangle / T \tag{4}$$

that reveals a peak versus frequency ω in case of present oscillations.
We can also study the statistics of the pulse number (spike count) $n(T)$ in a time window $(0,T)$. Of special interest for regularity of spiking is the growth of variance in time characterized by the diffusion coefficient of the pulse number

$$D_{\text{eff}} = \lim_{T\to\infty} \langle \Delta n^2(T) \rangle / (2T). \tag{5}$$

A slight growth of variance, i.e. a small value of D_{eff}, indicates clearly a regularity of the single spike count process $n(T)$ and thus serves as another measure to characterize coherence resonance.
The two-state approximation which permits us to calculate all those quantities analytically has been worked out in two recent papers [2, 3] (see also [4] for more details); here we just sketch this derivation. In case of a perfect time scale separation, i.e. for $\tau \to 0$ in eq. (1), the system stays on the two stable branches (the left and right ones) of $F(v)$, it may, however, switch between these branches (this is illustrated in Fig. 1, l.h.s.) In this limit we can set $\dot{v} = 0$ in eq. (1) which yields a two valued function $v = F_{L,R}^{-1}(u)$ and hence two different Langevin equations (second one of eqs. 1) corresponding to the motion on the left and right branch, respectively. Trajectories on the left (right) branch that reach the minimum $u_- = F(v_-)$ (maximum $u_+ = F(u_+)$) of $F(v)$ are absorbed and infinitely fast reinjected at the corresponding point on the right (left) branch (see Fig. 1, middle panel) The potentials that govern the dynamics on each branch can be read off from the respective Langevin equations:

$$U_{L,R}(u) = (u-b)^2/2 - \gamma \int^u du' F_{L,R}^{-1}(u'). \tag{6}$$

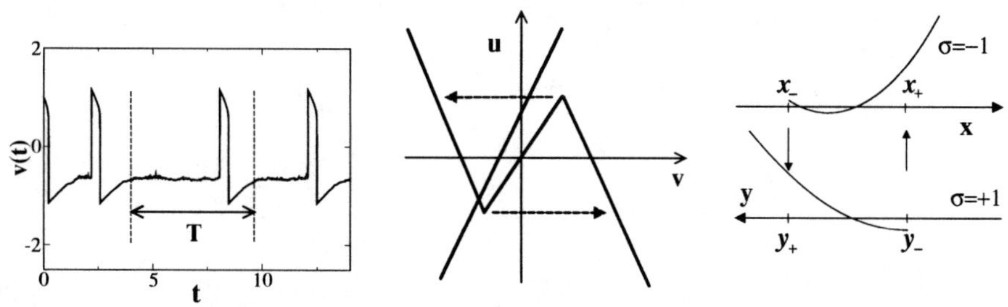

FIGURE 1. Left:sample trajectory of voltage variable with interspike interval indicated by the arrow. Middle: Piecewise linear null clines (eq. (2)) and possible transitions (arrows) for small τ. Right: two-state model with transition points, effective potentials (thin lines) and values attained by $\sigma(t)$.

Although their specific shapes depend on the function $F(v)$, there is an important difference between $U_L(u)$ and $U_R(u)$ (cf. Fig. 1,r.h.s.): while on the left branch we obtain a minimum in the potential that corresponds to the fixed point in the two-dimensional dynamics, for the potential on the right branch only a monotonous potential results. Under these conditions the system can escape from the left branch only by the assistance of noise while the right branch is left in a finite time ("downhill" motion until the absorbing boundary is reached) even in the absence of noise.
We may associate a discrete variable $\sigma(t) = \text{sgn}[v(t)]$ with the branch currently occupied, i.e. we approximate the excitable dynamics by a two-state model for which the waiting time in each state is governed by a Langevin dynamics and an absorbing boundary. The interspike interval consists of the sum of the subsequent passage times along the left and right branches. These times are statistically independent, therefore the spike train (the instants at which $\sigma(t) = -1 \to +1$) forms a *renewal process* [5]. For such a process, CV and diffusion coefficient of the pulse number can be expressed by the moments of the aforementioned passage times as follows

$$R = \sqrt{\langle \Delta T_L^2 \rangle + \langle \Delta T_R^2 \rangle} / (\langle T_L \rangle + \langle T_R \rangle) \,, \; 2D_{\text{eff}} = (\langle \Delta T_L^2 \rangle + \langle \Delta T_R^2 \rangle) / (\langle T_L \rangle + \langle T_R \rangle)^3 . \quad (7)$$

The moments of the passage times can be calculated using the classical quadrature formulae for the problem (see, e.g., [6, 4])

$$\langle T_{L,R} \rangle = \int_{u_-}^{u_+} dx\, I_{L,R}(x) \;, \quad I_{L,R}(x) = \frac{1}{D} e^{V_{L,R}(x)/D} \int_{-\infty}^{x} dy\, e^{-V_{L,R}(y)/D} , \quad (8)$$

$$\langle \Delta T_{L,R}^2 \rangle = 2 \int_{-\infty}^{u_+} dz\, [I_{L,R}(z)]^2 e^{-V_{L,R}(z)/D} \int_{z}^{u_+} dx\, \Theta(x-u_-) e^{V_{L,R}(x)/D} \quad (9)$$

with $V_R(x) = U_R(x)$, $V_L(x) = U_L(-x)$ and u_- and u_+ are the local minimum and maximum of the function $F(v)$, respectively. The quadrature results for CV and diffusion coefficient in the system with cubic null cline are shown in Fig. 2 and also compared

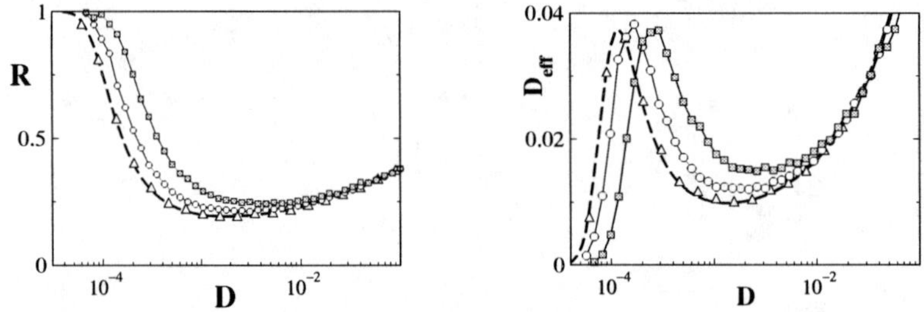

FIGURE 2. Coefficient of variation (left) and diffusion coefficient of spike count (right) versus noise intensity D for the FN model with cubic nonlinearity at $\gamma = 1.5, b = 0.6$. Approximation (dashed line) compared to simulations with $\tau = 10^{-3}$ (squares), $\tau = 10^{-4}$ (circles) and $\tau = 0$ (triangles).

to simulation results of the full dynamics at different values of the time scale separation constant τ. Both functions attain minimal values at a finite intensity of the applied noise indicating coherence resonance. Although there are only slight differences between approximation and numerical results, we would like to point out that CV and D_{eff} for small time scale separation reveal slightly deeper minima than the simulation results for finite τ. This means that a better time scale separation implies a more pronounced coherence. We note that while changing the systems parameters a minimum in CV is easy to obtain, the minimum in D_{eff} occurs only in case of strong excitability, i.e. just in the case where the CR effect is strongly pronounced. The diffusion coefficient of pulse number therefore constitutes a strong criterion of coherence resonance (see also [4], chapter 1.1).
The power spectrum of $\sigma(t)$ can be obtained applying Stratonovich's formula for a general two-state process [7] (we omit the δ peak at $\omega = 0$ due to a finite mean of $\sigma(t)$)

$$N(\omega) = \frac{8J_0}{\omega^2}\Re\left(\frac{(1-w_L(\omega))(1-w_R(\omega))}{1-w_L(\omega)w_R(\omega)}\right). \tag{10}$$

where $J_0 = 1/(\langle T_L + T_R \rangle)$ and $w_{L,R}(\omega)$ are the characteristic functions for the times T_L and T_R, respectively. For the piecewise linear nullclines eq. (2) these functions can be calculated

$$w_L(\omega) = e^{\frac{x_+^2 - x_-^2}{4\tilde{D}}} \frac{\mathscr{D}_{-i\omega}(x_+/\sqrt{\tilde{D}})}{\mathscr{D}_{-i\omega}(x_-/\sqrt{\tilde{D}})}, \quad w_R(\omega) = e^{\frac{y_+^2 - y_-^2}{4\tilde{D}}} \frac{\mathscr{D}_{-i\omega}(y_+/\sqrt{\tilde{D}})}{\mathscr{D}_{-i\omega}(y_-/\sqrt{\tilde{D}})}. \tag{11}$$

Here $\mathscr{D}_a(z)$ is the parabolic cylinder function [8] and we use rescaled time $\tilde{t} = (\gamma + 1)t$ and noise intensity $\tilde{D} = D/(1+\gamma)$ [3]. The $x_\pm, y_\pm$ are given by

$$x_\pm = \pm 1/2 - (b-\gamma)/(1+\gamma)\,,\; y_\pm = \pm 1/2 + (b+\gamma)/(1+\gamma) \tag{12}$$

The power spectrum resulting from these equations is shown for $\gamma = 1$, $b = 0.4$ and different noise levels in Fig. 3 (l.h.s.). While for small and large values of the noise intensity the spectrum falls off monotonously, at intermediate noise strength a peak versus frequency is obtained. This noise-induced eigenfrequency is another manifestation of coherence resonance. A comparison between this frequency and that given by the inverse mean interspike interval $\omega_m = 2\pi/\langle T\rangle$ is made in a counter plot in Fig. 3 (l.h.s., inset). Clearly, the noise-induced eigenfrequency is slightly larger than the mean frequency. In case of coherence resonance, however, both frequencies approach each other.

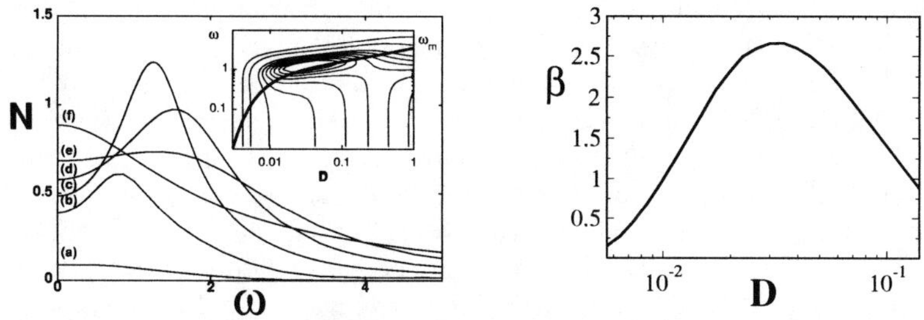

FIGURE 3. Left: output spectrum of the excitable system versus frequency and different noise levels (a)-(f): $\tilde{D} = 0.004, 0.009, 0.033, 0.094, 0.207, 0.769$. Inset: Contour plot of spectral density versus D and ω compared to mean frequency $\omega_m(\tilde{D})$ (thick line). Right: degree of coherence vs noise intensity.

We can measure the coherence of the oscillation by the so called degree of coherence

$$\beta = \frac{N(\omega_{max})}{(\omega_1 - \omega_2)/\omega_{max}} \quad \text{with} \quad N(\omega_1) = N(\omega_2) = N(\omega_{max})/a\,, \;\; \omega_1 < \omega_{max} < \omega_2 \quad (13)$$

Here, we choose $a = 1.2$, however, the value of a has no impact on the principal dependence of β on D. As can be seen this function (numerically calculated from eq. (10) and eq. (11)) attains a maximum versus noise intensity. A comparison to CV and D_{eff} curves for the system with piecewise linear null cline (not shown) reveals that all measures of CR attain their extremal value at roughly the same noise intensity if coherence resonance is strongly pronounced. For a weak CR effect, the minima of CV and D_{eff} and the maximum in the degree of coherence are attained at slightly different values that will depend on the specific definitions of the respective measure, e.g. the factor a in the above definition of β.

We note that the two-state description can be also used to calculate the linear response of the stochastic FN model with respect to periodic stimulation. In this case a possible functional role of coherence resonance becomes apparent - resonances with the noise-induced eigenfrequency might be useful for signal transmission through a neuron (see also [9, 10]).

CLUSTER OF STOCHASTIC CALCIUM CHANNEL

Next we consider as an application for an oscillating excitable dynamics a single cluster of calcium releasing ion-channels. Calcium plays the role of an important intra- and intercellular messenger in all types of cells and tissues. The elevation of free intracellular calcium concentration is due to the opening of calcium stores, e.g. the endoplasmatic reticulum (ER), which are gated by ion channels. Of importance for many intracellular processes is the family of inositol-(1,4,5)-triphosphate receptor (IP_3R) channels, for a review see [11], which is present in several types of tissues, e.g. neuronal tissue and smooth muscles.

We follow the idea that a certain number of ion channels in the cluster creates a necessary common level of noise which determines the stochastic dynamics of the whole cluster [12, 13, 14]. A mean field description of the calcium concentration, which is governed by fluctuations in the open probability of subunits, will show that the level of fluctuations, i.e. the number of channels in the cluster, plays the crucial role for the stochastic dynamics [15]. The power spectrum reveals collective stochastic calcium oscillations in the cluster which become most regular for an optimally selected noise level.

One channel is composed of four homolog subunits, each of which is activated by IP_3 and shows a bell shaped activation dependence on the calcium concentration [16], thus calcium induces and limits its own release. De Young and Keizer incorporated this to an eight-state model [17] which was later reduced to a two-state system (Li-Rinzel model) for this receptor [18, 19]. We apply the Li-Rinzel model for a single cluster of N_0 subunits, i.e. $N_0/4$ channels. The channel interaction is assumed to be instantaneous via the spatially homogeneous calcium concentration c=[Ca^{2+}] which is assumed to be realized by fast intracellular calcium diffusion. One channel is open if at least three out of four subunits are in the activated state [20],

$$P_{\text{open}} = x^4 + 4x^3(1-x)\,, \qquad x = \frac{pc(1-y)}{(p+K_1)(c+K_5)}, \tag{14}$$

where y is the probability of the subunit being calcium inactivated and p=[IP_3] is the concentration of the ligand IP_3.

If N of the N_0 subunits are in the inactivated state, a Master equation for the dynamics of inactivation can be formulated

$$\begin{aligned}\frac{\partial P(N,t)}{\partial t} &= -\left((N_0-N)K^+ + NK^-\right)P(N,t)\\ &+(N_0-N-1)K^+P(N-1,t) + (N+1)K^-P(N+1,t)\,.\end{aligned} \tag{15}$$

Therein the rate of inactivation $K^+(c)$ depends on the calcium concentration whereas the activation process, K^-, does not

$$K^+(c) = \frac{2c(K_1k_1k_4 + k_2k_4c + k_1k_2p)}{c(k_2+k_4) + 2k_1(K_1+p)}, \quad K^- = \frac{2(k_{-3}k_{-4} + k_{-2}(k_{-4}+k_3p))}{k_{-2}+k_{-4}+2k_3(K_3+p)}. \tag{16}$$

Eq. (15) can be expanded for $N \gg 1$ to give a Fokker-Planck Equation. The latter in turn is equivalent to the Langevin Equation

$$\dot{y} = (1-y)K^+ - yK^- + \sqrt{\frac{(1-y)K^+ + yK^-}{N_0}}\,\xi(t) \tag{17}$$

where $\xi(t)$ is zero-mean, Gaussian white noise, $\langle \xi(t)\xi(t+\tau)\rangle = \delta(\tau)$ and scales with an intensity inversely proportional to the number of clustering channels.

The intracellular calcium concentration is determined by

$$\dot{c} = (r_1 P_{\text{open}} + r_2)(c_{\text{ER}} - c) - r_3 \frac{c^2}{c^2 + K_p^2}, \qquad c_{\text{ER}} = (C_0 - c)/\alpha. \tag{18}$$

The first term models the gradient-dependent influx (Ca^{2+} source) while the second term represents the activity of the SERCA-pump (Ca^{2+} sink) which re-establishes this gradient; r_1, r_2 and r_3 are channel, leak and pump fluxes, respectively, c_{ER} is the ER calcium concentration, α is the ratio of ER volume to cell volume and C_0 is a constant, representing a local condition for a fixed amount of total cell calcium; our numeric standard parameters, including the dissociation constants $K_i = k_{-i}/k_i$, are given in [22]. Most of them are taken from [17, 21], some were adopted to new measurements [23] and some were slightly change by us to investigate other regimes.
To find an analytic description we have carried out further simplifications [15]: i) perfect time scale separation, ii) linear nullcline approximation, and iii) replacement of phase state dependent noise by additive noise with level taken in the fixed point $(c^s, y^s) = (0.076\,\mu M, 0.0802)$. This defines a two-state process switching between low and high intracellular calcium concentration. Shuai et al. [14] numerically described this process and computed its spectrum. We used the results of the piecewise linear FN model to find an analytic description [15].
The spectrum for the linearized system shows the same properties as Fig. 3, additionally the second harmonics of the peak are present. There exists an intermediate noise level for which the degree of coherence eq. (13) reaches a maximum value, for the corresponding number of channels calcium signalling can be considered most regular.
To prove the validity of our simplification we performed stochastic simulations of the unreduced Li-Rinzel model. Each of the four channel subunits was treated according to the Li-Rinzel model, i.e. the process $y(t) = 0$ or 1, to decide for a transition between these two states a uniformly distributed random number $\rho \in [0,1]$ was drawn and if $\rho\,\mathrm{d}t < K^{\pm}$ the subunit made the corresponding transition. For non-inhibited subunits the probability of being in the open-state is given by eq. (14) and the open state was set if $\rho' < x$, where ρ' is another. A channel opens if three or four subunits are in the open-state and the fraction N^{open}/N_0 of open channels substituted P_{open} in eq. (18). The evolution of $c(t)$ was sampled with $\mathrm{d}t = 0.01\,s$ for a time $T = 2621.44\,s$ and the time series was then zero averaged and fast Fourier transformed (with 2^{18} points). In order to obtain a smooth power spectrum we averaged over 300 runs. Results are displayed in Fig. 4(left). For a single channel (maximal noise level) the spectrum shows no peak but monotonously falls off for increasing frequencies. If the number of channels is increased, i.e. the system noise reduced, a peak emerges, reaches a maximum value and later starts to disappear again for very large clusters. Thus, the simulations show the same qualitative behavior as the reduced model shown on the r.h.s. in Fig. (4). To answer the key question – what is the optimal channel number per cluster with respect to signalling periodicity – we calculated the degree of coherence of the stochastic oscillations from eq. (13) with $a = 2$. The comparison of analytic and stochastic calculation is given in Fig. 4 (left) and shows excellent agreement.
In a numeric first approach similar results were obtained in [12, 13, 14] which showed the existence of a certain range of cluster size where signaling properties become most

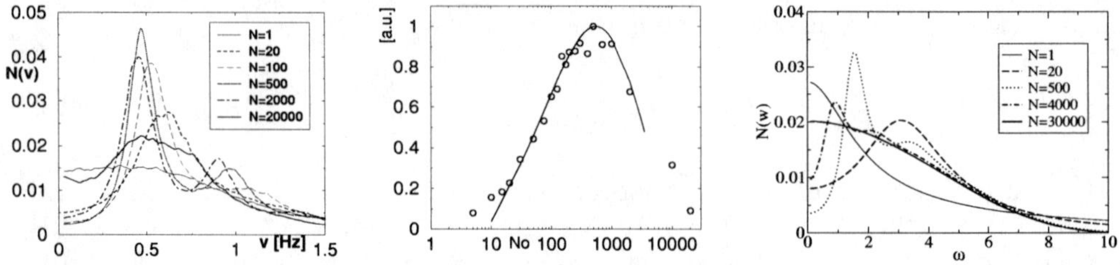

FIGURE 4. Left: Power spectra (left: simulations, right: analytic), degree of coherence [middle; stochastic simulation (circles) compared with analytic model (line)].

periodic and in this work we presented analytic evidence for that in a cluster of IP_3R-I calcium ion channels. We found a range of 20-750 channels per cluster optimal for signaling periodicity. The key parameter is the cluster size, i.e. the number of channels, governing the fluctuations in the clusters mean open probability.

This work was supported by DFG-Sfb 555 and Gk 268.

REFERENCES

1. A. Pikovsky and J. Kurths, *Phys. Rev. Lett.* **78**, 775 (1997).
2. B. Lindner and L. Schimansky-Geier, *Phys. Rev. E* **60**, 7270 (1999).
3. B. Lindner and L. Schimansky-Geier, *Phys. Rev. E* **61**, 6103 (2000).
4. B. Lindner, *Coherence and Stochastic Resonance in Nonlinear Dynamical Systems* (Logos-Verlag, Berlin, 2002).
5. D. R. Cox, *Renewal Theory* (Methuen, London, 1962).
6. C. W. Gardiner, *Handbook of Stochastic Methods* (Springer-Verlag, Berlin, 1985).
7. R. L. Stratonovich, *Topics in the theory of random noise*, vol.1, p.175 (Gordon and Breach, New York, 1963).
8. M. Abramowitz and I. A. Stegun, *Handbook of Mathematical Functions* (Dover, New York, 1970).
9. B. Lindner and L. Schimansky-Geier, *Phys. Rev. Lett.* **86**, 2934 (2001).
10. B. Lindner, L. Schimansky-Geier, and A. Longtin, *Phys. Rev. E* **66**, 031916 (2002).
11. Y. Yoshida, and S. Imai, *Jpn. J. Pharmacol.* **74**, 125 (1997).
12. P. Jung, and J. W. Shuai, *Europhys. Lett.* **56**, 29 (2001).
13. G. Schmidt, I. Goychuk, and P. Hänggi, *Europhys. Lett.* **56**, 22 (2001).
14. J.W. Shuai and P. Jung, *Phys. Rev. Lett.* **88**, 068102 (2002).
15. L. Meinhold and L. Schimansky-Geier, *Phys. Rev. E*, in print.
16. I. Bezprozvanny, J. Watras, and B.E. Ehrlich, *Nature* **351**, 751 (1991).
17. G.W. De Young and J. Keizer, *PNAS* **89**, 9895 (1992).
18. G.W. De Young and J. Keizer, *J. theor. Biol.* **166**, 431 (1994).
19. Y.X. Li and J. Rinzel, *J. theor. Biol.* **166**, 461 (1994).
20. J. Watras, I. Bezprozvanny, and B.E. Ehrlich, *J. Neurosci.* **11**, 3239 (1991).
21. M. Falcke, L. Tsimring, and H. Levine, *Phys. Rev. E* **62**, 2636 (2000).
22. Numeric model parameter: $K_1 = 0.0785\mu M$; $K_2 = 1.049\mu M$;$K_3 = 0.312\mu M$; $K_4 = 0.26393\mu M$; $K_5 = 0.0823\mu M$; k_1=400.0$(\mu M s)^{-1}$; k_2=0.3$(\mu M s)^{-1}$; k_3=400$(\mu M s)^{-1}$; k_4=0.3$(\mu M s)^{-1}$; k_5=80$(\mu M s)^{-1}$; $r_1 = 0.4\mathrm{s}^{-1}$; $r_2 = 0.02\mathrm{s}^{-1}$; $r_3 = 2.1(\mu M s)^{-1}$; $K_p = 0.08\mu M$; $C_0 = 2.0\mu M$; $p = 1.15\mu M$; $\alpha = 0.185$.
23. H. Yoneshima, A. Miyawaki, T. Michikawa, T. Furuichi, and K. Mikoshiba, *Biochem. J.* **322**, 591 (1997).

Singularities in Far-from-Equilibrium Distributions at Finite Noise Intensities

A. Bandrivskyy, S. Beri, D. G. Luchinsky, and P. V. E. McClintock

Department of Physics, Lancaster University, Lancaster LA1 4YB, UK

Abstract. How to find the (strongly non-Boltzmann) distribution in a far-from-equilibrium system is a problem of long standing. It appears in many different contexts, with topical examples including stochastic resonance and Brownian ratchets. One of the most promising approaches to the problem is through asymptotic analysis of the Fokker-Planck equation in the limit of small noise intensity. In simulations and experiments on real systems, however, the noise intensity is necessarily finite. Corrections to allow for finite noise intensity have recently been introduced for the particular case of escape. We are currently investigating the non-equilibrium distribution over the *whole* of phase space, for two model systems: the periodically driven, overdamped, Duffing oscillator and the inverted van der Pol oscillator. A modified Monte Carlo technique is being applied to investigate the limit of very small noise intensities. The next-to-leading order of approximation of the solution of the Fokker-Planck equation is used to compare the numerical results with the theory. We show, in particular, how changes in the non-equilibrium probability distribution induced by finite noise intensity are linked to an observable modification in the pattern of optimal paths. The numerical observations are in good agreement with theory.

INTRODUCTION

The analysis of fluctuations in systems away from thermal equilibrium represents an important, long-standing, problem [1] in statistical physics. It is one that has recently attracted renewed interest in the contexts of stochastic resonance [2, 3] and stochastic ratchets [4].

Away from thermal equilibrium, symmetries such as detailed balance and time-reversal symmetry are broken [5]. Although there are then no general methods available for calculation of the probability distribution, there is a promising approach involving asymptotic analysis of the Fokker-Plank equation. It is based on an approximation of the non-equilibrium distribution in the WKB-like form [6, 7]

$$\rho(x) = z(x)e^{-\frac{S(x)}{\varepsilon}}, \quad \varepsilon \to 0, \tag{1}$$

where the action function $S(x)$ plays the role of a non-equilibrium potential and $z(x)$ is a prefactor. Most earlier studies were restricted to calculations of the action function. For the *escape* problem, however, the prefactor has recently been considered in a number of publications. In [8, 9, 10, 11] the computation of the prefactor includes an analysis of the infinite discrete set of optimal trajectories that represent weak perturbations of the most probable escape path MPEP. In [12, 13] analytic results were obtained for escape rates at finite noise intensity. They utilized the logarithmic susceptibility (LS) theory

CP665, *Unsolved Problems of Noise and Fluctuations: UPoN 2002: Third International Conference,* edited by S. M. Bezrukov

[14, 15] to calculate the distribution away from the unstable limit cycle, and matched this distribution to the solution of the Fokker-Planck equation linearized near the cycle.

The methods of [8, 9, 10, 11, 12, 13] cannot, however, be applied to the analysis of noise-induced variations of the non-equilibrium distribution over the whole of phase space. In particular, they cannot address the problem of how singularities of the non-equilibrium potential [16, 17] result in singularities of the probability distribution in the presence of finite noise.

In this paper, we investigate the non-equilibrium distribution for two model systems: the periodically driven overdamped Duffing oscillator and the inverted van der Pol oscillator. In each case, the results are applicable over the whole of phase space. We use an asymptotic analysis of the Fokker-Planck equation that includes a calculation of the prefactor. The theoretical predictions are compared with the results of Monte Carlo simulations, using a modified technique that explores the limit of very small noise intensity, and with numerical integrations of the Fokker-Plank equation. We first outline the basis of the theory, and then apply it to the two nonlinear oscillators.

BASIS OF THE THEORY

We consider a 2-dimensional noise-driven dissipative dynamical system described by a Langevin equation of the form

$$\dot{x}_i = K_i(x) + \sqrt{\varepsilon}\sigma_{ij}\xi_j(t) \quad i,j = 1,2 \tag{2}$$

Here x is a coordinate in the state space of the system (2), $\xi_j(t)$ are Gaussian zero-mean uncorrelated sources of noise linearly mixed by the noise matrix σ_{ij}, and the noise intensity ε is the small parameter of the theory. For our two examples: (i) the periodically driven, bistable, overdamped Duffing oscillator is specified by $K_1 = x_1 - x_1^3 + A\sin(x_2)$, $K_2 = \Omega$, $\sigma_{ij} = \sqrt{2}\delta_{i1}\delta_{j1}$; and (ii) the inverted Van-der-Pol oscillator is specified by $K_1 = x_2$, $K_2 = -x_1 - 2\eta x_2\left(1 - x_1^2\right)$, $\sigma_{ij} = \sqrt{4\eta}\,\delta_{i2}\delta_{j2}$.

A stochastically equivalent Fokker-Planck equation corresponding to (2) describes the evolution of the probability density $\rho(x,t)$ for the system

$$\frac{\partial\rho}{\partial t} = \frac{\partial}{\partial x_i}\left(-\rho K_i + \frac{\varepsilon}{2}\frac{\partial}{\partial x_j}Q_{ij}\rho\right), \tag{3}$$

where the positively defined diffusion matrix $Q_{ij} = \sigma_{ik}\sigma_{jk}$. We are interested in a steady solution, so that $\partial\rho/\partial t = 0$.

Substituting (1) into the Fokker-Planck equation (3) and grouping the terms proportional to ε^{-1} one obtains a Hamilton-Jacobi equation in the form [6, 7]

$$H(p,x) = \frac{1}{2}Q_{ij}p_ip_j + K_i(x)p_i = 0, \tag{4}$$

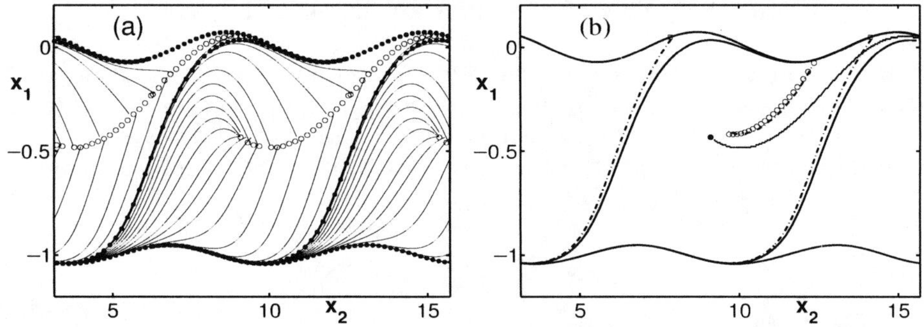

FIGURE 1. Effect of finite noise intensity on the escape paths and switching line for the periodically-driven overdamped Duffing oscillator with $A = 0.1$. (a) The limit $\varepsilon \to 0$. Optimal paths are shown as thin full lines and the switching line by open circles. The positions of the MPEP and of the stable and unstable states are shown by dotted lines. (b) Finite noise intensity, $\varepsilon = 0.01$. The switching line calculated via asymptotic analysis (dash-dotted line) is compared with the results of Monte Carlo simulations (open circles). Both it and the escape path (dash-dotted) are seen to differ from their zero noise equivalents (full curves). The stable and unstable solutions are also shown by full lines.

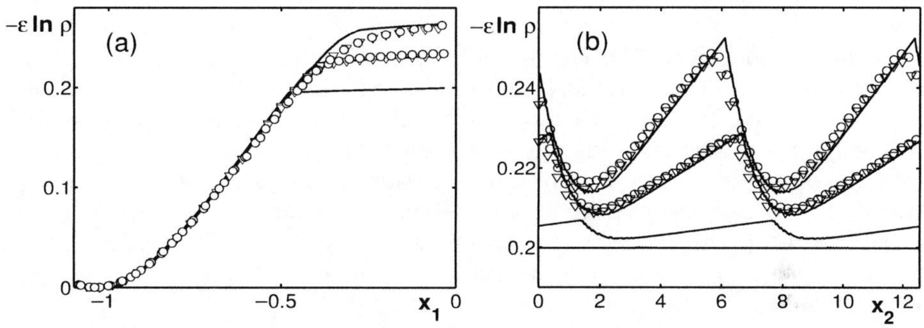

FIGURE 2. Cross-sections of the logarithm of the calculated probability distribution (full lines). (a) Results for $x_2 = 4.1$ and three different values of ε (from the top to the bottom): 0.02, 0.01, 0 are compared with Monte Carlo simulations (circles) and numerical integration of the Fokker Planck equation (triangles) for $\varepsilon = 0.02$ and 0.01 respectively. (b) Results calculated along the unstable limit cycle $x^{(u)}$ for four values of ε, from top to bottom, 0.01, 0.005, 0.001, and 0 are compared with Monte Carlo simulations and numerical integration of the Fokker Planck equation for $\varepsilon = 0.01$ and 0.005, respectively.

where $p_i \equiv \partial S/\partial x_i$ and $H(p,x)$ is the Hamiltonian function for an auxiliary Hamiltonian system. The solution of (4) can be found by the method of characteristics (see e.g. [6])

$$\begin{cases} \dot{x}_i = \frac{\partial H}{\partial p_i} = K_i + Q_{ij} p_j, \\ \dot{p}_i = -\frac{\partial K_j}{\partial x_i} p_j, \\ \dot{S} = \frac{1}{2} Q_{ij} p_i p_j. \end{cases} \tag{5}$$

Equations (5) describe a set of trajectories emanating from the stationary state of the system (2) and covering its state space.

APPLICATIONS TO NONLINEAR OSCILLATORS

Fig. 1(a) shows the trajectories described by equations (5) for the particular case of the Duffing system. Analysis of (4) shows that, in the general case of a non-equilibrium system, $S(x)$ is a multi-valued function of position x in the state space of system (2) [16, 17], i.e. more than one trajectory can arrive at the same point x, and the probability distribution takes the form [18]

$$\rho(x) = \sum_i z_i(x) e^{-\frac{S_i(x)}{\varepsilon}}, \tag{6}$$

where index i runs over the subset of trajectories that arrive to x. However, in the limit of $\varepsilon \to 0$ the only trajectories contributing to the sum are those that provide the global minimum of the action, known as *optimal paths*. The particular case of escape from the basin of attraction of a stationary state of the system (2) is governed by the properties of the so-called *most probable escape path* [19, 20] (MPEP) that emanates (at $t \to -\infty$) from the stationary state and arrives (at $t \to +\infty$) on the saddle cycle of (2).

As a result of global minimization [18] the state spaces of 2-dimensional systems such as (2) are separated by so-called *switching lines* into areas whither the system arrives along optimal paths of topologically different types. Along these lines the non-equilibrium potential is non-differentiable. Everywhere else, however, it is differentiable and it can therefore be treated as a generalized non-equilibrium thermodynamic potential for the system [16, 17]. The switching line and the most probable escape paths for the Duffing system are shown in Fig. 1. We note that this pattern is periodic in time with the period of the driving force T. In addition: (i) in each period there is only one path (MPEP) that approaches unstable limit cycle at $t \to +\infty$ (cf., however, a discussion of special cases [21]), (ii) all other paths are cut off by the switching line, (iii) the switching line also winds on the unstable limit cycle at $t \to +\infty$ always staying between two sequential turns of the MPEP.

With noise of finite intensity, this picture breaks down. In particular, as mentioned above, computation of the prefactor in the escape problem requires one to assume that, even for small noise intensity, more then one trajectory contributes to the probability of crossing the boundary [10]. The identification of the right set of trajectories that cross the boundary is a subtle problem [11] due to the fact [16] that optimal trajectories that are perturbations of the escape trajectory intersect one another wildly near the unstable limit cycle.

It is important to note that the switching lines and optimal paths are physically observable [22, 23] via measurements of the prehistory probability distribution [24] and that these observations are necessarily performed at finite noise intensities.

We now consider changes in the pattern of optimal paths, and in the logarithm of the probability distribution, induced by finite noise intensity, for the particular case of (2). Such an analysis can be conveniently performed in the next-to-leading order of approximation in ε of the solution of the Fokker-Plank equation, yielding [6] an equation for the prefactor in (1). Its calculation requires simultaneous integration of the equation for the second derivative of the action along the characteristics of (5). The equation for the second derivative of the action can be found by differentiating the Hamilton-Jacobi

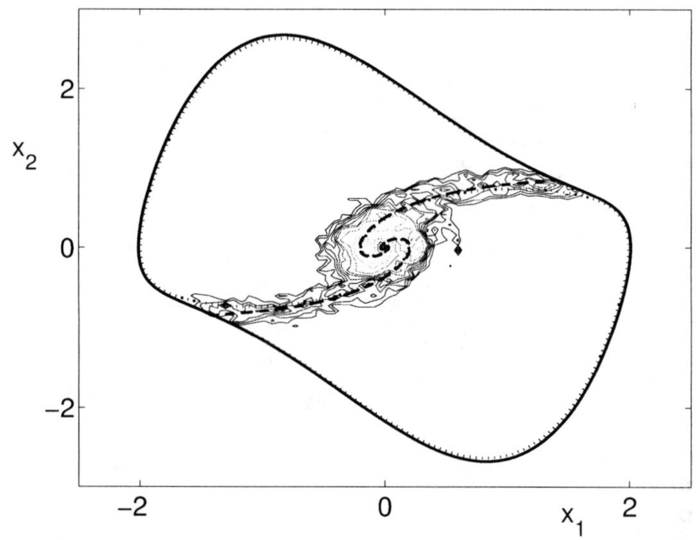

FIGURE 3. Monte Carlo simulations of escape events in the inverted van der Pol oscillator with $\eta = 0.5$, $\varepsilon = 0.05$. The contour plot of the probability distribution of simulated escape trajectories is compared with the zero-noise (dotted line) and finite noise (dashed line) most probable escape paths.

equation (4) twice [25] with respect to the position on the unstable Lagrangian manifold formed by trajectories emanating from the stationary state of (2). The resultant set of equations that have to be added to the system (5) is of form

$$\begin{cases} \frac{dz}{dt} = z\left(\frac{\partial K_i}{\partial x_i} + \frac{1}{2}Q_{ij}W_{ij}\right), \\ \dot{W}_{ij} = -p_m\frac{\partial^2 K_m}{\partial x_i \partial x_j} - W_{im}\frac{\partial K_m}{\partial x_j} - W_{jm}\frac{\partial K_m}{\partial x_i} - W_{jm}W_{ik}Q_{km}, \end{cases} \tag{7}$$

where $W_{ij} \equiv \frac{\partial^2 S}{\partial x_i \partial x_j}$.

The main results of the analysis of the extended system of equations (5),(7) are as follows. The position of the "switching line" is now defined, not by equality of the actions, but by the condition $z_1 \cdot \exp(-S_1/\varepsilon) = z_2 \cdot \exp(-S_2/\varepsilon)$ where z_i and S_i are respectively the prefactor and action calculated along trajectories that arrive to state x along two topologically different optimal paths. As a consequence, the switching line is shifted and crosses the saddle cycle (see Fig. 1(b)). The latter result is important because infinitely many optimal paths can now cross the saddle cycle. Moreover, for a finite noise intensity, the MPEP for the $\varepsilon \to 0$ limit ceases to be the most probable escape path, and its role is taken by a different path that crosses the saddle cycle (see Fig. 1(b)).

This picture is supported by direct measurements of the prehistory distribution of the optimal paths. The logarithm of the probability distribution calculated by asymptotic analysis of the Fokker-Planck equation is compared with the results of Monte-Carlo simulations and numerical integration of the Fokker-Plank equation in Fig. 2.

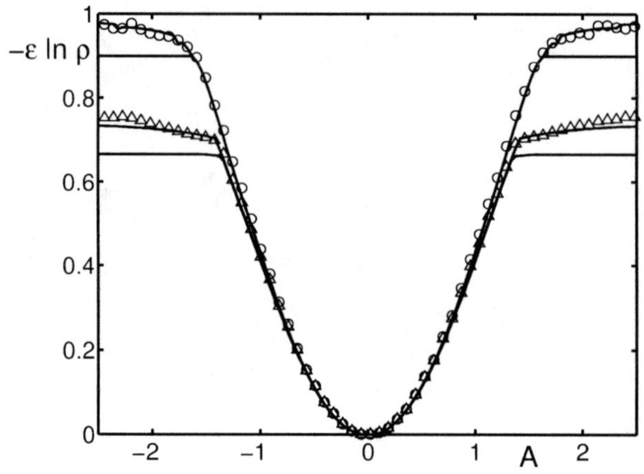

FIGURE 4. Sections $x_2 = x_1$ of $-\varepsilon \ln(\rho(x_1,x_2))$ for the calculated probability distribution (upper full line of each pair), for $\varepsilon = 0.01$, and η=0.25 (top) and $\eta = 0.5$ (bottom). The results are compared with Monte Carlo simulations for $\eta = 0.25$ (circles) and $\eta = 0.5$ (triangles). $A = \sqrt{x_1^2 + x_2^2}$ is the amplitude of oscillations. For comparison, we also show the theoretical results (lower full lines) with $\varepsilon = 0$ for both values of η.

Similar results are obtained for the inverted van der Pol oscillator. In Fig. 3, a contour plot of the probability distribution of the actual escape trajectories is compared with the zero-noise optimal escape trajectory (dotted line) and with the noise-shifted most probable escape path (dashed line). It can be seen from the figure that the distribution peaks along the corrected trajectory. Fig. 4 compares sections of $-\varepsilon \ln(\rho(x_1,x_2))$ obtained in Monte Carlo simulations with theoretical calculations of $-\varepsilon \ln(\rho(x_1,x_2))$ for two different values of η with noise intensity $\varepsilon = 0.01$. The theoretical curves obtained using equations (5) and (7) are in good quantitative agreement with the results of the numerical simulations.

CONCLUSION

In summary, we have investigated the important and hitherto unsolved problem of calculating the probability distribution in systems far away from thermal equilibrium in the presence of a small but *finite* noise intensity. The noise-induced variation of the probability distribution is described in terms of the topological changes of the modified non-equilibrium potential. In particular, for two archetypal non-equilibrium systems – the non-adiabatically driven overdamped Duffing oscillator and the inverted van der Pol oscillator – the noise induces shifts and smearing of the singularities of the non-equilibrium potential. As a result of the shift, the zero-noise most probable escape path ceases to reach the basin boundary. Instead, an infinite continuous set of trajectories can now cross the boundary even for very small noise strength. The theoretical predictions were verified by numerical solution of the Fokker-Planck equation and by Monte-Carlo simulations. The theoretical results are in good quantitative agreement with the numerical calculations for both systems over the whole of phase space. In the boundary region our results give a clear physical understanding of the origin of the set of trajectories crossing the boundary. This picture is consistent with the fundamental properties of the non-equilibrium potential which yield a more accurate calculation of the oscillation of the probability distribution at the boundary [8, 9, 10, 11, 12, 13]. In particular, these results suggest that the oscillations of the escape rate as a function of noise intensity predicted in [8, 11] cannot be observed. The method suggested in this paper can readily be extended to higher dimensional systems. Currently we are applying this technique to the calculations of the distribution of the mean-first passage time as a function of noise intensity in a Fitz-Hugh-Nagumo oscillator driven by a subthreshold periodic signal.

The work was supported by the Engineering and Physical Sciences Research Council (UK), the Joy Welch Trust (UK), the Russian Foundation for Fundamental Science, and INTAS.

REFERENCES

1. Onsager, L., *Phys. Rev.*, **37**, 405–426 (1931).
2. Dykman, M. I., Luchinsky, D. G., Mannella, R., McClintock, P. V. E., Stein, N. D., and Stocks, N. G., *Nuovo Cimento D*, **17**, 661–683 (1995).
3. Gammaitoni, L., Hänggi, P., Jung, P., and Marchesoni, F., *Rev. Mod. Phys.*, **70**, 223–287 (1998).
4. Julicher, F., Ajdari, A., and Prost, J., *Rev. Mod. Phys.*, **69**, 1269–1281 (1997).
5. Risken, H., *The Fokker-Plank Equation*, Springer, Berlin, ????, 2 edn.
6. Ludwig, D., *SIAM Rev.*, **17**, 605–640 (1975).
7. Freidlin, M., and Wentzel, A. D., *Random Perturbations in Dynamical Systems*, Springer, New-York, 1984.
8. Maier, R. S., and Stein, D. L., *Phys. Rev. Lett.*, **77**, 4860–4863 (1996).
9. Lehmann, J., Reimann, P., and Hänggi, P., *Phys. Rev. Lett.*, **84**, 1639–1642 (2000).
10. Lehmann, J., Reimann, P., and Hänggi, P., *Phys. Rev. E*, **62**, 6282–6303 (2000).
11. Maier, R. S., and Stein, D. L., *Phys. Rev. Lett.*, **86**, 3942–3945 (2001).
12. Smelyanskiy, V. N., Dykman, M. I., and Golding, B., *Phys. Rev. Lett.*, **82**, 3193–3197 (1999).
13. Smelyanskiy, V. N., Dykman, M. I., Rabitz, H., Vugmeister, B. E., Bernasek, S. L., and Bocarsly, A. B., *J Chem. Phys.*, **110**, 11488–11504 (1999).
14. Dykman, M. I., Rabitz, H., Smelyanskiy, V. N., and Vugmeister, B. E., *Phys. Rev. Lett.*, **79**, 1178–1181 (1997).
15. Smelyanskiy, V. N., Dykman, M. I., Rabitz, H., and Vugmeister, B. E., *Phys. Rev. Lett.*, **79**, 3113–3116 (1997).
16. Graham, R., and Tel, T., *Phys. Rev. Lett.*, **52**, 9–12 (1984).
17. Jauslin, H. R., *Physica A*, **144**, 179–191 (1987).
18. Dykman, M. I., Millonas, M. M., and Smelyanskiy, V. N., *Phys. Lett. A*, **195**, 53–58 (1994).
19. Dykman, M. I., and Krivoglaz, M. A., *Sov. Phys. - JETP*, **50**, 30–37 (1979).
20. Maier, R. S., and Stein, D. L., *Phys. Rev. Lett.*, **71**, 1783–1786 (1993).
21. Maier, R. S., and Stein, D. L., *Phys. Rev. Lett.*, **69**, 3691–3695 (1992).
22. Luchinsky, D. G., and McClintock, P. V. E., *Nature*, **389**, 463–466 (1997).
23. Luchinsky, D. G., McClintock, P. V. E., and Dykman, M. I., *Reports on Progress in Physics*, **61**, 889–997 (1998).
24. Dykman, M. I., McClintock, P. V. E., Smelyanskiy, V. N., Stein, N. D., and Stocks, N. G., *Phys. Rev. Let.*, **68**, 2718–2721 (1992).
25. Maier, R. S., and Stein, D. L., *SIAM J. Appl. Math.*, **57**, 752–790 (1997).

Stability Enhanced by the Noise in Nonlinear Systems with Metastable States

B. Spagnolo*, A. A. Dubkov† and N. V. Agudov†

*INFM, Unità di Palermo and Dipartimento di Fisica e Tecnologie Relative, Università di Palermo, Viale delle Scienze, I-90128 Palermo, Italy , Email: spagnolo@unipa.it, http://gip.dft.unipa.it
†Radiophysics Department, Nizhny Novgorod State University, Gagarin Ave.23, Nizhny Novgorod, 603950 Russia

Abstract. The problem of mean first-passage times (MFPT's) for Brownian particle in a fluctuating piecewise linear potential with metastable states is investigated. We find noise enhanced stability phenomenon (NES): The system remains in the metastable state for a longer time than in the deterministic case and the MFPT has a maximum at some noise intensity. We study the NES phenomenon in a system with metastable state in the presence of: (a) a periodical force, (b) a dichotomous random force. The conditions for the NES effect and the parameter region where the effect can be observed are obtained for all cases studied. In the presence of a dichotomous random force we obtain the analytical expression of MFPT as a function of the white noise intensity and the parameters of the dichotomous noise.

INTRODUCTION

The thermally activated escape of a Brownian particle over a potential barrier that is subjected to nonequilibrium fluctuations is central to many areas of physics, from diffusion in crystals to nucleation at phase transitions and has many applications in chemical and biological systems [1]-[3]. A particular challenging direction are systems far from thermal equilibrium, either due to non-thermal noise or external deterministic periodical forces [1]-[9]. Despite its experimental importance, the theory of oscillating barrier crossing in the regime of weak thermal noise is still at its beginning.

In present paper we will study the enhancement of stability by external noise in one-dimensional systems with a metastable state and with fluctuating potential. The fluctuations of the potential are due to: (i) a periodic driving force, (ii) a Markovian dichotomous random force. The noise enhanced stability (NES) phenomenon [4]-[7] has been observed in different physical systems. Among them there are the transient dynamics of an overdamped particle in a noisy cubic potential [9], the escape time in a tunnel diode periodically driven in a strong forcing regime [4, 6], the intermittent chaotic behaviour in one-dimensional maps [10], the noise induced stabilization [11] and the noise induced slowing down [12]. This effect implies that a system remains in the metastable state for a longer time then in deterministic case and the average escape time has a maximum at some noise intensity.

The paper is organized as follows. In the first section we consider the thermally activated escape of an overdamped Brownian particle over a potential barrier in the presence of a periodical dichotomous force. We analyse the role of different initial

CP665, *Unsolved Problems of Noise and Fluctuations: UPoN 2002: Third International Conference,* edited by S. M. Bezrukov

conditions of the particle on the average escape time from the metastable state and we give also a physical explanation of the NES effect. The case with a sinusoidal periodical force will be considered in the next section and the parameters region where the effect takes place will be given. The last section is devoted to the analysis of the thermally activated escape with randomly switching potential barrier. With this model we are able to obtain the exact expression of the average escape time as a function of the noise intensity and the switching mean frequency of the potential. We give also the conditions for the NES phenomenon.

DICHOTOMOUS DRIVING FORCE

We consider the model of overdamped Brownian motion described by the equation:

$$\frac{dx}{dt} = -\frac{dU(x)}{dx} + F(t) + \xi(t), \tag{1}$$

where $\xi(t)$ is the white Gaussian noise with zero mean, $< \xi(t)\xi(t+\tau) >= 2q\delta(\tau)$, $F(t)$ is the dichotomous driving force and $U(x)$ is a piecewise potential with the reflecting boundary and a metastable state at $x = 0$, and a maximum at $x = L$:

$$U(x) = \begin{cases} \infty, & x = 0 \\ hx, & 0 < x \leq L \\ E - k(x-L), & L < x \leq b, \end{cases} \tag{2}$$

where $h > 0$, $k > 0$ and $E = hL$ is the height of the potential barrier. For fixed potential $(F(t) = 0)$, when the particle is within the potential well, the decay time of metastable state increases exponentially in the limit $q \to 0$, according to Kramers formula [13]. If the starting position of the particle is between the maximum and the right boundary of the potential $(L < x_0 < b)$, then the initial state is unstable. Specifically if the initial position of the particle is between the maximum and the crossing point of the potential with the x-axis then the average escape time (MFPT) rises to infinity when $q \to 0$, while for zero noise we obtain a finite deterministic decay time. The average escape time has therefore a singularity at $q = 0$ for the following range of starting positions of the particle: $L < x_0 < L + E/k$. As a consequence a small quantity of noise can trap the particle into potential well and the NES effect appears [5, 7].

We consider now the initial state at $x(0) = 0$ and dichotomous driving force $F(t) = \pm a$ with period T. Potential $U(x)$ is defined by Eq.(2) where $h = 0$. We choose $F(t) = +a$, i.e. the potential barrier is absent for the first half of a period. In the absence of noise Eq.(1) has a periodical solution in the deterministic regime for $T < 2L/a$ and the particle always remains trapped in the metastable state at $x < L$. If the period of the driving force is:

$$\frac{T}{2} > \frac{L}{a} \tag{3}$$

the particle surmounts the region $[0, L]$ at time $t = L/a$ and the NES effect will occur if the position x of the particle at time $t = T/2$ is in the following range:

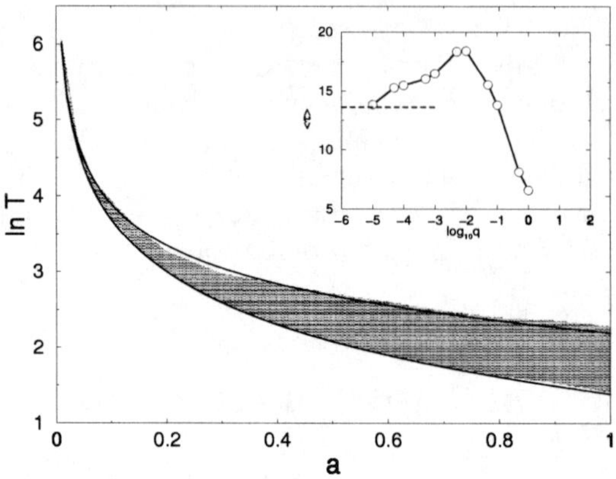

FIGURE 1. The shaded area is the region of the plane (lnT, a) where the NES effect is very strong: the average escape time is greater than 10% above the deterministic escape time. The parameters are: b=7, k=1, l=2. Inset: the average escape time versus the noise intensity for a=0.3 and T=13.5. The dashed line indicates the deterministic escape time.

$$L < x(T/2) < b. \tag{4}$$

This condition can be rewritten as follows [7]:

$$\frac{L}{a} < \frac{T}{2} < \frac{ab+kL}{a(a+k)}. \tag{5}$$

This inequality together with the condition $a < k$ gives the area on the parameters region (T,a) where the NES effect takes place for small noise intensity with respect to the barrier height (see Fig.1). Below the lower boundary we obtain Kramers-like behaviour. The magnitude of the NES effect decreases from the lower to the upper boundary. This is because in this parameters region the potential barrier is very small or absent during the noise induced escape process.

SINUSOIDAL DRIVING FORCE

In the case of sinusoidal force $F(t) = a \cdot \sin \omega t$ we consider the same fixed potential profile $U(x)$ of Eq.(1) with $h = 0$. The solution in the deterministic regime is $x(t) = (a/\omega) \cdot (1 - \cos \omega t)$ for $\omega > 2a/L$. In this case $x(t) < L$ for any t and the particle always remains in the metastable state. If the frequency is $\omega < 2a/L$ the particle surmounts the region $[0, L]$ and the solution reads:

$$x(t) = \begin{cases} (a/\omega) \cdot (1 - \cos \omega t), & 0 < t < \theta_1, \quad 0 < x(t) < L, \\ k(t - \theta_1) + (a/\omega) \cdot (1 - \cos \omega t), & t > \theta_1, \quad L < x(t) < b, \end{cases} \tag{6}$$

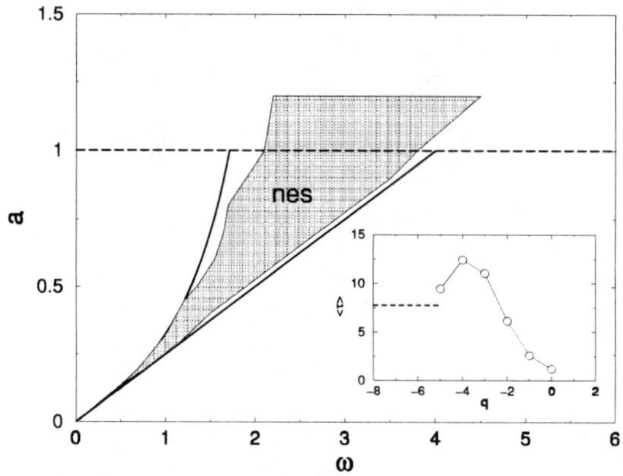

FIGURE 2. The shaded area is the region of the plane (a, ω) where the NES effect takes palce. The parameters are: b=1, k=1, l=0.5. Inset: the average escape time versus the noise intensity for a=0.1 and $\omega = 0.39$. The dashed line indicates the deterministic escape time.

where θ_1 is the time at which the particle crosses the point $x = L$. The condition for NES effect is again $L < x_1 < b$ where $x_1 = x(T/2)$. By adding a small quantity of noise into the system, the position of the particle at time $t = T/2$ is: $x_1(q) \approx x_1(0)$. The decay time for an initial position $x = 0$ is therefore $\tau(0,q) \approx T/2 + \tau(x_1,q)$, and $\tau(x_1,q) \gg \tau(x_1,0)$ because the potential barrier makes the average escape time very large just for $q \longrightarrow 0$. As a result the decay time $\tau(0,q)$ will increase with q and the NES phenomenon appears. The condition for NES effect now becomes:

$$\frac{k}{b}\left[\pi - \arccos\left(1 - \frac{\omega L}{a}\right)\right] < \omega < \frac{2a}{L}. \tag{7}$$

This inequality and the condition $a < k$ give the area on the parameters region (a, ω) where the NES effect takes place (see Fig.2). When the frequency ω and the amplitude a are chosen exactly on the left hand boundary of Eq.(7), the maximal height of induced potential barrier is zero, and the effect is very small. When we move to the right hand boundary of Eq.(7) the maximal barrier increases and the NES phenomenon too. After we cross the right boundary the deterministic decay time becomes infinite and the NES effect disappears. Then, in the presence of noise we get Kramers-like behaviour.

DICHOTOMOUS RANDOM FORCE

We consider now a randomly switching potential profile with reflecting boundary at $x = 0$ and absorbing boundary at $x = b$. In Eq.(1) therefore $U(x)$ is a fixed potential $(f(x) = -dU/dx)$ and $F(t) = a\eta(t)$, with $\eta(t)$ a Markovian dichotomous process which takes the values ± 1 with the switchings mean frequency ν. Exact results of MFPT for non-Markovian processes driven by two-state noise and without thermal diffusion $(D = 0)$ have been obtained in Ref.[14] and then generalized by various authors (see, for example, [15]-[17]). Exact equations for MFPT's for Brownian diffusion in switching potentials were first derived in [18].

Starting from backward Fokker-Plank equation and applying our boundary conditions

$$T'_{\pm}(0) = 0, \qquad T_{\pm}(b) = 0, \tag{8}$$

we obtain the following coupled differential equations [18]

$$\begin{aligned} DT''_{+} + [f(x) + a]T'_{+} + \nu(T_{-} - T_{+}) &= -1, &(9)\\ DT''_{-} + [f(x) - a]T'_{-} + \nu(T_{+} - T_{-}) &= -1. &(10) \end{aligned}$$

Here $T_{+}(x)$ and $T_{-}(x)$ are respectively the mean first-passage times for positive $\eta(0) = +1$ and negative $\eta(0) = -1$ initial velocity of Brownian particle. We introduce two auxiliary functions

$$T = \frac{T_{+} + T_{-}}{2}, \qquad \theta = \frac{T_{+} - T_{-}}{2}, \tag{11}$$

with boundary conditions: $T'(0) = \theta'(0) = 0, T(b) = \theta(b) = 0$, obtained by Eq. (8). So that from Eq.s (9)and (10) we obtain

$$\begin{aligned} DT'' + f(x)T' + a\theta' &= -1, &(12)\\ D\theta'' + f(x)\theta' + aT' - 2\nu\theta &= 0. &(13) \end{aligned}$$

After removing $T(x)$ from Eq.s (13) we obtain a third-order linear differential equation for the variable $\theta(x)$

$$\theta''' + \frac{2f(x)}{D}\theta'' + \left[\frac{f^2(x)}{D^2} + \frac{f'(x)}{D} - \gamma^2\right]\theta' - \frac{2\nu f(x)}{D^2}\theta = \frac{a}{D^2}, \tag{14}$$

where

$$\gamma = \sqrt{\frac{a^2}{D^2} + \frac{2\nu}{D}}. \tag{15}$$

We analyze here a piecewise linear potential, for which $f(x) = -dU/dx = k \cdot 1(x - L)$, where $1(x)$ is the step function and $0 < L < b$. We have a metastable state for $\eta(t) = -1$ and an unstable state for $\eta(t) = +1$. We will focus on MFPT T_{+}, with initial position of Brownian particle at point $x = 0$. We solve Eqs. (12)-(14) separately for regions $0 < x < L$ and $L < x < b$. Using the boundary conditions and the continuity conditions at $x = L$ we obtain for small noise intensity

$$T_{+}(D) \simeq T_{+}(0) + \frac{D}{a^2}g(\beta, \omega, s) + o(D), \tag{16}$$

where

$$g(\beta,\omega,s) = \frac{\beta^3\left[2+s(1+\beta^2)\right]}{(1+\beta)(1-\beta^2)}e^{-s}+\frac{\beta\left(1-\beta^2-2\beta^3\right)}{2(1-\beta^2)}\left(1-e^{-s}\right) - \frac{5+\beta}{2(1+\beta)}+2\omega\left(\frac{1}{1-\beta^2}-\frac{3}{\beta}\right)-\frac{2\omega^2}{\beta^2}, \quad (17)$$

with β, s and ω dimensionless parameters

$$\beta=\frac{a}{k}, \qquad \omega=\frac{\nu L}{k}, \qquad s=\frac{2\nu(b-L)}{k(1-\beta^2)}=\frac{2\omega}{1-\beta^2}\left(\frac{b}{L}-1\right), \quad (18)$$

and

$$T_+(0)=\frac{2L}{a}+\frac{\nu L^2}{a^2}+\frac{b-L}{k}\left[1-\frac{\beta}{s(1+\beta)}\left(1-e^{-s}\right)\right], \quad (19)$$

is the MFPT of initially unstable state in the absence of thermal diffusion.

The condition to observe the NES effect can be expressed by the following inequality

$$g(\beta,\omega,s)>0. \quad (20)$$

Let us analyze the structure of NES phenomenon region on the plane (β,ω). At very slow switchings $\nu\to 0$ ($\omega\to 0$, $s\to 0$) Eq. (20) takes the form

$$\beta>0{,}802; \qquad \omega<\frac{2\beta^2(1-\beta)-5\beta\left(1-\beta^2\right)^2/2}{6(1-\beta^2)^2-2\beta(1-\beta^2)+\beta^2(3\beta^2-1)(b/L-1)}. \quad (21)$$

In the case of $\beta\simeq 1$ we obtain from Eqs. (20) and (21)

$$\omega<\frac{1-\beta}{b/L-1}, \qquad \frac{1}{2}+\frac{5}{2}(1-\beta)<\omega<\frac{1}{2(1-\beta)}. \quad (22)$$

In Fig.3 are shown two NES regions (shaded area) on the plane (β,ω).

The NES effect occurs at the values of $\beta\simeq 1$, i.e. at very small steepness $k-a=k(1-\beta)$ of the reverse potential barrier for the metastable state. For our potential profile, a small noise intensity can move back Brownian particles into potential well, after they crossed the point $x=L$, thus increasing the MFPT.

CONCLUSIONS

We studied the noise enhanced stability (NES) phenomenon in one-dimensional systems with fluctuating metastable state in piecewise potential profiles with the reflecting boundary at the origin. The variations of the potential are due to: (i) a periodical force, (ii) a Markovian dichotomous noise. The analysis of MFPT's at small intensity of the thermal noise allows us to obtain analytically the region of NES phenomenon occurence in all cases studied. In contrast with the case of periodically driven metastable state, in

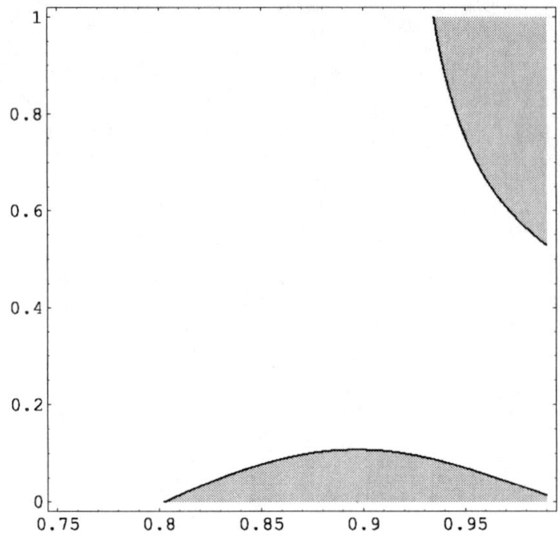

FIGURE 3. The shaded areas are the regions of the plane (ω,$\beta = a/k$) where the NES effect takes palce. The parameters are: L=1, b=1.2, k=1.

the presence of a dichotomous noise the NES effect can be observed only at very flattened potential profile beyond the potential well, i.e. in the absence of the reverse potential barrier for particles beyond the metastable state. Only in such a situation Brownian particles which are at large distances from the initial point can return to its original position by low noise intensity, producing an enhancement of stability of the metastable state of the system.

ACKNOWLEDGMENTS

This work has been supported by INFM (Istituto Nazionale di Fisica della Materia), by MIUR, by the INTAS Grant 01-0450, and Russian Foundation for Basic Research (Projects No.02-02-17517, No.00-15-96620).

REFERENCES

1. P. Reimann, *Phys. Rep.*, **361**, 57-265 (2002).
2. V. N. Smelyanskiy, M. I. Dykman and B. Golding, *Phys. Rev. Lett.*, **82**, 3193-3197 (1999).
3. J. Lehmann, P. Reimann and P. Hanggi, *Phys. Rev. Lett.*, **84**, 1639-1642 (2000).
4. R. N. Mantegna and B. Spagnolo, *Phys. Rev. Lett.*, **76**, 563-566 (1996); *Int. J. Bifurcation and Chaos*, **8** (4), 783-790 (1998).
5. N. V. Agudov and B. Spagnolo, "Decay Times of Unstable States Increased by the Noise", in *Stochastic and Chaotic Dynamics in the Lakes*, AIP, Eds. D. S. Broomhead et al., **502**, Melville,

New York, , 272-277 (2000).
6. R. N. Mantegna and B. Spagnolo, in *Stochastic Processes in Physics, Chemistry and Biology*, Lecture Notes in Physics, Eds. I. A. Freund and T. Poeschel, **557**, Springer-Verlag, Berlin, 327-337 (2000).
7. N. V. Agudov and B. Spagnolo, *Phys. Rev. E*, **64**, 035102 (R)(2001).
8. R. S. Maier and D. L. Stein, *Phys. Rev. Lett.*, **86**, 3942-3945 (2001).
9. I. Dayan, M. Gitterman and G. H. Weiss, *Phys. Rev. A*, **46**, 757-761 (1992).
10. R. Wackerbauer, *Phys. Rev. E*, **58**, 3036-3044 (1998).
11. A. Mielke, *Phys. Rev. Lett.*, **84**, 818-821 (2000).
12. D. Dan, M. C. Mahato and A. M. Jayannavar, *Phys. Rev. E*, **60**, 6421-6428 (1999).
13. H. A. Kramers, *Physica (Utrecht)*, **7**, 284-304 (1940).
14. P. Hänggi and P. Talkner, Phys. Rev. A **32**, 1934-1937 (1985).
15. J. Masoliver, K. Lindenberg, and B. J. West, Phys. Rev. A **33**, 2177-2180 (1986); M. A. Rodrigues and L. Pesquera, *ibid.* **34**, 4532-4534 (1986).
16. C. R. Doering, Phys. Rev. A **35**, 3166-3167 (1987).
17. U. Zürcher and C. R. Doering, Phys. Rev. E **47**, 3862-3869 (1993).
18. V. Balakrishnan, C. Van den Broek, and P. Hänggi, Phys. Rev. A **38**, 4213-4222 (1988).

MATERIALS, DEVICES

Thermal (Noise) Death of Moore's Law?

Laszlo B. Kish[1]

Department of Electrical Engineering, Texas A&M University, College Station TX 77843-3128, USA

Abstract. Physical limitations of Moore's Law of miniaturization would have a serious impact on technology and economy. According a new study [1] a thermodynamical effect, the increasing thermal noise voltage (Johnson-Nyquist noise) on decreasing characteristic capacitances, together with the constrain of using lower supply voltages to keep power dissipation manageable on the contrary of increasing clock frequency, has the potential to break abruptly Moore's Law within 6-8 years.

INTRODUCTION

In 1965 Moore observed that integrated circuit complexity evolved exponentially [2]. The silicon industry has been following Moore's law, and transistor sizes on chips have been shrinking exponentially [3-6]. The law has proven to be accurate enough that it is today's solid-state electronics industry target that must be met by semiconductor device manufacturers in order to remain competitive [3-6]. Today, characteristic sizes are around 100 nanometers [5], and the trend is believed to continue. Recent news claim MOS transistor sizes down at 9nm by leading research labs [7], and there are new integration technologies with a potential to follow Moore's Law for, at least, another two decades [8,9].

The successfully realized 9nm MOSFET indicates that there is no mesoscopic physical effect what would prohibit functioning of classical microelectric devices at that length scale. That means, as the technology of these silicon devices is the most advanced among all, that the possible length scale of exotic nanoelectronics, such as single electron devices and other exotic nanoelectronic devices, is confined to the range of 1-8 nm.

However, a recent study [1] indicates that increase of the integration density of computer chips, below 40nm characteristic size, may face another physical limit, in 6-8 years. If the present trend continues, abrupt and major complications due to false bit occurrences generated by thermal fluctuations (Johnson-Nyquist noise) can be expected. The study is based on CMOS technology which yields the most optimistic results. Bipolar transistors, single electron devices and other exotic nanoelectronic devices are supposed to be affected even more.

The present paper is the UPoN conference version of that study [1].

---

[1] Until 1999, his family name was Kiss

CP665, *Unsolved Problems of Noise and Fluctuations: UPoN 2002: Third International Conference,* edited by S. M. Bezrukov

SPEED, SIZE AND DISSIPATION VERSUS SIZE

In this section, we repeat the survey, in [1], about some well-established practical facts about speed, dissipation, capacitance, and size. At the following considerations, we focus on digital (C)MOS technology because it is the most widely used one. However, the same considerations can be used for other technologies with very similar outcome. During the considerations, we assume that we run the devices with the highest possible clock frequency they are fast enough to handle.

(C)MOS technology has the advantage that its power dissipation is the smallest among all. The reason for that is simple. When the devices are in a certain logical state, the current consumption is negligible. So, in the first approach, all dissipation takes place during transitions between logical states and the source of this dissipation is the need to charge or discharge the related capacitors. On Figure 1, a simple circuit model of the power dissipation related to the gate capacitance of a single transistor is shown. C is the capacitance of the gate, R is the drain-source resistance of the other (driving) transistor(s) which changes the logic state of the gate. U_0 is the supply voltage of the circuit.

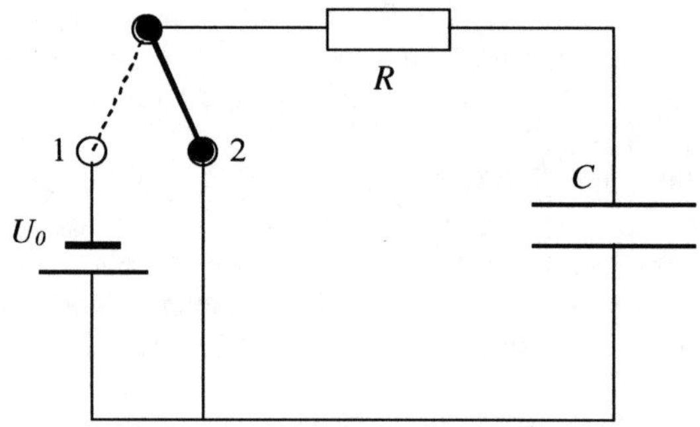

FIGURE 1. Circuit model of speed and dissipation [1].

From this picture, some well-known relations follow.

The upper limit of the clock frequency f_0, where the alternating voltage driving is still sufficient to switch the transistors between logical states, can be given as

$$f_0 \cong (RC)^{-1} . \tag{1}$$

The energy dissipated by this unit during one period of the clock frequency is scaling as CU_0^2 . The average power dissipation of this single unit (Fig. 1) is scaling as

$$P_1 \propto f_0 E_1 \propto (RC)^{-1} C U_0^2 \propto \frac{U_0^2}{R} . \tag{2}$$

Note, the clock frequency does not appear in this formula due to the assumption that we run the unit at the highest clock frequency f_0.

However, the total power dissipation P_N of a IC chip with fixed geometrical size containing N such units depends on N, R and C. To show the impact of miniaturization on the power dissipation and clock frequency of an *IC* chip with fixed size, first we have to determine the scaling of N, R and C versus the characteristic transistor size s. Obviously, N is scaling as $1/s^2$. Using the reasonable assumption that the depth of drain-source channel is constant, R can be approximated as constant. From these considerations, we obtain that P_N is scaling as

$$P_N \propto NU_0^2 / R \propto NU_0^2 \propto U_0^2 / s^2 \tag{3}$$

Relation 3 has an exact match with data describing the past and expected future of the evolution of microelectronic technology [5]. Therefore, if we want to keep the dissipation constant during miniaturization and also the dense packing of transistors, Eq. 3 requires the supply voltage to scale linearly with the size.

To calculate the thermal noise, we will need to know the capacitance. If the thickness of the gate oxide is constant then C is scaling as

$$C \propto s^2 \quad . \tag{4}$$

If, instead of constant gate thickness, we suppose that the gate oxide thickness is linearly scaling with s, we get

$$C \propto s \tag{5}$$

The practical situation has been between the two limits given by Eqs. 4 and 5 because the gate oxide thickness has gradually been shrinking, however not so fast as the characteristic size of the transistors on the chip [5].

3. THERMAL NOISE IN THE WHOLE BANDWIDTH

Thermal noise in resistors is a stationary Gaussian stochastic voltage fluctuation process with zero mean value. The well-known Johnson-Nyquist formula describes the power density spectrum S_u of the thermal noise voltage in the low frequency limit:

$$S_u(f) = 4kTR \quad , \tag{6}$$

where f is the frequency, k the Boltzmann constant, T the temperature and R the resistance. If the resistance is shunted by a capacitor then Eq. 6 and linear network theoretical calculations yield that the total root-mean-square (*RMS*) thermal noise voltage U_n on the capacitor is:

$$U_n = \sqrt{\frac{kT}{C}} \quad . \tag{7}$$

Eq. 7 can easily be obtained also from the energy equipartition theorem of equilibrium thermodynamics. As the capacitor has one thermodynamical degree of freedom, the mean energy in it should be $kT/2$ what requirement directly implies Eq. 7. This is the physical reason why the value of R does not appear in Eq. 7.

Eqs. 4, 5 and 7 imply that, when the full bandwidth is used like in digital circuits, the *RMS* thermal voltage is steadily increasing during miniaturization. This effect has not been widely realized yet, probably because of Eq. 6 what indicates a constant power density spectrum when the resistance is kept constant. The increasing bandwidth, $\approx(RC)^{-1}$, implies the increase of the *RMS* noise voltage even if its low-frequency spectrum remains unchanged. This increase evolves between the scaling limits of $1/s$ and $1/s^{0.5}$.

4. FALSE BIT-FLIPS DUE TO THERMAL NOISE

The mean frequency $\nu(U_{th})$ of crossing a threshold amplitude limit U_{th} by a Gaussian noise process of zero effective value is given by the Rice formula [10-12]:

$$\nu(U_{th}) = \frac{2}{U_n} \exp\left(\frac{-U_{th}^2}{2U_n^2}\right) \sqrt{\int_0^\infty f^2 S(f) df} \quad , \text{ where } U_n = \sqrt{\int_0^\infty S(f) df} \quad . \tag{8}$$

For band-limited white noise, $S(f)=S(0)$ for $f\leq f_c$ and $S(f)=0$ for $f_c<f$, the formula becomes:

$$\nu(U_{th}) = \frac{2}{\sqrt{3}} \exp\left(\frac{-U_{th}^2}{2U_n^2}\right) f_c \quad , \text{ where } U_n = \sqrt{S(0) f_c} \quad . \tag{9}$$

Whenever the thermal noise causes crossing of the value of the actual logic threshold voltage, which is the noise margin between the actual voltage difference between the two logic values, a false bit flip occurs. It is obvious from Eq. 9 that, in the case of low noise ($U_{th}/U_n<<1$), the explicit U_{th}/U_n ratio dominates the frequency of false bit-flips. This fact can also be seen on Fig. 2 where the frequency of false bit occurrences is plotted versus U_{th}/U_n . The clock frequency is assumed to be equal to f_c . Apparently, the practical difference between the f_c=2GHz and f_c=20GHz cases is negligible, especially, when we compare this dependence to the steep variation versus U_{th}/U_n . Similarly, the number of transistors in the computer has also a minor role at determining the necessary noise margin. It is obvious from Fig. 2 that U_t has to be greater by U_n about an order of magnitude for the reliability of the computer. Today's PCs have about $5*10^9$ transistors; 10^8 are in the processor and the rest in the RAM and the other ICs. In a decade, the clock frequency will approach 20 GHz and the number of transistors will grow further order of magnitude.

So in conclusion, it is reasonable to say that a safe thermal noise margin requires

$$U_{th}/U_n \geq 12 \tag{10}$$

in modern digital circuits.

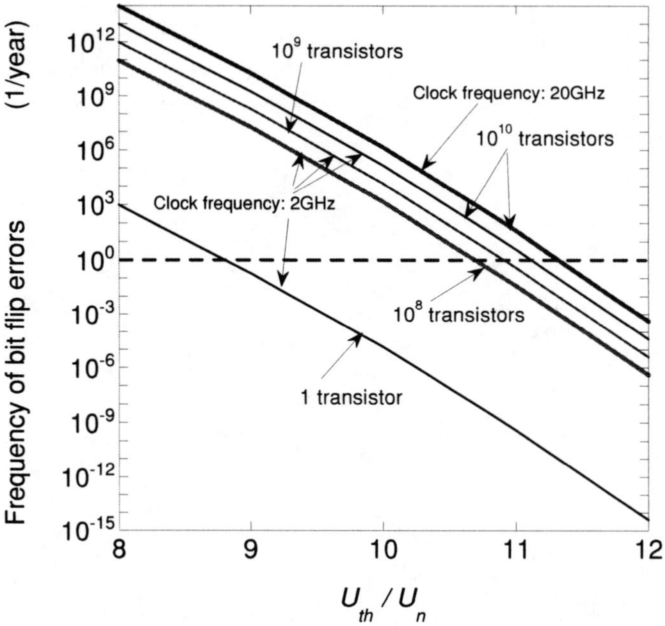

FIGURE 2. False bit frequency in microprocessors [1].

At the above considerations, error-correcting codes were not taken into account. One reason for that is the sake of simplicity, another reason is the trend toward more cache processing in complex logic integrated circuits, which reduces the advantages of error correcting codes [6].

5. LIMITATIONS OF MOORE'S LAW

In this section, we apply the results obtained in sections 2-4 for estimation of the practical situation with thermal noise. The logical threshold voltage $U_{th} = U_{highmin} - U_{lowmax}$, where $U_{highmin}$ is the minimal legal voltage value of the logic *high* state and U_{lowmax} is the maximal legal voltage value of the logic *low* state. Following the old 5V supply voltage practice, we assume that scales linearly with the supply voltage, $U_{th} = 0.6\ U_0$. Note, due to offset and crosstalk problems, the fraction 0.6 may be too optimistic for future practice, however, we have been unable to find any prediction how this fraction would diminish during miniaturization, so we stick to this optimistic value. The main results of this paper are shown on Figure 3.

The dissipation constrain curve is based on Rel. 3 and it shows how the U_{th} should be decreased versus the decreasing s in order to avoid increasing dissipation. The curve indicates the *allowed maximal* value of U_{th}. The right end of the curve represents today's data [5] (2002), s = 115 nm and $U_{th} = 0.6 * 1.5$ V = 0.9 V.

The noise constrain curves are based on Rels. 4,5,7 and 10 and they show how the U_{th} should be increased versus the decreasing s in order to avoid false bit-flips. The curves indicate the *required minimal* value of U_{th}, in the cases when Rel. 3 or 4 is valid, respectively. The real condition is between these two limits. The right end of the curves represent today's data [5] (2002), physical gate length of 53 nm, gate width/length aspect ratio of 2.5, gate thickness of 2 nm. A relative permittivity value of 3 was assumed for the gate oxide.

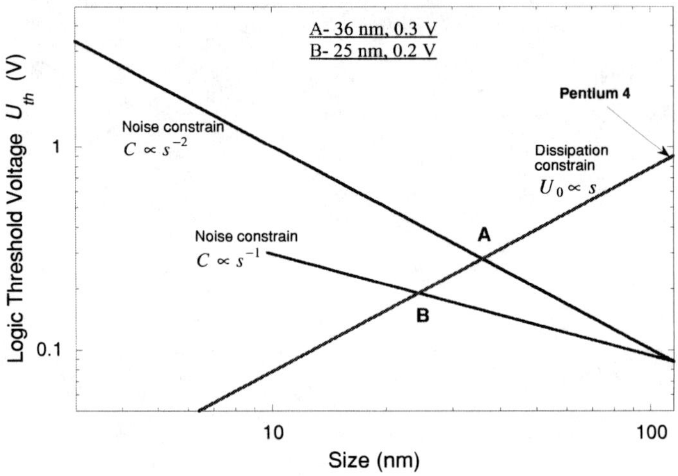

FIGURE 3. Moore's Law should break between the crossings A and B, or earlier [1].

The dissipation constrain curve intersects the noise constrain curves at points *A* (36 nm) and *B* (25 nm). Somewhere between these points, either the noise constrain or the dissipation constrain can't be satisfied, because the requirements are contradictory. Taking Moore's Law into account, a rough estimate shows that we will reach the danger zone in 6-8 years.

6. CONCLUSIONS

In this paper, we have studied potential technological difficulties of future miniaturization with high-density integration due to a fundamental an unavoidable noise process, the thermal noise. Our aim has been to take the attention to the problem of thermal noise, which has the potential of breaking Moore's Law.

It is important to emphasize here the nature of the problem. The arguments in this paper do not set any physical limit on transistor sizes. What is claimed here, the logical threshold voltage, therefore the supply voltage, cannot be reduced below a certain limit due to false bit-flips induced by thermal noise. The only way to get around this effect would be either giving up to increase the integration density, that is itself Moore's Law, or giving up to increase the clock frequency. However, that would be contradictory because increasing the clock frequency is one of the most important reasons for the miniaturization efforts.

For the sake of simplicity and generality, we used several strong approximations during these considerations. Most of the approximations were optimistic, so the real situation is probably worse than Fig. 3 shows. We used the following, optimistic simplifications.

1. Noises due to switching of devices (ground plane EMI), which are major factors in logic IC design, have been neglected.

2. Excess thermal noise due to hot electrons has not been taken into consideration.

3. The logical threshold voltage was supposed to be 60% of the supply voltage, which may be too high. Due to offset problems with the gate threshold voltage, this ratio will probably diminish while the supply voltage is decreased.

4. When applying the Rice formula for threshold crossing, we supposed that the clock frequency is the relevant bandwidth. However, in the nominator in the Rice formula, what contains the RMS velocity, the relevant bandwidth is always greater, it is determined by the second pole of the transfer function. So, the actual probability of threshold crossing is greater than supposed here.

5. Other noise phenomena (shot noise, $1/f$ noise) were neglected.

Some of the approximations were pessimistic:

i. Parasitic capacitors were neglected. These capacitors can increase gate capacitance by 20-30% [5].

ii. Error-correcting codes were not taken into consideration. As we mentioned above, there is a trend toward more cache processing in complex logic integrated circuits, which reduces the advantages of error-correcting codes [6]. Furthermore, note that a 10% increase of the U_n/U_{th} ratio causes a factor of ≈1000 increase in the false bit frequency. The very progressive growth of false bit-flip frequency at a small increase of the thermal noise or a small decrease of threshold voltage indicates that error-correcting codes can't save Moore's Law when the danger zone is reached.

Even the simplified considerations make it clear that the effect of thermal noise can become a serious issue of high-density integration at the beginning of the next decade, and the danger zone is approached already within this decade. The potential impact on future nanoelectronics technology is also evident.

It is important to note that, due to the optimistic sides of our estimation, there is a possibility that the described effects are already causing problems in the research and development of highly integrated logical ICs. As the false bit-flips occur at random locations and times, and the number of possible locations is great, it is not easy to identify the source of their origin.

ACKNOWLEDGMENTS

Valuable discussions with Dr. Bob Biard (Honeywell), Dr. Fred Strieter (Texas Instruments), Dr. Sergey Bezrukov (NIH), Prof. Mark Dykman (MSU) and Prof. Henry

Taylor (TAMU) are appreciated. This research has been supported by the Telecommunications and Informatics Task Force (TITF) of TAMU, by 3M Company and the startup found of the Department of Electrical Engineering of TAMU.

REFERENCES

1. Kish, L.B., "End of Moore's Law; Thermal (Noise) Death of Integration in Micro and Nano Electronics", Phys. Lett. A **305** 144–149 (2002); http://physicsweb.org/article/news/6/12/11
2. Moore, G.E., "Cramming more components onto integrated circuits", *Electronics* **38**, No. 8, April 19 (1965).
3. Fodor, I., *Physics Today* **53**, 106 (2000)
4. Brenner, A.E., *Physics Today* **54**, 84-84 (2001)
5. *International Technology Roadmap for Semiconductors*, Semiconductor Industry Association (2001), http://public.itrs.net
6. Condra, L., "The Impact of Semiconductor Device Trends on Aerospace Systems", Boeing report, unpublished
7. Fuller, B., "Moore's Law scales to at least 9nm", EE Times, (March 12, 2002) http://www.eetimes.com/semi/news/OEG20020312S0077
8. Editor, Global Sources (July 11, 2002) http://www.globalsources.com/MAGAZINE/EC/0209W2/TUTSMC.HTM
9. Neely, L. Electronic News, (July 10, 2002), http://email.electronicnews.com/cgi–bin2/flo?y=eHha0EjxX60DbD0Boi10Ag
10. Rice, S.O., Bell. Syst. Tech. J. 23, 282 (1944); 24, 46 (1945)
11. Kiss, L.B., in *Chaotic, Fractal, and Nonlinear Signal Processing*, ed. R. Katz, American Institute of Physics Press, 382 (1996)
12. Kedem, B., *Time Series Analysis by Higher Order Crossings, IEEE* (New York, 1994)

Origins of Non-Gaussian Noise in Metallic Manganites

M.B. Weissman, Akilan Palanisami, and J. N. Eckstein

Physics Department, University of Illinois at Urbana-Champaign
Urbana, Illinois 61801-3080

Abstract. Noise measurements on manganite colossal magnetoresistive materials show large non-Gaussian effects even in the metallic regime. Domain rotation effects are implicated, but the origin of the required current inhomogeneity is obscure.

Several manganite materials have drawn intensive study because their transitions into metallic phases are accompanied by an onset of ferromagnetism, giving rise to colossal magnetoresistance (CMR) [1]. Although early attempts to understand the CMR effect focussed on theories of homogeneous phase transitions, a variety of methods, including noise studies, show that the CMR effect often occurs at an inhomogeneous first-order transition. One of the most striking and accessible pieces of evidence is the presence of discrete fluctuations in the resistance even in relatively large samples [2-5]. Detailed thermodynamic measurements can be performed on these fluctuators [3-5], and their sizes (as determined from magnetic moments) are indeed tiny fractions of the sample size, as compared with the relatively large fractional effects on the resistance, $\delta R/R$. This effect makes at least qualitative sense because the highly inhomogeneous conductivity near the metal-insulator transition gives anomalously large current densities at some parts of a percolation-like network, enhancing the effect of fluctuations from those areas [2,3]. Here, in the spirit of UPON, we will we focus attention on a more mysterious aspect of the noise – the persistence of perceptible discrete fluctuations deep into the metallic regime, where there are no obvious reasons for extreme current inhomogeneity [6].

The materials we used were single-crystal films of $La_{0.7}Ca_{0.3}MnO_3$ (LCMO) prepared by atomic-layer-by-layer epitaxy (ALL) on $SrTio_3$ substrates. However, some closely related low-temperature results have been reported in laser-ablated films of the same material [4], and in the related material LSMO [7], containing Sr rather than Ca. Thus we believe that the phenomenon to be described is not a result of some accidental sample preparation detail, but may reflect a widespread property of the metallic regime in these CMR materials.

In ALL films whose temperature-dependent resistance, R(T), is a purely monotonic increasing function from 4K up to 150 K, we found numerous discrete fluctuators at low magnetic field, H, i.e. below the coercive field, for T < 20 K. (See Fig. 1.) In five

CP665, *Unsolved Problems of Noise and Fluctuations: UPoN 2002: Third International Conference,* edited by S. M. Bezrukov

cases, we were able to take data under enough conditions to clearly determine both the H and T derivatives of the occupation probablities for the high and low resistance states. From these derivatives and the probabilities themselves we unambiguously determine the energy, entropy, and magnetic moment differences between the two states of the fluctuator. For these low-T fluctuators, the ratios of entropy and energy changes to moment change are small compared to the ratios for fluctuations near the inhomogeneous first-order transition [6]. In fact, the sign of the energy and entropy changes seems to be random. These properties are exactly as expected for magnetic orientation fluctuations, as opposed to order/disorder fluctuations. Other types of data lead to a similar conclusion about noise in the same regime of LSMO [7].

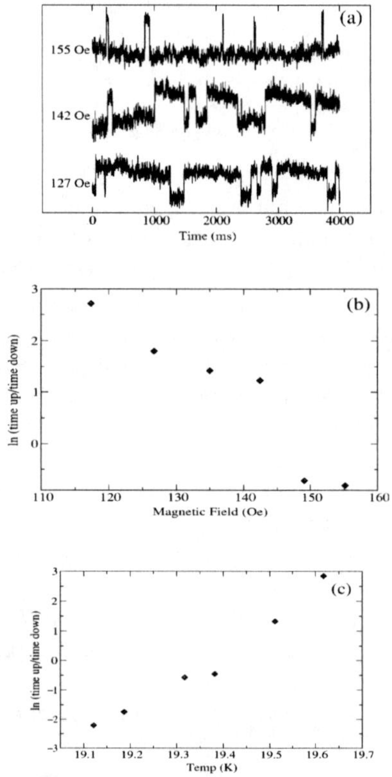

Figure 1

(a) Time traces of resistance vs. time are shown for a sample at T= 18.5 K.

(b) The field dependence of the ratio of the Boltzmann factor s of the discrete states is shown at T= 18.5 K.

(c) The temperature dependence of the ratio of the Boltzmann factors for the two states is shown at zero field.

From the size of the moment fluctuations, we know that the reorienting regions consist of roughly 3 $\times 10^5$ unit cells, out of a sample some 10^9 times larger. However, $\delta R/R \sim 10^{-4}$. The key unanswered question is why $\delta R/R$ is some 10^5 times larger than the fraction of the sample which fluctuates.

There are some serious constraints on the problem. The first is that the sample is far into a metallic regime, judged by its own R(T), quite aside from other evidence.

Therefore one cannot invoke extended areas of an insulating phase with small conducting links across them, unlike near the metal-insulator transition. The second is that the phenomenon is seen directly or indirectly in several samples from several groups, so that some accidental scratch, etc, cannot plausibly be invoked.

Although it is risky to trust seemingly logical trains of thought too far without detailed confirmation, we believe these results may point toward a specific mechanism. There are obviously magnetic domain walls below the coercive field. Although domain walls within the ferromagnetic metallic phase are not expected to contribute much to resistance, some calculations indicate that domain walls can lower their free energy by turning into thin layers of antiferromagnetic insulator [8]. If those layers are thin enough, conductance across them can occur primarily by tunneling, so that a rise in R(T) at low T would not be found. The topology of a network of line-like domain walls could require that all current cross those walls, even though they occupy a small fractional area of the sample. (Hence the distinction from the more ordinary percolation problem near the phase transition.) One challenge would be to see if any numerically consistent version of such a description actually could hold. Included in that would be some understanding of the soft percolation problem created by the domain wall pattern. Another would be to see what the role of disorder is in stabilizing the domain structure. How big is the parameter window within which domain walls tend to go insulating? Are extended defects required, or does the intrinsic point-like disorder suffice? Studies of the low-T noise in strain-free material, grown on a different substrate, would be particularly helpful. The role of the domain reorientation effects in understanding both noise and CMR in highly disordered material [2] is also open for study. In the absence of approximate local symmetry, do the fluctuations between phases and the domain orientation fluctuations become inextricably mixed, as in one model [8]?

ACKNOWLEDGMENTS

This work was supported by NSF DMR 99-81869.

REFERENCES

1. A. P. Ramirez, J. Phys. :Condensed Matter **9**, 8171 (1997).
2. H. T. Hardner, *et al.*, J. Appl. Phys. **81**, 272 (1997).
3. R. D. Merithew, *et al.*, Phys. Rev. Lett. **84**, 3442 (2000).
4. B. Raquet, A. Anane, S. Wirth, P. Xiong, and S. v. Molnar, Phys. Rev. Lett. **84**, 4485 (2000).
5. F. M. Hess, R. D. Merithew, M. B. Weissman, Y. Tokura, and Y. Tomioka, Phys. Rev. B **63**, 180408 (2001).
6. A. Palanisami, R. D. Merithew, M. B. Weissman, and J. N. Eckstein, Phys. Rev. B **64**, 132406/1 (2001).
7. R. Mathieu, P. Svedlindh, R. Gunnarsson, and Z. G. Ivanov, Phys. Rev. B **63**, 132407/1 (2001).
8. J. Burgy, M. Mayr, V. Martin-Mayor, A. Moreo, and E. Dagotto, Phys. Rev. Lett. **87**, 277202 (2001).

Steady State of Random Resistor Networks Under Biased Percolation: a Framework for Noise in Disordered Materials?

Cecilia Pennetta[1,2], Eleonora Alfinito[1] and Lino Reggiani[1,2]

[1] *INFM - National Nanotechnology Laboratory, Lecce, Italy,*
[2] *Dip. di Ingegneria dell'Innovazione, Università di Lecce, Via Arnesano, I-73100 Lecce, Italy,*
E-mail: cecilia.pennetta@unile.it

Abstract. Recently we have developed a new statistical approach which allows the study of the electrical noise and conduction of disordered materials over the full range of bias values, from the linear regime up to the electrical breakdown. In this approach, the resistance and the resistance fluctuations of disordered materials are determined by the competition of biased processes taking place in random resistor networks. In particular, the properties of the nonlinear regime are studied for different values of the model parameters and for different bias conditions. A strong nonlinearity of the relative variance of resistance fluctuations emerges in the pre-breakdown region which is also characterized by non-Gaussian noise. These results compare well with measurements of the electrical resistance and resistance noise performed in composites and in other conducting materials. The possibility of defining reliable failure precursors and other suitable quantities characterizing electrical breakdown phenomena in disordered materials is then discussed.

INTRODUCTION

The subject of the electrical stability in disordered systems has attracted a considerable interest in the recent literature [1, 2, 3, 4, 5, 6, 7, 8, 9]. Such phenomena have been successfully studied by using percolation theories [3, 4, 5, 6, 7, 8, 9]. In particular, large attention has been devoted to the critical exponents describing the resistance and its relative noise in terms of the conducting particle fraction, defect concentration, etc. [3, 4, 5, 6, 7, 8, 9]. However, few attempts [2, 3, 10, 11, 12] have been made so far to describe the behavior of a disordered medium over the full range of applied stress, i.e. by studying the response of the system to an external bias covering the full range of linear and nonlinear regimes. Relevant information is expected from these studies, like the identification of precursor phenomena, scaling relations, role of the disorder, etc. [1, 2, 3, 12]. Therefore, the understanding of breakdown properties of disordered materials, still represents an important unsolved problem, tightly connected to the study of the noise properties of these systems [13, 14]. Here we discuss some implications of these recent attempts, in particular of those described in Refs. [3, 10, 11, 12]. To this purpose, the electrical conduction of a disordered material has been studied, over the full range of bias values, from the linear regime up to the electrical breakdown, by considering the competition of two biased processes in a random resistor network (RRN) and by making use of Monte Carlo (MC) simulations. The results obtained [10, 11, 12]

CP665, *Unsolved Problems of Noise and Fluctuations: UPoN 2002: Third International Conference,* edited by S. M. Bezrukov

well agree with electrical measurements performed in composites [3] and in other conducting materials [15, 16]. As suggested by the authors of Ref. [3], the possibility of defining a universal quantity, characterizing the electrical breakdown of these materials, is considered. Furthermore, many other related and still open questions concerning the properties of resistance fluctuations near the breakdown and the predictability of electrical breakdown phenomena are also discussed.

MODEL

The RRN consists of a two-dimensional square-lattice network made of N_{tot} resistors of resistance r_n [9]. Different geometries can be studied; here we consider the simplest one which is the square $N \times N$, where N determines the linear size of the network. In practical applications, the value of N can be related to the ratio between the size of the sample and the average size of the grains composing the sample. An external bias, represented by a constant voltage V or by a constant current I, is applied to the RRN through electrical contacts realized by perfectly conducting bars at the left and right hand sides of the network. The RRN interacts with a thermal bath at temperature T_0 and the resistances r_n are taken linearly dependent on the local temperatures, T_n, by a temperature coefficient of the resistance (TCR), α. The local temperatures are calculated by taking into account Joule heating effects and thermal exchanges between neighbour resistors, according to the biased percolation model [7]:

$$T_n = T_0 + A\Big[r_n i_n^2 + \frac{3}{4N_{neig}} \sum_{l=1}^{N_{neig}} \left(r_l i_l^2 - r_n i_n^2\right)\Big] \tag{1}$$

where i_n is the current flowing in nth resistor and N_{neig} is the number of first neighbours around this resistor. The parameter A, measured in (K/W), describes the heat coupling of each resistor with the thermal bath and it determines the importance of Joule heating effects. Here, the initial state of the network is chosen to coincide with the perfect network ($r_n \equiv r_0$). We assume that two competing processes act to determine the network evolution. The first process consists of generating fully insulating defects (resistors with very high resistance, i.e. broken resistors) with probability: $W_D = \exp(-E_D/K_B T_n)$ [7], where E_D is an activation energy characteristic of the defect and K_B the Boltzmann constant. The second process consists of recovering the insulating defects with probability $W_R = \exp(-E_R/K_B T_n)$, where E_R is another activation energy, characteristic of this second process. For sufficiently high values of the bias, the probability of breaking (recovering) a resistor is higher in the so called "hot spots" of the RRN [9]. On the other hand, for vanishing small bias the Joule heating is negligible and $T_n \equiv T_0$, which gives rise to a homogeneous degradation [8, 9, 17]. As a result of the competition between generation and recovery processes, the RRN can reach a steady state or can evolve towards an irreversible breakdown, depending on the external conditions, T_0, I (or V), and on the values of the material parameters. In the first case, the network resistance fluctuates around the average value $< R >$. In the second case, a critical fraction of defects, p_c (percolation threshold), associated with a divergence of the network resistance, is reached [9]. We note that in the limit of a vanishing bias (random percolation) and infinite lattices ($N \to \infty$), the expression: $E_R < E_D + K_B T_0 \ln[1+\exp(-E_D/K_B T_0)]$ provides a sufficient condition for the existence of a steady state [17]. The evolution of the RRN is obtained by MC simulations carried out according to an iterative procedure described in detail in Refs. [10, 11, 12]. Each iteration step can be associated with an elementary time step on an appropriate time scale to be calibrated with experiments. In this manner it is possible to represent the simulated resistance evolution over either a time or a frequency domain, according to convenience. Except when differently specified, the simulations are performed by taking the following values for the parameters, which are chosen as plausible values: $N = 75$, r_0 ranges from $1 \div 10$ (Ω), $\alpha = 10^{-3}$ (K^{-1}), $A = 5 \times 10^5$ (K/W), $E_D = 0.170$ (eV), E_R in the range $0.026 \div 0.155$ (eV) and $T_0 = 300$ (K). The values of the external bias range from $0.001 \leq I \leq 3.0$ (A) under constant current conditions, and from $0.001 \leq V \leq 3.5$ (V) under constant voltage conditions.

RESULTS

The results are summarized in Figs. 1 to 6. Figure 1 shows the average resistance $< R >$ of the RRN as a function of the external current (each value has been calculated by averaging over time and over 20 independent realizations of the RRN). The data in this figure have been obtained by choosing $r_0 = 1$ (Ω) and $E_R = 0.043$ (eV). At the lowest biases the average resistance takes a value $< R >_0$ which represents the intrinsic linear response property of the network (Ohmic regime). When the external current overcomes a certain value, I_0, the average resistance starts to evidence the dependence on the bias. Thus, I_0 sets the current scale value for the onset of nonlinearity. For currents above I_0 the average resistance increases systematically with the current until a threshold current value, I_b, above which the RRN undergoes an irreversible breakdown. In the case of Fig. 1 it is $I_0 = 0.1$ (A) and $I_b = 2.1$ (A). By analysing the role played by the parameters r_0 and E_R on the dependence of $< R >$ upon the bias, we have found that the relative variation of the average resistance scales with the ratio I/I_0 as [10, 11, 12]:

$$\frac{< R >}{< R >_0} = g(I/I_0), \qquad g(I/I_0) \simeq 1 + a(I/I_0)^{\theta} \tag{2}$$

where the values of the dimensionless coefficient $a = (3.8 \pm 0.1) \times 10^{-4}$ and of the exponent $\theta = 2.1 \pm 0.1$ are found to be independent of the above cited parameters. It must be noticed that a superquadratic behavior of the average resistance emerges in the pre-breakdown region, where the relative variation of $< R >$ is characterized by a power law $(I/I_0)^{\theta_I}$ with an exponent $\theta_I = 3.7 \pm 0.3$ [11, 12]. However, this superquadratic behavior is present only for RRN sufficiently robust to breakdown. In fact, at increasing the recovery activation energy E_R, the ratio I_b/I_0 tends to decrease, i.e. the stability region is reduced and the dependence on the current of the average resistance can remain quadratic over the entire region [10, 12]. Within the numerical uncertainty, the value of the ratio $Y \equiv \frac{<R>_b}{<R>_0}$ is found to be independent of the initial resistance value [10, 12] and of the bias conditions (constant voltage or constant current) [11, 12], while it is found to be dependent on the recovery activation energy [10] and on the TCR [12]. In particular, it must be noticed that a zero value of the TCR gives rise to a bias dependence of $< R >$

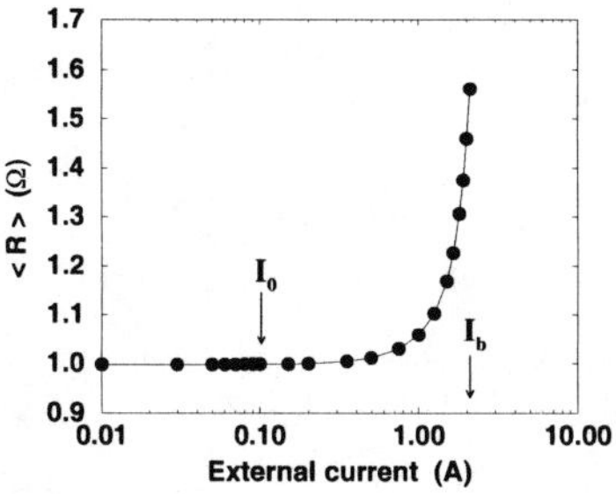

FIGURE 1. Steady state average resistance versus external current. The arrows on the left and on the right indicate the current values I_0 and I_b associated with the onset of the nonlinear regime and with the breakdown, respectively.

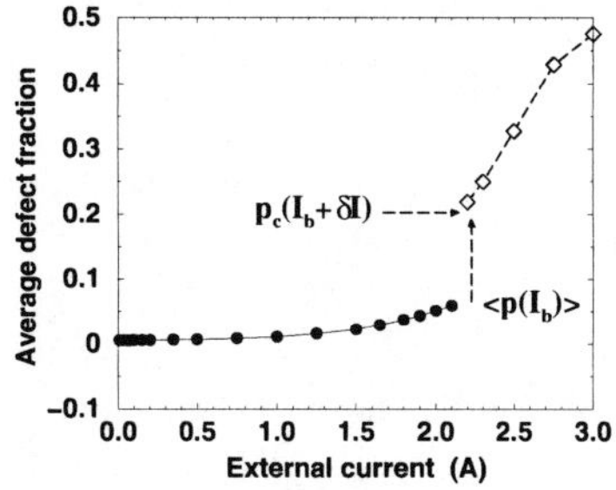

FIGURE 2. Steady-state average defect fraction versus current for $I < I_b$ (full circles) and percolation threshold versus current for $I > I_B$ (open diamonds). The percolation threshold is calculated by averaging over 20 breakdown realizations.

which belongs to a different universality class with respect to that described by Eq. (2) [12].

An important feature of the electrical breakdown described by the present approach, is that the threshold current I_b is associated with a first order transition. This features is illustrated in Fig. 2, where the steady-state average defect fraction $< p >$ and the percolation threshold p_C are reported as a function of the current for $I < I_b$ (full circles) and $I > I_b$ (open diamonds), respectively (the parameters are the same of Fig. 1). We can see that the value of $< p >$ for the last stable state of the RRN, at the current I_b, is significantly smaller than the percolation threshold at the current $I_b + \delta I$, only slightly higher than I_b. Indeed, it must be remarked that, as a consequence of the filamented growth of defects characteristic of biased percolation [7, 8], the percolation threshold depends on the value of the bias current [12] and it can take values lower than that corresponding to the random percolation ($p_C = 0.5$ on a square-lattice). We notice that all the above features, including the scaling behavior in Eq. (2) and the first order character of the transition, have been observed in resistance measurements of thin-films made of composites materials, performed over the full range of bias values up to breakdown [3].

In Fig. 4 we show the relative variance of resistance fluctuations, $\Sigma \equiv \frac{<\delta R^2>}{<R>^2}$, as a function of the external bias for two values of the TCR and under different bias conditions. Precisely, the curves with full and open triangles represent data obtained under constant current for $\alpha = 0$ and $\alpha = 10^{-3}$ (K^{-1}) respectively, while curves with full and open diamonds are obtained under constant voltage respectively for $\alpha = 0$ and $\alpha = 10^{-3}$ (K^{-1}). At low biases, $I \leq I_0$ (or $V \leq V_0$), the relative variance takes a constant value, Σ_0, which represents an intrinsic property of the system determined by the competition of two practically unbiased processes. Therefore, Σ_0 is independent of bias conditions and of α values (the scaling properties of Σ_0 as a function of $< R >_0$ have been studied in Ref. [17]). By contrast, a significantly different behavior of Σ is found in the nonlinear regime. To analyze this behavior, we report in Figs. 5 and 6, on a log-log plot, the relative variation of Σ as a function of the normalized bias under

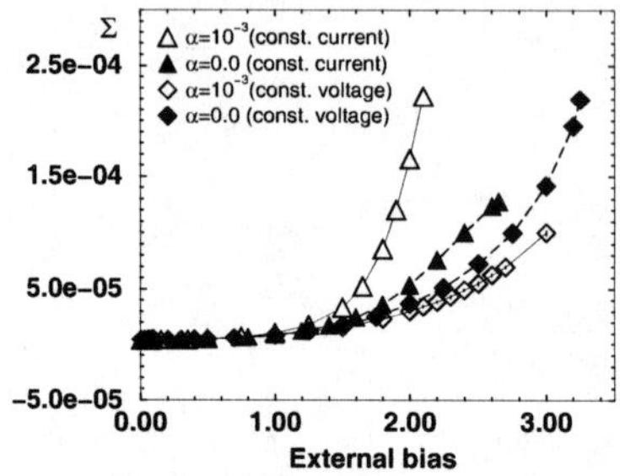

FIGURE 3. Relative variance of resistance fluctuations versus bias. Full and open triangles are obtained under constant current for $\alpha = 0$ and $\alpha = 10^{-3}$ (K^{-1}), respectively. Full and open diamonds are obtained under constant voltage for $\alpha = 0$ and $\alpha = 10^{-3}$ (K^{-1}), respectively.

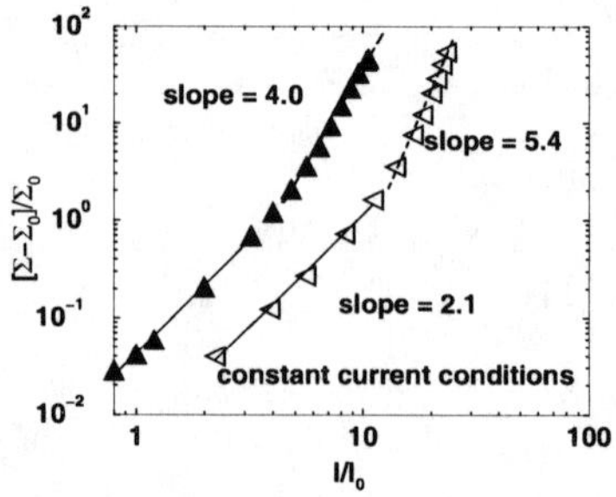

FIGURE 4. Relative variation of Σ versus the normalized bias under constant current conditions. Full triangles: $\alpha = 0$; open triangles: $\alpha = 10^{-3}$ (K^{-1}). The solid and dashed lines represent the best-fit of the data in the moderate bias and in the pre-breakdown regions, respectively.

constant current and constant voltage conditions, respectively. Only the nonlinear regime is shown in these figures. Two regions can be identified in this regime. The first region is at moderate bias, where Σ scales quadratically with the current, as [11]:

$$\frac{\Sigma}{\Sigma_0} = f(I/I_0), \qquad f(I/I_0) \simeq 1 + c(I/I_0)^\eta \tag{3}$$

with $\eta = 2.0 \pm 0.1$ and c a dimensionless coefficient. The second region is at high bias (pre-breakdown), where a strong superquadratic dependence appears. We notice that, under constant-current, the effect of the TCR strongly enhances the dependence of the resistance noise on the bias in this second region (the slope changes from 4.0 to 5.4). The opposite is found under constant-voltage conditions (in this case the slope decreases from 4.0 to 3.0). By comparing the effect of the two bias conditions for a fixed (and non zero) value of α, the growth of Σ in the pre-breakdown region is steeper under constant-current than under constant-voltage conditions. Remarkably, the slope in the pre-breakdown region is independent of bias conditions when the TCR is zero. Therefore the different behavior under the two bias conditions is due to the TCR effect. The role played by the parameters r_0 and E_R which determine the initial resistance and the intrinsic defect fraction, respectively, has been also analyzed [10, 12]. The slope of Σ in the pre-breakdown region is found to be independent of r_0 but dependent on E_R. This behavior of the relative variance of the resistance fluctuations is different from that obtained in the same region for the average resistance, as discussed in connection with Eq. (2). This strong nonlinearity and nonuniversality of Σ in the pre-breakdown region has been observed in electrical noise measurements in composites [4] and in other materials [13, 14]. Finally, it must be remarked that the quadratic dependence of Σ on the bias in the moderate bias region, described by Eq. (3), is valid independently of bias conditions and of the values of r_0, E_R and α. Further investigations, concerning the effects of the network structure and geometry are thus necessary to establish the

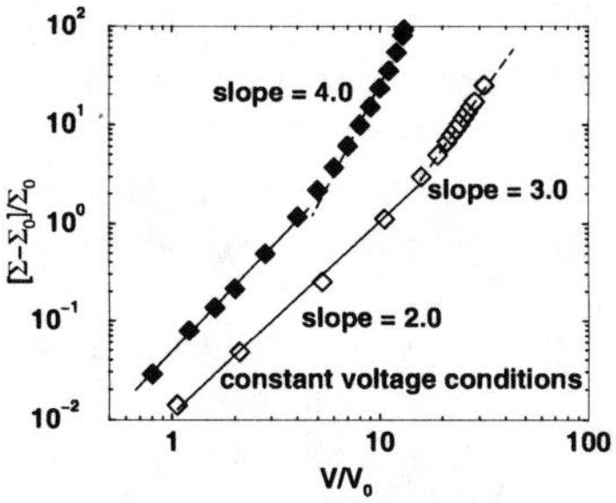

FIGURE 5. Relative variation of Σ versus the normalized bias under constant voltage conditions. Full diamonds: $\alpha = 0$; open diamond: $\alpha = 10^{-3}$ (K^{-1}). The solid and dashed lines represent the best-fit of the data in the moderate bias and in the pre-breakdown regions, respectively.

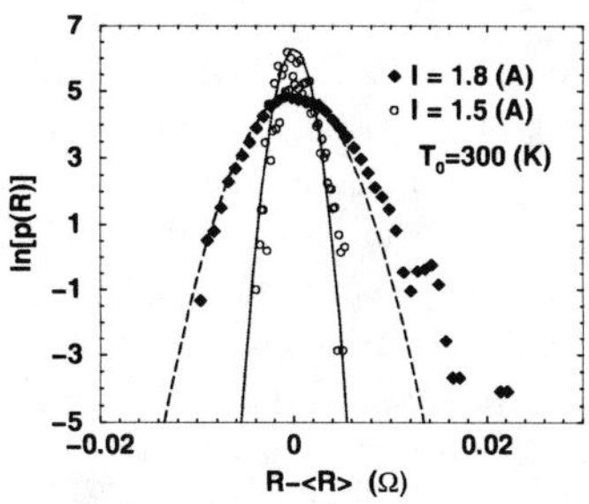

FIGURE 6. Distribution function of the resistance fluctuations for two bias values. Open circles correspond to $I = 1.5$ (A) and full diamonds to $I = 1.8$ (A). Since the plot is on a linear-log scale, the dotted and the long dashed curves represent the Gaussian fits to the two sets of data.

eventual universality of this behavior. The strong nonlinearity of Σ in the pre-breakdown region (Figs. 3-5) is found to be associated with the onset of a non-Gaussian noise, as shown in Fig. 7 where the distribution function of the resistance fluctuations around the average value is reported for two current values. From this figure we can see that the non-Gaussianity is significantly present at the highest bias. A similar behavior is found also at increasing values of the temperature for a fixed current [11]. A preliminary analysis shows that this non-Gaussianity seems sufficiently well described by the Nakagami distribution [18]. Finally, the power spectral density of resistance fluctuations is found to be Lorentzian and the corner frequency is found to be independent of bias but dependent on temperature [11, 12].

CONCLUSIONS AND OPEN QUESTIONS

All the results reported above are in good qualitative agreement with different measurements of the resistance and resistance noise performed in composites [3, 4], in metallic films [6, 16] and in nanostructured materials near the electrical breakdown [15]. Thus, we can conclude that the present approach catches the essential features of the nonlinear response to high currents (or voltages) and of breakdown phenomena in a wide class of disordered materials.

Now we briefly summarize some open questions connected to this study.
(i) To what extent the comparison with experiments can be made more quantitative?
(ii) Can we identify suitable parameters which act as reliable precursors of the electrical breakdown in real systems? In particular, Mukherjee et al. [3] suggested the definition of a dimensionless quantity, Λ, characterizing the breakdown properties of composites in the Joule regime. This quantity, function of Y, of the macroscopic temperature coefficient of the resistance, of the thermal conductivity and of the resistivity of the

conducting particles, would have a universal value. Our results for the dependence of the value of Y on the different model parameters follow the direction suggested by Mukherjee et al. Thus, can we confirm the existence of a universal value for Λ ?
(iii) Is it possible to generalize the scaling relations found in the linear regime (random percolation) to nonlinear regime (biased percolation) by introducing percolation thresholds and critical exponents that are functions of the applied bias?
(iv) How the dimensionality, the geometry, and the topology of the network would influence the results? In particular, is it possible to generalize the present approach to network with complex topology? Here we have studied a square lattice network. Actually, networks of complex topology [19] should be of great interest, particularly in the perspective of applying this approach to transport phenomena in self-organized structures.
(v) We have studied a RRN by assuming that all the elementary resistors have the same values of r_0, α, E_D and E_R. In other terms, in our approach the disorder inside the network arises from the fact that some resistors are broken and the temperature distribution is nonhomogeneous. On the other hand, it is possible to extend this approach by considering a RNN made by several "species" of resistors, i.e. resistors characterized by different values of the above parameters. Preliminary results, concerning the linear regime show a change in the power spectral density of the resistance fluctuations from Lorentzian to 1/f. Thus, we can ask what will happen for high biases, when the biased percolation can give rise to interesting self-organization phenomena.

ACKNOWLEDGEMENTS

This research is performed within the STATE project of INFM. Partial support is also provided by the MADESS II project of CNR (Italy). We thank Profs. L.B. Kish and Z. Gingl who introduced us to the biased percolation approach.

REFERENCES

1. Zapperi S., Ray P., Stanley H. E. and Vespignani A., *First-order transition in the breakdown of disordered media*. Phys. Rev. Letters, **78**, 1408-1411 (1997).
2. Lamaignère L., Carmona F. and Sornette D., *Experimental realization of critical thermal fuse rupture*. Phys. Rev. Letters, **77**, 2738-2741 (1996).
3. Mukherjee C. D., Bardhan K. K. and Heaney M. B., *Predictable electrical breakdown in composites*. Phys. Rev. Letters, **83**, 1215-1218 (1999).
4. Nandi U. N., Mukherjee C. D. and Bardhan K. K., *1/f noise in nonlinear inhomogeneous systems*. Phys. Rev. B, **54**, 12903-12914 (1996).
5. Dubson M. A., Hui Y. C., Weissman M. B. and Garland J. C., *Measurement of the fourth moment of the current distribution in two-dimensional random resistor networks*, Phys. Rev. B, **39**, 6807-6815 (1989).
6. Yagil Y., Deutscher G. and Bergman D. J., *Electrical breakdown measurements of semicontinuous metal films*, Phys. Rev. Letters, **69**, 1423-1426 (1992).
7. Gingl Z., Pennetta C., Kish L. B. and Reggiani L., *Biased percolation and abrupt failure of electronic devices*, Semic. Science Techn., **11**, 1770-1775 (1996).
8. Pennetta C, Reggiani L. and Trefan Gy, *Scaling and universality in electrical failure of thin films*, Phys. Rev. Letters, **84**, 5006-5009 (2000).

9. Stauffer D. and Aharony A., *Introduction to Percolation Theory*, Taylor & Francis, London (1992).
10. Pennetta C., Reggiani L., Alfinito E. and Trefan G., *Stationary regime of random resistor networks under biased percolation*, J. of Phys. C, **14**, 2371-2378 (2002).
11. Pennetta C., Reggiani L., Trefan G. and Alfinito E., *Resistance and resistance fluctuations in random resistor networks under biased percolation*, Phys. Rev. E, **65**, 066119-066119-10 (2002).
12. Pennetta C., *Resistance noise near to electrical breakdown*, Fluctuation and Noise Letters, **2**, R29-49 (2002).
13. Weissman M. B., *1/f noise and other slow, nonexponential kinetics in condensed matter*, Rev. Mod. Physics, **60**, 537-571 (1988).
14. *Unsolved Problems of Noise and Fluctuations*, Proc. 2nd Int. Conf. in Unsolved Problems of Noise, ed. by Abbott D. and Kish L. B., American Inst. of Physics, New York (1999).
15. L. B. Kish, C. Pennetta and Z. Gingl, *Biased Percolation Approach to Failure Propagation in Nanostructures and prediction of the Total Failure bt Electronic Noise Analysis*, Proc. SPIE's Annual Optical Science and Technology Meeting, Seattle, July, 2002, **in print**.
16. C. Pennetta, L. Reggiani, G. Trefan, F. Fantini, A. Scorzoni and I. De Munari, *A percolative approach to electromigration in metallic lines*, J. Phys. D: Appl. Phys., **34** 1421–1429 (2001).
17. Pennetta C., Trefan G. and Reggiani L., *Scaling law of resistance fluctuations in stationary random resistor networks*, Phys. Rev. Letters, **85**, 5238-5241 (2000).
18. Nakagami M., in *Statistical Methods in Radio Wave Propagation*, Hoffman W. G. (Ed), Pergamon, Oxford (1960).
19. Albert R. and Barabasi A. L., *Statistical mechanics of complex networks*, Rev. Mod. Physics, **74**, 47-97 (2002).

Instability of the Noise Level in Polymer Field-Effect Transistors with Non-Stationary Electrical Characteristics

Ognian Marinov[a)], M. Jamal Deen[a)], Jianfei Yu[b)], George Vamvounis[b)], Steven Holdcroft[b)], and William Woods[c)]

[a)] *Electrical and Computer Engineering Department, McMaster University, 1280 Main Street West, Hamilton, ON, L8S 4K1, Canada, E-mail: omarinov@yahoo.com*
[b)] *Department of Chemistry, Simon Fraser University,Vancouver, BC, Canada*
[c)] *School of Engineering Science, Simon Fraser University,Vancouver, BC, Canada*

Abstract. The low frequency noise (LFN) properties of field-effect transistors (FETs) using polymers as the semiconducting substrate material are investigated and explained in terms of the charge carrier transport in polymer thin-film structure. Three mechanisms contribute to the carrier transport – charge injection from source electrode into polymer, charge hopping between polymer molecules for drift transport toward the drain, and charge buildup, probably at polymer-oxide interface. Charge buildup is responsible for non-stationary electrical characteristics, but does not contribute significantly to the LFN. Charge hopping determines the maximum value of the mobility and the minimum value of mobility 1/f noise. The variations of the PFET characteristics are mainly due to dispersion in the injection barrier of source-to-polymer contact. High disorder in the polymer at the source contact can increase the leakage current in PFET and can introduce number fluctuation S_{G_N} in the polymer conduction on top of the mobility fluctuation. S_{G_N} is proportional to the DC power applied to the injection barrier and should be assumed as a voltage source, since carrier hopping in the polymer reduces the effect of the injection noise. At present, the physical origin of S_{G_N} is not fully understood.

INTRODUCTION

Organic- or polymer-based field-effect transistors (PFETs) have been increasingly investigated in the past two decades [1], [2]. This is in an effort to realize inexpensive thin-film transistors (TFTs) that can be used in some niche commercial applications, such as electronic tags, drivers in active matrix displays, or for sensing applications.

The low frequency noise (LFN) properties of the field-effect transistors (FETs) using polymers as the semiconducting substrate material were investigated and explained in terms of the non-stationary mobility μ in the semiconducting polymer. In our previous work [3], we found that the low-frequency noise on PFETs is of 1/f type for frequencies f<1kHz, as also reported in [4] and [5]. We developed a physically based quantitative description for the noise level in terms of the mobility fluctuation theory, relating the Hooge parameter to the non-stationary field-dependent mobility. This was a reasonable description since the spectral density S_I of the PFET drain current I_D was almost proportional to the DC power ($I_D \times V_{DS}$) applied onto the PFET's

CP665, *Unsolved Problems of Noise and Fluctuations: UPoN 2002: Third International Conference,* edited by S. M. Bezrukov

channel, with a slope of 9dB/dec. The slight decrease in the slope from the theoretical 10dB/dec was explained by the enhancement of the mobility at increased electrical field. Thus, the low-frequency noise could be assumed to be bulk effect in the polymer.

Our recent experiments on another set of samples confirmed that a relation between S_I and the DC power ($I_D \times V_{DS}$) does exist, as shown in Fig.1. However, in this case, S_I is not proportional to the DC power, since the slope in Fig.1 is almost 20dB/dec. Therefore, the Hooge equation cannot be used, assuming that the PFET's channel is a semiconductive bar. In addition, any normalization of the noise to the drain current, or drain voltage, does not show a constant value, as shown in Fig. 2 for the drain current over a range of temperatures.

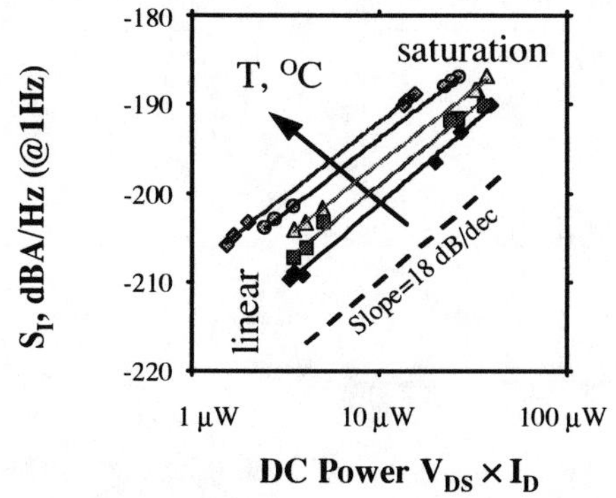

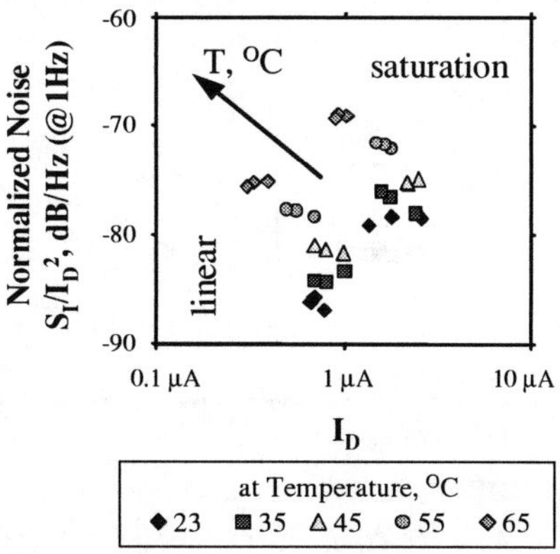

FIGURE 1. The noise S_I is a function of the DC power, applied on the polymer, for temperatures 23, 35, 45, 55, and 65°C. The PFET operates in both linear or saturation modes.

FIGURE 2. PFET performance degrades with increasing temperature – the noise increases and the field-effect (gate-bias dependent) current decreases.

In this paper, we report our new results for the LFN in PFET. Our investigation is focused on a possible explanation of noise in PFET with non-stationary electrical characteristics, which might be attributed by the polymer structure [6], [7], [8], interface and contact barrier effects [9], [10], [11], or the current transport mechanism in the structure [9], [12]. Since the variation in PFET electrical characteristics is significant, this is discussed first in the next section in terms of our findings for the charge carrier transport in polymer thin-film structures [13]. Then, the significance of the LFN noise sources in the PFET is analyzed in the section "1/f Noise Sources in PFET". The important findings of our work are summarized in the conclusions, and the unsolved problem of injection barrier noise is introduced as an open question at the end of the paper.

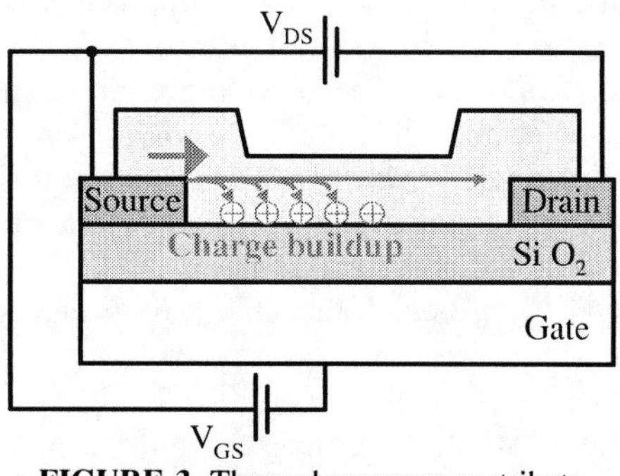

FIGURE 3. Three phenomena contribute to the charge carrier transport in PFET – injection (➔), drift (⟶) and buildup ( ).

ELECTRICAL CHARACTERISTICS OF PFET

The extensive electrical characterization in [13] showed that three phenomena contribute to the current flow in PFETs, and this is illustrated in Fig. 3. The first is charge injection from the source electrode into the polymer. In "good" PFETs, the rate of injection is dependent on the gate bias, and is almost independent of the drain bias. The second transport phenomenon is the drift due to the drain bias of carriers hopping between polymer molecules toward the drain. However, not all injected carriers are completely mobile. Some of them are immobilized for longer times in potential valleys. Thus, a charge buildup occurs at the polymer-oxide interface due to the electric field from the gate bias.

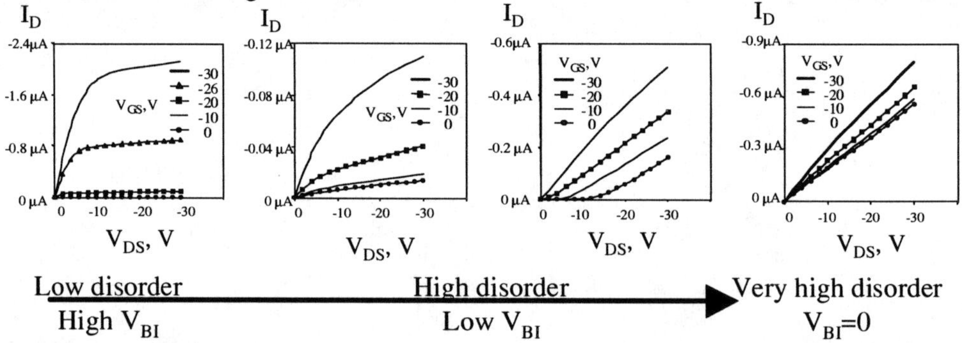

FIGURE 4. The disorder in the polymer at the source contact reduces the injection barrier V_{BI} of the contact. This causes variation in the shape of PFET electrical characteristics, since the charge carrier transport in polymer TFT is injection limited.

Overall, the electrical characterization in [13] showed that the charge transport in polymer field-effect transistors is limited by the injection from the source electrode into the polymer. This is because the shape of the characteristics depends on the quality of the polymer deposition. As illustrated in Fig. 4, the polymer TFT behaves as good field-effect transistor if the disorder in the polymer at the source contact is low. In this case, the potential barrier V_{BI} is high and uniform. However, if the disorder in the polymer at the source contact is higher, then the potential barrier V_{BI} decreases, and the drain bias can contribute to the carrier injection. As result, a leakage current occurs and this leakage can dominate the total drain current.

A very high disorder in polymer at the source contact can cause the injection barrier V_{BI} to vanish. In this case, the current flow through the TFT structure is determined by the electrical properties of the bulk polymer, and the structure behaves as a resistor, not as a field-effect transistor.

The effect of the charge buildup is complex. For example, it can cause a large variation in the measurement of the I-V characteristics of the polymer field-effect transistor, as shown in Fig. 5. The I-V curves are almost perfect, the drain current is high, when the measurement is performed for a short time of several seconds. However, if the sweep time is increased to 3 or 10 minutes, then the drain current decreases, indicating that the injection rate from the source electrode decreases. This is because the built-up space charge partially compensates the gate field, and the effective threshold voltage V_T of PFET increases with time.

Further increase of the measurement duration to 30 or 100 minutes provides enough time in order large number of carriers to be collected in polymer. Thus, the charge buildup modulates the polymer conductivity, and results in an increase of the leakage current with time. Since the charge buildup continues, the field-effect (gate bias dependent) current continues to decrease. In fact, the non-stationary effects due to charge buildup always appear during low-frequency noise measurements of PFET, because the measurement time is obviously about 10 to 30 minutes.

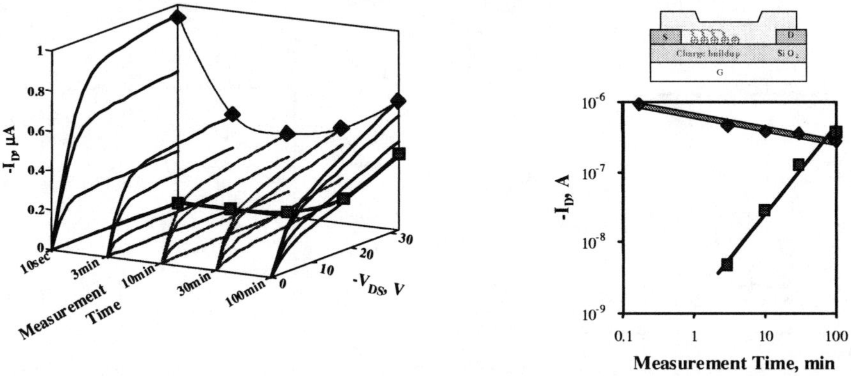

FIGURE 5. Charge buildup increases the number (N) of carriers in polymer. Two undesirable effects occur: increase of PFET leakage current (—■—), due to modulation of the polymer conductivity G_{DS}, and decrease of the gate-bias dependent field-effect current (◆), due to increase of the effective treshold voltage V_T of PFET.

1/F NOISE SOURCES IN PFET

The charge transport in PFET is relatively complicated. Three mechanisms, illustrated in Fig. 6, contribute to the transport, and any of them can create 1/f noise. Carrier hopping in polymer can be related to mobility noise (S_{G_μ}). Charge injection from the source electrode into the polymer modulates the polymer conductivity, since the number (N) of injected carriers may vary, and this can cause carrier number fluctuation in polymer conduction (S_{G_N}). Since the charge buildup can create a conduction layer at the polymer-oxide interface, the trapping-detrapping processes of interface states can introduce noise in the conduction layer. This will result in gate referred noise (S_{V_G}).

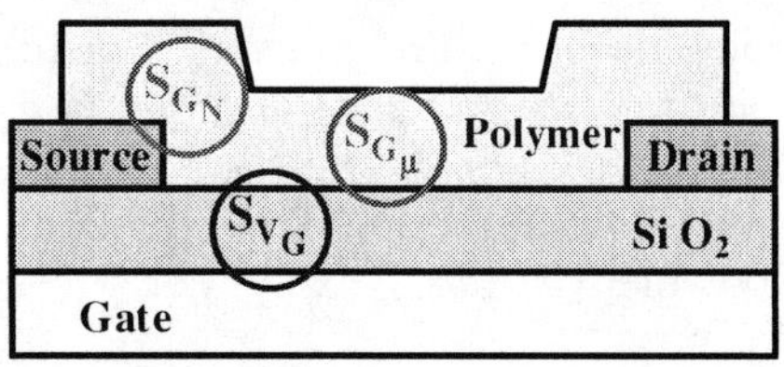

FIGURE 6. Low frequency (1/f) noise sources, relevant to the charge transport in PFET – mobility noise source (S_{G_μ}), injection barrier noise surce (S_{G_N}), and gate-referred noise source (S_{V_G})

The origin of low frequency noise in PFETs is studied experimentally to determine which noise source is dominant and how the noise level can be physically described.

First, we use the extraction technique [14] for the gate referred noise in MOSFETs. In this technique, the voltage noise (S_{V_G}) of interface charge is assumed constant. Using the PFET's transconductance g_m, S_{V_G} is converted into a current noise S_I in the drain terminal. As shown with eq. (1), the normalized noise becomes a function of the ratio g_m/I_D, and a slope of +10dB/dec is expected in Fig.7. However, the experimental data gave a negative slope. Therefore, S_{V_G} cannot explain the noise in PFETs.

$$\frac{S_I}{I_D^2} = \frac{g_m^2}{I_D^2} S_{V_G} \tag{1}$$

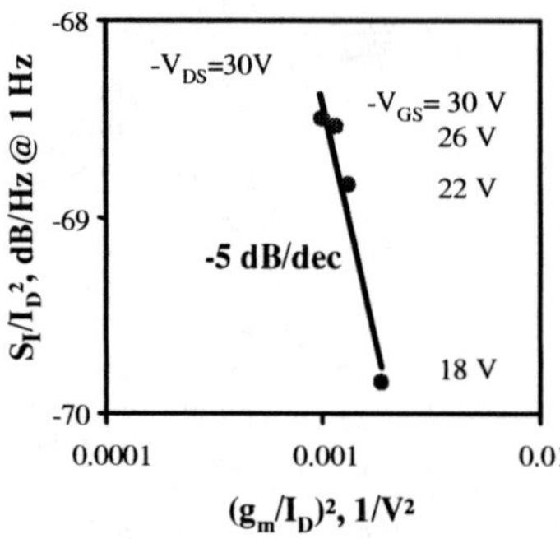

FIGURE 7. The gate referred noise model, given by eq. (1), cannot explain the noise in PFETs, because the normalized noise S_I/I_D^2 decreases when g_m^2/I_D^2 increases.

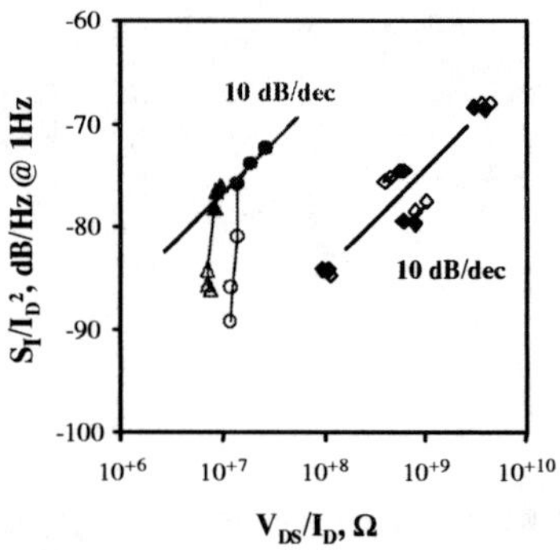

FIGURE 8. The proportionality $S_I/I_D^2 \propto V_{DS}/I_D$ indicates mobility noise S_{G_μ} in PFET, when the polymer is depleted.

Second, we use the extraction technique in [3] to describe the noise level in terms of the mobility fluctuation theory. It is assumed in this technique that the Hooge parameter α_H and the mobility μ in polymer are constant, and the number N of charge carriers is proportional to I_D. With these assumptions, a proportionality between normalized noise S_I/I_D^2 and resistance V_{DS}/I_D indicates mobility noise S_{G_μ} due to the hopping mechanism in polymer. Experimentally, as shown in Fig. 8, $S_I/I_D^2 \propto V_{DS}/I_D$ is observed, when the polymer layer is depleted and drain-source resistance V_{DS}/I_D in polymer TFT is high. These are for the cases of high impedance PFETs (diamonds in Fig. 8), or when the PFET is in saturation (closed symbols in Fig. 8). Thus, the noise in polymer is most likely due to mobility fluctuation in the carrier hopping mechanism.

In addition, the carrier hopping mechanism can suppress other noise sources, which is similar to the transition from dispersive to non-dispersive carrier transport in long polymer devices [9]. In polymer layers of lengths >1μm, after several hops, the carriers attain equilibrium waiting times between successive hops. Actually, since the PFET samples have 10μm channel length, it seems that this is the case when the polymer is depleted.

So far, we have shown that the gate referred noise (S_{V_G}) does not apply to PFET, and the mobility fluctuation (S_{G_μ}) in polymer can explain the 1/f noise in high impedance devices and in saturation mode of operation. However, S_{G_μ} cannot explain the noise in linear mode of operation, since S_I/I_D^2 is independent of PFET resistance V_{DS}/I_D, as shown with open triangles and open circles in Fig. 8 for two samples. For the case of linear mode of PFET operation, we assume that the fluctuation of the charge injection from the source electrode into the polymer modulates the polymer conductivity, and this causes the number fluctuation in polymer conduction (S_{G_N}).

S_{G_N} is located near the source electrode, as shown in Fig. 9. The carrier hopping mechanism in polymer conductance G_{DS} between drain and source terminals reduces the effect of injection barrier noise (S_{G_N}) on the drain side of polymer TFT. Therefore, it is reasonable to expect that 1/f noise level on the drain side is less than the level on the source side. It is experimentally observed that the noise on the drain side is about 3 to 5 dB less than the noise at the source side at same bias conditions in linear operation mode of PFET, as shown with open symbols in Fig. 9. The difference is even larger in saturation mode (closed symbols), confirming the effect of noise suppression when the polymer is depleted.

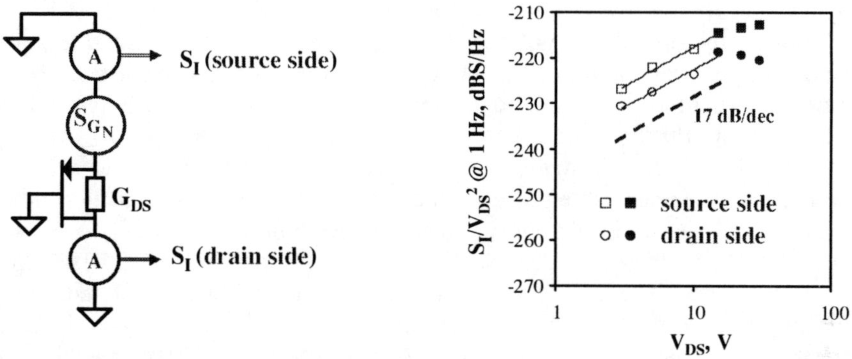

FIGURE 9. The effect of injection barrier noise (S_{G_N}) is reduced at the drain terminal, since the hopping mechanism in polymer conduction G_{DS} suppresses other noise sources.

Note that S_I/V_{DS}^2 is noise in the conductance G_{DS} caused by number fluctuation of injected carriers from source electrode. This is because

$$I_D = V_{DS} \times G_{DS} \,. \tag{2}$$

Assume that the noise is due to fluctuations of drain-source conductance (G_{DS}), then

$$\partial I_D = V_{DS} \times \partial G_{DS} \tag{3}$$

and the noise current is proportional to the conductance noise $S_{G_{DS}}$, scaled by the square of the drain bias voltage V_{GS}, given by

$$S_I = V_{DS}^2 \times S_{G_{DS}} \tag{4}$$

Generally, $S_{G_{DS}} \propto S_{G_N}$, if the injection barrier noise prevails in PFET, but S_{G_N} can be expressed also as a voltage or current source.

CONCLUSIONS

Three phenomena contribute to the charge carrier transport in polymer field-effect transistors – charge injection from source electrode into polymer, carrier hopping between polymer molecules and charge buildup.

The charge carrier hopping mechanism between polymer molecules depends on the type of the polymer, and it influences the maximum value of the mobility and the minimum value of mobility 1/f noise which is observed in the PFET.

The non-stationary electrical effects in PFETs are due to charge buildup, probably at polymer-oxide interface. The charge buildup is responsible for the increase of the leakage current and the decrease of the field-effect (gate bias dependent) current within a time scale of minutes.

The variations of the PFET characteristics are mainly due to variations in the injection barrier of the source-to-polymer contact. Therefore, the quality of the polymer deposition should be carefully investigated. Low disorder in the polymer at the source contact is beneficial for obtaining PFETs with repeatable characteristics, and low 1/f noise down to the limit of the mobility noise in the polymer.

In contrast, high disorder in polymer at the source contact can increase the leakage current in the polymer transistor. In this case, the injection barrier also becomes the dominant noise source, which introduces number fluctuation in the polymer layer on top of the mobility fluctuation.

Although the number fluctuation in the polymer conductance due to processes at injection barrier is a reasonable noise source in these PFETs, the physical origin of this source is an open question. It seems that the charge injection modulates the number (N) of carriers, and thus, the conduction (G_{DS}) in the polymer layer. The injection barrier noise source is sensitive to the DC drain bias when the polymer transistor operates in the linear mode. Therefore, it seems reasonable to assume that the injection barrier noise (S_{GN}) is proportional to the DC power applied on the injection barrier. Since the carrier hopping in polymer reduces the effect of the injection noise, S_{GN} should be assumed as a voltage source. However, to the best of our knowledge, strong evidence, which confirms this assumption is not available.

ACKNOWLEDGMENTS

The authors would like to thank the School of Engineering Science, Simon Fraser University, for the use of the clean room facilities and to several of their colleagues at both McMaster University and Simon Fraser University for assistance and cooperation. This work was supported in part by a strategic grant and a research grant from the Natural Sciences and Engineering Research Council (NSERC) of Canada.

REFERENCES

1. Deen, M.J., "Organic Semiconductor Devices," in *Wiley Encyclopedia of Electrical and Electronics Engineering (Vol. **15**)*, edited by J.G. Webster, John Wiley & Sons, New York, 1999, pp. 419-429.

2. Lodha, A., and Singh, R., *IEEE Transactions on Semiconductor Manufacturing* **14**(3), 281-296 (2001).
3. Deen, M. J., Marinov, O., Yu, J., Holdcroft, S., and Woods, W., *IEEE Transactions on Electron Devices* **48**(8), 1688-1695 (2001).
4. Martin, S., Dodabalapur, A., Bao, Z., Crone, B., Katz, H. E., Li, W., Passner, A., and Rogers, J. A., *Journal of Applied Physics* **87**(7), 3381-3385 (2000).
5. Necliudov, P. V., Rumyantsev, S. L., and Shur, M. S., *Journal of Applied Physics* **88**(9), 5395-5399 (2000).
6. Blom, P. W., de Jong, M. J. M., and Vleggaar, J. J. M., *Applied Physics Letters* **68**(23), 3308-3310 (1996).
7. Blom, P. W. M., and Vissenberg, M. C. J. M., *Physical Review Letters* **80**(17), 3819-3822 (1998).
8. Blom, P. W. M., Schoo, H. F. M., and Matters, M., *Applied Physics Letters* **73**(26), 3914-3916 (1998).
9. Martens, H. C. F., Brom, H. B., and Blom, P. W. M., *Physical Review B* **60**(12), R8489-R8492 (1999).
10. Nikitenko, V. R., Tak, Y.-H., and Bassler, H., *Journal of Applied Physics* **84**(4), 2334-2340 (1998).
11. Silvestre, G. C. M., Johnson, M. T., Giraldo, A., and Shannon, J. M., *Applied Physics Letters* **78**(11), 1619-1621 (2001).
12. Martens, H. C. F., Pasveer, W. F., Brom, H. B., Huiberts, J. N., and Blom, P. W. M., *Physical Review* B **63**(12), 125328-1 125328-7 (2001).
13. Marinov, O., Deen, M. J., Yu, J., Vamvounis, G., Steven Holdcroft, S., and Woods, W., "Variations of the Characteristics and Non-Stationary Current Transport in Polymer Thin Film Transistors," unpublished (2002).
14. Ghibaudo, G., Roux, O., Nguen-Duc, Ch., Balestra, F., and Brini J., *Phys. Stat. Sol (a)* **124**, 571 (1991).

Monte Carlo Simulation of Noise in Electronic Devices: Limitations and Perspectives

T. González*, J. Mateos, M. J. Martín-Martínez, S. Pérez, R. Rengel, B. G. Vasallo and D. Pardo

Departamento de Física Aplicada, Universidad de Salamanca
Plaza de la Merced s/n, 37008 Salamanca, Spain
**E-mail: tomasg@usal.es*

Abstract. We review the main problems met by the Monte Carlo technique when dealing with the study of noise in electronic devices. We also summarize the major challenges in the noise analysis associated with the continued evolution of electronic devices that could be addressed in the future through Monte Carlo simulations.

INTRODUCTION

Computer simulation has become a fundamental tool for the analysis of noise in electronic devices in the last two decades. Different numerical techniques, based on several transport models, have been developed to this end [1-4]. Among them, the Monte Carlo (MC) method has been shown to be very useful for the investigation of noise in electronic devices, specially at high frequencies (in the GHz range and beyond) and when a microscopic interpretation of the mechanisms at the origin of fluctuations is required. By coupling an ensemble MC simulator with a Poisson solver, the details of band structure and scattering mechanisms, as well as the self-consistent electric field, are intrinsically accounted for, thus incorporating in a natural way all the ingredients necessary for the analysis of noise (sources and transfer fields) in electronic devices [1]. Very successful implementations of this technique have been developed to analyze a variety of structures, where high-frequency noise has been satisfactorily investigated and interpreted, like in the case of n^+nn^+ [5,6], Schottky [7,8], planar-doped [9] and p-n [10] diodes; MESFETs [11,12], HEMTs [13,14], MOSFETs [15], BJTs and HBTs [16]; and mesoscopic structures [17-19]. In some cases, a remarkable agreement with experiments in real devices has been achieved [9,12,14].

However, despite of the potentiality and wide spectrum of application of MC simulations for the noise analysis, the limitations and problems met by this technique are manifold, mainly related to the very long computation times required to achieve enough statistical resolution (SR). On the other hand, many challenges have to be addressed in the future to deal with the new generations of electronic devices. The aim of this paper is to summarize all these topics.

CP665, *Unsolved Problems of Noise and Fluctuations: UPoN 2002: Third International Conference,* edited by S. M. Bezrukov

PROBLEMS/LIMITATIONS

In this section we review some of the main problems and limitations of the MC technique when dealing with the analysis of noise electronic in electronic devices. Most of these problems are finally related to the fundamental drawback of this technique: many implementations of MC simulators involve very large computation times, which are usually not affordable in technology development.

Technical problems

The above limitation is particularly pronounced in the case of noise analysis, since to achieve enough SR the simulations of the device history must be much longer and involve more simulated particles than static calculations [20]. From this point of view, much more efficient techniques exist for the analysis of electronic noise, like those based on transfer fields [3,4,21], but lacking in the detailed physical description of the system achieved with the MC method, which always constitutes the final reference to check the validity of any other approaches [3,21].

Thus, from a technical standpoint it is necessary to develop new algorithms able to reduce the statistical uncertainty of MC calculations without increasing the computation time. To this purpose, further work is necessary to improve the techniques used to:

(i) *Evaluate the instantaneous currents at the device terminals*. Initially, the currents were evaluated by counting the number of absorbed/injected particles and deriving the electric field at the contacts [22], technique that provides poor SR [7]. At present, the current estimators are typically based on the Ramo-Shockley theorem [23], aiming at involving as many particles as possible in the current evaluation to enhance the SR [13,24,25] and provide reliable noise results.

(ii) *Determine the noise characteristics*. Here, the correlation function approach [1] or the direct Fourier transform [26] of the device history (eventually divided in sub-histories) are systematically used to obtain the spectral densities of the different quantities of interest. New approaches requiring shorter computation times to achieve the same SR, like that reported in [27] based on the maximum entropy method (even if of reduced applicability), should be developed.

(iii) *Enhance the SR of the simulations in low-populated regions* (both in energy and space) by the use of particle expansion/compression algorithms. These variance-reduction techniques [28], which are specially necessary in barrier-controlled devices like p-n, Schottky-barrier diodes [29] and bipolar transistors, provide physically correct results for static characteristics, but meet problems in the noise analysis due to the different weight of the particles in the correlations induced by Coulomb interaction via the self-consistent field fluctuations. Thus, in the static case these techniques allow the MC analysis of barrier-controlled devices with enough SR in the full range of applied voltages, while the noise investigation must be restricted to voltages higher than a few k_BT/q below flat-band conditions [8,29].

Fast/slow processes

The concurrent presence of fast and slow processes having an influence on the noise in a given system makes very difficult its MC analysis. For example, the simultaneous consideration of scattering mechanisms (with typical times $\tau_S \approx 10^{-14}$-10^{-13} s) and generation-recombination (GR) processes (characterized by realistic lifetimes, $\tau_\ell \approx 10^{-5}$-$10^{-9}$ s) as sources of noise in a given semiconductor structure is nearly impossible from a computational point of view due to the extremely long computation time that would be required. Furthermore, slow processes (as compared with lattice scattering), like GR or number fluctuations associated with diffusive movement of carriers between terminals, can lead to the appearance of very long time tails in the correlation functions. These tails, firstly, may be very difficult to detect and, secondly, require again very long simulation times to achieve enough SR in their calculation [18,21].

In fact, to get enough significance of GR processes on the noise, the simulation time should be at least about $1000\tau_\ell$, which means that for every particle one must simulate the huge amount of 10^7-10^{12} scattering mechanisms. Only by artificially decreasing the characteristic times of GR processes some theoretical speculations can be done [30,31], as in the case reported in Fig. 1, corresponding to an homogenous semiconductor sample in the presence of both scattering and GR processes with characteristic times $\tau_s=10^{-14}$ s and $\tau_\ell=1.8\times10^{-12}$ s, respectively [30]. In Fig. 1(a) the current autocorrelation function $C_I(t)$ exhibits a two-time behavior related to the scattering mechanisms (evidenced in the velocity contribution) and GR processes (evidenced in the number contribution) at short and long time scales, respectively. Here, the tail associated with GR processes is relatively short due to the artificially low value adopted for τ_ℓ. The corresponding spectral density $S_I(f)$ [Fig. 1(b)] exhibits the expected double-Lorentzian shape.

Another example of these tails is reported in Fig. 2, corresponding to an homogenous 200 nm mesoscopic sample under a voltage of 40 k_BT/q in the presence of elastic scattering with characteristic time $\tau_s=5\times10^{-15}$ s, and in the absence of

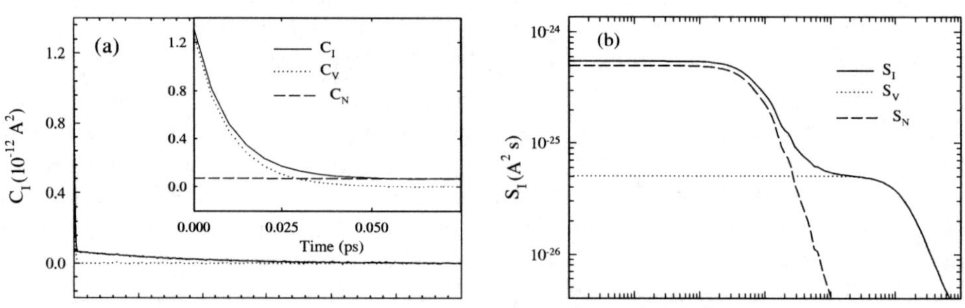

FIGURE 1. (a) Current autocorrelation function (including velocity and number contributions, with the short time behavior detailed in the inset), and (b) corresponding spectral density for the case of an homogenous semiconductor sample under far-from-equilibrium conditions in the presence of both scattering and GR processes with characteristic times $\tau_s=10^{-14}$ s and $\tau_\ell=1.8\times10^{-12}$ s, respectively [30].

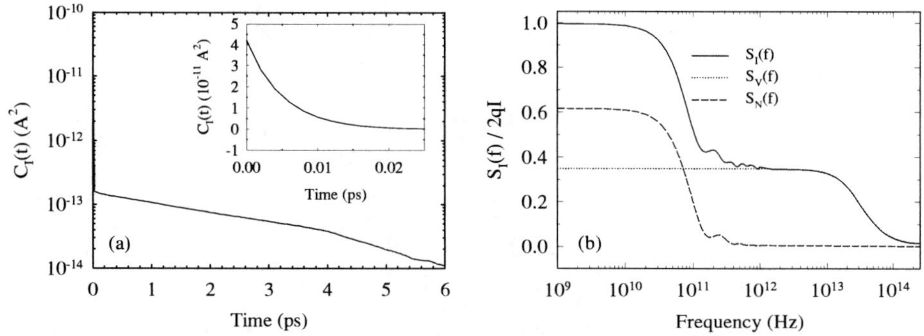

FIGURE 2. (a) Current autocorrelation function (short time behavior detailed in the inset), and (b) corresponding spectral density (velocity and number contributions are also shown) for the case of an homogenous 200 nm mesoscopic sample under a voltage of 40 k_BT/q in the presence of elastic scattering with characteristic time τ_s=5×10^{-15} s and in the absence of Coulomb correlations [18]. The carrier transit time is τ_T=5×10^{-12} s.

Coulomb correlations [18]. In this case the long time tail, associated with the number fluctuations controlled by the carrier transit time $\tau_T \approx 5\times10^{-12}$ s, has an amplitude about 2-3 orders of magnitude lower than the velocity contribution at the shortest times, which makes very difficult its detection and correct estimation. In fact, at first sight $C_I(t)$ seems to vanish at about 0.02 ps (see inset), while the tail is significant for more than 6 ps. This tail is crucial to evaluate correctly the spectral density, as observed in Fig. 2(b), since it provides an important contribution to the low-frequency value (about 60% of the total value).

Problems in specific devices

Some problems arise in the investigation of specific devices. Due to their importance for high-frequency applications, we report here those related to HBTs, HEMTs and RF MOSFETs.

The analysis of noise in bipolar devices, like BJTs and HBTs, involves the simultaneous self-consistent consideration of electrons and holes, and the presence of energy barriers controlling carrier transport. All these factors lead to a very poor SR of the MC currents at the terminals and to a high uncertainty in the noise results. This can be specially pronounced in the case of the base current in HBTs due to its very low value. Moreover, in the absence of variance-reduction techniques valid for the noise analysis, this is only affordable in a limited range of bias conditions, basically near high injection [16]. On the other hand, long time tails are found in the correlation functions, which makes even more difficult the study of current fluctuations in these devices, requiring very long, nearly prohibitive [21], simulation times to achieve a significant SR. To illustrate these problems, Fig. 3(a) shows the collector and base current autocorrelation functions for a Si/SiGe HBT (with a Ge trapezoidal shape in the base) working near high-injection conditions [16]. To obtain the results shown in Fig. 3, the calculations involve about 35000 simulated particles during 10 ns of device history (quite long time). Apparently, these correlation functions vanish at about 1 ps.

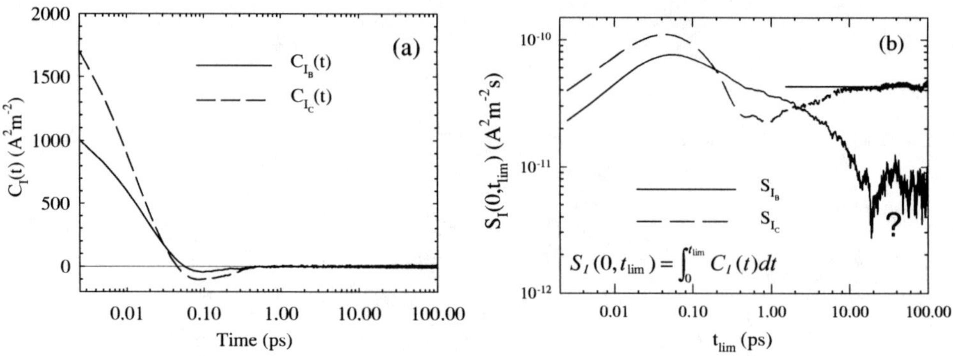

FIGURE 3. (a) Base and collector current autocorrelations functions vs. time, and (b) corresponding low-frequency spectral densities as a function of the integration time for the case of a Si/SiGe HBT operating near high-injection conditions [16].

However, even if taking very low absolute values around zero, there is a long tail which provides a significant contribution to the spectral density, as observed in Fig. 3(b), where the low frequency value of the spectral density $S_I(0)$ is shown as a function of the integration time t_{lim} used to perform the Fourier transform of $C_I(t)$. Indeed, after 1 ps the values of $S_I(0)$ still depend on t_{lim}, as pointed out in [21]. In the case of the collector current, beyond 20-30 ps a nearly constant value is reached and $S_{IC}(0)$ can be determined with enough accuracy. However, the base spectral density does not reach such a constant value, and it is extremely difficult to assign a reliable value to $S_{IB}(0)$.

In the case of FETs (HEMTs and MOSFETs), the calculation of the noise sources related to the gate terminal (S_{ig}, S_{igid}) at low frequency, where they are expected to vanish, is particularly delicate, since S_{ig} and S_{igid} are crucial in the evaluation of key figures of merit like the minimum noise figure ($F_{\min}$) [32]. Again, the poor SR due to their low values is the main problem. As an example, Fig. 4 shows the drain, gate and drain-gate correlation functions and low-frequency spectral densities (as a function of the integration time) for the case of a 100 nm T-gate AlInAs/GaInAs HEMT operating under saturation conditions [14]. The long time tails found in bipolar transistors are absent in the case of these devices provided that holes have no influence on transport. Indeed, the correlation functions vanish in the ps scale, though some oscillations around zero related to plasma effects are still present for longer times. In the case of the drain spectral density, the reliability of the MC results for $S_{id}(0)$ is high, even if some small dependence on t_{lim} is found. However, for the gate and gate-drain spectra, expected to vanish at low frequency in the absence of gate-leakage current, the uncertainty is higher, taking values oscillating significantly around zero. In order to obtain a reliable frequency dependence of $S_{ig}(f)$ and $S_{igid}(f)$, required to evaluate $F_{\min}(f)$, a common practice is to subtract the low-frequency value to the calculated spectrum [11].

The complete MC analysis of noise in these devices should include some additional effects like those related to the 2D electron gas present in the channel or the gate-leakage current associated with electron tunneling. These are also open problems never addressed so far.

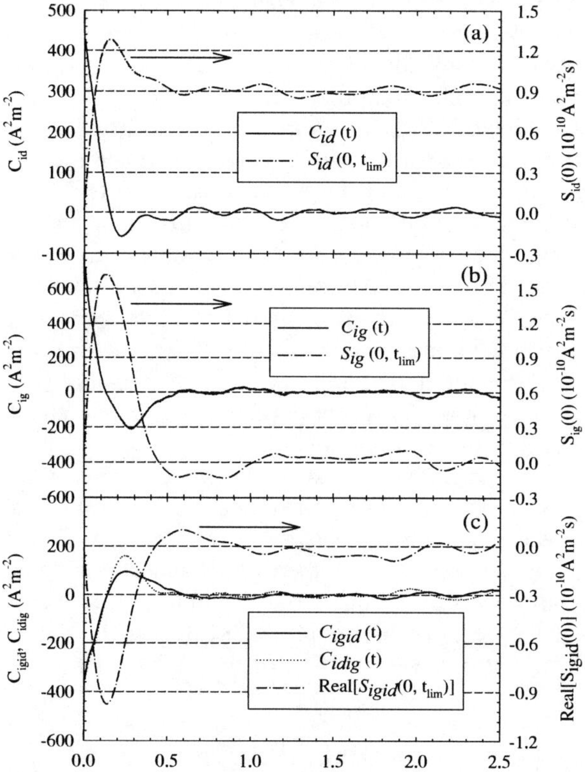

FIGURE 4. Correlation functions vs. time (left axis) and corresponding low-frequency spectral densities (right axis) as a function of the integration time for the case of (a) drain, (b) gate and (c) gate-drain current fluctuations in a 100 nm T-gate AlInAs/GaInAs HEMT operating under saturation conditions [14].

CHALLENGES/PERSPECTIVES

As future challenges and perspectives of application of the MC technique related to the noise analysis in electronic devices we mention the following:

Noise in devices operating under periodic large-signal conditions. This is a subject of growing activity in recent years [4,33]. The noise behavior under these working conditions, when the device response becomes nonlinear, is expected to change significantly with respect to that found under steady-state operation. Effects like noise upconversion [30] and harmonic-generation/noise competition [34] can take place under large-signal conditions. Some efforts are currently been made on this topic, both in bulk semiconductor [34] and diodes [35].

Noise in short-channel RF bulk and SOI MOSFETs. In the case of these devices, it is necessary to achieve an accurate evaluation of the noise terms related to the gate terminal, S_{ig} and S_{igid}, and the determination of the physical origin of excess noise as compared to long-channel predictions. An initial attempt can be found in [15].

Noise in devices in the presence of impact ionization. The MC technique can be specially appropriate to analyze the noise related to the hole gate-leakage current, the noise behavior just before breakdown or in the presence of kink effect in FETs, etc., subjects related to impact ionization processes not studied so far by MC simulations.

Noise in devices requiring a 3D simulation. To date, only 2D MC simulations have been used to study noise in electronic devices. However, some modern structures, like ultrashort MOSFETs and novel ballistic devices based on HEMT channels (T and Y branches) [36], require a 3D simulation which complicates the noise analysis.

Contact modeling. The boundary conditions adopted in the simulations can be crucial in the noise behavior of the devices, specially in the case of ballistic structures, where the noise comes from the injection statistics at the contacts. At this point, the development of models for momentum distributions and time statistics of injected carriers, taking into account the peculiarities of the contacts (level of degeneracy, presence of heterojunctions, non-uniform potentials, fractional statistics, etc.), are necessary. Some attempts can be found in [17,37].

Noise in quantum devices. So far, most of the work done by the MC method in the field of noise is performed within a semiclassical description of carrier transport. To deal with devices where quantum effects are determining, it is necessary to develop phase-coherent models compatible with the use of random MC techniques and able to provide instantaneous values of the quantities of interest. A first approach on this line, based on Bohm trajectories, can be found in [38], where the noise in a single-tunnel barrier is analyzed, still within a non self-consistent description.

CONCLUSIONS

As main conclusion we point out that the MC technique, recognized as a very useful tool when dealing with noise problems in electronic devices, meets as principal limitation the very long computation times required to obtain reliable results. Most of the problems reviewed in this paper are somehow linked to this limitation.

Anyway, with the continuous advent of more powerful computer facilities, these limitations are expected to be overcome, so that the application of this technique seems feasible for the analysis not only of dc transport but also of noise in more and more complex devices, and under new working conditions, like those listed in the challenges/perspectives section.

ACKNOWLEDGMENTS

Helpful discussions with E. Starikov, P. Shiktorov (Semiconductor Physics Institute, Vilnius, Lithuania) and L. Reggiani (Universitá di Lecce, Italy) on the subject of this paper are gratefully acknowledged. This work has been partially supported by the Dirección General de Investigación (Ministerio de Ciencia y Tecnología) and FEDER through the project TIC2001-1754, by the Consejería de Cultura de la Junta de Castilla y León through the project SA057/02, and by the European Commission through the NANOTERA project IST-2001-32517.

REFERENCES

1. L. Varani, L. Reggiani, T. Khun, T. González, and D. Pardo, *IEEE Trans. Electron Devices* **41**,1916 (1994).
2. H. L. Hartnagel, R. Katilius, and A. Matulionis, *Microwave Noise in Semiconductor Devices*, Wiley-Interscience, New York, 2001.
3. P. Shiktorov, E. Starikov, V. Gruzinskis, T. González, J. Mateos, D. Pardo, L. Reggiani, L. Varani, and J. C. Vaissiere, *Riv. Nuovo Cimento* **24**, 1 (2001).
4. F. Bonani and G. Ghione, *Noise in Semiconductor Devices: Modelling and Simulation*, Springer, Berlin, 2001.
5. L. Varani, T. Kuhn, L. Reggiani, and Y. Perlès, *Solid-State Electron.* **36**, 251 (1993).
6. T. González, D. Pardo, L. Varani, and L. Reggiani, *Appl. Phys. Lett.* **63**, 84 (1993).
7. J. G. Adams, T. W. Tang, and L. Kay, *IEEE Trans. Electron Devices* **41**, 575 (1994).
8. T. González, D. Pardo, L. Varani, and L. Reggiani, *J Appl. Phys.* **82**, 2349 (1997).
9. V. Gruzinskis, J. Liberis, A. Matulionis, P. Sakalas, E. Starikov, P. Shiktorov, B. Szentpali, V. Van Tuyen, and H. L. Hartnagel, *Appl. Phys. Lett.* **73**, 2488 (1998).
10. M. J. Martín-Martínez, D. Pardo, and J. E. Velázquez, *Appl. Phys. Lett.* **71**, 3382 (1997); *J. Appl. Phys.* **84**, 5012 (1998).
11. T. González, D. Pardo, L. Varani, and L. Reggiani, *IEEE Trans. Electron Devices* **42**, 991 (1995).
12. J. Mateos, T. González, D. Pardo, P. Tadyszak, F. Daneville, and A. Cappy, *Solid-State Electron.* **39**, 1629 (1996).
13. S. Babiker, A. Asenov, N. Cameron, S. Beaumont, and J. R. Barker, *IEEE Trans. Electron Devices* **45**, 1644 (1998).
14. J. Mateos, T. González, D. Pardo, V. Hoel, and A. Cappy, *IEEE Trans. Electron Devices* **47**, 1950 (2000).
15. R. Rengel, J. Mateos, D. Pardo, T. González, and M. J. Martín, *Semicond. Sci. Technol.* **16**, 939 (2001).
16. M. J. Martín-Martínez, S. Pérez, D. Pardo, and T. González, *J. Appl. Phys.* **90**, 1582 (2001); *Physica B* **314**, 381 (2002).
17. R. C. Liu, P. Eastman, and Y. Yamamoto, *Solid State Commun.* **102**, 785 (1997).
18. T. González, C. González, J. Mateos, D. Pardo, L. Reggiani, O. M. Bulashenko, and J. M. Rubí, *Phys. Rev. Lett.* **80**, 2901 (1998).
19. T. González, in *Proc. of the 16th Int. Conf. on Noise in Physical Systems and 1/f fluctuations*, edited by G. Bosman, World Scientific, Singapore, 2001, pp. 431-436.
20. C. Jungemann and B. Meinerzhagen, *IEEE Trans. Electron Devices* **48**, 985 (2001).
21. C. Jungemann, B. Neinhüs, S. Decker, and B. Meinerzhagen, *IEEE Trans. Electron Devices* **49**, 1250 (2002); ibid.**49**, 1258 (2002).
22. C. Moglestue, *IEEE Trans. Electron Devices* **32**, 2092 (1985).
23. H. Kim, H. S. Min, T. W. Tang, and Y. J. Park, *Solid-State Electron.* **34**, 1251 (1991).
24. V. Gruzinskis, S. Kersulis, and A. Reklaitis, *Semicond. Sci. Technol.* **6**, 602 (1991).
25. P. D. Yoder, K. Gartner, U. Krumbein, and U. Fichtner, *IEEE Trans. Comput.-Aided Des.* **16**, 1082 (1997).
26. W. H. Press, B. P. Flannery, S. A. Teulosky, and W. T. Vetterling, *Numerical Recipes*, Cambridge University Press, New York, 1992.
27. J. M. Miranda, J. L. Sebastián, and S. Muñoz, *Appl. Phys. Lett.* **72**, 238 (1998).
28. A. Pacelli and U. Ravaioli, *Solid-State Electron.* **41**, 599 (1997).
29. M. J. Martín-Martínez, T. González, D. Pardo, and J. E. Velázquez, *Semicond. Sci. Technol.* **11**, 380 (1996).
30. S. Pérez, T. González, S. Delage, and J. Obregon, *J. Appl. Phys.* **88**, 800 (2000).
31. B. G. Vasallo, J. Mateos, D. Pardo, and T. González, *Semicond. Sci. Technol.* **17**, 440 (2002).
32. S. D. Greaves and R. T. Unwin, *Microwave Opt. Technol. Lett.* **6** ,60 (1993).
33. F. Danneville, B. Tamen, A. Cappy, J. B. Juraver, O. Llopis, and J Grafeuil, *Fluctuation Noise Lett.* **1,** L189 (2001).
34. P. Shiktorov, E. Starikov, V. Gruzinskis, L. Reggiani, L. Varani, and J. C. Vaissière, *Appl. Phys. Lett.* **80**, 4759 (2002).
35. S. Pérez and T. González, *Semicond. Sci. Technol.* **17**, 696 (2002).
36. I. Shorubalko, H. Q. Xu, I Maximov, P. Omling, L. Samuelson, and W. Seifert, *Appl. Phys. Lett.* **79**, 1384 (2001).
37. T. González, J. Mateos, D. Pardo, L. Varani, and L. Reggiani, *Semicond. Sci. Technol.* **44**, L37 (1999).
38. X. Oriols, F. Martín, and J. Suñé, *Appl. Phys. Lett.* **79**, 1703 (2001); ibid. **80**, 4048 (2002).

Monte Carlo Simulation of Electronic Noise under Large-Signal Operation

E. Starikov*, P. Shiktorov*, V. Gružinskis*, S. Pérez†, T. González†, L. Reggiani**, L. Varani‡ and J.-C. Vaissière‡

*Semiconductor Physics Institute, Goshtauto 11, LT2600 Vilnius, Lithuania
†Departamento de Física Aplicada, Universidad de Salamanca, Plaza de la Merced s/n, E-37008 Salamanca, Spain
**INFM - National Nanotechnology Laboratory, Dipartimento di Ingegneria dell' Innovazione, Università di Lecce, Via Arnesano s/n, 73100 Lecce, Italy
‡Centre d'Electronique et de Micro-optoélectronique de Montpellier (CNRS UMR 5507), Université Montpellier II, 34095 Montpellier, France

Abstract. We present Monte Carlo calculations of electronic noise in bulk materials and submicron semiconductor structures subject to periodic electric fields or voltage/current excitations of high frequency and large amplitude. The modifications of the noise spectra with respect to the static case are analyzed. Under appropriate conditions the noise spectra exhibit a resonant peak which is interpreted as upconversion of hot-carrier intergroup noise.

INTRODUCTION

The development of semiconductor devices for frequency mixing and multiplication, especially in the THz frequency range [1,2], depends crucially on the intrinsic noise of the given device [1-4]. Indeed, under large-signal operation, the spectra of the regular response and of the noise overlap in the whole frequency range of interest. This fact has stimulated the development of theoretical analyses of electronic noise under large-signal conditions. Such investigations started recently in the framework of the impedance field method, usually implemented within the drift-diffusion model where the noise source is given by a modulated stationary diffusion coefficient [3,4]. However, under periodic large-signal operation the diffusion coefficient can differ significantly from its stationary value. This is the case of applied signals with frequencies high enough so that the cutoff of various relaxation processes and the appearance of some anomalies, such as noise upconversion, are expected [5,6]. To account for these effects, a microscopic approach to the problem based on most general principles is mandatory.

The main goal of this work is to address this issue by means of Monte Carlo (MC) simulations which investigate the behavior of hot-carrier noise in bulk materials and submicron semiconductor structures operating under high-frequency large-signal conditions. The MC procedure for noise calculations is based on the symmetric two-time correlation function of fluctuations and it is similar to that described in [7].

CP665, *Unsolved Problems of Noise and Fluctuations: UPoN 2002: Third International Conference,* edited by S. M. Bezrukov

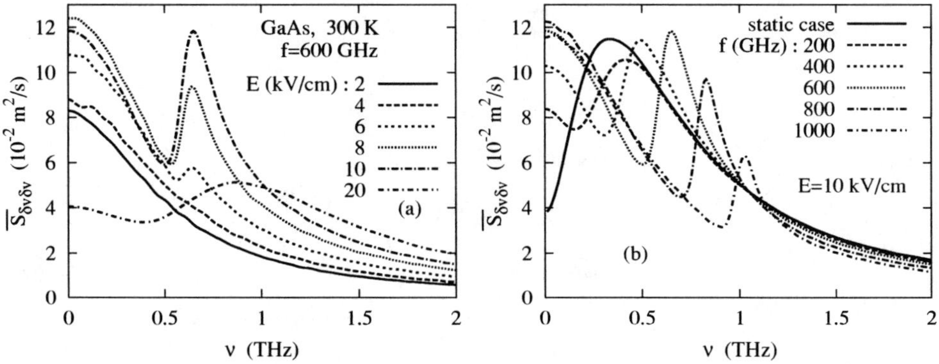

FIGURE 1. Noise spectra calculated by the MC method in bulk n-GaAs with doping level $N_D = 10^{16}\ cm^{-3}$ at room temperature under the action of a MWEF with different values of (a) amplitude E and (b) frequency f.

MODIFICATIONS OF HOT-CARRIER VELOCITY NOISE SPECTRA

Figure 1 illustrates the main modifications, with respect to the static case, exhibited by the stationary part of the spectral density of velocity fluctuations $\overline{S}_{\delta v \delta v}(\nu)$ under high-frequency large-signal operation [7]. Here, a microwave electric field (MWEF) $Ecos(2\pi ft)$ with different amplitudes E and frequencies f [Figs. 1(a) and 1(b), respectively] is applied to bulk n-GaAs. As follows from Fig. 1(a), at low amplitudes of the MWEF the spectrum shows the usual Lorentzian shape. When the amplitude becomes high enough for intervalley transfer to start (i.e., at $E > 5\ kV/cm$), a peak at the MWEF frequency appears, initially more pronounced the higher the field amplitude. Furthermore, for E above 10 kV/cm the peak begins to decrease and eventually shifts to higher frequencies. The behavior of the peak at increasing values of the excitation frequency is reported in Fig. 1(b) under hot-carrier conditions (at $E = 10\ kV/cm$). For comparison, the static case is also reported as solid line. To keep a similar heating of carriers, the static calculations were performed for a field $E_0 = 7\ kV/cm \approx E/\sqrt{2}$. In the static case $\overline{S}_{\delta v \delta v}(\nu)$ exhibits a peak at $\nu \approx 300\ GHz$ caused by intervalley transfer due to hot-carrier effects. As follows from Fig. 1(b), under large-signal conditions and fixed MWEF amplitude, when the excitation frequency f is increased, the low-frequency noise increases too, the hot-carrier peak of $S_{\delta v \delta v}(\nu)$ shifts to higher frequencies and finally transforms into a resonant-like peak centered at $\nu \approx f$.

Remarkably, a similar peak in the noise spectrum is also observed at low lattice temperatures when the MWEF amplitude is considerably lower than the threshold value for intervalley transfer, so that carrier heating is governed by optical phonon emission. Figure 2 shows the noise spectra in bulk n-InN at $T = 80\ K$ for several amplitudes and frequencies of the MWEF. At low values of the MWEF amplitude, carrier heating is insufficient for the electron energy ε to reach the value of the optical phonon energy $\hbar\omega_0$. Accordingly, all the carriers remain in the low energy region $\varepsilon < \hbar\omega_0$ (the so called

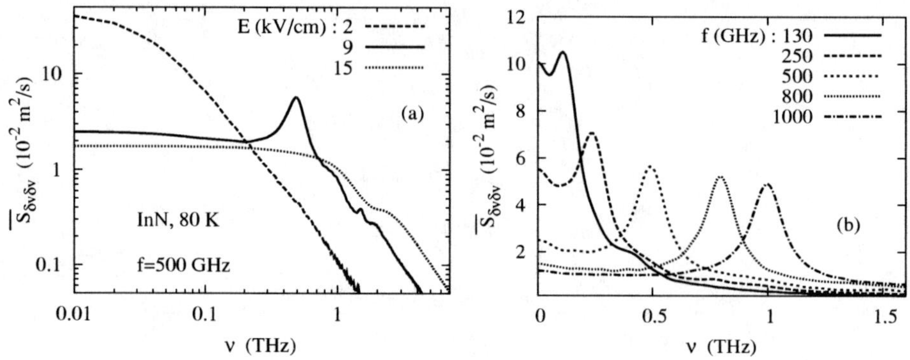

FIGURE 2. Noise spectra calculated by the MC method in bulk n-InN with doping level $N_D = 10^{16}\ cm^{-3}$ at 80 K under the action of a MWEF with different values of (a) amplitude E and (b) frequency f, at a nearly constant ratio E/f.

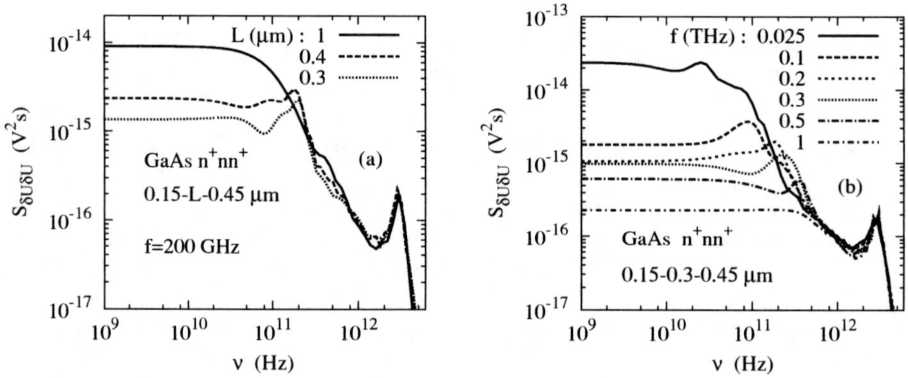

FIGURE 3. Voltage noise spectra in a GaAs n^+nn^+ structure under the action of a large-amplitude harmonic current $J_0 sin(2\pi ft)$ for different: (a) length of the n-region L and (b) frequency f of the excitation signal. $n^+ = 10^{17}\ cm^{-3}$, $n = 10^{16}\ cm^{-3}$.

passive region), where the main sources of scattering are acoustic phonons and ionized impurities because of the low lattice temperature. As a consequence, the spectrum of velocity fluctuations is found to take the usual Lorentzian shape [dashed curve in Fig. 2(a)]. At sufficiently high values of E all carriers reach a high energy $\varepsilon > \hbar\omega_0$ (the so called active region) during every half a period of the MWEF, then, they quickly emit an optical phonon and return back to the passive region. Again, the noise spectrum shows a nearly-Lorentzian shape [dotted line in Fig. 2(a)]. At an intermediate MWEF amplitude [solid line in Fig. 2(a)], the spectrum is found to exhibit a significant peak at the MWEF frequency, $\nu = f$, and minor peaks at higher harmonics which are superimposed to the Lorentzian-type spectrum. With the increase of the MWEF frequency f, the peak shifts to higher frequencies, and even becomes more pronounced, as reported in Fig. 2(b).

Here, calculations are performed keeping the ratio E/f nearly constant to have similar trajectories of the carrier motion in the passive region at different frequencies f.

An analogous behavior of the noise spectra is observed also in submicron n^+nn^+ GaAs structures. As an example, Fig. 3 shows the modifications of the voltage noise spectrum when an harmonic current of large amplitude is forced to flow through the structure [8]. In Fig. 3(a) the current density amplitude J_0 was varied to give an average electric field in the n-region of about 30 kV/cm, in Fig. 3(b) $J_0 = 6.48\ A/m^2$. We notice that the resonant peak is found to appear in sufficiently short structures, what indicates its dynamical origin.

Of course, in all the cases considered above, the resonant-like peaks in the noise spectra are associated with oscillations of the corresponding correlation functions at the large-signal frequency. It should be underlined that the presence and the properties of such a resonant-like peak is quite similar to what is observed in noise upconversion phenomena. Therefore, the following section will be devoted to a deeper investigation of these phenomena.

THE PROBLEM OF NOISE UPCONVERSION

The phenomenon of noise upconversion consists in the transfer of an excess noise component from the proper low-frequency region to a high-frequency region of the spectrum under the presence of a periodic signal of given amplitude. Typical examples are the transfer of $1/f$, trapping-detrapping or generation-recombination noise from the low-frequency to the GHz frequency region in voltage controlled oscillators [9] and systems driven by periodic large-amplitude excitations [4,6].

The simplest mathematical model to describe the upconversion process consists in representing some fluctuating quantity $x(t)$ in the form: $\{x(t)\} = cos(2\pi ft)\{u(t)\}$, which corresponds to the amplitude modulation of a high-frequency harmonic signal by some stationary random process $\{u(t)\}$. It is assumed that the random process $\{u(t)\}$ has zero average value and is characterized by the correlation function $C_{uu}(\tau) =< u(t-\tau/2)u(t+\tau/2) >$, with the corresponding spectral density $S_{uu}(\nu)$. The two-time symmetric correlation function of the random process $\{x(t)\}$, $C_{xx}(t,\tau) =< x(t-\tau/2)x(t+\tau/2) >$, can then be expressed as [5]:

$$C_{xx}(t,\tau) = \frac{1}{2}[cos(2\pi f\tau) + cos(4\pi ft)]C_{uu}(\tau). \tag{1}$$

According to this expression, the instantaneous spectral density of $x(t)$ can be decomposed into two terms, $S_{xx}(t,\nu) = \overline{S}_{xx}(\nu) + S'_{xx}(t,\nu)$, corresponding, respectively, to the stationary component

$$\overline{S}_{xx}(\nu) = \frac{1}{4}[S_{uu}(\nu+f) + S_{uu}(\nu-f)] \tag{2}$$

and to the nonstationary component

$$S'_{xx}(t,\nu) = \frac{1}{2}cos(4\pi ft)S_{uu}(\nu) \tag{3}$$

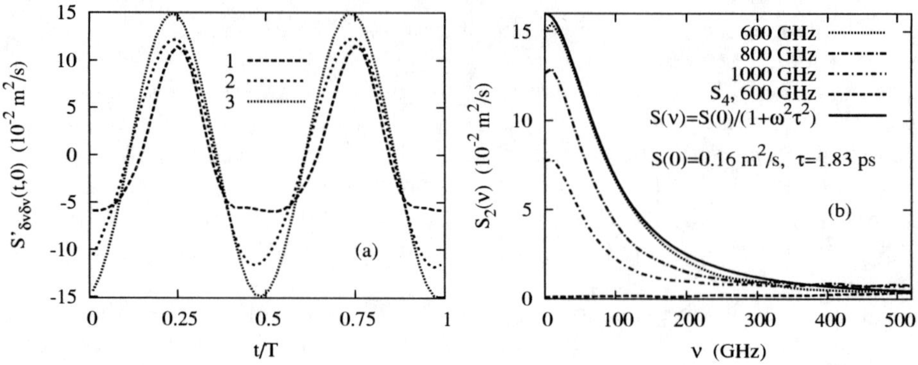

FIGURE 4. (a) Zero-frequency nonstationary part of the spectral density of velocity fluctuations $S'_{\delta v \delta v}(t,0)$ at different frequencies of the applied field $f = 200$, 400, 600 GHz (curves 1 to 3) and (b) upconverted spectral density represented by the second harmonic of $S_{\delta v \delta v}(t,\nu)$. The solid line shows the fitting of the MC data at $f = 600$ GHz to a Lorentzian shape. The system and the working conditions are the same as those in Fig. 1(b).

According to Eq. (2), due to the stochastic amplitude modulation of the harmonic signal the noise spectrum of the stationary random process $\{u(t)\}$ is shifted to the high-frequency region centered around the excitation frequency f. As a signature of the upconversion process, Eq. (3) indicates that the low-frequency nonstationary component of the instantaneous spectral density, $S'_{xx}(t,\nu)$, should exhibit an harmonic behavior with frequency $2f$. Since the instantaneous spectral density is always a periodic function with respect to the phase time t, it can be decomposed into the Fourier series as:

$$S_{xx}(t,\nu) = \overline{S}_{xx}(\nu) + \sum_{n=1}^{\infty} S_n(\nu) cos(2\pi fnt + \phi_n) \tag{4}$$

In the presence of a resonant-like peak in $\overline{S}_{xx}(\nu)$ centered around f and caused by the upconversion, the amplitude of the second harmonic of the Fourier decomposition given by Eq. (4), in accordance with Eq. (3), should be the upconverted spectrum itself, $S_2(\nu) = \frac{1}{2}S_{uu}(\nu)$, and the contribution to $S_{xx}(t,\nu)$ of all other harmonics $S_n(\nu)$ with $n \neq 2$ should be negligible.

The above features of noise upconversion are now applied to analyze the large-signal hot-carrier noise spectra presented in Figs. 1 and 2. In the case of n-GaAs [Fig. 1(b)], the time variation of the zero-frequency nonstationary part $S'_{\delta v \delta v}(t,0)$ of the spectral density of velocity fluctuations during one period of the microwave field is presented in Fig. 4(a) for different frequency values $f = 200$, 400, 600 GHz (curves 1 to 3). It is clearly seen, that starting from a frequency f of about 400 GHz the nonstationary part exhibits a nearly pure harmonic behavior at a frequency $2f$, in full agreement with Eq. (3). We conclude that here we deal with a noise upconversion process. The second harmonic of the instantaneous spectral density of velocity fluctuations, $S_2(\nu)$, which, in accordance with Eq. (4) gives an estimation of the upconverted spectrum, is shown in Fig. 4(b). Here, we consider the frequency values $f = 600$, 800, 1000 GHz, for which a pure harmonic behavior of $S'_{\delta v \delta v}(t,0)$ was observed. The upconverted spectra presented in Fig. 4(b) are well described by a Lorentzian function $S(\omega) = S(0)/(1+\omega^2\tau^2)$, with $S(0) = 0.16,\ 0.13,\ 0.08\ m^2/s$ and $\tau = 1.85,\ 2.1,\ 2.3\ ps$, at frequencies $f = 600,\ 800,\ 1000\ GHz$, respectively (see solid line corresponding to $f = 600\ GHz$). By comparing Fig. 4(b) with Fig. 1(b) one can see that, in full agreement with Eqs. (2) and (3), these values of $S(0)$ are about twice higher than the magnitude of the peaks in Fig. 1(b), when we assume that they are superimposed to a quasi-Lorentzian spectrum. (related to other contributions to the stationary spectrum different from the upconverted component). This fact confirms that the resonant-like peaks of the velocity noise spectra reported in Fig. 1 are caused by a noise upconversion process. A similar behavior of $S'_{\delta v \delta v}(t,0)$ and $S_2(v)$ was obtained for the case of bulk n-InN at $T = 80\ K$ presented in Fig. 2, what leads us to suggest the presence of an upconversion process also in this case.

OPEN QUESTIONS

The crucial open question we pose is: what kind of low-frequency noise is upconverted in these situations ?

For the case of GaAs, by considering that:

(i) at a given electric field amplitude E the main features of the upconverted spectrum strongly depend on frequency f (and viceversa);

(ii) as proven by MC simulations, the resonant-like peak is absent when intervalley transfer is absent;

(iii) with the increase of f the resonant peak appears as some transformation of the static hot-carrier peak;

we conclude that the upconverted noise is related to intervalley transfer and to the quasi-ballistic motion of carriers that cross the Γ valley. In accordance with the simple mathematical model considered above, the modulation of carrier velocity fluctuations under a strong microwave field is caused by random transitions of carriers between two groups with different dynamics, namely: (i) quasi-ballistic Γ electrons with insufficient energy to make intervalley transfer, (ii) the remaining high-energy electrons which suffer very quick randomizing scatterings due to intense intervally transfer. Thus, the physical mechanism responsible for the peak in the noise spectral density can be related to an upconversion of the so called intergroup or intervalley noise [10].

To support this conjecture, Fig. 5(a) reports the distribution function in velocity space along the field direction $f(v_x)$ at different time moments during half a period of the MWEF. The coexistance of at least two groups of electrons is evidenced by the following properties of the distribution profile. The central peak is caused mainly by slowly moving electrons populating the upper valleys. The wings are caused mainly by fast Γ-electrons, most of which perform quasiballistic motion. This scenario is further illustrated in Fig. 5(b), which shows the velocity time history of a single electron (solid line). For comparison, the dashed line plots the periodic function $Asin(2\pi f t)$ with a proper amplitude. One can clearly detect two kinds of velocity sequences, which correspond to quasi-ballistic and highly-randomizing motions of electrons.

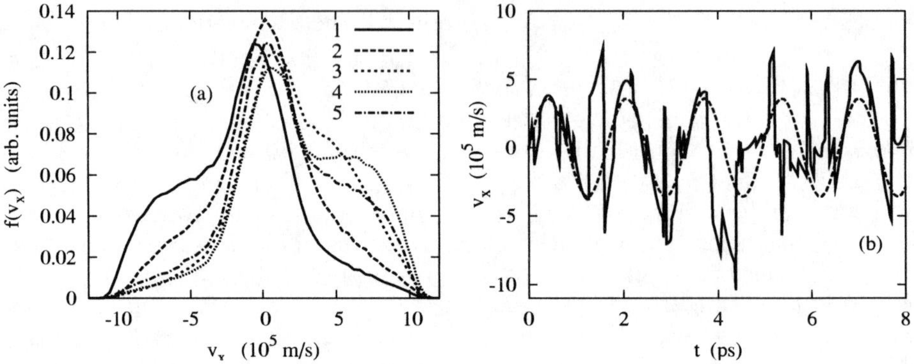

FIGURE 5. (a) Parallel velocity distribution function $f(v_x)$ calculated at different time moments corresponding to different phases (0, $\pi/4$, $\pi/2$, $3\pi/4$, π, curves 1 to 5, respectively) of the MWEF $E sin(2\pi f t)$, and (b) excerpt of time history of the parallel velocity (solid line). GaAs, 300 K, $E = 10\ kV/cm$, $f = 600\ GHz$.

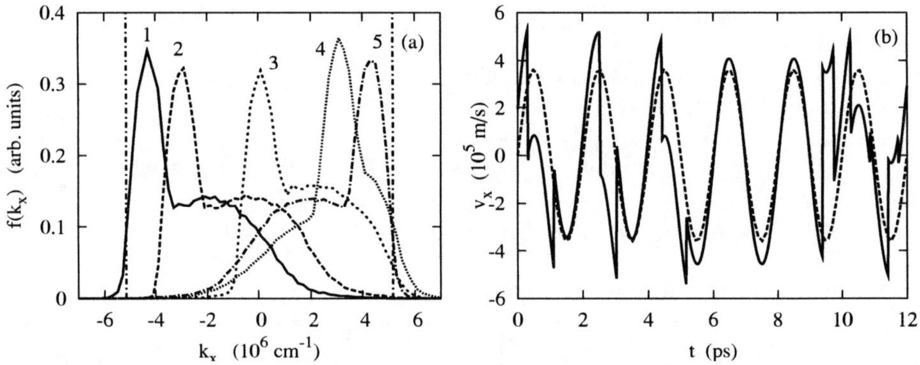

FIGURE 6. (a) Parallel wave-vector distribution function $f(k_x)$ calculated at different time moments corresponding to different phases (0, $\pi/4$, $\pi/2$, $3\pi/4$, π, curves 1 to 5, respectively, vertical lines indicate the passive region boundaries) of the MWEF $E sin(2\pi f t)$, and (b) excerpt of time history of the parallel velocity (solid line). n-InN, $80K$, $E = 9\ kV/cm$, $f = 500\ GHz$.

In the case of bulk n-InN at 80 K, again at least two groups of carriers can be detected in the presence of the resonant peak [11]. One group consists of carriers which cannot reach the optical phonon energy during half a period of the MWEF, so that their free motion always takes place inside the passive region of **k**-space. The second group consists of carriers which synchronize their free motion with the field phase in such a way that each carrier emits at least one optical phonon during every half a period, i.e. they reach the boundary of the active region $\varepsilon = \hbar\omega_0$. The transfer of electrons between the two groups is a stochastic process driven by strongly randomizing scattering events occurring inside the passive region. The formation and evolution of these two groups of carriers is illustrated in Fig. 6(a), which reports the time behavior of the distribution function along the field direction $f(k_x,t)$ during half a period of the MWEF. Here, the peak of the distribution corresponds to the group of carriers that cannot reach the active region boundary, and the lower plateau corresponds to the group of carriers that emit an optical phonon twice per period. An example of the resulting modulation of velocity fluctuations is reported in the velocity time history shown Fig. 6(b).

Thus, the random character of the intergroup transfer is responsible for the hot-carrier noise upconversion in bulk materials. Moreover, it is worthwhile to underline that, in contrast to the usual $1/f$ noise, generation-recombination noise, etc., which are resistance fluctuations present in the absence of a applied field, the "low-frequency" noise which is upconverted to higher frequencies is here originated by stochastic electron transitions between different trajectories created dynamically by the high-frequency large-amplitude alternating electric field.

It is well known that the spectral density of velocity fluctuations in bulk materials represents, in turn, the so called diffusion noise source that forms the basis of electronic noise calculations in semiconductor structures and devices by using the impedance field method [3,4]. Thus, since in the case of GaAs the noise source exhibits the resonant-like behavior under hot-carrier conditions, a similar behavior can be expected in inhomogeneous devices. Therefore, in principle, the appearance of the resonant peak in Fig. 3 could be of the same nature as in bulk GaAs.

ACKNOWLEDGMENTS

The authors wish to acknowledge the financial support of the NATO collaborative-linkage grant PST.CLG.977520, the Spanish *Secretaría de Estado de Educación y Universidades* through the grant SAB2000-0164, the *Dirección General de Investigación (MCyT)* and FEDER through the project TIC2001-1754, the Consejería de Educación y Cultura de la Junta de Castilla y León through the project SA057/02, the Italy-Spain Joint Action of the MIUR Italy (IT109) and MCyT Spain (HI2000-0138) and the french-lithuanian bilateral cooperation n. 12864 of french CNRS.

REFERENCES

1. R. Brazis, R. Raguotis, and M.R. Siegrist. *J. Appl. Phys.* **84**, 3474 (1998).
2. T.W. Crowe, J.L. Hesler, R.W. Weikle, and S.H. Jones, *Infrared Phys. Technol.* **40**, 175 (1999).
3. F. Danneville, G. Dambrine, and A. Cappy, *IEEE Trans. Electron Devices* **45**, 2207 (1998).
4. F. Bonani and G.Ghione, *Noise in Semiconductor Devices. Modeling and Simulation*, Springer, Berlin, 2001.
5. J.S. Bendat and A.G. Piersol. *Random data*. John Wiley & Sons, Inc., 1986.
6. S. Pérez, T. González, S.L. Delage, and J. Obregon, *J. Appl. Phys.* **88**, 800 (2000); *Semicond. Sci. Technol.* **16**, L8 (2001).
7. P. Shiktorov, E. Starikov, V. Gružinskis, L. Reggiani, L. Varani, and J.-C. Vaissière, *Appl. Phys. Lett.* **80**, 4759 (2002)
8. S. Pérez and T. González, *Semicond. Sci. Technol.* **17**, 696 (2002).
9. A. Hajimiri and T. H. Lee, *IEEE J. Solid-State Circuits* **33**, 179 (1998).
10. P. J. Price, *J. Appl. Phys* **31**, 949 (1960)
11. N. Ishida and T. Kurosawa, *Jap. J. Appl. Phys.* **64**, 2994 (1995).

To Be or not to Be a Source of Shot Noise: an Open Problem of Generation Recombination Noise

G. Gomila* and L. Reggiani†

*Centre de Recerca en Bioelectronica i Nanobiociencia and Departament d'Electronica-Universitat de Barcelona, Laboratori de Nanobioenginyeria- Parc Cientific de Barcelona, C/ Josep Samitier 1-5, 08028 Barcelona, Spain email-ggomila@pcb.ub.es
†INFM-National Nanotechnology Laboratory and Dipartimento di Ingegneria dell'Innovazione, Universitá di Lecce, Via Arnesano s/n, I-73100 Lecce, Italy. email-lino.reggiani@unile.it

Abstract. We address the fundamental question wether generation recombination noise can transform into shot noise at increasing values of the applied bias (current or voltage). From the seminal work of Van Vliet and Fassett (1965) the analytical answer to this question is affirmative. By contrast, recent numerical results by Bonani and Ghione (1999) provided a negative answer. For a simple semiconductor two level model consisting of a conducting band and an impurity level which provides charge carriers we solve exactly the appropriate second order stochastic differential equation self-consistently coupled with the Poisson solver. The effect of the diffusion current, neglected in the first analytical approach, is found to inhibit the cross-over between generation-recombination and shot noise, thus confirming more recent numerical results. Open problems related to the microscopic interpretation of these new results are pointed out.

INTRODUCTION

Since the seminal review of Van Vliet and Fassett [1] it is common belief that shot noise can appear in a homogeneous semiconductor sample due to the presence of generation-recombination processes in the presence of transport under the condition that the drift transit time becomes the shortest time scale when compared with carrier lifetime and dielectric relaxation time. Howevere, recently Gomila and Reggiani [2] have shown that, provided the stochastic equation includes the diffusion current, shot noise can appear due to the presence of diffusion noise alone. At this point a fundamental question arises: in a given two-terminal conductor where shot noise is observed how can we identify the origin of this shot noise? In order to provide an answer to this question one needs to solve the noise equations with the presence of generation-recombination noise source by accounting for the diffusion current and try to ascertain under which circumstances shot noise can originate from generation-recombination noise. This is the purpose of this work. We advance here the main result we have found, namely that, contrary to the previous findings of [1] generation-recombination noise does not give rise to shot noise. Instead, from a quadratic behavior with applied bias it tends to saturate to a constant value at the highest bias as a consequence of the finite size of the sample. This saturation regime (and the consequent absence of shot noise) was recently observed in a similar system by means of numerical simulations by Bonani and Ghione [3].

CP665, *Unsolved Problems of Noise and Fluctuations: UPoN 2002: Third International Conference,* edited by S. M. Bezrukov

Shot noise

Introduced by Schottky (1918) [4], shot noise is the current fluctuation associated with the discreteness of electrical charge. When being the only source of noise, the low-frequency spectral density of current fluctuations of shot noise is given by:

$$S_I^{shot}(0) = 2qI\,\gamma \tag{1}$$

where q is the unit charge responsible of transport, I the stationary current flowing in the two-terminal sample, and $\gamma \geq 0$ the Fano factor which probes the correlation between current pulses. The condition $\gamma = 1$ corresponds to full shot-noise (absence of correlations), while $\gamma < 1$ refers to suppressed shot-noise (negative correlations), and $\gamma > 1$ to enhanced shot-noise (positive correlations). Shot-noise is a rather ubiquoutus phenomenon and it has been observed in a large variety of systems like: vacuum diode and vacuum tubes in geneneral, tunnel diodes, Schottky and pn junctions, degenerate and nondegenerate mesoscopic devices, etc. (for a review see Refs. [5, 6]).

In the presence of thermal noise, the standard cross-over between thermal and shot noise for a linear systems in the absence of dielectric relaxation is given by:

$$S_I(0) = \frac{4k_BT}{R}[\frac{qV}{2k_BT}coth(\frac{qV}{2k_BT})] \tag{2}$$

where k_B is the Boltzmann constant, T the bath temperature , V the applied voltage and R the Ohmic resistance of the sample. We note the asymptotic linear behavior of the spectral density with applied bias.

The presence of this cross-over stems from the fact that shot noise is not an additive source of noise but follows from thermal noise when the applied voltage is sufficiently greater than the thermal value k_BT/q. Anomalies of this cross-over are a subject of present interest as reported in Ref. [2].

Generation recombination (GR) noise

Introduce by Van Vliet and Blok in (1956) [7], GR noise is the current fluctuations associated with the stochastic transfer of carriers between energy states belonging to different bands. The most simple example is the two level model of an extrinsic semiconductor (n or p type). In this case, a simple Lorentzian model gives for the current spectral density of GR noise the form

$$S_I^{gr}(0) = I^2 4\,\frac{\langle \Delta N^2 \rangle}{\langle N \rangle^2}\,\tau_l \tag{3}$$

with N the instantaneous number of free carriers, τ_l their lifetime and brakets denoting average with respect to the steady state. The proportionality to the square of the current evidences that GR is a kind of resistance noise (as the $1/f$ noise and others) and as such it is classified as an excess noise. GR noise is observed typically in bulk semiconductors

[8, 9]. The cross-over between thermal and GR noise is obtained by adding the two sources as

$$S_I(0) = \frac{4k_BT}{R} + I^2 4\frac{\langle \Delta N^2 \rangle}{\langle N \rangle^2}\tau_l \tag{4}$$

We notice the asymptotic quadratic behavior of the total spectral density with applied bias.

TRANSPORT MODEL OF GR NOISE

A more detailed calculation of GR noise requires a self-consistent calculation. Accordingly, we consider a two terminal structure consisting of a uniformly $n-$doped semiconductor sample of length L terminated by (metallic) Ohmic contacts with a single trap (donor) level controlling the electron concentration. We assume that the transversal dimensions of the sample are much greater than the longitudinal dimension, thus allowing for a one dimensional electrostatic treatment. Without loss of generality, only the GR noise source is considered. The physical model necessary to describe the noise properties of this system consists of the continuity equations for the free electrons and the ionized traps, the current equation for the electrons, the Poisson equation and appropriate boundary conditions. The continuity equations for free carriers and ionized traps densities read [1, 10]

$$\frac{\partial n(x,t)}{\partial t} = \frac{1}{q}\frac{\partial J_n(x,t)}{\partial x} - r_n(x,t) + g_n(x,t) + \gamma(x,t)\ , \tag{5}$$

$$\frac{\partial N_t^+(x,t)}{\partial t} = -r_n(x,t) + g_n(x,t) + \gamma(x,t)\ , \tag{6}$$

where $n(x,t)$ and $N_t^+(x,t)$ refers to the free electron and ionized trap densities at point x and time t, respectively, $J_n(x,t)$ the conduction electron current, $r_n(x,t)$ the recombination rate and $g_n(x,t)$ the generation rate. Furthermore, $\gamma(x,t)$ is the Langevin noise source related to GR processes which has zero mean and low frequency spectral density given by [1, 10]

$$2\int_{-\infty}^{+\infty} dt\,\overline{\gamma(x,t)\gamma(x,t')} = \frac{2}{A}\left[\bar{g}_n(x) + \bar{r}_n(x)\right]\delta(x-x')\ , \tag{7}$$

with A being the sample cross-sectional area and the bar denoting average with respect to fluctuations. In the drift-diffusion approach, the conduction current writes:

$$J_n(x,t) = q\mu_n n(x,t)E(x,t) + qD_n\frac{\partial n(x,t)}{\partial x}\ , \tag{8}$$

where μ_n is the electron mobility, $E(x,t)$ the electric field and D_n the diffusion coefficient satisfying Einstein relation, $D_n = \mu_n k_BT/q$, where non-degenerate statistics is assumed.

To perform a self-consistent calculation, we take into account the Poisson equation

$$\frac{\partial E(x,t)}{\partial x} = \frac{q}{\varepsilon}\left[N_t^+(x,t) - n(x,t)\right] , \tag{9}$$

where ε is the semiconductor dielectric permittivity. The total electric current is thus given by

$$I(t) = \frac{A}{L}\int_0^L dx \left[J_n(x,t) + \varepsilon \frac{\partial E(x,t)}{\partial t}\right] . \tag{10}$$

Finally, as appropriate boundary conditions we take Ohmic contacts made of ideal metal-semiconductor interfaces in the diffusion approximation, i.e. we assume that the diffusion velocity is much smaller than the contact recombination velocity, thus allowing to neglect any contribution from thermionic emission processes [11]. Accordingly, we take

$$\overline{n}(0) = \overline{n}(L) = n_c \; ; \;\; \delta n(0,t) = \delta n(0,t) = 0 , \tag{11}$$

where the carrier density at the contact is

$$n_c = N_C \exp\left(-\frac{q\phi_{bn}}{k_B T}\right) , \tag{12}$$

with N_C being the effective density of states in the conduction band and ϕ_{bn} the contact barrier height. Here, $\delta n(x,t)$ refers to the fluctuations of free electron density.

The model presented above admits a homogeneous stationary solution in the form

$$\overline{n}(x) = \overline{N}_t^+(x) = n_c \; ; \; \overline{E}(x) = \frac{\overline{I}}{qA\mu_n \overline{n}} ; \tag{13}$$

We note, that being the density of traps N_t and the contact density n_c two independent variables, one can always find for them values which satisfy simultaneously Eq. (13) and $\overline{g}_n(\overline{n},\overline{N}_t^+) = \overline{r}_n(\overline{n},\overline{N}_t^+)$. Under these homogeneous conditions the system behaves as a resistor with current-voltage characteristics satisfying Ohm's law, $\overline{I} = V/R$, with resistance $R = L/(qA\mu_n\overline{n})$.

To calculate the number fluctuations at low frequency

$$S_N(0) = 2\int_0^\infty \langle \Delta N(0)\Delta N(t)\rangle dt , \tag{14}$$

we use that

$$\Delta N(t) = A\int_0^L dx \delta n(x,t) , \tag{15}$$

where $\delta n(x,t)$ for low frequencies satisfies

$$\frac{\partial^2 \delta n}{\partial x^2} + \frac{1}{L_E}\frac{\partial \delta n}{\partial x} - \frac{1}{\widetilde{L}_D^2}\delta n = -\frac{\tau_l}{\widetilde{L}_D^2}\gamma . \tag{16}$$

Equation (16) can be derived by combining Eqs. (5), (6), (8) and (9) after linearization around the homogeneous stationary state and by neglecting the time derivatives due to the low frequency assumption. The parameters in Eq. (16) are defined as

$$\frac{L_E}{L} = \frac{k_B T}{qV} \; ; \; \widetilde{L}_D = \sqrt{\frac{k_B T \varepsilon}{q^2 \overline{n}(1 + \frac{\tau_n}{\tau^+})}} \; ; \; \frac{1}{\tau_l} = \frac{1}{\tau_n} + \frac{1}{\tau^+} \, . \tag{17}$$

Here, $\widetilde{L}_D$ is a renormalized Debye screening length and

$$\frac{1}{\tau_n} = \frac{\partial r_n(n, N_t^+)}{\partial n} - \frac{\partial g_n(n, N_t^+)}{\partial n} \; ; \tag{18}$$

$$\frac{1}{\tau^+} = \frac{\partial r_n(n, N_t^+)}{\partial N_t^+} - \frac{\partial g_n(n, N_t^+)}{\partial N_t^+} \, . \tag{19}$$

Below we will consider the two relevant cases when the diffusion current is neglected (to reproduce the results of Ref. [1]) and when it is accounted for.

Solution neglecting diffusion current

In this approximate case one obtains:

$$S_I^{gr}(0) = I^2 4 \langle \Delta N^2 \rangle \, \tau_l \left[1 + \frac{(e^{\lambda_2 L} - 1)}{(e^{\lambda_1 L} - 1)} (e^{\lambda_2 L} - e^{\lambda_1 L}) \left(\frac{1}{\lambda_2 L} - \frac{1}{\lambda_1 L} \right) \right] \tag{20}$$

with $\langle \Delta N^2 \rangle = AL\overline{g}_n \tau_l$, and

$$\lambda_{1,2} = \frac{1}{2L_E} \left(-1 \pm \sqrt{1 + \frac{4L_E^2}{\widetilde{L}_D^2}} \right) \tag{21}$$

We notice that there are three length scales of interest for the system, i.e. L (sample), L_E (field), L_D (Debye), which have the time scale counterpart $\tau_{diff} = L^2/D$ (diffusion transit time), $\tau_{tr} = L^2/(\mu V)$ (drift transit time) and $\tau_{dr} = \varepsilon/(q\mu_n n)$ (dielectric relaxation time). In the limit $I \to \infty$ it is found

$$S_I^{gr}(0) = 2q\beta \overline{I} \tag{22}$$

From the above equation we see that GR noise becomes shot noise with a value of the Fano Factor given by $\beta = \langle \Delta N^2 \rangle L^2 \tau_l / (\langle N \rangle \widetilde{L}_D^2 \tau_{diff})$

Solution including diffusion current

For the exact case when diffusion is accounted for we found [12]

$$S_I^{gr}(0) = I^2 4 \frac{\langle \Delta N^2 \rangle}{\langle N \rangle^2} \tau_l - \frac{4k_B T}{R} \beta s_I^{gr}(\lambda_1, \lambda_2) \tag{23}$$

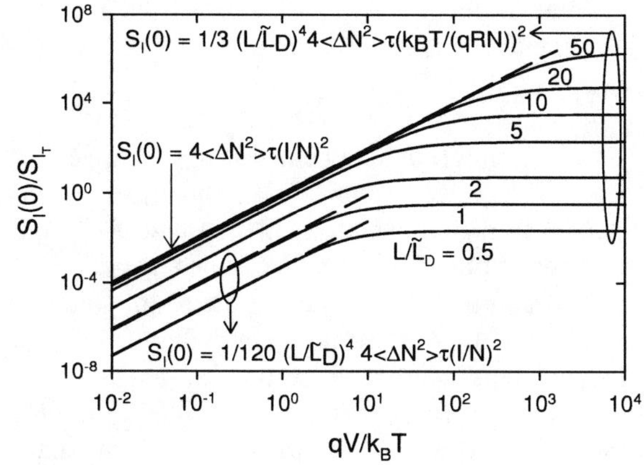

FIGURE 1. Spectral density of current fluctuations associated with GR noise vs applied voltage at different sample lengths

with

$$
\begin{aligned}
s_I^{gr}(\lambda_1,\lambda_2) \;=\; & \frac{\left(e^{\lambda_1 L}-1\right)\left(e^{\lambda_2 L}-1\right)(\lambda_1-\lambda_2)}{2L\lambda_1\lambda_2(\lambda_2+\lambda_1)\left(e^{\lambda_2 L}-e^{\lambda_1 L}\right)^2}\times \\
& \left[\lambda_1\left(1-e^{(\lambda_1+\lambda_2)L}-3e^{\lambda_2 L}+3e^{\lambda_1 L}\right)\right. \\
& \left. -\lambda_2\left(1-e^{(\lambda_1+\lambda_2)L}-3e^{\lambda_1 L}+3e^{\lambda_2 L}\right)\right] .
\end{aligned} \tag{24}
$$

The type of results obatined from the previous equations are summarized in Fig. 1 where we find convenient to normalize the current spectral density $S_I^{gr}(0)$ to the spectral density corresponding to an infinite sample (i.e. $L \gg |\lambda_{1,2}|$) with thermal current $I_T = k_BT/(qR)$, i.e. $S_{I_T}^{gr} = 4\langle\Delta N^2\rangle\,\tau_l\,(I_T/\langle N\rangle)^2$, and the applied voltage V to the thermal voltage $V_T = k_BT/q$. With these renormalizing factors, $S_I^{gr}(0)/S_{I_T}^{gr}(0)$ as a function of $qV/(k_BT)$ depends only on the dimensionless length, $L/\widetilde{L}_D$. The essential result we want to stress from Fig. 1 is that the spectral density saturates for high applied voltages to the value

$$
S_I^{gr,sat}(0) = \frac{1}{3}\left(\frac{L}{\widetilde{L}_D}\right)^4 4\langle\Delta N^2\rangle\tau_l\left(\frac{k_BT}{qR\langle N\rangle}\right)^2 . \tag{25}
$$

The critical voltage for saturation V_{sat} satisfies

$$
V_{sat} = \begin{cases} \sqrt{40}\,\frac{k_BT}{q} \;;\; for \;\; \frac{L}{\widetilde{L}_D} < 1 \\ \left(\frac{L}{\widetilde{L}_D}\right)^2 \frac{k_BT}{q} \;;\; for \;\; \frac{L}{\widetilde{L}_D} \geq 10 \end{cases} , \tag{26}
$$

(and an intermediate behavior for $1 < L/\widetilde{L}_D < 10$). The low bias behavior of $S_I^{gr}(0)$ displays also interesting features: it equals $I^2 4\langle\Delta N^2\rangle\tau_l/\langle N\rangle^2$ for $L/\widetilde{L}_D > 10$ and $0 < V < \left(L/\widetilde{L}_D\right)^2 k_BT/q$, but it is suppressed by a factor $(1/120)\left(L/\widetilde{L}_D\right)^4$ for $L/\widetilde{L}_D < 1$ and $0 < V < \sqrt{40}\frac{k_BT}{q}$ (and an intermediate behavior in the range $1 < L/\widetilde{L}_D < 10$). It is worth noting that the results displayed in Fig. 1 are qualitatively similar to the numerical results recently reported by Bonani and Ghione [3] for a more complex GR model. This fact indicates that saturation at high applied bias can be a general feature of GR noise.

The saturation of $S_I^{gr}(0)$ and the dependence of the results on the ratio $L/\widetilde{L}_D$ can be explained under the assumption that both space charge fluctuations and diffusion current play a relevant role. Indeed, although the system is neutral in average, i.e. $\overline{n}(x) = \overline{N}_t^+(x)$, instantaneously space charge fluctuations for which $\delta n(x,t) \neq \delta N_t^+(x,t)$ can appear. The relevance of these fluctuations is determined by their interaction with the rest of charges in the system through the long range Coulomb interaction. As a consequence, for $L < \widetilde{L}_D$ the long range Coulomb interaction does not play any role at all, and hence, as soon as the applied voltage is higher than the thermal value, the low frequency current spectral density saturates. On the other hand, when $L > \widetilde{L}_D$ the long range Coulomb interaction is able to restor space charge neutrality for low applied bias. However, as long as the applied bias is high enough as to make the transit time $\tau_{tr} = L^2/(\mu_n V)$ shorter than the renormalized dielectric relaxation time $\widetilde{\tau}_{dr} = \varepsilon/[q\mu_n\overline{n}(1+\tau_n/\tau^+)]$, the long range Coulomb correlations vanish and the low frequency spectral density saturates [2]. Furthermore, the fact that the diffusion current is playing a relevant role is seen on the fact that no shot noise is obtained in the high bias regime as it is obtained when the diffusion current is neglected [1]. Thus, the present work offers new perspectives in what concerns finite size effects on generation-recombination noise in semiconductor devices.

CONCLUSIONS

We have studied the nonequilibrium noise properties in unipolar homogeneous semiconductors of finite length in the presence of generation-recombination noise. It has been shown that the generation recombination contribution can exhibit quadratic or constant current dependences, but in no case proportional to the current (shot noise). An explanation for the erroneous predictions of existing theoretical works is related to the neglect of the diffusion current in the current equation. Indeed, when the diffusion current is included the finite size effect does the job of coupling different time scales. Independent results on the same direction obtained by means of numerical resolution of the noise equations have been reported for bipolar systems [3], thus providing an indirect confirmation of present findings. The microscopic interpretation of the saturation of the generation recombination noise as finite size effect remains an open problem which needs further investigation.

ACKNOWLEDGMENTS

Partial support from the MCyT-Spain through the Ramon y Cajal program and project No. BFM2001-2159 and from the MADESS II is gratefully acknowledged. Prof. F. Bonani of Torino University is acknowledged for the stimulating discussions carried out on the subject.

REFERENCES

1. K.M. van Vliet and J.R. Fasset in: *Fluctuation Phenomena in Solids*, Ed. R.E. Burgess (Academic Press, New York, 1965), pp. 267.
2. G. Gomila and L. Reggiani, Phys. Rev. B **62**, 8068 (2000).
3. F. Bonani and G. Ghione, Solid State Electr. **43**, 285 (1999).
4. W. Schottky, Ann. Phys. **57**, (1918).
5. A. van der Ziel, *Noise in Solid-State Devices and circuits*, (Wiley, New York, 1986).
6. Ya. M. Blanter and M. Büttiker, Physics Reports, **336**, 1 (2000).
7. K.M. van Vliet and J. Blok, Phyisica **22**, 231 (1956), ibidem **22**, 525 (1956).
8. N. Lukyanchikova, *Noise research in semiconductor physics*, Ed. B. Jones (Gordon and Breach Science Publishers, Amsterdam, The Netherlands, 1996).
9. V. Mitin, L. Reggiani and L. Varani in: *Noise and fluctuations control in electronic devices*, Ed. A.A. Balandin, (ASP, Stevenson Ranch, 2002).
10. F. Bonani and G. Ghione, *Noise in semiconductors devices, modeling and simulation*, (Springer Verlag, Berlin, 2001).
11. G. Gomila, O.M. Bulashenko and J.M. Rubí, J. Appl. Phys. **83**, (1998).
12. G. Gomila and L. Reggiani, Appl. Phys. Lett. **81**, 4380 (2002).

WAVES, HYDRODYNAMICS, AND MECHANICS

Noisy Wavefront Propagation in the Fisher-Kolmogorov-Petrovsky-Piscounov Equation

Charles R. Doering*†, Carl Mueller** and Peter Smereka*

*Department of Mathematics, University of Michigan, Ann Arbor, MI 48109-1109
†Michigan Center for Theoretical Physics, University of Michigan, Ann Arbor, MI 48109-1120
**Department of Mathematics, University of Rochester, Rochester, NY 14627

Abstract. We discuss some conjectures and open questions regarding the velocity of front propagation in the stochastic Fisher-Kolmogorov-Petrovsky-Piscunov equation

INTRODUCTION

The Fisher-Kolmogorov-Petrovsky-Piscunov (FKPP) equation [1, 2] is one of the most fundamental models in mathematical biology and ecology [3]. It describes a population $u(x,t)$ that evolves under the combined effects of spatial diffusion and local logistic growth and saturation. In one space dimension the FKPP equation may be written

$$\partial_t u = D\partial_{xx}u + \gamma u(1-u) \tag{1}$$

with diffusion coefficient D, low-density growth rate γ, and where the population has been normalized so that the stable saturation level is $u = 1$. A particularly interesting aspect of this nonlinear partial differential equation is that it gives a simple description of the invasion of the stable saturated state ($u = 1$) into regions of space occupied by the unstable "extinct" state ($u = 0$). It turns out that after transients, this invasion proceeds via propagation of a front at a constant velocity.

To find these travelling wave fronts we look for solutions of the form

$$u(x,t) = f(x-ct), \tag{2}$$

where the speed c is to be determined. Inserting this into the FKPP equation (1) yields an ordinary differential equation for the front shape $f(z)$:

$$Df'' + cf' + \gamma f(1-f) = 0. \tag{3}$$

The boundary conditions for the front shape function are determined by the setup for u. If we consider the setup for a right-moving front where $u \to 1$ as $x \to -\infty$ and $u \to 0$ as $x \to +\infty$, then the conditions for f are $f \to 1$ as the "time" $z \to -\infty$ and $f \to 0$ as $z \to +\infty$. In this case we look for positive velocities, $c > 0$, and the front is identified as the heteroclinic orbit connecting the unstable fixed point $(f, f') = (1,0)$ and the stable

CP665, *Unsolved Problems of Noise and Fluctuations: UPoN 2002: Third International Conference*, edited by S. M. Bezrukov

equilibrium state $(f, f') = (0,0)$ of the two-dimensional dynamical system defined by (3). Because $0 \le u(x,t) \le 1$ for any initial condition $0 \le u(x,0) \le 1$, we require that $0 \le f(z) \le 1$. Then it is easy to see that the determining factor in the analysis is the need to prevent oscillations around 0 near $f = 0$. This means that the "friction coefficient" c in the dynamical system for f must be sufficiently high that the stable fixed point $(f, f') = (0,0)$ is a node and not a spiral. Hence we find acceptable solutions for any speed $c \ge 2\sqrt{D\gamma}$.

It turns out that for sufficiently sharp initial fronts in $u(x,0)$, the minimum speed $c_{min} = 2\sqrt{D\gamma}$ is selected [1, 2], albeit very slowly [4, 5]. That is, for sharp initial fronts any reasonable definition of the instantaneous velocity $v(t)$ satisfies $v(t) = c_{min} + \mathcal{O}(t^{-1})$. This algebraic relaxation of the speed, combined with the fact that the speed is ultimately selected by the properties of the system far ahead of the bulk of the population front, leads to the terminology that the FKPP equation has a *pulled front* with *weak velocity selection*. The situation here is to be contrasted with reaction-diffusion fronts connecting metastable states. The velocity of such a *pushed front* is determined by the nonlinear dynamics in transition region defining the front and the speed adjusts itself exponentially fast, termed *strong velocity selection*.

Upon reflection it is clear that the extreme sensitivity of the pulled front to the details of the process near the unstable state cannot be robust or, really, physical. Indeed, any modification of the dynamics at the lowest population level where $u \approx 0$—for example due to population discreteness effects—would invalidate the velocity selection mechanism described above.

To fix ideas, consider the spatially homogeneous chemical reaction scheme

$$A + B \to A + A \text{ at rate } k_1 \quad \text{and} \quad A + B \to B + B \text{ at rate } k_2, \tag{4}$$

and let $N_A(t)$ and $N_B(t)$ be the number of A and B particles in the system. The reaction conserves the total number of particles so $N = N_A + N_B$ is a constant of the dynamics. Neglecting discreteness and fluctuation effects altogether leads to a simple description of the kinetics for the fraction of A particles, $U(t) = N_A(t)/N$, the logistic equation

$$\frac{dU}{dt} = \gamma U(1-U) \tag{5}$$

with $\gamma = k_1 - k_2 > 0$ (without loss of generality). This level of modeling predicts that any nonzero initial number of A particles will eventually grow to completely dominate the population.

Discreteness and the accompanying "noise" may qualitatively change this picture. If we model the dynamics of (4) by the appropriate Markovian random walk in particle number space [6] and, for convenience here, keep the just leading Markov diffusion process description in the $N \to \infty$ limit, the deterministic logistic differential equation (5) is replaced by the Itô stochastic differential equation

$$\frac{dU}{dt} = \gamma U(1-U) + \sigma\sqrt{U(1-U)}\,\xi(t) \tag{6}$$

where $\sigma^2 = \frac{k_1+k_2}{N}$ and $\xi(t)$ is a Gaussian white noise process with $\langle \xi(t) \rangle = 0$ and $\langle \xi(t)\xi(s) \rangle = \delta(t-s)$. The boundary conditions for the state variable U are absorbing at both extremes, $U = 0$ and $U = 1$, as is obvious from the reaction scheme in (4).

It is straightforward to compute the probability that an initial population of A particles will eventually become extinct—an eventuality that is simply impossible in the absence of noise. If $U(t)$ starts from (nonrandom) U_0, this extinction probability is

$$\mathrm{Prob}(U(t) \to 0 \text{ as } t \to \infty | U(0) = U_0) = \frac{e^{2\gamma(1-U_0)/\sigma^2} - 1}{e^{2\gamma/\sigma^2} - 1} \tag{7}$$

which is strictly positive if the initial population contains any B particles at all (i.e., if $0 \leq U_0 < 1$). There is a significant probability of extinction at low population: the probability of extinction is $1 - \mathcal{O}(U_0)$ as $U_0 \to 0$. Hence we would expect low population levels—such as those in the leading edge of a traveling wavefront in the FKPP equation—to naturally extinguish themselves. It is thus apparent that no mechanism which relies on the exponentially small population out in front of the wave should be physically relevant in the dynamics.

This lack of "structural stability" of the pulled front has been explored in quite alot of detail recently. In 1997 and 1998 Brunet and Derrida [7] and Kessler, Ner and Sander [8] considered the influence of discreteness in the population variable u on the front speed by replacing the $u(1-u)$ growth and saturation dynamics in the FKPP equation with $u(1-u)\Theta(u-N^{-1})$, where $\Theta(\zeta)$ is the step function vanishing when $\zeta < 0$. Here $N^{-1} \ll 1$ represents the discreteness level. Repeating the dynamical systems argument for the velocity of a traveling wavefront, they concluded that the selected speed is unique, strongly selected, and given for large N by

$$c \sim \sqrt{D\gamma}\left[2 - \frac{\pi^2}{(\log N)^2}\right] = c_{min} - \mathcal{O}((\log N)^{-2}). \tag{8}$$

This is slower than the minimum speed previously available. Moreover, it displays extremely slow convergence to the "continuum" limit for u as $N \to \infty$. Soon thereafter Pechenik and Levine [9] computationally verified this kind of velocity dependence on N in another stochastic model. Most recently Brunet and Derrida conjectured as well that in a discrete stochastic model the wavefront should diffuse with an $\mathcal{O}((\log N)^{-3})$ diffusion coefficient [10]. See also the contribution of Panja to this volume [11].

THE WAVEFRONT IN THE STOCHASTIC FKPP EQUATION

In view of the considerations above, we are led to study the stochastic Fisher-Kolmogorov-Petrovsky-Piscounov (sFKPP) equation

$$\partial_t u = D\partial_{xx} u + \gamma\, u(1-u) + \epsilon\sqrt{u(1-u)}\,\eta(x,t) \tag{9}$$

where $\eta(x,t)$ is a gaussian white noise process in space and time satisfying $\langle\eta(x,t)\rangle = 0$ and $\langle\eta(x,t)\eta(y,s)\rangle = \delta(x-y)\delta(t-s)$. This stochastic partial differential equation with multiplicative noise is to be interpreted as the continuum limit of a spatially discretized set of Itô equations; more on this below. We will refer to the coefficient ϵ as the noise strength, which one may think of as being proportional to $1/\sqrt{N}$ where N is

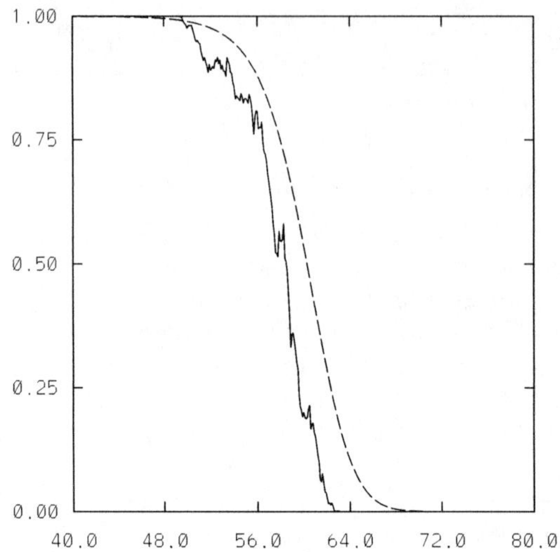

FIGURE 1. Snapshot of a traveling wave for the FKPP (dashed line) and the stochastic sFKPP equation (solid line). The results illustrate that the stochastic wavefront has compact support.

the saturation population at a lattice site or in an appropriately defined correlation volume. There have been many studies of the effects of spatial inhomogeneities and environmental randomness of the FKPP front dynamics, too [14], but we focus on this (statistically) spatially homogeneous model.

Although it is tempting to identify the sFKPP equation (9) as the fluctuating hydrodynamic description of a spatially extended reaction-diffusion model of (4), the extra noise introduced by the microscopic diffusive hopping complicates that connection. The sFKPP has, however, been rigorously derived as just such a description of an appropriate hydrodynamic limit of the contact process with long-range interactions [12].

The multiplicative noise in the sFKPP has profound qualitative effects on the front propagation. In 1995 Mueller and Sowers [13] proved that the sFKPP equation (9) has the so-called *compact support property*. This means, for example, that if the initial data $u(x,0)$ for the stochastic partial differential equation satisfies $u(x,0) = 1$ for $x \leq a$ and $u(x,0) = 0$ for $x \geq b$ where $-\infty < a < b < \infty$, then for all subsequent $t > 0$ the solution will satisfy $u(x,t) = 1$ for $x \leq a(t)$ and $u(x,t) = 0$ for $x \geq b(t)$ for some values $a(t)$ and $b(t)$ satisfying $-\infty < a(t) < b(t) < \infty$. That is, a front with sharp leading and trailing edges will always have sharp leading and trailing edges. This is despite the tendency for the diffusion to propagate u into the vacuous regions in front of the wave, and to decrease the population from the saturation level at the back of the front. The noise overwhelms the diffusive transport near the boundaries of the state variable at $u = 0$ and $u = 1$.

The compact support property of this process can be seen in numerical simulations. In Figure 1 we plot the results of direct numerical simulations of the FKPP equation (1) and stochastic sFKPP equation (9) explicitly showing the qualitative difference in the leading and trailing edges.

DUALITY AND INTERACTING PARTICLE SYSTEMS

The FKPP equation (1) is the mean-field hydrodynamic description of many microscopic reaction-diffusion systems including the single-species growth-coagulation process

$$A \to A+A \ \text{ at rate } \gamma \quad \text{and} \quad A+A \to A \ \text{ at rate } \chi, \tag{10}$$

where u would be a measure of the local density of A particles normalized by the equilibrium density $\sim \gamma/\chi$. It is tempting to interpret the stochastic sFKPP equation as a fluctuating hydrodynamic description of this process, as well, but a moment's reflection shows that this cannot be. Indeed, fluctuations around the equilibrium level of (10) corresponding to $u=1$ do not vanish for the $A \rightleftharpoons A+A$ process as they do (by definition) in the sFKPP equation. For (10) one would expect instead that the fluctuation strength should be proportional to at least $\sqrt{u}$ up to and beyond the $u=1$ level.

Nevertheless there is an exact rigorous relationship between the process in (10) and the stochastic Fisher-Kolmogorov-Petrovsky-Piscounov equation (9). The connection is by way of the notion of *duality*, an idea from the branch of probability theory concerned with interacting particle systems [15]. The duality relation between the growth-coagulation reaction and the sFKPP equation was discovered in 1986 by Shiga and Uchiyama [16]. It is most easily described in the discrete space setting; see [17] for a complete—and, we hope, accessible—derivation of this duality relation.

Impose a lattice on the x-axis with lattice spacing h and consider the discretized sFKPP equation

$$dU_i(t) = \left[\frac{D}{h^2}(U_{i+1} - 2U_i + U_{i-1}) + \gamma U_i(1-U_i)\right] dt + \sigma\sqrt{U_i(1-U_i)}dW_i \tag{11}$$

where the $W_i(t)$ are independent Brownian motions at each lattice site. The continuum limit of this system—taking $h \to 0$ with $x = ih$—is exactly the sFKPP equation (9) when we also send $\sigma \to \infty$ so that $\epsilon^2 = \sigma^2 h = \mathcal{O}(1)$.

On the other hand consider the spatially extended $A \rightleftharpoons A+A$ process (10) on the same spatial lattice where particles at site i hop to neighboring sites $i \pm 1$ at rate $\frac{D}{h^2}$ independently of all other processes. The rate coefficients γ and χ of the Markov process for the system are defined precisely as follows: If there are n_i particles at site i at time t, then they produce another particle there ($n_i \to n_i + 1$) at rate $n_i\gamma$, and the pair coagulation process ($n_i \to n_i - 1$) proceeds at rate $\frac{1}{2}n_i(n_i-1)\chi$. It is straightforward to write down the master equation for the joint probability distribution of all the occupation numbers $\{N_i(t)\}$ at all the lattice sites.

Duality between these systems holds when the noise amplitude σ^2 equals the coagulation rate χ. It is a connection between one processes evolving forward in time and the other evolving backwards in time: let $0 \le t \le T$, then $\sigma^2 = \chi \Rightarrow$

$$\left\langle \prod_i (1-U_i(t))^{N_i(T-t)} \right\rangle = \left\langle \prod_i (1-U_i(T-t))^{N_i(t)} \right\rangle \tag{12}$$

where the expectation $\langle\cdot\rangle$ is over both (independent) processes $\{U_i(t)\}$ and $\{N_i(t)\}$.

The duality relation allows us to express moments of the solution of the (discrete) sFKPP equation (11) at time t in terms of its initial data and the solution of the reaction-diffusion system. In particular, if we choose the initial condition for the particles so that there is just one particle at site i and no others anywhere else (i.e., $N_j(0) = \delta_{ij}$), then

$$\langle U_i(t) \rangle = 1 - \left\langle \prod_j (1 - U_j(0))^{N_j(t)} \right\rangle . \tag{13}$$

And if the initial data for $\{U_j\}$ is $U_j(0) = 1$ if $j \leq 0$ and $U_j(0) = 0$ if $j > 0$ (a setup for a wavefront propagating to the right), then duality produces the representation

$\langle U_i(t) \rangle =$ Prob(a site $j \leq 0$ has a particle at time t | only one particle at site i at time 0).

Note that the growth-coagulation process never becomes extinct: if we start with one (or more) particles we are ensured that there will always be at least one particle somewhere on the line. In fact, we expect that starting with only one singly occupied site i, the growth process will dominate until the unique equilibrium steady state with average occupation number $2\gamma/\chi$ at all lattice sites is achieved.

This connection implies that the asymptotic front speed for the sFPKK equation should be the same as the asymptotic front speed for the $A \rightleftharpoons A+A$ reaction-diffusion system—when the coagulation rate χ is the same as the square of the noise amplitude σ^2. In general we do not know an exact expression for the velocity of a front in the interacting particle model. However, we do know many things about a certain limit of the single-species growth-coagulation reaction-diffusion system, the so-called *diffusion-controlled* limit.

The diffusion-controlled $A \rightleftharpoons A+A$ process is the reaction-diffusion dynamics in (10) in the limit where $\chi \to \infty$ so that $A+A \to A$ instantaneously when two particles land on the same lattice site [18]. The equilibrium density ρ_{eq} of particles on the lattice is then nonzero only if the growth rate γ is also scaled up appropriately as $\chi \to \infty$. (By "density" we mean that $\rho_{eq}h$ is the equilibrium occupation probablility—equivalently the average occupation number—of a site.) Noting the correspondence of χ and σ^2 in the duality relation, this means that the strong-noise limit of the sFKPP equation is dual to the diffusion-controlled limit of the interacting particle process. We can make a quantitative connection between the system near the limit and the diffusion-controlled limit itself if we identify the equilibrium occupation probabilities $\rho_{eq}h$ and $2\gamma/\chi = 2\gamma/\sigma^2$.

The front propagation problem for the diffusion-controlled single-species growth-coagulation reaction-diffusion system has been solved exactly and in great detail [18, 19, 20]. In particular we know that the speed of the wavefront for the diffusion-limited process is $c = D\rho_{eq}$ [19, 20]. From this and the duality connection we can conjecture the behavior of the velocity of the sFKPP wave in the strong noise limit:

$$c = D\rho_{eq} \sim D\frac{2\gamma}{h\sigma^2} = \frac{2D\gamma}{\epsilon^2}. \tag{14}$$

In Figure 2 we plot the front velocities normalized by the minimum no-noise speed ($2\sqrt{D\gamma}$) as measured in direct numerical simulations of the (discrete) sFKPP equation (11) versus the dimensionless noise strength $\epsilon/(D\gamma)^{\frac{1}{4}}$. The solid lines are the asymptotic conjectures (8) for weak noise (using $N = 2\gamma/\sigma^2$) and (14) for strong noise.

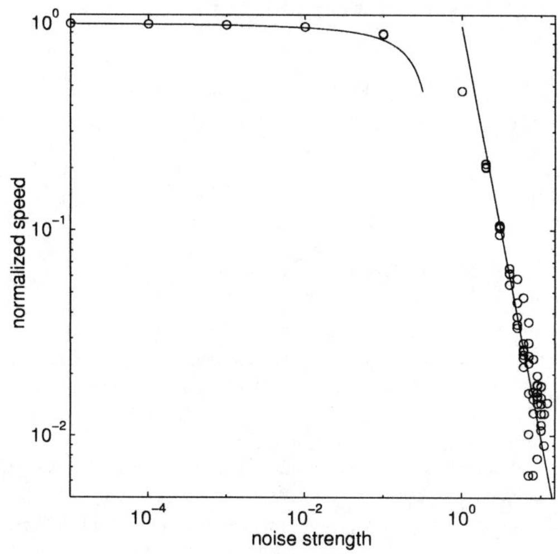

FIGURE 2. Wave speed (normalized by the no-noise speed) vs. dimensionless noise strength from the simulations (data), and the asymptotic conjectures for weak and strong noise (lines).

CONCLUSIONS AND UNSOLVED PROBLEMS

It is clear from Figure 2 that both the weak and strong noise conjectures for the front propagation speed are quantitatively accurate. Not unexpectedly, some fundamental equations remain:

1. Can these weak and strong noise asymptotic expressions for the front velocity be *proven* by analysis of the stochastic partial differential equation (with or without the use of duality)?
2. What can we discover about the statistics of the leading and trailing points of the front? (Mueller and Sowers proved that the fluctuating width of the front attains a stationary distribution as $t \to \infty$ [13]. What about front diffusion?)
3. In the other direction, what does the stochastic partial differential equation tell us about the statistical mechanics of the $A \rightleftharpoons A + A$ growth-coagulation reaction-diffusion process?
4. What can we deduce about front motion in higher spatial dimensions?

This last question in particular is of great interest for a variety of growth processes. In higher spatial dimensions even an initially smooth front can develop structure in the transverse direction(s) and fluctuations may play an even more dramatic role in the dynamics. Much recent effort has gone into studying particle systems whose mean-field descriptions are the higher dimensional FKPP equation [21, 22, 23, 24]. In these systems the effects of discreteness, fluctuations and noise may be very subtle and the quantitative behavior is difficult to discern.

ACKNOWLEDGMENTS

CRD thanks Profs. Lutz Shimansky-Geier (Humboldt University) and Peter Jung (Ohio University) for thoughtful comments. We acknowledge research support from NSF and NSA.

REFERENCES

1. R. A. FISHER, The wave of advance of advantageous genes, *Annals of Eugenics* **7**, 353 (1937).
2. A. KOLMOGOROV, I. PETROVSKY, AND N. PISCOUNOV, Etude de l'equation de la diffusion avec croissance de la quantité de matière et son application a un probleme biologique, *Moscow University Bulletin of Mathematics* **1**, 1 (1937).
3. J. A. MURRAY, *Mathematical Biology* (Springer-Verlag, Berlin, 1998).
4. M. BRAMSON, Convergence of Solutions of the Kolmogorov Equations to Traveling waves, *Memoirs of the American Mathematical Society* **44** (1983).
5. U. EBERT AND W. VAN SAARLOOS, Front propagation into unstable states: universal algebraic convergence towards uniformly translating pulled fronts, *Physica D* **146**, 1 (2000).
6. N. G. VAN KAMPEN, *Stochastic Processes in Physics and Chemistry* (North-Holland, Amsterdam, 1981).
7. E. BRUNET AND B. DERRIDA, Shift in the velocity of a front due to a cutoff, *Physical Review E* **56**, 2597 (1997).
8. D. A. KESSLER, Z. NER, AND L. M. SANDER, Front propagation: Precursors, cutoffs and structural stability, *Physical Review E* **58**, 107 (1998).
9. L. PECHENIK AND H. LEVINE, Interfacial velocity corrections due to multiplicative noise, *Physical Review E* **59**, 3893 (1999).
10. E. BRUNET AND B. DERRIDA, Effect of microscopic noise on front propagation, *Journal of Statistical Physics* **103**, 269 (2001).
11. D. PANJA, Surprising Aspects of Fluctuating "Pulled" Fronts, *UPoN 2002* (this volume).
12. C. MUELLER AND R. TRIBE, Stochastic PDEs arising from the long range contact and long range voter processes, *Probability Theory & Related Fields* **102**, 519 (1995).
13. C. MUELLER AND R. B. SOWERS, Random travelling waves for the KPP equation with noise, *Journal of Functional Analysis* **128**, 439 (1995).
14. J. XIN, Front propagation in heterogeneous media, *SIAM Review* **42**, 161 (2000).
15. T.M. LIGGETT, *Interacting Particle Systems* (Springer-Verlag, Berlin, 1985).
16. T. SHIGA AND K. UCHIYAMA, Stationary states and the stability of the stepping stone model involving mutation and selection, *Probability Theory & Related Fields* **73**, 87 (1986).
17. C. R. DOERING, C. MUELLER AND P. SMEREKA, *Physica A* (submitted, 2002).
18. M. A. BURSCHKA, C. R. DOERING, AND D. BEN AVRAHAM, Statics and Dynamics of a Diffusion-Limited Reaction: Anomalous Kinetics, Nonequilibrium Self-Ordering, and a Dynamic Transition, *Journal of Statistical Physics* **60**, 695 (1990).
19. C. R. DOERING, M. A. BURSCHKA, AND W. HORSTHEMKE, Fluctuations and correlations in a diffusion-reaction systems: Exact hydrodynamics, *Journal of Statistical Physics* **65**, 953 (1991).
20. D. BEN AVRAHAM, Fisher waves in the diffusion-limited coalescence process $A + A \rightleftharpoons A$, *Physics Letters A* **247**, 53 (1998).
21. J. RIORDAN, C. DOERING, AND D. BEN-AVRAHAM, Fluctuations and stability of Fisher waves, *Physical Review Letters* **75**, 565 (1995).
22. G. TRIPATHY AND W. VAN SAARLOOS, Fluctuation and relaxation properties of pulled fronts: A scenario for nonstandard Kardar-Parisi-Zhang scaling, *Physical Review Letters* **85**, 3556 (2000).
23. G. TRIPATHY, A. ROCCO, J. CASADEMUNT, AND W. VAN SAARLOOS, The universality class of fluctuating pulled fronts, *Physical Review Letters* **86**, 5215 (2001).
24. E. MORO, Internal fluctuations effects on Fisher waves, *Physical Review Letters* **87**, 238303.1-4 (2001).

Synthetizing Hydrodynamic Turbulence from Noise: Formalism and Applications to Plankton Dynamics

F. Sagués and R. Reigada*, J.M. Sancho†, R.M. Hillary** and M.A. Bees‡

*Departament de Química Física, Universitat de Barcelona, Diagonal 647, 08028 Barcelona, Spain
†Departament d'Estructura i Constituents de la Matèria, Facultat de Física, Universitat de Barcelona, Diagonal 647, 08028 Barcelona, Spain
**Department of Mathematics and Statistics, University of Surrey, GU2 7HX, UK
‡Department of Mathematics, University of Glasgow, G12 8QW, UK

Abstract. We present an analytical scheme, easily implemented numerically, to generate synthetic Gaussian 2*D* turbulent flows by using linear stochastic partial differential equations, where the noise term acts as a random force of well-prescribed statistics. This methodology leads to a divergence-free, isotropic, stationary and homogeneous velocity field, whose characteristic parameters are well reproduced, in particular the kinematic viscosity and energy spectrum. This practical approach to tailor a turbulent flow is justified by its versatility when analizing different physical processes occurring in advectely mixed systems. Here, we focuss on an application to study the dynamics of Planktonic populations in the ocean.

INTRODUCTION

Ideally, the statistical properties [1, 2] of a turbulent flow should come as the output of a first principles, Navier-Stokes based, formulation of the problem. However, we will adopt here a somewhat reversed perspective aimed at developing a methodology to construct what would be a sort of synthetic turbulence [3]. Rather than to retain the nonlinear coupling which makes possible the redistribution of energy from the largest length scales down into the smaller ones, we assume that the energy is incorporated into the system in an individual wave number basis, to reproduce the desired energy spectrum distribution of the turbulent flow in steady state.

The generation of the flow field proceeds from the two-dimensional simulation of a Langevin equation for the stream function $\eta(\mathbf{r},t)$,

$$\frac{\partial\eta(\mathbf{r},t)}{\partial t} = \nu\nabla^2\eta(\mathbf{r},t) + Q[\lambda^2\nabla^2]\nabla\cdot\zeta(\mathbf{r},t), \tag{1}$$

where ν stands for the kinematic viscosity and $Q[\lambda^2\nabla^2]$ denotes an operator which controls the spatial structure of the flow, with λ standing for its typical correlation length. Furthermore, $\zeta(\mathbf{r},t)$ represents a Gaussian white-noise field with zero mean value and whose covariance is given by

$$\langle\zeta^i(\mathbf{r}',t')\zeta^j(\mathbf{r}'',t'')\rangle = 2\varepsilon_0\nu\delta(t'-t'')\delta(\mathbf{r}'-\mathbf{r}'')\delta^{ij} \tag{2}$$

CP665, *Unsolved Problems of Noise and Fluctuations: UPoN 2002: Third International Conference*, edited by S. M. Bezrukov

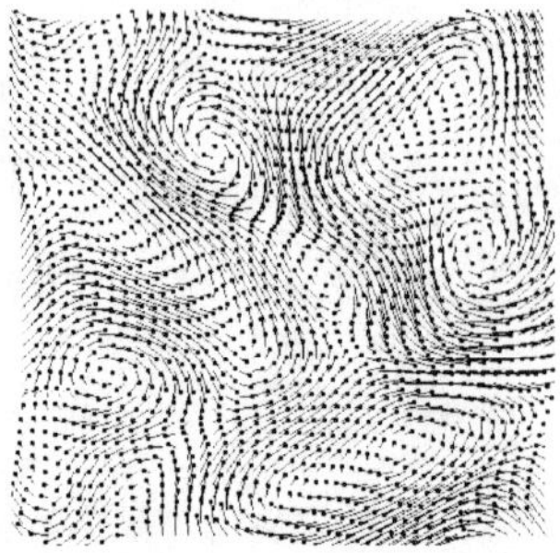

FIGURE 1. Snapshot of the "frozen" turbulent velocity field employed in the simulations in this study.

where ε_0 is the parameter that determines the intensity of the noise and further on that of the mimicked turbulent flow. This Langevin equation can be formally integrated in Fourier space to get the temporal evolution of the stream function. The incompressible two-dimensional discretized velocity field $\mathbf{U}_{i,j}$ follows from the stream function

$$\mathbf{U}_{i,j}(t) = \mathbf{U}(\mathbf{r}_{i,j},t) = \left(-\frac{\partial \eta(\mathbf{r}_{i,j},t)}{\partial y}, \frac{\partial \eta(\mathbf{r}_{i,j},t)}{\partial x} \right). \tag{3}$$

This synthetic fluid flow is characterized by three basic well-defined statistical properties; u_0^2 as the intensity of the flow, l_0 as the length correlation of the flow and t_0 as the time correlation of the flow [3]. These flow parameters are defined in terms of the velocity correlation function, which in turn, is formally associated to the energy spectrum (that depends on the form of the operator Q), and explicitly on the input parameters ν, ε_0, and λ. In this contribution we consider for most of the applications a flow with the Kraichnan spectrum [4]. In this case $E(k)$ in $2D$ is given by $E(k) \propto k^2 exp\left(-k^2/k_0^2\right)$ and correspondingly the diferential operator Q adopts the form $Q[\lambda^2\nabla^2] = exp(\lambda^2\nabla^2/2)$.

Although the general method of flow generation allows us to work with a non-stationary field, we shall not make use of the possibility in this study. For the sake of simplicity, and in order to avoid the existence of a great number of temporal scales, we take a stationary flow, and so fix $t_0 = \infty$ in the flow simulations. As an example, in Fig. 1 we show the flow realization used in this work. This flow has been generated using a square lattice of 128×128 points, and l_0 corresponds to a tenth of the linear size of the grid.

This model of synthetic turbulence has been used to study a large number of physical processes. Examples to be considered include dispersion of passive[5] and non-passive[6] particles, either reactive[7, 8] or non-reactive, front propagation phenomena[9] or phase-separating mixtures[10]. In this work we present an application to the study of planktonic populations in the ocean.

TURBULENT ADVECTION:INERTIAL EFFECTS

When describing the motion of advected plankton organisms or, in general any kind of non-passive particles, inertial effects due to their viscous and inertial properties have to be considered. In a Lagrangian prescription (the most visual and direct approach to study particle advection), the equation of motion for a spherical particle was derived by Maxey & Riley [11],

$$m_p\frac{d\mathbf{V}_n}{dt} = m_f\frac{D\mathbf{U}(\mathbf{X}_n)}{Dt} + 6\pi a\mu(\mathbf{U}(\mathbf{X}_n) - \mathbf{V}_n)$$

$$-\frac{m_f}{2}\left(\frac{d\mathbf{V}_n}{dt} - \frac{d\mathbf{U}(\mathbf{X}_n)}{dt}\right) - 6\pi a^2\mu\int_{-\infty}^{t}\frac{d(\mathbf{V}_n - \mathbf{U}(\mathbf{X}_n))/d\tau}{\sqrt{\pi\nu(t-\tau)}}d\tau, \tag{4}$$

being $\mathbf{X}_n$ the position of the particle n with velocity $\mathbf{V}_n$, in a velocity field, $\mathbf{U}(\mathbf{X}_n)$. m_p and m_f are the particle and fluid masses respectively, μ and ν are the dynamic and kinematic viscosities associated with the surrounding fluid and a is the radius of the particle. In Eq.(4) we have distinguished between two different derivative terms, D/Dt and d/dt. The first corresponds to the material derivative following a fluid element whereas the second denotes the derivative following the particle in the fluid such that $d\mathrm{U}/dt = \partial\mathrm{U}/\partial t + [\mathbf{V}_n \cdot \nabla]\mathbf{U}$. The first term in Eq. (4) is the Bernoulli term which is the force from the undisturbed flow, the second term is the Stokes viscous drag, the third is the added mass term and the final term is the Basset history force. This equation assumes that the particle, associated Reynolds number and fluid gradients around the particle surface are small [11]. Hereafter we consider that the drag and inertia terms are dominant and the particle is small, so that the history term can be neglected. We simplify even more the equation of motion assuming that the work done displacing a fluid element must be dependent on the forces on the fluid at that point. According to this idea we exchange the particle derivative for the fluid derivative in the added mass term to obtain a reduced version of Eq. (4) that reads [6, 11]

$$\frac{d\mathbf{V}_n}{dt} = A(\mathbf{U}(\mathbf{X}_n) - \mathbf{V}_n) + R\frac{D\mathbf{U}(\mathbf{X}_n)}{Dt}. \tag{5}$$

In this equation we have non-dimensionalized the time by rescaling it using the flow timescale l_0/u_0. We have defined $A = \alpha l_0/u_0$ as the dimensionless version of the inertia parameter $\alpha = (12\pi a\mu)/(2m_p + m_f)$, that corresponds to the balance between the viscous and inertial forces. The Bernouilli parameter $R = 3m_f/(2m_p + m_f)$ is the ratio between masses.

Eq. (5) can not be solved analytically, but it is possible to obtain reliable information from it in the limit where the viscous term dominates ($A > 1$). This asymptotic analysis was introduced by Maxey [12] for the singular case of infinitely heavy particles and extended to the general case for some of us [6]. It amounts to formally integrate the particle motion according to Eq. (5) under the assumption that A is large, so that exponential transients can be eliminated, and retain only terms up to the first order in A^{-1}. If we assume that the ambient flow is stationary, we have the following expression for the effective velocity field, $\mathbf{V}(\mathbf{r})$:

$$\mathbf{V}(\mathbf{r}) = \mathbf{U}(\mathbf{r}) + \frac{R-1}{A}[\mathbf{U}(\mathbf{r})\cdot\nabla]\mathbf{U}(\mathbf{r}) + \mathscr{O}(A^{-2}). \tag{6}$$

Actually, the above approximation assumes that the particle velocities adapt instantaneously to that of the surrounding fluid flow, and are only determined by the particle positions, neglecting the impact of trajectory memory. This is guaranteed by assuming that the viscous forces dominate the other ones. This allows the construction of an effective particle velocity field $\mathbf{V}(\mathbf{r})$ that closely matches the physical particle dynamics, and that can be easily implemented in a Eulerian formalism as we shall do later.

The most relevant signature of nonpassive advection is that particles drift from the flow trajectories due to inertia and tend to aggregate in various zones depending on their material characteristics [6, 12]. The aggregation of particles occurs in regions of negative divergence, $\nabla\cdot\mathbf{V} < 0$, in contrast to neutrally buoyant, passively advected particles. The divergence of the velocity $\mathbf{V}$ in Eq. (6) can be written in terms of the squares of the magnitudes of the local strain rate and the local vorticity, $\mathscr{S}^2$ and $|\Omega|^2$, respectively, of the original turbulent flow, $\mathbf{U}$, as follows

$$\nabla\cdot\mathbf{V} = \frac{R-1}{A}\left(2\mathscr{S}^2 - \frac{|\Omega|^2}{2}\right), \tag{7}$$

which leads to two different possibilities for aggregation. If the particle is heavier than the fluid ($R < 1$) then accumulation occurs in regions where $\mathscr{S}^2 > |\Omega|^2/4$, regions of high strain and low vorticity. Conversely, if a particle is lighter than the surrounding medium ($R > 1$) accumulation occurs in regions where $\mathscr{S}^2 < |\Omega|^2/4$, areas with high vorticity and low strain.

KINETIC MODEL FOR PLANKTON DYNAMICS

The versatility of our synthetic turbulence allowed thus to explore several physical systems in which the advection plays a crucial role. As anticipated, here we focus on the study of the influence of turbulent advection in Planktonic populations. Plankton patchiness has been observed on a wide range of spatial and temporal scales [13, 14, 15] and has been attributed to a range of physical and biological mechanisms. It is important to understand both the mechanisms that result in patchiness and the effect of patchiness on food-web interactions, which can have a major impact on fisheries policy.

Besides, at moderate spatial scales phytoplankton "blooms" can occur, whereby the population of phytoplankton rapidly increases in number and remains at this level for some period of time before returning to normal. This is the hallmark of an excitable system. Truscott & Brindley [16] investigated a two component phytoplankton-zooplankton (PZ) model which has the characteristics of an excitable system whose excitability is robust over a realistic parameter range. Normally, however, a large driving perturbation is required to initiate the bloom. Many mechanisms such as variations in salinity, temperature and nutrient mixing have been proposed to explain this initiation. In this work, we explore the effects of inertia in complex flows on excitable PZ dynamics. In particular, we shall address the question of whether the inertial separation of phytoplankton and zooplankton is sufficient to seed a bloom.

We propose the following system of reaction-advection-diffusion equations to represent the effects of planktonic interactions and diffusion in an Eulerian frame:

$$\frac{\partial P}{\partial t} = -\nabla \cdot (\mathbf{V}^P P - D_P \nabla P) + rP\left(1 - \frac{P}{K}\right) - \frac{\gamma Z P^2}{P^2 + \kappa^2},$$
$$\frac{\partial Z}{\partial t} = -\nabla \cdot (\mathbf{V}^Z Z - D_Z \nabla Z) + \frac{e\gamma Z P^2}{P^2 + \kappa^2} - \delta Z, \tag{8}$$

where $\mathbf{V}^i$ represents the velocity, with inertial and viscous corrections, of species i subject to advection by the flow obtained from Eq. (5), at lengthscales above the grid size, and grid-scale diffusive effects due to turbulent diffusion, with diffusivities D_P and D_Z. The last two terms in Eq. (8) correspond to the phytoplankton and zooplankton interactions following the excitable system proposed by Truscott & Brindley [16], which in essence models logistic phytoplankton growth, Holling type III grazing by zooplankton and a linear higher-predatory response with regard to the zooplankton mortality.

We non-dimensionalize Eq. (8) such that $P = Kp$, $Z = Kz$, $x = l_0\tilde{x}$, $y = l_0\tilde{y}$, $t = \tau l_0/u_0$, $\mathbf{V}^i = u_0\mathbf{v}^i$. After dropping the tildes this gives us the following non-dimensional version of Eq. (8),

$$\frac{\partial p}{\partial \tau} = -\nabla \cdot (\mathbf{v}^p p - d_p \nabla p) + \beta p(1-p) - \frac{\bar{\gamma} z p^2}{p^2 + \chi^2},$$
$$\frac{\partial z}{\partial \tau} = -\nabla \cdot (\mathbf{v}^z z - d_z \nabla z) + \varepsilon\left(\frac{\bar{\gamma} z p^2}{p^2 + \chi^2} - \sigma z\right), \tag{9}$$

where the dimensionless parameters are defined to be $\beta = rl_0/u_0$, $\varepsilon = e$, $\chi = \kappa/K$, $\sigma = \delta l_0/(\varepsilon u_0)$, $\bar{\gamma} = \gamma l_0/u_0$, $d_i = D_i/(u_0 l_0)$.

We choose L, the simulation box size, to be 100 km, the length scale of the eddies $l_0 \approx L/10$), and the characteristic velocity to be $u_0 = 0.075$ m s^{-1}, similar to that used in both Abraham [14] and Neufeld *et al.* [17]. From Truscott & Brindley [16], we take $\beta = 0.43$, $\chi = 0.053$, $\sigma = 0.34$ and $\varepsilon = 0.04$ (hence, the zooplankton population is the slow-recovery variable). Our choice of l_0 and u_o are such that we can realistically fix $\bar{\gamma} = 1$ and, hence, the reaction effects are comparable to advective effects. Kraichnan's spectrum has a maximum occurring at $k = 0.75$ and drops to a very small value at $k = 2$. Hence, the lengthscale associated with the highest energy is equal to $2\ \pi\ /\ 0.75 = 8.4$, or about a tenth of the length of the simulation region. The lengthscale associated with a wavenumber of 2 (residual energy) is given by $2\ \pi\ /\ 2 = \pi$, or about one 20th of the simulation region. If the simulation region is 100 km then the dimensional lengthscale associated with low spectral energy is $L/20 = 5 \times 10^5$ cm. This is the lengthscale for which we wish to implement a turbulent diffusivity in the reaction-advection-diffusion system. Hence, the diffusivity $D = 0.01 * l^{1.15} = 4 \times 10^4$ cm^2 s^{-1} $= 4$ m^2 s^{-1} (according to the empirical relationship extracted by Okubo [18] from experimental data). Therefore, the synthetic turbulence accounts for the large scale flow centered around a wavenumber k_0.

The system is spatially discretized using a square lattice of 128×128 cells, with a grid spacing of 0.5, and the time step used in the numerical scheme is $\Delta t = 0.001$. The integration of Eq. (9) was achieved by using a two-step *Lax-Wendroff* scheme which gave good numerical convergence for the parameter values employed. Simultaneously, we integrate Eq. (5) to obtain the velocities $\mathbf{V}^i$. We use the same $A = 6$ ($\gg 1$) for both species, but different Bernouilli parameters, R_p and R_z.

We begin from a homogeneous initial distribution of both species in their unexcited, equilibrium concentration values $p^* = 0.04589$ and $z^* = 0.04379$. The inertial separation under turbulent advection is enough to perturb the equilibrium distribution and, analogous to the threshold nature of perturbations in excitable systems, we find that there exist critical values of the Bernoulli parameters which separate regimes where no bloom occurs from regions where localized excitation and subsequent bloom propagation results.

First, we consider the extreme inertial regime, $R_p = 1.2$ and $R_z = 0.8$; namely, very light phytoplankton and very heavy zooplankton. As the phytoplankton accumulate in the eddies, the population locally exceeds the threshold value and excitation begins. Diffusion allows the excitation to move out from the eddies but the flow eventually forces this wave back into the vortices. At the same time the excitation decays and these three processes reach a steady, heterogeneous distribution of plankton, with phytoplankton at the excited state inside the eddies and below it outside. The pulse propagation is also tempered by the dilutory effects of the non-zero flow divergence enforced by inertial advective effects coupled with the geometry of the domain.

The most interesting scenario occurs for the regime where we have neutrally buoyant phytoplankton ($R_p = 1$) and slightly heavy zooplankton ($R_z = 0.95$). Here, we see the emergence, again without need of an inertial excitation, of an oscillatory bloom in the average phytoplankton population with a period of around one year, which is interesting in view of the annual nature of the observed blooms (although seasonal forcing is likely to have a major role). Fig. 2 presents the spatial evolution of the distribution of the phytoplankton population during one bloom cycle, whereas in Fig. 3 one may observe the nature of the oscillatory bloom in the mean population values of phytoplankton and zooplankton. The initial excitation is due to the absence of zooplankton in the eddies and, as before, the bloom propagates via diffusion throughout the domain. The main difference with the former scenario is that the expanding phytoplankton wave is not maintained in the vortices so the bloom decays and returns to the quiescent determination before the process is repeated.

What is fascinating is that the oscillatory regime is the most physically realistic scenario, as phytoplankton species are usually close to neutrally buoyant and zooplankton are slightly more dense than the surrounding ocean. However, notice that in the results shown here the value of the inertia parameter α from its dimensionless value ($A = 6$) is much smaller than what would be considered a realistic situation[19]. In a further work presented elsewhere [20] we adjust the value of the inertia parameter, obtaining qualitatively similar results that those ones presented here. Regardless of whether we use one or another value for α, we have corroborated that there are measurable qualitative differences between blooms involving zooplankton species of various sizes/densities, as it is reported in the experimental studies. However, a complete description of the problem would lead to a very complex model where, probably, the values of the inertial parameters and the material characteristics of the organisms would depend on time: $\alpha_i(t), R_i(t)$.

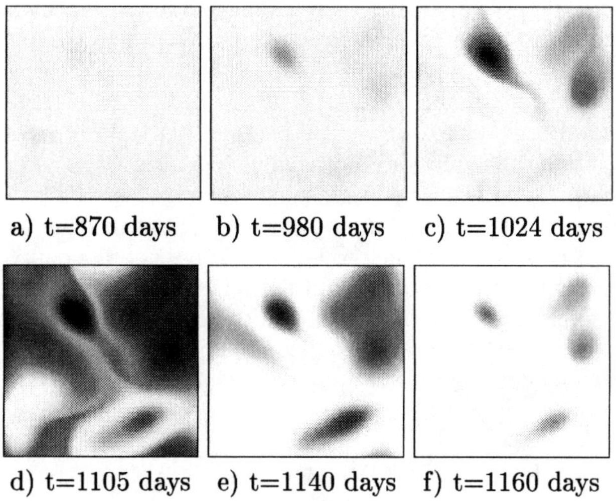

FIGURE 2. Snapshots of the phytoplankton population during the oscillatory bloom. Gray-scales as before.

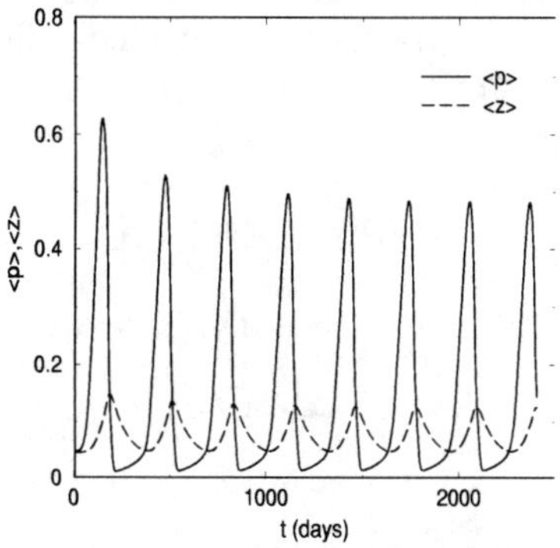

FIGURE 3. Oscillatory bloom in the mean concentrations of phytoplankton and zooplankton.

CONCLUSIONS

In this work, we have presented a methodology to tailor synthetic turbulent flows from white noise that can be easily implemented numerically. As a result we obtain a stochastic velocity field which enters in the equations as a multiplicative noise with a spatio-temporal structure. This makes the problem non Markovian which deserves new mathematical techniques [21]. As an example of the possible applications of this approach we have shown here the main features of the dispersion of nonpassive (inertial) particles under turbulent advection.

Applying these results and this metodology to the study of plankton dynamics, we have demonstrated how inertial separation of phytoplankton and zooplankton, coupled with excitable reaction dynamics, may contribute towards oceanic phytoplankton blooms and patchiness. Of particular interest was the oscillatory excitation, which occured in the most realistic inertial regime (neutrally buoyant phytoplankton and slightly heavy zooplankton). The period of this oscillation was approximately one year and occurred for a range of parameter values. The result is particularly intriguing due to the resulting annual period of the blooms, although we would emphasize that seasonal forcing is likely to play a great role. Such seasonal forcing would consist of variations in temperature, light, nutrients (due to mixing) or fluctuations in other components of the food-chain. However, with a minimal model and using experimentally determined values for the parameters, we obtain a natural period of almost one year, which may be further synchronized by the forcing. This self-initiating, periodic patch forming mechanism differs from other related work on advection enhanced blooms [17] as it requires no initial perturbation and is self-sustaining.

ACKNOWLEDGMENTS

The authors acknowledge the support of this research by the Ministerio de Educación y Cultura through Projects No. BFM2000-0624 and BXX2000-0638, and by the Comissionat per Universitats i Recerca de la Generalitat de Catalunya under Project No. 1999SGR00041. RMH and MAB gratefully acknowledge financial support from the EU Access to Research Infrastructure action of the *Improving Human Potential Programme*.

REFERENCES

1. A.S. Monin and A.M. Yaglom, *Statistical Fluid Mechanics* (MIT, Cambridge, MA, 1975).
2. W.D. McComb, *The Physics of Fluid Turbulence* (Oxford University, Oxford, 1990).
3. A.C. Martí, J.M. Sancho, F. Sagués and A. Careta, *Phys. Fluids* **9**, 1078 (1997).
4. R.H. Kraichnan, *Phys. Fluids* **13**, 22 (1970).
5. R. Reigada, A.C. Martí, F. Sagués, I.M. Sokolov and J.M. Sancho, *Phys. Rev. E* **62**, 4997 (2000).
6. R. Reigada, F. Sagués and J.M. Sancho, *Phys. Rev. E* **64**, 026307 (2001).
7. R. Reigada, F. Sagués, I.M. Sokolov, J.M. Sancho and A. Blumen, *Phys. Rev. Lett.* **78**, 741 (1997).
8. R. Reigada, F. Sagués and J.M. Sancho, *J. Chem. Phys.* **117**, 258 (1998).
9. A.C. Martí, F. Sagués and J. M. Sancho, *Phys. Fluids* **9**, 3851 (1997).
10. A. M. Lacasta, J. M. Sancho and F. Sagués, *Phys. Rev. Lett.* **75**, 1791 (1995).
11. M.R. Maxey and J.J. Riley, *Phys. Fluids* **26**, 883 (1983).
12. M.R. Maxey, *J. Fluid Mech.* **174**, 441 (1987).
13. P. J. S. Franks, *Limnol. Oceanogr.* **42**, 1297 (1997).
14. E.R. Abraham, *Nature* **39**, 577 (1998).
15. C.L. Folt and C.W. Burns, *TREE* **14** 300 (1999).
16. J.E. Truscott and J. Brindley, *Bull. Math. Biol.* **56**, 981 (1994).
17. Z. Neufeld, P.H. Haynes, V.C. Garçon and J.Sudre, *Geophys. Res. Lett.* **29**, 1029 (2002).
18. A. Okubo, *Deep-Sea Research* **18**, 789 (1971).
19. K.D. Squires and H. Yamazaki, *Deep-Sea Research I* **42**, 1989 (1995).
20. R. Reigada, R.M. Hillary, M.A. Bees, J.M. Sancho and F. Sagués, (accepted in *Proc. Roy. Soc. B*).
21. J.M. Sancho, M.A. Santos, S. Alonso and F. Sagués, (*in this volume*).

Surprising Aspects of Fluctuating "Pulled" Fronts

Debabrata Panja

Instituut-Lorentz, Universiteit Leiden, Postbus 9506, 2300 RA Leiden, The Netherlands[1]

Abstract. Recently it has been shown that when an equation that allows so-called pulled fronts in the mean-field limit is modelled with a stochastic model with a finite number N of particles per correlation volume, the convergence to the speed v^* for $N \to \infty$ is extremely slow — going only as $\ln^{-2} N$. However, this convergence is seen only for very high values of N, while there can be significant deviations from it when N is not too large. Pulled fronts are fronts that propagate into an unstable state, and the asymptotic front speed is equal to the linear spreading speed v^* of infinitesimal perturbations around the unstable state. In this paper, we consider front propagation in a simple stochastic lattice model. The microscopic picture of the front dynamics shows that for the description of the far tip of the front, one has to abandon the idea of a uniformly translating front solution. The lattice and finite particle effects lead to a "halt-and-go" type dynamics at the far tip of the front, while the average front behind it "crosses over" to a uniformly translating solution. In this formulation, the effect of stochasticity on the asymptotic front speed is coded in the probability distribution of the times required for the advancement of the "foremost occupied lattice site". These probability distributions are obtained by matching the solution of the far tip with the uniformly translating solution behind in a mean-field type approximation, and the results for the probability distributions compare well to the results of stochastic numerical simulations. This approach allows one to deal with much smaller values of N than it is required to have the $\ln^{-2} N$ asymptotics to be valid.

1. THE BASICS OF FRONT PROPAGATION: PULLED FRONTS

In pattern forming systems quite often situations occur where patches of different bulk phases occur which are separated by fronts or interfaces. In such cases, the relevant dynamics is usually dominated by the dynamics of these fronts. When the interface separates two thermodynamically stable phases, as in crystal-melt interfacial growth problems, the width of the interfacial zone is usually of atomic dimensions. For such systems, one often has to resort to a moving boundary description in which the boundary conditions at the interface are determined phenomenologically or by microscopic considerations. A question that naturally arises for such interfaces is the influence of stochastic fluctuations on the motion and scaling properties of such interfaces.

At the other extreme is a class of fronts which arise in systems that form patterns, and in which the occurrence of fronts or transition zones is fundamentally related to their nonequilibrium nature, as they do not connect two thermodynamic equilibrium phases

[1] Present Address: Institute for Theoretical Physics, Universiteit van Amsterdam, Valckenierstraat 65, 1018 XE Amsterdam, The Netherlands

CP665, *Unsolved Problems of Noise and Fluctuations: UPoN 2002: Third International Conference,* edited by S. M. Bezrukov

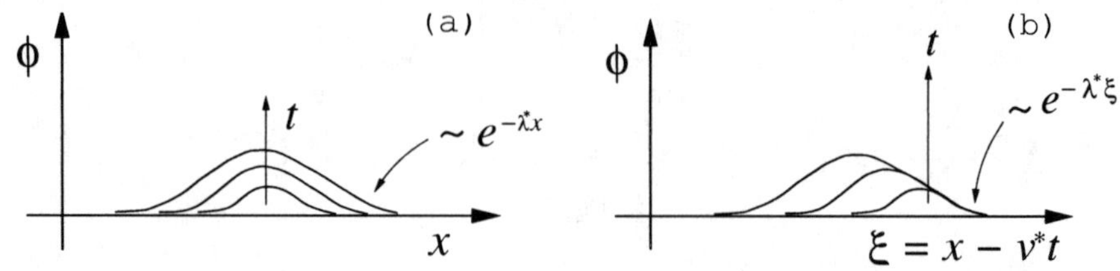

FIGURE 1. Illustration of v^* as the linear spreading speed of infinitesimal perturbations around the unstable state.

which are separated by a first order phase transition. In such cases — for example, chemical fronts [1], the temperature and density transition zones in thermal plumes [2], the domain walls separating domains of different orientation in in rotating Rayleigh-Bénard convection [3], or streamer fronts in discharges [4] — the fronts are relatively wide and therefore described by the same continuum equations that describe nonequilibrium bulk patterns. The lore in nonequilibrium pattern formation is that when the relevant length scales are large, (thermal) fluctuation effects are relatively small [5]. For this reason, the dynamics of many pattern forming systems can be understood in terms of the deterministic dynamics of the basic patterns and coherent structures. For fronts, the first questions to study are therefore properties like existence and speed of propagation of the front solutions of the deterministic equations. A class of fronts, for which these questions can be answered theoretically, are the so-called pulled fronts. Pulled fronts are the fronts that propagate into a linearly unstable state, and whose asymptotic front speed is equal to the linear spreading speed v^* of infinitesimal perturbations around the unstable state [6, 7, 8]. The name pulled front refers to the picture that in the leading edge of these fronts, the perturbation around the unstable state grows and spreads with speed v^*, while the rest of the front gets "pulled along" by the leading edge. The intuitive idea is captured in Fig. 1, where we consider the (exaggerated) growth and spreading of an infinitesimal perturbation around the linearly unstable state. Three snapshots of this perturbation, taken at three different time instants in the laboratory frame are plotted in Fig. 1(a). The initial perturbation is chosen in such a way that it decays as $\exp[-\lambda^* x]$ for $x \to \infty$, where λ^* is the exponent associated with v^*. The special status that the quantities λ^* and v^* have in the growth and spreading of this perturbation is coded in the fact that the $\exp[-\lambda^* x]$ decay of the perturbation for $x \to \infty$ remains preserved at *all* stages of its development. This fact is further illustrated in Fig. 1(b), where the same three snapshots [as in Fig. 1(a)] have been plotted in the comoving frame, moving with speed v^* w.r.t. the laboratory frame. That the notion of the leading edge ahead of the front grows and spreads with speed v^* and thereby pulls the rest of the front along with it is not merely an intuitive picture but can be turned into a mathematically precise analysis is illustrated by the recent derivation of exact results for the general power law convergence of the front speed to the asymptotic value v^* [8]. On the other hand, fronts which propagate into a linearly unstable state and whose asymptotic speed is $> v^*$ are referred to as pushed, as it is the nonlinear growth in the region behind the leading edge that pushes their front speed to higher values. If the state is not linearly unstable, then v^* is trivially zero; in such cases front propagation is always dominated by the nonlinear growth in

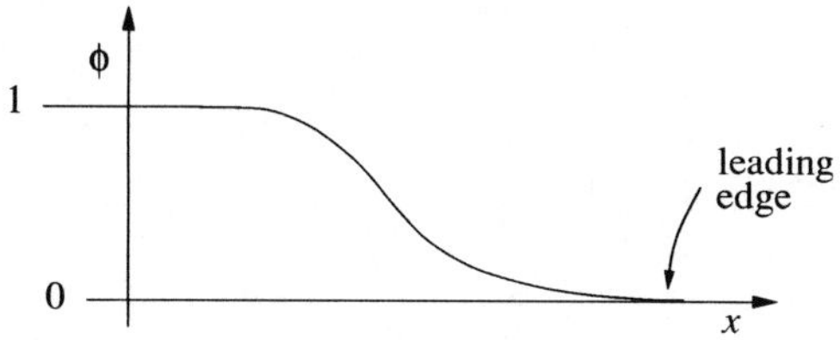

FIGURE 2. An instantaneous snapshot of the front configuration of Eq. (1).

the front region itself, and hence fronts in this case are in a sense pushed too. To obtain the asymptotic speed of a pushed front, one has to solve the full nonlinear equation; in general it is not possible to do so except for a special set of parameter values. To provide a more quantitative flavour of how the growth and spreading of infinitesimal perturbation around the unstable state play a very important role for pulled fronts, let us now consider and examine an example of a deterministic equation that admits pulled fronts. The equation that we choose for this purpose is the so-called Fisher equation, which was at first used to model the spreading of advantageous genes in a population [9]. In this model, the density of the advantageous genes is denoted by $\phi(x,t)$, whose dynamics is described by means of the equation

$$\frac{\partial \phi}{\partial t} = D\frac{\partial^2 \phi}{\partial t^2} + \phi - \phi^n. \qquad n > 1, \quad \text{for example 2 or 3} \tag{1}$$

Equation (1) has two stationary states, of which $\phi(x,t) = 0$ is (linearly) unstable and $\phi(x,t) = 1$ is stable. Therefore, if the system is prepared in a way such that these two states coexist in a certain region of space, then the stable state invades the unstable one and propagates into it. An instantaneous configuration of the resulting front is shown in Fig. 2.

To obtain the front solution admitted by Eq. (1), we rewrite it in a frame that moves w.r.t. the laboratory frame at a constant speed v, and look for a stationary solution of ϕ in this comoving frame. In terms of the comoving co-ordinate $\xi = x - vt$, Eq. (1) can be simply rewritten by means of a change of variables from (x,t) to (ξ,t), as

$$\frac{\partial \phi}{\partial t} - v\frac{\partial \phi}{\partial \xi} = D\frac{\partial^2 \phi}{\partial \xi^2} + \phi - \phi^n. \tag{2}$$

The crucial relevance of the growth and spreading of infinitesimal perturbations enters naturally in this front solution, as the propagating infinitesimal perturbations around the unstable state in the leading edge ahead of the front sets on the instability making way for further growth. At the leading edge of the front, the ϕ-values are very close to the unstable state value, i.e., $\phi \ll 1$, and one can neglect the nonlinear term ϕ^n compared to ϕ in Eq. (2). The stationary solution of the resulting linear equation can then be solved by using $\phi \sim \exp[-\lambda\xi]$, yielding the following relation between v and λ:

$$v(\lambda) = D\lambda + \frac{1}{\lambda}. \tag{3}$$

The curve for the dispersion relation between $v(\lambda)$ and λ is schematically shown in Fig. 3. It has a minimum at λ^*, and $v^* = v(\lambda^*) = 2\sqrt{D}$. The actual dispersion relation

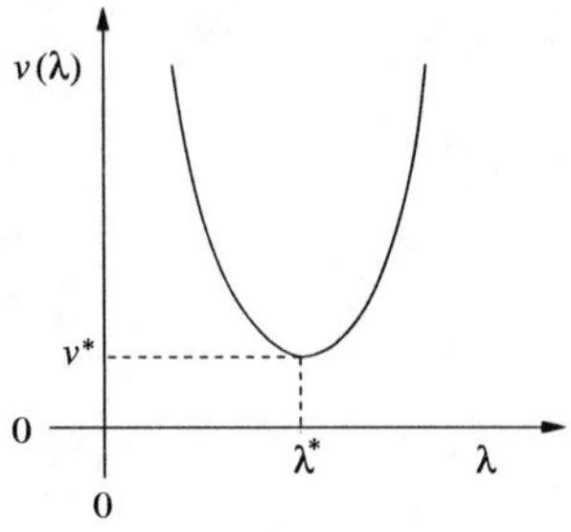

FIGURE 3. The dispersion relation (3) is schematically shown above.

depends on the model that one studies, but the U-shape is a characteristic of fronts that propagate into a linearly unstable state. Although Eq. (3) indicates that Eq. (1) has a front solution for all values of λ (and correspondingly all possible front speeds), from which it might a priori seem that the quantities λ^* and v^* are not special in any way, the actual selection of the asymptotic front speed is obtained only after a proper stability analysis of the front profile in the comoving frame. Such a stability analysis yields the result that with the initial condition that $\phi(x,t)|_{t=0}$ that decays faster than $\exp[-\lambda^* x]$ for $x \longrightarrow \infty$, the front speed converges uniformly[2] to v^* as [8, 10]

$$v(t) = v^* - \frac{3}{2\lambda^* t} + \mathcal{O}(t^{-3/2}), \tag{4}$$

as the front shape relaxes to its asymptotic steady state configuration $\phi^*(x - v^* t)$.

2. FLUCTUATING "PULLED" FRONTS: A DIFFERENT PARADIGM

From the above perspective, it is maybe less of a surprise that the detailed questions concerning the stochastic properties of inherently nonequilibrium fronts have been addressed, to some extent, only relatively recently [11, 12, 13, 14, 15, 16, 17, 18], and that it has taken a while for researchers to become fully aware of the fact that the so-called pulled fronts [6, 7, 8, 19] that propagate into an unstable state, do *not* fit into the common mold: they have anomalous sensitivity to particle effects [14, 15, 16, 20, 21], and have been argued to display uncommon scaling behaviour [18, 22, 23, 24, 25].

All these effects have one origin in common, namely the fact that the dynamics of pulled fronts, by its very nature, is not determined by the nonlinear front region itself, but by the region *at the leading edge of the front*, where deviations from the unstable state are small. To a large degree, this semi-infinite region alone determines the universal relaxation of the speed of a deterministic pulled front to its asymptotic value [7, 8, 14], as well as the anomalous scaling behaviour of stochastic fronts [22, 23, 24] in continuum

[2] Uniform convergence means that the convergence behaviour (4) of the front speed is the same irrespective of the value of ϕ at which the speed is being measured.

equations with multiplicative noise. The crucial importance of the region, where the deviations from the unstable state are small, also implies that if one builds a lattice model version of a front propagating from a stable into a linearly unstable state, the front speed is surprisingly sensitive to the dynamics of the tip (the far end) of the front where only one or a few particles per lattice site are present. This is the main subject of this paper, and we will demonstrate the effect of discreteness on front propagation by means of considering a discrete particle model of Fisher equation (1).

Following Refs. [11, 26], the discrete particle model of Fisher equation (1) that we will consider in this paper is that of the reaction-diffusion system X $\leftrightarrow$ 2X on a lattice. In this system, an X particle on any lattice site can diffuse to one of its nearest neighbour lattice sites with a diffusion rate D. The rate of the forward reaction X $\longrightarrow$ 2X is $N/2$, while the rate of the backward reaction is normalized to unity. A forward reaction on a lattice site k increases the number of X particles on lattice site k by one, and a backward 2X $\longrightarrow$ X reaction on lattice site k reduces the number of X particles on lattice site k by one. For large N values, the microscopic interaction indicates that an initial conglomeration of X particles grows both in size and spread. The microscopic rules also dictate that the growth saturates when the number of particles on a lattice site reaches approximately N — at that stage, on average, the amount of new X particles generated due to the forward reaction equals the amount of X particles annihilated due to the backward reaction.

The deterministic mean-field limit of the this reaction-diffusion system yields the equation [11, 26]

$$\frac{\partial \phi_k}{\partial t} = D\left[\phi_{k+1} + \phi_{k-1} - 2\phi_k\right] + \phi_k - \phi_k^2 . \tag{5}$$

Here, $\phi_k = \langle N_k \rangle / N$, where $\langle N_k \rangle$ is the conditionally (ensemble) averaged[3] number of X particles on the k-th lattice site. Equation (5) is the lattice version of Eq. (1), and it admits a pulled front solution, which can be obtained by using the uniformly translating solution $\phi_k(t) \equiv \phi(k - vt)$. From the analysis and discussion of the previous section, therefore, one can expect that the asymptotic front speed is v^*, where the corresponding front solution $\phi^*(k - v^*t)$ behaves $\sim \exp[-\lambda^*\xi]$ for $\xi \longrightarrow \infty$. Here $\xi = k - v^*t$ and λ^* and v^* correspond to the minimum of the dispersion relation $v(\lambda)$ vs. λ for Eq. (5) (similar to Fig. 3). In actuality, however, the asymptotic speed turns out to be $< v^*$. This was first observed numerically in Ref. [11].

The reason behind having an asymptotic speed $< v^*$ stems from the discreteness of the particles and the lattice effects, and is understood quite simply when one takes an instantaneous snapshot of the resulting front, moving from the left to the right, for one single realization at time $t \longrightarrow \infty$[4]. Such a snapshot is shown in Fig. 4. The important

---

[3] In view of the stochasticity in the system, one has to be careful about taking averages. A simple ensemble averaging does not yield a steady front shape due to diffusive wandering of individual front realizations. This notion is best understood by means of Fig. 5 of Ref. [11], which shows the ensemble average front shape by means of hashed lines. One needs to filter out the diffusive wandering of fronts before taking the average, and this is what is meant by conditional (ensemble) averaging. For a detailed discussion, see Sec. III.B of Ref. [26]

[4] The limit $t \longrightarrow \infty$ is necessary to ensure that the front relaxes to a steady shape in an average sense, so that there are no residual large scale transient effects.

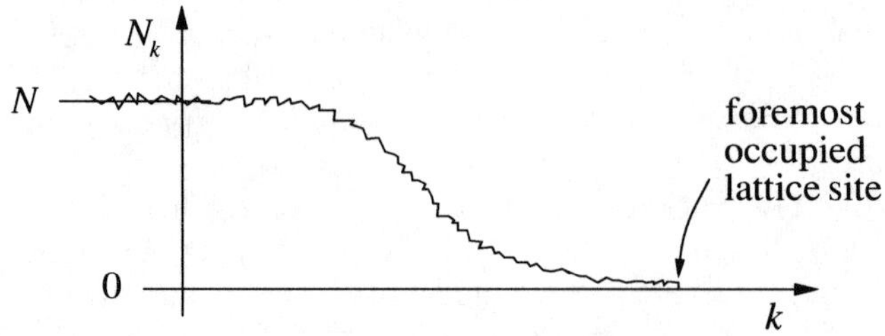

FIGURE 4. An instantaneous snapshot of a front realization (travelling from the left to the right) for the reaction diffusion system X $\leftrightarrow$ 2X, at time $t \to \infty$. Note the existence of the foremost occupied lattice site, on the right of which no lattice site has been previously occupied.

aspect to notice in Fig. 4 is that unlike the semi-infinite region that the leading edge occupies for a deterministic front (as shown in Fig. 2), there exists a foremost occupied lattice site (f.o.l.s.) in this snapshot, on the right of which no lattice site has ever been occupied before. In the language of the mathematicians, this means that the actual front region has a compact support. The crucial role played by the discreteness effects of the particles and the lattice is reflected in the mechanism of the front propagation at the f.o.l.s., where the mechanism is *not* that of a uniform translation, but instead, is of *halt-and-go, which in this reaction-diffusion model is diffusion dominated*, as we demonstrate below. First of all, it is clear from the very nature of the reaction-diffusion equation that unless there is at least one X particle on a lattice site, the growth in the number of particles on that lattice site does not take place. This means that the position of the f.o.l.s. moves towards the right *only* when a particle from the f.o.l.s. makes a diffusive hop towards the right. The movement of the f.o.l.s. is therefore diffusion dominated, as opposed to being also driven by growth; i.e., the phenomenon of leading edge ahead of the front setting on the instability making way for further growth that takes place for a deterministic mean field equation of a pulled front, does not take place for the corresponding discrete particle realization — this makes the fronts made of discrete particles on a lattice "pulled" as opposed to being pulled. Secondly, these diffusive hops are not continuous in time, i.e., there is a finite time difference between any two successive forward movements of the f.o.l.s. (this is what is meant by halt-and-go; the f.o.l.s. halts for the times between the successive hops), and these time differences are stochastic in nature. We stress here that the halt-and-go mechanism of the dynamics of the f.o.l.s. is a generic consequence of having discrete particles on a discrete lattice; in the present model that we consider, the movement is diffusion dominated, but there can be other models where the f.o.l.s. moves by means of some other mechanism (see for example, clock model [20]).

The subtlety that the movement of the f.o.l.s. is not driven by growth for "pulled" fronts made of discrete particles on a lattice was first realized by Brunet and Derrida [14], who chose to implement it by having a growth cutoff at a ϕ_k value $1/N$ in the deterministic mean-field Eq. (5). The choice of $1/N$ was motivated by the fact that the number of particles on the f.o.l.s. is $\mathcal{O}(1)$, which corresponds to a field value $\phi_k \sim 1/N$. The uniformly translating solution of the corresponding mean-field front equation for the reaction-diffusion process X $\leftrightarrow$ 2X

$$\frac{\partial \phi_k}{\partial t} = D\left[\phi_{k+1} + \phi_{k-1} - 2\phi_k\right] + \left[\phi_k - \phi_k^2\right]\Theta\left(\phi_k - \frac{1}{N}\right), \qquad (6)$$

where the Θ term is the heavyside theta function[5], yields an asymptotic speed v_{cutoff}, given by

$$v_{cutoff} = v^* - \frac{\pi^2 D \cosh \lambda^*}{\ln^2 N}. \qquad (7)$$

The essential correctness of Eq. (7), that the convergence of the asymptotic front speed to v^* behaves $\sim 1/\ln^2 N$ has been observed elsewhere in systems of physical interest such as clock model [20], in field-theoretical approach [16], and as well as in studies on noisy front propagation in Fisher equation in the limit of "weak noise" by mathematicians [28]. However, it has also been noted by many theorists working in this field that one often has to go to very high values of N to obtain the $1/\ln^2 N$ convergence of the asymptotic front speed to v^*. For large but more reasonable values of N, there are often significant deviations from Eq. (7) (see for example, Ref. [15]). A question which then naturally arises is the following: how does one bridge the gap between Eq. (7), which is valid for asymptotically large N, and the corresponding results for finite N?

Based on the discussion three paragraphs above, in what follows below, we will appeal to the reader for a more comprehensive description of fluctuating "pulled" fronts on a lattice that combines the stochastic halt-and-go nature of front propagation at the f.o.l.s. with a uniformly translating front solution few lattice sites behind the f.o.l.s. Such a description leads one to a concrete mathematical formalism developed elsewhere [26], of which we will provide only the flavour in this paper. This description and the associated formalism serves to bridge this gap between the results for asymptotically large N and that for reasonably large values of N. Afterwards, we will argue why Eq. (7) is essentially correct for asymptotically large N values. The mathematical formalism of Ref. [26], however, has open questions, which we will leave out for the next section.

We start with the fact that the very definition of the f.o.l.s. means that all the lattice sites on the right of it are empty (and they have never been occupied before). Naturally, a lattice site, which has never been occupied before attains the status of the f.o.l.s. as soon as one particle hops into it from the left. In reference to the lattice, the position of the f.o.l.s. remains fixed at this site for some time, i.e., after its creation, an f.o.l.s. remains the f.o.l.s. for some time. During this time, however, the number of particles on and behind the f.o.l.s. continues to grow. As the number of particles grows on the f.o.l.s., the chance of one of them making a diffusive hop on to the right also increases. At some instant, a particle from the f.o.l.s. hops over to the right: as a result of this hop, the position of the f.o.l.s. advances by one unit on the lattice, or, viewed from another angle, a new f.o.l.s. is created which is one lattice distance away on the right of the previous one. Microscopically, the selection process for the length of the time span

---

[5] The "pulled" nature can be made mathematically very precise for Eq. (6), see for example Ref. [27].

between two consecutive f.o.l.s. creations is stochastic, and the inverse of the long time average of this time span defines the front speed. Simultaneously, the amount of growth of particle numbers on and behind the f.o.l.s. itself depends on the time span between two consecutive f.o.l.s. creations (the longer the time span, the longer the amount of growth). As a consequence, on average, the selection mechanism for the length of the time span between two consecutive f.o.l.s. movements, which determines the asymptotic front speed, is nonlinear.

This inherent nonlinearity makes the prediction of the asymptotic front speed difficult. One might recall the difficulties associated with the prediction of pushed fronts due to nonlinear terms in this context, although the nature of the nonlinearities in these two cases is *completely different*. In the case of pushed fronts, the asymptotic front speed is determined by the mean-field dynamics of the fronts, and the nonlinearities originate from the *nonlinear growth terms in the partial differential equations* that describe the mean-field dynamics. On the other hand, for fronts consisting of discrete particles on a discrete lattice, the corresponding mean-field growth terms are *linear*, but since the asymptotic front speed is determined from the probability distribution of the time span between two consecutive f.o.l.s. movements, on average, it is the relation between this probability distribution and the effect of the linear growth terms that the nonlinearities stem from. As one can now clearly see, Brunet and Derrida's cutoff picture, as described above, does not take the stochasticity of the halt-and-go mechanism into account.

Our approach is to discuss a separate probabilistic theory for the hops to create the new f.o.l.s., and then to demonstrate that by matching the description of the behaviour in this region to the more standard one (of growth and roughly speaking, uniform translation) behind it, one obtains a consistent and more complete description of the stochastic and discreteness effects on front propagation. In the simplest approximation, the theory provides a very good fit to the data, but our approach can be systematically improved by incorporating the effect of fluctuations as well [26]. Besides providing insight into how a stochastic front propagates at the far tip of the leading edge, our analysis naturally leads to a more complete description that allows one to interpret the finite N corrections to the front speed for much smaller values of N than that are necessary to see the asymptotic result of Brunet and Derrida [14]. We stress here that such a procedure can be carried out for any fluctuating "pulled" front, although in the present context, we will confine ourselves only to the reaction-diffusion system $X \leftrightarrow 2X$.

In the resulting mathematical formalism [26], we follow the movement of the f.o.l.s. of *one single front realization* over a long time at large times. Let us denote the j successive values of the duration of halts of the f.o.l.s. by $\Delta t_1, \Delta t_2, \ldots, \Delta t_j$. One can then define the front speed as

$$v_N = \lim_{j\to\infty} j \left[\sum_{j'=1}^{j} \Delta t_{j'}\right]^{-1} . \tag{8}$$

Put in a different way, if we denote the probability that a f.o.l.s. remains the f.o.l.s. for time Δt by $\mathscr{P}(\Delta t)$, the asymptotic front speed, according to Eq. (8), is given by

$$v_N = \left[\int_0^\infty d(\Delta t)\, \Delta t\, \mathscr{P}(\Delta t)\right]^{-1} . \tag{9}$$

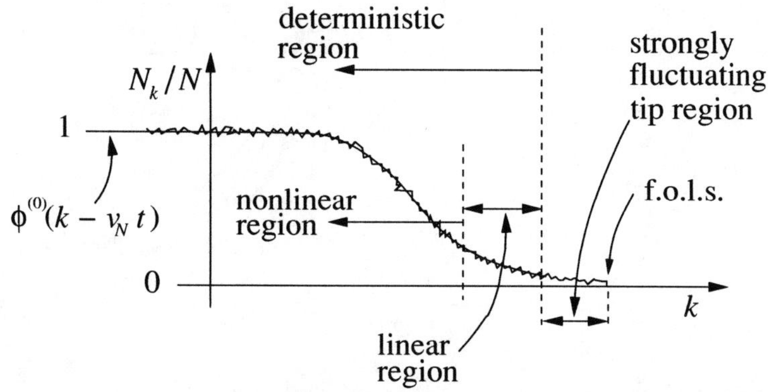

FIGURE 5. A snapshot of a realization of the reaction-diffusion system X $\leftrightarrow$ 2X (shown by the jaggered line) as $t \to \infty$, and how a theorist would picture such a front. In this picture, as the smooth line shows, a uniformly translating solution travelling with speed v_N, given by $\phi_k(t) = \phi^{(0)}(k - v_N t)$ and obeying Eq. (5), is suitable (in an average sense) everywhere but few lattice sites at the tip of the front leading up to the f.o.l.s. The region where such a description holds is denoted above by "deterministic region". The "deterministic region" is further subdivided into two parts — in the "linear region", the nonlinear term $\left[\phi^{(0)}\right]^2$ of Eq. (5) can be neglected. In the "nonlinear region", however, all the terms of Eq. (5) have to be taken into account.

Our goal is to obtain a theoretical expression of $\mathscr{P}(\Delta t)$, but to do so, we draw the reader's attention to Fig. 5. As described above, we assume that a uniformly translating (with speed v_N) deterministic mean-field description, given by $\phi_k(t) = \phi^{(0)}(k - v_N t)$, is valid everywhere *but a few lattice sites leading up to the f.o.l.s.*. The region where such a description [namely that its dynamics is given by Eq. (5)] holds is denoted by "deterministic region" in Fig. 5. This region can be further subdivided into two parts, a "linear region" and a "nonlinear region", where the nonlinear $\left[\phi^{(0)}\right]^2$ term can and cannot be neglected respectively. Although the amount of fluctuations in the number of particles on individual lattice sites at the strongly fluctuating tip region is of the same order of magnitude as the particle numbers themselves, following Ref. [26], we assume that the strongly fluctuating tip region can be modelled by a time dependent mean-field type description without uniform translation. In this description, at the tip region of the front, we express the front solution as $\phi_k(t) = \phi^{(0)}(k - v_N t) + \delta\phi_k(t)$. The quantities $\delta\phi_k(t)$ are non-zero due to the halt-and-go motion of the f.o.l.s., but at the boundary between the tip region and the "linear region" $\delta\phi_k(t)$ vanishes. As explained previously, to obtain the expression of $\mathscr{P}(\Delta t)$, one needs the expressions of $\phi^{(0)}(k - v_N t)$ and $\delta\phi_k(t)$. Moreover, the expression of v_N itself is needed to solve for $\phi^{(0)}(k - v_N t)$, and v_N can be determined only from $\mathscr{P}(\Delta t)$ as Eq. (9) shows. This indicates that the only way to obtain the expression of $\mathscr{P}(\Delta t)$ is to solve a whole system of equations self-consistently [26]. We also note here that in this self-consistent theory, there is an effective parameter.

This self-consistent set of equations are highly nonlinear and complicated, but due to the presence of the effective parameter in our theory, our procedure to obtain the $\mathscr{P}(\Delta t)$

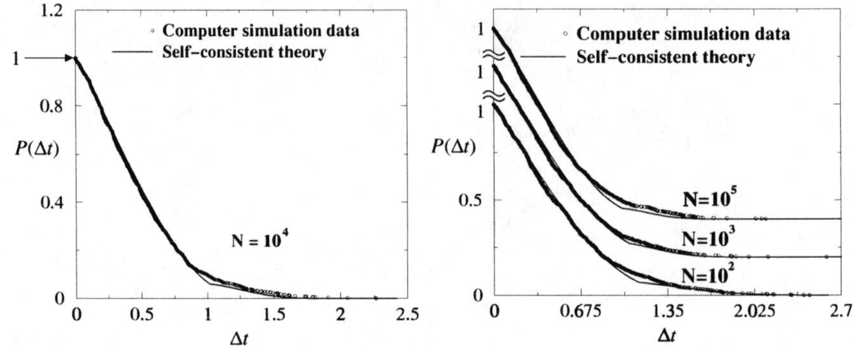

FIGURE 6. The agreement between the self-consistent theoretical curve of $P(\Delta t)$ [26] and computer simulation results.

is *not predictive*. However, the fact that it the theory generates a probability distribution that agrees so well with numerical simulations is indicative of the essential correctness of such a description [26] of a fluctuating "pulled" front. The self-consistent theoretical curves of $P(\Delta t) = \int_{\Delta t}^{\infty} dt' \mathscr{P}(t')$ for $D = 1$ and $N = 10^4, 10^2, 10^3$ and 10^5 (in that order) are shown in Fig. 6. The corresponding numerical comparison of front speeds are shown in Table I. First, we notice that in the graphs of Fig. 6, the theoretical curves lie below the simulation histograms at $\Delta t \approx 2/v_N$ — this is an artifact of the matching that we had to do for the expressions of $P(\Delta t)$ below and above $\Delta t \sim 2/v_N$. This difference occurs due to certain fluctuation and correlation effects [26]. Secondly, the agreement between the simulation data and the self-consistent theory is not very good for $N = 10^5$ — at this value of N, the simulation gets very slow and one has to continuously remove particles from the saturation region of the front to gain program speed, which affects the $P(\Delta t)$ histograms significantly for large times.

N	v_N(simulation)	v_N(theoretical)	v_N[Eq. (7)]
10^2	1.778	1.808	1.465
10^3	1.901	1.899	1.803
10^4	1.964	1.988	1.925
10^5	2.001	2.057	1.976

Table I: Comparison of v_N values, simulation, self consistent theory [26] [indicated by v_N(theoretical)], and that of Eq. (7).

In addition to such good agreements between our self-consistent theory and simulations for the $P(\Delta t)$ curves, a significant observation is the following: as the value of N is increased in our self-consistent theory, we find that the quantity $\delta\phi_k/\phi_k^{(0)}$ values at the tip of the front gradually reduces [26]. The stochastic halt-and-go character of the movement of the f.o.l.s., which is usually occupied by $\mathscr{O}(1)$ number of particles, however, continues to remain valid for any value of N. This implies that for very large N, one approaches the picture of a fluctuating "pulled" front, where a uniformly translating mean-field description (5) holds all the way up to one lattice site behind the f.o.l.s., while only the dynamics of the f.o.l.s. is a stochastic halt-and-go process. Such a simplified picture has been studied numerically in the last paper of Ref. [14].

We finally end this section with a short note arguing why the expression (7) is correct for asymptotically large N. In fact, it can be understood very simply when one observes that the length of the "linear region" in Fig. 5 increases as $\ln N$, and combines it with the expectation that the asymptotic front speed should be $< v^*$ (as motivated in the origin of the terminology "pulled" front). For a deterministic pulled front equation, such as Eqs. (1) or (5), the length of the linear region is infinite, and in such cases, in the absence of any length scale, the stability criterion of the front solution in the comoving frame dictates that for the selected front speed, it is good enough to consider exponentially decaying solution $\phi(\xi) \sim \exp[-\lambda\xi]$ with $\lambda \equiv \text{Re}(\lambda)$. On the other hand, for "pulled fronts" consisted of discrete particles on a lattice, the positivity of $\phi(\xi)$ demands that one has to consider an oscillatory function of wavelength $\pi/\ln N$ as a multiplicative factor to the exponentially decaying $\exp[-\lambda\xi]$ profile for the front solution $\phi(\xi)$ to obtain an asymptotic speed $< v^*$. Just from this argument alone, it is possible to derive the asymptotic front speed of Eq. (7).

3. OUTLOOK AND UNSOLVED PROBLEMS

When an equation that allows so-called pulled fronts in the mean-field limit is modelled with a stochastic model with a finite number N of particles per correlation volume, the convergence to the speed v^* for asymptotically large N, as obtained by Brunet and Derrida, behaves as $\ln^{-2} N$, and this behaviour is model independent. However, for large but more reasonable values of N, there are significant deviations from their result, and these deviations stem from the complicated stochastic halt-and-go dynamics of the foremost occupied lattice site, where the actual microscopic rules of the system under consideration play a crucial role, and therefore the deviations from the results of Brunet and Derrida are model dependent.

The message of this paper is as follows: to obtain the deviations from the $1/\ln^2 N$ convergence to v^* one really has to focus at the stochastic halt-and-go movement of the foremost occupied lattice site. From the mean-field limit of this fluctuating front, we know that the tip region is very important for its dynamics; as a result, the fluctuating tip plays a very significant role in deciding the asymptotic front speed, in which two very important aspects come to play a role — discrete nature of particles and discrete nature of the lattice indices. In this paper, we have outlined the formulation of a self-consistent theory, which we developed in Ref. [26], to model the fluctuating tip by means of a mean-field type description. The mean-field type description of the fluctuating tip is then matched to a uniformly translating solution behind. Due to the presence of an effective parameter in this self-consistent theory, it is *not predictive*. How to obtain a predictive theory for moderately large values of N, and demonstrate analytically how the corresponding front speed approaches to the expression (7) still remain unsolved problems. However, one has to remember that in actuality, the tip is strongly fluctuating and there are time-correlation effects involved (for a more detailed discussion, see Sec. IV.C of Ref. [26]). Any alternative predictive theory, that one might think at this point, must be able to take these into account, in addition to the mean-field type self-consistent theory of Ref. [26].

The prospect of such a theory however, looks grim at this point. Not only the problem becomes highly nonlinear, but also one must realize that the fluctuations in the number of X particles on the lattice sites at the tip of the front is of the same order as the number of X particles in them (~ 1), and there does not exist any small parameter that one can do perturbation theory with.

REFERENCES

1. E. Meron, Phys. Rep. **218**, 1 (1992).
2. G. Zocchi, P. Tabeling and M. Ben Amar, Phys. Rev. Lett. **69**, 601 (1992).
3. Y. Tu and M. C. Cross, Phys. Rev. E **69**, 2515 (1992).
4. U. Ebert, W. van Saarloos and C. Caroli, Phys. Rev. E **55**, 1530 (1997).
5. M. C. Cross and P. C. Hohenberg, Rev. Mod. Phys. **65**, 851 (1993).
6. E. Ben-Jacob, H. R. Brand, G. Dee, L. Kramer, and J. S. Langer, Physica D **14**, 348 (1985).
7. W. van Saarloos, Phys. Rev. A **39**, 6367 (1989).
8. U. Ebert and W. van Saarloos, Physica D **146**, 1 (2000).
9. R. A. Fisher, Ann. Eugenics **7**, 355 (1937).
10. M. Bramson, Mem. Am. Math. Soc. **44**, No. 285 (1983).
11. H. P. Breuer, W. Huber, and F. Petruccione, Physica D **73**, 259 (1994).
12. A. Lemarchand, A. Lesne, and M. Mareschal, Phys. Rev. E **51**, 4457 (1995).
13. J. Armero, J. M. Sancho, J. Casademunt, A. M. Lacasta, L. Ramírez-Piscina, F. Sagués, Phys. Rev. Lett. **76**, 3045 (1996).
14. E. Brunet and B. Derrida, Phys. Rev. E **56**, 2597 (1997). E. Brunet and B. Derrida, Comp. Phys. Comm. **122**, 376 (1999). E. Brunet and B. Derrida, J. Stat. Phys., **103**, 269 (2001).
15. D. A. Kessler, Z. Ner, and L. M. Sander, Phys. Rev. E. **58**, 107 (1998).
16. L. Pechenik and H. Levine, Phys. Rev. E **59**, 3893 (1999).
17. J. Armero, J. Casademunt, L. Ramírez-Piscina, and J.M. Sancho, Phys. Rev. E **58**, 5494 (1998).
18. J. Riordan, C. R. Doering, and D. ben-Avraham, Phys. Rev. Lett. **75**, 565 (1995).
19. G. Dee and J. S. Langer, Phys. Rev. Lett. **50**, 383 (1983).
20. R. van Zon, H. van Beijeren and Ch. Dellago, Phys. Rev. Lett. **80**, 2035 (1998).
21. D. Panja and W. van Saarloos, Phys. Rev. E **66**, 015206(R) (2002).
22. A. Rocco, U. Ebert, W. van Saarloos, Phys. Rev. E. **62**, R13 (2000).
23. G. Tripathy and W. van Saarloos, Phys. Rev. Lett. **85**, 3556 (2000).
24. G. Tripathy, A. Rocco, J. Casademunt, and W. van Saarloos, Phys. Rev. Lett. **86**, 5215 (2001).
25. E. Moro, Phys. Rev. Lett. **87**, 238303 (2001).
26. D. Panja and W. van Saarloos, Phys. Rev. E **66**, 036206 (2002).
27. D. Panja and W. van Saarloos; Phys. Rev. E **65**, 057202 (2002).
28. C. R. Doering, C. Mueller, and P. Smereka, *"Noisy Wavefront Propagation in the Fisher-Kolmogorov-Petrovsky-Piscunov Equation"*, UPoN 2002 (this volume); *"Interacting Particles, the Stochastic Fisher-Kolmogorov-Petrovsky-Piscunov Equation, and Duality"*, Physica A, submitted (2002).

GENERAL THEORY AND EXPERIMENT

Broomhead-King Hydrophobic Modes in Receptor-Targeted Peptide Design: How to Find Modes in a Short Data Sequence

Arnold J. Mandell*&, Karen A. Selz*&, Michael J.Owens&, and Michael F. Shlesinger#

*Cielo Institute, 486 Sunset Dr., Asheville NC 28804

&Dept. of Psychiatry and Behavioral Sciences and the Neuroscience Curricular Group, Emory University School of Medicine, Atlanta GA 30322

#Physical Sciences Division, Office of Naval Research, 800 N. Quincy St. Arlington VA 22217

Abstract. Broomhead-King modes are a lagged vector version of the Karhunen-Loeve orthogonal mode decomposition. We employ the Broomhead-King method to seek modes in short but evolutionarily stationary, protein and peptide, hydrophobically transformed, amino acid data sequences. We then use these modes to design 12-20 mer peptides whose amino acid series possess hydrophobic modes that match the leading modes of the targeted G-protein coupled, seven transmembrane, receptor proteins and act as indirect agonists and/or modulators of their *in vitro* and *in vivo* physiological activity. An example is briefly described involving peptide design targeting the human brain D_2 dopamine receptor, thought to be involved in Parkinson's disease and the action of some antipsychotic drugs.

INTRODUCTION

A pure random signal has a flat "white noise" power spectrum up to a cut-off frequency. Below the cut-off, all frequency modes have the same weight and no characteristic mode appears. For a signal with long range correlations and self-similar scaling, a 1/f frequency power spectrum appears, but again no characteristic frequency can be identified. A noisy oscillator can have a power spectrum with modal peaks rising above a noise floor, but this works best for long series of stationary data. What methods are available for analyzing short data series? The answer will depend on several factors and the information being sought. If one is looking for a particular frequency hidden in noise, perhaps a linear filter will be a good choice. If one is searching for chaos, perhaps discretizing the flow via a Poincare′ section followed by a return map works best. A wavelet transform is useful in locating sequence positions that are lost in spectral information. All-poles power spectra may work well for selecting leading modal peaks in short data sequences. We are interested mainly in sequences of data points that are biologically stable with respect to evolutionary time scales, even though their relatively short length would otherwise raise issues of transience and nonstationarity. These data points represent sequences of amino acid hydrophobic values in receptor proteins of 400-500 residue lengths and peptides up to

CP665, *Unsolved Problems of Noise and Fluctuations: UPoN 2002: Third International Conference*, edited by S. M. Bezrukov

30 amino acids long. We seek modal patterns that we can use to characterize membrane receptor protein data sequences [1,2]. These, in turn, can serve to direct the design of peptides with the potential for hydrophobic mode matched aggregation with them inducing changes in receptor-mediated physiological activity [3]. Our method will involve a decomposition of the data sequence into modes derived directly from the data.

ORTHOGONAL MODE DECOMPOSITION

One can always represent a function in a Fourier expansion, but for some data, many Fourier terms will be needed to represent it faithfully. An ingenious method exists, due separately to Karhunen [4] and Loeve [5], to decompose a data series into orthogonal modes. The sum of these modes converges rapidly to the function. The modes are not fixed as in sines, cosines, Bessel functions, or other known orthogonal functions, but are generated directly from the data. Different data sets will produce different Karhunen-Loeve modes.

Given that there are N data points x_i, ($i = 1, \cdots N$), we will assume that we can decompose the data series into N orthogonal functions ψ_n

$$x_i = \sum_{n=1}^{N} \lambda_n \psi_n(i) \tag{1}$$

We will assume that the ψ_n functions are orthogonal, i.e.,

$$\sum_{i=1}^{N} \psi_n(i)\psi_m(i) = \delta_{nm} \tag{2}$$

and then will show that they satisfy a certain eigenvalue problem. We could also provide the opposite proof, that if the ψ_n satisfy the eigenvalue problem then they constitute an orthogonal mode decomposition of the data. We want the ψ_n to appear as eigenvectors and the λ_n to be related eigenvalues. To accomplish this we form A_{ij} the autocorrelation matrix of the data,

$$A_{ij} = \left\langle x_i x_j \right\rangle = \sum_{n=1}^{N} \sum_{m=1}^{N} \lambda_n \lambda_m \left\langle \psi_n(i)\psi_m(j) \right\rangle \tag{3}$$

where the average is over an ensemble of data, and we assume that the average $\left\langle \psi_n(i)\psi_m(j) \right\rangle = \delta_{nm}$, giving

$$A_{ij} = \sum_{n=1}^{N} \lambda^2{}_n \psi_n(i)\psi_m(j) \tag{4}$$

Treating A as a matrix operator we will show that the ψ_n 's are its eigenvectors and the $\lambda^2{}_n$'s its eigenvalues. We perform the matrix multiplication,

$$\begin{aligned}\sum_{j=1}^{N} A_{ij}\psi_m(j) &= \sum_{j=1}^{N}\sum_{n=1}^{N}\lambda^2{}_n\psi_n(i)\psi_n(j)\psi_m(j)\\ &= \sum_{n=1}^{N}\lambda^2{}_n\psi_n(i)\delta_{nm} \\ &= \lambda^2{}_m\psi_m(i)\end{aligned} \tag{5}$$

If the data were considered to be continuous, the above equation would become an integral equation,

$$\int_0^\infty A(x,y)\psi_m(y)dy = \lambda_m^2\psi_m(x) \tag{6}$$

Both cases can be represented in matrix notation as,

$$\underline{\underline{A}}\,\overline{\psi} = \lambda^2\,\overline{\psi} \tag{7}$$

Generally, we form the autocorrelation ($x_i x_j$) matrix of the data and calculate the eigenvectors. These eigenvectors form an orthogonal mode decomposition of the data weighted in importance by their eigenvalues. We have used this method to explicate hydrophobic mode matches between shorter known neuropeptides and their longer membrane protein receptors based on matches in their modal decomposition [1,2]. We also use the calculated modes to design hydrophobic mode matched novel peptides to aggregate with and physiologically activate and/or modulate the indigenous activity of specific receptors [3].

BROOMHEAD-KING DECOMPOSITION

For N data points, determining the N orthogonal Karhuenen-Loeve mode eigenvectors involves inverting the NXN matrix in eqn (7). In practice, fewer than N modes need to be determined to have the sum in eqn (1) converge to the data. Initially it was thought that the number of dominant modes should somehow reflect the number of degrees of freedom or dimensionality of the signal. Later work suggested that only those dominant modes which looked like normal polynomial functions when their leading eigenvectors were composed with the original series should be counted [6]. The products of these compositions are called hydrophobic eigenfunctions in our studies [1,2]. Modes can be considered as unimportant if their eigenvalues are near zero and well separated from the larger eigenvalues. Broomhead and King [6] suggested embedding data into M-dimensional lag vectors where M << N. One would then only need a new matrix inversion of a small MXM matrix rather than the larger original NXN matrix, and one only need consider these M eigenvectors, rather than the larger set of N eigenvectors. This M-lag vector approach is particularly relevant to the number of leading hydrophobic modes found in protein amino acid sequences which in our studies of over two hundred proteins, these modes have been found to range from ≈ 2.0 to ≈ 15.0 amino acids in length [7], with usually one but at most three distinct eigenmodes for any single protein. The method is described below.

One forms M dimensional vectors from lagged data points vectors from the data $H_1, \cdots, H_N$

$$V_1 = (H_1, H_2, \cdots, H_M),$$
$$V_2 = (H_2, H_3, \cdots, H_{M+1}),$$
$$\cdots,$$
$$V_{N-M+1} = (H_{N-M+1}, H_{N-M+2}, \cdots, H_N).$$

In our case, the data will be the sequence of hydrophobic values of the associated amino acids composing the linear chain protein

Following Broomhead and King, consider the (N-M+1)XM matrix

$$V = \begin{pmatrix} V_1^T \\ V_2^T \\ \vdots \\ V_{N-M+1}^T \end{pmatrix} = \begin{pmatrix} H_1 & H_2 & \cdots & H_M \\ H_2 & H_3 & \cdots & H_{M+1} \\ \vdots & & & \vdots \\ H_{N-M+!} & H_{N-M+2} & \cdots & H_N \end{pmatrix} \tag{8}$$

We can now form $C=V^TV$ which is a (MX(N-M+1)) X ((N-M+1)XM) = MXM matrix.

$$C = \frac{1}{K}\begin{pmatrix} \sum_{i=1}^{K} H_i H_i & \cdots & \sum_{i=1}^{K} H_i H_{i+K-1} \\ & \vdots & \\ \sum_{i=1}^{K} H_{i+K-1} H_i & \cdots & \sum_{i=1}^{K} H_{i+K-1} H_{i+K-1} \end{pmatrix} \tag{9}$$

where K = N-M+1. This matrix has M eigenvectors $\psi_i(j)$ where both i and j = 1, …M, and M << N. We will use these M eigenvectors to capture the dominant spectral mode behavior in a sequence, even when the sequence is short, on the order of M=15.

ANALYSIS OF TARGET MEMBRANE RECEPTOR PROTEINS

The data series in which we are interested is generated from membrane receptor proteins. A protein is composed of a single, unbranched chain of amino acids. In these studies, each amino acid has a measure of hydrophobicity derived from its relative miscibility in a binary organic solvent/water partition in equilibrium at a standard temperature and pressure. These values as "free energies of transfer" from solvent to water can be treated like chemical potentials in kcal/mol that are normalized with respect to the amino acid glycine, G, which is arbitrarily set to zero [8]. These real numbers reflect the relative attraction of the hydrogen bonded water molecules of the aqueous lattice for themselves relative to each of the amino acids [9]. For a protein composed of N amino acids we have the series $H_1, \cdots, H_N$ representing the sequence of hydrophobicity values of the protein.

Peptides are short proteins (typically 10-30 amino acids long) and act through hydrophobic and polar interactions with specific receptors, triggering second messenger cascades that then activate a variety of physiological processes. We have an interest in transmembrane receptors that snake into and out of the cell membrane seven times, called G-protein coupled, 7 transmembrane receptors, GPCRs [10]. Over 50% of all of the drugs currently on the market are thought to act through GPCRs. As noted above, we have found that the all poles power spectra of the hydrophobically transformed peptides and those of the Broomhead-King data matrix-derived,

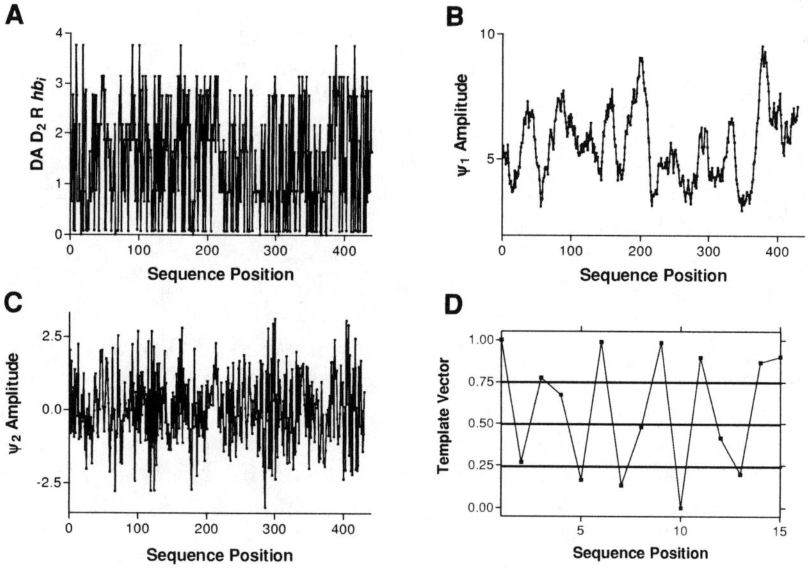

FIGURE 1 graphically summarizes the steps in the derivation of the eigenvector template used in human long form, D_2 dopamine receptor-targeted peptide design. The **Figure 1A** is a graph of the series of hydrophobic values (hb_i) of the amino acid sequence of the receptor. The **Figure 1B** is a plot of the results of the composition of the leading Broomhead-King eigenvector with the original hb_i series generating the leading hydrophobic eigenfunction which is labeled ψ_1. The $\approx$ 7 maxima correspond to the 7 highly hydrophobic transmembrane segments of the dopamine GPCR [12]. The **Figure 1C** is a graph of the second Brooomhead-King eigenfunction labeled ψ_2, (the second Broomhead-King eigenvector composed with the hb_i data series). An all poles power spectral transformation of ψ_2 would reveal two dopamine receptor hydrophobic variational frequencies that as wavenumbers are $\approx$ 8.12 and 2.61 amino acids in length [3]. The **Figure 1D** is a graph of the normalized second leading Broomhead-King eigenvector used to construct ψ_2 and which, when 4-partitioned, served as the template for random amino acid assignment by group as weighted by the distribution of that group's amino acids in human spinal fluid available for indigenous peptide synthesis [13].

autocovariance matrix's leading "eigenfunctions" (leading eigenvectors composed with the receptor's original hydrophobic series as noted above) have dominant peaks located at similar frequencies [1].

RECEPTOR-TARGETED NEW PEPTIDE DESIGN AND TESTING

The M-component Broomhead-King eigenvectors were designed to capture the essence of a signal in a much shorter length than the Karhunen-Loeve modes. Instead of, say a 500-component eigenvector (N=500) to represent a 500 amino acid length protein receptor, for the reasons given above, we use the much smaller (M=15) 15-component eigenvector. Plotting the 15 components against their sequence position, we arrive at a new series with 15 points. We then pick a new amino acid series such that the variations in sequential hydrophobic values match those in the eigenvector series. We accomplish this by dividing the eigenvector series of values into 4 levels of hydrophobicity which we match to Tanford's 20 amino acid hydrophobicity scale which has a natural 4-partition, 5 amino acids each. In this manner, we algorithmically design novel, membrane receptor-targeted peptides [3]. We briefly describe the design and testing of a new 15 amino acid peptide targeting the human D_2 dopamine receptor. Recent studies have shown that peptides of this length can cross the blood-brain barrier and manifest physiological activity [11], thus this process constitutes an entirely new approach to the rational design of brain peptide drugs. Our success rate using this approach has approximated 50%, which is considerably better than the 5 per 100,000 candidates yielded by more conventional high throughput screening of randomly varied candidate agents [3].

When ERNR... was infused in one microgram amounts in the dopamine cell body cluster in the nucleus accumbens, NA, in awake behaving rats via bilaterally implanted canulae, a typical dose-responsive dopaminergic behavioral activation, hyperactivity, was observed that resembled that seen following the administration of amphetamine or amphetamine-like stimulants [3]. ERNR...'s influence on more subtle brain dopamine cell dynamics was also reflected in its ability, when infused in the NA, to reverse the amphetamine-induced blockade of the rat's habituation to repeated administration of loud sounds called prepulse inhibition [15,16].

Broomhead-King hydrophobic modes may come to play important roles in the analyses of protein-protein interactions and the rational design of membrane-receptor targeted peptide drugs.

ACKNOWLEDGEMENTS

AJM and KAS are supported by grants from the National Institute of Mental Health, MH58026, and the National Cancer Institute, CA91384.

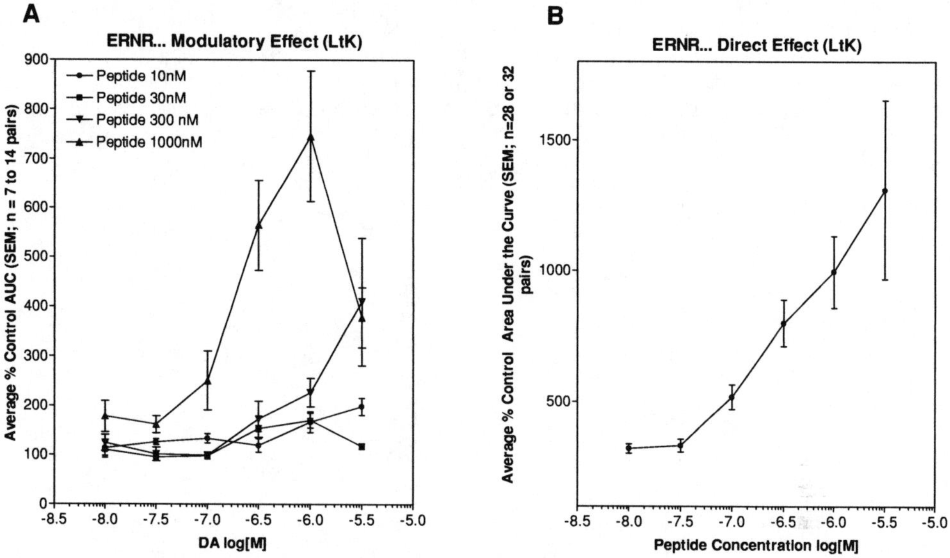

FIGURE 2A summarizes the effect of increasing levels of the eigenvector template designed peptide, H-ERNRKPYRPKNKYLL-0H, ERNR..., preceding the infusion of increasing levels of dopamine in a human long form D_2 dopamine receptor, stably transfected, LtK fibroblastic cell system. The receptor-mediated increases in cell metabolic activity reflected in the integrated (area under the curve, AUC), extracellular acidification (proton production) rates were monitored by Cytosensor® microphysiometry [14]. A peptide induced, dose-sensitive increment of the system's response to dopamine is evidenced. **FIGURE 2B** is a graph of a dose-response curve of the perfusion of peptide alone in this cell system using microphysiometry that demonstrates a dose dependent, indirect agonist, "direct" receptor-mediated, cellular activation effect in addition to the positive modulatory influence on the dopamine induced activation seen in Figure 2A

REFERENCES

1. Mandell, A. J., K. A. Selz and M. F. Shlesinger, *Proc. Natl. Acad. Sci. (USA)* **94** 13576-13581 (1997).
2. Mandell, A.J., Selz, K.A. and Shlesinger, M.F, *J. Physical Chem. B.* **104**, 3953-3959, (2000).
3. Mandell, A.J., Selz, K.A., Owens, M.J., Shlesinger, M.F., Kinkead, B., Gutman, D.A. and Arguragi, V. *Neuropsychopharm.* **26** (2002) In press.
4. Karhunen, K., Ann. Acad. Sci. Fennicae **37** (1946).
5. Loeve, M.M. *Probability Theory*, Princeton, Van Nostrand , NJ, (1955).
6. Broomhead D, and D. S. King, *J. Phys. A*, **20** L563-L65(1987).
7. Mandell, A.J., Selz, K.A., Owens, M., Shlesinger, M.F., Gutman, D.A. and Arguragi, V. In (ed. Raffa, R.) *Drug Receptor Thermodynamics: Introduction and Applications.* Wiley, NY. pp. 655-700 (2001).
8. Manavalen, P. and Ponnswamy, P.K. *Nature* **275** 273-274 (1978)
9. Tanford, C. *The Hydrophobic Effect* Wiley, NY.
10. Henderson, R., Baldwin, J.M., Ceska, T.A., Semlin, F., Beckmann and Downing, K.H. *J. Mol. Biol.* **213** 899-929 (1990)
11. Lu, A.G., Otero, D.A., Hiraiwa,M., O'Brien, J. *Neuroreport.* **11** 1791-1794 (2000).

12. Selbie, L.A., Hayes, G. and Shine, J. *DNA* **8** 683-689 (1989)
13. Perry, T.L., Hansen, S. and Kennedy, J. *J. Neurochem.* **24** 587-599.
14. Neve, K.A., Kozlowski, M.R. and Rosser, M.P. *J. Biol. Chem.* **267** 25748-25753 (1992)
15. Caine, S.B., Geyer, M.A. and Swerdlow, N.R. *Neuropsychopharm.* **12** 139-145 (1995)
16. Binder, E.B., Kinkead, B., Owens, M.J., Kilts, C.D. and Nemeroff, C.B. *J. Neurosci.* **21** 601-608 (2001).

System Size Stochastic and Coherence Resonance

A. Pikovsky* and A. Zaikin*

*Department of Physics, University of Potsdam, Postfach 601553, D-14415 Potsdam, Germany

Abstract. We demonstrate that in ensembles of noise-driven dynamical systems a maximum order can appear at a certain system size, while the noise and coupling remain constant. We discuss two situations with system size resonance. In the first one an ensemble of coupled nonlinear noisy oscillators demonstrates in the thermodynamic limit an Ising-type transition. When a small periodic force acts on the ensemble in the ordered phase, the linear response of the system has a maximum at a certain system size. Next, we show that in an ensemble of coupled noise-driven FitzHigh-Nagumo oscillators a maximum coherence of the output mean field occurs at a definite ensemble size.

INTRODUCTION

Constructive role of noise in nonlinear systems has been attracted large interest recently. One extremely popular example is ***Stochastic Resonance*** [1, 2]. As was demonstrated in [3], a response of a noisy nonlinear system to a periodic forcing can exhibit a resonance-like dependence on the noise intensity. In other words, there exists a "resonant" noise intensity at which the response to a periodic force is maximally ordered. Stochastic resonance has been observed in numerous experiments.

Being first discussed in the context of a simple bistable model, stochastic resonance has been also studied in complex systems consisting of many elementary bistable cells [4]. Again, one observes a resonance-like dependence on the noise intensity, moreover, the resonance may be enhanced due to coupling [5, 6]. Here in the next section we discuss another type of resonance in such systems, namely the ***System Size Stochastic Resonance***, when the dynamics is maximally ordered at a certain number of interacting subsystems [7]. Contrary to previous reports of array-enhanced stochastic resonance phenomena (cf. also [8, 9]), here we fix the noise strength, coupling, and other parameters; only the size of the ensemble changes.

Noteworthy, the order in a noise-driven system can have a maximum at a certain noise level even in the absence of periodic forcing, this phenomenon being called ***Coherence Resonance*** [10, 11, 12]. Below we demonstrate, taking an ensemble of coupled noise-driven FitzHugh–Nagumo systems, that it also can appear in the form of ***System Size Coherence Resonance***.

CP665, *Unsolved Problems of Noise and Fluctuations: UPoN 2002: Third International Conference*, edited by S. M. Bezrukov

SYSTEM SIZE STOCHASTIC RESONANCE

The basic model to be considered below is the ensemble of noise-driven bistable overdamped oscillators, governed by the Langevin equations

$$\dot{x}_i = x_i - x_i^3 + \frac{g}{N}\sum_{j=1}^{N}(x_j - x_i) + \sqrt{2D}\xi_i(t) + f(t)\,. \tag{1}$$

Here $\xi_i(t)$ is a Gaussian white noise with zero mean: $\langle \xi_i(t)\xi_j(t')\rangle = \delta_{ij}\delta(t-t')$; g is the coupling constant; N is the number of elements in the ensemble, and $f(t)$ is a periodic force to be specified later. In the absence of periodic force the model (1) has been extensively studied in the thermodynamic limit $N \to \infty$. It demonstrates an Ising-type phase transition at $g = g_c$ from the disordered state with vanishing mean field $X = N^{-1}\sum_i x_i$ to the "ferromagnetic" state with a nonzero mean field $X = \pm X_0$. A theory of this transition, based on the nonlinear Fokker-Planck equation, was developed in [13] where also the expressions for the critical coupling g_c are given.

While in the thermodynamic limit the full description of the dynamics is possible, for finite system sizes we have mainly a qualitative picture: in the ordered phase the mean field X switches between the values $\pm X_0$ and its average vanishes for all couplings. The rate of switchings depends on the system size and tends to zero as $N \to \infty$. The asymptotic dynamics in this limit has been discussed in [14].

For us, of the main importance is the fact that qualitatively the behavior of the mean field can be represented as the noise-induced dynamics in a potential with one minimum in the disordered phase (at $X = 0$) and two symmetric minima (at $X = \pm X_0$) in the ordered phase. Now applying the ideas of the stochastic resonance, one can expect in the bistable case (i.e. in the ordered phase for small enough noise or for large enough coupling) a resonant-like behavior of the response to a periodic external force when the intensity of the effective noise is changed. Because this intensity is inverse proportional to N, we obtain the resonance-like curve of the response in dependence of the system size. The main idea behind the system size resonance is that in finite ensembles of noise-driven or chaotic systems the dynamics of the mean field can be represented as driven by the effective noise whose variance is inverse proportional to the system size [14, 15, 16]. This idea has been applied to description of a transition to collective behavior in [17]. In [18] it was demonstrated that the finite-size fluctuations can cause a transition that disappears in the thermodynamic limit. The description of finite-size effects in deterministic chaotic systems using the effective noise concept has been suggested in [15]. We emphasize that noise plays an essential role in this picture: with $D = 0$ (1) is a deterministic oscillator (double or single well, depending on g), whose response to a periodic force does not depend on N.

Before proceeding to a quantitative analytic description of the phenomenon, we illustrate it with direct numerical simulations of the model (1), with a sinusoidal forcing term $f(t) = A\cos(\Omega t)$. Figure 1 shows the linear response function, i.e. the ratio of the spectral component in the mean field at frequency Ω and the amplitude of forcing A, in the limit $A \to 0$. For a given frequency Ω the dependence on the system size is a bell-shaped curve, with a pronounced maximum. The dynamics of the mean field $X(t)$ is illustrated in Fig. 2, for three different system sizes and for a particular frequency. The resonant dynamics (Fig. 2b) demonstrates a typical for stochastic resonance synchrony between the driving periodic force and the switchings of the field between the two stable positions. For non-resonant conditions (Fig. 2a,c) the switchings are either too frequent or too rare, as a result the response is small.

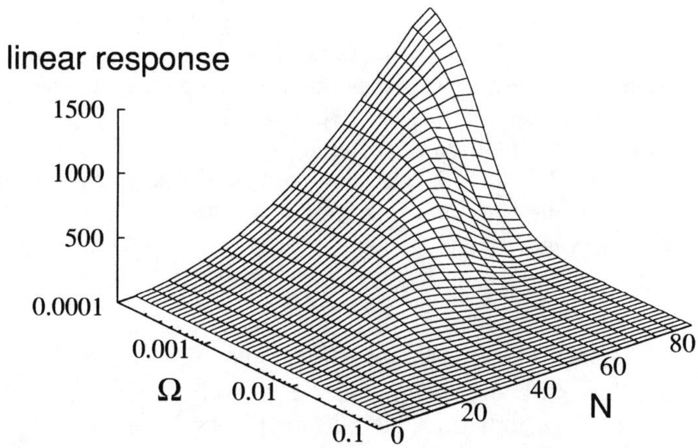

FIGURE 1. Linear response of the ensemble (1) ($D = 0.5$, $\varepsilon = 2$) in dependence on the frequency Ω and the system size N. Reprinted from [7].

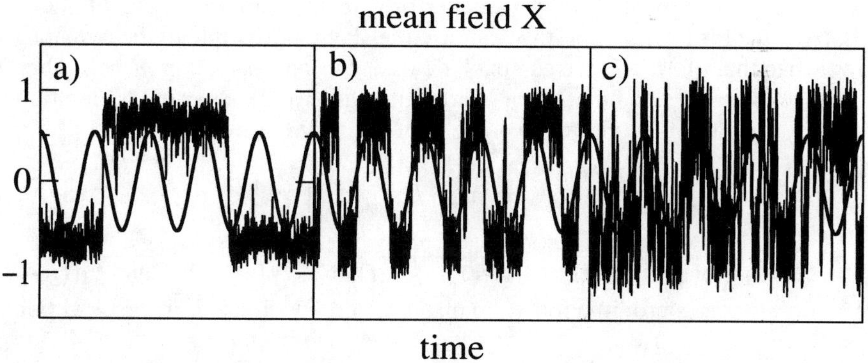

FIGURE 2. The time dependence of the mean field in the ensemble (1) for $D = 0.5$, $g = 2$, $A = 0.02$, $\Omega = \pi/300$, and different sizes of the ensemble: (a) $N = 80$, (b) $N = 35$, and (c) $N = 15$. We also depict the periodic force (its amplitude is not in scale) to demonstrate the synchrony of the switchings with the forcing in (b).

To describe the system size resonance analytically, we use, following [13], the Gaussian approximation. In this approximation one writes $x_i = X + \delta_i$ and assumes that δ_i are independent Gaussian random variables with zero mean and the variance M. Assuming furthermore that $N^{-1}\sum_i \delta_i^2 = M$ and neglecting the odd moments $N^{-1}\sum_i \delta_i$, $N^{-1}\sum_i \delta_i^3$, as well as the correlations between δ_i and δ_j, we obtain from (1) the equations for X and M:

$$\dot{X} = X - X^3 - 3MX + \sqrt{\frac{2D}{N}}\eta(t) + f(t), \quad (2)$$

$$\frac{1}{2}\dot{M} = M - 3X^2M - 3M^2 - gM + D, \quad (3)$$

where η is the Gaussian white noise having the same properties as $\xi_i(t)$. In the thermodynamic limit $N \to \infty$ the noisy term η vanishes. If the forcing term is absent ($f = 0$), the equations coincide with those derived in [13]. This system of coupled nonlinear equations exhibits a pitchfork bifurcation of the equilibrium $X = 0$, $M > 0$ at $g_c = 3D$. This bifurcation is supercritical for $D > 2/3$ in accordance with the exact solution of (1) given in [13], below we consider only this case. For $g > g_c$ the system is bistable with two symmetric stable fixed points

$$X_0^2 = (2 - g + S)/4, \quad M_0 = (2 + g - S)/12, \quad (4)$$

(here $S = \sqrt{(2+g)^2 - 24D}$) and the unstable point $X = 0$, $M = (1 - g + \sqrt{(1-g)^2 + 12D})/6$. Now, with the external noise η and with the periodic force $f(t)$ the problem reduces to a standard problem in the theory of stochastic resonance, i.e. to the problem of the response of a noise-driven nonlinear bistable system to an external periodic force (because the noise affects only the variable X, it does not lead to unphysical negative values of variance M, since $\dot{M}$ is strictly positive at $M = 0$). This response has a maximum at a certain noise intensity, which according to (2) is directly related to the system size.

To obtain an analytical formula, we perform further simplification of the system (2),(3). Near the bifurcation point we can use the slaving principle to the dynamics of X is slower than that of M, and we can exclude the latter one assuming $\dot{M} \approx 0$. Then from (3) we can express M as a function of X and substitute to (2). Near the bifurcation point we obtain a standard noise-driven bistable system

$$\dot{X} = aX - bX^3 + \sqrt{\frac{2D}{N}}\eta(t) + f(t), \quad (5)$$

where $a = 1 + 0.5(g-1) - 0.5\sqrt{(g-1)^2 + 12D}$, $b = -0.5 + 1.5(g-1)((g-1)^2 + 12D)^{-1/2}$. A better approximation valid also beyond a vicinity of the critical point can be constructed if we use $\bar{b} = aX_0^{-2}$ instead of b, where the fixed point X_0 is given by (4). Having written the ensemble dynamics as a standard noise-driven double-well system (5) (cf. [1, 2, 19]), we can use the analytic formula for the linear response R derived in [19]. It reads

$$R = \frac{NX_0^2}{2Da}\left(\frac{\mathscr{D}_{-3/2}(-\sqrt{s})}{\mathscr{D}_{-1/2}(-\sqrt{s})}\right)^2 \left[1 + \frac{\pi^2\Omega^2}{2a^2}\exp(s)\right]^{-1} \quad (6)$$

where $s = aNX_0^2/(2D)$, and $\mathscr{D}$ are the parabolic cylinder functions. We compare the theoretical linear response function with the numerically obtained one in Fig. 3. The qualitative correspondence is good, moreover, the maxima of the curves are rather good reproduced with the formula (6). This shows that the resonant system size is quite good quantitatively described by the Gaussian approximation, see Fig. 3.

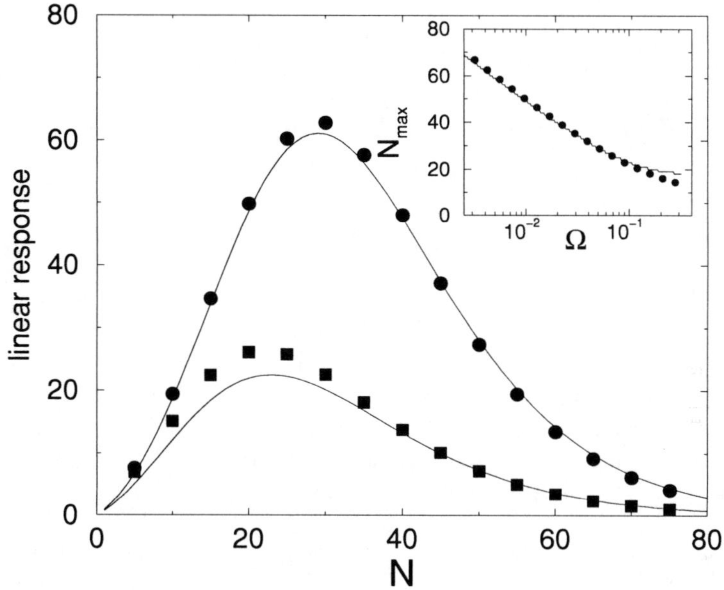

FIGURE 3. Comparison of the system-size dependencies of the linear response function for frequencies $\Omega = 0.05$ (circles) and $\Omega = 0.1$ (squares) with theory (6). The parameters are $D = 1$ and $g - g_c = 2.5$ (where the the exact g_c and the approximate $g_c = 3D$ are used for the ensemble and the Gaussian approximation, respectively). Inlet: Dependence of the system size yielding maximal linear response on the driving frequency Ω (circles: simulations of the ensemble (1), line is obtained by maximizing the expression (6)). Reprinted from [7].

As the last example of the system size resonance we consider a lattice where each individual element does not exhibit bistable noisy dynamics, but such a behavior appears due to interaction and multiplicative noise. This model is described by the set of Langevin equations [20, 21, 22]

$$\dot{x}_i = -x_i(1+x_i^2)^2 + \frac{g}{K}\sum_j (x_j - x_i) + \sqrt{2D}\xi_i(t)(1+x_i^2) + f(t)\,. \tag{7}$$

The difference to the model (1) is that the noise is multiplicative and the on-site potential has only one minimum. K is the number of elements to which the oscillator i is coupled, for global coupling $K = N$ and for a lattice of dimension d with nearest-neighbors coupling $K = 2d$. As has been demonstrated in [20, 21], in some region of couplings g system (7) exhibits the Ising-type transition. If an additional additive noise is added to (7), then one observes transitions between these states and the so-called double stochastic resonance in the presence of the periodic forcing [23, 24]. As is evident from the considerations above, such transitions occur even in the absence of the additive noise if the system is finite. Thus, the system size resonance should be observed in the lattice (7) as well. This is confirmed by numerical calculations presented in [7]. Note that in [24] a realistic electronic circuit modeling the ensemble similar to (7) is described, providing a possible experimental realization of the effect.

SYSTEM SIZE COHERENCE RESONANCE

Here we consider a one-dimensional lattice of N diffusively coupled FitzHugh–Nagumo systems

$$\varepsilon \frac{dx_k}{dt} = x_k - \frac{x_k^3}{3} - y_k + g(x_{k-1} + x_{k+1} - 2x_k)\,, \tag{8}$$

$$\frac{dy_k}{dt} = x_k + a + D\xi_k(t)\,, \qquad 1 \le k \le N\,. \tag{9}$$

We assume periodic boundary conditions and identity of the parameters of the oscillators, only the noise terms ξ_k are different ($\langle \xi_k \rangle = 0$, $\langle \xi_k(t)\xi_m(t')\rangle = \delta_{km}\delta(t-t')$). In [10] it has been numerically and analytically demonstrated, that pulses, generated by noise in a single FitzHugh–Nagumo oscillator, have maximal coherence at a certain noise level. In particular, given the interpulse intervals τ_i the ratio

$$R = \frac{\sqrt{\langle \tau^2 \rangle - \langle \tau \rangle^2}}{\langle \tau \rangle}$$

has been calculated, this quantity measures the closeness of the pulse train to a periodic one (where R obviously vanishes), and as a function of noise intensity D the factor R has a minimum.

Here we investigate the dependence of the factor R on the system size N, keeping the noise and other parameters constant. The results are presented in Fig. 4. The coherence factor R is minimal at $N \approx 10$. We emphasize that we have considered the system (8,9) under very strong coupling and for moderate system length N. This ensured that for considered system sizes (up to 100) all oscillators fired nearly synchronously. Thus the factor R does not depend on which oscillator is chosen for analysis. We expect that the dynamics in the lattice will become much more complex if either the coupling is reduced or the system size is increased, or both (cf. numerical experiments with weakly coupled lattice of FitzHugh–Nagumo oscillators [25]).

CONCLUSIONS

We have demonstrated that in coupled noise-driven oscillators a maximal order can occur at a certain system size. The effect is explained by reduction to the usual phenomena of stochastic and coherence resonances with an effective noise depending on the number of coupled elements. Similar effects have been recently reported for ion channels [26, 27]. One possible field of application of the system size resonance is the

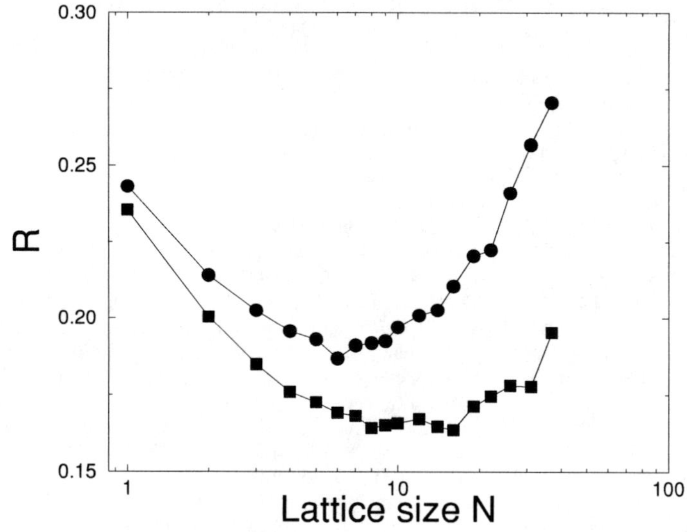

FIGURE 4. Coherence factor R vs. lattice size for the FitzHugh–Nagumo model (8,9) for $\varepsilon = 0.01$, $D = 0.2$, $g = 30$. Circles: $a = 1.05$, squares: $a = 1.03$.

neuronal dynamics. Individual neurons have been demonstrated to exhibit stochastic resonance [28]. While in experiments one can easily adjust noise to achieve the maximal sensitivity to an external signal, it may be not obvious how this adjustment takes place in nature. The above consideration shows, that changing the number of elements in a small ensemble of coupled bistable elements to the optimum can significantly improve the sensitivity. Moreover, changing its connectivity and/or coupling strength, a neuronal system can tune itself to signals with different frequencies.

ACKNOWLEDGMENTS

The work was supported by SFB 555 "Complex Nonlinear Processes".

REFERENCES

1. Gammaitoni, L., Hänggi, P., Jung, P., and Marchesoni, F., *Rev. Mod. Phys.*, **70**, 223–288 (1998).
2. Jung, P., *Phys. Reports*, **234**, 175–295 (1993).
3. Benzi, R., Sutera, A., and Vulpiani, A., *J. Phys. A: Math., Gen.*, **14**, L453–L457 (1981).
4. Jung, P., Behn, U., Pantazelou, E., and Moss, F., *Phys. Rev. A*, **46**, R1709–R1712 (1992).
5. Lindner, J. F., Medows, B. K., Ditto, W. L., Inchiosa, M. E., and Bulsara, A. R., *Phys. Rev. Lett.*, **75**, 3–6 (1995).
6. Lindner, J. F., Medows, B. K., Ditto, W. L., Inchiosa, M. E., and Bulsara, A. R., *Phys. Rev. E*, **53**, 2081–2086 (1996).
7. Pikovsky, A., Zaikin, A., and de la Casa, M. A., *Phys. Rev. Lett.*, **88**, 050601 (2002).

8. Neiman, A., Schimansky-Geier, L., and Moss, F., *Phys. Rev. E*, **56**, R9–R12 (1997).
9. Hu, B., and Zhou, C., *Phys. Rev. E*, **61**, R1001–R1004 (2000).
10. Pikovsky, A., and Kurths, J., *Phys. Rev. Lett.*, **78**, 775–778 (1997).
11. Neiman, A., Saparin, P. I., and Stone, L., *Phys. Rev. E*, **56**, 270–273 (1997).
12. Longtin, A., *Phys. Rev. E*, **55**, 868–876 (1997).
13. Desai, R. C., and Zwanzig, R., *J. Stat. Phys.*, **19**, 1 (1978).
14. Dawson, D., and Gärtner, J., *Stochastics*, **20**, 247–308 (1987).
15. Pikovsky, A. S., and Kurths, J., *Physica D*, **76**, 411–419 (1994).
16. Hamm, A., *Physica D*, **142**, 41–69 (2000).
17. Pikovsky, A., and Ruffo, S., *Phys. Rev. E*, **59**, 1633–1636 (1999).
18. Pikovsky, A. S., Rateitschak, K., and Kurths, J., *Z. Physik B*, **95**, 541–544 (1994).
19. Jung, P., and Hänggi, P., *Phys. Rev. A*, **44**, 8032–8042 (1991).
20. Van der Broeck, C., Parrondo, J. M. R., and Toral, R., *Phys. Rev. Lett.*, **73**, 3395 (1994).
21. Van der Broeck, C., Parrondo, J. M. R., Toral, R., and Kawai, R., *Phys. Rev. E*, **55**, 4084–4094 (1997).
22. Landa, P., Zaikin, A., and Schimansky-Geier, L., *Chaos, Solitons and Fractals*, **9**, 1367 (1998).
23. Zaikin, A., Kurths, J., and Schimansky-Geier, L., *Phys. Rev. Lett.*, **85**, 227 (2000).
24. Zaikin, A., Murali, K., and Kurths, J., *Phys. Rev. E*, **63**, 020103(R) (2001).
25. Sosnovtseva, O. V., Postnov, D. E., and Fomin, A. I., *Applied Nonlinear Dynamics*, **10**, 125–136 (2002).
26. Jung, P., and Shuai, J. W., *Europhys. Lett.*, **56**, 29–35 (2001).
27. Schmid, G., Goychuk, I., and Hänggi, P., *Europhys. Lett.*, **56**, 22–28 (2001).
28. Russell, D. F., Wilkens, L. A., and Moss, F., *Nature*, **402**, 291–294 (1999).

Noise-Induced Walking Patterns on Ratchets

José L. Mateos[§], Alexander Neiman[‡] and Frank Moss[‡]

§ Instituto de Física, Universidad Nacional Autónoma de México, México D. F. 01000 México

‡Center for Neurodynamics, University of Missouri at St. Louis, St. Louis, MO 63121, USA

Abstract. We propose a model for a walker moving on an asymmetric periodic ratchet potential mediated by noise. The walker has two feet that make steps in patterns. These are represented as two finite size particles coupled through a double-well potential. The system contains three noise sources: two independent white noises, modeling thermal noise, and exponentially correlated common noise. In the case where only white noise is present, we perform a bifurcation analysis which reveals different walking patterns available for various parameter settings. Numerical simulations showed the existence of current reversals and significant changes in the effective diffusion constant and in the synchronization index. We obtained an optimal transport, characterized by a maximum dimensionless ratio of the current and effective diffusion (Péclet number), when the periodicity of the ratchet potential coincides with the equilibrium distance between the two feet.

INTRODUCTION

There is an increasing interest in recent years in the study of nonlinear systems that can generate directed transport from unbiased nonequilibrium fluctuations. These, so called ratchet systems, can be modeled, for instance, by considering a Brownian particle in a periodic asymmetric potential and acted upon by external time-dependent forces of zero average. These forces can be deterministic or random in nature. This recent burst of work is motivated in part by the challenge to explain the unidirectional transport of molecular motors in the biological realm. For recent reviews see [1-4].

One particular motor protein, kinesin, has attracted considerable attention, motivated by experimental results in which the dynamical details of its motion can be measured [2,5,6]. Kinesin is a protein with two heads that perform a walking on microtubules inside cells. Motivated by these experimental results, several authors [7-10] have introduced diverse models in order to understand the particular walking of kinesin. Usually, these models consider two coupled particles subject to a ratchet potential that models the periodic asymmetric structure of microtubules. In [7,8], the authors consider two point particles on a ratchet potential and coupled elastically

CP665, *Unsolved Problems of Noise and Fluctuations: UPoN 2002: Third International Conference,* edited by S. M. Bezrukov

(linearly); the rest length between particles alternates in time between two values: 0 and 8 nm. They consider the presence of statistically independent Gaussian white noises acting on each particle. In [9], the authors studied the motion of two point particles coupled elastically moving in a ratchet potential and subject to independent white noises and temporally correlated multiplicative potential fluctuations. A similar model was consider in [10], where again two point particles are coupled linearly in a ratchet potential that alternates randomly in time between two configurations: on and off.

We introduced in this paper a model inspired by the walking of kinesin on microtubules, but that is not restricted to walking of motor proteins. The model can be useful to describe as well the walking of macroscopic objects (robots or animals) in the presence of fluctuations. The model consists of two *extended* particles coupled through a *non-linear* bistable potential and subject to correlated and white noises. The main question that this model tries to answer is the following: is it possible to walk without an external or internal periodic driving (pattern generator) using solely noise sources? Usually, walking involves the action of a pattern generator that introduces a characteristic period of the motion [11,12].

THE MODEL

We consider a walker moving on an asymmetric ratchet potential mediated by noise. The walker has two legs and feet. These are represented as two finite size particles coupled nonlinearly through a bistable potential. The walker moves on an asymmetric ratchet potential under the influence of three noises: two independent white noises acting on the two particles and a common colored noise acting on the system. The stochastic differential equations for the two particles, x and y, in the overdamped regime, are as follows:

$$\dot{x} = f_l(x) + r\,F_b(x-y) + \sqrt{2D_1}\,\xi_1(t) + \eta(t)\,,$$

$$\dot{y} = f_l(y) + r\,F_b(y-x) + \sqrt{2D_1}\,\xi_2(t) + \eta(t)\,.$$

These dimensionless equations represent two coupled rigid dimers or rods of finite size, given by l, on a ratchet potential given by

$$V(x) = C - (1/4\pi^2\delta)\left[\sin 2\pi(x-x_0) + (1/4)\sin 4\pi(x-x_0)\right],$$

where the constants C, x_0 and δ are introduced in order to center the minima of the periodic potential on the integers, [13,14]. The force due to this potential is

$$f(x) = (1/2\pi\delta)\left[\cos 2\pi(x-x_0) + (1/2)\cos 4\pi(x-x_0)\right].$$

In order to account for the finite size l of the rigid dimers we introduced the effective force

$$f_l(x) = (1/2)\left[f(x) + f(x+l)\right].$$

This means that we are representing not a continuous rod, but two point particles kept rigidly at a fixed dimensionless distance l.

The dimers are coupled through the *nonlinear* force $F_b(x-y) = 4(x-y)[\varepsilon - (x-y)^2]$, where ε is a parameter that determines the distance between the minima of the associated double-well potential

$$V_b(x-y) = (x-y)^4 - 2\varepsilon(x-y)^2,$$

with two minima separated by a distance $2\sqrt{\varepsilon}$ and a barrier of height ε^2. The parameter r represents the coupling strength between dimers. The parameter D_1 is the intensity of the zero-mean statistically independent Gaussian white "thermal" noises $\xi_1(t)$ and $\xi_2(t)$ acting on particles x and y respectively:

$$\langle \xi_i(t)\xi_j(s)\rangle = \delta_{ij}\delta(t-s).$$

The common noise source $\eta(t)$ is represented by the Ornstein-Uhlenbeck process and is Gaussian and exponentially correlated. It satisfies the equations

$$\dot{\eta} = -\frac{1}{\tau}\eta(t) + \frac{\sqrt{2D_2}}{\tau}w(t), \quad \langle \eta(t)\eta(s)\rangle = \frac{1}{\tau}e^{-|t-s|/\tau},$$

where τ is the correlation time and $w(t)$ is Gaussian white noise with zero mean and intensity D_2.

This model is different from previous ones described in the literature in several respects. First, we consider two coupled *extended* dimers, instead of just two point particles. Second, the model incorporates a *nonlinear* coupling between the two dimers through the double-well potential. It is important to stress the following: the distance between dimers, $x-y$, can be positive, negative or zero, depending on the location on the double-well potential. The potential minima are located at $\pm\sqrt{\varepsilon}$ and correspond to positive ($x>y$) and negative ($y>x$) distances between dimers, respectively. On the other hand, the local maximum at the origin $(x-y=0)$, when the two dimers are in the same position, corresponds to an unstable situation. So, we can think of a particle oscillating in the bistable potential back and forth between the two minima, as the walking of kinesin alternating its two heads, or a walker alternating its two feet. This nonlinear coupling allow us then to consider a very important aspect of a real walking that was lacking in previous models: the possibility of alternating the two feet. In models [7-10], the coupling between the particles is linear (a harmonic spring) and thus the particles cannot alternate positions. They simply can approach to each other, but once you have an ordering, say $x > y$, the ordering remains for the rest of the dynamics. In our case, on the other hand, we can have several types of walking: *alternating random* walking, where the two dimers alternate their order randomly when the walker moves through the ratchet; *rigid random* walking, where the two dimers move on the ratchet without exchanging their order.

Coming back to the stochastic equations that define our model, we see that we have the following parameters: the length of the dimers l, the coupling strength between dimers *r*, the distance $2\sqrt{\varepsilon}$ between the minima of the bistable potential, the intensity D_1 of white noise and the correlation time τ and intensity D_2 of the colored noise. The combination $r\varepsilon^2$ is proportional to the ratio of two barrier heights: the ratchet and the bistable. Both are energetic barriers that have to be overcome in order to perform a particular walking: alternating or rigid. If we overcome only the ratchet barrier, then we have a rigid walking.

An example of a typical walking that involves several types of walking is shown in Fig.1.

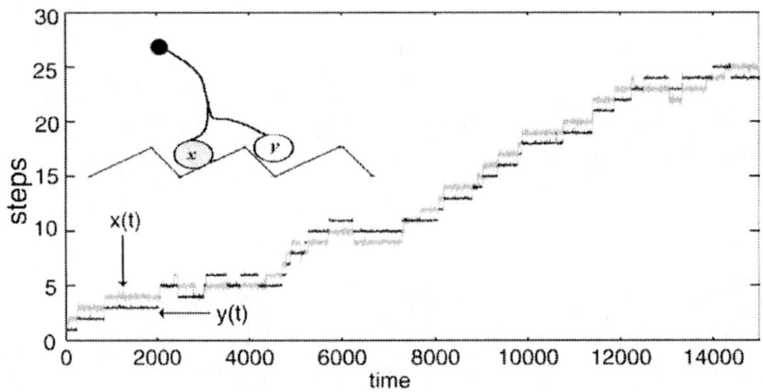

FIGURE 1. A typical trajectory of a walker moving along the ratchet. Vertical direction shows steps of *x* particle (grey) and *y* particle (black).

In this paper we present a preliminary study of the model with two parameters ε and *r* varied and with other parameters fixed.

BIFURCATION ANALYSIS

In the absence of common colored noise ($D_2 = 0$) the system satisfies the potential conditions [15] and is in equilibrium with a thermal bath. Thus, the stationary probability density $P_{st}(x, y)$ is given by the Boltzmann distribution [15]:

$$P_{st}(x,y) = N\exp[-\Psi(x,y)/D_1],$$

where

$$\Psi(x,y) = V(x) + V(y) + rV_b(x-y)$$

is the effective two-dimensional potential and N is the normalization constant. Indeed, there is no directed motion in this case, however, changes in the probability density structure correspond directly to the bifurcations of the correspondent deterministic system ($D_1 = 0$) [16,17]. Thus, various types of random walking on the potential $\Psi(x,y)$ landscape can be predicted just from the stationary probability density. In Fig. 2 we show contour lines of $P_{st}(x,y)$ for different values of the parameters r and a fixed value $\varepsilon = 0.25$ and $l = 0$. In the absence of coupling ($r = 0$) the system possesses 4 stable nodes, 4 unstable nodes and 8 saddles in a given frame ($0 \le x \le 2$, $0 \le y \le 2$). With the increase of r unstable nodes collides with saddles and disappear due to a saddle-node bifurcation. For $\varepsilon = 0.25$ this bifurcation occurs at $r = 0.014$. After this bifurcation there are only two unstable nodes. Further increase of the coupling strength leads to another saddle-node bifurcation, which results in the appearance of nodes and saddles near the stable node at $x = y = 1$. This bifurcation occurs at $r \approx 0.26$ for $\varepsilon = 0.25$ (see Fig.3).

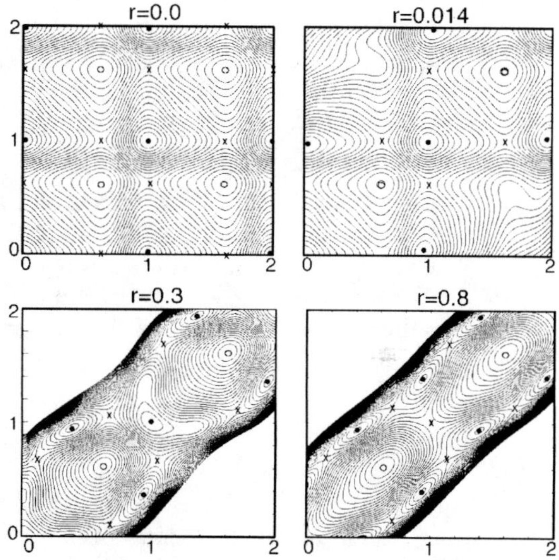

FIGURE 2. Contour lines of the stationary probability density as a function of x and y, for ε=0.25, l=0 and indicated values of the coupling strength r. The filled circles indicate stable nodes, the open circles indicate unstable nodes and crosses stands for saddles.

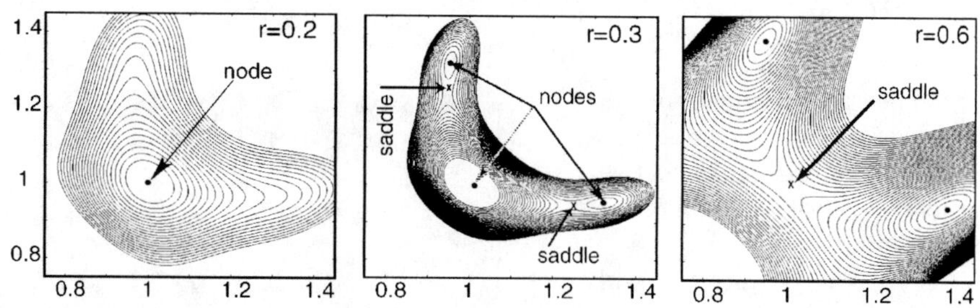

FIGURE 3. Contour lines of the stationary probability density in the magnified area near x=y=1 showing saddle-node bifurcation which results in formation of a saddle at x=y=1.

Finally, at $r = 0.5$ saddles and nodes born due to previous bifurcation collide with the node at $x = y = 1$, resulting in a saddle.

From Fig.3, we can see that alternating random walking occurs for large coupling strength, when the stable node at $x = y = 1$ is turned to a saddle. Otherwise, for smaller values of the coupling strength, the two dimers tend to remain together, which corresponds to the existence of stable node at $x = y = 1$.

CURRENT, EFFECTIVE DIFFUSION AND SYNCRONIZATION INDEX

In this section we will calculate numerically the main quantities of interest: the current, the effective diffusion coefficient and the synchronization index. The current is simply the average velocity of the center of mass of the walker. The effective diffusion constant is given by [16,18-20]

$$D_{eff} = \frac{1}{2}\frac{d}{dt}\left[\langle z^2(t)\rangle - \langle z(t)\rangle^2\right],$$

where $z(t) = (x + y)/2$ is the center of mass of the walker. Recently, there have been important results where the effective diffusion coefficient can be obtained analytically for the case of tilted periodic potentials [21-23]. Another important quantity that can be studied is the ratio between the current and the effective diffusion coefficient; the proper dimensionless ratio is refer to as the Péclet number [23].

A quantity that measures the degree of synchronization is the synchronization index [24], defined as $\gamma^2 = \langle\cos 2\pi(x - y)\rangle^2 + \langle\sin 2\pi(x - y)\rangle^2$.

The synchronization index changes from 0 to 1 and shows whether the difference between particle position is constant or varies randomly in time.

These quantities involve averaging over time and over 100 different realizations of noise sources.

The simulation results for the current are shown in Fig.4 as a function of two parameters. The current demonstrates reversals as a function of distance for the coupling strength larger than 0.2.

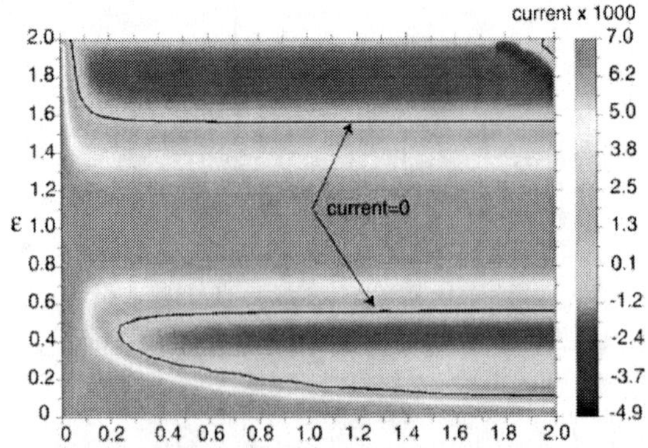

FIGURE 4. Contour lines of the probability current of the center of mass $z(t)$ calculated as a function of parameters r and ε. Other parameters values are: $D_1 = 0.01$, $D_2 = 0.1$, $\tau = 10$ and $l = 0$.

We can expect that the current is maximal for the distance ε corresponding to the distance between potential minima, that is, $\varepsilon = 1$. On the other hand, synchronization index is minimal in the case when the system performs random alternating motion. We note that in the case when common colored noise is switched on, there are two competitive tendencies, which affect walking patterns and, consequently, the synchronization index. The common noise tends to synchronize dimers even in the absence of coupling. On the other hand, coupling through the double-well potential tends to separate dimers. In Fig.5 we show the probability current, the effective diffusion and the synchronization index versus the distance ε for different values of the coupling strength. As we predicted, the current is maximal when the distance between dimers equals the distance between the wells of the ratchet potential $(\varepsilon = 1)$. In this case, we notice that the effective diffusion coefficient is minimum and then the ratio between current and effective diffusion (Péclet number) is maximum. The calculation of the synchronization index shows the region of ε when random alternation of dimers positions is available $(0.2 < \varepsilon < 0.7)$. The effective diffusion (middle panel) also shows non-monotonous behavior, indicating that the fastest diffusion was achieved at $\varepsilon = 0.25$.

We stress that these numerical simulations represent preliminary results and thus more extensive studies are necessary in order to understand the mechanisms behind current reversals and diffusion optimization.

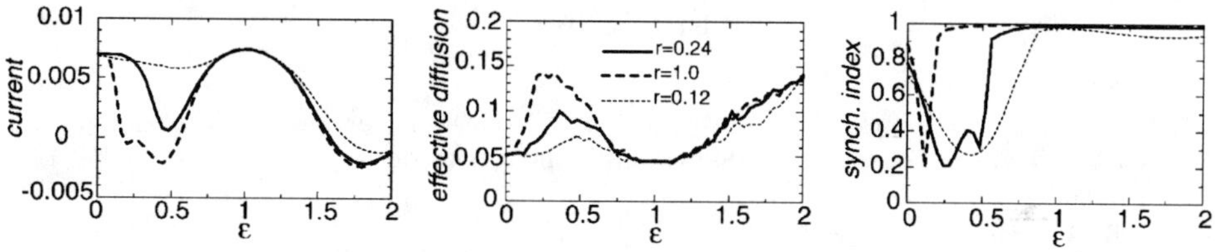

FIGURE 5. Current, effective diffusion and synchronization against ε for different values of r. Other parameters are the same as in Fig.4.

CONCLUDING REMARKS

In summary, we have introduced a model for a random walker that consists of two extended dimers coupled nonlinearly through a bistable potential, steeping on an asymmetric periodic ratchet potential. In contrast to linear coupling, the bistable forcing admits a richer dynamics where the ordering of the dimers can alternate. The dynamics then include two typical stepping patterns: alternating and/or rigid random walking. In our model, each "feet" is acted upon by white noise and additionally we have a common colored noise that drives the systems out of equilibrium and therefore allows a directed nonzero current. We have calculated the current, defined as the average velocity of the center of mass, the synchronization index and the effective diffusion coefficient, as a function of two parameters related with the coupling and the distance between particles. While there are time scales associated with our problem (the Kramers rates in the two potentials and the correlation time of the colored noise), there is no external pacemaker or pattern generator that imposes periodicity on the motion.

Numerical simulations showed the existence of current reversals and significant changes in the effective diffusion coefficient and in the synchronization index. In the case where the equilibrium distance between dimers is equal to the periodicity of the ratchet potential, we obtained an optimal transport, characterized by a maximum dimensionless ratio of the current and the effective diffusion coefficient (Péclet number). This is equivalent to a minimum *randomness* (inverse of Péclet number), as defined in the biological literature [25]. This coherent form of transport is also characterized by a maximum synchronization index between the two feet. Thus, we have a new model that may shed light on a number of currently interesting problems, ranging from noisy locomotion to transport of motor proteins, that establish a connection between transport in ratchets, bistability and synchronization.

ACKNOWLEDGMENTS

We gratefully acknowledge financial support from UNAM through project DGAPA-IN-111000. AN and FM acknowledge support from ONR Physics Division.

REFERENCES

1. Hänggi, P., and Bartussek, R., in *``Nonlinear Physics of Complex Systems",* Lecture Notes in Physics, Vol. **476**, edited by Parisi, J., Müller, S. C., and Zimmermann, W., Springer, Berlin, 1996, pp. 294-308.
2. Astumian, R. D., *Science* **276,** 917-922 (1997).
3. Jülicher F., Ajdari, A., and Prost, J., *Rev. Mod. Phys.* **69,** 1269-1282 (1997).
4. Reimann, P. *Phys. Rep.* **361,** 57-265 (2002).
5. Vale, R. D., and Milligan, R. A., *Science* **288,** 88-95 (2000).
6. Howard, J., *Mechanics of Motor Proteins and the Cytoskeleton,* Sinauer Associated, Inc. Sunderland, Massachusetts, 2001.
7. Derényi, I., and Vicsek, T., *Proc. Natl. Acad. Sci. USA* **93,** 6775-6779 (1996).
8. Stratopoulos, G. N., Dialynas, T. E., and Tsironis, G. P., *Phys. Lett. A* **252,** 151-156 (1999).
9. Klumpp, S., Mielke, A., and Wald, C., *Phys. Rev. E* **63,** 031914 (2001).
10. Elston, T. C., You, D., and Peskin, C. S., *SIAM J. Appl. Math.* **61,** 776-791 (2000).
11. Collins, J. J., and Stewart, I. N., *J. Nonlinear Science* **3,** 349-392 (1993).
12. Dickinson, M. H., Farley, C. T., Full, R. J., Koehl, M. A. R., Kram, R., and Lehman, S., *Science* **288,** 100-106 (2000).
13. Mateos, J. L., *Phys. Rev. Lett.* **84,** 258-261 (2000).
14. Mateos, J. L., *Physica D* **168-169,** 205-219 (2002).
15. Gardiner C.W., *Handbook of Stochastic Methods,* Springer-Verlag, Berlin, 1985.
16. Malchow, H. and Schimansky-Geier, L. *Noise and Diffusion in Bistable Nonequilibrium Systems*, Teubner-Texte zur Physik, Leipzig, 1985.
17. Neiman, A., *Phys. Rev. E* **49,** 3484-3487 (1994).
18. Stratonovich, R.L. *Topics in the Theory of Random Noise*, Vol.2, Gordon and Breach, New York-London-Paris, 1967.

19. Anishchencko, V. S., Astakhov, V. V., Neiman, A. B., Vadivasova, T. E., and Schimansky-Geier, L., *Nonlinear Dynamics of Chaotic and Stochastic Systems,* Springer-Verlag, Berlin, 2002.
20. Schimansky-Geier, L., Anishchenko, V. S., and Neiman, A., *Phase Synchronization: From Periodic to Chaotic and Noisy,* in *Neuroinformatics and Neural Modelling,* Chapter 2. Handbook of Biological Physics, Vol. **4,** edited by Moss, F., and Gielen, S., Elsevier, Amsterdam, 2001, pp. 23-82.

21. Reimann, P., Van den Broeck, C., Linke, H., Hänggi, P., Rubí, J. M., and Pérez-Madrid, A., *Phys. Rev. Lett.* **87**, 010602 (2001).
22. Reimann, P., Van den Broeck, C., Linke, H., Hänggi, P., Rubí, J. M., and Pérez-Madrid, A., *Phys. Rev. E* **65**, 031104 (2002).
23. Lindner, B., Kostur, M., and Schimansky-Geier, L., *Fluctuation and Noise Letters* **1,** R25-R39 (2001).
24. Rosenblum, M., Pikovsky, A., Kurths, J., Schäfer, C., and Tass, P. A., *Phase Synchronization: From Theory to Data Analysis,* in *Neuroinformatics and Neural Modelling,* Chapter 9. Handbook of Biological Physics, Vol. **4,** edited by Moss, F., and Gielen, S., Elsevier, Amsterdam, 2001, pp. 279-321.
25. Svoboda, K., Mitra, P. M., and Block, S. M., *Proc. Natl. Acad. Sci. USA* **91,** 11782-11786 (1994).

On the Amplitude and Time–structure Properties of $1/f^{\alpha}$ Noises

Z. Gingl[1], R. Mingesz and P. Makra

Department of Experimental Physics, University of Szeged, Hungary
Dom ter 9., Szeged, H-6720 Hungary

Abstract. An interesting invariant property of 1/f noises was found experimentally a few years ago: amplitude truncation does not change the shape of the power spectral density (PSD) under rather general conditions. Later the phenomenon was explained theoretically including extensions to $1/f^{\alpha}$ noises with $0<\alpha<2$ and associated unsolved problems were also presented [1-7]. In this work we present a brief review of our recent results and also report investigations of the amplitude–truncation and level–crossing properties of $1/f^{\alpha}$ noises in more detail. Particularly, we analyze the dependence of the spectral density and successive level–crossing interval statistics on the level of truncation. From the results of our numerical simulation, it turns out that the case $\alpha\approx1$ is unique in the range from 0 to 2 in two respects. First, 1/f noise is the most sensitive to the level of amplitude truncation when close levels are used. Second, the successive level–crossing intervals exhibit a correlation maximum for $\alpha\approx1$. These amplitude and time–structure related additions to our knowledge about 1/f noise further emphasize the uniqueness of this kind of noise. Several open problems arise including a possible explanation of the general presence of 1/f noise, theoretical justification of the above mentioned properties and more precise understanding of the role of the manifold exceptionality for the case $\alpha\approx1$, i.e. 1/f noise.

INTRODUCTION

1/f noise is very common in several natural and even artificial systems. Although 1/f noise was discovered several decades ago, the general occurrence of this phenomenon has not been understood yet, and many problems associated with the models and properties of 1/f noise have remained unsolved. One of the most challenging problems in noise research is to understand the generality and special features of 1/f noise and to introduce new models and tools to handle this kind of fluctuation. This requires extensive research into the properties of this noise, and new results in this field can help to understand it better. A strange invariant property of Gaussian 1/f noise was found experimentally a few years ago: the power spectral density (PSD) remains 1/f if the amplitude of the noise is truncated using two levels under rather general conditions. This nonlinear transformation can easily occur in overdriven measurements, saturating physical systems and quantities with limited range. Results for amplitude truncation were obtained for $1/f^{\alpha}$ noises with $0<\alpha<2$ by experimental investigations and numerical simulations; theoretical derivation was found for close truncation levels [1-7].

[1] e-mail: gingl@physx.u-szeged.hu

CP665, *Unsolved Problems of Noise and Fluctuations: UPoN 2002: Third International Conference,*
edited by S. M. Bezrukov

In this work we investigate the amplitude–truncation and level–crossing properties of $1/f^{\alpha}$ noises as functions of the truncation level. Interesting features were found using numerical techniques, showing that the case α close to 1 is unique in the range from 0 to 2. In the following, we present in brief an overview of amplitude truncation, the summary of our previous experimental and theoretical results, our new results and the open problems related.

OVERVIEW OF THE AMPLITUDE TRUNCATION INVESTIGATIONS

Many researchers might find a strange behavior during accidentally overdriven measurements of 1/f noise: the instrument showed the same 1/f power spectral density. Since the measurements were unreliable in this case probably nobody thought that it could tell something important about 1/f noise. In 1989 our group built a simple $1/f^2$ noise driven comparator circuit to generate $1/f^{1.5}$ noise [1], and also used 1/f noise driving sources – the dichotomous output noise seemed to have 1/f-like PSD, but this again we left without drawing any conclusions. Somewhat later we have investigated 1/f noise driven stochastic resonance in a Schmitt trigger and concluded that the output PSD has 1/f shape in the low frequency range and this was the first important sign for us that 1/f noise may have a special invariant property against amplitude saturation [2].

It was a real breakthrough in this field when we experimentally explored the amplitude saturation properties of 1/f noises for many different truncation levels [3]. These results were extended to $1/f^{\alpha}$ noises ($0<\alpha<2$) also and were presented at the first conference on unsolved problems of noise in 1996 [4]. A theoretical proof was found in 1999 and additional results were published in the subsequent years [5-7].

In the following we summarize the abovementioned results and show new achievements concerning the exceptionality of 1/f fluctuations and correlation properties of level-crossing intervals.

AMPLITUDE TRUNCATION OF 1/F NOISES

Amplitude truncation is defined in the following way:

$$y(t)=\begin{cases} x_{\min}, & \text{if } x(t)\le x_{\min} \\ x_{\max}, & \text{if } x(t)\ge x_{\max} \\ x(t), & \text{otherwise} \end{cases} \tag{1}$$

Here x(t) is the time-dependent amplitude of the zero-mean Gaussian noise, y(t) is the amplitude of the truncated noise, x_{min} and x_{max} are the truncation levels. For single level truncation, we can use the following formula:

$$y(t) = \begin{cases} -1, & \text{if } x(t) \le x_0 \\ +1, & \text{if } x(t) > x_0 \end{cases} \qquad (2)$$

where x_0 is the truncation level. Fig. 1. illustrates these transformations on a typical sample of 1/f noise.

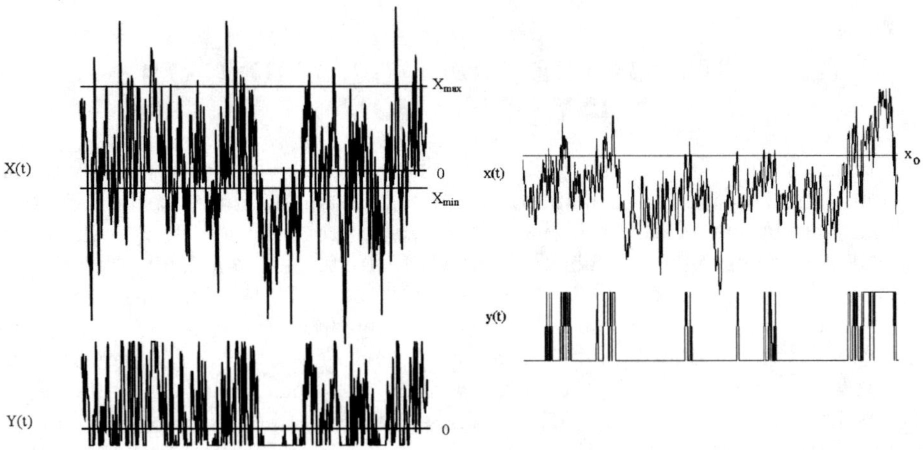

FIGURE 1. Illustration of the amplitude truncation of 1/f noise. The panel on the right shows a single level truncation.

For x_0=0, theoretical results show that the PSD of y(t) is proportional to $1/f^{\beta}$, where β is a function of α (detailed derivation can be found in ref. [5, 6]):

$$\beta = \begin{cases} \alpha, & \text{if } \alpha \le 1 \\ \dfrac{\alpha+1}{2}, & \text{if } \alpha > 1 \end{cases} \qquad (3)$$

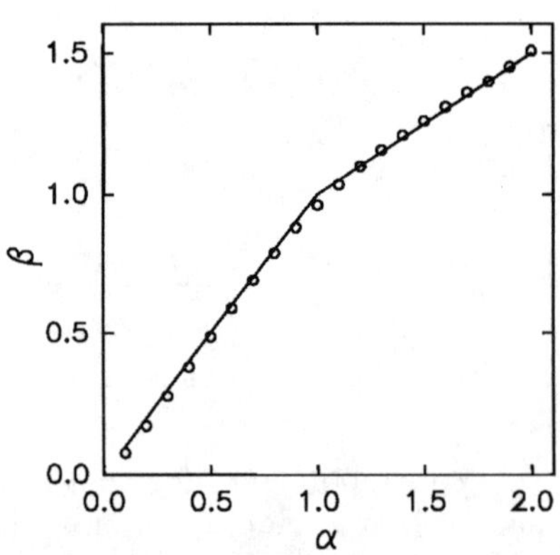

FIGURE 2. β versus α. The solid line represents the theoretical result, circles show numerical verification. The invariance (β=α) can be seen for α≤1.

Equation (3), however, is not valid for $x_0 \neq 0$, or in general, when x_0 differs from the mean value of the noise. Theoretical extensions are given for this case, valid in rather low frequency part [7], but real band-limited cases may differ significantly. Here we present numerical simulations using sample lengths of 65536 and the PSDs were calculated as an average of 1000 samples. The level x_0 was varied between 0 and 1.5σ, where σ was the standard deviation of the input noise. The spectral exponent α of the input noise was varied from 0 to 2 by steps of 0.25 and the results were plotted on Fig. 3. For $\alpha=2$ ($1/f^2$ noise) and $\alpha=0$ (white noise) the value of β does not depend on the level x_0, as it is expected ($1/f^2$ noise does not have definite mean value to relate to x_0 and white noise remains uncorrelated after truncation). Second order polynomials were fitted to the $\beta(x_0)$ curves and the coefficient A_2 of the second order term versus α is plotted on the right of Fig. 3. to characterize the curvature. One can see that this quantity has a maximum at 1/f noise ($\alpha \approx 1$), which means that 1/f noise is the most sensitive to the level of amplitude truncation.

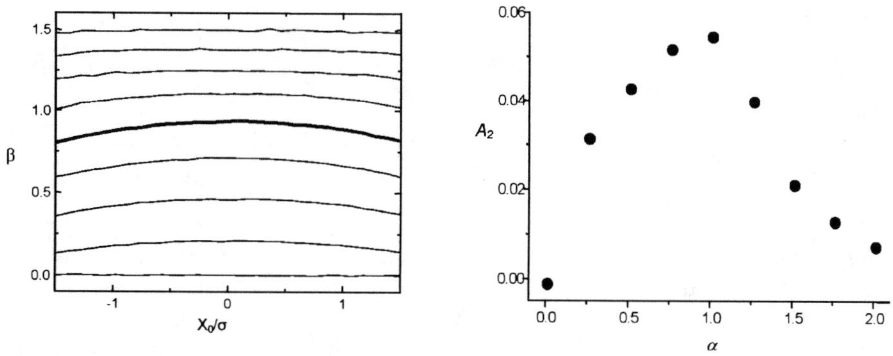

FIGURE 3. On the left β as a function of x_0/σ is shown. From bottom to top α is varied from 0 to 2 by steps of 0.25, the thick line corresponds to $\alpha=1$, i.e.1/f input noise. Coefficient A_2 of a second order polynomial fit on these curves is plotted on the right to characterize the curvature as a function of α. Note, that A_2 has a maximum at $\alpha \approx 1$.

We also investigated the level-crossing properties of $1/f^{\alpha}$ noises by analyzing the statistics of the time elapsed between neighboring level crossings. Fig. 4. illustrates the level-crossings of a signal and defines the T_U and T_L time intervals. We explored the correlation between successive T_U-T_U, T_L-T_L and T_U-T_L/T_L-T_U level-crossing intervals by numerical simulations. We found that the correlation is the strongest around α close to 1 again, namely for 1/f noise (see Fig. 5.). We have noticed also that the PSD of the time series of T_U-T_L/T_L-T_U type level-crossing intervals can be approximated as $1/f^{\gamma}$ and Fig. 6 shows how γ depends on α. Note that the highest correlation at α close to 1 can be observed again.

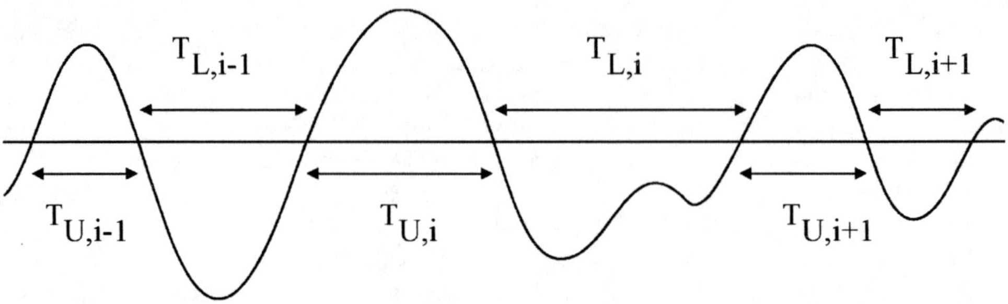

FIGURE 4. Level-crossing time intervals of a signal. $T_{U,i}$ and $T_{L,i}$ denote the time the signal spends above and below the level between neighboring level crossings, respectively.

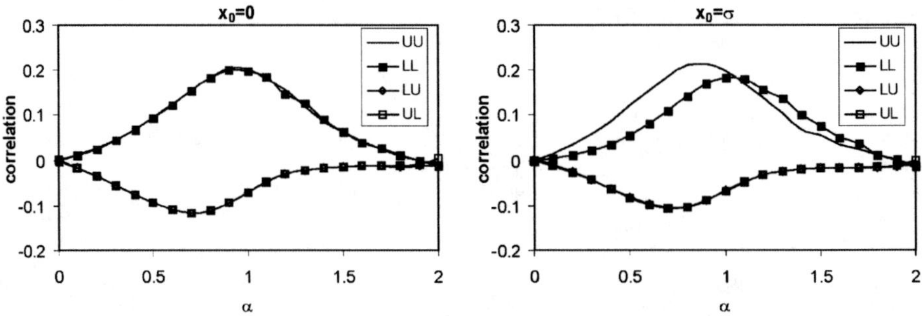

FIGURE 5. Correlation coefficients as a function of the power exponent of the input noise for T_U-T_U, T_L-T_L and T_U-T_L/T_L-T_U type level crossings. The two plots show results for levels equal to the mean value (zero) and to the standard deviation of the noise.

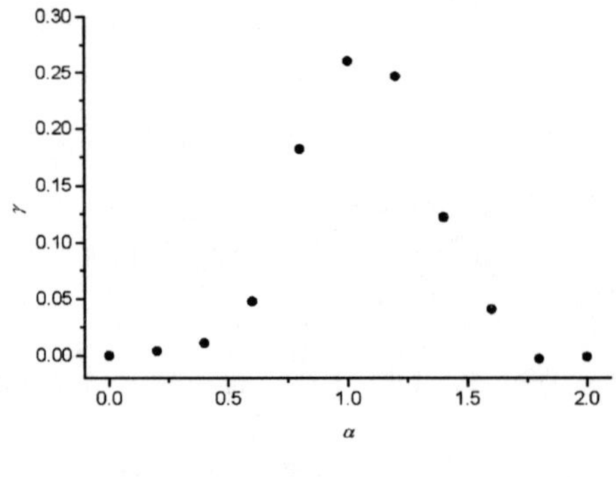

FIGURE 6. Spectral exponent γ versus α. The strongest correlation is shown at $\alpha \approx 1$.

CONCLUSION AND OPEN QUESTIONS

In this publication we briefly summarized our results about the amplitude truncation and time-structure related properties of $1/f^{\alpha}$ noises. We can conclude that the power spectral density remains $1/f^{\alpha}$ for $\alpha \leq 1$ if the amplitude of the noise is truncated using two levels under rather general conditions. This nonlinear transformation can easily occur in overdriven measurements, saturating physical systems and quantities with limited range. One may find related phenomena in:

- systems wherein amplitude clipping occurs,
- threshold crossing systems (like stochastic resonance),
- random dichotomous noises with 1/f spectrum,
- switching or stepping driven by 1/f noise.

Corresponding to these points the most important application seems to be the explanation of 1/f noise in ion channel currents [8, 9].

Furthermore, our new numerical simulations show that the amplitude and time-structure properties of $1/f^{\alpha}$ noises exhibit extreme values at $\alpha \approx 1$. Extreme features at $\alpha \approx 1$, i.e. for 1/f noise has been observed widely since its discovery, but their function often emerges as an open question. Note here, that it has also been reported recently that 1/f noise can play a prominently constructive role in some systems showing stochastic resonance [10, 11].

Although these facts seem to help to understand the mechanism and origin of 1/f noise, the possible role of the extreme features and a potential relation to the general occurrence are not yet clarified.

ACKNOWLEDGMENTS

The authors thank L.B. Kish for helpful discussion. This work has been supported by OTKA (Hungarian Academy of Sciences), grant no. T37664.

REFERENCES

1. Gy. Trefan, L.B. Kiss, Z. Gingl, I. Hevesi and M.I. Torok, "Diffusion noise generators realized by analog circuits", *Proc. 10th International Conference on Noise in Physical Systems*, ed. A. Ambrózy, Akadémia, Budapest, 1990, pp. 569-572.
2. L.B. Kiss, Z. Gingl, Zs. Marton, J. Kertesz, F. Moss, G. Schmera and A. Bulsara, *J.Stat.Phys.* **70** 451-462. (1993)
3. Z. Gingl and L.B. Kiss, "A Crucial Property of Gaussian 1/f Fluctuations" *Proc. Fluctuation Phenomena in Physical Systems*, Ed. V. Palenskis, Vilnius university Press, Palanga, Lithuania, 1994, pp. 274-279
4. Z.Gingl, L.B. Kiss, "Amplitude-saturation nonlinearity of Gaussian $1/f^k$ fluctuations", *Proceedings of the First International Conference on Unsolved Problems of Noise*, edited by Doering,Ch.R., Kiss, L.B., and Schlesinger,M.F., World Scientific, 1996, pp. 337-342.
5. Sh. Ishioka, Z. Gingl, D. Choi and N. Fuchikami, "Amplitude truncation of Gaussian 1/f^alpha noises", *Physics Letters A* **269** 7-12 (2000)
6. Gingl Z., Ishioka S.,Choi D., and Fuchikami N., *Chaos* **11** 3, 619-623 (2001).
7. D. Choi and N. Fuchikami, *J Phys Soc Jpn* **70** 1, 297-304 (2001).
8. S. M. Bezrukov, M. Winterhalter, *Physical Review Letters* **85** 202-205 (2000).
9. Z. Siwy, A. Fulinski, "Origins of $1/f^{\alpha}$ Noise in Membrane Channel Currents", *in this volume*
10. D. Nozaki, Y. Yamamoto, *Physics Letters A* **243**, 281-287 (1998).
11. L.B. Kish, S.M. Bezrukov, *Physics Letters A* **266**, 271-274 (2000).

Shot Noise as a Tool to Probe Microscopic Interactions

G. Gomila

Centre de Recerca en Bioelectrònica i Nanobiociència and Departament d'Electrònica-Universitat de Barcelona, Laboratori de Nanobioenginyeria-Parc Científic de Barcelona, C/ Josep Samitier 1-5, 08028 Barcelona, Spain

Abstract. In recent years it has been often argued that the shot noise that remains at zero temperature is the only one that can be used to probe microscopic interactions. Here, we present two examples where shot noise has a thermal origin and can be equally well used for this purpose. These two examples show that there is no essential difference between thermal and non-thermal shot noise

INTRODUCTION

The many degrees of freedom existing at a microscopic level are the responsible for the presence of spontaneous current fluctuations in electrically conducting systems. These current fluctuations can be characterized by means of the current spectral density defined as the real part of the Fourier transform of the current fluctuations correlation function, i. e.

$$S_I(f) = 4\int_0^\infty \overline{\delta I(t)\delta I(t+\tau)}\cos(2\pi f\tau) \tag{1}$$

where $\delta I(t) = I(t) - \bar{I}$ refers to the current fluctuations around the mean value, $\bar{I}$. The dependence of $S_I(f)$ on frequency, mean electric current and system parameters depends on the thermodynamic state of the conductor. At thermal equilibrium any electrically conducting systems displays Nyquist noise [1], whose current spectral density takes the form

$$S_I^{Nyquist}(f) = 4k_BT / R^{eq} \tag{2}$$

where k_B is the Boltzmann constant, T the temperature and R^{eq} the equilibrium resistance of the sample.

Out of equilibrium the relation between current noise and resistance breaks down and the current spectral density $S_I(f)$ takes values not related to the resistance of the sample. Typical examples of non-equilibrium noise types (often called *excess* noise) include *1/f* noise, generation-recombination noise and shot noise [2].

Among these excess noises, shot noise has received considerable attention in recent years in connection with the study of quantum phenomena in mesoscopic conductors

CP665, *Unsolved Problems of Noise and Fluctuations: UPoN 2002: Third International Conference*, edited by S. M. Bezrukov

[3]. Mesoscopic conductors are characterized by having, at least, one spatial dimension of the order of the wave length of carriers. Therefore, in these conductors the effects related to the quantum wave nature of the carriers become relevant. Examples of mesoscopic conductors include quantum point contacts or mesoscopic metallic conductors with a sample length shorter than the inelastic mean free path. It has been shown that in many instances mesoscopic conductors display shot noise with a current spectral density of the form [3],

$$S_I(f) = \gamma 2q\bar{I} \tag{3}$$

where q is the carrier charge and γ a dimensionless parameter referred to as the Fano factor. The value of the Fano factor provides information on the correlations existing among different current pulses. When $\gamma = 1$ no correlations exist among the different current pulses, while for $\gamma \neq 1$ they are present. The sensitivity of shot noise to the correlations among current fluctuations, and hence to the microscopic interactions, has made of shot noise a useful tool to probe microscopic interactions in mesoscopic conductors. For instance, in quantum point contacts with a single transmitting mode it has been found that $\gamma = 1 - t$, with t being the transmission probability of the mode, thus revealing the random transmission of the electron wave through the constriction [4]. Moreover, in mesoscopic metallic conductors it is found $\gamma = 1/3$ in the case that electron-electron interaction is absent and $\gamma = \sqrt{3}/4$ in the case that this interaction is present [5], thus probing directly the type of interaction existing in the conductor.

A particularity of the shot noise in mesoscopic conductors is that, having a quantum origin, it does not vanish at zero temperature. This property of shot noise in mesoscopic conductors has let to the belief that the *true* shot noise is the only one present at zero temperature, and that only this one can be used as a tool to probe microscopic interactions.

In what follows we question this statement and show that there is no essential difference between the shot noise generated by thermal fluctuations and the shot noise generated by quantum phenomena. Therefore, in both cases one can use shot noise as a tool to probe microscopic interactions in electrically conducting systems. To this end we will study two situations where shot noise has a thermal origin, and hence it would vanish at zero temperature, but in which shot noise can be used as a tool to probe microscopic interactions.

SYSTEM UNDER STUDY

The model system under study consists of a conducting sample of given length L in which two ohmic contacts are implemented. The cross sectional area of the sample is assumed to be large enough so as to allow for a one-dimensional electrostatic treatment. An external bias V is applied between the two contacts. We assume that the systems is under such a conditions as to make the quantum wave length of carriers much shorter than the sample length, thus making negligible any quantum effect related to the wave nature of the carriers. We consider two opposite transport regimes: (a) the ballistic transport regime, in which the sample length is much shorter than the inelastic and elastic mean free paths and (b) the macroscopic diffusive transport regime, in which the sample length is much larger than both the inelastic and elastic mean free paths. In the former case carriers experience no scattering while traversing the active sample region and the origin of the current fluctuations should be traced back to the thermionic emission of carriers from the contacts into the active region of the sample. In the later case carriers experience many inelastic scatterings with the lattice phonons, with which they are in local thermal equilibrium, and the origin of the current fluctuations is related to the random nature of these inelastic scattering processes. In both situations, current fluctuations have clearly a thermal origin, since they would vanish at zero temperature. In what follows, we will show that in both cases shot noise appears and can be used to probe microscopic interactions.

BALLISTIC TRANSPORT REGIME

In the case of ballistic transport, due to the absence of scattering processes inside the sample, the single source of fluctuations is the random thermionic emission of carriers from the contacts. The statistics of the injection of carriers depends on the energy range on which carriers are injected and on the nature of the quantum exclusion principle satisfied by the carriers. If carriers are injected with energies belonging to the tail of the contact distribution function (which can be approximated by a Maxwell-Boltzmann distribution function), then they follow a poissonian distribution. On the contrary if they are injected with energies around or below the chemical potential of the contacts, then they follow an injection statistics related to the exclusion principle they satisfy: a binomial distribution if they satisfy the Pauli Exclusion Principle [6], and more complicated statistics if they satisfy other quantum exclusion principle, as for instance a fractional exclusion principle.

A detailed analysis of the non-equilibrium current noise properties of these ballistic systems has been recently developed. The analysis has been based on the following Vlasov-Poisson-Langevin model [7],

$$\left[\frac{\partial}{\partial t}+v_x\frac{\partial}{\partial x}-\frac{\partial U(x,t)}{\partial x}\frac{\partial}{\partial v_x}\right]F(x,v_x,t)=0 \tag{4}$$

$$\frac{\partial U(x,t)}{\partial x^2}=-\int_{-\infty}^{+\infty}dv_xF(x,v_x,t) \tag{5}$$

subject to the boundary conditions [7],

$$\begin{aligned}F(0,v_x,t)&=\overline{F}_0(v_x)+\delta F_0(v_x,t); \quad v_x>0,\\ F(L,v_x,t)&=\overline{F}_L(v_x)+\delta F_L(v_x,t); \quad v_x<0,\end{aligned} \tag{6}$$

$$U(0,t)=0; \qquad U(L,t)=V \tag{7}$$

where v_x is the carrier velocity, $U(x,t)$ the potential energy and $F(v_x,x,t)$ the velocity distribution function. Moreover, $\overline{F}_{0,L}(v_x)$ and $\delta F_{0,L}(v_x,t)$ are the mean value and the fluctuations of the distribution function at the contacts of the sample. Finally, V is the applied bias. Dimensionless variables are used to simplify the notation (see Ref. [7]).

For this model a closed analytical solution has been found [7]. From it, it has been shown that when the system is driven to current saturation conditions by a high enough applied voltage bias (in analogy with vacuum tubes) then the system displays shot noise with a value of the Fano factor that depends on the nature of the quantum exclusion principle satisfied by the carriers [8,9]. This fact is illustrated in Fig. 1 where the Fano factor under saturation conditions is plotted as a function of the chemical potential of the contacts for different injection statistics related to the Pauli exclusion principle (*g=1*) and to Fractional Exclusion Statistics (*g<1*) (in the Figure the values of the chemical potential are taken with respect to the potential energy at the contact).

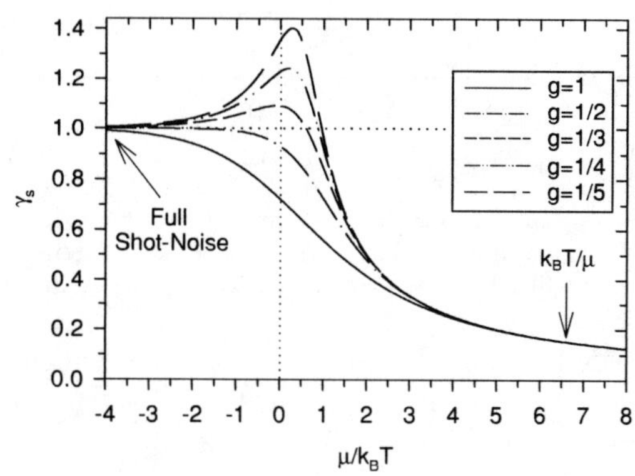

FIGURE 1. Fano factor under current saturation conditions for a one dimensional ballistic conductor for different injection statistics.

As can be seen in Fig. 1, when the chemical potential lies below (or above) the contact potential energy by an amount greater than $3k_BT$, there is no difference in the value of the Fano factor for the different exclusion principles. However, in the intermediate range of values, for which $-3k_BT < \mu < 3k_BT$, relevant differences are observed between the different curves, which can be used to identify the quantum exclusion principle satisfied by carriers [8,9].

Besides the correlations associated to the quantum exclusion principle, there can appear other correlations among current fluctuations in ballistic systems, as for instance long range Coulomb correlations. These correlations appear due to the modulation of the self-consistent potential energy barrier performed by the long range Coulomb interaction. These correlations are present when the system is not under current saturation conditions and give rise to a suppression of the current noise level below the shot noise level [7]. This fact is illustrated in Fig. 2 where we plot the low frequency current spectral density calculated for a 0.3 μm long GaAs n^+-i-n^+ ballistic diode at 77 K (solid line). In the Figure the shot noise region corresponds to the high bias voltage range in which the current spectral density is independent on bias and it is proportional to the current saturation value. For lower applied bias, the current spectral density lies below the shot noise level due to the presence of both Pauli and Coulomb correlations. By comparing the solid line with the dotted and dashed lines, corresponding to the absence of long range Coulomb correlations and to the absence of both Pauli and Coulomb correlations, respectively, one can appreciate the relevance of each of these correlating mechanisms.

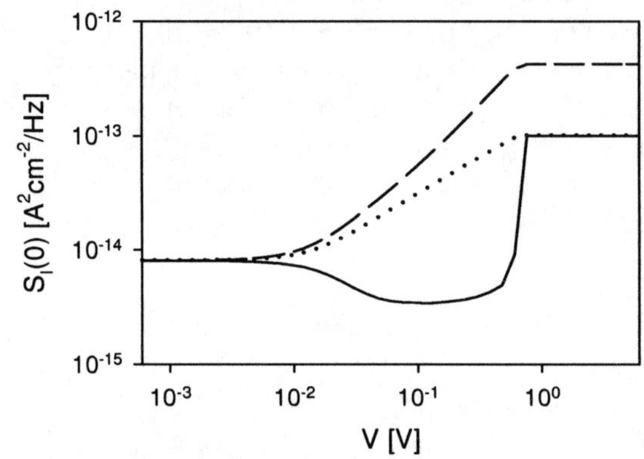

FIGURE 2. Low frequency current spectral density for a GaAs ballistic n^+-i-n^+ diode (solid line). Dotted line corresponds to the case in which the long range Coulomb correlations are neglected. Dashed line corresponds to the case in which both the long range Coulomb and Pauli correlations are neglected.

We have then provided a first example of a system in which shot noise generated by thermal fluctuations can give direct information about the relevance of different microscopic interactions.

MACROSCOPIC DIFFUSIVE TRANSPORT REGIME

In the macroscopic diffusive transport regime the source of current fluctuations is the random inelastic scattering of carriers with lattice phonons. In this transport regime the fluctuations generated by the thermoionic emission of carriers from the contacts can be neglected, since a few inelastic scattering processes suffice to completely randomize the carriers velocity. Therefore the single source of fluctuations is the one related to the diffusive motion of carriers.

The calculation of the non-equilibrium current noise properties of an homogeneous conductor in the macroscopic diffusive transport regime was addressed recently in Ref.[10] by means of the following drift-diffusion-Poisson-Langevin model,

$$J(x,t) = qn(x,t)\mu E(x,t) + qD\frac{\partial n(x,t)}{\partial x} + q\xi(x,t) \tag{8}$$

$$\frac{\partial E(x)}{\partial x} = \frac{q}{\varepsilon}\left(N_D - n(x,t)\right) \tag{9}$$

subject to the boundary conditions

$$n(0,t) = n(L,t) = n_c \tag{10}$$

Here, $J(x,t)$ is the current density, q the carriers charge, μ the mobility, $E(x,t)$ the electric field, D the diffusion coefficient, $n(x,t)$ the carrier density, N_D the donor density, ε the dielectric permittivity and n_c the contact density. Moreover, $\xi(x,t)$ is a Langevin noise source related to the diffusion of carriers [10].

An exact solution of this model for low frequencies can be derived in closed analytical form. One obtains for the low frequency current spectral density [10]

$$S_I(0) = S_I^{Nyquist}(0) + S_I^{ex}(0) \tag{11}$$

where the first term corresponds to the Nyquist noise given in Eq. (2) and the second term represents an excess noise given through [10]

$$S_I^{ex}(0) = \frac{\left(\lambda_2^2 - \lambda_1^2\right)\left(e^{\lambda_1} - 1\right)\left(e^{\lambda_2} - 1\right)}{2\lambda_2^2\lambda_1^2\left(e^{\lambda_1} - e^{\lambda_2}\right)^2}\left[\lambda_2\left(e^{\lambda_2} - 1\right)\left(e^{\lambda_1} + 1\right) - \lambda_1\left(e^{\lambda_1} - 1\right)\left(e^{\lambda_2} + 1\right)\right] \tag{12}$$

In the previous equation,

$$\lambda_{1,2} = -\frac{1}{2}\frac{\tau_D}{\tau_T}\left(1 \pm \sqrt{1 + 4\frac{\tau_T^2}{\tau_d \tau_D}}\right) \tag{13}$$

where $\tau_T = L^2/\mu V$ is the drift transit time, $\tau_D = L^2/D$ the diffusion transit time and $\tau_d = \varepsilon/\sigma$ the dielectric relaxation time, with σ being the sample conductivity. The additional contribution in Eq. (11) can be shown to give rise for high applied bias to shot noise, as illustrated in Fig. 3, where we plot the low frequency current spectral density as a function of current for different sample lengths.

We note two three interesting properties in Fig. 3: (i) the bias voltage at which the current spectral density departs from the Nyquist value depends on the ratio between the sample length and the Debye screening length, (ii) the Fano factor equals *1* for a non-degenerate conductor and k_BT/μ for a degenerate conductor, with μ being the chemical potential and (iii) shot noise appears when the transit time due to drift becomes shorter than the dielectric relaxation time, thus not giving time to the Coulomb correlations to set in [10].

These results allow monitoring in this system by means of shot noise both the Pauli and the long range Coulomb correlations. We have then provided a second example where shot noise generated by thermal fluctuations can be used to probe the relevance of microscopic interactions.

Interestingly enough, recent experiments performed on highly resistive CdTe seem to give support to the above theoretical predictions, as shown in Fig. 4, where we compare the experimental results to the theoretical predictions [11]

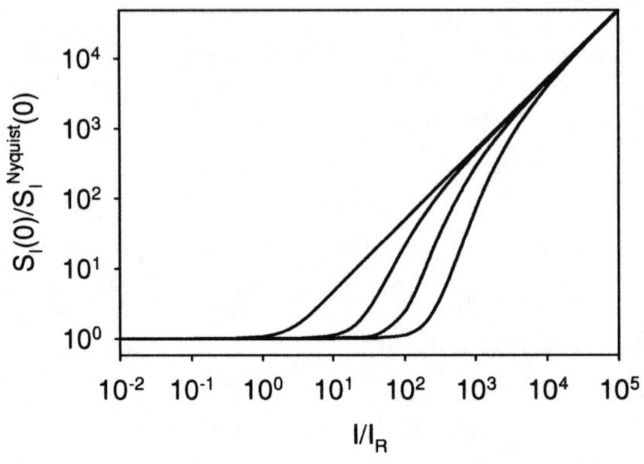

FIGURE 3. Low frequency current spectral density normalized to the Nyquist value as a function of the electric current normalized to a reference current for a homogeneous conductor in the macroscopic diffusive transport regime. Different values of the sample length are considered (curves from left to right L/L_D=*1*, *10*, *25* and *50*, where L_D is the Debye screening length).

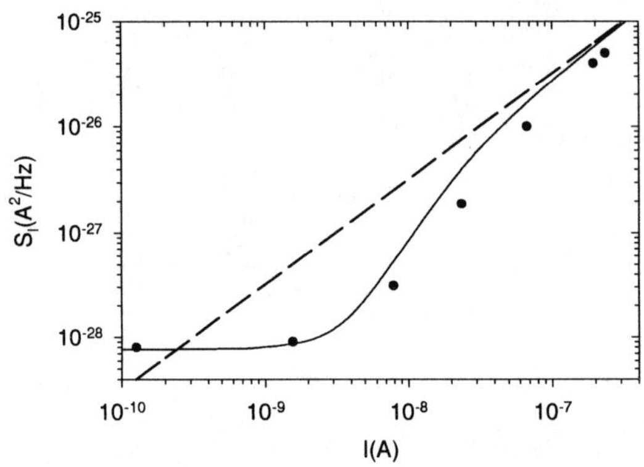

FIGURE 4. Calculated low frequency current spectral density as a function of current for a CdTe resistor (continuous line). Dashed line refers to the shot noise limit. Dots refer to the experimental results.

CONCLUSIONS

We have presented two examples where shot noise generated by thermal fluctuations can be used to probe microscopic interactions like the long-range Coulomb interaction and/or the Pauli (or Fractional Exclusion) Principle. The first example refers to the ballistic transport regime where current fluctuations are generated by the thermionic emission of carrier from the contacts. The second example concerns the macroscopic diffusive transport regime where current fluctuations are generated by the random inelastic scattering of carriers with lattice phonons. In both examples shot noise has a thermal origin, but it can be equally well used as a tool to probe the relevance of microscopic interactions. The broad generality of the physical arguments presented here suggests that the use of shot noise as a tool to probe microscopic interactions can be general for any electrically conducting system, and in particular, for those ones like biological systems (proteins, DNA, ionic channels, etc.) in which microscopic interactions are scarcely known at present.

ACKNOWLEDGMENTS

Most of the results presented in the present work have been obtained in close collaboration with Dr. Tomás González from Universidad de Salamanca (Spain), Prof. Lino Reggiani from Università di Lecce (Italy) and Inma R. Cantalapiedra from Universitat Politècnica de Catlunya (Spain). All of them are gratefully acknowledged. Financial support from the MCyT-Spain through the Ramon y Cajal Program and project No. BFM2001-2159 is gratefully acknowledged.

REFERENCES

1. Nyquist, H, *Phys. Rev. B* **32**, 110 (1928).
2. van der Ziel, A., *Noise in Solid State Devices and Circuits*, Wiley, New York, 1985.
3. Blanter, Ya. M. and Buttiker, M., *Phys. Rep.* **336**, 1 (2000).
4. Reznikov, M., Heiblum, M., Shtrikman, H. and Mahalu, D., *Phys. Rev. Lett.* **75**, 3340 (1995).
5. Henny, M., Oberholzer, S., Strunk, C. and Schonenber, C., *Phys. Rev. B* **59**, 13221 (1999).
6. González, T., Mateos, J., Pardo, D., Varani, L., and Reggiani, L., *Semicond. Sci. Technol.* **14**, L37 (1999).
7. Gomila, G., Cantalapiedra, I. R., González, T., and Reggiani, L., *Phys. Rev. B* **66**, 075302 (2002).
8. Gomila, G., and Reggiani, L., *Phys. Rev. B* **63**, 165404 (2001).
9. Gomila, G., and Reggiani, L., *Physica B* **314**, 141 (2002).
10. Gomila, G., and Reggiani, L. *Phys. Rev. B* **62** 8068 (2000).
11. Gomila, G., Pennetta, C., Reggiani, L., Sampietro, M., Ferrari, G., Bertuccio, G. in: *Noise in Physical Systems and 1/f fluctuations*, edited by G. Bosman, World Scientific, Singapur, 2001. pp.289

Can Quantum Regression Theorem Be Reconciled with Quantum Fluctuation Dissipation Theorem ?

P. Shiktorov*, E. Starikov*, V. Gružinskis* and L. Reggiani†

*Semiconductor Physics Institute, Goshtauto 11, LT2600 Vilnius, Lithuania
†INFM - National Nanotechnology Laboratory, Dipartimento di Ingegneria dell' Innovazione, Università di Lecce, Via Arnesano s/n, 73100 Lecce, Italy

Abstract. In the attempt to derive the regression theorem from the fluctuation dissipation theorem several authors claim the violation of the former theorem in the quantum case. Here we pose the question: does it exisist a quantum fluctuation dissipation theorem (QFDT) in its conventional interpretation ? It is shown that the relation usually called as the QFDT is the condition of detailed macroscopic energetic balance. Following this interpretation the existing conflict between the two theorems in the quantum case is removed.

INTRODUCTION

Under thermal equilibrium conditions the behavior of fluctuations of macroscopic observables of a physical system is governed by relationships which are formulated usually in terms of the regression theorem (the so called Onsager hypothesis [1]) and the fluctuation-dissipation theorem [2-6] (also known as the Nyquist relation). The former pertains to the time domain and states that the relaxation of a correlation of fluctuations is described by the same law governing the irreversible processes of the observable quantity itself. The latter pertains to the frequency domain and interrelates in some universal way the spectral characteristics of fluctuations and linear response (i.e. dissipation) of an observable of the physical system. Often, the fluctuation dissipation theorem is written as:

$$S_{xx}(\omega) = g^s(\omega) Im\{\alpha_x(\omega)\} \tag{1}$$

where $S_{xx}(\omega)$ is the spectral density of fluctuations of the observable x, $Im\{\alpha_x(\omega)\}$ the imaginary part of the generalized susceptibility which is responsible for the relaxation, and

$$g^s(\omega) = \hbar coth(\frac{\hbar\omega}{2kT}) = 2\hbar(\overline{n} + \frac{1}{2}) \tag{2}$$

is the Planck relating factor describing the average energy of the thermal radiation of the field mode with frequency ω.

In classical case $\hbar \to 0$, the two theorems give the same description of the spectrum of thermal fluctuations. By contrast, within the more general quantum case there appears a conflict between these two theorems since they predict different behavior of $S_{xx}(\omega)$. The quantum regression theorem (QRT) claims that the spectrum of fluctuations is

CP665, *Unsolved Problems of Noise and Fluctuations: UPoN 2002: Third International Conference,* edited by S. M. Bezrukov

determined only by the permitted transitions between energetic states of the system. In turn, the quantum fluctuation dissipation theorem (QFDT) states, that besides the eigenfrequencies of the system (as it required by the QRT) in the spectrum of the fluctuations there exist additionally the so called Matsubara frequencies, $\Omega_n = i\frac{2\pi kT}{\hbar}n$, where $n = \pm 1, \pm 2, ...$ The origin of these frequencies is related with the poles of the Planck factor $g(\omega)$.

This conflict is usually interpreted as a violation of the QRT (see, for example, Refs. [7-9]). In its most evident form such a violation is demonstrated in Ref. [9], where the conclusion statement announced that "there is no quantum regression theorem". The proof of the general character of such a statement is based on the fact that the violation of QRT follows from QFDT. However, a proof that QRT is valid independently of QFDT was given by Lax [10] on the basis of the general principles of quantum statistics (see also Refs. [11-13]). Since in Refs. [8,9] it is claimed that QFDT contradicts the validity of QRT, we argue that the origin of such a conflict is related with QFDT and its interpretation (see also Ref. [14]).

The aim of this article is to address this issue by considering the origin of such a conflict from a formal mathematical point of view.

FORMAL DERIVATION OF THE QFDT

In the operator representation, the symmetrized correlation function of the fluctuations of an observable x is written as [4-6]:

$$C_{xx}(\tau) = \frac{1}{2}Tr\{\hat{\rho}_s[\hat{x}(\tau)\hat{x}(0) + \hat{x}(0)\hat{x}(\tau)]\} \tag{3}$$

while the corresponding linear response function is given by Kubo formula [4-6]:

$$\alpha_x(\tau) = \frac{i}{\hbar}\Theta(\tau)Tr\{\hat{\rho}_s[\hat{x}(\tau)\hat{x}(0) - \hat{x}(0)\hat{x}(\tau)]\} \tag{4}$$

where $\hat{\rho}_s$ is the density operator which describes some stationary state of the physical system under test characterized by Hamiltonian $\hat{H}_s$, and $\Theta(\tau)$ is the unit step function.

The linear response described by Eq. (4) implies that the interaction between the system and the radiation is determined by the semi-classical Hamiltonian $\hat{V} = -\hat{x}f(t)$, where $f(t)$ is a classical force. By performing the Fourier transform of Eqs. (3) and (4), in accordance with ref. [6] one obtains:

$$\begin{bmatrix} S_{xx}(\omega) \\ \hbar Im\{\alpha_x(\omega)\} \end{bmatrix} = \frac{1}{2}\begin{bmatrix} J_{xx}(\omega) + J_{xx}(-\omega) \\ J_{xx}(\omega) - J_{xx}(-\omega) \end{bmatrix} \tag{5}$$

where

$$J_{xx}(\omega) = \int_{-\infty}^{\infty} Tr\{\hat{\rho}_s\hat{x}(\tau)\hat{x}(0)\}e^{i\omega\tau}d\tau = 2\pi\sum_{m,n}\rho_n|x_{mn}|^2\delta(\omega_{mn} - \omega) \tag{6}$$

is the spectral density corresponding to the one-directional in time correlation function $Tr\{\hat{\rho}_s\hat{x}(\tau)\hat{x}(0)\}$, $J_{xx}(-\omega)$ corresponds to $Tr\{\hat{\rho}_s\hat{x}(0)\hat{x}(\tau)\}$, ρ_n is the probability to find the system in the eigenstate with energy E_n, x_{mn} is the matrix representation of the operator $\hat{x}$ and $\omega_{mn} = (E_m - E_n)/\hbar$ is the frequency associated with the transition between the energetic states E_m and E_n.

Under thermal equilibrium $\hat{\rho}_s = exp(-\beta\hat{H}_s)/Tr\{exp(-\beta\hat{H}_s)\}$ with $\beta = 1/(kT)$. For the derivation of the QFDT an explicit expression for $J_{xx}(\omega)$ is not necessary, it is sufficient the existence of the quantum spectral relation [6]:

$$J_{xx}(-\omega) = e^{-\beta\hbar\omega} J_{xx}(\omega) \tag{7}$$

By substituting Eq. (7) into Eq. (5) one obtains:

$$\begin{bmatrix} S_{xx}(\omega) \\ \hbar Im\{\alpha_x(\omega)\} \end{bmatrix} = \frac{1}{2}\begin{bmatrix} 1 + p(\omega) \\ 1 - p(\omega) \end{bmatrix} J_{xx}(\omega) \tag{8}$$

where $p(\omega) = exp(-\beta\hbar\omega)$. According to Ref. [3-6], the QFDT is then obtained by excluding the factor $J_{xx}(\omega)$ which is common to both $S_{xx}(\omega)$ and $Im\{\alpha_x(\omega)\}$. As a consequence of such a derivation, the zeros of $Im\{\alpha_x(\omega)\}$ determined by the factor $1 - p(\omega) = 0$ (see Eq. (8)) become the poles of the Planck factor in Eq. (1), i.e., they originate the Matsubara frequencies and, in turn, the QRT-QFDT conflict. Thus, the poles of the Planck factor at the Matsubara frequencies can not be considered as independent of the frequency dependence of $Im\{\alpha_x(\omega)\}$ which, in accordance with Eq. (8), has zeros at the same frequencies:

$$Im\{\alpha_x(\omega)\}\Big|_{\omega=\Omega_n} = 0 \tag{9}$$

Therefore, from a mathematical point of view, in this case the right-hand side of Eq. (1) contains an indefinite form of $\frac{0}{0}$-type.

Now we pose the following open question: does the constraint given by Eq. (9) merely represent the formal requirement necessary for a rigorous derivation of the QFDT, or bring a proper physical meaning ?

An attempt to answer this question is detailed in the following section.

INTERRELATION BETWEEN THE QFDT AND THE PRINCIPLES OF ENERGY BALANCE

As physical model we shall consider a sufficiently large isolated system subdivided into two subsystems. The first corresponds to some physical system under test, the second represents the surrounding world. In this case the total Hamiltonian can be written as: $\hat{H} = \hat{H}_S + \hat{H}_T + \hat{V}$, where $\hat{H}_S$ and $\hat{H}_T$ are Hamiltonians of subsystems, and $\hat{V}$ describes the interaction between these subsystems. By using the standard procedure [15] to construct the master equations for the statistical operators of each subsystem, $\hat{\rho}_i$ $(i = S, T)$, and assuming that the interaction is weak one obtains the following equation for time variations of the average energy $< \hat{H}_i >$ in the i-th subsystem:

$$\frac{d}{dt} < \hat{H}_i > = \frac{1}{\hbar^2}\int_0^{\infty} Tr_{S+T}\{\hat{\rho}_S\hat{\rho}_T[[\hat{H}_i, \hat{V}], \hat{V}(-\tau)]\} d\tau \tag{10}$$

where $< \hat{H}_i >= Tr_i\{\hat{\rho}_i \hat{H}_i\}$, $\hat{V}(-\tau) = exp[-\frac{i}{\hbar}(\hat{H}_S + \hat{H}_T)\tau]\hat{V} exp[\frac{i}{\hbar}(\hat{H}_S + \hat{H}_T)\tau]$.

The energy exchange described by Eq. (10) satisfies the conservation law for the total energy of an isolated system in the form corresponding to the assumption of a weak interaction: $\frac{d}{dt} < \hat{H}_S + \hat{H}_T >= 0$. By assuming that the interaction Hamiltonian $\hat{V} = -\hat{x}\hat{f}$ is linear and factorized with respect to the variables of both subsystems the matrix representation in the right-hand side of Eq. (10) takes the form:

$$\frac{d}{dt}\begin{bmatrix} < \hat{H}_S > \\ < \hat{H}_T > \end{bmatrix} = \frac{\pi}{\hbar}\sum_{m,n}^{S}\sum_{M,N}^{T} |x_{mn}|^2 |f_{MN}|^2 \begin{bmatrix} \omega^S_{mn} \\ \omega^T_{MN} \end{bmatrix} (\rho^S_m \rho^T_M - \rho^S_n \rho^T_N)\delta(\omega^S_{mn} + \omega^T_{MN}) \quad (11)$$

From Eq. (11) one directly obtains the condition of the microscopic detailed energetic balance (MiDEB):

$$\rho^S_m \rho^T_M = \rho^S_n \rho^T_N \quad or \quad \frac{\rho^S_m}{\rho^S_n} = \frac{\rho^T_N}{\rho^T_M} \quad (12)$$

which must be satisfied with respect to only those energy states of both subsystems that are directly involved in the interaction, i.e., when $\omega^S_{mn} = \omega^T_{MN}$, $|x_{mn}|^2 \neq 0$, $|f_{MN}|^2 \neq 0$.

Now, let us formulate the conditions of the energy balance at the macroscopic level of description. For this sake in Eq. (11) we replace the term $\delta(\omega^S_{mn} + \omega^T_{MN})$ by $\int \delta(\omega^S_{mn} - \omega)\delta(\omega^T_{MN} + \omega)d\omega$ and rewrite Eq. (11) by using the matrix representation of the asymmetric spectral density $J(\omega)$ given by Eq. (6):

$$\frac{d}{dt} < \hat{H}_S >= -\frac{d}{dt} < \hat{H}_T >= \frac{1}{4\pi\hbar}\int \omega[J_{xx}(-\omega)J_{ff}(\omega) - J_{xx}(\omega)J_{ff}(-\omega)]d\omega \quad (13)$$

From Eq. (13) we obtain the condition of macroscopic detailed energy balance (MaDEB) as:

$$J_{xx}(-\omega)J_{ff}(\omega) = J_{xx}(\omega)J_{ff}(-\omega) \quad (14)$$

which requires to be fulfilled for any value of the current frequency ω. We notice that the condition given by Eq. (14) is not the only form which can be used to express such a detailed balance. For example, by using the definitions of $Im\{\alpha_x(\omega)\}$ and $S_{xx}(\omega)$ given by Eq. (5) it is easy to show that Eq. (14) can be rewritten in an equivalent form as:

$$Y_f(\omega)Im\{\alpha_x(\omega)\} = Y_x(\omega)Im\{\alpha_f(\omega)\} \quad (15)$$

where for $Y_i(\omega)$ one can use any of the three spectral densities, namely: $J_{ii}(\omega)$, $J_{ii}(-\omega)$ and $S_{ii}(\omega)$ $(i = x, f)$. The equivalence of Eqs. (14) and (15) from the point of view of the formal determination of $J(\omega)$, $S(\omega)$ and $Im\{\alpha(\omega)\}$ given by Eqs. (5) and (6) does not mean their entire macroscopic equivalence. This is related to the fact, that with respect to Eq. (14), Eq. (15) involves additionally the notion of the macroscopic function of linear response, $\alpha_i(\omega)$, all properties of which cannot be obtained from Eqs. (5) and (6) only.

Let us consider the conditions when the MiDEB given by Eq. (12) and the MaDEB given by Eqs. (14) and (15) can be considered as equivalent descriptions of the energy balance under steady state. In the following we shall use the formulation of balance conditions through some ratio of characteristics of a subsystem [see the right-hand side version of Eq. (12)] since such ratios sometimes are universal functions which are independent from the internal properties of the interacting subsystems. Let us rewrite Eq. (14) in the form:

$$\frac{J_{xx}(-\omega)}{J_{xx}(\omega)} = \frac{J_{ff}(-\omega)}{J_{ff}(\omega)} \equiv p(\omega) \quad or \quad J_{ii}(-\omega) = p(\omega) J_{ii}(\omega) \tag{16}$$

where $p(\omega)$ is common for both subsystems factor which is some single-valued function of the current frequency and it satisfies the condition $p(-\omega) = p^{-1}(\omega)$. It is easy to show that the equivalence of the energy balance description given by Eq. (12) and (16) is satisfied if $p(\omega)$ can be defined from the microscopic level as:

$$\frac{\rho_m^S}{\rho_n^S} = \frac{\rho_N^T}{\rho_M^T} = p(\omega)\Big|_{\omega=\omega_{mn}=\omega_{NM}} \tag{17}$$

Let us rewrite the MaDEB condition given by Eq. (15) in a form analogous to Eq. (16)

$$\frac{Y_x(\omega)}{Im\{\alpha_x(\omega)\}} = \frac{Y_f(\omega)}{Im\{\alpha_f(\omega)\}} \equiv g^Y(\omega) \quad or \quad Y_i(\omega) = g^Y(\omega) Im\{\alpha_i(\omega)\} \tag{18}$$

Here the function of current frequency $g^Y(\omega)$ will depend on which spectral density [i.e., the symmetric $S(\omega)$ or asymmetric $J(\pm\omega)$] is used to formulate the balance conditions. The fulfillment of Eq. (16) allows us to represent the frequency dependence of $S_{ii}(\omega)$ and $Im\{\alpha_i(\omega)\}$ of both subsystems in a form entirely analogous to thermal equilibrium [see Eq. (8)] with the only difference that now $p(\omega)$ is not necessarily given by the thermal value. From the above it is easy to see that all the functions $g^Y(\omega)$ in Eq. (18) are determined by by the frequency dependence of $p(\omega)$ only:

$$g^Y(\omega) = \frac{\hbar}{1-p(\omega)} \begin{cases} 1 & , for \quad J(\omega) \\ p(\omega) & , for \quad J(-\omega) \\ 1+p(\omega) & , for \quad S(\omega) \end{cases} \tag{19}$$

When $p(\omega)$ is a universal function of frequency [e.g., in thermal equilibrium when $p(\omega) = exp(-\beta\hbar\omega)$] there is the possibility to formulate the MaDEB conditions in terms of expressions which relate in some universal way the macro-characteristics of only one of the interacting subsystem [see the right-hand side expressions in Eqs. (16) and (18)]. However, in so doing, it is easy to loose the physical meaning of these expressions.

In thermal equilibrium, Eq. (18) allows the QFDT to be given an alternative physical interpretation with respect to the conventional one. Indeed, by replacing in Eq. (18) $Y_i(\omega)$ with the symmetrical spectral density $S_{ii}(\omega)$ one obtains the MaDEB condition [right-hand side expression in Eq. (18)] in a form which is identical for both the subsystems and which coincides with the conventional form of the QFDT given by Eq. (1). This allows us to conclude that the QFDT describes the detailed energetic balance between the interacting physical systems under thermal equilibrium. As a consequence, the usual interpretation that the frequency dependences of the Planck factor $g^s(\omega)$ [see Eq. (2)] and of the imaginary part of the generalized susceptibility $Im\{\alpha_i(\omega)\}$ are independent (the source of the QRT-QFDT conflict) is in contradiction with both the MiDEB and MaDEB principles. When these principles are fulfilled, the poles of $g^s(\omega)$ and the zeros of $Im\{\alpha_i(\omega)\}$ are determined by the same factor $1-p(\omega)$ [see Eqs. (19) and (8), respectively]. The neglect of this property when treating fluctuation phenomena corresponds in essence to the violation of the MiDEB and MaDEB principles and, hence, it implies the violation of the condition of stationarity at least at the Matsubara frequencies $\omega = \Omega_n$ where $p(\Omega_n) = 1$.

The formulation of the balance conditions represented in terms of a ratio of macro-characteristics of one or of both subsystems can serve as a source of incorrect interpretation of the MaDEB principle and, as a consequence, can lead to the violation of the energy conservation law for subsystems with partially overlapping energetic spectra. The typical example is the interaction between the harmonic oscillator with eigenfrequency ω_s and the thermal bath characterized by a quasi-continuous spectrum. By using for the matrix elements of oscillator the relations [15]: $|x_{mn}|^2 = (1/\omega_s)\gamma_s^2\hbar(n+1)\delta_{m,n+1}$ and $\omega_{mn}^S = \omega_s(m-n)$, from Eq. (6) one obtains asymmetric spectral densities $J(\pm\omega)$ represented as:

$$\begin{bmatrix} J(-\omega) \\ J(\omega) \end{bmatrix} = 2\pi\frac{\gamma^2(\omega)\hbar}{\omega}\begin{bmatrix} \overline{N}(\omega) \\ \overline{N}(\omega)+1 \end{bmatrix}\delta(\omega_s-\omega) \tag{20}$$

where $\overline{N}(\omega_s) = \sum_{n=0}^{\infty} np_n = [exp(\beta\hbar\omega_s)-1]^{-1}$, $\gamma^2(\omega_s)$ is the electro-dipole matrix element square which can depend upon the oscillator frequency. In Eq.(20), due to the properties of the δ-function we have performed the replacement usually used in such situations when dependence of $\gamma^2(\omega_s)$ and $\overline{N}(\omega_s)$ on ω_s is replaced by their dependence on the current frequency ω. In this way, from Eq. (20) one obtains:

$$\frac{J(-\omega)}{J(\omega)} = p(\omega) = \frac{\overline{N}(\omega)}{\overline{N}(\omega)+1} \tag{21}$$

which defines in explicit form $p(\omega)$ in the whole frequency range, while the spectrum of a single oscillator is defined at the frequency $\omega = \omega_s$ only, i.e. it is described by a δ-function which in Eq. (21) disappeared from explicit consideration.

By considering the thermal bath, also in this case it is used the harmonic oscillator model which describes one mode of the radiation field. To obtain the spectral densities $J(\pm\omega)$ for the thermal bath it is sufficient to perform in the right-hand side of Eq. (20) a summation over all the possible modes, that is over subindex s. In so doing, the dependence of $\gamma^2(\omega_s)$ and $\overline{N}(\omega_s)$ on ω_s but not on ω can remain. By using the standard procedure for systems with a quasi-continuous spectrum, i.e., by replacing $\sum_s$ by $\int G(\omega_s)d\omega_s$, where $G(\omega_s)$ is the density of the radiation mode with frequency ω_s in the interval $d\omega_s$, one obtains:

$$\begin{bmatrix} J(-\omega) \\ J(\omega) \end{bmatrix} = 2\pi\frac{G(\omega)\gamma^2(\omega)\hbar}{\omega}\begin{bmatrix} \overline{N}(\omega) \\ \overline{N}(\omega)+1 \end{bmatrix} \tag{22}$$

From Eq. (22) one can again obtain Eq. (21), which is now valid in the whole frequency range. When a single oscillator interacts with the thermal bath, the interaction takes place not in the whole frequency range, as it would follow from Eqs. (21) or (18), but only at the oscillator eigen-frequency, where the energy exchange is only possible.

CONCLUSIONS AND OPEN QUESTIONS

Here we have shown that the QFDT expressed in its usual form of Eq. (1) in essence represents the macroscopic principle of detailed energy balance between a physical system and the surrounding world interacting with it. Such an interpretation, and the related restrictions concerning the frequency dependence of the imaginary part of the generalized susceptibility of the system [see Eq. (9)], leads to two main consequences. First, the announced conflict between QRT and QFDT related with the Matsubara frequencies is reconciled. Second, some conventional notions as macroscopic dissipation and relaxation closely related to the QFDT should be revisited. Indeed, these notions cannot be treated as internal characteristics of one of the two interacting subsystems which compose the total isolated system.

In the framework of the macroscopic approach where the energy dissipation is described by $Im\{\alpha(\omega)\}$, the energy conservation law under the energy balance implies that the power dissipated by one of subsystems from another must be returned back, i.e. it is equal to the power dissipated by another subsystem from the first. The form of the MaDEB given by Eq. (15) just express this mathematically, by claiming that the spectra of the power dissipated by each subsystem from another are equal. Therefore, with respect to one subsystem, one side of Eq. (15) can be treated as the dissipated power and another side as the power returned back, i.e. as emitted power. This is illustrated by the following example, which considers the system interaction with thermal bath. By using Eq. (22), the MaDEB conditions given by Eq. (15) take the form:

$$2\hbar\omega\begin{bmatrix}\overline{N}(\omega)+\frac{1}{2}\pm\frac{1}{2}\\ \overline{N}(\omega)+\frac{1}{2}\end{bmatrix}Im\{\alpha_x(\omega)\}=\omega\begin{bmatrix}J_{xx}(\pm\omega)\\ S_{xx}(\omega)\end{bmatrix} \tag{23}$$

The left-hand side of Eq. (23) describes the power dissipated by the system when this power is provided by the thermal bath. The right-hand side of Eq. (23) describes the power which the system returns back to the thermal bath. If the only photon part, $\hbar\omega\overline{N}(\omega)$, of the full energy of the field is involved into the absorption process, then the returning spectrum, $J_{xx}(-\omega)$, corresponds to the spectrum of the spontaneous emission processes occurring in the system. This is easy shown by using the explicit form of $J(-\omega)$ given by Eq. (6). If now one reverses time, i.e. replace ω by $-\omega$, the energy of the thermal bath which is involved into dissipation process will be $\hbar\omega[\overline{N}(\omega)+1]$. This means that with respect to one of the interacting subsystems the processes of energy exchange in equilibrium conditions shows an asymmetry under time reversal. If the system has absorbed the full energy of the field together with the zero-field value, $\hbar\omega[\overline{N}(\omega)+\frac{1}{2}]$ (the case of the QFDT), to fulfill MaDEB it would be necessary that the system returns back the same full energy. In this case, the emitted spectrum which returns back would be described by the symmetric spectral density $S_{xx}(\omega)$. Here the following question arises, which spectrum of fluctuations and correspondingly which correlation function are experimentally measured: the asymmetric or symmetric ones ?

In applications, the QFDT is often used to describe the relaxation phenomena in systems interacting with a thermal bath. In this interpretation as one of the formulation of the MaDEB conditions there appears a question about the correctness of such a use. Indeed, if the relaxation process implies that the system is approaching the stationary state, it means that the system is not in a stationary state. It is evident that under nonstationary conditions all the relations considered above loose their meaning, and any attempt to use them will imply a violation of the energy conservation law. Thus, the QFDT can be applied for the description of only those relaxation phenomena which occur under energy equilibrium conditions, as described by the QFDT itself.

The MaDEB conditions lead to another important property of the relaxation processes: the description of the relaxation in the interacting subsystems which compose an isolated system cannot be considered in independent way. From the definition of the linear response function $\alpha_i(\omega)$ it follows that the spectral densities of the fluctuations in each of the subsystems must be interrelated as $S_{ii}(\omega) = |\alpha_i(\omega)|^2 S_{jj}(\omega)$ where $i \neq j$. Together with the MaDEB in the form given by Eq. (18) this leads to additional restrictions put on the frequency dependence of $Im\{\alpha_i(\omega)\}$ pertaining to each of the two subsystems:

$$Im\{\alpha_i^{-1}(\omega)\} = Im\{\alpha_j(\omega)\} \quad (24)$$

where $Im\{\alpha_i^{-1}(\omega)\} = Im\{\alpha_i(\omega)\}/|Im\{\alpha_i(\omega)\}|^2$ is the relaxation law in one of the subsystems, which in accordance with Eq. (24) is determined by the imaginary part of the generalized susceptibility of the other subsystem. For example, in the presence of an interaction of some systems with a thermal bath characterized by the radiation mode density $G(\omega) = \omega^2\pi^{-2}c^{-3}$ [15] the relaxation in the system must be described by the law:

$$Im\{\alpha_i^{-1}(\omega)\} = \frac{1}{2\hbar}[J_{ff}(\omega) - J_{ff}(-\omega)] = \frac{\gamma^2(\omega)}{\pi c^3}\omega \quad (25)$$

If the electrodipole matrix element square, γ^2, is independent of ω, one obtains the usual law of viscous friction.

The financial support of the NATO collaborative-linkage grant PST.CLG.977520, and the french-lithuanian bilateral cooperation n. 12864 of french CNRS is acknowledged.

REFERENCES

1. L. Onsager, *Phys. Rev.* **38**, 2265 (1931).
2. H. Nyquist, *Phys. Rev.* **32**, 110 (1928).
3. H.B. Callen and T.A. Welton, *Phys. Rev.* **83**, 34 (1951).
4. L. Landau and E. Lifshitz, *Statistical Physics* (Addison Wesley Reading, Mass., 1974)
5. R. Kubo, M. Toda, N. Hashitsume, *Statistical Physics II* (Spriger-Verlag, Berlin, 1985)
6. D.N. Zubarev, *Nonequilibium Statistical Thermodinamics*, (Nauka, Moscow, 1971)
7. H. Grabert, *Z. Phys.* **B 49**, 161 (1982).
8. P. Tolkner, *Ann. Phys. (N.Y.)* **167**, 380 (1986).
9. G.W. Ford and R.F. O'Connell, *Phys. Rev. Lett.* **77**, 798 (1996).
10. M. Lax, *Phys. Rev.* **129**, 2342 (1963).
11. V.H. Louisell, *Quantum Statistical Properties of Radiation* (Wiley, New York, 1973), Sec. 6.6.
12. C. Cohen-Tannoudji, J. Dupont-Roc and G. Grinberg, *Atom-Photon Interactions* (Wiley, New York, 1992), Chap. IV, p. 350.
13. A. Mandel and E. Wolf, *Optical Coherence and Quantum Optics* (Cambridge University Press, Cambridge, 1995), Sec. 17.1.
14. P. Shiktorov, E. Starikov, V. Gružinskis, L. Reggiani L. Varani and J.C. Vaissière, *cond-matt/0011420*, Nov. 24, 2000.
15. P. Meystre, M. Sargent, *Elements of Quantum Optics* (Springer-Verlag, 1991).

Excess Noise In The Output Of Linear Fluctuating Systems

Michael H.W. Hoffmann

University of Ulm, Department of Microwave Techniques
Albert-Einstein-Allee 41, D-89069 Ulm, Germany
Fax: +49-7301-5026359, Email: Michael.Hoffmann@ieee.org

Abstract. The impulse-response of many linear systems in biophysics, medicine, electronics, economics and other fields fluctuates randomly; i.e.; it depends not only upon the time-*difference* from the excitation of the impulse; but it depends also randomly upon the absolute point in time the impulse was applied to the system. These systems might be described, therefore, as linear time-variant (LTV) systems. Application of the theory of LTV-systems shows the power-spectrum of the output of such a system, driven by input-noise, contains additional terms as compared to linear time-invariant systems. In this paper a very simple LTV-system is considered that contains four different terms at its output; one term is seen to be proportional to the input-power-spectrum; a second term is proportional to the power spectrum of the system's fluctuations; a third term describes additional equilibrium output-noise. While these three terms are expected to appear in the output-spectrum, the fourth term is less known: It is due to the interactions between the input-noise and the system's fluctuations. It will be shown that this excess noise might contain power at frequencies that did not appear in the input-spectrum, nor in the spectrum of the fluctuations. Its occurrence might potentially explain so-far inexplicable spectra following a $1/f^{\alpha}$-law, and raises, furthermore, numerous new questions.

INTRODUCTION

Linear, noise-generating systems are known from many fields of research as for instance biology, physics, medicine, electronics and economics. The system properties of many of them may fluctuate randomly with time. Examples of such systems are (under certain constraints) ion-channels in biological cells or resistors in semi-conducting material, or flow of goods between two regions with differing prices for those goods.

In this paper, a simple model is considered that takes into consideration possible fluctuations of the transfer behavior. In contrast to the usually assumed time-invariance of the transfer function, a linear time-variance of the transfer function is considered. From a mathematical point of view, time-variant parameters of the describing equations are allowed rather than constant parameters. There exists a well-proven theory of linear time-varying systems, see for example [1, 2], which will be applied to describe the model.

CP665, *Unsolved Problems of Noise and Fluctuations: UPoN 2002: Third International Conference,* edited by S. M. Bezrukov

METHODS

The methods applied in this paper are demonstrated with the model depicted in Fig. 1. It has a single, scalar input-signal proportionally linked to the output-signal, where the factor of proportionality is fluctuating with time. An additional internal source of noise exists that describes equilibrium noise (i.e., noise when the system is not driven by an input-signal). In this model, $s(t)$ is the input to the system and $g(t)$ its output.

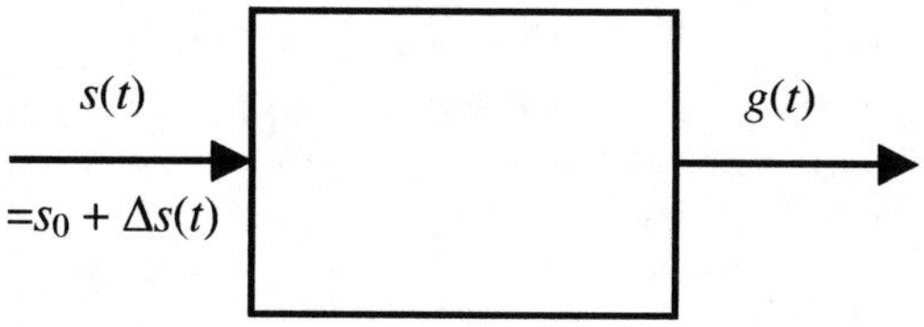

FIGURE 1. Simple linear, time-variant system with fluctuating input

The input is described as a constant term s_0 with a randomly varying term $\Delta s(t)$ superimposed upon it.

$$s(t) = s_0 + \Delta s(t) \quad . \tag{1}$$

The impulse-response of the system under idealized conditions and with respect to an impulse that is excited at point ξ in time is assumed to be:

$$h(t,\xi) = \{h_0 + \Delta h(t)\}\delta(t-\xi) \tag{2}$$

The output of the signal is consequently assumed as

$$g(t) = \int_{-\infty}^{\infty} h(t,\xi)s(\xi)d\xi = \{h_0 + \Delta h(t)\}\{s_0 + \Delta s(t)\} + \Delta g(t) \quad , \tag{3}$$

where $\Delta g(t)$ takes into account fluctuations of the output that are also present, when the input is identically zero (equilibrium fluctuations).

As an example, consider an ohmic resistor with fluctuating resistance $r_0 + \Delta r(t)$ that is driven by a current with average value i_0, and that is subject to the shot effect and thus fluctuates by a value $\Delta i(t)$. This resistance will produce an additional thermal noise voltage $\Delta u(t)$. Thus, the complete output voltage of the resistor is

$$u(t) = \{r_0 + \Delta r(t)\}\{i_0 + \Delta i(t)\} + \Delta u(t) \quad . \tag{4}$$

The aim of this paper is to compare the power spectrum of the input $s(t)$ with that of the output $g(t)$, or in the case of the resistor, to compare the power spectrum of $i(t)$ with that of the output $u(t)$.

In the case of the above example, it is also of interest to compare the spectrum of the power dissipated in the resistor, which is the spectrum evolving from the cross-correlation between input and output. Therefore, the spectra evolving from auto-correlation functions as well as that evolving from the cross-correlation function of a system as described by equation (3) are analyzed in the following.

In order to be able to compute correlation functions and from them spectra, further assumptions on the statistical properties of the processes $\Delta s(t)$, $\Delta g(t)$ and $\Delta h(t)$ are necessary. It will be assumed that all three processes have zero-mean and that they are wide-sense stationary (WSS) and mutually uncorrelated.

Under these conditions, it will then be easy to compute statistical average values and correlation functions:

$$\langle g(t)\rangle = h_0\, s_0 \quad ; \tag{5}$$

$$R_s(\tau) = s_0^2 + R_{\Delta s}(\tau) \quad ; \tag{6}$$

$$R_g(\tau) = h_0^2\left\{s_0^2 + R_{\Delta s}(\tau)\right\} + R_{\Delta g}(\tau) + s_0^2 R_{\Delta h}(\tau) + R_{\Delta h}(\tau)\, R_{\Delta s}(\tau) \quad ; \tag{7}$$

$$R_{g,s}(\tau) = h_0\left\{s_0^2 + R_{\Delta s}(\tau)\right\} \quad ; \tag{8}$$

where $R_X(\tau)$ is the auto-correlation function of the process X and where $R_{X,Y}(\tau)$ is the cross-correlation function of the processes X and Y. A Fourier transform of the correlation-functions then yields the studied spectra.

RESULTS AND DISCUSSION

The Output Spectrum

Applying a Fourier transform to the above correlation functions results in:

$$S_s(\omega) = s_0^2\, 2\pi\, \delta(\omega) + S_{\Delta s}(\omega) \quad ; \tag{9}$$

$$S_g(\omega) = \underbrace{h_0^2\, S_s(\omega)}_{\text{spectrum transferred from input}} + \underbrace{S_{\Delta g}(\omega)}_{\text{equilibrium noise}} + \underbrace{s_0^2 S_{\Delta h}(\omega)}_{\text{spectrum due to fluctuations of impulse-response}} + \underbrace{\frac{1}{2\pi}\int_{-\infty}^{\infty} S_{\Delta s}(\eta)\, S_{\Delta h}(\omega-\eta)\, d\eta}_{\textbf{excess spectrum}} \quad , \tag{10}$$

$$S_{g,s}(\omega) = h_0 S_s(\omega) \quad . \tag{11}$$

While the first term in the output spectrum S_g is not a surprise at all, and while the next two terms might also be expected, the last term seams to be less well known (Kish has already treated a special case [3], where this term appeared in a different notation that did not lead to a convolution integral).

It is emphasized that the excess-noise spectrum is *not a product of two spectra* as it is often assumed in the literature when dealing with fluctuating systems.

It is pointed out in addition that for the output power-spectrum

$$S_g(\omega) \neq h_0^2\, S_s(\omega) \quad ; \quad S_g(\omega) \neq h_0^2\, S_s(\omega) + S_{\Delta g}(\omega) \quad ; \tag{12}$$

as opposed to assumptions that are found frequently in the literature, although the system is supposed to fluctuate!

As an example, consider an ohmic resistor with fluctuating resistance $r_0 + \Delta r(t)$ that is driven by a current with average value $i_0 + \Delta i(t)$, and which generates a thermal noise voltage $\Delta u(t)$. Then the voltage spectrum is *not* proportional to the current power spectrum. This proportionality might potentially apply as an approach, but it is certainly not the whole truth. For that same example,

$$S_{u,i}(\omega) = r_0 S_i(\omega) \quad ; \tag{13}$$

indicates that the (true) power spectrum follows another rule other than the power-spectrum found from the voltage.

The most important message from equation (10), however, is the fact that there is an excess noise term that might change the spectrum in a very interesting way.

Spectrum-Spreading By An LTV-System

In order to get a first impression of how the output spectrum might be influenced, some simple examples are shown in Figures 2 to 4.

In Figure 2, the power-spectra of the input and the fluctuations of the system are shown together with the excess (!) power-spectrum of the output.

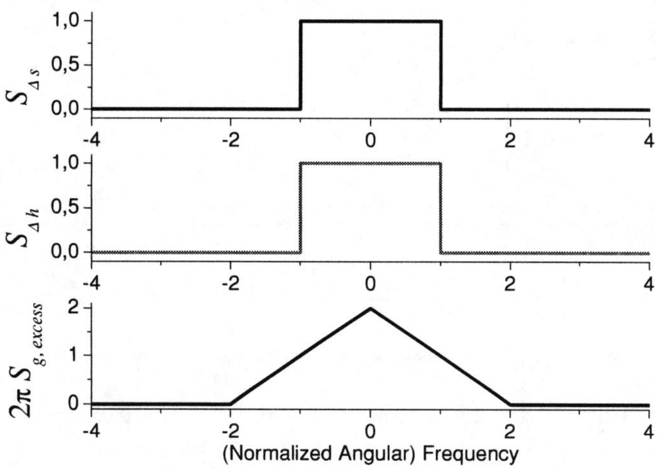

FIGURE 2. Excess spectrum for a simple case

Note that even at frequencies where the power-spectrum of the input as well as that of the fluctuations vanished, the output power-spectrum differs from zero: The output-spectrum is spread as compared to both the input spectrum and the spectrum of the fluctuations. That is definitely a behavior, which is completely unexpected in linear time-*invariant* systems. It is, however, a natural behavior of a linear time-*variant* system.

In a next step, a cascade of two systems each with a behavior identical to that of Fig.2 and with the same input as before is considered. The excess-part of its output power-spectrum is shown in Fig.3.

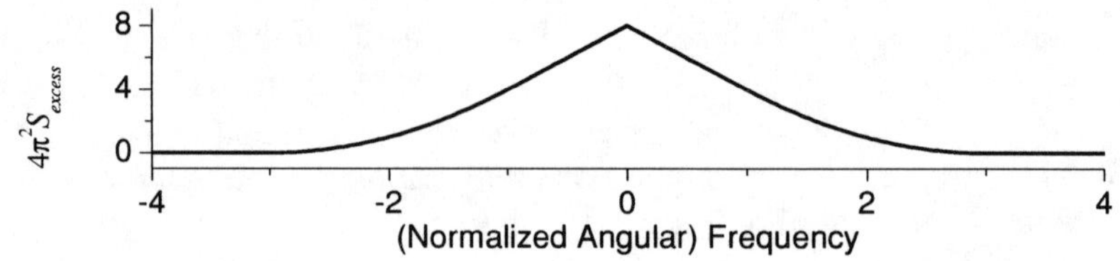

FIGURE 3. Excess spectrum for a simple case

Note that the spectrum is not only spread again, but, additionally, it is smoothed. This behavior develops more and more with an increasing number of stages in the cascade. This is demonstrated in Figure 4, where excess spectra at the output terminals of several stages in a cascade of similar linear fluctuating systems are shown:

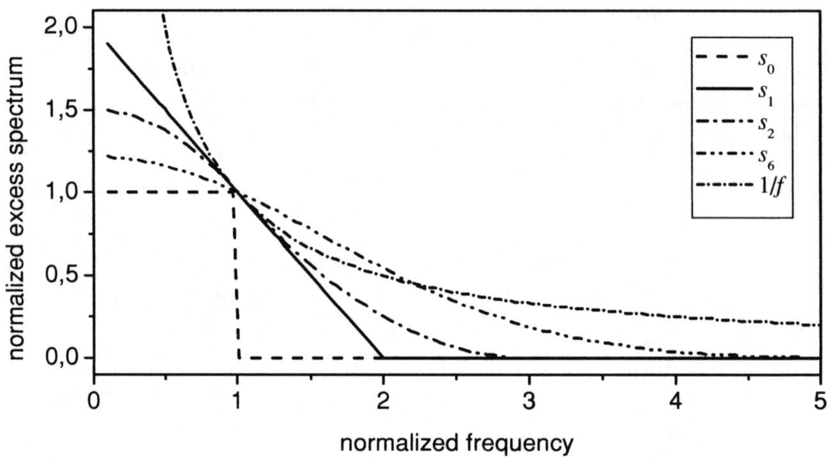

FIGURE 4. Excess spectrum for a simple case

s_0 is the spectrum of the fluctuations in the input of the cascade, s_1 is the excess-spectrum at the output of the first stage, s_2 is the excess spectrum at the output of the second stage and s_6 is the excess spectrum at the output of the sixth stage. For comparison, a $1/f$ curve is shown. All curves are normalized to the value 1 at normalized frequency 1.

Curve s_n vanishes for absolute values of the normalized frequency greater than $n+1$. All curves s_n are monotonously decreasing. For sufficiently large value of the normalized frequency, $1/f$ is an upper bound for s_n.

The displayed results apply for a very special and not too realistic assumption of the input variations and for the fluctuations of the impulse response. They demonstrate, however, that there might be alterations of the output- spectrum in principle, which have not been taken into consideration so far.

Whether these properties of linear, fluctuating systems might be the key to explain $1/f^{\alpha}$-noise in such systems is an unsolved problem.

OPEN QUESTIONS

There remain numerous other open questions.

Since, traditionally, excess-spectra have not been expected, established practice needs to be re-checked with respect to linear, time-*variant* systems. This is not a triviality, since one of the typical underlying assumptions has been (with vanishing equilibrium fluctuations) S_g be $h_0^2\ S_s$, which has been shown *not* to apply in all cases. Therefore, the following questions are open for discussion again:

1. Can evidence be given for the fluctuations of the parameters of the transfer function in concrete examples? If so,
   i) What are the statistical properties of these parameters?
   ii) Can causes be identified for these properties?
2. Based on these findings, will it be possible to design systems with minimized excess noise?
3. With these findings, will it be possible to use that noise as piece of information rather than as disturbance?

Another group of questions is more specific to the basic theory of time-varying systems:

4. How will output spectra be influenced, if there are multi-dimensional input signals and multi-dimensional output signals?
5. How will output spectra be influenced, if the system has memory effects?
6. Will non-linear effects hide or emphasize this behavior?

The most exciting question, however, (at least for this author) is:

7. Might $1/f^{\alpha}$-noise find its explanation as a fundamental property of any distributed (cascaded) time-varying system?

REFERENCES

1. Zadeh, L. A., *Circuit Analysis of Linear Varying-Parameter Networks.* J. Appl. Phys. **21**, 1171-1177 (1950).
2. Bello, P. A., *Characterization of Randomly Time-Variant Linear Channels. IEEE* Trans. Com. **11**, 360-393 (1963).
3. Kish, L. B., *An excess Johnson-noise due to 1/f and other bulk conductivity noises.* Solid State Communications **60**, 125-127 (1986).

Electronic Delocalization in Finite One-Dimensional Correlated–Disordered Binary Solids

Plamen Ch. Ivanov*, Pedro Carpena†, Pedro Bernaola–Galván† and H.Eugene Stanley*

*Center for Polymer Studies and Department of Physics, Boston University, Boston, MA 02215
†Center for Polymer Studies and Department of Physics, Boston University, Boston, MA 02215 and Departamento de Física Aplicada II. E.T.S.I. de Telecomunicación, Universidad de Málaga, 29071 Málaga, Spain

Abstract. We report a strong electronic delocalization when long-range power-law correlations are introduced in 1-D disordered binary solids. We perform numerical simulations using a 1-D tight-binding model, and we find that the long-range correlations in the system lead to a considerable extension of the localization length for a large fraction of the electrons in the system within a broad energy band. We discuss the possible relevance of this result for DNA, which displays long-range correlations and was recently reported to be a 1-D disordered conductor.

The behavior of electron wave functions in perfectly ordered crystals is governed by Bloch's Theorem, which states that the probability of finding an electron is the same over the entire crystal [1], so electron wave functions are extended. The extended states of ordered systems can carry current, and thus the system behaves as a metal (conductor). However, electron states can become localized when disorder is introduced in the solid, and the system can become an insulator. This phenomenon is called Anderson localization [2], and the metal-insulator transition driven by the disorder is known as "Anderson transition", which has attracted considerable interest [3, 4].

Here we describe the solid with a tight-binding model characterized by a Hamiltonian in which one orbital (single electron) and a single random energy ε_i are assigned to each lattice site i

$$H = \sum_i \varepsilon_i \,|i><i| + \sum_{<i,j>} t\,|i><j|, \tag{1}$$

where t is the electronic overlap (hopping term) between the wave functions of electrons centered at two nearest neighbor sites i, j. For a perfectly ordered crystal, the site energies have the same value or follow a periodic pattern. For disordered solids, the energies ε_i can be randomly chosen from a certain probability distribution, e.g. uniform distribution or Gaussian. We choose t to be unity, thus fixing the energy scale of the system. The probability distribution generating the site energies ε_i is characterized by a parameter W which accounts for the spread of the distribution and thus for the degree of disorder in the system. In the case of a uniform distribution, W can be the distribution width, while in the case of Gaussian distribution W can be the standard deviation.

CP665, *Unsolved Problems of Noise and Fluctuations: UPoN 2002: Third International Conference*, edited by S. M. Bezrukov

In the $W \to 0$ limit, when the coupling energy t between two neighbors dominates, we recover a perfect lattice, and Bloch's Theorem applies: there are extended electron states and the system can conduct [1]. For sufficiently large W, the order is lost, and the electron wave functions become localized. This localization phenomenon strongly depends on the system dimension. For 3-D systems at zero temperature, there is a critical value W_c for which a metal–insulator transition occurs [5, 6]. In the case of the uniform distribution $W_c = 16.5$. For $W < W_c$, in spite of some degree of disorder, the electron wave functions are sufficiently extended, and the system behaves as a metal, while for $W > W_c$, the electron wave functions become localized, and the system behaves as an insulator. For 2-D and 1-D disordered systems, it is commonly accepted that even infinitesimal disorder will produce localized states and thus at zero temperature the system behaves as an insulator in the thermodynamic limit, although in 2-D it is possible to find a metal-insulator transition for strongly interacting systems [7], and also if a magnetic field is applied [8] or if spin-orbit coupling is taken into account [9].

Here we consider the case of 1-D system. We show that randomly chosen but *long-range correlated* site energies ε_i can lead to extended wave functions for large but finite systems and nearly extended wave functions for infinitely large systems and thus can lead to conductivity and enhanced electron transport, in contrast to the expectation (based on the assumption of uncorrelated disorder) that no extended states can be found in 1-D disordered systems [1] at zero temperature. Traditionally, localization of the electron wave function in 1-D is considered to be unaffected by the form and width of the distribution from which the site energies ε_i are chosen. Although all proofs establishing particle localization in 1-D are model–dependent, exponential localization of all eigenstates in 1-D is believed to occur at zero temperature [10, 11, 12]. In this case, electron wave functions are of the form

$$\Psi(x) = f(x)e^{-|x-x_0|/\lambda}, \tag{2}$$

where $f(x)$ is a random function which depends on the particular realization of the disordered chain, and λ is the localization length of the state, which intuitively is a measure of the size of the wave function.

The simplest 1-D model exhibiting Anderson localization is the random binary alloy [13], where there are two types of atoms, A and B, and correspondingly two different diagonal energies, ε_A and ε_B. To build the series of diagonal energies of the Hamiltonian describing the system (1), ε_A and ε_B are assigned at random without correlation to each lattice site with probabilities p and $1-p$ respectively. An example of a possible segment of an infinite lattice is $\ldots\varepsilon_B\varepsilon_A\varepsilon_A\varepsilon_B\varepsilon_A\varepsilon_B\varepsilon_B\varepsilon_A\varepsilon_B\varepsilon_A\ldots$. For the random binary alloy, the corresponding wave functions are exponentially localized and the system behaves as an insulator (see Fig. 1).

Next, we show that introducing correlations in the disorder can change the physics dramatically. The first attempt to introduce correlations into the random binary alloy model was the random dimer model [14, 15]. In this model, *short-range* correlations are introduced as follows: a dimer $\varepsilon_A\varepsilon_A$ is assigned to each lattice site with probability p and a monomer ε_B with probability $1-p$. A possible realization could be $\ldots\varepsilon_B\varepsilon_A\varepsilon_A\varepsilon_B\varepsilon_A\varepsilon_A\varepsilon_A\varepsilon_A\varepsilon_B\varepsilon_B\ldots$ The main property of the random dimer model is that there exists a particular energy (the resonant energy of the dimer $\varepsilon_A\varepsilon_A$) for which there is a

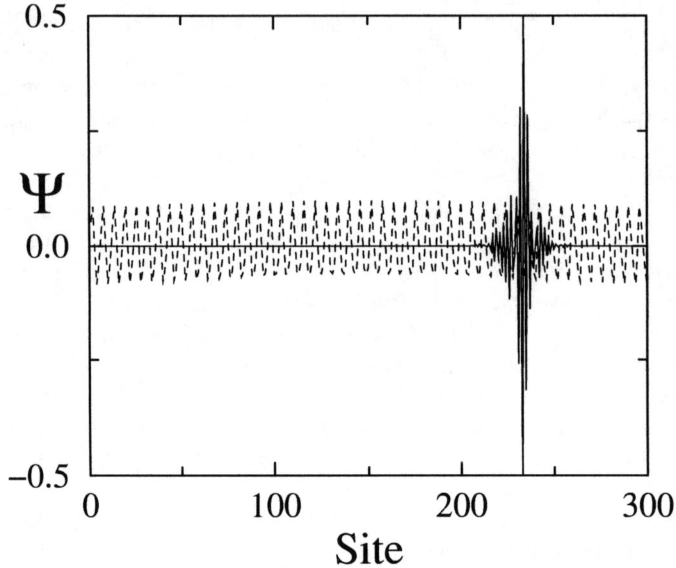

FIGURE 1. Typical wave functions for a periodic binary chain (dashed line) and for the random binary alloy (solid line). Both are calculated in a system of 300 atoms, and with $\varepsilon_A = 1/2$ and $\varepsilon_B = -1/2$.

perfect electron transmission through the system. Wave functions corresponding to energies close to this resonant energy behave as extended states. However, they do not form an energy band of extended states, since the number of extended wave functions with energy values close to this resonant energy is of the order of $\sqrt{N}$, where N is the system size [14, 15]. Thus the fraction of extended states is negligible in the limit of large systems. The random dimer model has been applied to explain the conduction properties of some conducting polymers [14, 15]. A recent attempt to introduce long-range correlations in 1-D systems utilizes site energies ε_i chosen to form a long-range correlated continuous function [16]. However, to maintain the spread of the density of states, a normalization is needed which ultimately eliminates the correlations in the site energies and produces a series of energies that tend to be constant [17]. If such a normalization is not performed, even stronger localization than in the purely random case is observed [17].

We next propose a new 1-D model of correlated disorder which generates extended states and thus leads to conductivity in finite but very large system sizes of $\approx 1\mu m$ which are of interest in nanotechnology. We generate binary sequences with long–range correlations using the modified Fourier filter method [19], and we use these sequences as the site energies in the binary chain. In essence, the algorithm generates a random noise in the frequency domain, multiplies this noise by a power–law with the desired exponent, and then Fourier transforms the signal back into real space. Since we must generate a binary sequence, we consider a transformation which maps any positive value of the

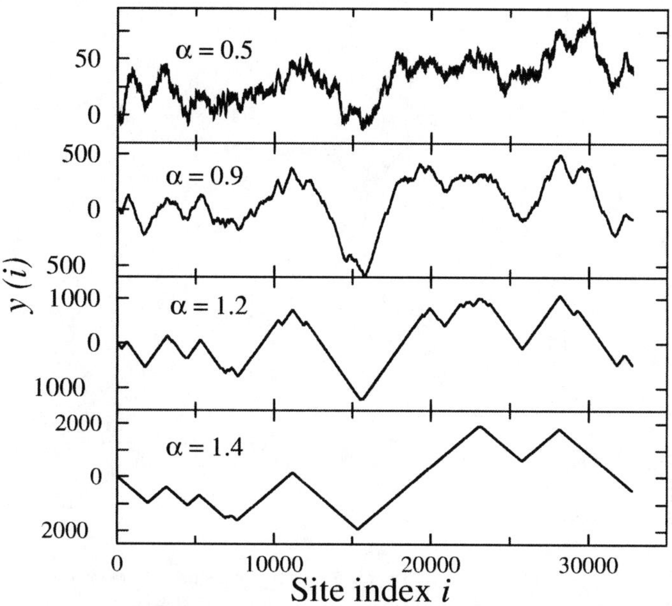

FIGURE 2. Examples of long-range correlated binary series. For simplicity, we plot the series $y(i) \equiv \sum_{j=1}^{i} \varepsilon_j$, where $\varepsilon_j = \varepsilon_A$ or ε_B, is a binary sequence characterized by different correlations. Here we choose $\varepsilon_A = 1/2$ and $\varepsilon_B = -1/2$, so a region of positive slope indicates abundance of ε_A, and conversely, a region with negative slope indicates abundance of ε_B. Note that the two types of atoms tend to form clusters of any size which are power-law distributed with a higher probability for large sizes as the correlations increase. The power–law correlations in the binary sequence are characterized by the scaling exponent α obtained from DFA.

correlated series into ε_A, and any negative value into ε_B. Such mapping can change the correlation properties of the series, and therefore the correlations are not properly quantified by the power–law exponent in the original correlated series.

To quantify the correlations in the final binary sequence, we use the scaling exponent α of the detrended fluctuation analysis (DFA) method [20, 21, 22, 23]. Briefly, the DFA method consists of the following steps: (i) divide the series into boxes of size n; (ii) calculate the fluctuations of the series around the local polynomial trend inside each box; (iii) average these fluctuations for all the boxes and calculate the root mean squared fluctuation function $F(n)$. Iterate the procedure for a broad range of scales n. A power–law relation between $F(n)$ and n indicates the presence of scaling: $F(n) \propto n^{\alpha}$. The scaling (or correlation) exponent α quantifies the correlation properties of the binary sequence: if $\alpha = 0.5$ there is no correlation (white noise); if $\alpha < 0.5$ the binary sequence is anticorrelated while if $\alpha > 0.5$ there are positive long-range correlations in the signal (Fig. 2). DNA sequences can be cited as an example of real chains displaying such long–range correlations [18].

Next, we study the localization properties of correlated-disordered binary chains. We characterize the size of the states (spread of the electron wave functions) by the localization length λ, which we obtain using the transfer matrix method [5]. The Schrödinger

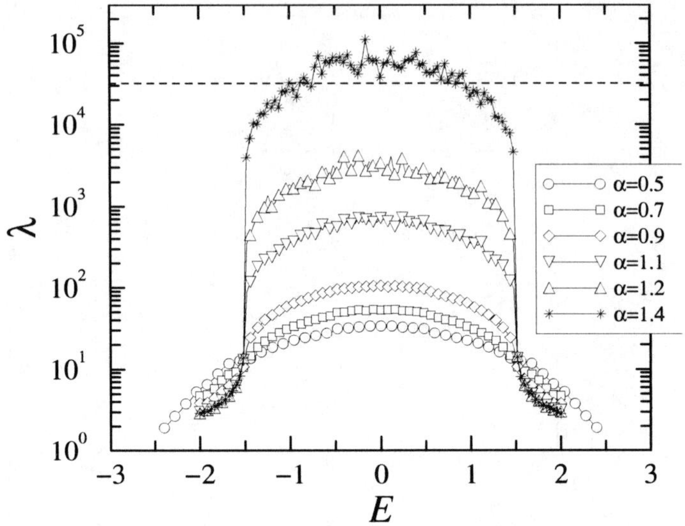

FIGURE 3. Behavior of λ as a function of energy for several values of the correlations. The horizontal dashed line corresponds to the fixed system size $N = 2^{15} \approx 3x10^4$. Note that for $\alpha \approx 1.4$ we find a threshold in α above which extended states exist in the central region of the energy band. This energy interval is given by $[\varepsilon_A - 2t, \varepsilon_A + 2t] \cap [\varepsilon_B - 2t, \varepsilon_B + 2t]$. For the particular case shown here, $t = 1$, $\varepsilon_A = -\varepsilon_B = 1/2$, the energy interval in which the increase of λ occurs is [-1.5,1.5].

equation for the Hamiltonian (1) becomes

$$E\psi_n = t\psi_{n-1} + \varepsilon_n \psi_n + t\psi_{n+1}, \tag{3}$$

where E is the energy corresponding to the electron wave function, $|\psi_n|^2$ is the probability to find an electron at site n in the chain, and $\varepsilon_n = \varepsilon_A$ or ε_B. Using the transfer matrix method, we can write Eq. (3) in the recursive form

$$M_n \begin{pmatrix} \psi_n \\ \psi_{n-1} \end{pmatrix} = \begin{pmatrix} \psi_{n+1} \\ \psi_n \end{pmatrix}, \quad M_n = \begin{pmatrix} \frac{E-\varepsilon_n}{t} & -1 \\ 1 & 0 \end{pmatrix} \tag{4}$$

The localization length $\lambda(E)$ is then defined by

$$\frac{1}{\lambda(E)} = \lim_{N\to\infty} \frac{1}{N} \ln \left| \frac{\psi_N}{\psi_0} \right| \tag{5}$$

where N is the chain length. We choose $\psi_0 = \psi_1 = 1/\sqrt{2}$, and for every realization of the potential we apply Eq. (4) to obtain ψ_N and we calculate $\lambda(E)$ from Eq. (5). Finally, we average λ over a large ensemble of chain realizations for a fixed system size N and fixed value of the energy E. We repeat this procedure for different values of N and E.

Our results for the localization length λ show that correlations in the sequence of site energies in the binary chain strongly modify the localization properties of the

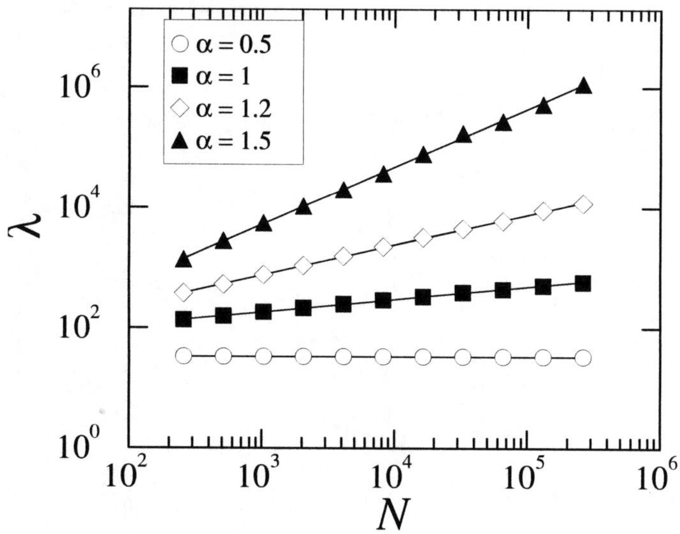

FIGURE 4. Behavior of λ averaged around $E = 0$ as a function of the system size N and for different values of the correlations. The system size N ranges from 2^8 atoms to 2^{18} atoms. The solid lines correspond to fits of the type $\lambda \propto N^\gamma$, from where the exponent γ has been obtained.

electronic wave functions (Fig. 3). For a fixed system size N, we obtain the shortest localization length as a function of electron energy E for the disordered chain ($\alpha = 0.5$), which corresponds to the result for random binary alloy [13]. We find, surprisingly, that long-range correlations ($\alpha > 0.5$) change drastically the localization properties of the electronic states. When we increase the long-range correlations in the chain, λ increases in the central region of the energy band, and eventually, for a fixed system size N, we find a critical value α_c above which $\lambda > N$, which corresponds to extended wave functions in the energy region which we consider (see Fig. 3). The energy region in which this increase of λ takes place is precisely the energy region where the bands corresponding to the ordered chains formed only by A atoms or only by B atoms overlap. Thus, we obtain a broad band of extended states (for a fixed system size N), in contrast to the random dimer model case, where only a singular value of the energy E corresponds to an extended wave function.

To speak properly of extended states, we need the following condition:

$$\lim_{N\to\infty} \frac{\lambda}{N} = \text{constant} \neq 0. \tag{6}$$

If this condition is not fulfilled the localization length λ becomes negligible in comparison to the system size N in the thermodynamic limit, and the wave function is therefore localized.

Hence, we next study the behavior of λ as a function of N for different values of the correlations. Since λ is also a function of the electron energy E, we choose

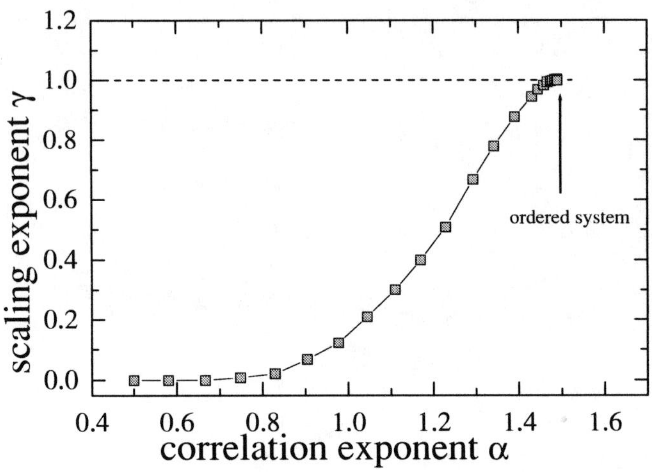

FIGURE 5. The calculated exponent γ as a function of the correlation exponent α characterizing the binary chain.

$E = 0$ — i.e. the central energy of the band. Our results in Fig. 3 suggest that a similar behavior can be expected for every other value of E in the energy interval $[\varepsilon_A - 2t, \varepsilon_A + 2t] \cap [\varepsilon_B - 2t, \varepsilon_B + 2t]$. To avoid excessive numerical fluctuations, we average λ in a small energy window around $E = 0$, i.e. the interval [-0.05,0.05]. We find a power–law relation between λ and N:

$$\lambda \propto N^{\gamma} \tag{7}$$

where the exponent γ depends on the scaling exponent α characterizing the long–range correlations in the site energies ε_A and ε_B in the chain (Fig. 4). For α close to 0.5 we find that $\gamma \approx 0$, and thus we have localized states. As α increases, γ also increases, and for $\alpha \to 1.5$ we obtain, within the error bars of our calculations, that $\gamma = 1$ which indicates that the system behaves as a conductor in the thermodynamic limit (Fig. 5). However, at $\alpha = 1.5$ the number of segments of type A or B does not increase with the system size but remains constant even in the thermodynamic limit of infinite system size. Thus at this limiting value of the correlation exponent α the size of the segments increases with the system size, and in the thermodynamic limit the binary chain undergoes a transition from a disordered (insulating) system to an ordered (conducting) system [24].

In summary, we have found that long-range correlations change the localization properties of 1–D disordered binary solids. We show that the localization length of the electron wave function is greatly increased by the long-range correlations leading to conductivity within a broad energy band for systems of large but finite size. We note that recent work [25] studies how temperature can lead to conductivity in 1-D uncorrelated disordered binary systems used to model DNA chains, and shows that DNA, considered an insulator at zero temperature, can become a conductor when the temperature increases. Our results, however, suggest that DNA conductivity could exist independently of the temperature, provided there are sufficiently strong long-range correlations. Indeed, recent studies report the presence of long-range correlations in DNA [18]. Our work suggests that conductivity in short DNA chains (small N) could exist even for weaker correlations than those discussed above [26]. Moreover, since the localization length exceeds the size of the segments with correlated disorder, the electron wave function overlaps with parts of the neighboring uncorrelated disordered segments and can affect their conducting properties, thus facilitating important genomic functions, especially those related to the electronic transport through the chain, as the repair of certain mutations [27]. Finally, our finding of conductivity arising from long-range correlated disorder may shed light on the conduction mechanisms at non-zero temperature, when both the correlations and the temperature could contribute.

ACKNOWLEDGMENTS

We thank the Spanish Ministerio de Ciencia y Tecnologia/Grant No. BFM2002-00183 and NIH/National Center for Research Resources (P41RR13622) for support.

REFERENCES

1. N.W. Ashcroft and N.D. Mermin, *Solid State Physics* (Holt–Saunders, London, 1976).
2. Anderson, P.W. *Phys. Rev.* **109**, 1492-1505 (1958).
3. Kramer, B. and MacKinnon, A. *Rep. Prog. Phys.* **56**, 1469-1564 (1993).
4. Janssen, M. *Phys. Rep.* **295**, 1-91 (1998).
5. Mackinnon, A. and Kramer B. *Phys. Rev. Lett.* **47**, 1546-1549 (1981).
6. Hofstetter, E. and Schreiber M. *Phys. Rev. B* **48**, 16979-16985 (1993).
7. Kravchenko, S.V., Simonian, D., Sarachik, M.P., Mason, W., and Furneaux, J.E. *Phys. Rev. Lett.* **77**, 4398-4941 (1996).
8. Weymer, A. and Janssen M. *Ann. Physik (Leipzig)* **7**, 159-173 (1998).
9. Schweitzer, L. and Zharekeshev, I.Kh. *J. Phys.: Condens. Matter* **9**, L441-L445 (1997).
10. Mott, N.F. and Twose, W. *Adv. Phys.* **10**, 107-163 (1961).
11. R. Abou-Chacra, P.W. Anderson, and D.J. Thouless. *J. Phys. C* **6**, 1734-1752 (1973).
12. Ishii, K. and Matsuda, H. *Prog. Theor. Phys. Suppl.* **45**, 56-86 (1970).
13. Davids, P.S. *Phys. Rev. B* **52**, 4146 (1995).
14. Dunlap, D.H., Wu, H.-L. and Phillips, P. *Phys. Rev. Lett.* **65**, 88-91 (1990).
15. Phillips, P. and Wu, H.-L. *Science* **252**, 1805-1812 (1991).
16. de Moura, F.A.B.F. and Lyra, M.L. *Phys. Rev. Lett.* **81**, 3735-3738 (1998).
17. Kantelhardt, J.W., Russ, S., Bunde, A., Havlin, S. and Webman, I. *Phys. Rev. Lett.* **84**, 198 (2000).
18. Peng, C.-K. et al. *Nature* **356**, 168-170 (1992).
19. Makse, H.A., Havlin, S., Schwartz, M. and Stanley, H.E. *Phys. Rev. E* **53**, 5445-5449 (1996).
20. Peng C.-K. et al. *Phys. Rev. E* **49**, 1685-1689 (1994).
21. Kantelhardt, J.W., Koscielny-Bunde, E., Rego, H.H.A, Havlin, S. and Bunde, A. *Physica A* **294**, 441-454 (2001).
22. Hu, K., Ivanov, P.Ch., Chen, Z., Carpena, P. and Stanley, H.E. *Phys. Rev. E* **64**, 011114(19) (2001).
23. Chen Z., Ivanov P.Ch., Hu K., and Stanley, H.E. *Physical Review E* **65**, 041107(15) (2002).
24. Carpena P., Bernaola-Galván P., Ivanov P.Ch. and Stanley, H.E.(in preparation).
25. Yu, Z.G. & Song, X. *Phys. Rev. Lett.* **86**, 6018-6021 (2001).
26. Carpena P., Bernaola–Galván P., Ivanov P.Ch., and Stanley, H.E. *Nature* **418**, 955-959 (2001).
27. Dandliker, P.J., Holmlin, R.E. and Barton, J.K. *Science* **275**, 1465-1468 (1997).

List of Participants

Derek Abbott, *Adelaide University, Australia*
Abdulrahman Alfuraih, *George Washington University*
David Allingham, *University of Warwick, Coventry, UK*
Andrew Allison, *Adelaide University, Australia*
Dean Astumian, *University of Maine*
Sonya Bahar, *Weill-Cornell Medical College*
Gábor Balázsi, *Northwestern University*
Alexander Berezhkovskii, *CIT, NIH*
Sergey Bezrukov, *NICHD, NIH*
Martin Bier, *East Carolina University*
Gijs Bosman, *University of Florida*
Juan Cabrera, *University of Chicago*
Dante Chialvo, *University of California*
John Clay, *NINDS, NIH*
James Collins, *Boston University*
Leonardo Dagdug, *CIT, NIH*
Emmanuele DiBenedetto, *Vanderbilt University*
Edmund Di Marzio, *NIST*
Mingzhou Ding, *Florida Atlantic University*
Charles Doering, *University of Michigan*
Mark Dykman, *Michigan State University*
Robert Eisenberg, *Rush Medical Center*
Jianfeng Feng, *Sussex University, UK*
Jordi Garcia-Ojalvo, *Technical University of Catalonia, Spain*
Oleg Gerashchenko, *St. Petersburg Nuclear Physics Institute, Russia*
Dirk Gillespie, *University of Miami School of Medicine*
Zoltan Gingl, *University of Szeged, Hungary*
Brage Golding, *Michigan State University*
Gabriel Gomila, *University of Barcelona, Spain*
Tomas Gonzalez, *University of Salamanca, Spain*
Irina Gopich, *NIDDK, NIH*
Ewa Gudowska-Nowak, *Smoluchowski Institute of Physics, Poland*
Michael Hoffmann, *University of Ulm, Germany*
Gerhard Hummer, *NIDDK, NIH*
Nathan Israeloff, *Northeastern University*
Plamen Ivanov, *Boston University*
Kuni Iwasa, *NIDCD, NIH*
Fernan Jaramillo, *Carleton College*
Jorge Jose, *Northeastern University*
Peter Jung, *Ohio University*
Frank Jülicher, *Max Planck Institute, Germany*
Ad. Kalmijn, *Scripps Institution of Oceanography*
John Kasianowicz, *NIST*
Laszlo Kish, *Texas A&M University*

Samuel Kou, *Harvard University*
Vladimir Kuznetsov, *NICHD, NIH*
Hans Liljenström, *Agora for Biosystems, Sweden*
Katja Lindenberg, *University of California*
Benjamin Lindner, *University of Ottawa, Canada*
Andre Longtin, *University of Ottawa, Canada*
David Lubensky, *Bell Labs*
Dmitri Luchinskii, *Lancaster University, UK*
Riccardo Mannella, *University of Pisa, Italy*
Abraham Mara, *R.W. Johnson PRD*
Ognian Marinov, *McMaster University, Canada*
Jose Mateos, *Institute of Physics, Mexico*
Peter McClintock, *Lancaster University, UK*
Mark McDonnell, *Adelaide University, Australia*
Jason Middleton, *University of Ottawa, Canada*
John Milton, *University of Chicago*
Toshio Mori, *Kyushu University, Japan*
Rob Morse, *Keele University, UK*
Frank Moss, *University of Missouri*
M. Muthukumar, *University of Massachusetts*
Alexander Neiman, *University of Missouri*
Ekaterina Nestorovich, *NICHD, NIH*
Ralph Nossal, *NICHD, NIH*
Toru Ohira, *Sony Computer Science Labs, Japan*
Debabrata Panja, *University of Leiden, Netherlands*
Adrian Parsegian, *NICHD, NIH*
Johan Paulsson, *Princeton University*
Cecilia Pennetta, *University of Lecce, Italy*
Arkady Pikovsky, *University of Potsdam, Germany*
Mark Pustovoit, *St. Petersburg Nuclear Physics Institute, Russia*
Lino Reggiani, *University of Lecce, Italy*
Tatiana Rostovtseva, *NICHD, NIH*
David Russell, *University of Missouri*
Francesc Sagues, *University of Barcelona, Spain*
Jose Sancho, *University of Barcelona, Spain*
Lutz Schimansky-Geier, *Humboldt University, Germany*
Pavel Shiktorov, *Semiconductor Physics Institute, Lithuania*
Michael Shlesinger, *Office of Naval Research*
Amit Singer, *Tel-Aviv University, Israel*
Zuzanna Siwy, *Gesellschaft fuer Schwerionenforschung, Germany*
Alexander Smirnov, *NHLBI, NIH*
Bernardo Spagnolo, *University of Palermo, Italy*
Mark Spano, *Carderock Laboratory*
Eugene Stanley, *Boston University*
Evgenij Starikov, *Semiconductor Physics Institute, Lithuania*
Aneta Stefanovska, *University of Ljubljana, Slovenia*

Nigel Stocks, *University of Warwick, Coventry, UK*
Bela Suki, *Boston University*
Attila Szabo, *NIDDK, NIH*
Peter Talkner, *University of Augsburg, Germany*
Peter Tass, *University of Cologne, Germany*
Mukund Thattai, *Massachusetts Institute of Technology*
Peter Tieleman, *University of Calgary, Canada*
Lawrence Ward, *University of British Columbia, Canada*
George Weiss, *CIT, NIH*
Michael Weissman, *University of Illinois*
Kurt Wiesenfeld, *Georgia Institute of Technology*
Stephen Winters-Hilt, *University of California*
Sunney Xie, *Harvard University*
Yoshi Yamamoto, *University of Tokyo, Japan*
Renat Yulmetyev, *Kazan State Pedagogical University, Russia*
Robert Zwanzig, *NIDDK, NIH*

AUTHOR INDEX